Clinical Virology Manual

Second Edition

Steven Specter, Ph.D.
Professor

Gerald Lancz, Ph.D.
Professor

Department of Medical Microbiology and Immunology
University of South Florida College of Medicine
Tampa, Florida

Elsevier

New York • Amsterdam • London • Tokyo

Sole distributors outside the United States and Canada:
Elsevier Science Publishers B.V.
P.O. Box 1000 AE Amsterdam, the Netherlands

© 1992 by Elsevier Science Publishing Company, Inc.

This book has been registered with the Copyright Clearance Center, Inc. For further information, please contact the Copyright Clearance Center, Salem, Massachusetts. OCLC 26928665

All inquiries regarding copyrighted material from this publication, other than reproduction through the Copyright Clearance Center, should be directed to: Rights and Permissions Department, Elsevier Science Publishing Company, Inc., 655 Avenue of the Americas, New York, New York 10010. FAX 212-633-3977.

Elsevier Science Publishing Company, Inc.
655 Avenue of the Americas, New York, New York 10010

Library of Congress Cataloging-in-Publication Data

Clinical virology manual / edited by Stephen Specter, Gerald Lancz
  2nd. ed.

  Includes bibliographical references and index.
  ISBN 0-444-01662-7 (hardcover : alk. paper)
  1. Diagnostic virology—Handbooks, manuals, etc.
I. Specter, Steven. II. Lancz, Gerald
[DNLM: 1. Virology—methods.   2. Virus Diseases—diagnosis. QW
160 C641]
QR387.C48   1992
616'.0194-dc20
DNLM/DLC
for Library of Congress

Current printing (last digit)
10 9 8 7 6 5 4 3 2 1

Manufactured in the United States of America

*We dedicate this book to our children and our
wives. It is their individual expressions,
which, in reflection, present a collage of their love,
that provide us the encouragement, energy, and
vigor with which we undertake this task.*

*To Rachel and Brian; Daniel, Ross, and David;
and to Sharon and Randie.*

Contents

Preface to the Second Edition   xiii
Preface to the First Edition   xv
Contributors   xvii

SECTION 1
LABORATORY PROCEDURES FOR
DETECTING VIRUSES   1

**1 Quality Control in Clinical
  Virology**   3
Ann L. Warford
  1.1   Introduction   3
  1.2   Facility Design and
        Maintenance   3
  1.3   Personnel   5
  1.4   Procedure Manual   5
  1.5   Quality Control   5
  1.6   Viral Transport Media (VTM)   6
  1.7   Submission of Smears   7
  1.8   Submission of Sera   7
  1.9   Cell Culture   8
  1.10 Media and Solutions   10
  1.11 Water Quality   11
  1.12 Reference Virus Stocks   12
  1.13 Reagents, Stains, Antisera, and
        Kits   12
  1.14 Instruments   13
  1.15 Statistics and Backup
        Procedures   15
  1.16 Proficiency Testing   16
  References   17

2 Specimen Requirements   19
Thomas F. Smith
  2.1   Introduction   19
  2.2   Specimen Selection   20
  2.3   Specimen Collection   20
  2.4   Transport   31
  2.5   Processing Specimens   34
  2.6   Serologic Determination   35
  References   36

3 Primary Isolation of Viruses   43
*Marie L. Landry and
G. D. Hsiung*
  3.1   Introduction   43
  3.2   Virus Isolation in Cell Cultures   44
  3.3   Appendix: Additional Methods
        for Virus Isolation   61
  References   65

**4 The Cytopathology of Virus
  Infections**   71
*Roger D. Smith and
Karen Sutherland*
  4.1   Introduction   71
  4.2   Preparation and Staining   71
  4.3   Virus Cytopathology   73
  References   87

**5 Electron Microscopy and
  Immunoelectron Microscopy**   89
Frances W. Doane
  5.1   Introduction   89

5.2 Clinical Specimens Suitable for Virus Detection By Electron Microscopy   91
5.3 Negative Staining Methods   94
5.4 Immunoelectron Microscopy Methods   97
5.5 Thin Sectioning Methods   101
5.6 Processing Cell Culture Isolates for Electron Microscopy   102
References   105

6 **The Interference Assay**   111
Charles A. Reed
6.1 History   111
6.2 Specimen Collection, Transport, and Storage   112
6.3 Specimen Preparation and Inoculation   112
6.4 ECHO-11 Challenge and Neutralization   113
6.5 Pros and Cons   114
References   114

7 **Immunofluorescence**   117
Linda L. Minnich and C. George Ray
7.1 History   117
7.2 Materials   117
7.3 Methods   120
7.4 Practical Details for Indirect Immunofluorescence   121
7.5 Application of Fluorescent Antibody Technique in Detection of Antigens in Cell Culture   124
7.6 Resolution of Technical Problems   125
7.7 Application of Fluorescent Antibody Technique and Its Future   126
References   127

8 **Radioimmunoassay**   129
Isa K. Mushahwar and Thomas A. Brawner
8.1 Introduction   129
8.2 Competitive Binding Radioimmunoassay   129
8.3 Solid Phase Radioimmunoassays   130
8.4 Hepatitis B Serologic Markers   130

8.5 Direct Solid Phase Sandwich Assays   131
8.6 Indirect (Competitive) Radiometric Assays for Hepatitis B Virus Antibodies   134
8.7 Assay for IgM Class Antibodies   135
8.8 Other Viral Markers   138
References   145

9 **Enzyme Immunoassay**   153
Andrew J. O'Beirne and John L. Sever
9.1 Introduction   153
9.2 Materials   157
9.3 ELISA Test Methods   165
9.4 Quantitation   168
9.5 Instruments   169
9.6 Comments and Sources of Error   169
References   170

10 **Immunoperoxidase Detection of Viral Antigens in Cells**   189
Helen Gay, John J. Docherty, and Joseph E. Kall, Jr.
10.1 Introduction   189
10.2 Staining Considerations   192
10.3 Materials for Immunoperoxidase Staining   193
10.4 Methods   193
10.5 Results   194
10.6 Discussion   195
References   199

11 **Complement Fixation Test**   203
Lillian M. Stark and Arthur L. Lewis
11.1 Introduction   203
11.2 General Considerations   206
11.3 Test Setup   209
11.4 Test Performance   211
References   223

12 **Neutralization**   229
Harold C. Ballew
12.1 Introduction   229
12.2 Standardization of Test Materials   230
12.3 Test Procedures   233
12.4 Calculations of 50 Percent Endpoints   236
12.5 Preparation of Serum Pools   236

12.6 Interpretation of Serologic
Results 239
References 240
Bibliography 241

13 Hemagglutination Inhibition and Hemadsorption **243**
Leroy C. McLaren
13.1 Hemagglutination Inhibition Test 243
13.2 Hemadsorption 247
13.3 Hemadsorption Inhibition Test 248
Appendix 248
References 249

14 Immune Adherence Hemagglutination **251**
Evelyne T. Lennette and David A. Lennette
14.1 Introduction 251
14.2 Historic Background 252
14.3 IAHA Microtiter Procedure 252
14.4 Red Blood Cells 255
14.5 Hemagglutination Patterns 256
14.6 Detection of IgM Antibodies 256
14.7 Sensitivity and Specificity 257
14.8 Advantages and Disadvantages 258
14.9 Appendix 259
References 259

15 IgM Determinations **263**
Kenneth L. Herrmann and Dean D. Erdman
15.1 Introduction 263
15.2 Methods Used for IgM Antibody Determination 263
15.3 Interpretation of Assay to Detect IgM Antibodies 270
References 273

16 Antiviral Drug Susceptibility Testing **277**
Edgar L. Hill and M. Nixon Ellis
16.1 History 277
16.2 Materials 277
16.3 Dye Uptake Assay Methods 278
16.4 Conclusion 283
References 283

17 Nucleic Acid Hybridization **285**
Attila T. Lörincz
17.1 Introduction 285
17.2 Specimen Handling 285

17.3 Hybridization 287
17.4 Specific Hybridization Tests 291
17.5 Problems in Test Performance and Interpretation: Choosing the Correct Test 292
17.6 Application of Hybridization to Viral Testing 293
17.7 Pros and Cons of Nucleic Acid Testing 295
References 296

18 Application of Western Blotting for the Diagnosis of Viral Infections **299**
Peter G. Medveczky
18.1 Introduction 299
18.2 History and Principle of Western Blotting 299
18.3 The Western Blot Procedure 301
18.4 Advantages and Disadvantages of Western Blot Assay 304
18.5 Commercial Kits and Nitrocellulose Strips with Blotted Proteins 305
References 306

19 Polymerase Chain Reaction **309**
Carleton T. Garrett, Kathleen Porter-Jordan, and Suhail Nasim
19.1 General Considerations 309
19.2 Polymerase Chain Reaction— Basic Principles 310
19.3 Advantages Deriving From In Vitro Amplification 311
19.4 Limitations of PCR 313
19.5 Summary 317
References 317

SECTION 2
VIRAL PATHOGENS 319

20 Respiratory Viruses **321**
Britt Åkerlind-Stopner and Maurice A. Mufson
20.1 Introduction 321
20.2 Influenza Virus 323
20.3 Parainfluenza Virus 327
20.4 Respiratory Syncytial Virus 329
20.5 Adenovirus 332
20.6 Coronavirus 334
20.7 Rhinovirus 335
References 337

21 Enteroviruses Including Hepatitis A Virus 341
Heinz Zeichhardt
21.1 Introduction 341
21.2 History of Virus Discovery 341
21.3 Structural, Biophysical, Biochemical, & Biological Characteristics 342
21.4 Epidemiology 345
21.5 Pathogenesis and Clinical Syndromes 349
21.6 Incubation Times 353
21.7 Immune Response 353
21.8 Vaccination 354
21.9 Laboratory Diagnosis 355
References 357

22 Rotavirus, Enteric Adenoviruses, Norwalk Virus, and Other Gastroenteritis Tract Viruses 361
Miguel O'Ryan, David O. Matson, and Larry K. Pickering
22.1 Introduction 361
22.2 Rotaviruses 361
22.3 Enteric Adenoviruses 375
22.4 Norwalk Viruses 376
22.5 Astroviruses 379
22.6 Caliciviruses 379
22.7 Other Viruses 380
22.8 Treatment and Prevention of Viral Gastroenteritis 381
22.9 Summary 381
References 381

23 Viral Hepatitis 397
Mario R. Escobar
23.1 Introduction 397
23.2 Brief Description of the Characteristics of the Viruses 400
23.3 Pathogenesis of Viral Hepatitis 403
23.4 Pathology of Viral Hepatitis 407
23.5 Laboratory Diagnosis of Viral Hepatitis 409
References 417

24 Rabies 425
George R. Anderson
24.1 History 425
24.2 Characteristics of the Virus 426
24.3 Pathogenesis 427
24.4 The Isolation and Cultivation of Rabies Virus 430

24.5 Virus Identification 431
24.6 Considerations for the Clinical Virologist 434
24.7 Future Approaches to Rabies Prevention 436
References 438

25 Arboviruses 443
Robert E. Shope
25.1 History 443
25.2 Biochemical, Serologic, and Epidemiologic Characteristics of Arboviruses 444
25.3 Pathogenesis 447
25.4 Clinical Description 448
25.5 Laboratory Procedures 449
25.6 Trends 451
References 451

26 Papovaviruses 455
Keerti V. Shah
26.1 Introduction 455
26.2 Human Papillomaviruses 455
26.3 Human Polyomaviruses 464
References 469

27 Herpes Simplex Viruses 473
Laure Aurelian
27.1 Introduction 473
27.2 Characteristics of Herpes Simplex Viruses 473
27.3 Replicative Cycle 476
27.4 Pathogenesis of Herpes Simplex Virus Infection 477
27.5 Epidemiology of Herpes Simplex Virus Infections 479
27.6 Herpes Simplex Virus-Induced Latency 480
27.7 Diagnosis of Herpes Simplex Virus Infections 482
27.8 Diagnosis of Encephalitis Due to Herpes Simplex Virus 489
27.9 Serologic Diagnosis 489
27.10 Immunity and Vaccine Development 491
27.11 Unusual Features of Herpes Simplex Viruses 493
27.12 Treatment of HSV Infections 494
References 496

28 Cytomegalovirus, Varicella-Zoster Virus, Epstein-Barr Virus, and Human Herpesvirus Type 6   **501**
David Paar and Stephen E. Straus
28.1   Introduction   501
28.2   Cytomegalovirus   501
28.3   Varicella-Zoster Virus   507
28.4   Epstein-Barr Virus   511
28.5   Human Herpesvirus-6 (HHV-6)   515
References   518

29 Poxviruses   **527**
James H. Nakano
29.1   Description of Diseases   527
29.2   Description of Viruses   533
29.3   Collection and Handling of Specimens   535
29.4   Methods for Isolation   535
29.5   Methods for Identification   537
29.6   Serologic Methods for Antibody Assay   540
References   544

30 Parvoviruses   **547**
Stanley J. Naides
30.1   Introduction   547
30.2   Characteristics of Parvovirus B19   548
30.3   Pathogenesis of Virus Infection   549
30.4   Clinical Manifestations of Parvovirus B19 Infection   551
30.5   Presence of B19 in Various Tissues   555
30.6   Preferred Sites for Virus Isolation   556
30.7   Virus Stability   556
30.8   Propagation of B19 in Vitro   556
30.9   Laboratory Diagnosis of B19 Infection   557
30.10   Unusual Features   561
30.11   Prevention of and Therapy for Diseases Due to Human Parvovirus   561
References   562

31 Measles, Mumps, and Rubella   **571**
David A. Fuccillo and John L. Sever
31.1   Measles Virus   571
31.2   Mumps Virus   574

31.3   Rubella   576
References   581

32 Human Retroviruses   **585**
Fulvia di Marzo Veronese, Paolo Lusso, Jorg Schüpbach, and Robert C. Gallo
32.1   Introduction   585
32.2   History and Characterization   585
32.3   Genome Organization and Protein Products   588
32.4   Pathogenesis of Human T-Cell Leukemia Virus Type I Infection   591
32.5   Pathogenesis of Human Immunodeficiency Virus (HIV) Infection   597
32.6   Laboratory Diagnosis of Human Retrovirus Infections   602
32.7   Stability and Safety Precautions   614
References   615

33 Chlamydia   **627**
Julius Schachter
33.1   Introduction   627
33.2   Characteristics of the Organism   628
33.3   Pathogenesis   629
33.4   Infected Sites and Methods of Collection   630
33.5   Stability, Storage, and Transport   630
33.6   Nonculture Methods for Detecting Chlamydiae   631
33.7   Direct Cytologic Examination   632
33.8   Enzyme Immunoassay   632
33.9   Methods for Isolation   633
33.10   Identification   634
33.11   Serodiagnosis   634
References   637

SECTION 3
REFERENCE LABORATORIES   **641**

34 Virology Services Offered by the Federal Reference Laboratory Centers for Disease Control   **643**
Kenneth L. Herrmann
34.1   Introduction   643
34.2   NCID Organization   644
34.3   Reference and Disease Surveillance   644

34.4 Requirements for All
Specimens 644
34.5 Special Requirements for Viral
Diseases 650
34.6 Additional Suggestions for
Viral Specimens 651
34.7 Requirements for Specimens
for DVRD and DVBID 652
34.8 Reference and Disease
Surveillance Request Form 653

**35 State Laboratory Virology
Services 659**
Steven Specter and Gerald Lancz
35.1 Introduction 659

35.2 Submission of Specimens 659
35.3 Scope of Services 664
35.4 Turnaround Time for Result
Reporting 665

**36 Laboratories Offering Viral
Diagnostic Services 671**
Steven Specter and Gerald Lancz
36.1 Introduction 671

Index 681

Preface to the Second Edition

The aims of the Clinical Virology Manual have not changed since the first edition. Employing the scientific principle "If it ain't broke, don't fix it" we have left the original preface, which describes the goals of the book, intact. The second edition of the Clinical Virology Manual is an updated and expanded version of the first edition. Some of the chapters were expanded to include new viral agents that were recently discovered to be clinically significant to man or because there was too little information about these agents at the time of the first edition, e.g., parvoviruses, human herpesvirus 6, hepatitis C virus, etc. The chapter on human retroviruses has been expanded and changed considerably from the first edition considering the veritable explosion of information regarding these infectious agents and their significance in human disease. There are also two new chapters in the second edition which describe techniques that were more recently developed and which are having a major impact upon the clinical virology labora-

tory, i.e., the polymerase chain reaction and western blotting. We have also significantly expanded the section of the book that lists the various installations, from private to public and federal, that provide virologic services as part of their mission to the community and/or nation. It is our hope that this compilation will be used as a reference and a resource by laboratorians, to make them aware of colleagues in their locale who perform virologic services and as a source with whom they might interact and collaborate to improve diagnostic virology services in their area. Our overall goal is to help keep clinical virologists and clinicians current with the discipline. Finally, we gratefully acknowledge the guidance and support we have received for both editions of the Clinical Virology Manual from our friend and chairman, Herman Friedman.

Steven Spector
Gerald Lancz

Preface to the First Edition

Clinical virology is an area that is undergoing rapid expansion. As a service for patient care, the utility of the clinical virology laboratory has increased significantly in the past decade. Due to the availability of commercial test kits, sophisticated yet simple diagnostic reagents, and the standardization of laboratory assays, accurate, reliable and, in many instances, rapid protocols are currently available for the diagnosis of a variety of viral agents producing human infections. Thus, the demands (on both the physician and the clinical laboratory virologist) for the diagnosis of viral infections will continue to increase. With this in mind, this volume is written as both an aid to the clinician and as a guide for the clinical laboratory.

This manual has three sections. The first describes laboratory procedures to detect viruses. The initial chapters deal with quality control in the laboratory and specimen handling, areas that are critical for an effective diagnostic laboratory. This is followed by individual chapters that provide information or a detailed protocol on how to set up and test samples for viral diagnosis using this technique. Both classical and the newer, more experimental techniques are described in detail.

The second section focuses on the viral agents. Viruses are grouped into chapters based on a target organ-system categorization. In this way, viruses producing infection in a particular organ or tissue are discussed and compared in a single chapter. This approach more accurately reflects the problems and choices faced by the attending physician and clinical technician for the diagnosis of a viral infection. Each chapter includes information relating basic, pathogenic, immunologic, and protective measures concerning each virus group, as well as information on its isolation, propagation, and diagnosis. This section also includes a chapter on *Chlamydia*. There are two reasons for including this family: The clinical laboratory often isolates and diagnoses *Chlamydia*, and the techniques used in its isolation and diagnosis are used in other instances.

The third section is designed to be used for reference. Here we supply information about Federal Reference Laboratories at the Centers for Disease Control and their role in the diagnosis of viral infection. The diagnostic and regulatory activities of state health laboratories and services available at individual hospital laboratories are provided in survey form. This listing is somewhat incomplete in that it contains information provided in response to an initial questionnaire and follow-up.

The aim and scope of this volume is service: to the physician, as a source of basic and clinical information regarding viruses and viral diseases, and to the laboratories, as a reference source to aid in the diagnosis of virus infection by providing detailed information on individual techniques and the impetus to expand services offered.

Steven Specter
Gerald Lancz

Contributors

Britt Åkerlind-Stopner, M.D.
Department of Virology
Karolinska Institute
School of Medicine
Stockholm, Sweden

George R. Anderson, D.V.M.
Acting Bureau Chief
Bureau of Laboratory and Epidemiological
* Services*
Michigan Department of Public Health
Lansing, Michigan

Laure Aurelian, Ph.D.
Professor and Director of Laboratories
Virology/Immunology Laboratories
University of Maryland School of Medicine
Baltimore, Maryland

Harold C. Ballew, M.D.
Retired Virologist
Centers for Disease Control
Atlanta, Georgia

Thomas A. Brawner, Ph.D.
Senior Scientist
Department of Infectious Diseases and
* Immunology*
Hepatitis Laboratory
Abbott Laboratories
North Chicago, Illinois

Frances W. Doane, M.D.
Professor
Department of Microbiology
Faculty of Medicine
University of Toronto
Toronto, Ontario, Canada

John J. Docherty, Ph.D.
Professor and Chairman
Department of Microbiology and
* Immunology*
Northeastern Ohio Universities
College of Medicine
Rootstown, Ohio

M. Nixon Ellis, Ph.D.
Departments of Clinical Investigation and
* Virology*
Burroughs-Wellcome Company
Research Triangle Park, North Carolina

Dean D. Erdman, Ph.D.
Research Microbiologist
Respiratory and Enteric Viruses Branch
Division of Viral and Rickettsial Diseases
National Center for Infectious Diseases
Centers for Disease Control
Atlanta, Georgia

Mario R. Escobar, Ph.D.
Professor of Pathology
Scientific Director, Clinical Immunopathology/
Virology Section
Department of Pathology
Medical College of Virginia
Virginia Commonwealth University
Richmond, Virginia

David A. Fuccillo, Ph.D.
Director
Shady Grove Laboratory
Rockville, Maryland

Robert C. Gallo, M.D.
Chief, Laboratory of Tumor Cell Biology
National Cancer Institute
Bethesda, Maryland

Carleton T. Garrett, M.D., Ph.D.
Professor of Pathology and Genetics
George Washington University Medical Center
Department of Pathology
Washington, D.C.

Helen Gay, M.S.
Retired Research Assistant
Northshore University Hospital
Manhasset, New York

Kenneth L. Herrmann, M.D.
Deputy Director
Division of Viral and Rickettsial Diseases
National Center for Infectious Diseases
Centers for Disease Control
Atlanta, Georgia

Edgar L. Hill, Ph.D.
Departments of Clinical Investigation and
Virology
Burroughs-Wellcome Company
Research Triangle Park, North Carolina

G.D. Hsiung, Ph.D.
Professor Emeritus, Department of Laboratory
Medicine
Yale University School of Medicine
New Haven, Connecticut
Director Emeritus
Virology Laboraotry
Veterans Administration Hospital
West Haven, Connecticut

Joseph E. Kall, Jr., M.S.
Department of Microbiology and Immunology
Northeastern Ohio Universities
College of Medicine
Rootstown, Ohio

Gerald Lancz, Ph.D.
Professor
Department of Medical Microbiology and
Immunology
University of South Florida College of
Medicine
Tampa, Florida

Marie L. Landry, M.D.
Associate Professor, Laboratory and Internal
Medicine
Yale University School of Medicine
New Haven, Connecticut
Director
Virology Reference Laboratory
Vetrans Affairs Medical Center
West Haven, Connecticut

David A. Lennette, Ph.D.
Co-Director
Virolab Incorporated
Emeryville, California

Evelyne T. Lennette, Ph.D.
Co-Director
Virolab Incorporated
Emeryville, California

Arthur L. Lewis, D.V.M., M.P.H.
Epidemiology Research Director
Epidemiology Research Center
Office of Laboratory Services
Florida Department of Health and
Rehabilitative Services
Tampa, Florida

Attila T. Lorincz, Ph.D.
Vice President and Scientific Director
Digene Diagnostics, Incorporated
Silver Spring, Maryland

Paolo Lusso, M.D.
Laboratory of Tumor Cell Biology
National Cancer Institute
Bethesda, Maryland

David O. Matson, M.D., Ph.D.
Department of Pediatrics and Division of
   Molecular Virology
Baylor College of Medicine
Houston, Texas

LeRoy McLaren, Ph.D.
Professor, Department of Microbiology
College of Medicine
Albuquerque, New Mexico

Peter G. Medveczky, M.D.
Department of Medical Microbiology and
   Immunology
University of South Florida College of
   Medicine
Tampa, Florida

Linda L. Minnich, M.S.SM(AAM)
Section Supervisor
Virology Laboratory
Charleston Area Medical Center
Charleston, West Virginia

Maurice A. Mufson, M.D.
Professor and Chairman
Department of Medicine
Marshall University
School of Medicine
Huntington, West Virginia

Isa K. Mushahwar, Ph.D.
Head, Department of Infectious Diseases and
   Immunology
Hepatitis Laboratory
Abbott Laboratories
North Chicago, Illinois

Stanley J. Naides, M.D.
Assistant Professor
Division of Rheumatology
University of Iowa
College of Medicine
Iowa City, Iowa

James H. Nakano, Ph.D.
(deceased February 9, 1990)
Formerly Chief, Poxvirus Laboratory and
   World Health Organization Collaborating
   Center for Smallpox and Other Poxvirus
   Infections
Virus Exanthems and Herpes Virus Branch
Division of Viral Diseases
Centers for Disease Control
Atlanta, Georgia

Suhail Nasim, M.D.
Associate Professor
George Washington University Medical Center
Department of Pathology
Washington, D.C.

Andrew J. O'Beirne, Dr.P.H.
General Manager, Diagnostics Division
BioWhittaker, Incorporated
Walkersville, Maryland

Miguel O'Ryan, M.D.
Department of Pediatrics
University of Texas Medical School
Houston, Texas

David Paar, M.D.
Assistant Professor of Medicine
Department of Infectious Diseases
University of Texas Medical Branch
Galveston, Texas

Larry K. Pickering, M.D.
Professor and Vice Chairman for Research
Director, Center for Pediatric Research
Department of Pediatrics
Eastern Virginia Medical School
Children's Hospital of the King's Daughters
Norfolk, Virginia

Kathleen Porter-Jordan, M.D., Ph.D.
Molecular Oncology, Incorporated
Gaithersburg, Maryland

C. George Ray, M.D.
Fred Hutchison Cancer Center
Division of Infectious Disease
Seattle, Washington

Charles A. Reed, A.B.
Research Associate
Supervisor of Clinical Virology
Department of Pediatrics
Washington University School of Medicine
St. Louis, Missouri

Julius Schachter, Ph.D.
Professor of Epidemiology
Department of Laboratory Medicine
San Francisco General Hospital
San Francisco
Director, World Health Organziation
  Collaborating Center for Reference on
  Chlamydia
San Francisco, California

Jorg Schüpbach, M.D.
Institute for Immunology and Virology
University of Zurich
Zurich, Switzerland

John L. Sever, M.D., Ph.D.
Senior Vice President of Medical and
  Academic Affairs
Chairman, Department of Child Health and
  Development
Children's Hospital National Medical Center
Washington, D.C.

Keerti V. Shah, M.D., Dr.P.H.
Department of Immunology and Infectious
  Diseases
The Johns Hopkins University
School of Hygiene and Public Health
Baltimore, Maryland

Robert E. Shope, M.D.
Professor of Epidemiology
Yale Arbovirus Research Unit
Department of Epidemiology and Public
  Health
Yale University School of Medicine
New Haven, Connecticut

Roger D. Smith, M.D.
Professor and Director
Department of Pathology and Laboratory
  Medicine
University of Cincinnati Medical Center
Cincinnati, Ohio

Thomas F. Smith, Ph.D.
Head, Section of Clinical Microbiology
Department of Laboratory Medicine
Mayo Clinic and Mayo Foundation
Rochester, Minnesota

Steven Specter, Ph.D.
Professor
Department of Medical Microbiology and
  Immunology
University of South Florida College of
  Medicine
Tampa, Florida

Lillian M. Stark, Ph.D., M.P.H.
Epidemiology Research Center
Office of Laboratory Services
Florida Department of Health and
  Rehabilitative Services
Tampa, Florida

Steven S. Straus, M.D.
Chief, Medical Virology Section
Laboratory of Clinical Investigation
NIAID
National Institute of Health
Bethesda, Maryland

Karen Sutherland, C.T., (A.S.C.P.)
Diagnostic Cytology Section
Department of Pathology and Laboratory
  Medicine
University of Cincinnati Medical Center
Cincinnati, Ohio

Fulvia di Marzo Veronese, Ph.D.
Department of Cell Biology
Advanced BioScience Laboratories,
  Incorporated
Kensington, Maryland

Ann L. Warford, Dr.P.H.
Microbiology
Children's Hospital
San Diego, California

Heinz Zeichhardt
Private Lecturer
Institute for Clinical and Experimental
  Virology
The Free University of Berlin
Hindenburgdamm, Berlin, Germany

Clinical Virology Manual

Section 1

Laboratory Procedures
for Detecting Viruses

1

Quality Control in Clinical Virology

Ann L. Warford

1.1 INTRODUCTION

Quality control in the clinical virology laboratory has traditionally consisted of procedures designed to ensure accurate and reproducible laboratory results. In the 1990s this definition has been expanded to include the assurance of test results that are sensitive, specific, and clinically relevant, now called quality assurance.

In order to provide clinically relevant information, the test turnaround time must allow for rapid reporting of results to effectively modify therapy on the patient tested. A thorough and timely quality control program can help improve both turnaround time and cost effectiveness by reducing the need for repeated testing, recollecting specimens, and by eliminating errors resulting in misdiagnosis or inappropriate therapy.

Minimally, quality control procedures must be performed and documented as required for laboratory accreditation and licensure. The quality control standards for accreditation by the College of American Pathologists (CAP), Joint Committee for Accreditation of Healthcare Organizations (JCAHO), and licensure for the federal medicare program have been detailed re-

cently (August et al, 1990). Decisions to check performance of each lot or each usage of reagents or equipment must also be made with consideration for local, state, and federal requirements.

Ideally, quality control should encompass all activities, from test selection and specimen collection through the interpretation presented in the final report. Good communication between the clinical virologist, the pathologist, and the medical staff aids greatly in achieving the goal of the quality assurance.

1.2 FACILITY DESIGN AND MAINTENANCE

The physical plant for a clinical virology lab should be designed with three basic concepts in mind:

1. Biohazard control, to minimize risk to laboratory personnel and the general public
2. Protection of cell cultures and media used in isolation and detection of viruses from contamination; use of separate ar-

eas to isolate uninfected and infected cell cultures

3. Provision of space adequate for current and future functions without compromising the quality of work and quality control activities

The amount of space necessary for virology has been variously reported as ranging from 135 to 175 net square feet/person or at least 10 net linear feet of bench space/person (Bartlett, 1985; Blank and Schrunk, 1985). Other concepts, such as facility design to maximize work efficiency and maintain a pleasant work environment conducive to personnel satisfaction and retention are also increasing in importance.

The level and type of service provided by virology laboratories will vary and will therefore impact on specific laboratory design specifications. Critical factors include:

1. Cell culture production for viral isolation
2. Viral isolation using commercially produced cell cultures
3. Viral antigen detection
4. Viral serology
5. Referral service to off-site virology laboratory

The full service virology laboratory that performs viral isolation in cell culture will require a facility designed specifically for clinical virology with the following features:

1. The facility should be physically separated from the microbiology laboratory and not share common air returns or equipment, such as a biological safety cabinet or incubator.
2. The environment should be controlled so that the ambient temperature is in the range of 22 °C to 26 °C, and relative humidity is 30% to 50%.
3. The facility should be under negative pressure with respect to the rest of the

laboratory if required for the type or concentration of viruses that are cultured (Richardson and Barkley, 1984).

4. Positive air pressure is preferred for areas involved in cell culture production and media preparation, whereas the negative pressure areas are best suited for viral isolation.
5. Areas separate from the main laboratory should be provided for nucleic acid amplification and preparation of high titer stocks of viruses and viral antigen controls.
6. Good aseptic technique must be observed. Some of these are the use of universal safety precautions, daily decontamination of all work surfaces with an effective disinfectant such as 2% alkaline glutaraldehyde (Favero, 1985), use of lab coats and latex gloves, and minimization of aerosal generation. (Warren, 1981; Richardson and Barkley, 1984).
7. Class II biological safety cabinets with HEPA filters and external venting should be used for all cell culture and viral specimen work (Richardson and Barkley, 1984). Separate biological safety cabinets are preferable for cell culture production and viral isolation. If only one biological safety cabinet is available, uninoculated cell culture work should be scheduled before inoculation of cell cultures for virus isolation and the cabinet should be decontaminated with an effective disinfectant between these operations.
8. The facility must be properly maintained. Laboratory safety procedures such as annual certification of biological safety cabinets, procedures for electrical safety, fire control, and chemical and physical hazards control must be observed. Centrifuges with both aerosol containment and temperature control are vital for viral isolation work. Biohazardous and radioactive waste must be

properly handled and disposed of according to local, state, and federal regulations.

9. An ultra-low freezer (preferably with temperature recorder) capable of maintaining −60 °C or lower is essential for storage of virus stocks and clinical specimens until the results of viral isolation are complete.

1.3 PERSONNEL

The key to a quality viral diagnostic service is the personnel recruitment, training, continuing education, and retention of a qualified staff. The availability of laboratorians with the appropriate education and license or certification as required by local and state regulations is decreasing. With the decline in the number of experienced technologists, the need for improved technical training and continuing education programs after employment together with supervisory review of work increases. Another aspect of maintaining quality of service is appropriate workload levels. A range of 500 to 1000 specimens/year/person was found in a survey of virology laboratories (Bartlett, 1985). Excessive workloads are not consistent with quality, particularly with subjective tasks requiring judgment, such as microscopy. The need for adequate time for accurate microscopic screening has been established in cytology where a maximum number of specimens per cytologist per day has been recommended by a number of State Departments of Health.

Lastly, one of the main objectives of any quality assurance program should be to provide feedback to personnel on the errors commonly detected or unusual isolates as an educational process to increase knowledge and therefore, the level of performance in diagnostic testing.

1.4 PROCEDURE MANUAL

An essential tool for the laboratory staff is a complete and current procedure manual available at the bench. The manual should contain a detailed, stepwise procedure for all tasks performed in the laboratory written according to the guidelines established by the National Committee on Clinical Laboratory Standards (NCCLS, 1984) in "Clinical Laboratory Procedure Manual Guideline GP2-A," and as also required by the College of American Pathologists. Each written procedure must include all of the following items:

Title
Principle of test
Specimen collection and handling
Reagent preparation and storage
Calibration
Quality control materials and corrective
    action
Step-by-step procedure
Calculations
Result reporting with procedure for abnormals
Source of error
Limitations
References

The procedure should include which cell lines provide optimal isolation of viruses for each specimen source, instructions on the interpretation of cytopathic effect (CPE) in cell culture, and differential identification procedures for each cell line and specimen source. The manual must be reviewed at least annually by the laboratory supervisor and pathologist, but should be updated promptly as improved techniques become available and are used.

1.5 QUALITY CONTROL

A set of written instructions for the use of the viral diagnostic service should be provided to patient care providers who use the service and should include all of the following:

1. Test purpose and limitations
2. Patient selection
3. Time of specimen collection
4. Specimen collection sites and methods
5. Specimen transport medium
6. Specimen transport and holding instructions
7. Availability of test and hours performed
8. Test turnaround time
9. Result reporting procedures

The criteria for rejecting specimens should also be provided to the physicians as well as an indication of those specimens which will be processed with a disclaimer as to the validity of the test as follows:

1. Unlabelled or mislabelled specimen— recollect
2. Improper transport medium, wood or calcium alginate swab—process if possible with disclaimer or recollect
3. Excessive transport time—process with disclaimer
4. Excessive temperature during transport—recollect frozen or warm specimens and transport at proper temperature, (1 °C to 10 °C)
5. Improper site for test requested (i.e., stool for respiratory syncytial virus, RSV)—call for clarification
6. Quantity of specimen not sufficient (QNS)—call for test priority
7. Inadequate cellular material for immunofluorescence—recollect

Specimen collection information is included in the laboratory manual, but to promote quality assurance, these instructions must be provided to medical and nursing staffs, preferably in pocket handbooks and as part of continuing education programs. No procedure in the laboratory

can compensate for erroneous specimen collection and handling. The quality of the specimen may ultimately determine the quality of the result.

The types of viral specimens required for the diagnosis of viral syndromes are catalogued both in this text and elsewhere (Lennette, 1985; Schmidt and Emmons, 1989). The important principles are collection of viral specimens early in the disease course from sites with high diagnostic significance and prompt transport to the laboratory in viral transport medium (VTM) at 1 °C to 5 °C. Freezing is no longer recommended since enveloped viruses such as cytomegalovirus (CMV) and RSV lose infectivity with a freeze-thaw cycle particularly at −20 °C and without the addition of cryoprotectant, e.g., room temperature storage also may result in a dramatic loss of viral titer. Transport at 1 °C to 5 °C has been shown to result in the best recovery rates (Stagno et al, 1980; Lennette, 1985; Johnson, 1990).

1.6 VIRAL TRANSPORT MEDIA (VTM)

The composition and type of VTM can profoundly affect viral isolation rates (Johnson, 1990). In general, the VTM should:

1. Be buffered, e.g., Hank's balanced salt solution or sucrose phosphate
2. Contain a substance, usually protein, such as gelatin, fetal bovine serum, or bovine serum albumin that stabilizes viruses
3. Employ nontoxic swabs—swabs of calcium alginate or those on wood sticks have been reported to be toxic to herpes simplex virus (HSV) and *Chlamydia trachomatis* although toxicity can also occur with certain lots of cotton and dacron swabs (Crane et al, 1980; Mahony and Chernesky, 1980; Johnson,

1990). Pretesting cotton or dacron-plastic swabs and the prompt removal of the swab at the earliest opportunity are recommended procedures.

4. Contain antibiotics, particularly for specimens from nonsterile body sites or when it is anticipated there will be a long delay before cell inoculation. Gentamicin is stable at room temperature for several months but vancomycin and fungizone degrade rapidly at this temperature making storage temperature for VTM a consideration. Penicillin should not be included in any transport medium used for chlamydial isolation since chlamydiae are susceptible to this antibiotic in vitro.

5. The VTM must be nontoxic to cell culture and noninhibitory to the antigen detection systems.

1.7 SUBMISSION OF SMEARS

The submission of smears for fluorescent (IF) or immunoperoxidase (IP) staining must also be quality controlled. Recent availability of highly specific and sensitive monoclonal antibodies has made direct immunostaining an attractive alternate to viral and chlamydial cultures particularly when delays in transit are common. However, the rapidity of staining techniques is still offset by the subjectivity of test interpretation. Criteria for acceptance of a smear include:

1. The smear should be from an anatomic site appropriate for the pathogen. Smears should be made by rolling the swab across a cleaned glass slide.

2. The amount of blood or purulent discharge in a smear preparation needs to be evaluated for acceptability due to the problems associated with nonspecific staining of background debris.

3. The smear diameter should be limited to

a reasonable size, such as 5 to 15 mm, as larger smears waste reagents, and require excessive time to examine.

4. The smear should contain a representative number of cells on a total smear or per field basis and the appropriate type of cells such as columnar epithelial cells from the endocervix for *C. trachomatis* and basal rather than squamous epithelial cells for the detection of HSV from skin lesions.

5. Tissue imprint smears should include a separate imprint area or slide to permit staining both with an antiviral conjugate and normal serum to be used as an autofluorescence control.

1.8 SUBMISSION OF SERA

Sera submitted for viral serology usually present less of a problem in quality control than do specimens for culture or antigen detection. However, the following should be considered:

1. Sera that are excessively hemolyzed, lipemic, or bacterially contaminated should be rejected.

2. Sera should be heat treated to inactivate complement (56 °C, 30 minutes) depending on the specific test(s) to be performed.

3. Paired sera should be submitted for assay of antiviral IgG levels. A single serum sample may be of little diagnostic value, as it often only reflects immune status rather than identifying an acute illness. The physician and clinical virologist or serologist should discuss the times when blood samples should be drawn during the disease course.

4. IgG assays for TORCH (toxoplasma, rubella, CMV, and HSV) organisms should compare antibody levels in both the infant's and the mother's serum. The virologist and the ordering physician should discuss the timing of specimens

for serodiagnosis or when viral isolation is preferred.

In the event that a specimen is rejected for any of the criteria cited above, the ward or attending physician must be informed. This is preferably done with an oral report followed by a written report, stating the reason(s) for the rejection of the specimen. Again, extenuating circumstances may warrant acceptance of a substandard specimen, but this is done only after consultation among the clinical virologist, pathologists, and attending physician. Acceptance of a substandard specimen should be documented and the attending physician must be informed of the possible ramifications of an improperly submitted specimen. Routine submission of substandard specimens indicates the need for a continuing education program on use of the viral diagnostic service.

1.9 CELL CULTURE

Cell culture represents the current "gold standard" for the isolation and identification of viruses and chlamydiae, but are dependent upon biological systems that may be of variable quality and availability. Testing each lot of commercially obtained cell cultures *before use* is precluded even when cell cultures are delivered weekly because maximal isolation rates are best achieved with low passage cell monolayers—those under 11 days old (Thiele et al, 1987; Fedorko et al, 1989). Therefore, quality control for cell cultures is performed in parallel with cultures inoculated with patient specimens. Because of demonstrated one hundred-fold or greater differences in sensitivity among commercial cell lines for detection of known viral samples, when lower virus yields or isolation rates are suspected the particular cell supplier or cell line can be excluded or replaced.

Procedures for the quality control of

in-house cell culture production are well described elsewhere (Schmidt and Emmons, 1989) and the following paragraphs refer to the quality control of commercially-prepared monolayers.

Weekly records of all cell lines and lots received should be maintained to establish the need for repeating attempts to isolate virus from a specimen, to obtain replacement shipments of cells, and for changing vendors or cell lines used. Because commercially-prepared cell cultures may be back ordered, contaminated, demonstrate exposure to toxic products, damaged or lost in transit, it is not advisable to rely on a single vendor or cell line for any one test performed. The files maintained should record sensitivity, evidence of toxicity to the cells, sterility, and confluence at the time of arrival as shown in Figure 1.1, Cell Line QC. These data should be documented weekly for each lot of cells used along with the corrective action instituted for any problems associated with cell culture.

Cells purchased from a commercial supplier should be certified to be free from mycoplasmal, fungal, and bacterial contamination. All cells should be examined on receipt. Both fed and unfed uninoculated tubes and vials of each lot should be retained for 14 to 35 days as appropriate, to pinpoint possible contamination problems and serve as negative controls for CPE and hemadsorption testing.

Primary monkey kidney (PMK) cells may be contaminated with endogenous simian viruses such as SV-5, SV-40, or even herpesvirus B (herpesvirus simiae). Because herpesvirus simiae infections can be fatal to humans, cells with viral contamination should be autoclaved and then discarded. A negative control (uninoculated culture) from each lot of PMK must be tested with every hemadsorption test on samples inoculated with patient material for possible hemadsorption due to SV-5 contamination (See Figure 1.2).

Figure 1.1 Sample chart for recording cell line quality control.

CELL LINE: _____
VENDOR/SOURCE:

DATE REC'D	PLANT DATE	PASS NO.	pH	APPEARANCE		STERILITY	REFEED MEDIA LOT(S)	POSITIVE CONTROL TITER	DISCARD DATE	COMMENTS	TECH	SUPV	CORRECTIVE ACTION
				INITIAL	FINAL								

Figure 1.2 Sample chart for recording hemadsorption quality control.

HEMADSORPTION QC

HAD DATE	DATE INOC	PMK CELLS		GPRBC LOT DATE REC'D	POSITIVE CONTROLS STOCK#/DATE		NEGATIVE CONTROLS		CONTAMINATION	TECH/SUPV	CORRECTIVE ACTION
		VENDOR #1 LOT/PLANT DATE	VENDOR #2 LOT/PLANT DATE		VENDOR #1	VENDOR #2	VENDOR #1	VENDOR #2			

1.10 MEDIA AND SOLUTIONS

All media, solutions, sera, and glassware should be pretested for sterility, pH, toxicity, endotoxin level, and growth promotion or interference with detection of virus before contact with essential cell cultures or patient specimens (See Figure 1.3). Medium QC is used to document and trace origins of any toxicity, contamination, or interference noted to the lot of medium or additives. If single lots of medium or solutions of volumes are purchased that can be used within the shelf life, the burden of quality control testing can be reduced. Purchase of large lots of fetal bovine serum (FBS) is particularly cost-effective since a twelve-month supply can be obtained with a discount and stored frozen at $-20\,°C$ for more than a year. Each lot of FBS must be carefully checked in parallel with the previous lot for endotoxin level, the ability to support the maintenance of fibroblasts for 35 days, and the absence of any inhibitory activity on the detection of viruses and chlamydiae. Commercial lots of FBS causing inhibition in the isolation of both myxoviruses and *C. trachomatis* have been reported (Schmidt and Emmons, 1989).

The level of antimicrobial agents in all culture media and solutions should be assayed before use. Final concentration greater than 15 μg gentamicin sulfate/mL, 2.5 μg amphotericin B/mL, and 10 μg vancomycin hydrochloride/mL can be toxic to cell culture. Conversely, with low antimicrobial levels, bacteria and fungi from nonsterile patient samples can overgrow the cell cultures and interfere with virus isolation.

Contamination is almost inevitable at some time when using cell cultures either from lapses in technique, nonsterile patient specimens, or indigenous agents in primary cell cultures. However, a good quality control program, adequate staff training, and procedure documentation provide efficient

Figure 1.3 Chart for recording medium/reagent quality control.

MEDIUM QC														TYPE:
DATE PREP/ REC'D	SOURCE	LOT #	EXP. DATE	pH	TOXICITY CELL LINE	STERILITY	POSITIVE CONTROL TITER	ADDITIVES (LOT NO.)	ANTIBIOTIC(S) LEVELS (EXP. DATES)	DISCARD DATE	TECH	SUPV	CORRECTIVE ACTION

Table 1.1 Cell Culture Contamination Checklist: Bacterial, Fungal, and Viral

1. Cells:	One lot or several
	Primary cell line—Simian virus
2. Technique:	Inoculation vs refeeding
3. Media or Serum:	
4. Antibiotic mix:	Nonsterile or incorrect concentration
5. Specimen:	Cross contamination
	Can original specimen be repeated?

From Warford, 1990.

postcrisis intervention and minimize the number of repeated cultures and recollected specimens. Quality control that may aid in minimizing risk of contamination include:

1. *Pretesting* of all media, sera, and solutions used in cell culture maintenance to ensure that it is sterile (Bowdre, 1985).
2. Use of separate biological safety cabinets when possible, one for refeeding uninoculated cell cultures and one for manipulating inoculated cultures. If a single cabinet is used, uninoculated cell culture feeding should precede specimen or stock virus work and the biological safety cabinet should be thoroughly decontaminated with an effective disinfectant after each use, allowing adequate time for contact disinfection.
3. Each technologist should have his or her own set of media and solutions initialed and dated with each use. Small bottles of media are preferable to reduce the number of uses and the risk of contamination.
4. Separate sterile pipets should be used for each specimen. Avoid contamination of safety pipetting devices.
5. Cell culture refeeding and specimen inoculation should be performed first in human fibroblasts, followed by continuous cell lines with primary simian cell lines opened last to avoid possible cross-contamination of cell cultures with endogenous simian viruses.
6. The technologist should initial, date, and keep records of each passage, feeding, and other manipulation performed on each cell tube.

When contamination occurs, the Table 1.1 contamination checklist may be helpful in tracing its source. If the source cannot be identified, it may be cost-effective to destroy all suspicious media and retrain any new personnel in aseptic technique. Personnel employed by cell-culture vendors also experience lapses in quality assurance and ship contaminated cells; however, not all vendors will replace cell cultures in a timely fashion. A partnership with another virology laboratory is the only recourse against significant losses due to a limited cell supply.

1.11 WATER QUALITY

Use of water with the highest possible quality, monitored daily for pH and electrical resistance and weekly for bacterial colony counts is advisable because early cell degeneration can result when low levels of endotoxin and other substances are present. A resistivity reading of 10 megaohms and bacterial count of fewer than 10 colonies per milliliter should be maintained. Pyrogen-free water for cell culture is best obtained by an initial deionization step yielding "Type I water" (NCCLS, 1985) followed by a second ultra-filtration step. Storage of purified water for cell culture media should be avoided since growth of pseudomonads can result. Filtration of water with 0.2 micron filters is most desirable with the filter placed close to the final outlet.

1.12 REFERENCE VIRUS STOCKS

Reference virus stock cultures may be obtained from American Type Culture Collection (ATCC, Rockville, MD) or other reference institutions such as the Centers for Disease Control or the National Institutes of Health. These reference stocks or pooled patients' isolates of the same virus may serve several purposes in the clinical virology laboratory quality control program:

1. The reference stocks may be used to demonstrate typical CPE in susceptible host cell lines. These stocks may be used to train medical technology and medical students. Fixed and stained infected cell cultures (Gurtler et al, 1982) serve as a ready supply of material to demonstrate typical CPE while reducing biohazard risk to students.

2. Cell culture tubes and shell vials infected with reference stocks can be used as positive controls for all viral methods—direct specimen IF, immunoperoxidase (IP), enzyme immunoassay (EIA), shell vial pre-CPE antigen detection, hemadsorption, and interference assays. These cell cultures, infected with reference virus stocks, are superior to fixed control slides which only provide a check on the final staining technique and reagents and are no assurance that the cells, centrifugation, incubation time and temperature, and fixation are adequate.

3. Using titered reference virus and chlamydia stocks, especially low titer posi-

tive controls, and quantifying the number of foci or inclusions can help detect errors that produce a partial loss of culture sensitivity before the isolation rate reaches zero. An example of this is shown in the Chlamydia QC (Figure 1.4), where a decrease in the number of inclusions was evident and upon trouble-shooting with the Chlamydia Culture Checklist (Table 1.2) the problem was traced to a lot of commercially-obtained cycloheximide that was of suboptimal concentration.

Reference virus stocks can be stored indefinitely at $-70\,°C$ in a solution containing a cryoprotectant, 10% dimethyl sulfoxide and FBS (Schmidt and Emmons, 1989) with the exception of CMV and varicella zoster virus (VZV). Alternatively, CMV and VZV tube cultures with more than 2+ CPE can be scraped and the fluid-cell mixture stored for at least a week at $-1\,°C$ to $4\,°C$ and used to inoculate cell cultures as infectious positive controls.

1.13 REAGENTS, STAINS, ANTISERA, AND KITS

The same basic principles apply to all shipments of reagents, stains, antisera, and kits received:

1. All shipments should be entered into an inventory log with the lot number, date received, transport temperature and condition, and expiration date noted.

2. Reagents transported or stored at the wrong temperature or condition must be discarded.

3. Each kit and container must be initialed and dated upon use.

4. New lots must be tested in parallel with previously tested lots with both patient specimens as well as positive and negative controls. Positive and negative controls are also required for each run in

Figure 1.4 Sample chart for recording *Chlamydia* culture quality controls.

CHLAMYDIA CULTURE QC

DATE INOC	DATE STAIN	MCCOY CELL LOT/PLANT DATE	CYCLOHEXIMIDE LOT/EXP DATE	FA LOT/SOURCE	# INCLUSIONS/HPF (AVG 10 FIELDS)		NEGATIVE CONTROL	TECH	SUPV	CORRECTIVE ACTION
					HIGH POS	LOW POS				
8/9	8/11	8/7	#206 10/90	#V26 SYVA	5.1	2.2	0	MJ	ALW	NONE
8/11	8/13	8/7	#206 10/90	#V26 SYVA	4.5	1.7	0	PBV	ALW	NONE
8/12	8/14	8/7	#206 10/90	#V26 SYVA	1.6	0	0	YKS	ALW	REPEAT

most cases. A single parallel run may not be adequate to detect subtle lot-to-lot variations, particularly an increase in nonspecific reactions with polyclonal antisera. In California, both high and low titer positive controls are required by the Department of Health. It is a good procedure to include heterologous as well as homologous positive viral controls in testing new lots of antisera for cross reactions among related viruses such as the herpesvirus group.

5. Antisera must be titered for determination of optimal dilutions for neutralization, IF, and IP staining with the microscope(s) to be used. Box (checkerboard) titration of primary and secondary antisera used in IF or IP staining provides the most satisfactory results for microscopists but is practical only when the lot size is sufficient for more than a month.

1.14 INSTRUMENTS

Clinical laboratory instruments should be subjected to routine preventive maintenance and also surveyed on a regular basis for satisfactory performance. Consistent test results cannot be expected with unreliable equipment. Furthermore, improperly maintained and defective instruments can pose a threat to personnel safety.

The minimum preventive maintenance performed should follow the manufacturer's recommendations. Note that these recommendations assume normal usage and applications for the instrument. Excessive use or special applications may warrant additional preventive maintenance. This work can be performed by in-house instrument specialists. Logs should be kept of the routine preventive maintenance, and service manuals should be readily available to perform routine trouble-shooting. Work beyond the scope of in-house capabilities can be performed under manufacturer's service

Table 1.2 Chlamydia Culture Checklist

1. Confluent McCoy cell monolayer: young cells (24 to 72 hours old)
2. Pretested medium and serum (*no* penicillin)
3. Incubator CO_2 and temperature monitored daily
4. Cycloheximide: optimal concentration checked with inclusion counts of pooled patient control specimen and expiration date checked
5. Centrifugation speed: $3000 \times g$
6. Centrifugation temperature monitored at $33\,°C$ to $36\,°C$
7. Incubation period: 48 hours (± 6 hours)
8. Fixation and staining technique
9. Inclusion counts of controls checked:
   Positive: ATCC strain and pooled patient specimen
   Negative: Viral transport medium with antibiotics cell control (no cycloheximide)
10. Subpassage all tissue and toxic specimens

From Warford, 1990.

contracts. These service contracts should specify both the cost and the response time for service calls. Test turnaround time can be negatively impacted by excessive instrument downtime awaiting service.

Equipment logs should be maintained with the following information including the date work is done and the technologist's initials:

1. Instrument name, serial number, and date put in use
2. Routine performance check procedures
3. Acceptable performance ranges
4. Routine preventive maintenance
5. Instrument performance failure including specific details of steps preceding failure, type of problems and lot numbers and specimen numbers involved in occurrence
6. Date and time of service request and response
7. Corrective action taken to minimize future downtime and repeat testing of the equipment to show proper operation: initialed by supervisor

Each laboratory must meet the requirements of licensing or certifying agencies and comply with local, state, and federal requirements for equipment checks and maintenance, some of which have been detailed recently (August et al, 1990). The following are some general recommendations for routine laboratory maintenance and performance checks on instruments commonly found in the clinical virology laboratory:

Autoclaves: Daily temperature check and recording thermometer graph; monthly spore strip testing.

Biological safety cabinets: Daily cleaning of ultraviolet lamp and check of air intake flow of 50 ft/min or more across face opening; decontamination of work surfaces before and after each use; annual or semi-annual checks for air velocity and filter integrity; paraldehyde decontamination before filter replacement.

Centrifuges: Quarterly speed calibrations with tachometer; annual inspection of motor and drive system; weekly decontamination.

Incubators: Daily temperature, CO_2, and humidity checks; weekly decontamination of interior.

Microscopes: Daily cleaning of objectives and stage; log of lamp usage with mercury vapor bulbs; annual overhaul by service contractor.

Microdiluters and pipettors: Gravimetric or spectrophotometric check before first use and monthly thereafter; annual overhaul.

pH meters: Multiple point check with reference buffers before each use.

Refrigerators and freezers: Daily temperature check; annual check of compressor and refrigerant levels; emergency power and alarm systems preferable.

Rotators: Daily rpm check.

Spectrophotometers: Absorbance and linearity check with each run.

Water baths and heat blocks: Daily temperature check; weekly decontamination.

1.15 STATISTICS AND BACKUP PROCEDURES

With the current demand for rapid turn-around time and cost-effective procedures, antigen detection methods have replaced virus and chlamydia isolation in many clinical laboratories. However, it is essential to perform culture isolation as a backup procedure to antigen detection methods, at least during the initial evaluation period of new rapid methods for the following reasons:

1. Cell culture remains the more sensitive diagnostic method assuming appropriate specimen collection and transport.
2. Cell culture remains the more specific diagnostic method, as compared to antigen detection, particularly in low prevalence populations.
3. Culture provides an accurate standard for judging subjective methods such as antigen detection by direct IF.
4. More than one agent may be present in a specimen and clinically significant, such

as RSV and *C. trachomatis* isolated from a respiratory specimen.
5. Cell culture is a more all encompassing screening method when multiple etiologic agents are possible such as in viral pneumonia due to infection with influenza A and B viruses, parainfluenza viruses 1–4, RSV, adenoviruses, CMV, and other agents.
6. Lastly, cell culture, when performed with antigen detection, provides a safety net for errors or problems in rapid methods just as antigen detection may serve as a backup for those specimens which result in cytotoxic contamination or rapid degeneration.

Statistical methods can be useful in virology although this field has been subjective and qualitative rather than quantitative in the past. Traditional Levey Jennings quality control methods are now applicable for automated and quantitative optical density readings from EIA equipment. (Stewart and Koepke, 1987). EIA runs are considered out-of-control according to the "CAP modified" approach of Haven, et al, 1980, when

1. Patient results appear incorrect, regardless of the control results; an example of this is an unusual number of positive or borderline results on a single run.
2. Any single control value deviates more than 3 standard deviations (SD) from the mean of the control values.
3. One control value is greater than 2 SD from the mean on two consecutive runs.
4. Two controls are used and both control values are greater than 2 SD from the mean.

Isolation rates tallied by month, cell line, specimen type, and even technologist provide valuable data when compared with prior rates for determining sources of error. Some of the common errors encountered in clinical virology are listed in Table 1.3. A

Table 1.3 Sources of Error in Clinical Virology

1. Clerical: Mislabeling, data transcription errors, mismatches
2. Collection: Poor technique or timing, inadequate cells, wrong swab
3. Specimen: Inappropriate site, QNS
4. Media and reagents: Wrong type or concentration, expired or contaminated
5. Cells: Insensitive due to age, passage level, or selection inappropriate for agent
6. Water: Contaminated, toxic, acidic or alkaline, high particle count
7. Equipment: Malfunction or nonfunction, uncalibrated
8. Volume: Specimen or reagent, pipetting error
9. Time: Incorrect incubation periods, expired storage limits
10. Temperature: Excessive transport or incubation temperatures
11. Technique: Poor test performance, i.e., asepsis during refeeding, inadequate washing on ELISA, or IF staining
12. Interpretation: Microscopic or macroscopic, i.e., CPE, specific fluorescence, EIA color end point
13. Results reporting: Provision of inadequate information on test limitations or need for recollection of specimens or failure to notify patient care providers in a timely manner of critical findings

decrease in the monthly CMV isolation rate from bloods but not urines might implicate the leukocyte separation technique. A dramatic rise in the isolation rate of a new technologist by IF might suggest the need for further training. Decreases in isolation rates associated with cell lines and reagents may indicate a need for improved quality control testing and corrective action.

1.16 PROFICIENCY TESTING

Formal proficiency testing is now mandatory for all clinical virology laboratories as part of the Clinical Laboratory Improvements Amendments (CLIA, 1988). Quarterly sets of five challenges, such as provided by CAP, must be tested in each laboratory annually.

Sanctions, including suspension of certification for Medicare and Medicaid reimbursement, may be imposed on laboratories scoring less than 80% in two of three successive quarterly sets of samples.

Supervisors may also implement internal proficiency testing on a more frequent basis to permit all members of the staff to gain experience with unknown samples. However, both of these types of controls, CAP samples and internal unknowns, are likely to be accorded favored treatments with the staff consulting and testing in duplicate. Only "blind controls" indistinguishable from patient specimens serve to establish the accuracy and reproducibility of the laboratory's routine performance.

Regardless of the type of proficiency tests employed, the benefit of these samples is to check the procedures, reagents, equipment, and personnel of the laboratory. Proficiency test failures may provide opportunities for improvement if the appropriate corrective action is taken. It is useful to document errors as follows:

1. Type of failure
2. Time, date, and day of failure
3. Name(s) of technologists performing test(s)
4. Reagent lot(s) associated with failure
5. Possible reason(s) for failure
6. Recommended changes to technique, training, procedure, reagents, equipment, and supervisory review process

7. Inservice and/or discussion date(s) with pathologist, virologist, and technologist(s)

Quality control or proficiency test failures which are detected and discussed can serve to educate the staff. These errors can also improve the overall performance of the laboratory when errant procedures are corrected, preventing diagnostic test errors and providing quality assurance for the patient.

REFERENCES

August, M.J., Hindler, J.A., Huber, T.W., and Sewell, D.L. 1990. A.S. Weissfeld, coordinated. Cumitech 3A: Quality control and Quality Assurance Practices in Clinical Microbiology. Washington, D.C.: American Society for Microbiology.

Bartlett, R.C. 1985. Quality control in clinical microbiology. In E.H. Lennette, A. Balows, W.J. Hausler, Jr., and H.J. Shadomy (eds.), Manual of Clinical Microbiology. 4th ed, Washington, D.C.: American Society for Microbiology, pp. 14–23.

Blank, C.H. and Schrunk, R.L. 1985. Human resources factor. In J.M. Miller and B.B. Wentworth (eds.), Methods or Quality Control in Diagnostic Microbiology. Washington, D.C.: American Public Health Association, pp. 263–269.

Bowdre, J.H. 1985. Viral isolation and antigen detection. In J.M. Miller and B.B. Wentworth (eds.), Methods for Quality Control in Diagnostic Microbiology. Washington, D.C.: American Public Health Association, pp. 211–233.

Crane, L.R., Gutterman, P.A., Chapel, T., and Lerner, M. 1980. Incubation of swab materials with herpes simplex virus. J. Infect. Dis. 141:531.

Favero, M.S. 1985. Sterilization, disinfection, and antisepsis in the hospital. In E.H. Lennette, A. Balows, W.J. Hausler, Jr. and H.J. Shadomy (eds), Manual of Clinical Microbiology, 4th Ed, Washington, D.C.: American Society for Microbiology, pp. 129–137.

Fedorko, D.P., Ilstrup, D.M., and Smith, T.F. 1989. Effect of age of shell vial monolayers on detection of cytomegalovirus from urine specimens. J. Clin. Microbiol. 27:2107–2109.

Gurtler, J., Ballew, H., and Smith, T. 1982. Cell culture medium for cytopathic effects in cell cultures. Lab. Med. 13:244–245.

Haven, G.T., Lawson, N.S., and Ross, J.W. 1980. Quality control outline. Pathologist 34:619–624.

Johnson, F.B. 1990. Transport of viral specimens. Clin. Micro. Rev. 3(2):120–131.

Lennette, D.A. 1985. Collection and preparation of specimens for virological examination. In E.H. Lennette, A. Balows, W.J. Hausler, Jr., and H.J. Shadomy (eds.), Manual of Clinical Microbiology, 4th ed. Washington, D.C.: American Society for Microbiology, p. 687–693.

Mahony, J.B. and Chernesky, M.A. 1985. Effect of swab type and storage temperature on the isolation of *Chlamydia trachomatis* from clinical specimens. J. Clin. Micro. 22:865–867.

National Committee for Clinical Laboratory Standards, 1985. Approved standard (C3-P2): Specifications for reagent water used in the clinical laboratory. Villanova, PA.

Richardson, H.J. and Barkley, W.E. (eds.) 1984. Biosafety in Microbiological and Biomedical Laboratories. Washington, D.C.: Centers for Disease Control and National Institutes of Health. U.S. Government Printing Office, pp. 10–23.

Schmidt, N.J. and Emmons, R.W. 1989. Diagnostic Procedures for Viral, Rickettsial, and Chlamydial Infections. 6th ed. Washington, D.C.: American Public Heart Association.

Stagno, S., Pass, R.F., Reynolds, D.W., Moore, M.A., Nahmias, A.J., and Alford, C.A. 1980. Comparative study of diagnostic procedures for congenital cytomegalovirus infection. Pediatrics 65:251–257.

Stewart, C.E. and Koepke, J.A. 1987. Basic Quality Assurance Practices for Clinical Laboratories. Philadelphia, PA: J.B. Lippincott.

Thiele, G.M., Bicak, M.S., Younk, A., Kinsey, J., White, R.J., and Purtilo, D.J. 1987. Rapid detection of cytomegalovirus by tissue culture, centrifugation, and immunofluorescence with monoclonal antibody to an early antigen. J. Virol. Meth. 16:327–338.

Warford, A.L. 1990. Troubleshooting in the clinical virology laboratory. Clin. Micro. Newsletter. 12:41–44.

Warren, E. 1981. Laboratory safety. In J.A. Washington (ed.), Laboratory Procedures in Clinical Microbiology. N.Y.: Springer-Verlag, pp. 729–744.

2

Specimen Requirements
Selection, Collection, Transport, and Processing

Thomas F. Smith

2.1 INTRODUCTION

Historical perspective suggests that the diagnosis of most viral infections has been based on clinical grounds rather than by specific laboratory tests. The extensive time lapse between submission of specimens and the availability of results and the lack of effective forms of treatment or prevention of viral diseases have been major reasons for this situation. However, this trend has changed dramatically, especially in the last few years. For example, herpes simplex virus (HSV), once thought to produce only benign lip lesions, is now recognized to be a major cause of sexually transmitted disease that can be transmitted to newborns (Stagno and Whitley, 1985; Growdon et al, 1987). Further, our understanding of the medical importance of HSV has increased due to its ability to produce a wide spectrum of clinical manifestations, including keratitis, genital infections, gingivostomatitis, encephalitis, and oftentimes disseminated disease in immunosuppressed individuals (Kohl, 1988; Bergmann et al, 1990).

Because of the availability of specific monoclonal reagents, together with creative commercial innovations, rapid laboratory diagnosis of many viral infections [adenovirus, HSV, respiratory syncytial virus (RSV), rotavirus] can be achieved by microbiology laboratories that lack preexisting virologic capabilities. These assays, generally based on the immunologic detection of viral antigens are best known for the diagnosis of HSV infections. For example, HSV antigens could be detected 48 hours postinoculation using a combination of "spin amplification" of the virus in a shell viral assay and subsequent antigen detection using an enzyme immunoassay (Michalski et al, 1986). More recently, this virus has been detected directly in clinical specimens using four hour (DuPont Herpchek; Dascal et al, 1989) or 15 minute (Kodak Surecell; Dorian et al, 1990) procedures. It is important to recognize, however, that these are not exclusive (stand alone) systems for detecting HSV, i.e., conventional cell cultures are also used. Nevertheless, 80% to 95% of HSV infections can now be diagnosed in the laboratory and reported to the clinical service within the same day the specimen is submitted. Once the laboratory diagnosis of

HSV infection has been established, specific antiviral drug therapy can be instituted in many clinical situations even at the primary level of medical care (Whitley et al, 1986; Gäbel et al, 1988; Krusinski, 1988; Stone and Whittington, 1990). This trend of rapid diagnosis followed by institution of antiviral therapy has followed for other viral infections as well, e.g., varicella-zoster virus (VZV)-acyclovir, RSV-ribavirin, and cytomegalovirus (CMV)-ganciclovir.

The success rate of the diagnostic laboratory is dependent on frequent communication between the laboratory and the physician. This is especially true in this era of evolving technology. No technique, however, regardless of how rapid, is useful unless the quality of the specimen is adequate regarding the source, method of transport, and means of processing. Finally, the ever increasing cost of operating a clinical laboratory obligates the laboratory director to suggest to clinicians the selection of the best specimen(s) that would yield a definitive laboratory diagnosis of a viral infection.

2.2 SPECIMEN SELECTION

Comprehensive tables listing the selection of viral specimens based on clinical infections and the suspected viral agents have been published (Smith, 1984; Greenberg and Krilov, 1986; Lennette et al, 1991). A specific guide for most clinical situations is presented (Table 2.1).

Several viruses may cause respiratory tract infections. In almost all cases, a throat or nasal swab/washing should provide a suitable specimen. With other viral infections, such as those involving the central nervous system (CNS) or those associated with congenital disease, several types of specimens should be submitted.

2.3 SPECIMEN COLLECTION

2.3.1 Timing

Specimens should be collected early in the acute phase of infection. However, the duration of viral shedding depends on the type of virus and systemic involvement, as well as other factors. For example, the duration of RSV shedding is usually 3 to 7 days, but with a range of 1 to 36 days (Hall et al, 1976a;b). Further, viral excretion was of longer duration in patients with lower respiratory tract disease than in those with clinical manifestations limited to the upper respiratory tract.

The presence of a virus in a specimen may depend on the source of the sample. Generally, VZV can be recovered from lesions for up to seven days after the initial vesicles appear, but from blood only during the late incubation period or days 1 to 4 of the acute illness (Feldman and Epp, 1979).

Interpretation of the significance of enterovirus isolates from stool specimens is confounded by the prolonged shedding (6 to 8 weeks) of these viruses, especially in children (Modlin, 1986). Since virus is commonly present for a shorter time in the oropharynx, and often yields virus from specimens collected during the first 5 to 7 days of illness, the interpretation of a positive virus isolation is less ambiguous.

The immunocompetence of a patient with a viral illness has a significant effect on the time and duration of virus excretion. For example, high concentrations of HSV ($>10^4$ plaque forming units) are detected in specimens from lesions of immunocompromised patients for more than three weeks, while the mean duration of HSV shedding from immunocompetent men and women with genital infection is 11.4 days (Corey et al, 1983). Echoviruses were recovered from the cerebrospinal fluid (CSF) of children with agammaglobulinemia for periods varying from two months to three years (Wilfert et al, 1977). Varicella-zoster virus could be

Table 2.1 Specimen Information for Diagnostic Virology Services

General Disease Categories	Virus	Serology	Specimens to be Submitted for Culture	Container or transport device — Culturette® Swab	Sterile Screw-capped Container	Vacutainer (Serology)	Volume	Other Considerations
Respiratory pharyngitis, croup, bronchitis, pneumonia	Cytomegalovirus	Yes[a]	Urine	X[b]			5–10 mL urine	
			Throat swab		X			
			Sputum		X	X	5 mL sterile clotted blood for serology	
			Blood (5 mL of heparinized blood in green capped tube)					
			Bronchoalveolar lavage (BAL)		X			
	Enterovirus	No	Throat swab	X				
	Herpes simplex virus	Yes	Throat swab	X		X	5 mL sterile clotted blood for serology	
			Sputum		X			
	Influenza virus	Yes	Throat swab	X		X	5 mL sterile clotted blood for serology	
			Sputum		X			
			BAL		X			
	Mumps virus	Yes	Throat swab (Stensen's duct)	X		X	5–10 mL urine; 5 mL sterile clotted blood for serology	
			Urine		X			
	Parainfluenza virus	No	Throat swab	X				
			BAL		X			
	Respiratory syncytial virus	Yes	Nasopharyngeal wash or aspirate; nasopharyngeal swab (Calgi-swab®)		X	X	5 mL sterile clotted blood for serology	
	Adenovirus	Yes	Throat swab	X		X	5 mL sterile clotted blood for serology	
Exanthem (maculopapular)	Adenovirus	Yes	Throat swab	X		X	5 mL sterile clotted blood for serology	
	Enterovirus	No	Throat swab	X				
			Rectal swab	X				

Table 2.1 (*continued*)

General Disease Categories	Virus	Serology	Specimens to be Submitted for Culture	Culturette® Swab	Sterile Screw-capped Container	Vacutainer (Serology)	Volume	Other Considerations
Exanthem (maculopapular) continued	Rubella virus	Yes				X	5 mL sterile clotted blood for serology	Single serum only required for determination of immune status to rubella virus. For serologic evidence of acute phase infections (rubella or other agents listed), 2 or more serum specimens are needed taken 2–3 weeks apart—**unless IgM** class antibody can be demonstrated with the first serum specimen.
	Measles (rubeola)	Yes				X		
	Less frequently:							
	Parainfluenza virus	No	Throat swab	X				
	Respiratory syncytial virus	Yes	Nasopharyngeal wash or aspirate; nasopharyngeal swab	X (Calgi-swab)®	X	X	5 mL sterile clotted blood for serology	
Exanthem (vesicular)	Herpes simplex virus	Yes	Vesicle swab; Vesicle scrapings on slide for HSV direct FA test (See Other Considerations)	X		X	5 mL sterile clotted blood for serology	Vesicle scrapings for Direct FA test: Place small drop of saline in each of 2 separate areas of a glass slide 5 mm to 10 mm apart. Transfer skin scrapings from a scalpel blade to the saline and spread the cells over a small circular area (5 mm to 10 mm in diameter). Dry the slide at room temperature and transport the slides to the laboratory in a cardboard mailer.
	Varicella-zoster virus	Yes	Vesicle swab	X		X	5 mL sterile clotted blood for serology	

Table 2.1 (*continued*)

General Disease Categories	Virus	Serology	Specimens to be Submitted for Culture	Culturette® Swab	Sterile Screw-capped Container	Vacutainer (Serology)	Volume	Other Considerations
Central nervous system (aseptic meningitis and encephalitis)	LaCrosse (California virus)	Yes				X	5 mL sterile clotted blood for serology	Viral isolation not attempted.
	Enterovirus	No	CSF		X		1–2 mL	
			Throat swab	X				
			Rectal swab	X				
			Serum (infants)		X		1–2 mL	
			Brain biopsy, if available		X			
	Herpes simplex virus	Yes (see Other Considerations)	CSF		X (1–2 mL CSF for serology)	X	5 mL sterile clotted blood for serology	CSF may contain antibodies to HSV indicating CNS infection due to this virus.
			Vesicle	X				
			Brain biopsy					
	Measles (rubeola) virus	Yes				X	5 mL sterile clotted blood for serology	Viral isolation generally not successful.
	St. Louis encephalitis	Yes				X	5 mL sterile clotted blood for serology	Viral isolation not attempted.
	Western equine encephalitis	Yes				X	5 mL sterile clotted blood for serology	Viral isolation not attempted.
	Rabies	Yes				X	5 mL sterile clotted blood for serology	Antibody to rabies virus determined by the Centers for Disease Control with sera from individuals exposed to a possible rabid animal. CDC will not perform rabies antibody titers in sera from persons administered human diploid vaccine. CDC will test sera from individuals who have received duck embryo vaccine and/or from vaccinees who are immunosuppressed.
	Mumps	Yes	CSF		X			
			Urine		X		5–10 mL urine	
			Throat swab (Stensen's duct)	X		X	5 mL sterile clotted blood for serology	

Table 2.1 (*continued*)

General Disease Categories	Virus	Serology	Specimens to be Submitted for Culture	Culturette® Swab	Sterile Screw-capped Container	Vacutainer (Serology)	Volume	Other Considerations
Infectious mononucleosis	Epstein-Barr virus	Yes				X	5 mL sterile clotted blood for serology	Viral isolation not attempted. Immunofluorescence test for antibodies to EBV indicated in those patients with heterophile-negative determinations.
	Cytomegalovirus	Yes	Urine / Throat swab	X	X	X	5–10 mL urine / 5 mL sterile clotted blood for serology	
	Hepatitis viruses, HAV, HBV, HCV	Antigen and/or antibody test				X	5 mL sterile clotted blood for antigen or antibody determinations	
Gastroenteritis	Rotavirus	Antigen detection	Stool (5 g)		X			Rotavirus antigen in stool specimens detected by enzyme immunoassay procedure.
	Norwalk-like agents	Yes				X	5 mL sterile clotted blood for serology	Assays for antibodies to Norwalk-like agents are performed in only a few research laboratories.
	Adenovirus	Yes	Stool (5 g) or rectal swab			X	5 mL sterile clotted blood for serology	Few "high numbered" serotypes of adenoviruses have been associated with gastroenteritis.
Genital infections	HSV	Yes - (See Other Considerations)	Vesicle swab	X		X	5 mL sterile clotted blood for serology	Serology may be useful only for primary genital infections due to HSV. Specimens for culture are recommended.
			Vesicle scrapings on slide for HSV direct FA test (See Other Considerations)					Vesicle scrapings for direct FA test: Place small drop of saline in each of 2 separate areas of a glass slide 5 mm to 10 mm apart. Transfer skin scrapings from a scalpel blade to the saline and spread the cells over a small circular area (5 mm to 10 mm in diameter). Dry the slide at room temperature and transport the slides to the laboratory in a cardboard mailer.

Table 2.1 (*continued*)

General Disease Categories	Virus	Serology	Specimens to be Submitted for Culture	Container or transport device			Volume	Other Considerations
				Culturette® Swab	Sterile Screw-capped Container	Vacutainer (Serology)		
Genital infections continued	Human papilloma-virus (HPV)		Endocervical swab biopsy tissue paraffin-embedded tissue	Specific collection kits required				HPV can be detected in Papanicolaou-stained cells; alternatively, certain biotypes (6,11; 16,18,31,33,35) can be detected by commercial kits using nucleic acid detection methods.
Congenital	Cytomegalo-virus	Yes	Urine Throat swab	X	X	X	5–10 mL urine 5 mL sterile clotted blood for serology	
	Herpes simplex virus	Yes	Vesicle swab Throat swab	X X		X	5 mL sterile clotted blood for serology	
	Rubella virus	Yes				X	5 mL sterile clotted blood for serology	Viral isolation not attempted. IgM antibody to rubella virus should be assayed using serum from babies up to 6 months of age. IgG class antibody should not be determined since its presence reflects only passive transfer from the mother.
	Enterovirus	No	Throat swab Rectal swab Serum (see other considerations)	X X	X			Cord blood is a useful and productive specimen for recovering enteroviruses in congenital infections.

Table 2.1 (*continued*)

General Disease Categories	Virus	Serology[a]	Specimens to be Submitted for Culture	Container or transport device[b]			Volume	Other Considerations
				Culturette® Swab	Sterile Screw-capped Container	Vacutainer (Serology)		
Ocular Infections	HSV	Yes (See Other Considerations)	Swab (eye) Vesicle swab	X X		X	5 mL sterile clotted blood for serology	Serology may be useful only for the primary ocular infections due to HSV. Specimens for culture are recommended. Vesicle scrapings for direct FA test: Place small drop of saline in each of 2 separate areas of a glass slide 5 mm to 10 mm apart. Transfer skin scrapings from a scalpel blade to the saline and spread the cells over a small circular area (5 mm to 10 mm in diameter). Dry the slide at room temperature and transport the slides to the laboratory in a cardboard mailer.
	Adenovirus	Yes	Swab (eye)	X		X	5 mL sterile clotted blood for serology	

[a] Availability of serology for laboratory diagnosis of infection.

[b] "X" indicates the type of transport device to be used for transport of the specimen submitted to the laboratory.

recovered from the buffy coat of blood specimens from patients with malignant disease for at least eight days after the onset of cutaneous lesions, but could not be isolated from patients with typical varicella and no underlying malignancy (Myers, 1979). New lesions due to VZV infection in immunologically normal children develop over a four-day period while new lesions form in most immunocompromised children for more than five days. Immunocompromised adults with zoster shed virus for longer (7.0 days) than otherwise normal adults (5.3 days). In addition, zoster is much more likely to disseminate cutaneously in immunocompromised than in immunocompetent hosts (Balfour, 1988).

Some viruses, such as CMV, rubella, adeno-, and enteroviruses can be excreted from various sites in the asymptomatic individual for months or years. Therefore, the timing of specimen collection is critical with regard to being able to associate the virus with the current disease. For example, a CMV-positive urine specimen collected during the first three weeks of life from an infant with congenital disease is strong evidence supporting the viral etiology of the congenital anomalies, presumably since the agent was acquired in vitro and active viral infection and replication of the virus is present. Conversely, isolation of CMV from the same source after three weeks would not discriminate between congenitally or postnatally acquired CMV infection (Spector, 1983). Thus, CMV infection in an infant obtained as a result of passage of the fetus through the birth canal (postnatal infection) would likely require at least three weeks for viral replication to reach detectable levels in the urine.

Adeno- and enteroviruses can be excreted in stool specimens for several days to weeks, even though the ability of these viruses to cause gastroenteritis is debated by some (Johansson et al, 1990). Nevertheless, both viruses are recognized pathogens causing acute upper respiratory tract disease (Kepfer et al, 1974). The isolation of either agent from a stool specimen, but not from other sources (respiratory tract) would not have etiologic significance. Importantly, disseminated adenovirus infections can occur in immunocompromised hosts with the virus being recovered from the same body sites as CMV and even cytopathic effects in cell cultures can be similar to those of this virus (Landry et al, 1987). Interestingly, both enteroviruses and adenoviruses were recovered over a period of several weeks from a patient with Bare Lymphocyte Syndrome, a form of partial combined immunodeficiency (Arens et al, 1987). Thus, persistent or chronic excretion of these viruses may merely reflect the compromised ability of the host to eliminate viral replication, but, on the other hand, adeno- and enteroviruses may produce disease in these patients with limited antiviral defenses.

2.3.2 Source

2.3.2a Throat, Nasopharyngeal Swab, Nasal Washing

Throat or nasal washing may be more productive for viral isolation than throat or nasal swabs, but few comparisons have been reported. Generally, nasopharyngeal aspirates are more productive than nasopharyngeal swabs for the diagnosis of RSV (Ahluwalia et al, 1987). The convenience of using a swab by medical personnel and the willingness of the patient to allow collection of this specimen (compared with washings) are important factors in this choice. Swabs are considered superior to nasal wash because several problems are associated with the latter. Nasal wash specimens submitted for fluorescent antibody detection of viral antigen often contain debris, such as mucus, squamous cells, leukocytes, and erythrocytes. Also, the number of cells obtained by nasal wash was smaller than that obtained by nasal and oropharyngeal swabs combined (Kim et al, 1983; Blumenfeld et al, 1984). Frayha et al (1989) demonstrated that nasopharyngeal swabs and nasopharyngeal aspirates were equally effective for the diagnosing of RSV, parainfluenza virus, and influenza virus in children.

2.3.2.b Sputum

Recovery of a virus from this source does not necessarily reflect lower respiratory tract infection, but may be due to "contamination" of the specimen with the agent present in the throat. Alternatively, the recovery or cytologic detection of a virus from a transtracheal aspirate or transbronchial biopsy indicates lower respiratory tract involvement (Blumenfeld et al, 1984). Although Kimball et al (1983) obtained an overall isolation rate of 20% from patients diagnosed as having radiologically confirmed pneumonia, they suggested that not all of the viruses recovered from this source may be of clinical importance as the etiologic agents of lower respiratory tract disease in their study population. Thus, the type and frequency of viral isolates was influenza virus (H_3N_2), six; RSV, two; HSV, nine; and rhinovirus, three, causing the investigators to conclude that only the influenza and RSV (total, 8%) caused lower respiratory tract disease in their patients. This conclusion was based on the known recognition of these viruses as lower respiratory tract pathogens and the presence of radiographic evidence of pneumonia in these eight patients, but not in the others from whom other viruses (HSV, rhinovirus) were detected. On this basis, they speculated that sputum specimens may be of particular value in the laboratory diagnosis of lower respiratory tract disease due to viruses. Unfortunately, because of the severity of illness in their study population, these investigators were unable to obtain throat washings for comparison with the sputum specimens.

Interestingly, HSV and CMV are capable of growth in human alveolar macrophages, thus, bronchopulmonary lavage specimens yielding these cells may be useful as a relatively simple technique for laboratory diagnosis of lower respiratory tract infection due to these viruses without resorting to an open-lung biopsy procedure (Kahn and Jones, 1988).

2.3.2.c Bronchoalveolar Lavage

Bronchoscopy, transbronchial biopsy, and bronchoalveolar lavage (BAL) have been useful procedures for obtaining specimens, especially for the diagnosis of *Pneumocystis carinii* infections. Of these, BAL has provided an alternative to open lung biopsy for the rapid diagnosis of viral infections, especially CMV (Martin and Smith, 1986). With this procedure, the entire tracheobronchial tree is inspected with a fiber-optic bronchoscope. After removing the scope, cleaning it of secretions, and reinserting it into the involved segment of the lung, saline is instilled and then removed by vacuum to obtain the lavage specimen of suspended cells. The cells are washed and inoculated onto cell cultures, or stained and assayed directly for virus using immunologic or nucleic acid detection methods. In one year at the Mayo Clinic, 80 viral isolates were obtained from BAL specimens. Seventy-one (89%) were CMV; four, parainfluenza virus; three, influenza virus; and two, enterovirus. Woods et al (1990) indicated that the presence of alveolar lymphopenia in a patient and laboratory diagnosis of CMV using antibodies to early antigens of the virus was highly suggestive of disease although correlation of these results with positive cytologic findings (inclusion bodies) would increase the specificity of tests regarding CMV etiology.

2.3.2.d Rectal Swabs and Stool Specimens

The use of feces as a specimen for virus isolation has been reduced for cases of gastroenteritis with the realization that viruses that are noncultivatable in cell cultures (rotavirus, Norwalk-like agent, and perhaps some adenoviruses) are responsible for most cases of viral gastroenteritis. Enteroviruses can be isolated as commonly from the feces of individuals without disease compared to patients with gastroenteritis. However, it is frequently useful to submit stool or rectal swabs in addition to other specimens (particularly CSF) for the laboratory diagnosis of CNS disease presumed to be of enteroviral etiology. In this situation, an enterovirus may be excreted in the gastrointestinal tract, but the agent may not be recovered from CSF. Importantly, the etiologic association of such isolates from the gastrointestinal tract becomes somewhat more tenuous than from CSF, however, owing to the common excretion of these agents in stool subsequent to respiratory tract infection (Bowen et al, 1983). There is a higher rate of virus isolation from stool specimens but these are less convenient than rectal swabs for the diagnosis of viral gastroenteritis (Mintz and Drew, 1980). Detection of the fastidious adenovirus types (40 and 41) by nucleic acid probes may help to clarify the role of these agents in gastrointestinal diseases (Kottoff et al, 1989; Krajden et al, 1990).

Viruses surrounded by a lipid envelope generally are not found in an active form in stool specimens, although CMV is an important cause of intestinal disease and has been recovered from that source (Drew, 1988).

2.3.2.e Urine

Many viruses are excreted in the urine during the incubation period. Mumps, adenovirus, and CMV are commonly recovered from urine after symptoms develop. It is important to recognize that mumps virus can be isolated from urine when specimens from other sites are negative, as in the case of CNS disease. Papillomavirus also is excreted in urine, but detection requires highly sensitive techniques such as nucleic acid hybridization (Melchers et al, 1989). Similarly, the polyomaviruses, JC and BK, can be detected by immunologic methods in shell vial cell cultures, but detection by in situ hybridization or nucleic acid amplification is more sensitive than the isolation of the virus in cell culture (Marshall et al, 1990). The clinical significance of polyomaviruses excretion in urine needs to be determined.

Lee and Balfour (1977) reported that urine specimens submitted for the laboratory diagnosis of CMV infection can be inoculated directly into cell cultures because the virus is not concentrated in urine sediment after low-speed centrifugation. However, in one study, low speed centrifugation (500 g for 10 minutes) of urine specimens prior to inoculation presumably removed toxic materials and thus increased viral detection in shell vial cell cultures (Lipson et al, 1990). Because of the relatively low titer, CMV in urine must be concentrated first to be detectable by nucleic acid hybridization (Landini et al, 1990). Urine is the best single specimen for recovery of CMV, but in some cases the virus has been isolated using only a throat swab (Henson et al, 1972; Glenn, 1981). This demonstrates that CMV excretion can be sporadic from body sites and therefore multiple specimens should be processed. In third-trimester pregnant women, cervical excretions appreciably exceeded urinary shedding of virus, the rates being 11.6% (48 of 404) and 6.3% (29 of 463), respectively (Reynolds et al, 1973). As expected, HSV was recovered more frequently from specimens obtained from the cervical canal compared with urine from pregnant women with acute genital infections (Kawana et al, 1982).

2.3.2.f Dermal Lesions

HSV (70%), VZV (29%), coxsackievirus type A (1%), and perhaps some echoviruses are the principal agents that can be recovered on a routine basis from dermal lesions (Desada-Tous et al, 1977; Smith, 1983; Smith and Wold, 1991). The ability to detect or isolate HSV varies with the stage of the lesion. For example, HSV was recovered from 94% of vesicular lesions, 87% of pustular lesions, 70% of ulcers, and 27% of crusted lesions (Moseby et al, 1981). Similarly, smears prepared with cells obtained from vesicles for Papanicolaou, crystal violet, or immunofluorescence staining were superior to cells obtained from ulcers for the diagnosis of HSV infections. Skin biopsy of cutaneous lesions may be important in the diagnosis of systemic CMV infections of immunocompromised patients (Swanson and Feldman, 1987).

2.3.2.g Cerebrospinal Fluid

Nonpolio enteroviruses are the most common isolates from the CSF. HSV, an important cause of CNS disease, rarely has been isolated from the CSF except when recovered from this source in association with meningitis caused by HSV-2 (Rubin, 1983). Alternatively, detection of viral antigens (Bos et al, 1987) or of amplified nucleic acid sequences in CSF of patients with HSV and enterovirus infections should be investigated for possible routine use based on recent reports (Rotbart, 1990; Rowley et al, 1990). Although isolation rates of viruses from CSF specimens generally are low (<4%), this source has been particularly productive for the recovery of enterovirus (67% to 80%) (Rubin, 1983; Wildin and Chonmaitree, 1987); CMV, VZV, and adenoviruses are rarely isolated from CSF in immunocompromised hosts.

Togaviruses, although present in CSF, usually are not collected for testing because they do not replicate well in the cell cultures commonly used for routine viral diagnosis.

2.3.2.h Eye

HSV (66%) and adenovirus (34%) are the viruses commonly associated with eye infections (Smith, 1984; Chastel et al, 1988; Claoué et al, 1988); however, enteroviruses have been associated with hemorrhagic conjunctivitis (Pal et al, 1983). Cytomegalovirus retinitis is recognized as an initial manifestation of the acquired immunodeficiency syndrome (Henderly et al, 1987). Neonates infected with other agents such as rubella, VZV and HPV may exhibit ocular involvement (Naghashfar et al, 1986; Lambert et al, 1989).

2.3.2.i Blood

Isolation of viruses from the blood provides evidence of acute phase infection and symptomatic disease (Neiman et al, 1977). For example, CMV viremia in patients prior to bone marrow transplantation is predictive of both CMV pneumonia and gastrointestinal disease (Meyers et al, 1990). Specific separation procedures that allow the collection of different leukocyte fractions of the blood will yield higher isolation rates of viruses, such as CMV and VZV, compared with buffy coat preparations, especially with specimens from neutropenic patients (Howell et al, 1979; Paya et al, 1988). Detection of antigenemia using a mixture of antibodies to CMV may be the earliest indicator of systemic infection by this virus (van der Bij et al, 1988; Miller, 1991; van den Berg et al, 1991). On the other hand, enteroviruses were isolated from 14 of 31 frozen serum specimens obtained from hospitalized patients with enterovirus infection that had been documented previously by the recovery of virus from stool, throat, or CSF (Prather et al, 1984). Similarly, whole blood was used to document congenital infection due to echovirus type 11 (Jones et al, 1980). Viremia with human immunodeficiency virus (HIV) may predispose to coinfection with CMV in the blood (Shibata et al, 1988b).

2.3.2.j Tissue

Generally, lung and other tissue from the respiratory tract and brain tissue are the only tissue specimens that yield viruses in cell cultures. Occasionally, liver and, rarely, spleen tissue have yielded CMV or HSV. Of 95 viral isolates from tissue over an eight-year period (isolation rate, 3.6%), 82 (86%) of these were CMV, HSV, parainfluenza, influenza, rhino- and adenovirus from respiratory tissues (Smith, 1983). Obviously, selection of particular tissue specimens will improve the recovery rate of viruses from this source. For example, of 105 open-lung biopsy specimens, obtained mostly from immunosuppressed adults, CMV was recovered from 20 (19%), influenza virus type A from one. Generally, however, the recovery rate is less than 10% and the majority of isolates are CMV from lung specimens. The low rate of isolation may stem from the release of viral inhibitors in tissue after homogenization. Alternatively, enzymatic digestion of tissue fragments has provided higher rates of viral isolation compared with homogenized specimens (Shope et al, 1972). The ability to detect viruses such as CMV, HPV, and JCV from tissue fixed in formaldehyde and embedded in paraffin adds a new dimension to the laboratory diagnosis of viral infection (Shibata et al, 1988a; Jiwa et al, 1989; Telenti et al, 1990).

2.4 TRANSPORT
2.4.1 Swabs

Swabs are used for the collection of specimens from throat, dermal, rectal, and ocular sites. Of a total of almost 19,000 specimens submitted for viral culture during 1988 at the Mayo Clinic, 12,000 (63%) were collected with swabs (Smith and Wold, 1991). A variety of fibers have been used for the tip of the shaft of commercially available swabs including rayon, cotton, dacron, polyester, and calcium alginate. All materials, with the notable exception of calcium alginate-tipped swabs, have been acceptable for general use for collection of viral specimens. Calcium alginate has demonstrated toxicity for HSV and should not be used in diagnostic virology (Crane et al, 1980; Bettoli et al, 1982).

2.4.2 Stability of Viruses

In general, viruses that are enveloped, such as the herpesviruses and the myxo- and paramyxoviruses (especially RSV), are relatively labile compared with those without envelopes. Nevertheless, these viruses survive transit for at least 24 to 48 hours if maintained at 4 °C (Levin et al, 1984). HSV can survive for as long as 2 hours on the surface of skin, 3 hours on cloth, and 4 hours on plastic (Turner et al, 1982).

2.4.3 Transport System

Comprehensive studies of viral transport media are difficult to find in the scientific literature. The reason is likely the substantial expense of comparing two or more types of media for transport of swabs from the physician's office to the laboratory. As a result, viral transport devices are often selected on the basis of convenience (Coyle et al, 1987). For example, at the Mayo Clinic, Culturettes® (Becton-Dickinson Microbiology Systems, Cockeysville, MD) have been used for almost 20 years. This swab consists of a plastic tube containing a sterile rayon-tipped applicator and an ampule of modified Stuart's transport medium, which was originally formulated approximately 35 years ago specifically to prolong

Table 2.2 Recovery of Herpes Simplex Virus from Genital Specimens Transported in Several Media

Type of Laboratory	Transport Medium	Number of Isolates/Number of Specimens	Recovery (percent)
Private virology reference	Hanks' Balanced Salt Solution + 0.5% gelatin	33/124	26.6
Medical center and reference laboratory (Mayo Clinic)	Culturette® (Stuart's transport medium)	60/179	33.5
Community hospital laboratory	Eagle's Minimal Essential Medium +2% fetal bovine serum	32/110	29.1
Medical school hospital laboratory	Veal infusion broth + 0.5% gelatin	63/214	29.4

the viability of *Neisseria gonorrhoeae* in transit. These swabs have allowed recovery of commonly isolated viruses in clinical laboratories at rates ranging up to 50%. An advantage for our institution is that one swab, the Culturette®, may be used for specimen collection and transport of organisms appropriate for diagnostic use in bacteriology, mycology, parasitology, and virology.

Two other transport media [Hanks' balanced salt solution (HBSS) and Leibovitz-Emory medium (LEM)] were compared with Stuart's and incorporated into the ampules of Culturettes® by the manufacturer. The swabs were coded and provided to pediatricians with instructions to insert three swabs simultaneously into the oropharynx or on vesicles on the dermal surface, and then transported to the laboratory by routine means (transport time, 30 minutes to 21 hours). Of 80 isolates from 200 children (40% isolation rate), 72 (90%) were recovered in HBSS, 64 (80%) in Stuart's, and 63 (79%) in LEM. Although the greatest number of isolates were recovered in HBSS, the differences in isolation rates among the three media were not statistically significant (Huntoon et al, 1981). It would be necessary to use several hundred patients in such a study in order to demonstrate any substantial difference in the performance of the media.

The incorporation of protein, such as serum, albumin, or gelatin into transport media has been advocated as a means of stabilizing viruses during transit to the laboratory. We participated in a four-site evaluation of a laboratory test kit for rapid diagnosis of genital HSV infections. All four laboratories processed specimens sent to their laboratories with a request for HSV culture. Two laboratories used a transport medium containing 0.5% gelatin and another incorporated 2% fetal bovine serum as a stabilization agent. Our laboratory used the Culturette® during the study, and almost all specimens were submitted from another reference laboratory that required a minimum of 24 hours of transit time between collection and receipt at the Mayo Clinic laboratory. Transport time for specimens submitted to the other three laboratories was routinely shorter than 24 hours. The addition of protein to stabilize HSV during transit provided no apparent positive effect on the overall isolation rate of virus (Table 2.2). Admittedly, this is a comparison of apples with oranges, because the specimens are not common, however, the methodology used at all investigational sites was similar.

Comparative studies of viral transport media have been reviewed comprehensively by Johnson (1990). Most studies used laboratory-passaged strains of virus in assessment of different formulations of transport media (Chernesky et al, 1982; Smith, 1983). HSV survives equally well in Stuart's or HBSS compared with protein-containing medium (Rodin et al, 1971; Yeager et al, 1979). Thus, at least for HSV, it would seem that the type of transport medium is not of paramount importance for virus isolation. The presence of protein is believed to help stabilize viruses during a freeze-thaw cycle.

Virocult (Medical Wire and Equipment Co., Cleveland, OH) is another commercially available self-contained viral collection and transportation system that consists of a sterile pack containing a green color-coded collection swab and plastic transtube. The transtube contains a small sponge saturated with 1.0 mL of a buffered phosphate (pH 7.2) solution containing D-glucose, lactalbumin hydrolysate, chloramphenicol, and cycloheximide. At ambient temperature it has a shelf-life of two years. Surprisingly, the recovery rate of HSV from specimens submitted with the Virocult system was 15.0% and was 15.7% from samples transported in tryptose phosphate broth (TPB). The investigators concluded that Virocult was at least equal in efficiency to TPB and offered the advantage of commercial availability, extended shelf-life at ambient temperature, and ease of use (Perez et al, 1984). In other evaluations, this transport system provided an HSV isolation rate of 22.4% from 2,000 clinical specimens, a result consistent with most other transport media (Johnson et al, 1984; Mayo et al, 1985). Interestingly, however, HSV could be successfully isolated after a holding period of 12 days [2 °C (best) or ambient temperature, 22 °C] in the Virocult transport tube. Overall results from their study indicated that shipping times of up to

2 to 3 days from outlying areas usually result in satisfactory survival of the virus.

An unique commercial transport system for viruses, the Transporter Tube (Bartels Immunodiagnostics, Bellevue, WA), consists of a plastic centrifuge tube containing a monolayer of human diploid fibroblast monolayers around the lower conical portion of the centrifuge tube with 2 mL of medium. This system was tested in parallel with sucrose-phosphate-glutamate (SPG) for the recovery of viruses from specimens transported an average distance of 73 miles to the laboratory. The Transporter Tubes were kept at ambient temperature and the SPG extracts were refrigerated at 1 °C to 10 °C on cold packs. Ninety-two (91%) of 101 viral isolates were recovered in cell culture from the Transporter Tube compared with 82 (81%) from SPG. Twenty-five (24.7%) of the viral isolates were detected by cytopathic effects (CPE) that developed 1 to 4 days earlier with specimens transported in the Transporter Tube than specimens transported in SPG. Of the 101 total virus isolates in the study, however, only nine were recovered exclusively in SPG and only 19 in Transporter Tubes. Overall, the authors felt that the Transporter Tubes system was superior to SPG because of the earlier development of CPE and the system facilitated the transport of clinical material at ambient temperature.

2.4.4 Storage of Specimen

Viruses, such as adeno- and enteroviruses, that do not have structurally labile lipid envelopes survive freeze-thaw procedures with relatively little loss of viral titer. Conversely, a single freeze-thaw cycle may decrease the titer of HSV by 100-fold. Even storage at room temperature for 1 to 30 days significantly reduced the infectivity of this virus. In contrast, none of 65 samples stored for this period of time at 4 °C in HBSS or TPB media showed more than a tenfold loss in infectivity. For short-term

(<5 days) transit or storage of most viral suspensions, therefore, the specimen should be held at 4 °C, rather than frozen (Chernesky et al, 1982).

Preservation of the infectivity of labile viruses such as RSV and VZV for reference purposes can be accomplished without refrigeration by initially freeze-drying the preparation in the presence of a stabilizer containing sucrose, phosphate compounds, sodium glutamate, and bovine albumin. More practically, solutions of dimethyl sulfoxide (DMSO) in concentrations ≥5%, sucrose-phosphate-glutamate containing 1% bovine albumin, and other formulations containing serum, glycerol, skim milk, or other proteins have been shown to be useful cryoprotectants for viruses (Tannock et al, 1987; Gallo et al, 1989; Johnson, 1990).

2.5 PROCESSING SPECIMENS

2.5.1 Inoculation

Generally, 0.2 to 0.3 mL are inoculated onto human diploid fibroblast (HDF) cells, primary monkey kidney cells, and a continuous cell line, such as HeLa or HEp-2 cells. Primary rabbit kidney cells may be used in place of HDF cells for the recovery of HSV (Moore, 1984) and rhabdosarcoma (RD) and buffalo green monkey (BGM) kidney cells have been found to yield more enteroviruses compared with the usual cell systems.

Removal of liquid medium from cell monolayers before inoculation of a specimen, in order to allow for adsorption of viral particles to the cells (1 hour), probably enhances the rate of recovery of these viruses, although comparative data from clinical specimens have not been published. Based on studies that have demonstrated that low-speed centrifugation (2000–3000 × g) of the specimen inoculum onto cell culture monolayers increased the efficiency of

infection of *Chlamydia* spp. (Reeve et al, 1975), murine CMV (Osborn and Walker, 1969), and the AD169 strain of human CMV (Hudson et al, 1976), we applied this technique to urine specimens for the diagnosis of human strains of CMV (Gleaves et al, 1984). In this procedure, 0.2 mL urine is inoculated onto monolayers of MRC-5 cells seeded in 1-dram shell vials containing a circular cover slip. The inoculated shell vials are then centrifuged at 700 × g for 1 hour, medium is added back, and the vials incubated at 36 °C for 16 hours. At this time, the cover slip containing the cell monolayer is tested for the presence of early antigen of CMV by immunofluorescence. This rapid test reduced the detection time for CMV from an average of eight days in conventional tube cell cultures to one day (16 hours). The sensitivity of the rapid test was 100% when compared with conventional cell culture isolation. If the specimens were not centrifuged, however, the sensitivity was only 37.5%. This technique has been applied for the rapid diagnosis of HSV infections in the laboratory, and seems applicable to other viruses, including adenovirus, influenza A, and VZV (Gleaves et al, 1985; Espy et al, 1986; Espy et al, 1987; Gleaves et al, 1988). The only limitation of the methodology is the availability of specific monoclonal or polyclonal antibodies for use as a probe to detect early antigens synthesized by a virus. Application of nucleic acid detection methods for rapid laboratory diagnosis of viral infections will depend on their favorable comparison with shell vial assays in terms of sensitivity, specificity, simplicity, and cost.

Inoculation of specimens directly into cell cultures at the bedside has yielded 500-fold higher titers of certain agents, such as RSV (Hall and Douglas, 1975). However, a study of 135 samples inoculated at bedside or held for three hours at 4 °C before transport to the laboratory showed no difference in the rate of RSV recovery.

Of 51 positive specimens, 44 (86%) were positive by both inoculation procedures, three (6%) were recovered only from specimens inoculated at bedside, and four (8%) were positive for RSV only when the specimen was inoculated in the laboratory (Bromberg et al, 1984).

2.5.2 Examination

Conventional test tube cell cultures are examined at 40 × to 100 × magnifications for the presence of typical viral CPE. In the near future, the need for this type of visual examination could be reduced, as new rapid and sensitive techniques become available for each virus. For example, an ELISA test could detect HSV prior to the development of CPE by assaying lysates of infected cells (Morgan and Smith, 1984; Michalski et al, 1986). Other rapid techniques, such as the use of a biotinylated monoclonal antibody to HSV detected streptavidin-horseradish peroxidase (Herpchek™, DuPont Co.) and nucleic acid amplification by polymerase chain reaction (PCR), for the detection of viruses are not dependent on the use of cell cultures, and appear to be extremely sensitive (Dascal et al, 1989; Verano and Michalski, 1990; Persing, 1991). However, with most rapid techniques (except PCR), especially those performed directly on the clinical specimen, the sensitivity generally is between 80% and 100% compared with conventional cell cultures. Overall, the most successful and sensitive techniques have combined initial viral amplification in cell cultures with the subsequent application of a rapid detection assay a few hours later (Jespersen et al, 1989; Arens et al, 1991; Smith et al, 1991). Recently, membrane immunoassays (sensitivity, 80% to 90%) for the rapid (~15 minute) diagnosis of influenza virus, type A, HSV, RSV, and rotavirus have been developed and evaluated (Chernesky et al,

1988; Swierkosz et al, 1989; Dorian et al, 1990; Rothbarth et al, 1991; Waner et al, 1991; Zimmerman et al, 1991).

Even with the practical advent of rapid techniques in clinical virology, therefore, we must endeavor to preserve the maximal infectivity of viruses in specimens by giving careful attention to the principles of proper specimen selection, collection and transport, and inoculation of cell cultures.

2.6 SEROLOGIC DETERMINATION

Although techniques and high quality reagents have reduced the time required for the diagnosis of many viral infections to just a few hours, the need for some serologic tests is still apparent. Agents such as Epstein-Barr virus (EBV), rubella, measles, arboviruses (especially togaviruses), and hepatitis viruses do not replicate in the battery of cell cultures generally used in the clinical laboratory. As such, serology remains the test of choice for the laboratory diagnosis of these virus infections. Further, the assessment of immune status of individuals to rubella, CMV (renal transplant recipients), and VZV (children with neoplastic disease), has remained an important diagnostic function of viral laboratories. The detection of viral-specific IgM class antibodies in the acute phase serum of patients, in contrast to IgG antibodies, usually indicates a recent primary infection (with a few exceptions, e.g., HSV) with a particular agent. Inclusion of the proper controls and separation methods is necessary to provide specific, reliable results (Smith, 1983; see Chapter 15). Thus, sensitive techniques for IgM determination using a single specimen provide a necessary complement to rapid techniques for demonstrating viral antigens directly in specimens or after amplification in cell culture.

REFERENCES

Ahluwalia, G., Embree, J., McNicol, P., Law, B., and Hammond, G.W. 1987. Comparison of nasopharyngeal aspirate and nasopharyngeal swab specimens for respiratory syncytial virus diagnosis by cell culture, indirect immunofluorescence assay, and enzyme-linked immunosorbent assay. J. Clin. Microbiol. 25:763–767.

Arens, M.Q., Knutsen, A.P., Schwarz, K.B., Roodman, S.T., and Swierkosz, E.M. 1987. Multiple and persistent viral infections in a patient with Bare lymphocyte syndrome. J. Infect. Dis. 156:837–841.

Arens, M., Owen, J., Hagerty, C.M., Reed, C.A., and Storch, G.A. 1991. Optimizing recovery of cytomegalovirus in the shell vial culture procedure. Diagn. Microbiol. Infect. Dis. 14:125–130.

Balfour, H.H. Jr. 1988. Varicella-zoster virus infections in immunocompromised hosts: A review of the natural history and management. Am. J. Med. 85:68–73.

Bergman, O.J., Morgensen, S.C., and Ellegaard, J. 1990. Herpes simplex virus and intraoral ulcers in immunocompromised patients with haematologic malignancies. Eur. J. Clin. Microbiol. Infect. Dis. 9:184–190.

Bettoli, E.J., Brewer, P.M., Oxtoby, M.J., Zaidi, A.A., and Guinan, M.E. 1982. The role of temperature and swab materials in the recovery of herpes simplex virus from lesions. J. Infect. Dis. 145:399.

Blumenfeld, W., Wager, E., and Hadley, W.K. 1984. Use of transbronchial biopsy for diagnosis of opportunistic pulmonary infections in acquired immunodeficiency syndrome (AIDS). Am. J. Clin. Pathol. 81:1–5.

Bos, C.A., Olding-Stenkvist, E., Wilterdink, J.B., and Scheffer, A.J. 1987. Detection of viral antigens in cerebrospinal fluid of patients with herpes simplex virus encephalitis. J. Med. Virol. 21:169–178.

Bowen, G.S., Fisher, M.C., Deforest, A., Thompson, C.M., Jr., Kleger, B., and Friedman, H. 1983. Epidemic of meningitis and febrile illness in neonates caused by ECHO type 11 virus in Philadelphia. Pediatr. Infect. Dis. 2:359–363.

Bromberg, K., Daidone, B., Clarke, L., and

Sierra, M.F. 1984. Comparison of immediate and delayed inoculation of HEp-2 cells for isolation of respiratory syncytial virus. J. Clin. Microbiol. 20:123–124.

Chastel, C., Adrian, T., Demazure, M., Legrand-Quillien, M.C., Lejeune, B., Colin, J., and Wigand, R. 1988. Molecular epidemiology of two consecutive outbreaks of adenovirus 8 keratoconjunctivitis. J. Med. Virol. 24:199–204.

Chernesky, M.A., Ray, C.G., and Smith, T.F. 1982. Laboratory diagnosis of viral infections. In W.L. Drew (coordinating ed.), Cumitech 15. Washington, D.C.: American Society for Microbiologists.

Chernesky, M., Castriciano, S., Mahony, J., Spiewak, M., and Schaefer, L. 1988. Ability of TESTPACK ROTAVIRUS enzyme immunoassay to diagnose rotavirus gastroenteritis. J. Clin. Microbiol. 26:2459–2461.

Claoué, C.M.P., Ménage, M.J., and Easty, D.L. 1988. Severe herpetic keratitis. I: Prevalence of visual impairment in a clinic population. Br. J. Ophthalmol. 72:530–533.

Corey, L., Adams, H.G., Brown, Z.A., and Holmes, K.K. 1983. Genital herpes simplex virus infections: Clinical manifestations, course, and complications. Ann. Intern. Med. 98:958–972.

Coyle, M.B., Granato, P.A., Morello, J.A., and Zabransky, R.J. 1987. Results of the survey of virology laboratory methods. Clin. Microbiol. Newslett. 9:97–108.

Crane, L.R., Gutterman, P.A., Chapel, T., and Lerner, A.M. 1980. Incubation of swab materials with herpes simplex virus. J. Infect. Dis. 141:531.

Dascal, A., Chan-Thim, J., Morahan, M., Portnoy, J., and Mendelson, J. 1989. Diagnosis of herpes simplex virus infection in a clinical setting by a direct antigen detection enzyme immunoassay kit. J. Clin. Microbiol. 27:700–704.

Deseda-Tous, J., Byalt, P.H., and Cheny, J.D. 1977. Vesicular lesions in adults due to echovirus 11 infections. Arch. Dermatol. 113:1705–1706.

Dorian, K.J., Beatty, E., and Atterbury, K.E. 1990. Detection of herpes simplex virus by the Kodak SureCell herpes test. J. Clin. Microbiol. 28:2117–2119.

Drew, W.L. 1986. Controversies in viral diagnosis. Rev. Infect. Dis. 8:814–824.

Drew, W.L. 1988. Cytomegalovirus infection in patients with AIDS. J. Infect. Dis. 158:449–456.

Espy, M.J., Hierholzer, J.C., and Smith, T.F. 1987. The effect of centrifugation on the rapid detection of adenovirus in shell vials. Am. J. Clin. Pathol. 88:358–360.

Espy, M.J., Smith, T.F., Harmon, M.W., and Kendal, A.P. 1986. Rapid detection of influenza virus by shell vial assay with monoclonal antibodies. J. Clin. Microbiol. 24:677–679.

Feldman, S., and Epp, E. 1979. Detection of viremia during incubation of varicella. J. Pediatr. 94:746–748.

Frayha, H., Castriciano, S., Mahony, J., and Chernesky, M. 1989. nasopharyngeal swabs and nasopharyngeal aspirates equally effective for the diagnosis of viral respiratory disease in hospitalized children. J. Clin. Microbiol. 27:1387–1389.

Gäbel, H., Flamholc, L., and Ahlfors, K. 1988. Herpes simplex virus hepatitis in a renal transplant recipient: Successful treatment with acyclovir. Scand. J. Infect. Dis. 20:435–438.

Gallo, D., Kimpton, J.S., and Johnson, P.J. 1989. Isolation of human immunodeficiency virus from peripheral blood lymphocytes stored in various transport media and frozen at −60 °C. J. Clin. Microbiol. 27:88–90.

Gleaves, C.A., Lee, C.F., Bustamante, C.I., and Meyers, J.D. 1988. Use of murine monoclonal antibodies for laboratory diagnosis of varicella-zoster virus infection. J. Clin. Microbiol. 26:1623–1625.

Gleaves, C.A., Smith, T.F., Shuster, E.A., and Pearson, G.R. 1984. Rapid detection of cytomegalovirus in MRC-5 cells inoculated with urine specimens by using low-speed centrifugation and monoclonal antibody to an early antigen. J. Clin. Microbiol. 19:917–919.

Gleaves, C.A., Wilson, D.J., Wold, A.D., and Smith, T.F. 1985. Detection and serotyping of herpes simplex virus in MRC-5 cells using centrifugation and monoclonal antibodies 16 h postinoculation. J. Clin. Microbiol. 21:29–32.

Glenn, J. 1981. Cytomegalovirus infections following renal transplantation. Rev. Infect. Dis. 3:1151–1178.

Greenberg, S.B., and Krilov, L.R. 1986. Laboratory diagnosis of viral respiratory disease. In W. Drew and S. Rubin (eds.), Cumitech 21. American Society for Microbiology: Washington, DC.

Growdon, W.A., Apodaca, L., Cragun, J., Peterson, E.M., and de la Maza, L.M. 1987. Neonatal herpes simplex virus infection occurring in second twin of an asymptomatic mother: Failure of a modern protocol. J. Am. Med. Assoc. 257:508–511.

Hall, C.B. and Douglas, R.G. Jr. 1975. Clinically useful method for the isolation of respiratory syncytial virus. J. Infect. Dis. 131:1–5.

Hall, C.B., Douglas, R.G. Jr., and Geiman, J.M. 1976a. Respiratory syncytial virus infections in infants: Quantitations and duration of shedding. J. Pediatr. 89:11–15.

Hall, C.B., Geiman, J.M., Biggar, R., Kotok, D.I., Hogan, P.M., and Douglas, R.G. Jr. 1976b. Respiratory syncytial virus infections within families. N. Engl. J. Med. 294:414–419.

Hammond, G., Hannan, C., Yeh, T., Fischer, K., Mauthe, G., and Straus, S.E. 1987. DNA hybridization for diagnosis of enteric adenovirus infection from directly spotted human fecal specimens. J. Clin. Microbiol. 25:1881–1885.

Henderly, D.E., Freeman, W.R., Smith, R.E., Causey, D., and Rao, N.A. 1987. Cytomegalovirus retinitis as the initial manifestation of the acquired immune deficiency syndrome. Am. J. Ophthalmol. 103:316–320.

Henson, D., Siegel, S.E., Fuccillo, D.A., Matthew, E., and Levine, A.S. 1972. Cytomegalovirus infections during acute childhood leukemia. J. Infect. Dis. 126:469–481.

Howell, C.L., Miller, M.J., and Martin, W.J. 1979. Comparison of rates of virus isolation from leukocyte populations separated from blood by conventional and Ficoll-Paque/Macrodex methods. J. Clin. Microbiol. 10:533–537.

Hudson, J.B., Misra, V., and Mosmann, T.F. 1976. Cytomegalovirus infectivity: Analysis of the phenomenon of antifungal enhancement of infectivity. Virology 72:235–243.

Huntoon, C.J., House, R.F. Jr., and Smith, T.F. 1981. Recovery of viruses from three transport media incorporated into Culturettes. Arch. Pathol. Lab. Med. 105:436–437.

Jespersen, D.J., Drew, W.L., Gleaves, C.A., Meyers, J.D., Warford, A.L., and Smith, T.F. 1989. Multisite evaluation of a monoclonal antibody reagent (Syva) for rapid diagnosis of cytomegalovirus in the shell vial assay. J. Clin. Microbiol. 27:1502–1505.

Jiwa, N.M., Raap, A.K., van de Rijke, F.M., Mulder, A., Weening, J.J., Zwaan, F.E., The, T.H., and van der Ploeg, M. 1989. Detection of cytomegalovirus antigens and DNA in tissues fixed in formaldehyde. J. Clin. Pathol. 42:749–754.

Johansson, M.E., Zweygberg Wirgart, B., Grillner, L., and Björk, O. 1990. Severe gastroenteritis in an immunocompromised child caused by adenovirus type 5. Pediatr. Infect. Dis. J. 9:449–450.

Johnson, F.B. 1990. Transport of viral specimens. Clin. Microbiol. Rev. 3:120–131.

Johnson, F.B., Leavitt, R.W., and Richards, D.F. 1984. Evaluation of the virocult transport tube for isolation of herpes simplex virus from clinical specimens. J. Clin. Microbiol. 20:120–122.

Jones, M.J., Kolb, M., Votava, H.J., Johnson, R.L., and Smith, T.F. 1980. Intrauterine echovirus type 11 infection. Mayo Clin. Proc. 55:509–512.

Kahn, F.W. and Jones, J.M. 1988. Analysis of bronchoalveolar lavage specimens from immunocompromised patients with a protocol applicable in the microbiology laboratory. J. Clin. Microbiol. 26:1150–1155.

Kawana, T., Kawogoe, K., Takizawa, K., Chen, J.T., Kawaguchi, T., and Sakamoto, S. 1982. Clinical and virologic studies on female genital herpes. Obstet. Gynecol. 60:456–461.

Kepfer, P.D. Hable, K.A., and Smith, T.F. 1974. Viral isolation rates during summer from children with acute upper respiratory tract disease and healthy children. Am. J. Clin. Pathol. 61:1–5.

Kim, H.W., Wyatt, R.G., Fernie, B.F., Brandt, C.D., Arrobio, J.O., Jeffries, B.C., and Parrott, R.H. 1983. Respiratory syncytial

virus detection by immunofluorescence in nasal secretions with monoclonal antibodies against selected surface and internal proteins. J. Clin. Microbiol. 18:1399–1404.

Kimball, A.M., Foy, H.M., Cooney, M.K., Allan, I.D., Mattock, M., and Plorde, J.J. 1983. Isolation of respiratory syncytial and influenza viruses from the sputum of patients hospitalized with pneumonia. J. Infect. Dis. 147:181–184.

Kohl, S. 1988. Herpes simplex virus encephalitis in children. Pediatr. Clin. North Am. 35:465–483.

Kottoff, K.L., Lasonsky, G.A., Morris, J.G., Jr., Wasserman, S.S., Singh-Nag, N., and Levine, M.M. 1989. Enteric adenovirus infection and childhood diarrhea: An epidemiologic study in three clinical settings. Pediatrics 84:219–225.

Krajden, M., Brown, M., Petrasek, A., and Middleton, P.J. 1990. Clinical features of adenovirus enteritis: A review of 127 cases. Pediatr. Infect. Dis. J. 9:636–641.

Krusinski, P.A. 1988. Treatment of mucocutaneous herpes simplex infections with acyclovir. J. Am. Acad. Dermatol. 18:179–181.

Lambert, S.R., Taylor, D., Kriss, A., Holzel, H., and Heard, S. 1989. Ocular manifestations of the congenital varicella syndrome. Arch. Ophthalmol. 107:52–56.

Landini, M.P., Trevisani, B., Guan, M.X., Ripalti, A., Lazzarotto, T., and La Placa, M. 1990. A simple and rapid procedure for the direct detection of cytomegalovirus in urine samples. J. Clin. Lab. Analysis 4:161–164.

Landry, M.L., Fong, C.K.Y., Neddermann, K., Solomon, L., and Hsiung, G.D. 1987. Disseminated adenovirus infection in an immunocompromised host: Pitfalls in diagnosis. Am. J. Med. 83:555–559.

Lee, M.S. and Balfour, H.H. Jr. 1977. Optimal method for recovery of cytomegalovirus from urine of renal transplant patients. Transplantation 24:228–230.

Lennette, D.A. 1991. Preparation of specimens for virological examination, In A. Balows, W.J. Hausler, Jr., K.L. Herrmann, H.D. Isenberg, and H.J. Shadomy, (eds.) Manual of Clinical Microbiology, 5th ed. Washington, D.C.: American Society for Microbiology, pp. 818–821.

Levin, M.J., Leventhal, S., and Master, H.A. 1984. Factors influencing quantitative isolation of varicella-zoster virus. J. Clin. Microbiol. 19:880–883.

Lipson, S.M., Costello, P., Forlenza, S., Agins, B., and Szabo, K. 1990. Enhanced detection of cytomegalovirus in shell vial cell culture monolayers by preinoculation treatment of urine with low-speed centrifugation. Curr. Microbiol. 20:39–42.

Marshall, W.F., Telenti, A., Proper, J., Aksamit, A.J., and Smith, T.F. 1990. Rapid detection of polyomavirus BK by a shell vial cell culture assay. J. Clin. Microbiol. 28:1613–1615.

Martin, W.J. II and Smith, T.F. 1986. Rapid detection of cytomegalovirus in bronchoalveolar lavage specimens by a monoclonal antibody method. J. Clin. Microbiol. 23:1006–1008.

Mayo, D.R., Brennan, T., Egbertson, S.H., and Moore, D.F. 1985. Rapid herpes simplex virus detection in clinical samples submitted to a state virology laboratory. J. Clin. Microbiol. 21:768–771.

Melchers, W.J.G., Schift, R., Stolz, E., Lindeman, J., and Quint, W.G.V. 1989. Human papillomavirus detection in urine samples from male patients by the polymerase chain reaction. J. Clin. Microbiol. 27:1711–1714.

Meyers, J.D., Ljungman, P., and Fisher, L.D. 1990. Cytomegalovirus excretion as a predictor of cytomegalovirus disease after marrow transplantation: Importance of cytomegalovirus viremia. J. Infect. Dis. 162:373–380.

Michalski, F.J., Shaikh, M., Sahraie, F., Desai, S., Verano, L., and Vallabhaneni, J. 1986. Enzyme-linked immunosorbent assay spin amplification technique for herpes simplex virus detection. J. Clin. Microbiol. 24:310–311.

Miller, H., Rossier, E., Milk, R., and Thomas, C. 1991. Prospective study of cytomegalovirus antigenemia in allograft recipients. J. Clin. Microbiol. 29:1054–1055.

Mintz, L. and Drew, W.L. 1980. Relation of culture site to the recovery of nonpolio enteroviruses. Am. J. Clin. Pathol. 74:324–326.

Modlin, J.F. 1986. Perinatal echovirus infection: Insights from a literature review of 61 cases of serious infection and 16 outbreaks in nurseries. Rev. Infect. Dis. 8:918–926.

Moore, D.F. 1984. Comparison of human fibroblast cells and primary rabbit kidney cells for isolation of herpes simplex virus. J. Clin. Microbiol. 19:548–549.

Morgan, M.A. and Smith, T.F. 1984. Evaluation of an enzyme-linked immunosorbent assay for the detection of herpes simplex virus antigen. J. Clin. Microbiol. 19:730–732.

Moseby, R.C., Corey, L., Benjamin, D., Winter, C., and Remington, M.L. 1981. Comparison of viral isolation, direct immunofluorescence, and indirect immunoperoxidase techniques for detection of genital herpes simplex virus infection. J. Clin. Microbiol. 13:913–918.

Myers, M.G. 1979. Viremia caused by varicella-zoster virus: Association with malignant progressive varicella. J. Infect. Dis. 140:229–233.

Naghashfar, Z., McDonnell, P.J., McDonnell, J.M., Green, W.R., and Shah, K.V. 1986. Genital tract papillomavirus type 6 in recurrent conjunctival papilloma. Arch. Ophthalmol. 104:1814–1815.

Neiman, P.E., Reeves, W., Ray, G., Flourney, N., Lerner, K.G., Sale, G.E., and Thomas, E.D. 1977. A prospective analysis of interstitial pneumonia and opportunistic viral infection among recipients of allogeneic bone marrow grafts. J. Infect. Dis. 136:754–767.

Osborn, J.E. and Walker, D.L. 1969. Enhancement of infectivity of murine cytomegalovirus *in vitro* by centrifugal inoculation. J. Virol. 2:853–858.

Pal, S.R., Szucs, G.Y., and Melnick, J.L. 1983. Rapid immunofluorescence diagnosis of acute hemorrhagic conjunctivitis caused by enterovirus 70. Intervirology 20:19–22.

Paya, C.V., Wold, A.D., and Smith, T.F. 1988. Detection of cytomegalovirus from blood leukocytes separated by Sepracell-MN and Ficoll-Paque/Macrodex methods. J. Clin. Microbiol. 26:2031–2033.

Perez, T.R., Mosman, P.L., and Juchau, S.V. 1984. Experience with Virocult as a viral collection and transportation system. Diagn. Microbiol. Infect. Dis. 2:7–9.

Persing, D.H. 1991. Polymerase chain reaction: Trenches to benches. J. Clin. Microbiol. 29:1281–1285.

Prather, S.L., Jenista, J.A., and Menegus, M.A. 1984. The isolation of nonpolio enteroviruses from serum. Diagn. Microbiol. Infect. Dis. 2:353–357.

Reeve, P., Owen, J., and Oriel, J.D. 1975. Laboratory procedures for the isolation of *Chlamydia trachomatis* from the human genital tract. J. Clin. Pathol. 28:910–914.

Reynolds, D.W., Stagno, S., Hosty, T.S., Tiller, M., and Alford, C.A. Jr. 1973. Maternal cytomegalovirus excretion and perinatal infection. N. Engl. J. Med. 289:1–5.

Rodin, P., Hare, M.J., Barwell, C.F., and Withers, M.J. 1971. Transport of herpes simplex virus in Stuart's medium. Br. J. Vener. Dis. 47:198–199.

Rotbart, H.A. 1990. Diagnosis of enteroviral meningitis with the polymerase chain reaction. J. Pediatr. 117:85–89.

Rothbarth, P.H., Hermus, M.-C., and Schrijnemakers, P. 1991. Reliability of the new test kits for rapid diagnosis of respiratory syncytial virus infection. J. Clin. Microbiol. 29:824–826.

Rowley, A.H., Whitley, R.J., Lakeman, F.D., and Wolinsky, S.M. 1990. Rapid detection of herpes-simplex-virus DNA in cerebrospinal fluid of patients with herpes simplex encephalitis. Lancet 1:440–441.

Rubin, S.J. 1983. Detection of viruses in spinal fluid. Am. J. Med. 75:124–128.

Shibata, D.K., Arnheim, N., and Martin, W.J. 1988a. Detection of human papilloma virus in paraffin-embedded tissue using the polymerase chain reaction. J. Exp. Med. 167: 225–230.

Shibata, D., Martin, W.J., Appleman, M.D., Causey, D.M., Leedom, J.M., and Arnheim, N. 1988b. Detection of cytomegalovirus DNA in peripheral blood of patients infected with human immunodeficiency virus. J. Infect. Dis. 158:1185–1192.

Shope, T.C., Klein-Robbenhaar, J., and Miller, G. 1972. Fatal encephalitis due to *Herpesvirus hominis*: Use of intact brain cells for isolation of virus. J. Infect. Dis. 125:542–544.

Smith, T.F. 1983. Clinical uses of the diagnostic virology laboratory. Med. Clin. North Am. 67:935–951.

Smith, T.F. 1984. Diagnostic virology in the community hospital. Postgrad. Med. 75: 215–223.

Smith, T.F. and Wold, A.D. 1991. Changing trends of diagnostic virology in a tertiary care medical center. *In* L.M. de la Maza and E.M. Peterson (eds), Medical Virology X. Elsevier, New York, pp. 1–16.

Smith, M.C., Creutz, C., and Huang, Y.T. 1991. Detection of respiratory syncytial virus in nasopharyngeal secretions by shell vial technique. J. Clin. Microbiol. 29:463–465.

Spector, S.A. 1983. Transmission of cytomegalovirus among infants in hospital documented by restriction-endonuclease-digestion analysis. Lancet 1:378–380.

Stagno, S., and Whitley, R.J. 1985. Herpesvirus infections of pregnancy. Part II: Herpes simplex virus and varicella-zoster virus infections. N. Engl. J. Med. 313:1327–1330.

Stone, K.M. and Whittington, W.L. 1990. Treatment of genital herpes. Rev. Infect. Dis. 12:S610–S619.

Swanson, S., and Feldman, P.S. 1987. Cytomegalovirus infection initially diagnosed by skin biopsy. Am. J. Clin. Pathol. 87:113–116.

Swierkosz, E.M., Flanders, R., Melvin, L., Miller, J.D., and Kline, M.W. 1989. Evaluation of the Abbott TESTPACK RSV enzyme immunoassay for detection of respiratory syncytial virus in nasopharyngeal swab specimens. J. Clin. Microbiol. 27:1151–1154.

Tannock, G.A., Hierholzer, J.C., Bryce, D.A., Chee, C.-F., and Paul, J.A. 1987. Freeze-drying of respiratory syncytial viruses for transportation and storage. J. Clin. Microbiol. 25:1769–1771.

Telenti, A., Aksamit, A.J. Jr., Proper, J., and Smith, T.F. 1990. Detection of JC virus DNA by polymerase chain reaction in patients with progressive multifocal leukoencephalopathy. J. Infect. Dis. 162:858–861.

Turner, R., Shehab, Z., Osborne, K., and Hendley, J.O. 1982. Shedding and survival of herpes simplex virus from "fever blisters." Pediatrics 70:547–549.

van den Berg, A.P., van Son, W.J., van der Giessen, M., The, T.G., and van der Bij, W. 1991. Detection of cytomegalovirus (CMV) immediate early antigens in peripheral blood leukocytes during active CMV infection. Nephron 57:115–116.

van der Bij, W., Schirm, J., Torensma, R., van Son, W.J., Tegzess, A.M., and The, T.H. 1988. Comparison between viremia and antigenemia for detection of cytomegalovirus in blood. J. Clin. Microbiol. 26:2531–2535.

Verano, L. and Michalski, F.J. 1990. Herpes simplex virus antigen direct detection in standard virus transport medium by DuPont Herpchek enzyme-linked immunosorbent assay. J. Clin. Microbiol. 28:2555–2558.

Waner, J.L., Todd, S.J., Shalaby, H., Murphy, P., and Wall, L.V. 1991. Comparison of Directigen Flu-A with viral isolation and direct immunofluorescence for the rapid detection and identification of influenza A virus. J. Clin. Microbiol. 29:479–482.

Whitley, R.J., Alford, C.A., Hirsch, M.S., Schooley, R.T., Luby, J.P., Aoki, F.Y., Hanley, D., Nahmias, A.J., Soong, S.-J., and the NIAID Collaborative Antiviral Study Group. 1986. Vidarabine versus acyclovir therapy in herpes simplex encephalitis. N. Engl. J. Med. 314:144–149.

Whitley, R., Arvin, A., Prober, C., Corey, L., Burchett, S., Plotkin, S., Starr, S., Jacobs, R., Powell, D., Nahmias, A., Sumaya, C., Edwards, K., Alford, C., Caddell, G., and Soong, S.-J. 1991. Predictors of morbidity and mortality in neonates with herpes simplex virus infections. N. Engl. J. Med. 24:450–454.

Wildin, S. and Chonmaitree, T. 1987. The importance of the virology laboratory in the diagnosis and management of viral meningitis. Am. J. Dis. Child. 141:454–457.

Wilfert, C.M., Buckley, R.H., Mohanakumar, T., Griffith, J.F., Katz, S.L., Whisnant, J.K., Eggleston, P.A., Moore, M., Treadwell, E., Oxman, M.N., and Rosen, F.S. 1977. Persistent and fatal central-nervous system echovirus infections in patients with agammaglobulinemia. N. Engl. J. Med. 296: 1485–1489.

Woods, G.L., Thompson, A.B., Rennard, S.L., and Linder, J. 1990. Detection of cytomegalovirus in bronchoalveolar lavage specimens: Spin amplification and staining with a monoclonal antibody to the early nuclear antigen for diagnosis of cytomegalovirus pneumonia. Chest 98:568–575.

Yeager, A.S., Morris, J.E., and Prober, C.G. 1979. Storage and transport of cultures for herpes simplex virus, type 2. Am. J. Clin. Pathol. 72:977–979.

Zimmerman, S., Moser, E., Sofat, N., Bartholomew, W.R., and Amsterdam, D. 1991. Evaluation of a visual, rapid, membrane, enzyme immunoassay for the detection of herpes simplex virus antigen. J. Clin. Microbiol. 29:842–845.

3

Primary Isolation of Viruses

Marie L. Landry and G. D. Hsiung

3.1 INTRODUCTION

Viruses are obligate intracellular parasites and, therefore, require living cells in which to replicate. This is very different from the cultivation of bacteria, for which nutrient broth or agar plates suffice. The "living cells" essential for virus isolation and assay can be in the form of cultured cells, embryonated eggs, or laboratory animals, most frequently newborn mice (Figure 3.1). The variety of methods and host systems employed for the isolation of different viruses from clinical specimens reflects the fact that the optimum growth conditions for each virus may differ tremendously. If an insensitive host system is inoculated with a specimen containing a particular virus, or if suboptimal growth conditions exist, the virus probably will not be isolated and a false-negative result will be obtained. Due to limitations in resources and personnel, not all available isolation systems can be maintained in every laboratory. Depending on the patient population the laboratory serves, a decision can be made to select the best host systems and methods needed to optimize the isolation of those viruses causing the most morbidity in that group of patients.

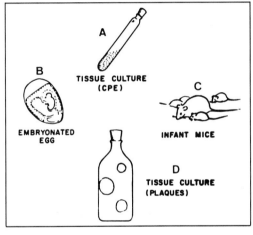

Figure 3.1 Host systems for virus isolation, (A) tissue culture method (CPE); (B) embryonated eggs; (C) newborn mice; (D) tissue culture method (plaques). (Reproduced with permission from Hsiung, G.D. 1982. Diagnostic Virology. New Haven: Yale University Press, p. 18.)

Although embryonated eggs and laboratory animals are very useful for the isolation of certain viruses, cell cultures are the sole isolation system utilized in most clinical virology laboratories and remain the mainstay of viral diagnosis. Thus, this chapter will focus primarily on cell culture techniques. However, virus isolation meth-

43

Table 3.1 Cell Cultures Commonly Used in a Clinical Virology Laboratory

Cell Culture	Origin	Number of Subpassages
Primary	Kidney tissues from monkeys, rabbits, etc. Embryos from chickens, guinea pigs, etc.	1 or 2
Diploid (limited passage)	Human embryonic lung/ or human newborn foreskin	20–50
Heteroploid	Human epidermoid carcinoma of larnyx (HEp-2), of cervix (HeLa), of lung (A549)	Indefinite

ods using other host systems are reviewed in the Appendix at the end of this chapter.

3.2 VIRUS ISOLATION IN CELL CULTURES

3.2.1 Background

The discovery by Enders, Weller, and Robbins in the late 1940s that poliovirus replicates in cultivated mammalian cells derived from nonnervous tissues revolutionized and simplified procedures for the isolation of viruses (Enders et al, 1949). Until that time intact animals or embryonated eggs were the common systems used. After that landmark discovery, cell cultures were prepared for virus studies from a wide variety of animal and human tissues and as a result, in the years following, most of the common viruses we are familiar with today were discovered.

3.2.2 Types of Cell Culture

Cell cultures are generally separated into three types (Table 3.1): Primary cells, which are prepared directly from animal or human tissues and can be subcultured only one or two passages; diploid cell cultures,

which usually are derived from human tissues, either fetal or newborn, and can be subcultured 20 to 50 times before senescence; and continuous cell lines, which can be established from human or animal tissues, from tumors, or following the spontaneous transformation of normal tissues. These have a heteroploid karyotype and can be subcultured an indefinite number of times. However, sensitivity to different viruses may change after serial passage.

3.2.3 Variation in Susceptibility to Different Viruses

Cell cultures vary greatly in their susceptibility to different viruses (Table 3.2). If a virus is inoculated into an insusceptible cell culture, the virus will not be able to replicate and a negative result will be obtained. When small amounts of virus are present in a clinical sample, a positive result may be obtained only when the most sensitive systems are used. Therefore, it is critical that those caring for the patients inform the laboratory of the clinical syndrome and/or virus(es) suspected, so that the most sensitive cell cultures can be used and appropriate detection methods employed.

Table 3.2 Variation in Sensitivity of Cell Cultures to Infection by Viruses Commonly Isolated in a Clinical Virology Laboratory

	Cell Culture[a]			
Virus	PMK	HDF	HEp-2	RK
RNA VIRUS				
Enterovirus	+++	++	+/−	−
Rhinovirus	+	+++	+	−
Myxovirus	+++	+	−	−
Respiratory syncytial	++	+	+++	−
DNA VIRUS				
Adenovirus	+	++	+++	+/−
HSV	+	++	++	+++
VZV	+	+++	−	−
CMV	−	+++	−	−

Abbreviations: PMK, primary monkey kidney; HDF, human diploid fibroblasts; HEp-2, human heteroploid cell line; RK, primary rabbit kidney.

[a] Degree of sensitivity: +++, highly sensitive; ++, moderately sensitive; +, low sensitivity; −, nonsensitive.

3.2.4 Virus Isolation Methods

3.2.4.a Obtaining and Processing Specimens

Although this area has been reviewed in the previous chapter, it should be reiterated that without appropriate specimens that are properly collected and promptly transported to the laboratory, the subsequent time and effort spent in isolation attempts will be wasted. Accomplishing this is an important task of the clinical virology laboratory and requires continuing communication with and education of the clinicians.

3.2.4.b Supplies and Equipment Needed

The materials needed for the isolation of viruses in cell culture (Table 3.3) are those necessary for the safe handling and processing of cell cultures, maintenance and observation of cell cultures, and preservation and storage of clinical specimens and virus isolates. Although the clinical virologist is primarily interested in virus isolation, maintaining different cell cultures in healthy condition is absolutely necessary in order to ensure good results. A wide variety of cell cultures are available commercially and can be purchased and delivered weekly or biweekly, depending on the needs of the laboratory. It may be elected to prepare some cell cultures in the laboratory from available animal or human tissues (e.g., rabbit kidney, guinea pig embryo, human newborn foreskin), or passage certain cell lines [e.g., HEp-2, human diploid fibroblast (HDF) cell strains] for reasons of availability, economy, or quality. The preparation and maintenance of cell cultures can be found in several reference books cited in the reference list (Hsiung and Green, 1978; Hsiung, 1982; Schmidt and Emmons, 1989).

3.2.4.c Inoculation and Incubation

To isolate a spectrum of viruses, cell cultures of several different types are inoculated, such as HDF, a human heteroploid cell line, (e.g., HEp-2, A549), and a primary monkey kidney cell culture. Alternatively, for specific indications (e.g. herpes simplex virus infection), limited cultures

Table 3.3 Supplies and Equipment Needed for Isolation of Viruses in Cell Culture

	Supplies and Equipment Needed
Processing of cell cultures	Laminar flow hood, pipettes, automatic pipetting device, pipette jar and discard can, disposable gloves, disinfectant, and sterile glass- and plasticware
Maintenance of cell cultures	Culture media, serum, antibiotics, 4 °C refrigerator, test tube racks, and/or rotating drum, 35 °C incubators, CO_2 incubator, waterbath, and upright and inverted microscopes
Preservation and storage of viruses	Freezer vials, ultralow temperature freezer (-70 °C), and DMSO as stabilizer

intended to detect only one or two virus types can be performed.

The cell type(s) most susceptible to the suspected viruses in the clinical specimen should be included. Ideally, only healthy, freshly prepared, young cell cultures should be used, because aged cells are less sensitive to virus infection. All cell cultures should be examined under the microscope before inoculation to ensure that the cells are in good condition. Although techniques may vary somewhat for different viruses, in general, the following procedures apply:

1. Pour off culture media and inoculate specimens, 0.1–0.3 mL, into each culture tube (Figure 3.2). Shell vial cell cultures, each containing cells grown on a coverslip, can be inoculated in a similar manner, centrifuged, then incubated. The coverslips are removed later for fixing and staining and the monolayers examined for virus-induced antigens. For further details, see section 3.2.4.e. Uninoculated cultures should be kept in parallel for comparison.

2. Allow specimen to adsorb in the incubator at 35 °C for 30 to 60 minutes. Then, 1.0–1.5 mL of maintenance medium should be added and the inoculated cultures returned to the incubator. Inocu-

lated cultures can be placed in a rotating drum if available, which is optimal for the isolation of respiratory viruses, especially rhinoviruses, and results in earlier appearance of cytopathic effect (CPE) for many viruses. More rapid rotation of cultures can also enhance the speed and sensitivity of virus recovery (Mavromoustakis et al; 1988). If stationary racks are used to conserve space, it is critical that culture tubes be positioned so that the cell monolayer is bathed in nutrient medium, otherwise the cells will degenerate, especially at the edge of the monolayer.

3. Check inoculated culture tubes daily for the first week, then every other day for virus-induced CPE. Compare with uninoculated control tubes from the same lot of cell cultures.

4. Certain specimens, such as urine and stool, frequently will be toxic to the cell cultures and this toxicity can be confused with virus-induced cytopathology. With such specimens it is a good practice to check inoculated tubes either after adsorption or within 24 hours of inoculation and refeed with fresh medium if necessary. If toxic effects are extensive, it may be necessary to sub-

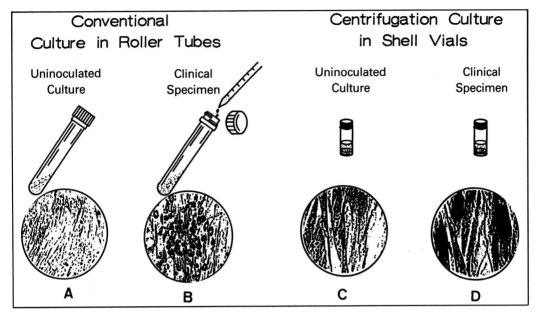

Figure 3.2 A and B: Conventional culture in roller tubes (10× objective). (A) Uninoculated culture showing fibroblast cell monolayer. (B) Similar cell culture inoculated with a clinical specimen showing virus-induced cytopathic effect (small rounded cells). C and D: Centrifugation culture in shell vials (40× objective). (C) Uninoculated culture as seen in (A) but fixed and stained. (D) Similar cell culture inoculated with a clinical specimen fixed and stained with immunoperoxidase staining, showing virus-induced antigen in the nuclei of infected cells (dark areas).

passage the inoculated cells in order to dilute toxic factors and provide viable cells for virus growth.

5. Inoculated cultures and the uninoculated controls should be kept for observation for virus-induced effects for at least two weeks, during which time cell cultures may need to be refed to maintain the cells in good condition. Some cultures, such as HEp-2 cells, may require refeeding or subculturing every few days. Great care must be taken when refeeding cultures that cross contamination from one specimen to another does not occur. Separate pipettes should be used for separate specimens.

6. When virus-induced effects occur (see below), passage infected cultures (especially in doubtful cases) into a fresh culture of the same cell type to ensure recovery of virus for further identification of the isolate. For certain cell-associated viruses, such as cytomegalovirus (CMV) or varicella-zoster virus (VZV), it is necessary to trypsinize and passage intact infected cells (Taylor-Robinson, 1959). Adenovirus can be subcultured after freezing and thawing infected cells, which disrupts the cells and releases intracellular virus (Rowe and Hartley, 1962).

7. For certain fastidious viruses, or when the amount of infectious virus in the specimen is low, blind passage (i.e., subculture of the inoculated culture in the absence of virus-induced effects) into a set of fresh culture tubes may be necessary before virus growth can be detected.

In order to enhance viral replication and provide a more rapid and sensitive viral diagnosis, herpesvirus-infected cell cultures have been treated with dimethyl sulfoxide (DMSO) and/or dexamethasone (DEX). The results reported have been mixed, with some authors reporting enhancement (West et al, 1989) and others reporting no effect (Espy et al, 1988). This discrepancy may be explained by the age of the treated cells. DMSO and DEX appear to result in more rapid and extensive viral CPE when confluent, static cells are used (West et al, 1988). Since actively growing subconfluent cell cultures are sensitive to virus infection, DMSO or DEX treatment does not enhance the induction of CPE.

3.2.4.d Detection of Virus-induced Effects

Cytopathic effects. Many viruses can be identified by the characteristic cellular changes they induce in susceptible cell cultures. These can be visualized under the light microscope. Examples of CPE characteristic for each particular virus are shown in Figure 3.3 and described in greater detail in the sections on individual viruses. Degree of CPE is usually graded from + to + + + + as it progresses to involve less than 25% of the cell monolayer (+), to 50% (+ +), 75% (+ + +), and finally 100% (+ + + +). There are two important points that should be emphasized regarding CPE induced by virus:

1. The *rate* at which the CPE progresses may help to distinguish similar viruses; for example, HSV progresses rapidly to involve the entire monolayer of several cell systems (Figure 3.4), whereas, two other herpesviruses, CMV and VZV, grow primarily in HDF cells and progress slowly over a number of days or weeks (Weller et al, 1958; Weller and Hanshaw, 1962).

2. The *type* of cell culture(s) in which the virus replicates is important; that is, although the CPE may be similar within a virus group, the susceptibility of different cell types to different viruses may differ greatly. For example, both polio

and echovirus induce similar CPE in primary rhesus monkey kidney (RhMK), however, echovirus does not induce CPE in HEp-2 cells, thus, allowing presumptive identification (Figure 3.5).

It should be cautioned that virus-induced CPE must be distinguished from "nonspecific" CPE caused by toxicity of specimens, contamination with bacteria or fungi, or old cells. A subculture into fresh cells should amplify virus effects and dilute toxic effects. With experience, the appearance of the cellular changes, taken together with the susceptible cell systems, the specimen source, and clinical disease, usually allow a presumptive diagnosis to be made as soon as the virus-induced cellular changes occur.

Hemadsorption. Parainfluenza and sometimes influenza viruses may not induce any distinctive cellular changes; however, they do possess hemagglutinins, which have an affinity for red blood cells (RBC). When a freshly obtained guinea pig RBC suspension is added to the infected cultures, the RBC adsorb onto the infected cells, resulting in a hemadsorption phenomenon as shown in Figure 3.6(2). When a culture shows positive hemadsorption, the culture fluid is subcultured into a fresh culture to confirm the virus isolation and to permit further identification. Caution should be taken when aged guinea pig RBC are used, however, because nonspecific hemadsorption often occurs in an uninoculated culture [Figure 3.6(3)] and should be distinguished from that resulting from a specific viral infection (Dowdle and Robinson, 1966). Furthermore, the hemadsorption test is usually performed at 4 °C or 22 °C because the RBC will elute when incubated at 37 °C. It should be noted that not all viruses that agglutinate RBC can adsorb them onto infected cell monolayers, because hemadsorption is a property of those viruses that bud from the host cell membrane during maturation and thus express viral hemag-

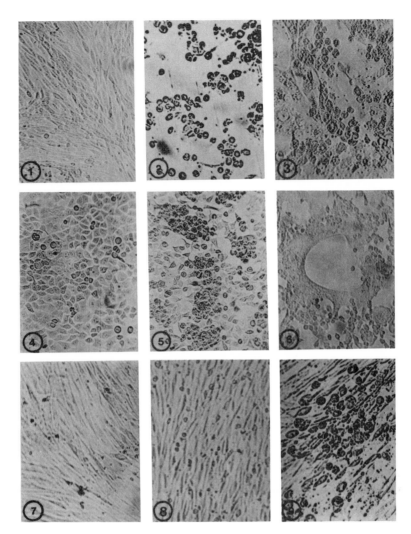

Figure 3.3 Examples of characteristic cytopathic effects (CPE) of different viruses: (1) Uninfected rhesus monkey kidney (RhMK) cells; (2) poliovirus CPE in RhMK cells; (3) influenza B virus CPE in RhMK cells; (4) uninfected HEp-2 cells; (5) adenovirus CPE in HEp-2 cells; (6) respiratory syncytial virus CPE in HEp-2 cells; (7) uninfected human diploid fibroblasts (HDF); (8) rhinovirus CPE in HDF cells; and (9) cytomegalovirus CPE in HDF cells.

glutinin on the surface of the infected cell. The technique is further described in Chapter 13.

Interference. Certain viruses that do not readily induce CPE in infected cultures can be detected by their ability to interfere with the growth of a second virus inoculated into the same culture. This test has been used most frequently in detection of rubella virus and is referred to as the "interference phe-nomenon." African green monkey kidney cell cultures infected with rubella virus commonly do not show CPE. After a 10-day incubation, a standard dose of a challenge virus, such as echovirus 11, is inoculated into the same tube, as well as into a control tube without rubella virus, and incubated for two or more additional days. The infection of these cells by rubella virus will inhibit the replication of a super-

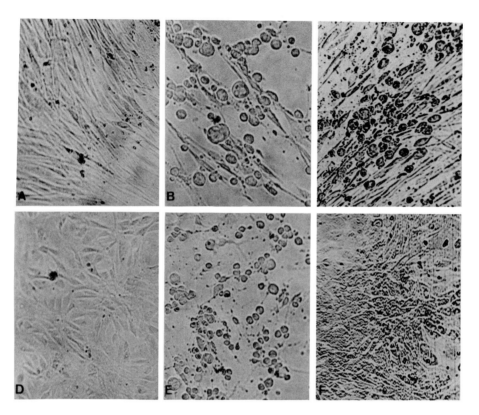

Figure 3.4 Cell susceptibility and rate of progression of CPE of two herpesviruses: herpes simplex virus type 1 (HSV-1) and human cytomegalovirus (CMV). (A) Uninfected WI-38 cells; (B) extensive HSV-1 CPE in WI-38 cells, two days postinoculation; (C) CMV in WI-38 cells, one week postinoculation; (D) uninfected rabbit kidney cells (RK); (E) extensive HSV-1 CPE in RK cells, one day postinoculation; (F) absence of CMV CPE in RK cells, two weeks postinoculation. (Modified from Hsiung, G.D. 1982. Diagnostic Virology, Figure 79. New Haven: Yale University Press, p. 206.)

infection by echovirus 11. Thus, echovirus CPE will not be observed in rubella virus-infected cultures, due to the presence of an interfering agent.

3.2.4.e Centrifugation Culture for Virus Isolation (Shell Vial Technique)

Conventional virus isolation requires observation of roller tube cultures for CPE, which can take weeks to appear. However, the rapid diagnosis of viral infections is increasingly important in patient management. One of the most significant contributions to rapid diagnosis in the clinical

laboratory has been the application of centrifugation cultures to viral diagnosis.

For a number of years, it has been recognized that low-speed centrifugation of specimens onto cell monolayers enhances infectivity of certain viruses as well as chlamydia, (Reeve et al, 1975; Hudson et al, 1976; Oefinger et al, 1990). The mechanism for this effect is unclear, and may involve centrifugation of virus aggregates, of virus attached to cell debris, or an effect on cell membranes to enhance virus entry.

In 1984 the use of centrifugation cultures followed by staining with a monoclonal antibody at 24 hours postinoculation

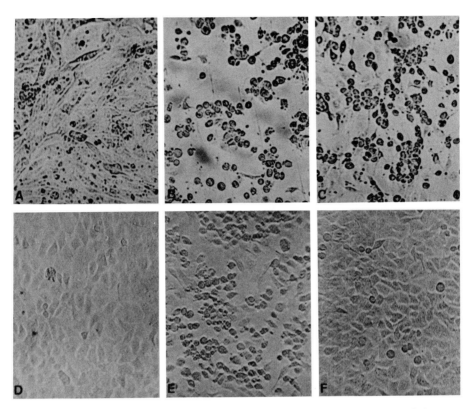

Figure 3.5 Differential susceptibility of cell cultures to enteroviruses. (A) Uninfected rhesus monkey kidney (RhMK); (B) poliovirus infected RhMK cells showing advanced CPE; (C) echovirus infected RhMK cells showing advanced CPE; (D) uninfected HEp-2 cells; (E) poliovirus infected HEp-2 cells showing advanced CPE; (F) echovirus infected HEp-2 cells showing absence of CPE. (Reproduced with permission from Hsiung, G.D. 1982. Diagnostic Virology, Figure 20. New Haven: Yale University Press, p. 94.)

was first reported for CMV (Gleaves et al, 1984) (Figure 3.7). Subsequent reports have documented its usefulness in rapid diagnosis of other viruses (Gleaves et al, 1985; Espy et al, 1987; Gleaves et al, 1988).

The shell vial technique combines 1) cell culture to amplify virus in the specimen; 2) centrifugation to enhance viral infectivity and 3) early detection of virus-induced antigen (before CPE) by the use of high quality specific antibodies. It can be used for any virus that replicates in cell culture and for which a specific antibody is available. For viruses with a long replication cycle, such as CMV, viral antigens produced early in the replication cycle can

be detected many days before CPE is apparent using light microscopy. For viruses that replicate faster, e.g., HSV, or if the antibodies available are to late rather than early replication products, less time is gained using the shell vial technique.

Although centrifugation culture enhances the rapidity of diagnosis of viral infections, numerous studies have demonstrated that for maximal sensitivity, both conventional culture and centrifugation cultures should be performed in parallel (Rabella and Drew, 1990). The overall sensitivity of the shell vial technique varies with the type of specimen (Paya et al, 1987), the length and temperature of cen-

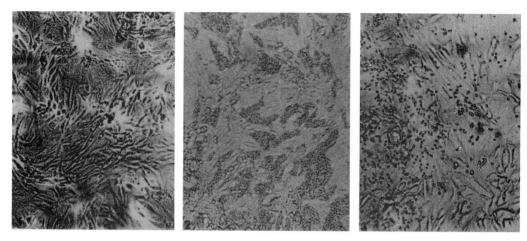

Figure 3.6 Hemadsorption of guinea pig red blood cells by parainfluenza virus in monkey kidney cells (MK). (1) Uninfected MK cells; (2) specific hemadsorption in parainfluenza infected MK cells; (3) nonspecific hemadsorption seen with aged red blood cells in uninfected cell cultures. (Modified from Hsiung, G.D. 1982. Diagnostic Virology, Figures 43 and 44. New Haven: Yale University Press, p. 145.)

trifugation (Shuster et al, 1985), the virus, the cell cultures, the antibody employed, and the time of fixation and staining. The use of young cell monolayers (Fedorko et al, 1989), with inoculation of multiple shell vials enhances the recovery rate (Paya et al, 1988). Toxicity, particularly problematic with blood and urine specimens, can lead to cell death and the loss of the monolayer, necessitating blind passage of the specimen or specimen reinoculation. Furthermore, with all rapid techniques that use specific antibodies or nucleic acid probes to detect viruses, only the virus sought will be de-

Figure 3.7 Centrifugation culture: Detection of CMV early antigens in infected nuclei at 24 hours postinoculation (immunoperoxidase stain).

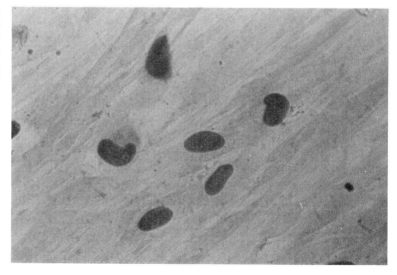

tected. Conventional isolation using a spectrum of cell cultures can detect a variety of virus types, including unanticipated agents.

Reagents and equipment

Antibodies to specific viral types, either fluorescein or peroxidase labeled

Cold acetone

Shell vials - 1 dram, 15 × 45 mm, with stoppers

Coverslips - 12 mm diameter

Cell cultures grown on coverslips in shell vials, sensitive to the suspected viruses

Low speed centrifuge with adapters for shell vials

Humidified chamber

Rotator or rocker

Suction flask and vacuum source

Test procedure. Inoculation of shell vials

1. Prepare three shell vials for blood specimens suspected of containing CMV; for all other specimens and viruses, prepare two shell vials.
2. Remove cap and aspirate medium from shell vial.
3. Inoculate prepared specimen onto monolayer, 0.2 to 0.3 mL per vial.
4. Replace cap and centrifuge (30 to 60 minutes at $700 \times g$).
5. Aspirate inoculum for blood, urine and stool samples then rinse with 1 mL of medium to reduce toxicity.
6. Add 1.0 mL of maintenance medium to each shell vial and incubate at 35 °C for 1 to 2 days.

Fixation of shell vials

1. Before fixation, inspect the coverslips for toxicity, contamination, etc. If necessary, passage the cell suspension to a new vial and repeat incubation before staining.
2. If monolayer is intact, aspirate medium from shell vials and rinse once with 1.0 mL of PBS. If monolayer appears fragile, do not rinse with PBS.
3. Aspirate medium completely, add 1.0 mL of cold acetone to each shell vial and allow cells to fix for 10 minutes.
4. Aspirate the acetone and allow the shell vial to dry completely.

Staining of coverslips

1. Add 1.0 mL of PBS to each coverslip, then aspirate the PBS.
2. Pipet 150 μL of antiserum (appropriately titrated and diluted) into the shell vial. Replace the cap.
3. Rock the tray holding shell vials to distribute the reagent, then check to see that no coverslips are floating above the reagent.
4. Place rack holding the shell vials in a humidified chamber in the 35 °C incubator.
5. Incubate for 30 minutes.
6. Add 1.0 mL of PBS to the shell vial, then aspirate. Pipet a second 1.0 mL of PBS into the shell vial and allow the monolayer to soak for 5 minutes, then aspirate. For *direct assays* (primary antibody is labeled), go directly to step "9". For *indirect assays* (primary antibody is not labeled):
7. Pipet 150 μL of labeled conjugate onto the monolayer.
8. Repeat steps "3–5."
9. Add distilled water to the shell vial. Using forceps and a wire probe, remove coverslip, and blot on tissue or absorbent paper, e.g., Kimwipe. (Exposure of slide to distilled water should be kept to less than 30 to 45 seconds).
10. Add one drop of mounting fluid to a properly labeled slide and place coverslip on mounting fluid with cell side down, being careful not to trap air bubbles.

Table 3.4 An Example of Determination of $TCID_{50}$

	Number of Cultures		Total Cultures			
Virus Dilution	With CPE	No CPE	With CPE	No CPE	CPE Ratio	% Cultures with CPE
10^{-3}	4	0	9	0	9/9	100
10^{-4}	3	1	5	1	5/6	83
10^{-5}	2	2	2	3	2/5	40
10^{-6}	0	4	0	7	0/7	0

Reading procedure. Coverslips are examined using 20× objective with a fluorescence microscope equipped with the appropriate filters to maximize detection of the FITC label (or a light microscope if a peroxidase label is used). A known positive control is run for each viral antigen with each assay. Non-infected monolayers are fixed and stained as negative antigen controls. For indirect IFA, normal goat serum, or PBS plus FITC conjugate, is used as a negative serum control.

The pattern of fluorescence varies depending upon the virus sought, the antiserum used, the cell culture, and the stage of virus replication. Even a single cell, characteristically stained, is considered a positive result.

The test should be repeated if,

1. The staining pattern is not typical for the virus sought,
2. Non-specific staining is observed, or
3. The staining color is more yellow than green.

3.2.5 Virus Assay and Identification

3.2.5.a Virus Infectivity Assay by the End Point of CPE

At times it is necessary to quantitate the amount of infectious virus present in a specimen or a cell culture. The specimen can be assayed by determining the highest dilution of the fluid that produces CPE (or hemadsorption or interference) in 50% of the cell cultures inoculated; this endpoint is the 50% tissue culture infectious dose ($TCID_{50}$), as follows:

1. Add 0.9 mL of HBSS to seven sterile test tubes.
2. Add 0.1 mL of virus suspension to the first dilution tube, mix thoroughly and transfer 0.1 mL of the mixture to the next tube.
3. With a separate 1-mL pipette, mix the suspension and transfer 0.1 mL to the next tube; continue this process with all seven tubes.
4. Inoculate 0.1 mL of each dilution of the virus suspension into a tube of a sensitive cell culture, four tubes per dilution (use a separate pipette for each dilution).
5. Determine the assay endpoint by the method of Reed and Muench (1938) (see Table 3.4).

Calculation: 50% end point

$$= \frac{\% \text{ with CPE} >50\% - 50\%}{\% \text{ with CPE} >50\% - <50\%}$$

$$= \frac{83 - 50}{83 - 40}$$

$$= 0.7$$

Therefore a virus dilution of $10^{4.7}$ per 0.1 mL represents one $TCID_{50}$, i.e., at that dilution 50% of the inoculated cultures will become infected. A dilution of $10^{2.7}$ per 0.1 mL of virus suspension will contain 100 $TCID_{50}$ in a volume of 0.1 mL.

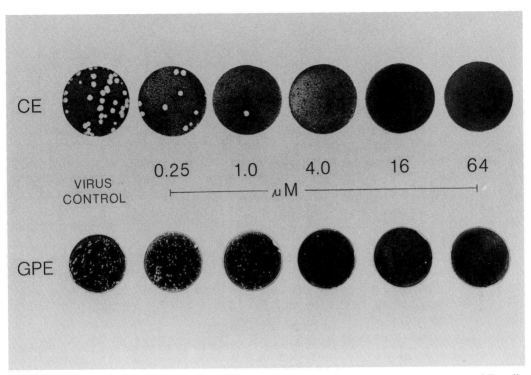

Figure 3.8 Acyclovir inhibition of HSV type 2 induced plaque formation in chick embryo (CE) cells and guinea pig embryo (GPE) cells. Note that in more susceptible cell systems (GPE cells), a higher concentration of ACV is required to inhibit viral plaques than in the less sensitive CE cells. (Reproduced with permission from Landry ML, Mayo D, Hsiung GD: Rapid and accurate viral diagnosis. Pharmacol. Ther. 40:287–328, 1989, Pergamon Press.)

3.2.5.b Plaque Formation

Many viruses that produce CPE, and also certain viruses that do not produce detectable CPE under fluid medium in cell cultures, may be detected by their ability to form plaques in cell monolayers under a solid medium, such as agar, agarose, or methyl cellulose (Porterfield, 1960; Hsiung, 1961; Hsiung, 1982). Virus plaques are colorless areas of infected cells, which do not take up the vital stain neutral red when it is incorporated into the overlay medium. Alternatively, monolayers can be stained with crystal violet after the medium is removed to visualize plaques (Figure 3.8). Virus particles from a focus of infection are localized by the solid overlay medium and the virus

spreads from infected cells to adjacent cells resulting in discrete foci of infection. Different enteroviruses induce plaques of varying size and shape in much the same manner that different bacteria produce characteristic colonies (Hsiung and Melnick, 1957). Characteristic plaques can be helpful in detecting mixed infections of two viruses within the same virus group, such as poliovirus and echovirus, which induce the same type of CPE but show two distinct sizes and shapes of plaques. Because one infectious unit is capable of initiating one plaque, this technique can be used both for accurate quantitative assay of virus infectivity and for purification of virus strains. Plaque assays are not commonly used in a clinical laboratory at the present time; therefore, detailed procedures are not included in this chapter and the reader is referred to the reference list at the end of this chapter.

It should be noted, however, that as demand for antiviral sensitivity testing of virus isolates increases, plaque reduction assays may be more frequently performed in the near future. Figure 3.8 shows the reduction in HSV-2 induced plaques in the presence of increasing concentrations of acyclovir (ACV). The cell system used in antiviral assays has a significant effect on the amount of drug required to inhibit the virus (Hu and Hsiung, 1989). Highly sensitive cell systems such as GPE cells require more ACV to effect a 50% reduction in plaques than less sensitive chick embryo (CE) and Vero cells. For HSV antiviral assays, Vero cells are the most commonly used.

3.2.5.c Identification of Virus Isolates

As mentioned above, presumptive identification usually can be made on the basis of characteristic virus induced effects (e.g., type of CPE or hemadsorption) and selective cell susceptibility. For final identification, staining infected cells with either fluorescein or peroxidase-labeled antibodies is most commonly used. Monolayers showing CPE or hemadsorption are trypsinized or dislodged, the cells transferred to a welled-slide and then stained with specific antiserum (Rossier et al, 1989). Nucleic acid hybridization, enzyme-linked immunosorbent assay, and latex agglutination can also be used for virus identification. When determination of specific type is requested for enteroviruses or adenoviruses, neutralization of virus-induced cytopathology in cell culture can be performed. These tests will be discussed in greater detail in subsequent chapters.

Occasionally a new isolate cannot be identified by the standard serologic tests,

and it may be necessary to study the more basic properties of the new agent (Hsiung, 1982). These include the nucleic acid type, size and shape of the isolate, as well as the presence or absence of a lipid envelope. Nucleic acid type can be determined by exposure to 5'-bromodeoxyuridine (BrdU), an inhibitor of DNA viruses, followed by assay of virus infectivity. Virus size can be estimated by passing infected culture fluids through a series of filters or by electron microscopy (EM), and the presence of a lipid envelope can be determined by exposure of the virus to ether, then reinoculation into culture to determine if infectivity has been lost. The morphologic properties of the infecting virus can also be determined by EM. These tests would generally be performed only by reference laboratories.

3.2.6 Advantages and Limitations

The advantages of cell culture for virus diagnosis include relative ease compared with animal inoculation, broad spectrum and sensitivity when compared with other available diagnostic methods, and the recovery of unknown or unexpected infectious virus(es) that may be present in the specimen. It is limited by the difficulty in maintaining cell cultures, by the sometimes variable quality of cultures, and by the decreased sensitivity of cell lines at higher passage levels. Contamination with adventitious agents, such as endogenous viral agents and mycoplasma, occurs which can inhibit the growth of viruses in clinical specimens (Hsiung, 1968; Smith, 1970; Stanbrige, 1971; Chu et al, 1973). Endogenous viruses that are latent in the tissue culture can be reactivated during cultivation and cause CPE or hemadsorption, and thus, be confused with virus isolated from the patient's specimen. In addition, some common viruses as yet do not produce identifiable effects in readily available cell cultures—for example, hepatitis viruses, rotavirus, Norwalk virus, some group A coxsackieviruses and togaviruses—so that other methods of detection are necessary. To get the best results from primary isolation in cell culture, it is most important to maintain healthy cell cultures and have a spectrum of cell types available. In general, a primary monkey kidney cell culture, an HDF cell strain, and a human heteroploid cell line (e.g., as HEp-2 or A549) constitute a satisfactory combination. If the isolation of a particular virus is a high priority, the most sensitive system available should be selected.

Another problem in isolating certain viruses, especially of the myxo- and paramyxovirus groups, is the presence of inhibitory substances and/or antibodies in calf serum used in the cell culture media (Krizanova and Rathova, 1969). Ideally, maintenance media for inoculated cultures should be serum-free; however, serum is required for long-term maintenance of cells. Using fetal or agamma calf serum reduces this problem, but adds to expense. To date, no completely satisfactory, chemically defined medium is available.

3.2.7 Viruses Commonly Isolated in a Clinical Laboratory

3.2.7.a Herpes Simplex Viruses Types 1 and 2

Both herpes-simplex virus type 1 and type 2 (HSV-1, HSV-2) infect a wide variety of cell cultures and animals. However, differences in sensitivity of cell cultures are evident, particularly when specimens containing low titers of virus are inoculated (Zhao et al, 1987). Primary cell cultures, such as rabbit kidney (RK), human embryonic kidney (HEK), and guinea pig embryo (GPE) cells, are all very sensitive to HSV infection (Landry et al, 1982). Three continuous cell lines, mink lung (ML), RD, and A549 cells, have also been found to be highly sensitive, more so than commercial

human diploid fibroblast (HDF) strains (Woods and Young, 1988; Johnston et al, 1990). Others have prepared HDF in their own laboratories from human placenta, newborn foreskin, or embryonic tissues and found them to be highly sensitive. Comparison of sensitivities of different cultures is shown in Table 3.5. Vesicular fluids, throat swabs, and genital lesions are the most common sources for virus isolation. HSV produces a rapid degeneration of cells, often appearing within 24 hours of inoculation of the cell culture (Figure 3.4). Over 90 percent of positives will be identified within three to five days (Herrmann, 1972). Occasionally, CPE develops later and, rarely, will be detected after blind passage.

The CPE begins as clusters of enlarged, rounded, refractile cells and spreads to involve the entire monolayer, usually within 48 hours. Another type of CPE also can be seen with HSV type 2: the formation of multinucleated giant cells. Subcultures are performed by passaging 0.2 mL of supernatant fluid to a fresh culture tube.

Identification of virus as HSV can be done by immunofluorescence or neutralization. Commercial kits are available that use immunoperoxidase stains to identify HSV infected cells (Fayram et al, 1983; Sewell et al, 1984). Isolates also can be differentiated as type 1 or 2 by several means, including immunofluorescence with monoclonal antibodies (Balkovic and Hsiung, 1985), ability to replicate in chick embryo cell monolayers (Nordlund et al, 1977), sensitivity to bromovinyldeoxyuridine (BVDU) (Mayo, 1982), or by restriction enzyme analysis (Lonsdale, 1978).

3.2.7.b Varicella-zoster Virus

HDF are the most sensitive cells for the isolation of VZV, although the virus also has been isolated using human epithelial cells, primary MK cells, and occasionally GPE cells. Vesicle fluid and lesion swabs

Table 3.5 Comparison of Sensitivity and Rapidity of CPE Induced by Herpes Simplex Virus in Different Cell Culture Systems

HSV dose (TCID$_{50}$)	Cell Culture/Days Postinoculation CPE Observed											
	RK			HDF			HEp-2			Vero		
	1[a]	2	4	1	2	4	1	2	4	1	2	4
10000	+++	+++	+++[b]	0	+	++	0	+	++	0	++	++
1000	++	++	+++	0	+	++	0	0	++	0	+	++
100	+	+	+++	0	0	+	0	0	0	0	0	0
10	0	+	++	0	0	0	0	0	0	0	0	0
1	0	0	+	0	0	0	0	0	0	0	0	0

Abbreviations: RK, primary rabbit kidney cells; HDF, human diploid fibroblast, WI-38/MRC-5; HEp-2, human heteroploid cell line derived from carcinoma of the larynx; Vero, African green monkey kidney cell line; TCID$_{50}$, tissue culture infectious dose, 50 percent.

[a] Days postinoculation.

[b] Degree of CPE: +, 25 percent of cells show CPE; ++, 50 percent of cells show CPE; +++, 75 percent of cells show CPE; ++++, 100 percent of cells show CPE.

are the usual sources for VZV isolation. The virus is quite labile; therefore, prompt inoculation into cell culture is desirable. In contrast with HSV, VZV does not cause disease in newborn mice or embryonated eggs.

Cytopathology starts as foci of rounded enlarged cells, as seen with HSV, however, the onset and progression are much slower and the foci of CPE tend to progress linearly along the axis of the cells similar to CMV. However, VZV infected foci degenerate more rapidly than CMV. CPE first appears four to seven days after inoculation but may take two or three weeks. The virus is cell-associated and subpassages are performed by trypsinization and passage of infected intact cells to fresh monolayers of cells. Stocks of VZV should be maintained as suspensions of viable cells frozen at or below −70 °C. Final identification is by immunofluorescence using monoclonal antibodies. The use of centrifugation cultures enhances the sensitivity of detection for VZV (Gleaves et al, 1988).

3.2.7.c Cytomegalovirus

HDF are the single most successful culture system for the isolation of CMV. The source of the fibroblasts can be either human embryonic tissues or newborn foreskin. The latter, however, lose their sensitivity after the tenth to 15th passage. Virus can be isolated from a variety of body secretions including urine, saliva, tears, milk, semen, stools, vaginal or cervical secretions, and leukocytes. Isolation of CMV from bronchoalveolar lavage specimens has been used quite successfully in recent years to diagnose pulmonary infections in lieu of lung biopsy (Broaddus et al, 1985). Clinical specimens can produce CPE within a few days to many weeks, depending on the amount of virus in the specimen. Characteristic CPE consists of foci of enlarged, refractile cells that slowly enlarge over weeks and often do not involve the entire monolayer [Figure 3.4(C)]. Thus, it is important that the monolayers be maintained in good condition for four weeks. On the other hand, when a large quantity of CMV is inoculated, one may see generalized rounding at 24 hours that can be confused with HSV. Incubation of cultures at 36 °C instead of 33 °C results in more rapid onset of CPE and higher isolation rates (Gregory and Menegus, 1983). For subculture, early passage of intact infected cells is essential. Alternatively, degenerating mono-

Table 3.6 Host Susceptibility for Enteroviruses

Virus Type	Serotypes	Cell Cultures			Newborn Mice
		RhMK	HDF	HEp-2	
Poliovirus	1–3	++[a]	++	++	−
Coxsackie B	1–6	++[a]	++	++	++
Coxsackie A	1–24	+/−	+/−	−	++
Echovirus	1–34	++	+	−	−

Abbreviations: RhMK, primary rhesus monkey kidney; HDF, human diploid fibroblast; HEp-2, human heteroploid cell line.

[a] Degree of sensitivity: ++, sensitive; +, less sensitive; −, not sensitive.

layers can be trypsinized and then dispensed onto fresh uninfected cells. To maintain stocks of CMV, viable cells should be frozen in modified Eagle's medium with 20% calf serum and 10% dimethyl sulfoxide (DMSO). Identification of isolates can be accomplished with immunofluorescence. As described in section 3.2.4.e, centrifugation cultures have had a major impact on the rapid diagnosis of CMV infections. However, it is still recommended that conventional cultures be performed in parallel (Rabella and Drew, 1990).

3.2.7.d Adenovirus

In general, human adenoviruses produce CPE in continuous human cell lines, such as HEp-2 and A549, in HEK and HDF cell cultures. Each of these cell systems has its disadvantages: the continuous cell lines may be difficult to maintain; HEK often are not readily available and are expensive; the HDF are less sensitive and the changes produced are not characteristic (Mahafzah and Landry, 1989). Non-human cells, such as rhesus monkey kidney (RhMK), are of variable sensitivity and virus growth is slower. Throat swabs, nasal swabs, eye swabs, and stool are good sources of virus, the choice depending on the clinical syndrome. Characteristic CPE consists of grape-like clusters of rounded cells [Figure 3.3(5)], which appear in two to seven days

with types 1, 2, 3, 5, 6, and 7. Other adenovirus types may require four weeks or blind passage. Adenovirus remains cell-associated, similar to VZV and CMV; however, adenovirus is nonenveloped and stable to freezing and thawing. Therefore, two to three cycles of freezing at −70 °C and thawing disrupts the cells and releases intranuclear infectious virus. A number of enteric adenovirus types associated with diarrhea have been detected by electron microscopy, but do not grow readily in routine cell culture (Gary et al, 1979).

Identification of isolates can be done by immunofluorescence using antihexon serum for the adenovirus group. Isolates can be separated into four subgroups by agglutination with rat and rhesus erythrocytes; hemagglutination inhibition (HAI) and neutralization tests will identify virus types.

3.2.7.e Enteroviruses

The first 67 types of enteroviruses have been divided into subgroups based on their growth characteristics in cell cultures and pathogenicity in newborn mice (Table 3.6). However, many exceptions occur and newly-recognized virus types are now classified simply as enteroviruses. In general, enteroviruses grow best in epithelial cells of primate origin. Polio- and coxsackie B viruses grow well in primary MK and HEp-2 cells, echovirus grows well in primary MK but not HEp-2 cells, and the universal host for coxsackie group A is the newborn mouse (Lipson et al, 1988); however, some strains grow in HDF, HEK, MK, RD (a rhabdomyosarcoma cell line), or GPE cells (Schmidt et al, 1975; Landry et al, 1981). Enteroviruses can be recovered from feces, throat swabs, CSF, blood, vesicle fluid, conjunctival swabs, and urine. Characteristically, infected monolayer cells round up, become refractile, shrink, degenerate, then detach from the surface of the culture vessel [Figure 3.3(2)]. Virus in the supernatant fluid can be subpassaged. Preliminary identification can be determined by characteristic CPE and differential cell susceptibility (Hsiung, 1982; Johnston and Siegel, 1990). Final identification by microneutralization tests in cell culture using antiserum pools is expensive and time-consuming, and is reserved for reference laboratories. In the future, identification may be performed using molecular techniques such as nucleic acid hybridization and polymerase chain reaction (Chapman et al, 1990).

3.2.7.f Rhinovirus

Rhinoviruses are classified as picornaviruses along with the enteroviruses, but can be separated from the latter by their sensitivity to low pH. Unlike enteroviruses, therefore, they cannot survive passage through the stomach and are not found in the gut. They can be isolated in cells of human origin (usually HDF), although some laboratories have used HEK, and occasionally primary MK. Varying sensitivity of different lots of cells can be a problem. Many rhinovirus types were originally isolated in organ cultures of human embryonic trachea. Sources of virus include nasal swabs or washes and throat swabs. Cultivation at 33 °C in a roller drum apparatus is optimal (Gwaltney and Jordan, 1964). CPE may occur from the first to the third week of incubation. The CPE is similar to the enteroviruses, starts as foci of rounded cells, and spreads gradually [Fig-

ure 3.3(8)]. CPE may not progress and may even disappear; if it is not progressing, subpassage of supernatant fluids from infected cells should be performed. Identification of isolates is by characteristic CPE and inactivation at pH 3. Typing by neutralization tests is reserved for research laboratories.

3.2.7.g Influenza

Primary MK is the most widely used cell culture for isolation of influenza, although the MDCK cells may be used if trypsin is added (Frank et al, 1979). Influenza A is reliably isolated in eggs and usually in MK as well, though strains differ (Smith and Reichrath, 1974). Influenza B is more readily isolated in MK than in eggs. Throat swabs, nasal swabs, and nasal washings are good sources for virus and should be collected early in illness. Serum components may inhibit influenza virus from replicating. Therefore, serum should be removed from cell cultures by rinsing with HBSS before inoculation and cultures should be maintained in serum-free media after inoculation. Incubation at 33 °C in a roller drum is optimal for isolation. The presence of virus is generally detected by hemadsorption of guinea pig RBC onto infected monolayers [Figure 3.6(2)]. CPE may be seen with influenza B [Figure 3.3(3)], and often with influenza A, but it occurs later than the detection of virus by hemadsorption. Subcultures can be performed by passaging the supernatant fluids. Isolates can be identified as influenza A or B by immunofluorescence using monoclonal antibodies. Strain differences are determined by hemagglutination inhibition.

3.2.7.h Parainfluenza

Primary MK is the most sensitive system for isolation of these viruses. Parainfluenza grows poorly or not at all in hens' eggs. HEK, HDF, and HEp-2 are less sensitive.

Throat and nasal swabs are good sources for virus. Cell cultures should be washed with HBSS before inoculation and refed with media without serum. Incubation at 33 °C to 36 °C in a roller drum is optimal. The presence of virus is detected by hemadsorption [Figure 3.6(2)], which occurs before CPE. Parainfluenza type 2 may produce syncytia, especially in HEp-2 cells. On subculture parainfluenza type 3 also may induce syncytia formation. In those instances when high levels of virus are present, hemadsorption may be detected in the infected cultures within a few days; with specimens containing less virus, ten days or more of incubation may be necessary. Identification is by hemadsorption inhibition or immunofluorescence.

3.2.7.i Respiratory Syncytial Virus

Respiratory syncytial virus (RSV) grows best in continuous cell lines, such as HEp-2, in which it produces characteristic syncytia [Figure 3.3(6)]. However, syncytium formation is variable and viral replication may be missed. If HEp-2 cells are confluent and five to seven days old when inoculated, syncytia may not form. Rather non-specific rounding may occur. Syncytia formation is also dependent upon the presence of adequate levels of glutamine and calcium in the medium (Marquez and Hsiung, 1967; Shahrabadi and Lee, 1988). Primary MK cells show CPE. Isolation in A549 cells is variable. HDF cells support its growth but are less sensitive and the cytopathology is not characteristic. RSV is found in respiratory secretions from the nose and oropharynx. However, RSV is more readily detected from nasopharyngeal aspirates than from swabs (Ahluwalia et al, 1987). Rapid diagnosis of RSV is very important in order to isolate infected infants in the hospital. Antigen detection tests, such as ELISA and IF, performed directly on clinical specimens, are extremely useful to achieve this. However, viral culture is still

necessary for optimum detection of RSV and for the recovery of other viral respiratory pathogens (Blanding et al, 1989). Identification is by immunofluorescence.

3.3 APPENDIX: ADDITIONAL METHODS FOR VIRUS ISOLATION

3.3.1 Explant Culture or Cocultivation

Clinical specimens, such as fresh tissue cells, can be grown out (explanted) or cocultivated with cells susceptible to the suspected agent in the specimen and observed for the development of CPE. The procedure is as follows:

1. Mince tissues finely in 0.25% trypsin.
2. Centrifuge at low speed to pellet the cells.
3. Remove the trypsin.
4. Resuspend in Hanks' balanced salt solution (HBSS) with antibiotics to a 10% or 20% suspension.
5. For cocultivation, inoculate 0.1 to 0.2 mL onto cell monolayers.
6. For explant culture, resuspend in growth medium with 20% calf serum and disperse into petri dishes or flasks.

3.3.2 Organ Culture

For certain fastidious viruses, such as rhinovirus, coronavirus, or rotavirus, isolation can be accomplished using whole or part of an organ in vitro, such as human embryonic trachea or intestine, allowing preservation of architecture and/or function (Tyrrell and Bynoe, 1965; McIntosh et al, 1967; Wyatt et al, 1974). Virus growth may be detected by subculture of nutrient fluids onto monolayers, or by observation of cessation of ciliary activity and eventual degeneration of epithelial cells. This technique is tedious and is not routinely used in a clinical laboratory. Detailed procedures can be found in the textbooks in the reference list.

3.3.3 Virus Isolation in Embryonated Eggs

3.3.3.a Background

The chick embryo is a highly sensitive host for the primary isolation of several virus types. Some influenza A virus strains are more readily isolated in embryonated eggs than in cell culture, although the reverse is true for influenza B virus strains. The chick embryo is also a sensitive host for the isolation of mumps virus, however, MK cells are a sensitive host for the rapid isolation of this agent. For primary isolation of influenza and mumps, inoculation of the amniotic sac is necessary. Subsequently, virus isolates can be adapted to grow in the allantoic sac.

The chorioallantoic membrane (CAM) is highly susceptible to pock formation by HSV and poxviruses. With the availability of numerous sensitive cell cultures for the isolation of HSV and vaccinia viruses, and the eradication of smallpox, CAM inoculation is now rarely employed. However, HSV and poxviruses can be differentiated by the morphologic characteristics of the pocks that they produce in this system.

3.3.3.b Maintenance and Source of Eggs

Fertile hens' eggs for virus isolation should be obtained from flocks that are free of infection (Newcastle disease virus and mycoplasma can be particularly troublesome). Eggs should be incubated at 37 °C in an atmosphere of 40% to 70% humidity to ensure proper development of the air sac. To prevent adhesions of the embryonic membranes and to keep the embryo centralized, the eggs should be turned two to four times each day. After four to five days of incubation the eggs are candled to determine which ones are fertile and contain developing embryos.

3.3.3.c Inoculation and Incubation[1]
Amniotic or Allantoic Cavity Inoculation

1. Use 7- to 13-day-old embryonated eggs (optimum for mumps, 7 to 8 days; for influenza, 10 to 13 days)
2. Candle the eggs to locate the embryo and detect movement; make a puncture through the shell over the air sac
3. Inoculate 0.1 to 0.2 mL of the specimen into the amniotic sac, which is entered with a quick stabbing motion, and/or allantoic cavity as the needle is withdrawn using a 1¾ inch-long, 23-gauge needle; use three or four eggs per specimen. When the needle is in the amniotic cavity, gentle pressure will move the embryo [Figure 3.9(A)]. (These procedures should be performed while the egg is illuminated on a candler)
4. Seal the hole in the shell with Scotch tape
5. Incubate eggs at 35 °C to 37 °C with air sac uppermost
6. Candle inoculated eggs daily. Discard those that die within 24 hours after inoculation

Chorioallantoic Membrane Inoculation

1. Candle eggs containing 9- to 12-day-old developing chicken embryos; mark an area free from large blood vessels on the side where the embryo is located and the area over the air sac
2. Drill two slits in the eggshell, one on the side and the other over the air sac
3. Puncture the shell membrane under the slits with a sterile needle; care should be taken not to damage the chorioallantoic membrane (CAM)
4. With a rubber bulb, gently apply suction at the hole over the air sac. If this procedure is carried out while the egg is being candled, one can see the CAM

---

[1] Parts excerpted from Diagnostic Virology by G.D. Hsiung, Yale University Press, 1982.

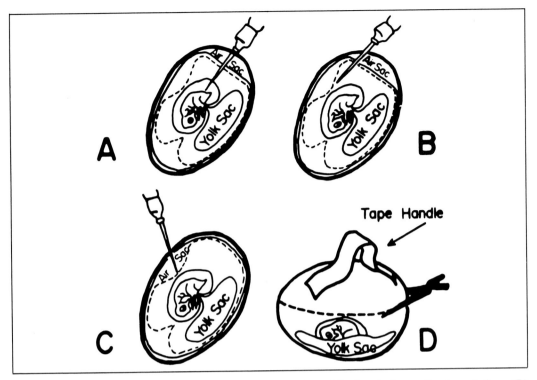

Figure 3.9 Embryonated egg inoculation and harvesting. (A) Amniotic cavity inoculation; (B) allantoic cavity inoculation. (C) chorioallantoic membrane inoculation (note artificial air sac); (D) chorioallantoic membrane harvesting. (Reproduced with permission from Hsiung, G.D. 1982. Diagnostic Virology, Figure 5. New Haven: Yale University Press, p. 22.)

drop and a new, artificial air sac form

5. Place 0.05 mL inoculum on the dropped membrane by inserting the needle through the side slit to a depth of about 5 mm [Figure 3.9(C)]. Withdraw the needle very slowly. Seal the opening with Scotch tape or wax and incubate at 35 °C to 37 °C with the artificial air sac uppermost

6. Candle the inoculated eggs daily; discard those that die within 24 hours after inoculation

3.3.3.d Harvesting, Assay, and Identification of Isolates

Amniotic and Allantoic Fluids

1. Harvest amniotic fluids and allantoic fluids separately, two to four days after inoculation for influenza and five to seven days for mumps. The procedure for harvesting egg fluids is as follows:

- Chill eggs at 4 °C overnight, or for 30 minutes at −20 °C then two to four hours in the refrigerator, in order to clot the blood
- Open the egg shell over the air sac
- Cut out and remove the overlying shell membrane and CAM with scissors
- Aspirate amniotic and allantoic fluids separately with sterile capillary pipettes

2. Carry out a spot hemagglutination test by mixing 0.025 mL allantoic or amniotic fluids, both undilute and a 1:5 dilution in phosphate buffered saline (PBS) with 0.025 mL of a 0.5% suspension of guinea pig or chick RBC. Allow this

mixture to stand at room temperature for 35 to 45 minutes before reading. If a hemagglutinating virus is present, the RBC will not settle to the bottom but will remain in suspension. Nonspecific hemagglutination can be caused by bacterial contamination

3. Blind passage of egg fluids in eggs may be necessary before virus is detected

4. Isolates are identified by the inhibition of hemagglutination using specific antisera (HI test)

Chorioallantoic Membrane

1. Place the inoculated egg on a holder so that the slit through which the inoculum was delivered faces upward

2. Make a tape handle over the area of inoculation [Figure 3.9(D)]

3. Cut off the top half of the eggshell, including the infected area, and gently remove the CAM, which is attached to the shell

4. Place the infected CAM in a Petri dish with a few milliliters of phosphate buffered saline; spread the membrane flat against the bottom of the dish; place the dish on a dark surface to facilitate counting of pocks

5. Variola, vaccinia, monkeypox, cowpox viruses, and HSV types 1 and 2 can be differentiated by the morphologic characteristics of the pocks they produce on the CAM. In doubtful cases, the pocks can be ground and examined by EM, or used as antigens for complement fixation or agar gel precipitation tests.

3.3.4 Virus Isolation in Mice

3.3.4.a Background

Although suckling mice are the definitive host for certain viruses, isolation of virus in mice is rarely performed in a clinical laboratory today. Group A and B coxsackieviruses were originally isolated in suckling mice and were differentiated by the pathologic lesions and illnesses that they produce

(Melnick, 1983). Coxsackie B viruses, however, are readily isolated in cell culture. Although no universal cell system for the isolation of group A viruses has been found, several cell systems such as RD cells (Schmidt et al, 1975), GPE cells (Landry et al, 1981), and HDF support the growth of a number of the coxsackie A serotypes.

The suckling mouse is also the universal host for togaviruses, many of which are etiologic agents of encephalitis. Several cell lines and cell cultures derived from insects are now available for many of these viruses (Shope and Sather, 1979).

3.3.4.b Inoculation and Observation for Illness

Pregnant mice are obtained and the entire litter of mice is inoculated with the clinical specimen within 24 to 48 hours of birth, as described below (Figure 3.10). Older mice are less susceptible to infection. Mice should be obtained from pretested virus-free mouse colonies (Sendai virus, coronavirus, and many others can be troublesome). Materials needed include animal facilities for housing noninfected and infected animals, needles, syringes, scissors, and forceps.

1. Inoculate newborn mice within 24 to 48 hours of birth, with 0.01 to 0.02 mL/mouse intracerebrally, and/or with 0.03 to 0.05 mL/mouse intraperitoneally, using a 27-gauge ⅜-in. long needle and a ½-cc syringe (Figure 3.10)

2. Check inoculated mice twice daily for signs of illness, paralysis, or death (Figure 3.10)

3.3.4.c Harvesting and Processing Infected Tissues[2]

1. Harvest mouse brain or skeletal muscle when animals are paralyzed or when other symptoms appear

---

[2] Parts excerpted from *Diagnostic Virology* by G.D. Hsiung, Yale University Press, 1982.

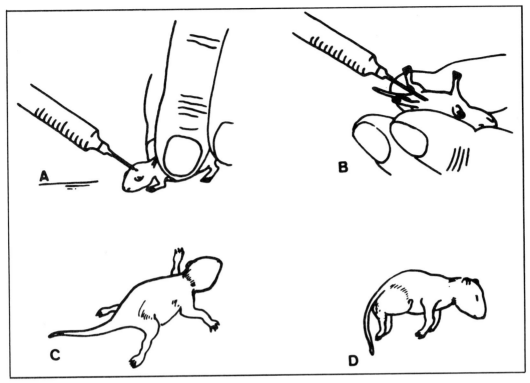

Figure 3.10 Newborn mouse inoculation. (A) Intracerebral inoculation; (B) intraperitoneal inoculation; (C) flaccid hindlimb paralysis as seen with coxsackie A virus infection; (D) spastic paralysis secondary to coxsackie B infection. (Modified from Hsiung, G.D. 1982. Diagnostic Virology, Figure 4. New Haven: Yale University Press, p. 22.)

2. Make 10% tissue suspension and inoculate into additional mice and/or appropriate cell cultures, if the latter are available, for further study

3.3.4.d Identification of Isolates

Mouse protection neutralization of virus-induced death or inhibition of CPE in cell culture can be performed for final identification. It is important to recognize that the mice may also become ill due to their own endogenous viruses and that cannibalism may occur.

When working with togaviruses, it should be appreciated that the isolation and identification of certain of these viruses can be associated with risks to the laboratory personnel (Hanson et al, 1967). Unless appropriate precautions can be taken, no attempts should be made to isolate these viruses outside of the appropriate reference facilities.

REFERENCES

Ahluwalia, G., Embree, J., McNicol, P., Law, B., and Hammond, G. 1987. Comparison of nasopharyngeal aspirate and nasopharyngeal swab specimens for respiratory syncytial virus diagnosis by cell culture, indirect immunofluorescence assay, and enzyme-linked immunosorbent assay. J. Clin. Microbiol. 25:763–767.

Balkovic, E.S. and Hsiung, G.D. 1985. Comparison of immunofluorescence with commercial monoclonal antibodies to biochemical and biological techniques for typing clinical herpes simplex virus isolates. J. Clin. Microbiol. 22:870–872.

Blanding, J.G., Hoshiko, M.G., and Stutman, H.R. 1989. Routine viral culture for pediatric respiratory specimens submitted for direct immunofluorescence testing. J. Clin. Microbiol. 27:1438–1440.

Broaddus, C., Dake, M.D., Stulbarg, M.S., Blumenfeld, W., Hadley, W.K., Golden, J., and Hopewell, P.C. 1985. Bronchoalveolar lavage and transbronchial biopsy for the diagnosis of pulmonary infections in the acquired immunodeficiency syndrome. Ann. Intern. Med. 102:747–752.

Chapman, N.M., Tracy, S., Gauntt, C.J., and Fortmueller, U. 1990. Molecular detection and identification of enteroviruses using enzymatic amplification and nucleic acid hybridization. J. Clin. Microbiol. 28:843–850.

Chu, F.C., Johnson, J.B., Orr, H.C., Probst, P.G., and Petricciani, J.C. 1973. Bacterial virus contamination of fetal bovine sera. In Vitro 9:31–34.

Dowdle, W.R. and Robinson, R.Q. 1966. Nonspecific hemadsorption by rhesus monkey kidney cells. Proc. Soc. Exp. Biol. Med. 121:193–198.

Enders, J.F., Weller, T.H., and Robbins, F.C. 1949. Cultivation of the Lansing strain of poliomyelitis virus in cultures of various human embryonic tissues. Science 109:85–87.

Espy, M.J., Smith, T.F., Harmon, M.W., and Kendall, A.P. 1986. Rapid detection of influenza virus by shell vial assay with monoclonal antibodies. J. Clin. Microbiol. 24:677–679.

Espy, M.J., Hierholzer, J.C., and Smith, T.F. 1987. The effect of centrifugation on the rapid detection of adenovirus in shell vial. Am. J. Clin. Pathol. 88:358–360.

Espy, M.J., Wold, A.D., Ilstrup, D.M., and Smith, T.F. 1988. Effect of treatment of shell vial cell cultures with dimethyl sulfoxide and dexamethasone for detection of cytomegalovirus. J. Clin. Microbiol. 26:1091–1093.

Fayram, S.L., Aarnaes, S., and de la Maza, L.M. 1983. Comparison of Cultureset to a conventional tissue culture-fluorescent antibody technique for isolation and identification of herpes simplex virus. J. Clin. Microbiol. 18:215–216.

Fedorko, D.P., Ilstrup, D.M., and Smith, T.F. 1989. Effect of age of shell vial monolayers on detection of cytomegalovirus from urine specimens. J. Clin. Microbiol. 27:2107–2109.

Frank, A.L., Couch, R.B., Griffis, C.A., and Baxter, B.D. 1979. Comparison of different tissue cultures for isolation and quantitation of influenza and parainfluenza viruses. J. Clin. Microbiol. 10:32–36.

Gary, G.W., Jr., Hierholzer, J.C., and Black, R.E. 1979. Characteristics of noncultivable adenoviruses associated with diarrhea in infants: A new subgroup of human adenoviruses. J. Clin. Microbiol. 10:96–103.

Gleaves, C.A., Smith, T.F., Shuster, E.A., and Pearson, G.R. 1984. Rapid detection of cytomegalovirus in MRC-5 cells inoculated with urine specimens by using low-speed centrifugation and monoclonal antibody to an early antigen. J. Clin. Microbiol. 19:917–919.

Gleaves, C.A., Wilson, D.J., Wold, A.D., and Smith, T.F. 1985. Detection and serotyping of herpes simplex virus in MRC-5 cells by use of centrifugation and monoclonal antibodies 16 hr postinoculation. J. Clin. Microbiol. 21:29–32.

Gleaves, C.A., Lee, C.F., Bustamante, C.I., and Meyers, J.D. 1988. Use of murine monoclonal antibodies for laboratory diagnosis of varicella-zoster virus infection. J. Clin. Microbiol. 26:1623–1625.

Gregory, W.W. and Menegus, M.A. 1983. Effect of incubation temperature on isolation of cytomegalovirus from fresh clinical specimens. J. Clin. Microbiol. 18:1003–1005.

Gwaltney, J.M., Jr. and Jordan, W.S., Jr. 1964. Rhinoviruses and respiratory disease. Bacteriol. Rev. 28:409–422.

Hanson, R.P., Sulkin, S.E., Buescher, E.L., Hammon, W.McD., McKinney, R.W., and Work, T.H. 1967. Arbovirus infections of laboratory workers. Science 158:1283–1286.

Herrmann, E.C., Jr. 1972. Rates of isolation of

viruses from a wide spectrum of clinical specimens. Am. J. Clin. Pathol. 57:188–194.

Hsiung, G.D. 1961. Production of an inhibitor substance by DA myxovirus. Auto-inhibition of plaque formation. Proc. Soc. Exp. Biol. Med. 108:357–360.

Hsiung, G.D. 1968. Latent virus infections in primate tissues with special reference to simian viruses. Bact. Rev. 32:185–205.

Hsiung, G.D. 1982. Diagnostic Virology, 3rd ed. New Haven: Yale University Press.

Hsiung, G.D. and Green, R.H., eds. 1978. Virology and Rickettsiology, Vol. 1, Parts 1 and 2. Handbook Series in Clinical Laboratory Science, Section H. West Palm Beach, FL: CRC Press.

Hsiung, G.D. and Melnick, J.L. 1957. Morphologic characteristics of plaques produced on monkey kidney monolayer cultures by enteric viruses (poliomyelitis, coxsackie and ECHO groups). J. Immunol. 78:128–136.

Hu, J.M. and Hsiung, G.D. 1989. Evaluation of antiviral agents: I. In vitro perspectives, Antiviral Research 11:217–232.

Hudson, J.B., Misra, V., and Mosmann, T.R. 1976. Cytomegalovirus infectivity: Analysis of the phenomenon of centrifugal enhancement of infectivity. Virology 72:235–243.

Johnston, S.L.G., Wellens, K., and Siegel, C.S. 1990. Rapid isolation of herpes simplex virus by using mink lung and rhabdomyosarcoma cell cultures. J. Clin. Microbiol. 28:2806–2807.

Johnston, S.L.G. and Siegel, C.S. 1990. Presumptive identification of enterovirus with RD, HEP-2, and RMK cell lines. J. Clin. Microbiol. 28:1049–1050.

Krizanova, O. and Rathova, V. 1969. Serum inhibitors of myxoviruses. Curr. Top. Microbiol. Immunol. 47:125–151.

Landry, M.L., Madore, H.P., Fong, C.K.Y., and Hsiung, G.D. 1981. Use of guinea pig embryo cell cultures for isolation and propagation of group A coxsackieviruses. J. Clin. Microbiol. 13:588–593.

Landry, M.L., Mayo, D.R., and Hsiung, G.D. 1982. Comparison of guinea pig embryo cells, rabbit kidney cells and human embryonic fibroblast cell strains for the isolation of herpes simplex virus. J. Clin. Microbiol. 15:842–847.

Landry, M.L., Mayo, D.R., and Hsiung, G.D. 1989. Rapid and accurate viral diagnosis. Pharmacol. Ther. 40:287–328.

Lipson, S.M., Walderman, R., Costello, P., and Szabo, K. 1988. Sensitivity of rhabdomyosarcoma and guinea pig embryo cell cultures to field isolates of difficult-to-cultivate group A coxsackieviruses. J. Clin. Microbiol. 26:1298–1303.

Lonsdale, D.M. 1978. A rapid technique for distinguishing herpes simplex virus type 1 from type 2 by restriction enzyme technology. Lancet 1:849–851.

Mahafzah, A.M. and Landry, M.L. 1989. Evaluation of immunoflourescent reagents, centrifugation, and conventional culture for the rapid diagnosis of adenovirus infections. Diagn. Microbiol. Infect. Dis. 12:407–411.

Marquez, A. and Hsiung, G.D. 1967. Influence of glutamine on multiplication and cytopathic effect of respiratory syncytial virus. Proc. Soc. Exp. Biol. Med. 124:95–99.

Mavromoustakis, C.T., Witiak, D.T., and Hughes, J.H. 1988. Effect of high-speed rolling on herpes simplex virus detection and replication. J. Clin. Microbiol. 26:2328–2331.

Mayo, D.R. 1982. Differentiation of herpes simplex virus types 1 and 2 by sensitivity to (E)5-(2-Bromovinyl)-2' deoxyuridine. J. Clin. Microbiol. 15:733–736.

McIntosh, K., Dees, J.H., Becker, W.B., Kapikian, A.Z., and Chanock, R.M. 1967. Recovery in tracheal organ culture of novel viruses from patients with respiratory disease. Proc. Natl. Acad. Sci. USA 57:933–940.

Melnick, J.L. 1983. Portraits of viruses: The picornaviruses. Intervirology 20:61–100.

Nordlund, J.J., Anderson, C., Hsiung, G.D., and Tenser, R.B. 1977. The use of temperature sensitivity and selective cell culture system for differentiation of herpes simplex virus types 1 and 2 in a clinical laboratory. Proc. Soc. Exp. Biol. Med. 155:188–123.

Oefinger, P.E., Shawar, R.M., Loo, S.H., Tsai, L.T., and Arnett, J.K., 1990. Enhanced recovery of cytomegalovirus in conventional tube cultures with a spin-amplified adsorption. J. Clin. Microbiol. 28:965–969.

Paya, C., Wold, A.D., and Smith, T.F. 1987. Detection of cytomegalovirus infections in specimens other than urine by shell vial assay and conventional tube cell cultures. J. Clin. Microbiol. 25:755–757.

Paya, C., Wold, A.D., Ilstrup, D.M., and Smith, T.F. 1988. Evaluation of number of shell vial cultures per clinical specimen for rapid diagnosis of cytomegalovirus infection. J. Clin. Microbiol. 26:198–200.

Porterfield, J.S. 1960. A simple plaque-inhibition test for the study of arthropod-borne viruses. Bull. W.H.O. 22:373.

Rabella, N. and Drew, W.L. 1990. Comparison of conventional and shell vial cultures for detecting cytomegalovirus infection. J. Clin. Microbiol. 28:806–807.

Reed, L.J. and Muench, H.A. 1938. A simple method of estimating fifty percent end points. Am. J. Hyg. 27:493–497.

Reeve, P., Owen, J., and Oriel, J.D. 1975. Laboratory procedures for the isolation of *Chlamydia trachomatis* from the human genital tract. J. Clin. Pathol. 28:910–914.

Rossier, E., Miller, H.R., and Phipps, P.H. 1989. Rapid Viral Diagnosis by Immunofluorescence. Ottawa, Canada: Univ. of Ottawa Press.

Rowe, W.P. and Hartley, J.W. 1962. A general review of adenoviruses. Ann. N.Y. Acad. Sci. 101:466–474.

Schimidt, N.H., Ho, H.H., and Lennette, E.H. 1975. Propagation and isolation of group A coxsackieviruses in RD cells. J. Clin. Microbiol. 2:183–185.

Schimidt, N.J. and Emmons, R.W., eds. 1989. Diagnostic Procedures for Viral, Rickettsial and Chlamydial Infections. Washington, D.C.: American Public Health Association.

Sewell, D.L., Horn, S.A., and Pilbeck, P.W. 1984. Comparison of Cultureset and Bartels Immunodiagnostics with conventional tissue culture for isolation and identification of herpes simplex virus. J. Clin. Microbiol. 19:705–706.

Shahrabadi, M.S. and Lee, P.W.K. 1988. Calcium requirement for syncytium formation in HEp-2 cells by respiratory syncytial virus. J. Clin. Microbiol. 26:139–141.

Weller, T.H. and Hanshaw, J.B. 1962. Virologic and clinical observation on cytomegalic inclusion disease. N. Engl. J. Med. 266:1233–1244.

Weller, T.H., Witton, M.M., and Bell, E.J. 1958. The etiologic agents of varicella and herpes zoster: Isolation, propagation, and cultural characteristics in vitro. J. Exp. Med. 108:843–852.

West, P.G., Aldrich, B., Hartwig, R., and Haller, G.J. 1988. Enhanced detection of cytomegalovirus in confluent MRC-5 cells treated with dexamethasone and dimethyl sulfoxide. J. Clin. Microbiol. 26:2510–2514.

West, P.G., Aldrich, B.A., Hartwig, R., and Haller, G.J. 1989. Increased detection of herpes simplex virus in MRC-5 cells treated with dimethyl sulfoxide and dexamethasone. J. Clin. Microbiol. 27:770–772.

Woods, G.L. and Young, A. 1988. Use of A-549 cells in a clinical virology laboratory. J. Clin. Microbiol. 26:1026–1028.

Shope, R.E. and Sather, G.E. 1979. Arboviruses. In E.H. Lennette and N.H. Schmidt (eds.), Diagnostic Procedures for Viral, Rickettsial and Chlamydial Infections. Washington, D.C.: American Public Health Association.

Smith, K.O. 1970. Adventitious viruses in cell cultures. Prog. Med. Virol. 12:302–336.

Smith, T.F. and Reichrath, L. 1974. Comparative recovery of 1972–1973 influenza virus isolates in embryonated eggs and primary rhesus monkey kidney cell cultures after one freeze-thaw cycle. Am. J. Clin. Pathol. 61:579–584.

Stanbridge, E. 1971. Mycoplasmas and cell cultures. Bact. Rev. 35:206–227.

Taylor-Robinson, D. 1959. Chicken pox and herpes zoster. III. Tissue culture studies. Br. J. Exp. Pathol. 40:521–532.

Tyrrell, D.A.J. and Bynoe, M.L. 1965. Cultivation of a novel type of common cold virus in organ cultures. Br. Med. J. 1:1467–1470.

Wyatt, R.D., Kapikian, A.Z., Thornhill, T.S., Sereno, M.M., Kim, H.W., and Chanock, R.M. 1974. In vitro cultivation in human fetal intestinal organ culture of a reovirus-like agent associated with nonbacterial gastroenteritis in infants and children. J. Infect. Dis. 130:523–528.

Zhao, L., Landry, M.L., Balkovic, E.S., and Hsiung, G.D. 1987. Impact of cell culture sensitivity and virus concentration on rapid detection of herpes simplex virus by cytopathic effects and immunoperoxidase staining. J. Clin. Microbiol. 25:1401–1405.

The Cytopathology of Virus Infections

Roger D. Smith and Karen Sutherland

4.1 INTRODUCTION

Virus-infected cells that exfoliate or are scraped from the skin or mucous membranes may contain readily identifiable morphologic changes that permit rapid diagnosis. In many instances, the cytologic alterations may be so distinctive as to be pathognomonic for infection with a specific agent. In others, the changes may point to a virus group or merely raise a suspicion of infection to be confirmed by other means. The purpose of this chapter is to describe the methods used to obtain and prepare cells for cytologic examination, and to illustrate characteristic changes encountered in common virus infections. Most of the methods described are used routinely by diagnostic cytology laboratories and are best applied by the cytotechnologist or pathologist who analyzes cytologic material on a daily basis.

4.2 PREPARATION AND STAINING

The proper collection, fixation, preparation, and staining of specimens for cytology is essential. Rapid fixation in 95% alcohol or with cytology spray fixative before the smear dries is imperative for accurate interpretation. The purpose of fixation is to maintain the existing form and structure of the cellular elements and to achieve consistent staining characteristics and identifiable structures. Improper specimen preparation will decrease the diagnostic accuracy and may lead to false-positive results. Air drying causes nuclear swelling and distortion, cytoplasmic vacuolization, and atypical staining. These changes can mimic nuclear and/or cytoplasmic alterations seen with some virus infections or can distort the characteristic details to such an extent that viral cytopathology cannot be identified. For example, the ground-glass nuclear appearance seen in early herpes simplex virus (HSV) infection can be confused with the smudgy nuclear detail seen in air-dried specimens. Aqueous fixatives (e.g., formalin) result in poor staining and irregular condensation of chromatin, which can be mistaken for a nuclear inclusion.

The preparatory methods utilized for the microscopic examination of cytologic specimens can be divided into four categories: direct smears, preparation by cytocentrifugation, membrane filter preparation, and cell block preparation (Bales and Durfee, 1979).

4.2.1 Direct Smears

Direct scraping of vesicular or bullous lesions of the skin and mucous membranes is the simplest method of cell collection for the identification of viral changes. The scrapings from the base and edges of the suspected lesion should be smeared evenly onto a clean glass slide and immediately fixed with 95% alcohol or cytology spray fixative. Once fixed, the smears are almost indefinitely stable at room temperature and can be stored, mailed, or transported to a cytology laboratory for further processing. The slides are then stained with a modified Papanicolaou staining technique. Proper sampling is important because the crusted or eczematous areas often fail to show the diagnostic cellular features. If a specific lesion, such as an ulcer or unroofed vesicle, is present, the base and edges of the lesion should be thoroughly scraped with a spatula, tongue blade, or endoscopic brush to insure proper sampling.

4.2.1.a Modified Papanicolaou Stain

The modified Papanicolaou stain technique is used to detect cytologic changes due to viral infection. The procedure for staining is as follows:

1. Ten dips in 95% ETOH
2. Ten dips in 70% ETOH
3. Ten dips in 50% ETOH
4. Ten dips in distilled H_2O
5. Two minutes in hematoxylin [Gill hematoxylin, consisting of 2190 mL distilled H_2O, 750 mL ethylene glycol, 6 g hematoxylin (C.I. #75290), 0.6 g sodium iodate, 528 g aluminum sulfate, and 60 mL glacial acetic acid]
6. One minute under running tap water
7. One minute in Scott's tap water substitute (consisting of 10 g anhydrous magnesium sulfate, 2 g sodium bicarbonate, and 1 L distilled water)
8. Ten dips in tap water
9. Ten dips in 50% ETOH
10. Ten dips in 95% ETOH
11. One and one-half minutes in OG-6[1]
12. Ten dips in 95% ETOH
13. Ten dips in 95% ETOH
14. Three minutes in EA-65[1]
15. Ten dips in 95% ETOH
16. Ten dips in 95% ETOH
17. Ten dips in 95% ETOH
18. Ten dips in 100% ETOH
19. Ten dips in 100% ETOH
20. Ten dips in 100% ETOH/Hemo-De[2] (equal amounts)
21. Ten dips in Hemo-De, in three consecutive dishes
22. Coverslip slide with Permount media[3]

4.2.2 Cytocentrifugation and Filtration

New cytocentrifugation and filtration techniques have been developed which concentrate small numbers of cells suspended in fluids and are the preferred method for preparation of samples from urine, cerebrospinal, and body cavity fluids. To determine which of the two techniques should be used, one needs to consider the expected number of cells in the fluid. Fluids containing many cells should be centrifuged. The specimen is placed in a centrifuge tube and spun at 1500 rpm for 10 minutes. The supernatant fluid is decanted leaving a volume of 2 mL in the tube and the sediment is resuspended. The specimen is then prepared using the standard cytocentrifugation technique (Barrett, 1976). If little or no sediment is present, then the filtration technique (Gill, 1976) is the method of choice.

---

[1] From Harleco Co., Gibbstown, New Jersey.

[2] Clearing Agent from Fisher Scientific Co., Philadelphia, Pennsylvania.

[3] From Fisher Scientific Co., Philadelphia, Pennsylvania.

The use of membrane filters should be limited to cases where the number of cells is low and where additional cell sampling presents a problem. Cytospin preparations offer several advantages. The cells are evenly dispersed on the slide (monolayer), little or no background artifact is present, and the cell preparation can be utilized easily for other diagnostic procedures, such as special stains, immunofluorescence, immunoperoxidase, and electron microscopy. Cell block preparations require a histopathology laboratory and are used when there is an abundance of cellular material that can be embedded in paraffin and sectioned for histologic examination. Appropriate precautions to insure against potential infection of laboratory personnel should be taken with all human tissues and body fluids. In processing unfixed specimens for cytology, the individual should wear gloves and other protective garments. Fluids that potentially can aerosolize should be processed in a class II biological safety cabinet.

4.3 VIRUS CYTOPATHOLOGY

The eye, skin, and respiratory, genital, and urinary tracts are locations that readily yield cytologic material for rapid viral diagnosis. Characteristic cytologic changes depend on the cytopathic effect of a virus in infected cells, which need not include all cells of the involved organ. In practice, however, it is most useful to consider cytologic alterations in the context of organ system affected and clinical presentation. Therefore, the following discussion and illustrations are organized according to organ system.

4.3.1 Viral Infections of the Respiratory Tract

Smears of cells obtained by nasal and throat swabs, tracheal aspirates, sputum, bronchial washings and brushings, and pulmonary lavage may exhibit cytologic alterations that are diagnostic of virus infection (Table 4.1). In adults, the most frequently encountered virus that is readily detectable by cytology is cytomegalovirus (CMV) (Warner, 1964). Particularly in patients with the acquired immunodeficiency syndrome (AIDS) and those who are receiving immunosuppressive therapy for transplantation or cancer chemotherapy, the rapid cytologic identification of characteristic inclusions may be a great asset in patient management, particularly because CMV isolation in tissue culture takes many days and often weeks. Because CMV infection involves the lungs, a deep specimen containing pulmonary macrophages is needed. Patients with CMV pneumonia rarely produce abundant sputum, thus requiring bronchial washings and brushings or pulmonary lavage to obtain an adequate specimen. The use of bronchoalveolar lavage in the evaluation of diffuse pulmonary infiltration in the immunocompromised patient yields large numbers of cells that can be used for rapid cytologic diagnosis, particularly of cytomegalovirus (Springmeyer et al, 1986). The characteristic cytologic changes are seen in pulmonary macrophages or in cells lining the alveoli. They most often exhibit a single nucleus but occasionally two or more nuclei that are four to six times their normal size are present (Figure 4.1). Early in the infection these nuclei contain amphoteric or basophilic inclusions that are granular, and the inclusions become condensed and surrounded by a halo in the later stages of infection. Smaller, more eosinophilic oval cytoplasmic inclusions that are periodic acid-Schiff reaction (PAS)-positive are often, but not invariably, present. At later stages of the infection, cytoplasmic inclusions may predominate with an empty or collapsed nucleus (Figure 4.6).

Characteristic inclusion-bearing cells may be observed in sputum and bronchial

Table 4.1 Cytopathology of Respiratory Viral Infections

Virus	Clinical Presentation	Cytologic Findings
Adeno	URI Pneumonia	Small multiple eosinophilic IN inclusions (early); large, single, dense basophilic IN inclusion (late)
CMV	Pneumonia	Cytomegaly; large, single, amphophilic IN inclusions; small PAP-positive IC inclusion
HSV	Tracheobronchitis	Large ground-glass nucleus (early): eosinophilic IN inclusions (late): multinuclearity with nuclear molding
Parainfluenza	Bronchitis pneumonia	Cytomegaly, single nucleus, small eosinophilic IC inclusions
RSV	Tracheobronchitis pneumonia	Large multinucleated cells; IC basophilic inclusions with prominent halos
Measles	Prodromal	Mulberry-like clusters of lymphocytic nuclei in nasal secretions
	Pneumonia	Multinucleated giant cells with IN and IC eosinophilic inclusions
Nonspecific (many viruses)	Bronchitis pneumonia	Ciliocytophthoria

Abbreviations: IN, intranuclear, IC, intracytoplasmic.

washings in HSV tracheobronchitis, which is encountered often in immunosuppressed and burn patients (Vernon, 1982). Inclusion-bearing cells tend to be multinucleated and contain either eosinophilic intranuclear inclusions that are centrally located and surrounded by a halo or, at an earlier stage of infection, these cells contain ground-glass inclusions that stain poorly (Figure 4.2). The chromatin often appears as a basophilic ring condensed at the periphery of the nuclear membrane. Cytoplasmic inclusions are not present. When there are multiple nuclei, they are frequently molded or indented by each other (Figure 4.3). Because HSV tends to produce cellular necrosis, the background of these smears usually contains an abundance of cellular debris.

Adenovirus infections of the upper respiratory tract may be identified by smears of secretions from the nasopharynx. Adenovirus pneumonia may be diagnosed by finding typical intranuclear inclusions in bronchial or epithelial cells obtained by bronchoscopy. At an early stage (Figure 4.16) the nucleus contains multiple small, rounded eosinophilic inclusions, each surrounded by a halo. At a later stage, a single larger, dense, basophilic intranuclear inclusion is seen (Figure 4.17).

Respiratory syncytial virus (RSV) and parainfluenza viruses frequently cause bronchitis and pneumonia in infants and young children. They can be rapidly diagnosed by identifying characteristic cytologic changes in respiratory epithelial cells obtained by nasopharyngeal swabs and tra-

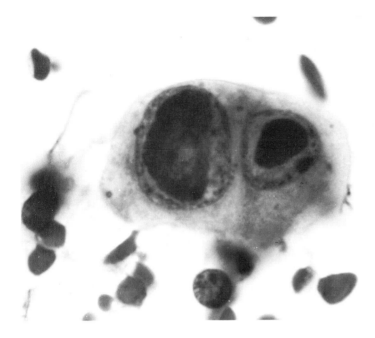

Figure 4.1 CMV in bronchial brushing. Large intranuclear inclusions in binucleated cell and small cytoplasmic inclusions. Pap stain, ×800

cheal aspirates (Naib et al, 1968). Parainfluenza can be differentiated from RSV by finding large cells containing a single nucleus and multiple small eosinophilic inclusions. In RSV infection, the epithelial cells are large and multinucleate and the cytoplasm contains multiple basophilic inclusions with prominent halos.

Measles can be detected during the prodrome by finding mulberry-like clusters of lymphocytes having up to 50 nuclei in smears of nasal secretions (Tomkins and

Figure 4.2 HSV in sputum. Epithelial cells showing the ground-glass nuclear appearance of the early stage. Pap stain, ×800

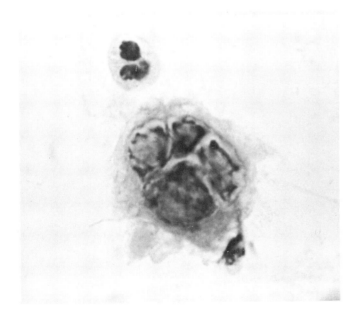

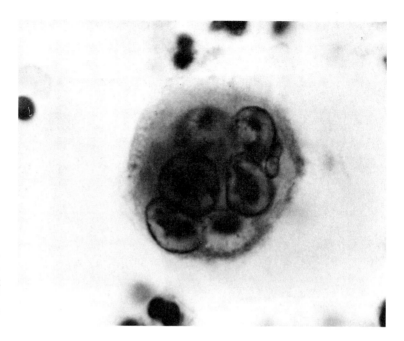

Figure 4.3 HSV in bronchial brushings. Multinucleated cell with molding of nuclei and typical HSV intranuclear inclusions. Pap stain, ×800

Macanlay, 1955). Measles giant cells (Figure 4.4) are multinucleate respiratory epithelial cells with intranuclear and cytoplasmic inclusions. These cells may appear in the sputum of patients with measles pneumonia.

A nonspecific change referred to as ciliocytophthoria is found in various inflammatory diseases of the respiratory tract and, in particular, virus infections (Pierce and Hirsch, 1958). The ciliated bronchial epithelial cells undergo a degenerative process in which a pinching off occurs between the cytoplasm and nucleus, resulting in detached tufts of cilia, and a degenerating nucleus and cytoplasm (Figure 4.5). The degenerated cytoplasm may contain small, round eosinophilic inclusion bodies (Takahashi, 1981). Ciliocytophthoria occurs most frequently with influenza, parainfluenza, and adenovirus infection, but may also occur in bronchiectasis and other nonviral

Figure 4.4 Measles in lung tissue. Paraffin embedded tissue of measles pneumonia showing a multinucleated giant cell with multiple intranuclear and intracytoplasmic inclusions. H&E stain, ×550

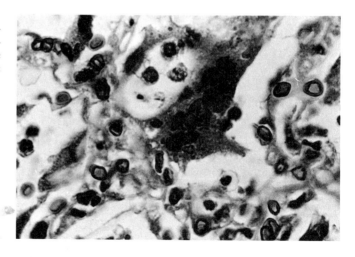

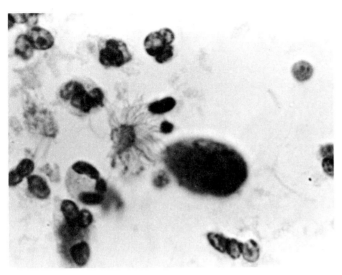

Figure 4.5 Ciliocytophthoria in sputum. Single karyorrhexic nucleus with degenerated cytoplasm containing small inclusion bodies and a tuft of detached cilia. Pap stain, ×550

inflammatory conditions. It is seen more frequently in sputum specimens than in those obtained by bronchoscopy.

4.3.2 Virus Infections of the Urinary Tract

Although many viruses that cause systemic infections have been isolated from the urine, those most readily diagnosed by urine cytology are CMV, HSV, and a member of the papovavirus group, designated the BK virus (BKV) (Table 4.2). In each of these infections, epithelial cells of the urinary tract from the renal tubules to the bladder and urethra, detach and enter the urine. These cells often contain characteristic inclusions. Because of the relatively

Table 4.2 Virus Infections of the Urinary Tract

Virus	Clinical Presentation	Cytologic Findings
Adeno	Hemorrhagic cystitis	Dense basophilic IN inclusions in transitional cells
BK (human papova)	Urethral stenosis in renal transplant and asymptomatic immunosuppressed patients	Large full mucoid IN inclusions (early): dense full basophilic inclusions bulging from cytoplasm (late)
CMV	Asymptomatic immunosuppressed patients	Large basophilic IN inclusions surrounded by halo; multiple eosinophilic IC inclusions
HSV	Generalized infection or local cystitis; may be contaminant from herpes genitalis	Ground-glass nuclei (early), eosinophilic IN inclusions (late); multinuclearity, may be part of tubular cast
Measles	Measles with exanthema	Multinucleated giant cells with eosinophilic IN and IC inclusions

Abbreviations: IN, intranuclear; IC, intracytoplasmic.

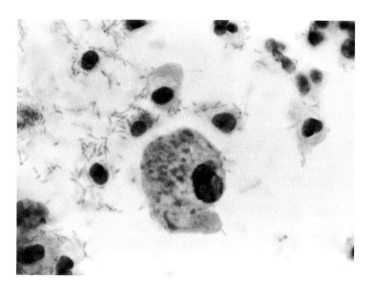

Figure 4.6 CMV in urine. Multiple intracytoplasmic inclusions with no significant intranuclear inclusions in late stage of infection. The nucleus is degenerated and appears empty. Pap stain, ×750

small number of cells in a large fluid volume, filtration or cytocentrifugation is necessary to obtain a suitable preparation (Schumann et al, 1977). Each of these infections occurs most often, but not exclusively, in immunosuppressed patients (commonly renal transplant recipients), and may coexist as mixed infections. CMV infected urothelial cells were first described in the urine of newborn infants with cytomegalic inclusion disease (Fetterman, 1952). Cytologic examination of the urinary sediment is indicated with any infant suspected of having neonatal CMV infection. Positive cytology is seen in approximately 50% of neonates that will subsequently have CMV-positive cultures (Hanshaw et al, 1968). CMV is the most frequently encountered viral infection in renal transplant recipients. In one study it was found in 31% of 2354 cytologically examined routine urine samples obtained from 91 patients (Traystman et al, 1980). Cellular changes include cytomegaly of the urothelial cells with typical large intranuclear inclusions surrounded by a clear halo and smaller eosinophilic cytoplasmic inclusions. This cytopathology most often involves single cells (Figure 4.1). Although the classical large, dense, intranuclear inclusion is the

easiest to identify, occasional cells may be binucleated and some cells may have large dense eosinophilic cytoplasmic inclusions with little or no evidence of an intranuclear inclusion (Figure 4.6).

BKV was first isolated from the urine of a 39-year-old man who developed ureteral stenosis 4 months after renal transplantation (Gardner et al, 1971). The urine sediment contained abnormal transitional cells with dense intranuclear inclusions composed of crystalline arrays of papova virions as revealed by electron microscopy. BKV was later isolated from many asymptomatic transplant recipients (Coleman, 1975) and two other patients with ureteral stenosis (Coleman et al, 1978). At an early stage, the most recognizable cytologic change due to BKV infection is the enlarged nucleus in an epithelial cell that contains a mucoid inclusion filling the nucleus (Figure 4.7). A more homogeneous, densely basophilic inclusion is detached at a later stage (Figure 4.8). This is sometimes referred to as a "decoy cell." Frequently, the nucleus appears to be bulging from the cytoplasm or thrusting from it, giving it a comet-like effect (Figure 4.9). Although most involved cells have single nuclei, occasional binucleated forms are seen with

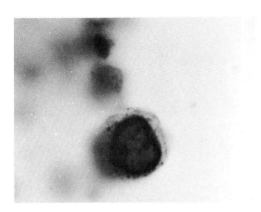

Figure 4.7 BKV in urine. Urothelial cell with the early changes of BKV infection. The enlarged nucleus has a condensed nuclear membrane with a mucoid inclusion. Pap stain, ×750

both types of inclusions present. At a later stage, the inclusion shrinks from the nuclear membrane leaving an incomplete thin halo. The intranuclear inclusions of BKV can be distinguished from those of CMV by the complete and consistent halo around the CMV inclusion and the lack of cytoplasmic inclusions in BKV (compare Figure 4.1 with 4.8).

Cytologic changes of HSV in the urinary sediment are similar to those described in the respiratory tract. They

Figure 4.8 BKV in urine. Binucleated transitional cell with a dense homogeneous intranuclear inclusions. Pap stain, ×550

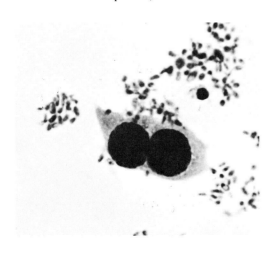

include multinuclear syncytial cells with enlarged ground-glass nuclei, seen at an early stage of infection, and typical eosinophilic intranuclear inclusions surrounded by a halo at a later stage. Elongated clumps of infected epithelial cells, probably of tubular origin, may contain inclusions in varying stages of development (Figure 4.10). Urinary sediment cells characteristic of HSV may occur in a generalized HSV infection involving the kidney, or a localized cystitis; they also may result from herpes genitalis, particularly when there is vaginal involvement during the infection (Masukawa et al, 1972). Other cytologic changes that may be observed in urinary sediment cells include intranuclear inclusions of adenovirus associated with acute hemorrhagic cystitis in children (Numazaki et al, 1973) and inclusion-bearing cells in the urine of patients with measles (Bolande, 1961).

4.3.3 Virus Infections of the Genital Tract

Cytologic recognition of typical viral changes in cells of routine ''pap'' smears is the most readily available and cost-effective method of detecting genital herpes infections (Table 4.3). This is important in abating the spread of genital HSV, as well as epidemiologic studies that indicate a putative etiologic involvement of HSV-2 in human neoplasia. Cytologic recognition of HSV is of critical importance in directing the management of pregnancy near term. The overall incidence of HSV in routine vaginal smears has been reported at approximately 0.3% (Naib, 1980), although this figure varies greatly depending on the patient population. The sensitivity of cytology for detecting HSV infection depends somewhat on the location of the herpetic lesions and the adequacy of the sample. In one study (Vontver et al, 1979), 41% of 69 cases with external lesions that were virus isolation-positive had positive smears, but the rate was 23% with women that had only

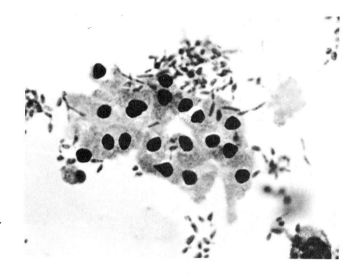

Figure 4.9 BKV in urine. Urothelial cells each containing a dense intranuclear inclusion, some of which have a comet-like configuration. Pap stain, ×550

cervical lesions. Similar results were reported in a study of 76 patients with genital HSV comparing virus isolation, immunofluorescence, immunoperoxidase, and cytology as means of making a diagnosis (Moseley et al, 1981). The overall positive cytology rate was 37.6%, but was 47.9% for cases with vaginal or cutaneous lesions. Significantly, there were a number of cases in which the cytology was positive but virus isolation in tissue culture from the same sample was negative. This discrepancy is repeatedly encountered in reported studies comparing virus isolation and cytopathology with various viruses at different sites, and indicates that the greatest diagnostic yield with virus infections amenable to cytologic diagnosis is from the combination of cytology and virus isolation.

The identifiable cytologic changes in genital HSV infection are identical to those described with infections of the respiratory and urinary tracts. At an early stage, the enlarged nuclei have a bland ground-glass appearance with the chromatin displaced to the periphery, resulting in an apparent thickening of the nuclear membrane (Figure 4.11). At a later stage, the nucleus contains an eosinophilic inclusion surrounded by a clear halo. Multinuclear cells are common with up to ten nuclei, which often exhibit molding. Inclusions in multinucleated cells may all be at the same stage (Figure 4.12) or may exhibit different stages of development (Figure 4.13). HSV-1 and -2 produce identical morphologic changes and cannot be differentiated on the basis of cytology. Although it has been reported that primary HSV infection can be differentiated from recurrent or secondary infection by a predominance of bland ground-glass nuclei in primary infection (Ng et al, 1970), this has not been confirmed in subsequent studies (Naib, 1980).

Figure 4.10 HSV in urine. Renal tubular cast showing characteristic HSV inclusions and nuclear molding. Pap stain, ×800

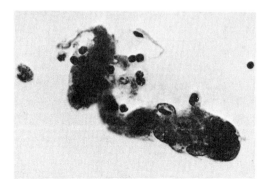

Table 4.3 Virus Infections of the Genital Tract

Virus	Clinical Presentation	Cytologic Findings
HSV (types I and II)	Herpes genitalis	Ground-glass nuclei (early; eosinophilic IN inclusions (late) with peripheral chromatin condensation; multinuclearity
Papilloma	Condyloma acuminatum; cervical dysplasia	Enlarged hyperchromatic nucleus; rare basophilic IN inclusions perivascular cytoplasmic clearing and vacuolar degeneration (koilocytotic change)
Molluscum contagiosum	Vaginal, penile, or perineal papule with central umbilication	Large dense staining IC inclusions displacing nucleus; squamous cell often bean-shaped

Abbreviations: IN, intranuclear; IC, intracytoplasmic.

Infection of the cervical or vaginal mucosa and the skin of the perineum with a human papilloma virus (HPV) may result in proliferation of epithelial cells forming a vegetative papillary growth known as a condyloma or venereal wart. Atypical cellular changes often accompany the proliferative process and may result in cellular alterations similar to those of malignant cells (Meisels et al, 1981). HPV has been associated with premalignant cervical dysplasia and carcinoma, and most recently identified by immunohistology and/or molecular probes in association with squamous cell carcinoma of the vulva (Zachow et al, 1982; Pilotti et al, 1984). At the present time, cytology is the only practical way of identifying an HPV infection of the genital tract when a characteristic gross condylomatous lesion is not observed with

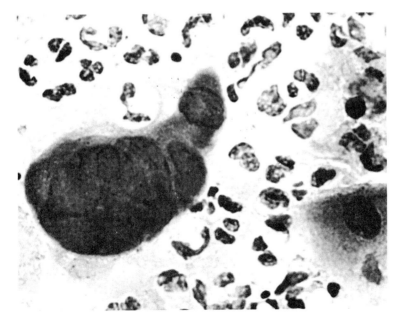

Figure 4.11 HSV in genital smear. Early stage of infection showing the molded nuclei with ground-glass nuclear appearance. Pap stain, ×800

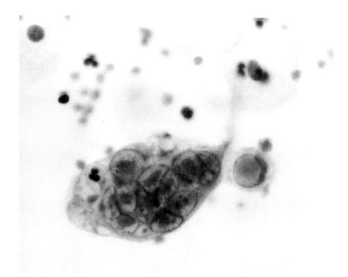

Figure 4.12 HSV in genital smear. Multinucleated giant cells with intranuclear inclusions surrounded by halos. Pap stain, ×720

molecular techniques used for confirmation. The changes attributable to virus infection involve squamous cells, which appear swollen and have a perinuclear halo with poor cytoplasmic keratinization resulting in irregular staining. This produces a picture referred to as koilocytotic change (Figure 4.14). The nucleus is frequently enlarged and occasionally contains a poorly defined basophilic inclusion which, by electron microscopy, is composed of virus particles and fibrillar material (Caras-Cordeo et al, 1981). Although the reported incidence of anogenital HPV infections in var-

Figure 4.13 HSV in genital smear. Note inclusions at different stages. Pap stain, ×720

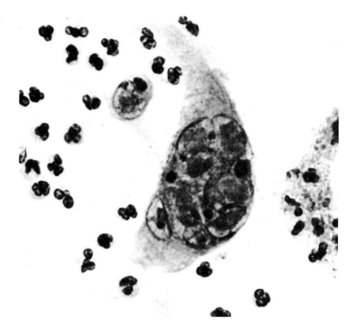

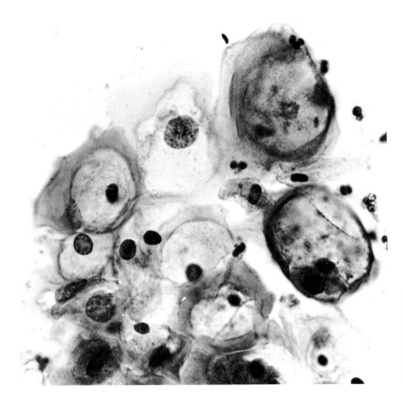

Figure 4.14 Papilloma virus in genital smear. Swollen squamous cells exhibiting koilocytotic change characteristic of a condylomatous lesions. Pap stain, ×720

ious populations varies greatly depending on the molecular method used to detect and type HPV, there is agreement that the majority of premalignant and malignant cervical and vulvar lesions contain certain types (mostly HPV 16 or 18) of HPV genomic material (Reeves et al, 1989). There is also general agreement that koilocytotic change reflects HPV infection although the accuracy of diagnosis based on cytologic changes alone has not yet been fully documented. Because some types of HPV, particularly HPV 18, have been associated with rapid malignant transformation and evolution from dysplasia to carcinoma, the presence of koilocytotic atypia indicates a need for cervical biopsy or at least close followup with repeat cytology. At present, there is no agreement as to the clinical usefulness of molecular identification and typing of the virus although these studies, using the polymerase chain reaction (PCR), are important in epidemiologic evaluation

of this sexually transmitted infection (Bauer et al, 1991).

Other viruses that have been identified by characteristic cytologic findings in vaginal smear include CMV and the poxvirus, which causes molluscum contagiosum (Brown et al, 1981). Molluscum contagiosum is a benign cutaneous infection most often observed in children and young adults, which is easily transmitted by direct contact. Circumstantial evidence suggests the infection is transmitted between young adults during sexual intercourse. Although the lesions have a characteristic appearance consisting of a small, firm papule with a centralized umbilication on the skin or vaginal mucosa, virus isolation techniques to confirm the diagnosis are not yet available. However, the histopathology necessitating biopsy and the morphologic changes in individual cells as observed in a pap stained smear are diagnostic. The cytologic changes consist of large, dense staining

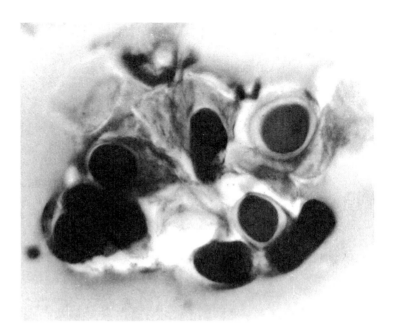

Figure 4.15 Molluscum contagiosum in genital smear. Molluscum bodies from a vaginal smear showing large intracytoplasmic inclusions displacing the nucleus. Pap stain, ×720

cytoplasmic inclusions occupying the entire squamous cell and resulting in peripheral displacement of a flattened nucleus (Figure 4.15). The cells frequently assume a bean shape with the nucleus displaced to the concave aspect.

4.3.4 Virus Infections of the Eye

In many common ocular lesions, and particularly in those involving the cornea, a rapid diagnosis is essential in order to initiate therapy and avoid progressive corneal damage. Although biopsy and virus isolation are the usual definitive procedures for the diagnosis of virus infections involving the cornea and conjunctiva, exfoliative cytology offers a simple, inexpensive, and rapid means of diagnosing adenovirus keratoconjunctivitis and keratitis due to herpes viruses (Naib et al, 1967; Schumann et al, 1980) (Table 4.4). In addition, characteristic cytologic changes consisting of multinucleated giant cells in measles keratoconjunctivitis and the large dense cytoplasmic inclusions of molluscum contagiosum in conjunctival and eyelid lesions may also yield a definitive diagnosis, although these

infections as isolated ophthalmic disease are rarely encountered. Finally, cytology is most useful for the diagnosis of chlamydial conjunctivitis, which may be clinically difficult to differentiate from the disease caused by adenovirus but can be diagnosed by the finding of characteristic cytoplasmic inclusions caused by the chlamydial infection (Gupta et al, 1979).

Specimens containing conjunctival or corneal cells should be collected by a physician, preferably an ophthalmologist, by swab or superficial conjunctival or corneal scraping. Corneal scraping requires examination with a slit lamp microscope for localization of the lesion. The collected material is immediately spread on an alcohol-moistened slide and, after partial evaporation, the slide is placed in 95% ethyl alcohol for proper fixation. Because there are usually very few cells and little fluid substrate, air-drying of the smears is a frequent problem, but can be avoided by immediate fixation of the specimen on the slide.

In adenovirus infections, the conjunc-

Table 4.4 Virus Infections of the Eye

Virus	Clinical Presentation	Cytologic Findings
Adeno	Acute (epidemic) keratoconjunctivitis and conjunctivitis with pharyngiitis	Multiple eosinophilic IN inclusions (early); dense central basophilic inclusions surrounded by halo (late)
HSV	Corneal vesicle or ulcer; may be isolated ophthalmic lesion or with other HSV vesicles	Multinucleated cells with eosinophilic IN inclusions surrounded by halo; nuclei has ground-glass appearance (early stage)
Molluscum contagiosum	Reddish papular 5-mm lesions of eyelid or conjunctiva	Large dense basophilic IC inclusions displacing nucleus
Varicella zoster	Vesicular eruptions in dermatome involving eye (shingles) or accompanying chicken pox	Multinucleated cells with IN eosinophilic inclusions
Chlamydia	Granular conjunctiva with corneal ulcerations (trachoma) or conjunctivitis only (inclusion conjunctivitis)	Enlarged corneal (trachoma only) and conjunctival cells with numerous IC basophilic inclusions surrounded by individual halos

Abbreviations: IN, intranuclear; IC, intracytoplasmic.

tival or corneal cells are mixed with lymphocytes and plasma cells, and contain distinctive intranuclear inclusions. In the early stages of infection (Figure 4.16), the intranuclear inclusions are multiple, small, and sometimes granular and eosinophilic. At a later stage, the small inclusions coalesce as a single dense basophilic body, usually centrally located and surrounded by a clear halo (Figure 4.17).

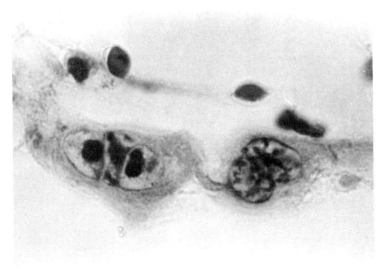

Figure 4.16 Adenovirus in conjunctival scraping. Multiple small intranuclear inclusions characteristic of the early stage of infection. Pap stain, ×800

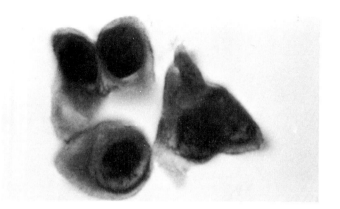

Figure 4.17 Adenovirus in conjunctival scraping. Conjunctival cells showing a centrally located intranuclear inclusion surrounded by a clear halo seen in late stages of the infection. Pap stain, ×800

Keratoconjunctivitis caused by HSV can occur as a primary lesion or as part of a systemic infection. Superficial scrapings from the margin of the ulcerated area will usually contain multinucleated cells with characteristic large eosinophilic intranuclear inclusions surrounded by prominent halos. Early in the infection, as with HSV at other sites, scrapings will reveal enlarged nuclei having a ground-glass appearance.

Herpes zoster keratitis is well recognized clinically and usually does not require additional diagnostic confirmation, such as cytology. However, scrapings of the lesions will yield cells similar to the ones found in HSV infection, although it is reported that syncytia and intranuclear inclusions are less prominent than with HSV

(Naib, 1967). The acute conjunctivitis that occurs, usually during the prodromal of measles, is also associated with characteristic cytologic findings seen in conjunctival smears or scrapings. The characteristic cells may contain up to 100 round nuclei surrounded by an abundant cytoplasm in which there are numerous eosinophilic inclusions. Occasionally, similar eosinophilic inclusions can be found within the nuclei. These findings can precede the appearance of the typical exanthem by 2 to 3 days.

The morphologic changes in conjunctival and corneal cells due to virus infection must be differentiated from the cytologic changes due to chlamydial infections causing trachoma and inclusion conjunctivitis. In both these diseases, the epithelial cells

Figure 4.18 Chlamydia in conjunctival scraping. Epithelial cells with intracytoplasmic inclusions surrounded by distinct halos. Pap stain, ×720

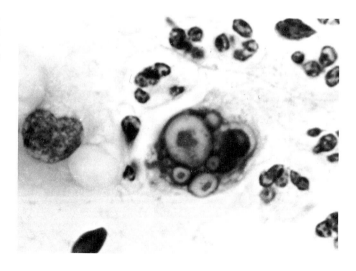

generally are enlarged and have abundant cytoplasm containing clusters of basophilic intracytoplasmic inclusions, each surrounded by a large individual halo (Figure 4.18). In trachoma, the corneal cells are involved, whereas, only conjunctival cells show the changes in the more benign chlamydial conjunctivitis. The presence of cells containing cytoplasmic inclusions in cytologic examination of specimens from the eye suggests chlamydial, rather than viral, infection.

REFERENCES

Bales, C. and Durfee, G. 1979. Cytologic techniques. In L. Koss (ed.), Diagnostic Cytology, Vol. 2, 3rd ed. Philadelphia: J.B. Lippincott Co., pp. 1187–1266.

Bauer, H.M., Ting, Y., Greer, C.E., Chambers, J.C., Tashiro, C.J., Chimera, J., Reingold, A., and Manos, M.M. 1991. Genital human papillomavirus infection in female university students as determined by a PCR-based method. JAMA 265:472–477.

Barrett, D. 1976. Cytocentrifugation technique. In C. Keebler, J. Reagan, and G.L. Wied (eds.), Compendium on Cytopreparatory Techniques, 4th ed. Chicago: Tutorials of Cytology, pp. 80–83.

Bolande, R.P. 1961. Significance and nature of inclusion-bearing cells in the urine of patients with measles. N. Engl. J. Med. 265: 919–923.

Brown, S.T., Nalley, J.F., and Kraus, S.N. 1981. Molluscum contagiosum. Sex. Transm. Dis. 8:227–234.

Caras-Cordero, M., Morin, C., Roy, M., Fortier, M., and Meisels, A. 1981. Origin of the koilocytes in condylomata of the human cervix. Ultrastructural study. Acta Cytol. 25:383–392.

Coleman, D.V. 1975. The cytodiagnosis of human polyoma infection. Acta Cytol. 9:93–96.

Coleman, D.V., MacKenzie, E.F.D., Gardner, S.D., Poulding, J.M. 1978. Human polyomavirus (BK) infection and ureteric stenosis in renal allograft recipients. J. Clin. Pathol. 31:338–347.

Fetterman, G.H. 1952. New laboratory aid in clinical diagnosis of inclusion disease of infancy. Am. J. Clin. Pathol. 22:424–427.

Gardner, S.D., Field, A.M., Coleman, D.V., and Hulme, B. 1971. New human papovavirus (BK) isolated from urine after renal transplantation. Lancet 1:1253–1257.

Gill, G. 1976. Methods of cell collection on membrane filters. In C. Keebler, J. Reagan, and G.L. Wied (eds.), Compendium on Cytopreparatory Techniques, 4th ed. Chicago: Tutorials of Cytology, pp. 34–44.

Gupta, P.A., Lee, E.F., Erozan, Y.S., Frost, J.K., Geddes, S.T., and Donovan, P.A. 1979. Cytologic investigations in chlamydia infection. Acta Cytol. 23:315–320.

Hanshaw, J.B., Steinfeld, H.J., and White, C.J. 1968. Fluorescent-antibody test for cytomegalovirus macroglobulin. N. Engl. J. Med. 279:566–570.

Masukawa, T., Jarancis, J.C., Rytel, M., and Mattingly, R.F. 1972. Herpes genitalis virus isolation from human bladder urine. Acta Cytol. 16:416–428.

Meisels, A., Ray, M., Fortier, M., Morin, C., Cassas-Cordero, M., Shah, K.V., and Turgeon, H. 1981. Human papillomavirus infection of the cervix. Acta Cytol. 25:7–16.

Moseley, R.C., Corey, L., Benjamin, D., Winter, C., and Remington, M.L. 1981. Comparison of viral isolation, direct immunofluorescence, and direct immunoperoxidase techniques for detection of genital herpes simplex virus infection. J. Clin. Microbiol. 13:913–918.

Naib, Z.M., Clepper, A.S., and Elliott, S.R. 1967. Exfoliative cytology as an aid in diagnosis of ophthalmic lesions. Acta Cytol. 11:295–303.

Naib, Z., Stewart, J., Dowdle, W., Casey, H., Marine, W., and Nahmias, A. 1968. Cytologic features of viral respiratory tract infections. Acta Cytol. 12:162–171.

Naib, Z.M. 1980. Exfoliative cytology in the

rapid diagnosis of herpes simplex infection. In A.J. Nahmias, W.R. Dowdle, and R.F. Schinaza (eds.), The Human Herpes Virus: An Interdisciplinary Perspective. Amsterdam: Elsevier/North Holland Biomedical Press, pp. 381–386.

Ng, A.B.P., Reagan, J.W., and Yen, S.S. 1970. Herpes genitalis: Clinical and cytopathological experience with 256 patients. Obstet. Gynecol. 36:645–651.

Numazaki, Y., Kumasaka, T., Yano, N., Yamanaka, M., Miyazawe, T., Takai, S., and Ishida, N. 1973. Further study on acute hemorrhagic cystitic due to adenovirus type 11. N. Engl. J. Med. 289:344–347.

Pierce, C.H. and Hirsch, J.G. 1958. Ciliocytophthoria relationship to viral respiratory infections in humans. Proc. Soc. Exp. Biol. Med. 98:489–492.

Pilotti, S., Rilke, F., Shah, K.V., Torre, G.D., and DePalo, G. 1984. Immunohistochemical and ultrastructural evidence of papilloma virus infection associated with in situ and microinvasive squamous cell carcinoma of the vulva. Am. J. Surg. Pathol. 8:751–761.

Reeves, W.C., Brinton, L.A., Garcia, M., Breenes, M.M., Herrero, R., Gaitan, E., Tenorio, F., De Britton, R.C., and Rawls, W.E. 1989. Human papillomavirus infection and cervical cancer in Latin America. N. Engl. J. Med. 320:1437–1441.

Schumann, G.B., Berring, S., and Hill, R. 1977. Use of the cytocentrifuge for the detection of cytomegalovirus inclusions in the urine of renal allograft patients. Acta Cytol. 21:168–172.

Schumann, G.B., O'Dowd, G.J., and Spinnler, P.A. 1980. Eye cytology. Lab. Med. 11:533–540.

Springer, S., Hackman, R.C., Holle, R., Greenberg, G.M., Weems, C.E., Myerson, D., Meyers, J.D., and Thomas, E.D. 1986. Use of bronchoalveolar lavage to diagnose acute diffuse pneumonia in the immunocompromised host. J. Infect. Dis. 154:604–610.

Takahashi, M. 1981. Color Atlas of Cancer Cytology. Tokyo-New York: Igaku-Shoin, p. 291.

Tomkins, V. and Macanlay, J.C. 1955. A characteristic cell in nasal secretions during prodromal measles. JAMA 157:711–712.

Traystman, M.D., Gupta, P.K., Shah, K.V., Reissig, M., Cowles, L.T., Hillis, W., and Frost, J.K. 1980. Identification of viruses in the urine of renal transplant recipients by cytomorphology. Acta Cytol. 24:501–510.

Vernon, S.E. 1982. Cytologic features on nonfatal herpes virus tracheobronchitis. Acta Cytol. 26:237–242.

Vontver, L.A., Reeves, W.C., Rathay, M., Corey, L., Remington, M.A., Tolentino, E., Schweid, A., and Holmes, K.K. 1979. Clinical course and diagnosis of genital herpes virus infection and evaluation of topical surfactant therapy. Am. J. Obstet. Gynecol. 133:548–554.

Warner, N.E., McGrew, E.A., and Nanos, S. 1964. Cytologic study of sputum in cytomegalic inclusion disease. Acta Cytol. 8:311–315.

Zachow, K.R., Ostrow, R.S., Bender, M., Watts, S., Okagaki, T., Pass, F., and Faras, A.J. 1982. Detection of human papillomavirus DNA in anogenital neoplasms. Nature 300:771–773.

5

Electron Microscopy and Immunoelectron Microscopy

Frances W. Doane

5.1 INTRODUCTION

Virus diagnosis by electron microscopy (EM) relies on the detection and identification of viruses on the basis of their characteristic morphology. As long ago as 1948 it was realized that the electron microscope could be used to distinguish one virus from another, and soon after the first commercial electron microscopes became available, they were being used to distinguish the large brick-shaped smallpox virus from the smaller, round varicella-zoster virus (Nagler and Rake, 1948; van Rooyen and Scott, 1948). With the introduction of the simple but effective negative staining technique for revealing viral ultrastructure (Brenner and Horne, 1959) it became possible to observe far greater structural detail than had been possible with the earlier shadowing technique. An increasing number of virologists ventured into this field when it became apparent that the electron microscope could be an important tool in rapid diagnosis of virus infections (Peters et al, 1962; Nagington, 1964; Cruickshank et al, 1966; Chambers et al, 1966; Doane et al, 1967, 1969; Joncas et al, 1969). In more recent years immunoelectron microscopy (IEM),

involving the visualization by EM of virus-antibody complexes, has increased the sensitivity of virus detection, providing a means of serotyping viruses directly on the EM specimen grid (Almeida et al, 1963; Almeida and Waterson, 1969; Doane 1974).

This chapter presents a number of EM methods that we and our colleagues have found to be useful in diagnostic virology. For more extensive dissertations the reader should consult other sources (Doane and Anderson, 1977, 1987; Kjeldsberg, 1980; Davies, 1982; Field, 1982; Hsiung, 1982; Palmer and Martin, 1982; Chernesky and Mahony, 1984; McLean and Wong, 1984; Oshiro, 1985; Miller, 1986).

5.1.1 Advantages

A major advantage of virus diagnosis by EM is the ability to visualize the virus. The virologist who employs other methods in the diagnostic armamentarium is denied this gratification, relying instead on indicator systems—indirect clues involving cytopathology, lysing blood cells, color changes or radioactivity counts—that signify the presence of a virus. By identifying viruses directly, on the basis of their morphology, it is possible to perform an examination without a preconceived concept of the etiological agent, in contrast with those assays that require a specific viral probe (e.g., a particular viral antibody or nucleic acid sequence). Speed is another major advantage provided by EM. With the commonly used negative staining technique, a clinical specimen can be processed within minutes after collection. In many cases, virus morphology alone is sufficiently characteristic to permit family identification. Finally, by depending on the morphology of a virus for identification, whether or not it is still infective is of no consequence in making that identification. Thus, EM methods are applicable to the detection of inactivated viruses, and to viruses that are difficult to culture, such as hepatitis B, Norwalk or papovaviruses.

5.1.2 Limitations

The increasing use of EM in virus laboratories attests to the value of this instrument in diagnostic virology. But it would be unwise to enter this field without a realistic understanding of its limitations. Probably the most serious limitation of the electron microscope as a tool in diagnostic virology is its inability to examine multiple specimens coincidentally. Thus, it is not possible for an active virus laboratory to examine each clinical specimen by EM. Consequently, only selected specimens will be examined by this technique.

A second limitation of the electron microscope is one shared by other detection methods, viz. there must be a minimum number of virus particles present in the specimen in order to be detected on the EM specimen grid. Some clinical specimens, such as vesicle fluid from herpetic or poxvirus lesions, or feces from rotavirus-infected patients, usually have a very high virus content and are ideal EM specimens. Others, such as throat washings and urine, may contain too little virus to be detected by EM. For this reason many laboratories

initially inoculate all specimens into cell culture systems. This serves to screen out the negatives, and amplify the virus content in the positives. The EM can then be introduced at this stage, to identify the cell culture isolates.

5.1.3 Special Facilities

A high-resolution transmission electron microscope (TEM) and appropriately trained personnel are essential elements for diagnostic virology by EM. One should consider the facility of operation and available service when purchasing a TEM. Alternatively, the virus laboratory can depend entirely on the services of an external EM facility, thereby avoiding maintenance responsibilities.

The principal EM methodology used in diagnostic virology is the negative staining technique, which requires little special equipment. The EM specimen grids can be coated with plastic alone, but it is preferable to stabilize the plastic with carbon, a procedure that requires a simple vacuum evaporator. Plastic/carbon grids can be prepared in large quantities, and stored for several weeks.

If thin sectioning techniques are to be employed, a fairly extensive collection of equipment and reagents must be available, including oven, ultramicrotome, glass or diamond knives, fixatives, embedding media, and stains. In addition, someone must be available to cut thin sections (1/10 to 1/20 μm)—a skill that is not easily acquired. For this reason, most virus laboratories will prefer to rely on a neighboring pathology facility for tissue fixation, embedding, and sectioning. It then remains for the virologist to examine prepared sections by EM.

Rarely will one perform electron microscopy without the need—or desire—to produce micrographs. Consequently, the operator must have a supply of photographic materials and possess a technical capability in photographic processing.

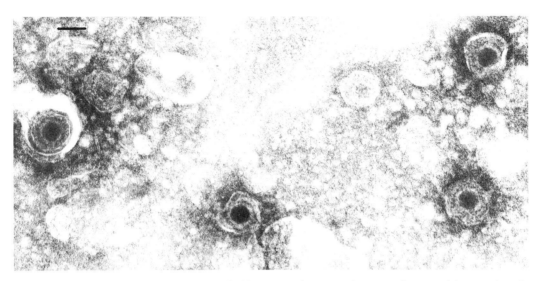

Figure 5.1 Negatively stained herpes varicella-zoster virus seen in smear from vesicle scraping. In this particular field all of the virus particles exhibit a stain-penetrated core surrounded by a hexagonal-shaped capsid and a tightly adhering envelope. Bar = 100 nm.

5.1.4 Safety Precautions

When preparing clinical specimens for electron microscopy, their potential pathogenicity should always be kept in mind. The commonly-used negative staining technique is almost always carried out on un-fixed specimens, and should preferably be performed in a laminar flow hood. If a hood is unavailable, a bench area protected with plastic- or foil-backed absorbent paper, which should be disposed of appropriately based on institutional requirements after use, can be used. Rubber gloves should be worn when working on an open bench. A good all-purpose disinfectant is 0.6% sodium hypochlorite (prepared from commercially available laundry bleach). Prepared specimen grids should be exposed to UV radiation (approximately 700 to 1000 μW/cm^2 at a distance of 4 to 6 inches) for at least ten minutes, prior to EM examination. Forceps used to handle grids should be wiped with an alcohol swab and flamed *briefly* immediately after use. Care should be taken in discarding examined grids, as some viruses are not readily inactivated by exposure to the electron beam. The 2.5%

glutaraldehyde solution used to fix tissues for histopathological examination has been shown to be effective in inactivating poliovirus and hepatitis B virus after 30 minutes (Boudouma et al, 1984; Kobayashi et al, 1984).

5.2 CLINICAL SPECIMENS SUITABLE FOR VIRUS DETECTION BY ELECTRON MICROSCOPY

5.2.1 Vesicle Fluid and Crusts

Vesicle fluid and scrapings from poxvirus and herpetic skin lesions are ideal EM specimens, as they often contain large quantities of virus that can be detected readily (Figure 5.1) (Peters et al, 1962; Nagington, 1964; Cruickshank et al, 1966; Macrae et al, 1969; Blank et al, 1970). Vesicle fluid can be collected in a capillary tube or a fine-bore needle attached to a small syringe. It is expelled onto a glass slide and mixed with a drop of filtered, distilled water.

A scalpel blade scraping of the base of

a vesicle can be smeared on a glass slide, and moistened with a small quantity of water. Crusts removed from dried vesicles can be ground in water on a slide, using a scalpel blade.

Specimens to be negatively stained are added to the EM grid either by touching a coated grid to the fluid on the slide or by processing the suspension by the agar diffusion method (see below).

5.2.2 Respiratory Tract Secretions

On occasion, these specimens contain a sufficient quantity of virus (e.g., paramyxoviruses) to be detected by direct EM (Doane et al, 1967; Joncas et al, 1969), although they usually require some form of amplification, such as ultracentrifugation, IEM, or passage in cell culture. A small amount of specimen is placed on a glass slide. The specimen, if it is viscous, can be diluted slightly with distilled water prior to negative staining.

5.2.3 Cerebrospinal Fluid

Evans and Melnick (1949) reported the EM detection of a herpesvirus in cerebrospinal fluid (CSF) from a patient with herpes zoster. One of the first clinical specimens we examined was CSF from a patient with mumps encephalitis; a paramyxovirus was detected by negative staining (Doane et al, 1967). In general, concentration of virus content should be attempted by, for example, Airfuge® ultracentrifugation (see below).

5.2.4 Feces

The negative staining technique can be used to detect a wide variety of viruses in feces (Tyrrell and Kapikian, 1982; Bock and Whelan, 1987), including rotaviruses (Flewett et al, 1973; Bishop et al, 1973; Middleton et al, 1974), coronaviruses (Caul et al, 1975; Mathan et al, 1975), adenovi-

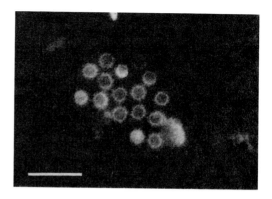

Figure 5.2 Negatively stained Norwalk virus in partially purified stool suspension. Bar = 100 nm. (Micrograph kindly provided by Dr. Fouad Bishai, Ontario Ministry of Health.)

ruses (Anderson and Doane, 1972; Flewett et al, 1974), astroviruses (Madeley and Cosgrove, 1975), caliciviruses (Madeley and Cosgrove, 1976; Flewett and Davies, 1976; Kjeldsberg, 1986), hepatitis A virus (Feinstone et al, 1973), and Norwalk virus (Figure 5.2) (Kapikian et al, 1972; Lin et al, 1991). The larger viruses found in feces (rotavirus, adenovirus, coronavirus) are usually readily identified by their distinct morphology (Figure 5.3). However, many of the "small round viruses" (SRV), which range in size from 20 nm to 40 nm, lack distinctive morphological features and may occur only in small numbers; consequently, they are often difficult to detect and identify. An abridged version of the classification scheme adopted by the UK Public Health Laboratory Service (Caul and Appleton, 1982; Appleton, 1987) is given at the top of page 93.

Some investigators routinely prepare 10% to 20% suspensions of feces, which they clarify free of bacteria and debris using light centrifugation, then ultracentrifuge to pellet any virus present (Flewett et al, 1974; Davies, 1982). A simple and effective method is to mix a small amount of crude feces with 1% ammonium acetate, and then transfer the mixture to a grid for negative staining (Middleton et al, 1977).

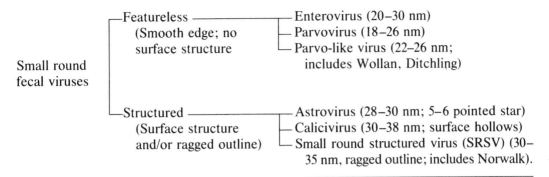

Small round
fecal viruses

┌─ Featureless ─────────── Enterovirus (20–30 nm)
│   (Smooth edge; no ───── Parvovirus (18–26 nm)
│   surface structure └──── Parvo-like virus (22–26 nm;
│   includes Wollan, Ditchling)
│
│
└─ Structured ─────────── Astrovirus (28–30 nm; 5–6 pointed star)
     (Surface structure ── Calicivirus (30–38 nm; surface hollows)
     and/or ragged outline) └── Small round structured virus (SRSV) (30–
                                35 nm, ragged outline; includes Norwalk).

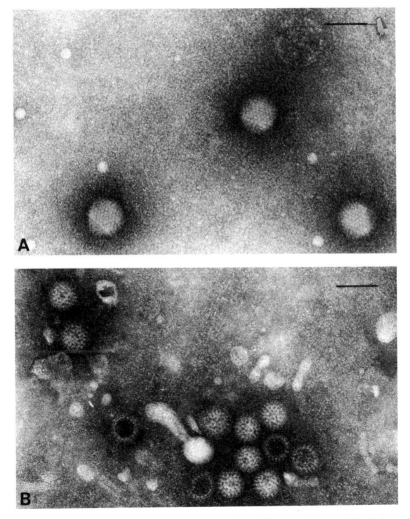

Figure 5.3 Isometric viruses found in negatively stained human feces. (A) Adenoviruses; (B) rotaviruses. Each family can be identified by morphological characteristics, such as virion size and the appearance of the capsid(s) and capsomers. Bar = 100 nm. (Micrographs kindly provided by M. Szymanski. Reproduced from Doane and Anderson 1977, with permission.)

5.2.5 Urine

Herpesvirus particles have been demonstrated by negative staining in the urine of patients excreting cytomegalovirus (Montplaiser et al, 1972; Lee et al, 1978). Papovaviruses of the polyomavirus genus have been detected in the urine of immunosuppressed renal transplant patients (Gardner et al, 1971; Field, 1982) and in the urine of pregnant women (Coleman et al, 1977).

5.2.6 Blood

Electron microscopy has provided a sensitive and rapid method for demonstrating hepatitis B virus in serum, both by direct EM and by IEM (Figure 5.4) (Bayer et al, 1968; Almeida et al, 1969; Hirschman et al, 1969; Dane et al, 1970).

5.2.7 Tissues

Any number of different tissues have been examined by EM for the presence of viruses. Brain biopsy tissue from patients with suspected herpes simplex virus infection can be ground with a tissue grinder or mortar and pestle in filtered distilled water to produce a suspension that is then negatively stained. By this method, herpesvirus particles have been identified within a few minutes after receipt of the specimen (Chia and Spence, personal communication; Szymanski and Middleton, personal communication). Although this method is undoubtedly the most rapid, it may not be as sensitive as thin sectioning (Field, 1982). Standard fixation and embedding procedures, or a rapid embedding method (described later) have been used to demonstrate virus in the brain of patients with herpes simplex encephalitis (Harland et al, 1967; Roy and Wolman, 1969), subacute sclerosing panencephalitis (SSPE) (Figure 5.5) (Bouteille et al, 1965; Herndon and Rubinstein, 1968), and progressive multifocal leukoencephalopathy (PML) (Zu Rhein and Chou, 1965).

5.3 NEGATIVE STAINING METHODS

Several commonly used negative staining methods are presented below. They vary in complexity and in sensitivity. Except where noted, they use essentially the same materials. An excellent review of the negative staining technique has been published by Hayat and Miller (1989).

5.3.1 Materials

300 mesh copper specimen grids, coated with a Formvar or Parlodion film (preferably stabilized with a thin layer of carbon)

Negative stain: 2% phosphotungstic acid (PTA) adjusted to pH 6.5 with 1N KOH; or 1% uranyl acetate, pH 4.0; or 3% ammonium molybdate, pH 7.2; stain must be prepared in distilled water sterilized by filtration to remove bacteria

Fine-bore pasteur pipettes

Filter paper

EM forceps

UV lamp

5.3.2 Direct Application Method

As the name implies, this is the simplest of the negative staining methods. It cannot be used for specimens that contain a high concentration of salt.

1. Place a small drop of specimen on a coated copper grid held with EM forceps.
2. Add a drop of negative stain.
3. Touch the fluid with the torn edge of filter paper, leaving only a moist layer on the grid.
4. Air-dry (one to two minutes).

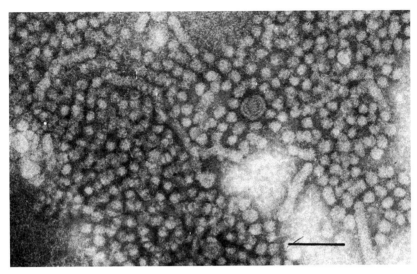

Figure 5.4 Hepatitis B virus particles from chronic carrier. Patient's serum was incubated for one hour with rabbit anti-HBs prior to negative staining. A single isometric virus particle is seen amidst the spherical and tubular forms of HBsAg. Bar = 100 nm. (Micrograph kindly provided by M. Fauvel. Reproduced from Doane and Anderson 1977, with permission.)

5.3.3 Water Drop Method

This is a simple method for removing salt from specimens that contain a high salt concentration which, if allowed to dry on the EM grid, will produce crystals that tend to hide virus particles. It yields a nicely stained preparation, but it requires a starting concentration of approximately 10^9 virus particles per milliliter (Doane et al, 1969).

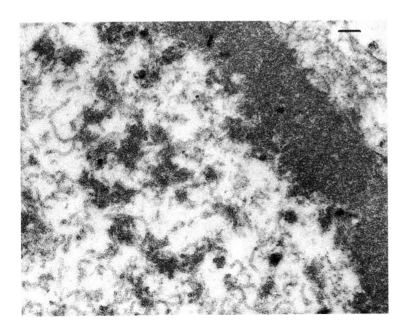

Figure 5.5 Thin section (in region of nucleus) of brain biopsy from patient with SSPE. Note margination of chromatin across upper right corner, and worm-like viral nucleocapsids scattered throughout the nucleoplasm. Bar = 100 nm. (Reproduced from Doane and Anderson 1977, with permission.)

1. On a waxed surface (e.g., Parafilm®) place a drop of filtered, distilled water.
2. Place a small drop of specimen on top of the water drop.
3. Briefly touch the coated surface of a specimen grid to the top of the drop.
4. Add negative stain and air-dry as described above in Direct Application Method.

5.3.4 Agar Diffusion Method

This is a modification of the method of Kelen et al, (1971). It is useful for specimens that contain a high salt concentration, especially when the virus concentration is low. The limit of detectability is approximately 10^7 virus particles per milliliter (Anderson and Doane, 1972).

Special materials required Flexible 96-well microtiter plates with wells filled approximately 3/4 with 1% agar or agarose; covered with transparent sealing tape and stored at 4 °C until needed.

1. For each specimen to be examined cut a pair of agar-filled cups from the plate; remove the sealing tape and dry the agar surface at room temperature for approximately five minutes.
2. Place a coated specimen grid on the surface of each cup.
3. Add a drop of specimen to each grid.
4. Air-dry at room temperature (30 to 60 minutes).
5. Add a drop of negative stain; remove the grid and air-dry as described in Direct Application Method.

5.3.5 Pseudoreplica Method

Although more involved than the Agar Diffusion Method, this method utilizes the same principles (Smith, 1967; Lee et al, 1978; Boerner et al, 1981).

Special materials required

10 mm × 10 mm × 5 mm squares of 2% agar (or agarose)
Glass slides
0.5% Formvar in ethylene dichloride

1. Place a cube of agar on a glass slide; place a drop of specimen in the center of the cube.
2. Air-dry the specimen (10 to 15 minutes).
3. When the surface appears dry, flood with 1 to 2 drops of Formvar solution; drain off any excess with absorbent paper.
4. When the Formvar has dried completely, trim the cube slightly on all four sides and move it to the very end of the slide.
5. Slowly dip the end of the slide into a container of negative stain, at a slight angle, until the Formvar film floats off.
6. Place a bare 300 mesh copper grid in the center of the floating film.
7. Turn up the corner of a piece of filter paper that is two to three times the size of the floating film; holding this corner, gently place the paper on top of the film; as soon as the paper becomes wet, quickly flip the paper (plus film and grid) 180° out onto paper toweling.

5.3.6 Airfuge® Ultracentrifugation

Hammond et al, (1981) routinely employ this method for all clinical specimens submitted to the virus laboratory. Detection sensitivity may be increased by as much as 1,000-fold.

Special materials required

Airfuge® ultracentrifuge with EM-90 rotor
Parafilm
Filtered distilled water
2.5% glutaraldehyde or 10% formalin

1. Place a coated specimen grid into the end of each of the six rotor sectors.

2. Add 90 μL of specimen into each sector.
3. Centrifuge at top speed (approximately 90,000 rpm) for 30 minutes.
4. To stain each grid, invert it briefly on a drop of negative stain resting on Parafilm.
5. Air-dry.
6. After use, decontaminate the rotor and rotor cover by immersing in glutaraldehyde or formalin for 15 to 30 minutes; rinse in tap water, brushing the sectors with a cotton-tipped swab; soak briefly in 90% alcohol; air-dry.

Alain et al, (1987) have reported increased sensitivity of detection by using the A-100 fixed angle rotor with the Airfuge® ultracentrifuge. Samples up to 240 μL could be processed, and required only five minutes centrifugation.

5.4 IMMUNOELECTRON MICROSCOPY METHODS

Immunoelectron microscopy (IEM), like other immunoassays, involves the use of specific antibody as a probe or detector of viral antigen. Unlike other immunoassays, however, IEM permits the direct visualization of the resultant immune complex itself. In reading IEM assays one looks for antibody-trapped negatively-stained virus particles or viral antigen.

IEM has a variety of applications in virology (Doane, 1974). It was used effectively to provide the first morphological characterization of rubella virus (Best et al, 1967), and was the means by which hepatitis B virus ("Australia antigen") was originally discovered (Bayer et al, 1968). IEM is a recommended technique for the detection of noncultivatable viruses such as Norwalk virus and calicivirus, using patient serum mixed with the clinical specimen to produce visible immune complexes (Kapikian et al, 1976; Kjeldsberg, 1986; Kjeldsberg et al, 1989; Bishai, 1990).

It can also be used to increase the sensitivity of EM detection. We have found that the presence of antibody in our EM detection system can increase the sensitivity for enteroviruses by 100-fold (Anderson and Doane, 1973).

IEM has been used to serotype several viruses, including enteroviruses (Anderson and Doane, 1973; Petrovicova and Juck, 1977), adenoviruses (Luton, 1973; Vassall and Ray, 1974; Svensson and von Bonsdorff, 1982), papovaviruses (Gardner et al, 1971; Penny et al, 1972; Penny and Narayan, 1973), myxoviruses (Kelen and McLeod, 1974; Edwards et al, 1975), and rotaviruses (Gerna et al, 1988).

IEM can also be used to assay specific antibody. This approach, which has been extensively studied by Kapikian et al (1975, 1976, 1979), uses a predetermined concentration of reference virus mixed with dilutions of acute and convalescent sera. The amount of antibody present is determined by measuring the width of the antibody halo surrounding the aggregated virus particles.

Whether using IEM for viral antigen or antibody detection, visualization and specific identification of immune complexes can be enhanced through the use of a colloidal gold label (Stannard et al, 1982; Beesley and Betts, 1985; Hopley and Doane, 1985; Kjeldsberg, 1986; Doane et al, 1990; Wu et al, 1990). This can be bound either to the primary antiviral antibody, or to a secondary immunoreagent such as antispecies antibody or protein A or G. Colloidal gold suspensions can be prepared in the laboratory (Roth et al, 1978) and complexed to immunoglobulin or protein A as required (Geoghegan and Ackerman, 1977), or the conjugate can be purchased commercially.

Like other immunoassays, IEM procedures require careful attention to controls. It is essential that all reactants first be titrated to determine optimum test concentrations, and each test must be run in parallel with the appropriate positive and negative controls.

5.4.1 Direct Immunoelectron Microscopy Method

Materials

Plastic/carbon-coated copper specimen grids, 300 mesh

Microtiter plate

Antibody (monoclonal, polyclonal)

Flexible microtiter plates (see agar diffusion method)

Negative stain

1. Prepare antiserum (or virus) dilutions in a microtiter plate; prepare 1:1 mixtures of virus and antiserum in cups; incubate at 37 °C for one hour.
2. Process the mixtures by the agar diffusion method.
3. Examine grid for the presence of virus-antibody aggregates.

The original method described by Almeida and Waterson (1969) recommended an overnight refrigeration of the virus-antiserum mixture after step 1, followed by centrifugation at 10,000 to 15,000 rpm for 30 minutes. The pellet is then resuspended in distilled water and negative stained.

5.4.2 Serum-in-Agar Method

This is a modification of the agar diffusion method, incorporating antiserum in the agar itself (Anderson and Doane, 1973). We have found it to be the most sensitive of the IEM methods described here (Pegg-Feige and Doane, 1984; Doane et al, 1990). Reference antisera should initially be titrated to determine the IEM endpoint (highest dilution producing a visible immune complex), and can then be used at 10 to 100 times this value.

Materials

Plastic/carbon-coated copper specimen grids, 300 mesh

Microtiter plates (flexible plastic)

Antibody (monoclonal, polyclonal)

Negative stain

1. Prepare a 1% molten solution of agar (or agarose) by heating to boiling point.
2. Allow the molten agar to cool to approximately 45 °C and add antiserum.
3. Pipette the mixture into microtiter cups (approximately 3/4 full); allow to solidify at room temperature (plates can now be covered with plastic sealing tape and stored at 4 °C).
4. For use, cut a pair of cups from the plate, remove the tape, and air-dry the agar surface (approximately five minutes at room temperature).
5. Place a coated specimen grid on the surface of each cup.
6. Add a drop of specimen to each grid.
7. Air-dry at room temperature (30 to 60 minutes).
8. Add a drop of negative stain; remove the grid and air-dry as described for direct application method.
9. Examine grid for the presence of virus-antibody aggregates (Figure 5.6).

5.4.3 Solid Phase Immunoelectron Microscopy

Originally described by Derrick (1973) for use with plant viruses, this method has also been used for detection, and for typing, of a number of human viruses (Wood and Bailey, 1987; Gerna et al, 1988; Doane et al, 1990). The presence of antibody on the grid has been shown to increase the sensitivity of detection of enteroviruses approximately 60-fold over direct EM (Pegg-Feige and Doane, 1984). For the detection of rotaviruses in stool, Svensson et al (1983) found solid phase immunoelectron microscopy (SPIEM) to be approximately 30 times more sensitive than direct EM and ten times more sensitive than enzyme-linked immunosorbent assays (ELISA). Wu et al (1990) reported even greater sensitivity when gold-labeled protein G was incorporated in their SPIEM assay for rotavirus.

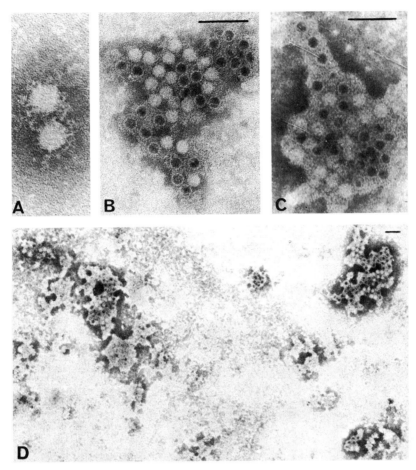

Figure 5.6 Immunoelectron microscopy of enterovirus, using the serum-in-agar method. (A) At high magnification antibody molecules can be seen bridging adjacent particles. (Courtesy of F. Lee) (B) Immune complex formed with dilute antiserum shows little surrounding antibody. (C) Complex formed with lower dilution of antiserum has a dense surrounding halo of antibody. (D) Immune complexes may be large enough to be detected at low magnifications. Bars = 100 nm. (Reproduced from Doane 1974, with permission.)

Materials

Formvar/carbon grids, 300 mesh, pre-treated with UV light (1700 uW/cm² for 30 minutes prior to use)

Antiserum

Protein A, 1 mg/mL

Parafilm

0.05 M Tris buffer (pH 7.2)

Negative stain

The procedure given below is one developed for rotavirus (Pegg-Feige and Doane, 1983). It is recommended that optimum test conditions be determined for individual viruses, prior to routine use of SPIEM.

1. Invert a freshly pretreated grid on a drop of protein A on Parafilm; leave at room temperature for ten minutes.

2. Drain grid briefly by touching the edge to filter paper; transfer across three separate drops of Tris buffer (total one to two minutes).

3. Invert on a drop of 1/100 dilution of rotavirus antiserum for ten minutes.

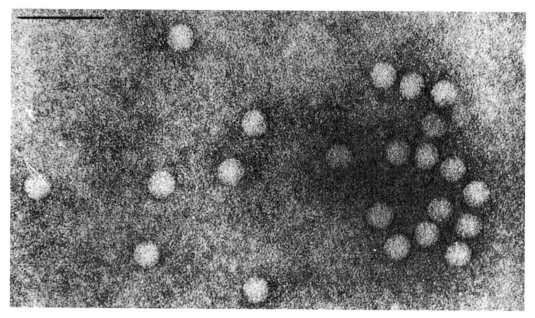

Figure 5.7 Immunoelectron microscopy of enterovirus, using SPIEM. A Formvar/carbon grid was exposed to protein A, then to poliovirus antiserum, prior to exposure to the specimen. Bar = 100 nm. (Kindly provided by K. Pegg-Feige.)

4. Drain briefly; transfer across three separate drops of Tris buffer (one to two minutes).

5. Invert on a drop of specimen at room temperature for 30 minutes.

6. Rinse on three drops of Tris buffer (one to two minutes).

7. Float briefly on a drop of negative stain; air-dry.

8. Examine grid for trapped virus (Figure 5.7).

Grids coated with protein A and/or antiserum are most effective when used shortly after preparation. They can be stored at 4°C, but with a resultant 50% reduction in trapping efficiency after four to five weeks.

5.4.4 Immunogold Method

The immunogold method given here was developed in our laboratory by Miss Nan Anderson.

Materials Serum-in-agar cups containing antiviral antibody (prepared as described above). 15 nm colloidal gold particles complexed to protein A (protein A-gold).

1. Mix equal volumes of protein A-gold and test specimen; leave at room temperature for five to ten minutes.

2. Place a drop of mixture on duplicate grids on a serum-in-agar cup. Allow the mixture to air-dry (30 to 60 minutes).

3. Rinse the grids by immersing sequentially in five separate drops on Parafilm; the first three drops consist of the buffer used to dilute the protein A-gold (0.5 M tris pH 7.0; 0.15% NaCl; 0.5 mg/mL polyethylene glycol 20,000; 0.1% NaN$_3$); the final two drops consist of filtered distilled water.

4. Place grids on a drop of negative stain for 30 seconds; remove, air-dry, and examine for the presence of gold label (Figure 5.8).

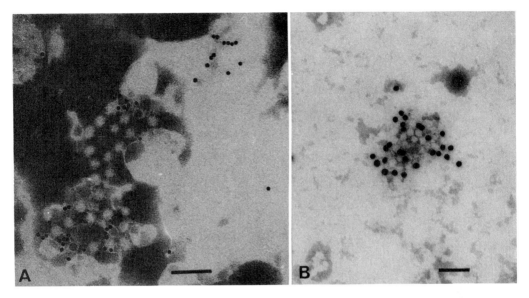

Figure 5.8 The dense colloidal gold particles conjugated to protein A improve visualization of immune complexes, and assist in specific virus identification. (A) Norwalk virus from stool, mixed with patient's serum, and incubated with a conjugate of protein A and 10 nm gold. (Micrograph kindly provided by Dr. Fouad Bishai, Ontario Ministry of Health.) (B) Poliovirus from an infected cell culture, mixed with poliovirus type 2 antiserum, and incubated with a conjugate of protein A and 15 nm gold. (Kindly provided by Nan Anderson.) Bars = 100 nm.

5.4.5 Routine Immunoelectron Microscopy Screening for Clinical Specimens or Cell Culture Isolates

By incorporating immune serum globulin in the serum-in-agar method, a broad non-selective virus detection system is produced (Juneau, 1979; Berthiaume et al, 1981). Enteroviruses detected by this method can be further identified by using the serum-in-agar method with pooled reference antisera (Doane and Anderson, 1987).

Materials Serum-in-agar cups containing pooled human gamma globulin, or pooled reference antisera, or individual antisera.

1. Process the specimen by the serum-in-agar method, using cups containing pooled gamma globulin at a final dilution

of 1/50; identify the virus family on the basis of virus particle morphology.

2. To serotype the identified virus, process by the serum-in-agar method, but use pooled or individual reference antisera; identify the virus type by carefully comparing test grids with control grids (virus alone, and virus plus normal serum) for the presence or absence of virus aggregates.

5.5 THIN SECTIONING METHODS

Any standard EM embedding procedure can be used to process tissue samples or cell pellets (Weakley 1981). However, these procedures usually take one to three days to complete and, for a rapid diagnosis, a shortened embedding method such as the one given below (Doane et al, 1974) is advantageous. Preservation of ultrastruc-

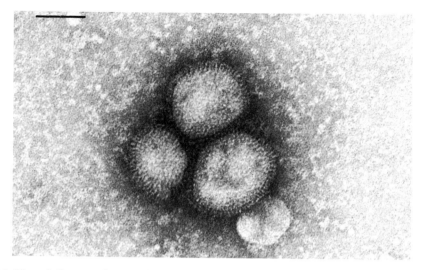

Figure 5.9 Three influenza virus particles from negatively stained cell culture lysate. Characteristic morphologic features include coarse spikes and an intact envelope (not easily penetrated by stain). Bar = 100 nm.

ture and specimen contrast are slightly reduced by this method, when compared with standard methods.

Materials

2.5% glutaraldehyde in 0.13 M phosphate buffer, pH 7.3

1% osmium tetroxide in 0.13 M phosphate buffer, pH 7.3

Acetone

Epoxy embedding medium

Bare 200-mesh copper grids

Miscellaneous equipment and reagents for preparing thin sections (e.g., oven, microtome, knives, stains)

1. Fix 1 mm cubes of tissue in glutaraldehyde for 15 minutes at 4 °C.
2. Rinse in three changes of phosphate buffer (one minute each).
3. Fix in osmium tetroxide for 15 minutes at room temperature.
4. Dehydrate through acetone as follows: two changes of 70% (total, five minutes);

three changes of 100% (total, five minutes).
5. Place in 1:1 mixture of 100% acetone and embedding medium for ten minutes at room temperature.
6. Transfer through two changes of 100% embedding medium (five minutes each).
7. Place in embedding capsule in fresh embedding medium and heat at 95 °C for 60 minutes.
8. Cool block; cut and stain sections as usual.

5.6 PROCESSING CELL CULTURE ISOLATES FOR ELECTRON MICROSCOPY

Because it is not possible to examine a large number of clinical specimens by EM on a daily basis, many virus laboratories routinely select only those specimens that have induced a cytopathic effect in inoculated cell cultures. The viral isolates can then be identified by EM, either by negative stain-

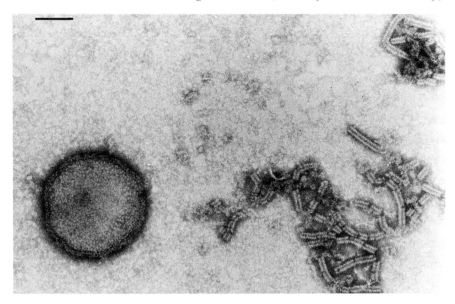

Figure 5.10 Simian paramyxovirus found in "normal" monkey cell culture (negatively stained). An intact virus particle with a fine fringed envelope is seen at left. At right are clusters of viral nucleocapsids exhibiting the herringbone configuration that is characteristic of paramyxoviruses. Bar = 100 nm.

Figure 5.11 Thin section of adenovirus-infected cell culture. Crystalline arrays of virus particles are seen in the nucleus. Bar = 100 nm.

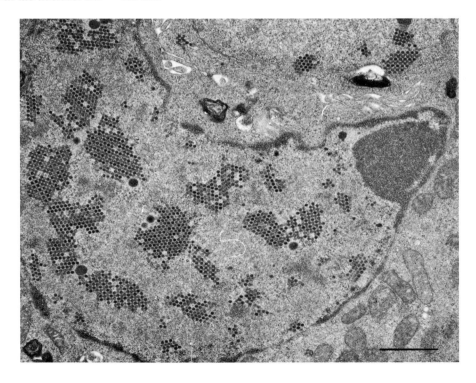

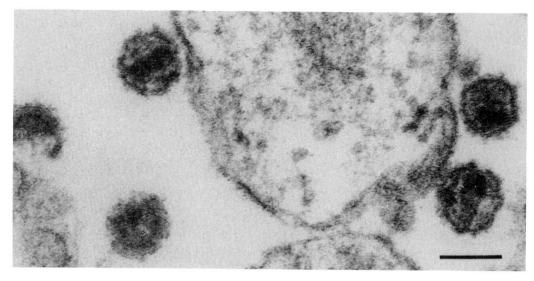

Figure 5.12 Thin section of H-9 cell culture infected with human immunodeficiency virus type 1 (HIV-1). Bar = 100 nm.

ing (Figures 5.9–5.10) or by thin sectioning (Figures 5.11–5.12).

5.6.1 Negative Staining

Materials

Filtered distilled water

Pasteur pipettes

Negative stain

Agar diffusion cups with specimen grids

1. Withdraw medium from culture to be examined; put medium aside temporarily.
2. Add two to three drops of water to culture; resuspend cells.
3. After two to three minutes, negatively stain cell lysate by the agar diffusion method or by the serum-in-agar IEM method.
4. If no virus is found in the lysate, process the temporarily-stored medium by Airfuge® ultracentrifugation and negative staining or by the serum-in-agar method.

5.6.2 Thin Sectioning

Materials

Hematocrit centrifuge

1.3 mm × 75 mm capillary tubes

Parafilm

Plasticine

Paper clip

2.5% glutaraldehyde in 0.13 M phosphate buffer, pH 7.3.

1. Gently resuspend cells from culture into overlying medium, using a rubber policeman or a Pasteur pipette.
2. Transfer suspension to a small conical-tipped centrifuge flask; centrifuge at 1500 rpm for three minutes.
3. Withdraw the medium and replace with two to three drops of glutaraldehyde; transfer to a sheet of Parafilm.
4. Draw the suspension into a capillary tube; seal one end with a small plug of plasticine.
5. Centrifuge in a hematocrit centrifuge for three minutes at 12,500 rpm; the cells now form a compact pellet immediately above the plasticine.
6. Score the glass tube and break at a distance 6 to 7 mm above the cell pellet.

7. Invert the tube so that the open end is directed toward a drop of glutaraldehyde on Parafilm; use a straightened paper clip to push against the plasticine plug, forcing the cell pellet into the fixative.

8. Transfer the cell pellet to a vial containing fresh fixative; process by a standard embedding method or by the rapid embedding method.

ACKNOWLEDGMENT

The assistance of Nan Anderson and Joan Stubberfield is gratefully acknowledged.

REFERENCES

Almeida, J., Cinader, B., and Howatson, A. 1963. The structure of antigen-antibody complexes. J. Exp. Med. 118:327–340.

Almeida, J.D., Zuckerman, A.J., Taylor, P.E., and Waterson, A.P. 1969. Immune electron microscopy of the Australia-SH (serum hepatitis) antigen. Microbios 2:695–698.

Almeida, J.D. and Waterson, A.P. 1969. The morphology of virus-antibody interaction. Adv. Virus Res. 15:307–338.

Anderson, N. and Doane, F.W. 1972. Agar diffusion method for negative staining of microbial suspensions in salt solutions. Appl. Microbiol. 24:495–496.

Anderson, N. and Doane, F.W. 1973. Specific identification of enteroviruses by immuno-electron microscopy using a serum-in-agar diffusion method. Can. J. Microbiol. 19:585–589.

Appleton, H. 1987. Small round viruses: Classification and role in food-borne infections. In G. Bock and J. Whelan (eds), Novel Diarrhoea Viruses. Ciba Foundation No. 128. Chichester: John Wiley & Sons, pp. 108–125.

Bayer, M.E., Blumberg, B.S., and Werner, B. 1968. Particles associated with Australia antigen in the sera of patients with leukemia, Down's syndrome and hepatitis. Nature (London) 218:1057–1059.

Beesley, J.E. and Betts, M.P. 1985. Virus diagnosis: A novel use for the protein A-gold probe. Med. Lab. Sci. 42:161–165.

Berthiaume, L., Alain, R., McLaughlin, B., Payment, P., and Trepanier, P. 1981. Rapid detection of human viruses in faeces by a simple and routine immune electron microscopy technique. J. Gen. Virol. 55:223–227.

Best, J.M., Banatvala, J.E., Almeida, J.D., and Waterson, A.P. 1967. Morphological characteristics of rubella virus. Lancet 2:237–239.

Bishai, F. 1990. Personal communication.

Bishop, R.F., Davidson, G.P., Holmes, I.H., and Ruck, B.J. 1973. Virus particles in epithelial cells of duodenal mucosa from children with acute non-bacterial gastroenteritis. Lancet 2:1281–1283.

Blank, H., Davis, C., and Collins, C. 1970. Electron microscopy for the diagnosis of cutaneous viral infections. Br. J. Dermatol. 83:69–80.

Bock, G. and Whelan, J. 1987. Novel Diarrhoea Viruses. Ciba Foundation Symposium No. 128. Chichester: John Wiley & Sons.

Boerner, C.F., Lee, F.K., Wickliffe, C.L., Nahmias, A.J., Cavanagh, H.D., and Strauss, S.E. 1981. Electron microscopy for the diagnosis of ocular viral infections. Ophthalmology 88:1377–1380.

Boudouma, M., Enjalbert, L., and Didier, J. 1984. A simple method for the evaluation of antiseptic and disinfectant virucidal activity. J. Virol. Meth. 9:271–276.

Bouteille, M., Fontaine, C., Verdrenne, C., and Delarue, J. 1965. Sur un cas d'encephalite subaigue a inclusions. Etude anatomo-clinique et ultrastructurale. Rev. Neurol. 118:454–458.

Brenner, S. and Horne, R.W. 1959. A negative staining method for high resolution electron microscopy of viruses. Biochim. Biophys. Acta 34:103–110.

Caul, E.O. and Appleton, H. 1982. The electron microscopical and physical characteristics of small round human fecal viruses: An interim scheme for classification. J. Med. Virol. 9:257–265.

Caul, E.O., Paver, W.K., and Clarke, S.K.R. 1975. Coronavirus particles in faeces from patients with gastroenteritis. Lancet 1:1192.

Chambers, V.C., Ito, Y., and Evans, C.A. 1966. Technique for visualizing papovaviruses in tumors and in tissue cultures. J. Bacteriol. 91:2090–2092.

Chernesky, M.A. and Mahony, J.B. 1984. Detection of viral antigens, particles, and early antibodies. Yale J. Biol. Med. 57:757–776.

Coleman, D.V., Russel, W.J.I., Hodgson, J., Tun P., and Mowbray, J.F. 1977. Human papovavirus in Papanicolaou smears of urinary sediment detected by transmission electron microscopy. J. Clin. Pathol. 30: 1015–1020.

Cruickshank, J.G., Bedson, H.S., and Watson, D.H. 1966. Electron microscopy in rapid diagnosis of smallpox. Lancet 2:527–530.

Dane, D.S., Cameron, C.H., and Briggs, M. 1970. Virus-like particles in serum of patients with Australia-antigen-associated hepatitis. Lancet 1:695–698.

Davies, H.A. 1982. Electron microscopy and immune electron microscopy for detection of gastroenteritis viruses. In D.A. Tyrrell, and A.Z. Kapikian (eds.), Virus Infections of the Gastrointestinal Tract. New York: Marcel Dekker, pp. 37–49.

Derrick, K.S. 1973. Quantitative assay for plant viruses using serologically specific electron microscopy. Virology 56:652–653.

Doane, F.W. 1974. Identification of viruses by immunoelectron microscopy. In E. Kurstak, and R. Morisset, (eds.), Viral Immunodiagnosis. New York: Academic Press, pp. 237–255.

Doane, F.W. and Anderson, N. 1977. Electron and immunoelectron microscopic procedures for diagnosis of viral infections. In E. Kurstak and C. Kurstak (eds), Comparative Diagnosis of Viral Diseases, Vol. II, part B. New York: Academic Press, pp. 505–539.

Doane, F.W. and Anderson, N. 1987. Electron Microscopy in Diagnostic Virology: A Practical Guide and Atlas. New York: Cambridge University Press.

Doane, F.W., Anderson, N., Chao, J., and Noonan, A. 1974. Two-hour embedding procedure for intracellular detection of vi-ruses by electron microscopy. Appl. Microbiol. 27:407–410.

Doane, F.W., Anderson, N., Chatiyanonda, K., Banatyne, R.M., McLean, D.M., and Rhodes, A.J. 1967. Rapid laboratory diagnosis of paramyxovirus infections by electron microscopy. Lancet 2:751–753.

Doane, F.W., Anderson, N., Lee, F., Pegg-Feige, K., and Hopley, J. 1990. Evaluation of immunoassays for electron microscopy. In L.M. de la Maza and E.M. Peterson (eds.), Medical Virology 9. New York: Plenum, pp. 1–16.

Doane, F.W., Anderson, N., Zbitnew, A., and Rhodes, A.J. 1969. Application of electron microscopy to the diagnosis of virus infections. Can. Med. Assoc. J. 100:1043–1049.

Edwards, E.A., Valters, W.A., Boehm, L.G., and Rosenbaum, M.J. 1975. Visualization by immune electron microscopy of viruses associated with acute respiratory disease. J. Immunol. Meth. 8:159–167.

Evans, A.S. and Melnick, J.L. 1949. Electron microscope studies of the vesicle and spinal fluids from a case of herpes zoster. Proc. Soc. Exp. Biol. Med. 71:283–286.

Feinstone, S.M., Kapikian, A.Z., and Purcell, R.H. 1973. Hepatitis A: Detection by immune electron microscopy of a virus-like antigen associated with acute illness. Science 182:1026–1028.

Field, A.M. 1982. Diagnostic virology using electron microscopic techniques. Adv. Virus Res. 27:1–69.

Flewett, T.H. and Davies, H. 1976. Caliciviruses of man. Lancet 1:311.

Flewett, T.H., Bryden, A.S., and Davies, H. 1973. Virus particles in gastroenteritis. Lancet 2:1497.

Flewett, T.H., Bryden, A.S., and Davies, H. 1974. Diagnostic electron microscopy of faeces. I. The viral flora of the faeces as seen by electron microscopy. J. Clin. Pathol. 27:603–608.

Gardner, S.D., Field, A.M., Coleman, D.V., and Hulme, B. 1971. A new human papovavirus (BK) isolated from urine after renal transplantation. Lancet 1:1253–1257.

Geoghegan, W.D. and Ackerman, G.A. 1977. Adsorption of horseradish peroxidase, ovomucoid and anti-immunoglobulin to colloidal gold for the indirect detection of concanavalin A, wheat germ agglutinin and goat anti-human immunoglobulin G on cell surfaces at the electron microscope level: A new method, theory and application. J. Histochem. Cytochem. 25:1187–1200.

Gerna, G., Sarasini, A., Coulson, B.S., Parea, M., Torsellini, M., Arbustini, E., and Battaglia, M. 1988. Comparative sensitivities of solid-phase immune electron microscopy and enzyme-linked immunosorbent assay for serotyping of human rotavirus strains with neutralizing monoclonal antibodies. J. Clin. Microbiol. 26:1143–1151.

Hammond, G.W., Hazelton, P.R., Chuang, I., and Klisko, B. 1981. Improved detection of viruses by electron microscopy after direct ultracentrifuge preparation of specimens. J. Clin. Microbiol. 14:210–221.

Harland, W.A., Adams, J.H., and McSeveney, D. 1967. Herpes simplex particles in acute necrotising encephalitis. Lancet 2:581–582.

Hayat, M.A. and Miller, S.E. 1989. Negative Staining: Methods and Applications. New York: McGraw Hill.

Herndon, R.M. and Rubinstein, L.J. 1968. Light and electron microscopy observations on the development of viral particles in the inclusions of Dawson's encephalitis (subacute sclerosing panencephalitis). Neurology 18 (part 2):8–20.

Hirschman, R.J., Schulman, N.R., Barker, L.F., and Smith, K.O. 1969. Virus-like particles in sera of patients with infectious and serum hepatitis. JAMA 208:1667–1670.

Hopley, J. and Doane, F.W. 1985. Development of a sensitive protein A-gold immunoelectron microscopy method for detecting viral antigens in fluid specimens. J. Virol. Meth. 12:135–147.

Hsiung, G.D. 1982. Diagnostic Virology. New Haven: Yale University.

Joncas, J.H., Berthiaume, L., Williams, R., and Beaudry, P. 1969. Diagnosis of viral respiratory infections by electron microscopy. Lancet 1:956–959.

Juneau, M.L. 1979. Role of the electron microscope in the clinical diagnosis of viral infections from patients' stools. Can. J. Med. Technol. 41:53–57.

Kapikian, A.Z., Dienstag, J.L., and Purcell, R.H. 1976. Immune electron microscopy as a method for the detection, identification, and characterization of agents not cultivable in an in vitro system. In N.R. Rose and H. Friedman (eds), Manual of Clinical Immunology. Washington: American Society of Microbiology pp. 467–480.

Kapikian, A.Z., Feinstone, S.M., Purcell, R.H., Wyatt, R.G., Thornhill, T.S., Kalica, A.R., and Chanock, R.M. 1975. Detection and identification by immune electron microscopy of fastidious agents associated with respiratory illness, acute nonbacterial gastroenteritis, and hepatitis A. Perspect. Virol. 9:9–47.

Kapikian, A.Z., Wyatt, R.G., Dolin, R., Thornhill, T.S., Kalica, A.R., and Chanock, R.M. 1972. Visualization by immune electron microscopy of a 27 nm particle associated with acute infectious non-bacterial gastroenteritis. J. Virol. 10:1075–1081.

Kapikian, A.Z., Yolken, R.H., Greenberg, H.B., Wyatt, R.G., Kalica, A.R., Chanock, R.M., and Kim, H.W. 1979. Gastroenteritis viruses. In E.H. Lennette and N.J. Schmidt (eds), Diagnostic Procedures for Viral, Rickettsial and Chlamydial Infections. Washington: American Public Health Association, pp. 927–995.

Kelen, A.E., Hathaway, A.E., and McLeod, D.A. 1971. Rapid detection of Australia/SH antigen and antibody by a simple and sensitive technique of immunoelectron microscopy. Can. J. Microbiol. 17:993–1000.

Kelen, A.E. and McLeod, D.A. 1974. Differentiation of myxoviruses by electronmicroscopy and immunoelectronmicroscopy. In E. Kurstak and R. Morisset (eds), Viral Immunodiagnosis. New York: Academic Press, pp. 257–275.

Kjeldsberg, E. 1980. Application of electron microscopy in viral diagnosis. Path. Res. Pract. 167:3–21.

Kjeldsberg, E. 1986. Demonstration of calicivirus in human faeces by immunosorbent and immunogold-labelling electron microscopy methods. J. Virol. Meth. 14:321–333.

Kjeldsberg, E., Ånestad, G., Greenberg, H., Ørstavik, I., Pedersen, R., and Slettebø. 1989. Norwalk virus in Norway: An outbreak of gastroenteritis studied by electron microscopy and radioimmunoassay. Scand. J. Infect. Dis. 21:521–526.

Kobayashi, H., Tsuzuki, M., Koshimizu, K., Toyama, H., Yoshihara, N., Shikata, T., Abe, K., Mizuno, K., Otomo, N., and Oda, T. 1984. Susceptibility of hepatitis B virus to disinfectants or heat. J. Clin. Microbiol. 20:214–216.

Lee, F.K., Nahmias, A.J., and Stagno, S. 1978. Rapid diagnosis of cytomegalovirus infection in infants by electron microscopy. N. Engl. J. Med. 299:1266–1270.

Lin, Y.P., Nicholas, K., Ball, F.R., McLaughlin, B., and Bishai, F.R. 1991. Detection of Norwalk-like virus and specific antibody by immune-electron microscopy with colloidal gold immune complexes. J. Virol. Meth. 35:237–253.

Luton, P. 1973. Rapid adenovirus typing by immunoelectron microscopy. J. Clin. Pathol. 26:914–917.

Macrae, A.D., Field, A.M., McDonald, J.R., Meurisse, E.V., and Porter, A.A. 1969. Laboratory differential diagnosis of vesicular skin rashes. Lancet 2:313–316.

Madeley, C.R. and Cosgrove, B.P. 1975. 28 nm particles in faeces in infantile gastroenteritis. Lancet 2:451–452.

Madeley, C.R. and Cosgrove, B.P. 1976. Calicivirus in man. Lancet 1:199.

Mathan, M., Mathan, V.I., Swaminathan, S.P., Yesudoss, S., and Baker, S.J. 1975. Pleomorphic virus-like particles in human faeces. Lancet 1:1068–1069.

McLean, D.M. and Wong, K.K. 1984. Same-Day Diagnosis of Human Virus Infections. Boca Raton: CRC Press.

Middleton, P.J., Szymanski, M.T., Abbott, G.D., Bortolussi, R., and Hamilton, J.R. 1974. Orbivirus acute gastroenteritis of infancy. Lancet 1:1241–1244.

Middleton, P.J., Szymanski, M.T., and Petric, M. 1977. Viruses associated with acute gastroenteritis in young children. Am. J. Dis. Child. 131:733–737.

Miller, S.E. 1986. Detection and identification of viruses by electron microscopy. J. Electron Micros. Tech. 4:265–301.

Montplaisir, S., Belloncik, S., Leduc, N.P.,

Onji, P.A., Martineau, B., and Kurstak, E. 1972. Electron microscopy in the rapid diagnosis of cytomegalovirus: Ultrastructural observations and comparison of methods of diagnosis. J. Infect. Dis. 125:533–538.

Nagington, J. 1964. Electron microscopy in differential diagnosis of variola, vaccinia and varicella. Br. Med. J. 2:1499–1500.

Nagler, F.P.O. and Rake, G. 1948. The use of the electron microscope in diagnosis of variola, vaccinia and varicella. J. Bacteriol. 55:45–51.

Oshiro, L.S. 1985. Application of electron microscopy to the diagnosis of viral infections. In E.H. Lennette (ed.), Laboratory Diagnosis of Viral Infections. New York: Marcel Dekker, pp. 55–72.

Palmer, E.L. and Martin, M.L. 1988. Electron Microscopy in Viral Diagnosis. Boca Raton: CRC Press.

Pegg-Feige, K. and Doane, F.W. 1983. Effects of specimen support film in solid phase immunoelectron microscopy. J. Virol. Meth. 7:315–319.

Pegg-Feige, K. and Doane, F.W. 1984. Solid phase immunoelectron microscopy for rapid diagnosis of enteroviruses. In G.W. Bailey (ed.), 42nd Annual Proceedings of the Electron Microscopy Society of America.

Penny, J.B. and Narayan, O. 1973. Studies of the antigenic relationships of the new human papovaviruses by electron microscopy agglutination. Infect. Immun. 8:299–300.

Penny, J.B., Weiner, L.P., Herndon, R.M., Narayan, O., and Johnson, R.T. 1972. Virions from progressive multifocal leukoencephalopathy: Rapid serological identification by electron microscopy. Science 178:60–62.

Peters, D., Nielsen, G., and Bayer, M.E. 1962. Variola. Dtsch. Med. Wschr. 87:2240–2246.

Petrovicova, A. and Juck, A.S. 1977. Serotyping of coxsackieviruses by immune electron microscopy. Acta Virol. 21:165–167.

Roth, J., Bendayan, M., and Orci, L. 1978. Ultrastructural localization of intracellular antigens by the use of protein A-gold complex. J. Histochem. Cytochem. 26:1074–1081.

Roy, S. and Wolman, L. 1969. Electron microscopic observations on the virus particles in herpes simplex encephalitis. J. Clin. Pathol. 22:51–59.

Smith, K.O. 1967. Identification of viruses by electron microscopy. In H. Busch (ed), Methods in Cancer Research, Vol. 1, New York: Academic Press, pp. 545–572.

Stannard, L.M., Lennon, M., Hodgkiss, M., and Smuts, H. 1982. An electron microscopic demonstration of immune complexes of hepatitis B e-antigen using colloidal gold as a marker. J. Med. Virol. 9:165–175.

Svensson, L. and von Bonsdorff, C.D. 1982. Solid-phase immune electron microscopy (SPIEM) by use of protein A and its application for characterization of selected adenovirus serotypes. J. Med. Virol. 10:243–253.

Svensson, L., Grandien, M., and Pettersson, C.A. 1983. Comparison of solid-phase immune electron microscopy by use of protein A with direct electron microscopy and enzyme-linked immunosorbent assay for detection of rotavirus in stool. J. Clin. Microbiol. 18:1244–1249.

Tyrell, D.A. and Kapikian, A.Z., eds. 1982. Virus Infections of the Gastrointestinal Tract. New York: Marcel Dekker.

van Rooyen, C.E. and Scott, G.D. 1948. Small-pox diagnosis with special reference to electron microscopy. Can. J. Pub. Health, 39: 467–477.

Vassall, J.H. and Ray, C.G. 1974. Serotyping of adenoviruses using immune electron microscopy. Appl. Microbiol. 28:623–627.

Weakley, B.S. 1981. A Beginner's Handbook in Biological Electron Microscopy. Edinburgh: Churchill Livingstone.

Wood, D.J. and Bailey, A.S. 1987. Detection of adenovirus types 40 and 41 in stool specimens by immune electron microscopy. J. Med. Virol. 21:191–199.

Wu, B., Mahony, J.B., Simon, G., and Chernesky, M. 1990. Sensitive solid-phase immune electron microscopy double-antibody technique with gold-immunoglobulin G complexes for detecting rotavirus in cell culture and feces. J. Clin. Microbiol. 28: 864–868.

Zu Rhein, G.M. and Chou, S.M. 1965. Particles resembling papovaviruses in human cerebral demyelinating disease. Science 148: 1447–1479.

The Interference Assay

Charles A. Reed

6.1 HISTORY

For many decades, rubella had been looked upon as a rather unimportant and mild disease of childhood. It was commonly referred to as German measles or the three-day measles, and seemed to cause few complications. Rubella had first been described in the German medical literature by DeBergen in 1752 and Orlow in 1758 (Emminghaus, 1870). But it was not until 1941 that the teratogenic effects of this virus, when acquired during pregnancy, became well documented. It was then that an Australian ophthalmologist, Gregg (1941), reported a high incidence of cataracts and other anomalies among offspring of mothers who had contracted rubella early in pregnancy during a rubella epidemic in Australia. A few years later, Swan et al (1943, 1946) reported more cases of congenital defects with later observations related especially to rubella. Since then, many reports have appeared in the literature confirming these findings and reporting many other congenital defects and complications of pregnancy following maternal rubella. Through the years the pattern of events with their resulting effects have become known as "the rubella syndrome." Many retrospective studies and other reports in the literature are in agreement with the earlier findings (Skinner, 1961; also see Chapter 31).

With the advent of newer virologic techniques, more precisely defined chemical media, and a multiplicity of cell lines developed in the late 1940s and early 1950s, the world of virology exploded with the isolation and identification of many viruses (Melnick et al, 1979). However, it was not until 1962 that two groups of investigators succeeded in isolating the rubella virus in two different systems. Weller and Neva were able to isolate the virus in primary human amnion cell cultures, which demonstrated a visible cytopathic effect (CPE) with time (Weller and Neva, 1962). Parkman et al (1962) were able to demonstrate the presence of the virus in primary African green monkey kidney (AGMK) cell cultures, which produced no rubella virus-associated CPE. However, it was shown that rubella virus infection interfered with the CPE production of a challenge virus (e.g., ECHO-11). This was the first demonstration of the interference assay that could be used to identify rubella virus subsequent to its isolation in cell culture. At about the

111

same time, Veronelli et al (1962), reported similar findings using a continuous Rhesus monkey cell line, LLC-MK$_2$, which had been shown to be useful for viral research (Hull et al, 1962; Veronelli et al, 1962).

Parkman et al continued their research with the rubella virus interference phenomenon and it was shown that infection in cell culture interfered with CPE production by many enteroviruses, myxoviruses, and arboviruses in AGMK cell cultures (Parkman et al, 1964a). Thirteen different cell lines were tested and it was found that viral replication occurred in 11 of these cell lines, but the interference of the CPE by a challenge virus varied with the different lines. In addition, their studies showed that rubella virus is heat labile and that specimens should be chilled or frozen after collection. In continuing studies, Parkman et al demonstrated that rubella viral replication and neutralization tests showed similar results when the AGMK and LLC-MK$_2$ cell lines were compared (Parkman et al, 1964b). All of the techniques and methods were put to the test when the United States was swept by a major epidemic of rubella in 1964–1965 and most virology laboratories started to culture for the rubella virus. With the licensure of a rubella vaccine in 1969, however, the demands on the clinical virus lab for culturing rubella virus have been significantly reduced.

6.2 SPECIMEN COLLECTION, TRANSPORT, AND STORAGE

As with all viral isolation, the process starts with the proper collection and handling of the specimen (see Chapter 2). Once inoculated onto tissue culture, the remaining specimen should be frozen and stored at −70 °C.

For rubella identification by the interference assay the following specimens are appropriate (Sanders, 1978; Herrman, 1979).

1. Respiratory: Nasal and pharyngeal swabs from children and adults. Throat washings may be suitable from adults. These should be collected no later than five days after the onset of the rash.

2. Blood: Heparinized blood should be collected as near the onset of the rash as possible and as early as seven days after possible exposure.

3. Cerebrospinal fluid: CSF is most useful in cases involving suspected congenital rubella.

4. Urine: Urine should be collected as aseptically as possible. Urine has been shown to contain virus for months after birth of a congenitally infected infant (Cooper and Krugman, 1966).

5. Tissues: All tissues and body fluids may be collected and tested when appropriate. These would include ocular tissues and fluids, autopsy or biopsy tissues, placenta, amniotic fluid, and any fetal tissues that are accessible.

6.3 SPECIMEN PREPARATION AND INOCULATION

Specimens collected on swabs should be expressed into 2 mL of maintenance medium if not collected directly into a transport broth. All specimens—with the exception of CSF—should be treated with the appropriate antimicrobial agents per each laboratory's preference. Tissue material should be ground and prepared as a 10% suspension in the same maintenance medium that is to be added to the final cell culture. Our laboratory utilizes Earle's basic salt solution (EBSS), Eagle's minimum essential medium (EMEM) and 2% fetal calf serum as maintenance medium. Adjust the pH of all specimens to 7.0 to 7.2.

Select the cell line to be inoculated. Primary AGMK is the most widely used, but it has drawbacks that will be discussed later. LLC-MK$_2$ is used in our laboratory if problems arise with the AGMK. Remove the medium from the selected tube cultures, using four tubes for each specimen. Inoculate 0.25 mL of the specimen directly onto each cell culture followed by a one-hour adsorption period at 36 °C (stationary rack). Add 1.5 mL of fresh maintenance medium to each tube and incubate in a roller drum. Observe each tube daily and record any changes noted.

6.4 ECHO-11 CHALLENGE AND NEUTRALIZATION

Rubella virus may not produce visible CPE, but may be detectable using the property of viral interference. For example, if rubella virus is growing in AGMK cells, these cells become resistant to ECHO-11 virus, to which they are ordinarily highly susceptible. As a result, no CPE is produced following infection by ECHO-11. The following procedure is our adaptation of the procedures of Parkman et al (1964b):

1. Prepare a stock virus pool of ECHO-11 in LLC-MK$_2$; aliquot into tubes (1 mL per tube) and stored at −70 °C
2. Titer this challenge virus (ECHO-11) in LLC-MK$_2$ cells to an endpoint using the Reed-Muench (1938) method
3. Inoculate specimens submitted that are suspected to contain rubella virus (including all specimens from newborns with congenital malformations) into four tubes of AGMK cells of the same lot
4. At ten days postinoculation remove the medium from two inoculated tubes and two control tubes of AGMK cells.
   - Add 100 TCID$_{50}$ in 0.1 mL of ECHO-11 challenge virus into one inoculated tube and one control tube each. Add 10 TCID$_{50}$ in 0.1 mL of ECHO-11 challenge virus into each of the remaining inoculated and control tubes. These dilutions of virus stock

are made in EMEM supplemented with 2% fetal calf serum
   - Incubate in a stationary rack for one hour at 36 °C; manually rotate the tubes every 10 to 15 minutes
   - Add 1.5 mL EMEM with 2% FCS at the end of this incubation period
   - Incubate in a stationary rack at 36 °C
5. Examine tubes daily; interference is present if extensive CPE develops in the control cell cultures but not in the cell cultures inoculated with the clinical specimen

Note: This method allows for detection of low titer viral replication. If the 10 TCID$_{50}$ challenge tube is negative for ECHO-11, the unchallenged tubes should be harvested and passed to four new tubes. Repeat steps 3 and 4 above.

1. Interference-neutralization test:
   - Harvest material (by one freeze-thaw cycle) from the two inoculated tubes that were not subjected to challenge virus
   - Place 0.15 mL of the harvested material in a sterile screw cap tube (16 × 75 mm) and add 0.15 mL rubella antiserum at a dilution to represent 50 neutralizing antibody units per 0.1 mL
   - As a control, incubate the harvest material without antiserum
   - Incubate samples in a stationary rack for one hour at 36 °C
   - Inoculate 0.2 mL of the virus antiserum mixture and 0.1 mL of the untreated virus into one tube each of AGMK cells; incubate at 36 °C in a stationary rack. Duplicate tubes may be used.
   - After five days inoculate each tube with 100 TCID$_{50}$ ECHO-11; if duplicate tubes were prepared, one set may be challenged with 10 TCID$_{50}$ of ECHO-11; the virus is confirmed as rubella virus if viral interference is

prevented by antiserum treatment, and is again demonstrated in the control tube

Note: Several different log dilutions of virus can be employed at step 6 if the virus titer is too high for the neutralizing capacity of the antibody.

6.5 PROS AND CONS

The "rubella-challenge-interference" system is probably the most rapid system of isolating the virus from a patient specimen in a single passage. It can be achieved within 14 days on primary AGMK. Other cell lines that are capable of replicating rubella virus usually require more than one blind passage before a visible CPE is noted. During the period of time required for virus replication the physical condition of the cell sheet is critical and must be visually compared very carefully with uninfected control tubes. Most laboratories today do not pass the specimen from tubes of a set that has shown a positive ECHO-11 CPE. The specimen is reported negative after 14 days.

However, because laboratories may experience contamination with indigenous resident simian viruses in purchased AGMK cell lines from time to time, these cell lines should be examined daily for simian virus associated CPE and tested for the presence of these simian viruses. If the cell cultures are positive for contaminating agents they should not be used because misleading results may occur. If specimens have already been inoculated into tubes of that lot, reinoculate a new lot of tested cells from the specimen sample that has been held in frozen storage. This problem is not encountered if the LLC-MK$_2$ line is used, however, LLC-MK$_2$ cells are not as sensitive for the primary isolation of rubella virus. The LLC-MK$_2$ line can be used for the neutralization test subsequent to the primary isolation of the virus.

Because of the expense of the cell cultures and laboratory personnel time that are required to isolate, challenge and then confirm rubella virus isolation, this protocol has been questioned especially if a serologic test will give the physician the necessary information. However, virus isolation and identification by interference assay is still the standard for proof of current rubella virus infection.

REFERENCES

Cooper, L.Z. and Krugman, S. 1966. Diagnosis and management: Congenital rubella. Pediatrics 37:335–338.

Emminghaus, H. 1870. Uber rubeolen. Jahrb. Kinderheilkd. 4:47–59.

Gregg, N.M. 1941. Congenital cataract following German measles in the mother. Trans. Ophthal. Soc. Aust. 3:35–46.

Herrmann, K.L. 1979. Rubella virus. In E.H. Lennette and J. Schmidt (eds.), Diagnostic Procedures for Viral, Rickettsial and Chlamydial Infections, 5th ed. Washington, D.C.: American Public Health Association, pp. 725–766.

Hull, R.H., Cherry, W.R., and Tritch, O.J. 1962. Growth characteristics of monkey kidney cell strains LLC-MK$_1$, LLC-MK$_2$ and LLC-MK$_2$ (NCTC-3196) and their utility in virus research. J. Exp. Med. 115:903–917.

Melnick, J.L., Wenner, H.A., and Phillips, C.A. 1979. Enteroviruses. In E.H. Lennette and N.J. Schmidt (eds.), Diagnostic Procedures for Viral, Rickettsial and Chlamydial Infections, 5th ed. Washington, D.C.: American Public Health Association, pp. 471–534.

Parkman, P.D., Buescher, E.L., and Artenstein, M.S. 1962. Recovery of rubella virus from army recruits. Proc. Soc. Exp. Biol. Med. 111:225–230.

Parkman, P.D., Buescher, E.L., Artenstein, M.S., McCown, J.M., Mundon, F.K., and Druzd, A.D. 1964a. Studies of rubella I. Properties of the virus. J. Immunol. 93:595–607.

Parkman, P.D., Mundon, F.K., McCown, J.M., and Buescher, E.L. 1964b. Studies of rubella II. Neutralization of the virus. J. Immunol. 93:608–617.

Reed, L.J., and Muench, H. 1938. A simple method of estimating fifty percent end points. Am. J. Hyg. 27:493–497.

Sanders, C.V., Jr. 1978. Diagnostic virology. In H. Rothschild, F. Allison, Jr., and C. Howe (eds.), Human Diseases Caused by Viruses. New York: Oxford University Press, pp. 259–281.

Skinner, C.W. 1961. The rubella problem. Am. J. Dis. Child. 101:78–86.

Swan, C., Tostevin, A.L., Moore, B., Mayo, H., and Black, G.H.B. 1943. Congenital defects in infants following infectious disease during pregnancy. Med. J. Australia 2:201–210.

Swan, C., Tostevin, A.L., and Black, G.H.B. 1946. Final observations on congenital defects in infants following infectious diseases during pregnancy with special reference to rubella. Med. J. Australia 2:889–908.

Veronelli, J.A., Maassab, H.F., and Hennessy A.V. 1962. Isolation in tissue culture of an interfering agent from patients with rubella. Proc. Soc. Exp. Biol. Med. 111:472–476.

Weller, T.H. and Neva, F.A. 1962. Propagation in tissue culture of cytopathic agents from patients with rubella-like illness. Proc. Soc. Exp. Biol. Med. 111:215–225.

7

Immunofluorescence

Linda L. Minnich and C. George Ray

7.1 HISTORY

Immunofluorescence (IF) was first developed in 1941 when Coons et al (1941) used immune serum globulin labeled with a fluorescent dye to detect corresponding antigens. By the late 1950s, the technique had been successfully applied for the rapid diagnosis of herpes simplex virus (HSV) in corneal scrapings and lesions (Biegeleisen et al, 1959), influenza A virus (Liu, 1956) and others. In 1966, the indirect IF test became the recommended test for the rapid diagnosis of rabies. The popularity of IF as a diagnostic tool for identifying virus infections has waxed and waned, but this technique has again become more widely used, in part due to the commercial availability of monoclonal antibodies to a wider variety of viruses than ever before. Most exhibit excellent sensitivity and specificity (Minnich et al, 1988).

7.2 MATERIALS

7.2.1 Specimens

Proper collection of specimens appropriate for the clinical syndrome are essential for successful utilization of immunofluores-cence in the rapid diagnosis of viral infections (Almeida et al, 1979). Appropriate specimens for diagnosis of respiratory infections include nasopharyngeal-throat swabs, nasal aspirates, bronchoalveolar lavages and lung biopsies (Chernesky et al, 1982; Gardner et al, 1968). Inappropriate specimens include sputum, throat swabs, and gargles which have significantly lower amounts of viruses. Collection methods are described below.

Nasopharyngeal-throat swab specimens are collected by inserting a type 1, flexible shaft cotton or rayon tipped swab into the nasopharynx and holding it in place for 60 seconds. This swab is removed and placed in viral transport medium. A vigorous swab of the peritonsillar area should be obtained and placed in the same vial, producing a pooled sample. Slides for direct examination may be prepared by firmly rolling a properly collected nasopharyngeal swab onto a welled slide (Minnich and Ray, 1980).

Nasal aspirates, as previously described by Gardner and McQuillin (1980), are collected by passing a number 8 French catheter through each nostril into the nasopharynx. This catheter is attached to a mucous trap that is connected to a source of suction, which may be as simple as a large syringe. The catheter is passed while applying suction and the sample is collected in the mucous trap.

Bronchoalveolar lavages and bronchial washes are collected in the same manner as for bacteriologic studies (Cordonnier et al, 1987; Emanual et al, 1986). Care should be taken to avoid using phosphate buffered saline (PBS) containing preservatives or contaminating the specimens with local anesthetics, e.g., xylocaine, as these may inactivate the virus or be toxic to cell cultures.

Lung biopsies and/or aspirates should be collected according to standard procedures. Touch preparations of the tissue can be made for direct antigen detection. Aspirates may be cytocentrifuged directly onto the glass slide. Alternatively, centrifugation of specimen at 1,500 to 2,500 × g produces a pellet of cells which may be transferred to a slide for antigen detection.

Lesions are sampled by scraping or firmly swabbing the affected site and transferring the cells either directly to a slide or placing them in viral transport medium. Local anesthetics, frequently used during the collection of samples from the eye, may inactivate viruses such as HSV, varicella-zoster virus (VZV), and even adenoviruses, and should be avoided if virus isolation in cell culture is also planned. Thus, in cases when the comfort, compliance, or safety of the patient requires utilization of local anesthetics, antigen detection may be more sensitive than detection of virus by cell culture. For best results, vesicular skin lesions should be selected and cells at the base of the lesion gently but firmly scraped or swabbed. Older, crusted lesions may produce false positive IF results.

7.2.2 Reagents

Although the number of commercially available reagents has dramatically increased in recent years, the need for each laboratory to perform its own quality control and adhere to strict criteria for reading and interpreting slides has not diminished. This section on quality control of reagents includes information which is necessary for using specific and anti-species antisera as well as other reagents such as PBS, fixatives, slides, and mounting fluids.

7.2.3 Quality Control of Specific Antisera

Specific antisera are initially tested to assure reactivity with the homologous agent. This is done by titrating serial dilutions of the serum in the test procedure. It is important that the same incubation environment, wash times, and other reagents to be used for the actual testing are applied for this titration. If an indirect system is being used, either a pretitrated optimal dilution of the conjugated secondary (anti-species) antibody is used or an arbitrary dilution (usually 1:10 or 1:20 if it is the laboratory's initial evaluation) is selected. A four- to eight-fold greater concentration than that which produces a 1+ (dim but definite) fluorescence result is used for testing to document nonreactivity with other heterologous antigens including other viruses, tissues, and cell cultures.

Evaluation of antisera using uninfected cells, normal tissues, and other irrelevant or unrelated viruses which may be present in the clinical specimens is then made using the dilutions determined above. For example, antisera to respiratory agents should be evaluated for reactivity to respiratory syncytial virus (RSV), influenza A viruses (both H_1N_1 and H_3N_2 strains), influenza B virus, parainfluenza virus types 1–4, mumps virus, measles virus, adenovirus, rhinovirus, poliovirus types 1–3, HSV, cytomegalovirus (CMV), and at least two enteroviruses, e.g., coxsackie B4 and echo 9. Antisera to be used on skin lesions should be evaluated for reactivity with HSV, VZV, and CMV. Each antiserum should react only with the homologous virus and exhibit no reactivity with uninfected cell culture, tissue, or heterologous viruses. Final evaluation of an antiserum prior to using it is made by testing positive and negative clinical samples as well as samples positive for heterologous viruses if possible. This will provide additional information, e.g., reactivity of the antisera to yeasts or other microbial antigens which may be found in patient specimens. After this initial, extensive evaluation, confirmation of the antiserum and conjugate reactivity is usually documented by including homologous virus as a control antigen on a use day basis.

7.2.4 Quality Control of Anti-species Conjugate

Dilutions of antiglobulin conjugate along with an appropriate primary antiserum should be tested on infected cell cultures and positive human specimens to determine the optimal dilution for use. A preliminary optimal dilution may be obtained using infected cell cultures, but usually must be modified for final use on clinical samples. In general, a four to eight-fold greater concentration of serum is required for use with clinical samples.

The optimal dilutions are then tested for nonspecific reactions on cell lines likely to be used in the laboratory as well as nonspecific reactivity on human material that does not contain the virus in question.

The need to develop immunization schedules for large animals, e.g., calves and horses, as well as utilization of yolk sacs from eggs produced by immunized hens has been diminished by the advent of monoclonal antibodies. Monoclonal antibodies have many potential advantages, including a theoretically unlimited supply, reproducibility, high titers, and strict antigenic specificity. This restricted reactivity can sometimes be a source of difficulty in clinical settings. Large numbers of wild-type strains must be tested to determine if slight or significant antigenic mutations lead to failure to detect the antigen by a particular monoclonal antibody. Alternatively, the use of a pool of monoclonal antibodies broadens the reactivity pattern against the variation of strains of viruses present in the human population. A monoclonal antibody that reacts with a highly conserved antigen or antigenic epitope may also be used. For screening, a pool of monoclonal antibodies to candidate viruses may be used (Stout et al, 1989).

The selection of particular monoclonal and/or polyclonal antibody preparations for use in direct antigen detection and cell culture confirmation is made by the individual laboratory. Regardless of which system is used, careful and continuous quality control is of paramount importance.

7.2.5 The Physics of Immunofluorescence and the Microscope

The basic principle of immunofluorescence is that certain dyes (fluorochromes) become excited when they are stimulated by light of short wave lengths, in the ultraviolet (UV) and violet-blue end of the spectrum. To revert to the resting state, the fluorochrome molecules emit light of a longer wavelength, which is visible by the human eye. The process is known as fluorescence. The fluorochrome most frequently used in immunofluorescence is fluorescein isothiocyanate, although other dyes (e.g., rhodamine isothiocyanate) can also be used. For simplicity, only fluorescein will be considered. Fluorescein has a peak wavelength for absorbing light, thereby, achieving maximum stimulation at 490 nm. The pre-emission wavelength is 517 nm. The successful utilization of the IF technique rests with the ability to separate these two wavelengths.

The art of producing good IF is compromise. A light source is needed, usually a mercury vapor or halogen bulb, which stimulates fluorescein near its optimum at 495 nm. An exciting filter system capable of blocking out all light above 500 nm is used so that maximum stimulation is achieved. Then a barrier filter is employed, which only allows through light over 517 nm that will be seen as an apple-green color. Ordinary glass filters cannot achieve sharp cutoff points, and therefore, cannot achieve maximum stimulation. Recently, more effective fluorescence has been achieved by using interference filters that can transmit excitation light up to 490 nm and should be used with a suitable barrier filter that transmits light over 500 nm. Interference filters are layers of glass in which thin layers of metallic salt compounds are deposited in a vacuum. If the vacuum coating process is repeated using different materials a multilayer filter is obtained. Light that does not pass through interference filters is reflected. Variation of the thickness and refractive index of the vacuum-deposited layers makes it possible to select specific wavelengths of transmission and reflection that are close to the wavelength of maximum absorption for the fluorochrome to be used. A further property of fluorochromes is their ability to be coupled with proteins. If they are coupled with antibodies they can be used to detect antigens by irradiating such stained specimens under the fluorescence microscope. There are two kinds of microscopes that can be used for fluorescence microscopy: transmitted light and incident light (epifluorescence). The first method is now rarely used and should be considered obsolete. A more detailed review of the theory of IF and fluorescence microscopy has been presented in Gardner and McQuillin (1980). It should also be noted that the titer of FITC-conjugated antisera as well as the primary antiserum in an indirect system varies with the bulb and microscope. Therefore, titrations may vary from laboratory to laboratory within two- to fifty-fold dilutions.

7.3 METHODS

Many debates have been raised on the advantages and disadvantages of direct vs indirect IF, and more recently, monoclonal vs polyclonal antibodies. Some laboratorians still hold the belief that the indirect is more sensitive because of the amplification possible with the second step, by providing additional fluorescein molecules. Others believe that the cleaner preparations of direct conjugates provide increased sensitivity and specificity due to ease of reading. Monoclonal antibodies in general provide the cleanest preparations but have limitations in their ranges of antigenic reactivity—a problem that often can be resolved by use of a pool of monoclonals (Minnich et al, 1987). Selection of both the method and antibody type should be determined by the individual laboratory after consideration of the reagents and technical expertise available.

7.3.1 Specimen Preparation

Nasopharyngeal secretions collected as previously described should be transported at 2 °C to 8 °C (wet ice) to the laboratory as soon as possible. Transportation times of less than 24 hours are acceptable for both direct antigen detection and virus isolation. Smears may be prepared directly at the bedside if welled slides are available. The use of regular glass slides without a hydrophobic barrier often results in spread of the smear over a large area, necessitating the use of excessive quantities of expensive reagent(s).

Swabs and aspirates submitted in transport medium require centrifugation prior to the preparation of slides for examination. Cells are pelleted at 350 to 500g at 2 °C to 8 °C for 10 to 15 minutes. The supernatant fluid is used to inoculate cell cultures for virus isolation, if desired. The cell pellet is washed in PBS if needed to remove excess mucus by resuspending the cells in PBS and recentrifuging. The PBS supernatant is discarded and the cells are resuspended in PBS with 1% to 2% bovine serum albumin or fetal bovine serum to make a cloudy suspension. Excessively thick suspensions can result in a loss of cells during the staining and washing procedures or result in difficulty the in microscopic assessment of the cells. Conversely, cell suspensions that are too thin may contain an inadequate number of cells to assess. Preparation of extra slides enables the laboratorian to build a specimen library for future evaluation of reagents, as well as for teaching and research purposes.

The type of slides to be used for antigen detection impacts upon the success of the test procedure. Hydrophobic, welled slides are commercially available in a variety of colors, well sizes, and number of wells. The hydrophobic character of the slides is cost effective when reagent savings are considered. While other slides are less expensive, they may require precleaning with acetone and/or alcohol to remove residual film which prevents the cells from properly adhering. Labeling should be done with a pencil to avoid dissolving inks with the acetone.

Fixation of the smears is the next consideration. Traditionally, cold acetone has been used. The frequent lack of availability of an explosion-proof refrigerator for storage of this organic solvent has stimulated the evaluation of fixation at room temperature. This has proven to work equally as well provided the acetone is changed when it becomes cloudy as a result of absorption of water from specimens or contamination. Fixation times do not seem to be as critical as once thought. Two to ten minutes are adequate, but longer fixation times are not

deleterious to most antigens. Some laboratorians simply flood the slide with acetone and allow it to evaporate. Remember that acetone is extremely flammable and care must be taken to avoid open flames or sparks in the immediate area. Formalin and alcohols destroy many viral antigens and are therefore not routinely used as fixatives. Fixed slides have traditionally been stored for long periods at −30 °C or colder, but many antigens survive quite well for longer than a year at room temperature. Removal from the freezer in high humidity or repeated freeze-thaw cycles may be deleterious to many antigens.

7.4 PRACTICAL DETAILS FOR INDIRECT IMMUNOFLUORESCENCE

The direct method, which involves the use of conjugated primary antiserum, consists of steps 1–4 and 7–8 of the indirect method, and will not be discussed separately.

1. Fixed and properly identified slides are placed in a humidified chamber which may simply consist of a plastic box in which a wet towel is overlaid with plastic or wooden sticks. This keeps the edge of the slide from direct contact with the towel and provides a moist atmosphere to prevent the antiserum and conjugate from drying on the wells.

2. A drop of the appropriate, properly diluted antiserum is placed onto the designated well and spread to cover all of the well. Avoid touching the cells when spreading the antiserum.

3. Incubate the slides at 35 °C to 37 °C for 30 to 60 minutes (time determined in initial titration) to allow the antigen-antibody reaction to take place.

4. After incubation, remove excess antiserum by gently washing with PBS. A wash bottle may be used if care is taken to prevent the stream of PBS from being

directed on the smear. Alternatively, the slide may be dipped into a jar of PBS. The number and length of time of the PBS washes are determined by the initial evaluation of the antiserum being used. In general, three ten-minute washes are essential to remove unbound antibodies when polyclonal antiserum is used. One to two one-minute washes are usually adequate for monoclonal antisera. Washes of longer duration may produce false negative results.

5. Excess PBS is blotted from the slide and the secondary, conjugated antispecies antibody is added to the wells and spread to cover the entire area of the smear. This conjugated anti-species antibody will react with primary antibody bound in the first reaction.

6. Repeat the incubation and wash steps as above. If the primary antibody is monoclonal, the second wash steps will be no more than two five-minute washes even if the secondary antibody is not monoclonal. Careful adherence to wash times developed in the initial evaluation of the system must be observed to avoid both non-specific fluorescence and false negative results.

7. The slide is then immersed in or flooded with distilled water for one to two minutes to remove PBS crystals. A counter stain such as Evans' blue may be added to the distilled water. The time for the distilled water rinse may be increased if PBS crystals interfere with reading of the slides. If crystals are noted, the coverslip may be removed by soaking in distilled water and the slide rinsed for additional time to remove the crystals.

8. A coverslip is then placed on the slide using a mounting medium such as Tris-buffered glycerol, pH 9.0 ± 0.5. Immersion oil should not be used due to inactivation of the fluorescein by a low pH. Some laboratorians recommend using a non-fluorescing immersion oil di-

rectly on the slide. The utilization of "high-dry" lenses makes it easier to examine the slide with a coverslip in place and thereby avoid the messiness of oil objectives. Distilled water may be used as a mounting medium if photographs are to be taken; however, slides treated in this manner may not be stored for long periods of time.

Reading and interpreting stained slides requires a combination of scientific criteria and technical expertise. Strict criteria for staining patterns must be defined and observed. These staining patterns will vary with the antiserum used and the cells in the specimen. For example, a pooled monoclonal antibody preparation for RSV produces a much more striking speckling appearance with irregular granules in cells from a nasopharyngeal specimen than the smaller, more uniform granules seen in an impression smear of lung tissue. (Figures 7.1 and 7.5). The staining seen with HSV-infected cells varies with the type of virus infecting the cells. Proper training of technologists in reading and interpreting specimens cannot be overemphasized.

False positive results can be due to a variety of factors. One of interest is the dull green fluorescence produced by certain infant formulas that contaminate specimens from the oral cavity, which persist despite repeated washing of the cells. Yeast and some bacteria which adhere to cells may produce fluorescence, especially if polyclonal antisera are used. When strict criteria for interpretation of results when using a particular cell type and seeing a particular staining pattern are observed, these "false positives" are not a problem. It should also be noted that the exact staining pattern observed in infected cell cultures may not be precisely replicated in clinical samples. Adequate, properly collected and processed samples are essential for successful utilization of the IF technique.

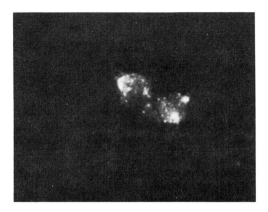

Figure 7.1 RSV in cells of a nasopharyngeal secretion. × 1100

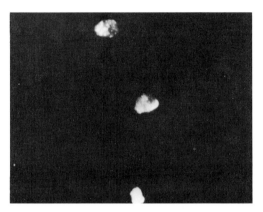

Figure 7.2 Influenza A in cells of a nasopharyngeal secretion. × 1100

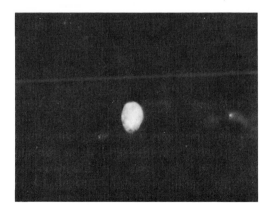

Figure 7.3 Adenovirus in cells of a nasopharyngeal secretion. × 1100

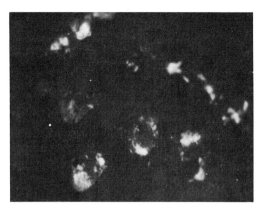

Figure 7.4 Parainfluenza 1 in cells of a nasopharyngeal secretion. × 1100

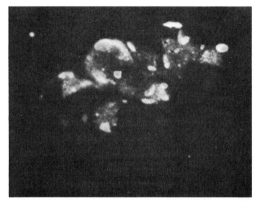

Figure 7.5 Impression smear of lung showing RS virus infected cells. × 1100

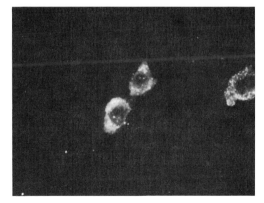

Figure 7.6 Impression smear of lung showing cells infected with parainfluenza virus type 3. × 1100

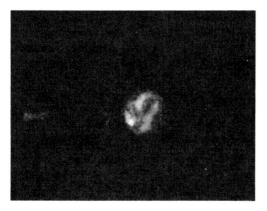

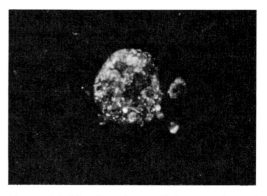

Figure 7.7 Herpes virus hominis in a skin scraping. × 1100

Figure 7.8 Measles giant cell in secretion of patient with acute lymphatic leukemia. × 1100

7.4.1 Viruses For Which the Fluorescent Antibody Technique Is Available

Fortunately for today's diagnostic virologist, polyclonal and monoclonal antibodies against a large number of viruses are commercially available. Monoclonal antibodies against measles virus, mumps virus, RSV, influenza virus types A and B, adenovirus group antigen, HSV, VZV, CMV, and parainfluenza virus types 1–4 are readily available (Minnich et al, 1988; Waner et al, 1985; McQuillan et al, 1985; Minnich et al, 1991). With the exceptions of influenza virus A and B, RSV, VZV, parainfluenza virus type 1–3, and HSV, they are primarily available for the indirect IF test system. Even with the use of "licensed kits," in-house quality control, as previously described, is required. With the reappearance of measles and mumps in developed countries, the potential cross reactivity of antisera to RSV and the parainfluenza viruses with measles and mumps viruses, respectively, must be carefully considered and quality controlled. Photographs of common respiratory virus infected cells, stained by IF, are presented in Figures 7.1 through 7.8. Descriptions of the reactions usually observed are presented in Table 7.1.

7.5 APPLICATION OF FLUORESCENT ANTIBODY TECHNIQUE IN DETECTION OF ANTIGENS IN CELL CULTURE

Direct detection of viral antigens in clinical samples has proven useful in rapid diagnosis of respiratory viruses HSV and VZV. In addition, detection of antigens in cell cultures before the appearance of cytopathic effects has further enhanced the speed of detection for agents such as CMV, adenoviruses, and measles virus. Tube cultures may be scraped and the cells placed on a slide; however, the efficiency has been markedly enhanced by the incorporation of spin-amplified vial cultures (Espy and Smith, 1987; Gleaves et al, 1985; 1987; Paya et al, 1987; Lucas et al, 1989; Marshall et al, 1990; Gleaves et al, 1987; Espy et al, 1987). This system has proved useful for HSV, VZV, adenovirus, influenza viruses, parainfluenza viruses, measles virus, and CMV (Espy and Smith, 1987; Espy et al, 1986; Minnich et al, 1988; Rabella and Drew, 1990.).

In this "shell vial" technique, the appropriate cells are grown on coverslips placed in vials. Specimens are inoculated directly onto the coverslips and centrifuged at 25 °C to 35 °C and 700 g for approxi-

Table 7.1 Expected Staining Patterns of Virus Infected Cells Stained by IF

Virus	Expected Staining Pattern
Adenovirus	Nuclear and/or cytoplasmic staining of ciliated respiratory epithelial and epithelial cells.
Cytomegalovirus	Nuclear and/or cytoplasmic in epithelial and other affected cells, including alveolar macrophages, gastric epithelium.
Enteroviruses	Cytoplasmic staining only of epithelial cells and neurons.
Herpes simplex virus	Nuclear and cytoplasmic staining of epithelial cells and/or neurons. (Note: Do not confuse non-specific stianing of polymorphonuclear cells with specific staining of epithelial cells.)
Influenza viruses	Nuclear and cytoplasmic staining of ciliated respiratory epithelial cells.
Measles virus	Cytoplasmic only staining of respiratory epithelium, epithelial cells; morphology of multinucleated giant cells may be observed.
Mumps virus	Cytoplasmic staining only of respiratory epithelial cells.
Parainfluenza viruses	Cytoplasmic staining only in ciliated respiratory epithelial cells.
Rabies virus	Cytoplasmic staining only in neurons.
Respiratory syncytial virus	Cytoplasmic staining only in ciliated respiratory epithelial cells.
Varicella-zoster virus	Cytoplasmic and/or nuclear staining of epithelial cells and neurons.

mately 40 minutes. Longer centrifugation times and increased speeds do not improve results and may be deleterious to the cells. After centrifugation, the vials are refed with appropriate maintenance medium or may be incubated for one hour at 35 °C to 37 °C prior to refeeding. After an incubation period of 18 hours to four to five days (see Table 7.2 for suggested incubation times),

Table 7.2 Recommended Incubation Times Prior to Immunofluorescent Staining for Detection of Viral Antigens in Cell Cultures

Virus	Incubation Time
Adenovirus	2–5 days
Cytomegalovirus	1–5 days
Herpes simplex virus	18 hours to 2 days
Influenza virus	1–2 days
Measles virus	1–3 days
Mumps virus	1–3 days
Parainfluenza viruses	1–3 days
Polyomavirus BK	1–3 days
Respiratory syncytial virus	3–5 days
Varicella-zoster virus	2–5 days

the cells in the vials are acetone-fixed, removed, and stained with appropriate antisera. HSV, measles virus, influenza viruses, parainfluenza viruses, and often CMV have frequently replicated enough to provide sufficient antigen for detection at 18 to 36 hours after inoculation. Other viruses such as adenovirus and sometimes slower replicating strains of CMV may require four to five days of incubation prior to staining (Espy et al, 1987). This system provides a more rapid approach to identification of these viruses than traditional roller tube cultures. See Chapter 3 in this volume for a more complete description.

7.6 RESOLUTION OF TECHNICAL PROBLEMS

As with any procedure, technical problems may be encountered and must be resolved. Table 7.3 lists those pitfalls that are more commonly encountered with considerations for their resolution.

Table 7.3 Potential Sources of Technical Problems with Immunofluorescence and Items to Consider for Their Resolution

REAGENTS

Primary antisera, conjugated or unconjugated
· specificity (too much cross reactivity or too specific for routine clinical use)
· optimal titers not properly determined or no longer valid (e.g., loss of reactivity with storage)
· loss of reactivity due to frequent warming during use or improper storage
· contamination with other reagents or bacteria
· disassociation of conjugate from antibody if direct conjugate

Diluents and wash solutions
· pH below 7.0 or above 10.0
· bacterial contamination
· crystallization

Mounting fluids
· pH below 7.0 or above 10.0
· bacterial contamination
· crystallization
· autofluorescence

Counterstain
· autofluorescence
· loss of specific fluorescence by improper concentration

TECHNIQUE
· improperly or inadequately collected samples
· improper transportation of specimens
· errors in handling specimens and/or reagents
· quality of specimens
· preparation of specimens
· improper fixation, especially with wrong fixative
· under or over washing
· wrong anti-species globulin if indirect method
· wrong primary antibody or conjugate
· incorrect illumination or improperly aligned bulb
· wrong or improperly functioning or aligned filter system

READING AND INTERPRETATION
· Nonspecific fluorescence due to contaminants, autofluorescence, mucus, reactions between specimens and antiglobulins, milk, or other specimen-related factors
· Failure to follow specific criteria for diagnosis:
    intracellular fluorescence with proper staining pattern
    color
    experience

7.7 APPLICATION OF FLUORESCENT ANTIBODY TECHNIQUE AND ITS FUTURE

The role of the diagnostic virology laboratory has changed over the years from that of primarily epidemiologic surveillance to that of providing rapid, accurate diagnosis of viral infections to facilitate cohorting of patients, treatment with anti-viral agents, and cost-containment by reduction of unnecessary additional diagnostic procedures and therapies. In the current health care environment, limitation of resources places additional pressure on the laboratory to provide a rapid diagnosis with the lowest possible cost to the patient, facility, and third-party payer. In this environment, IF techniques provide the diagnostic virologist with cost and time efficient tools. For example, although 9% or more of lower respiratory virus infections in infants and young children involve more than one etiologic agent (Ray and Minnich, 1987), many such infections can be accurately and rapidly diagnosed by IF. Once established and continuously monitored for quality control, the laboratory may be justified in not processing these specimens for further isolation and identification of homologous or additional viruses. This approach requires careful communication with the primary physician. Also it is useful in reducing the cost for both reagents and technical time. Experience with the IF technique is impor-tant in increasing its utility to the laboratory and reducing the time required for reading and interpreting the results.

The role of IF in the diagnostic virol-ogy laboratory is further assured by the ability to provide quality control for the system in a cost effective manner. After initial evaluation and titration, minimal quality control to assure consistent reactiv-ity of reagents and regular reviews of tech-nologist's performance are adequate to ensure a sensitive and specific system.

ACKNOWLEDGMENT

The photomicrographs and the text describ-ing immunofluorescence physics were orig-inally provided by Dr. P.S. Gardner when he authored this chapter for the first edi-tion. We gratefully acknowledge this ongo-ing contribution from him.

REFERENCES

Almeida, J.D., Atanasiu, P., Bradley, D.W., Gardner, P.S., Maynard, J., Schuurs, A.W., Voller, A., and Yolken, R.H. 1979. Manual for rapid viral diagnosis. WHO Off-set Publication No. 47.

Biegeleisen, J.Z. Jr., Scott, L.V., and Lewis, V. Jr. 1959. Rapid diagnosis of herpes simplex virus infection with fluorescent antibody. Science 129:640–641.

Chernesky, M.A., Ray, C.G., and Smith, T.F. 1982. Cumitech 15, Laboratory diagnosis of viral infections. W.L. Drew, (Coordinating ed.), Washington, D.C.: American Society for Microbiology.

Coons, A.H., Creech, H.J., and Jones, R.N. 1941. Immunological properties of antibody containing a fluorescent group. Proc. Soc. Exp. Biol. Med. 47:200–202.

Cordonnier, C., Escudier, E., Nicolas, J.-C., Fleury, J., Deforges, L., Ingrand, D., Bri-cout, F., and Bernaudin, J.-F. 1987. Evalu-ation of three assays on alveolar lavage fluid in the diagnosis of cytomegalovirus pneu-monitis after bone marrow transplantation. J. Infect. Dis. 155:495–500.

Emanuel, D., Peppard, J., Stover, D., Gold, J., Armstrong, D., and Hammerling, U. 1986. Rapid immunodiagnosis of cytomegalovirus pneumonia by bronchoalveolar lavage using human and murine monoclonal antibodies. Ann. Intern. Med. 104:476–481.

Espy, M.J., Hierholzer, J.C., and Smith, T.F. 1987. The effect of centrifugation on the rapid detection of adenovirus in shell vials. Am. J. Clin. Pathol. 88:358–360.

Espy, M.J. and Smith, T.F. 1987. Simultaneous seeding and infection of shell vials for rapid detection of cytomegalovirus infection. J. Clin. Microbiol. 25:940–941.

Espy, M.J., Smith, T.F., Harmon, M.W., and Kendal, A.P. 1986. Rapid detection of influ-enza virus by shell vial assay with monoclo-nal antibodies. J. Clin. Microbiol. 24:677–679.

Fulton, R.E. and Middleton, P.J. 1975. Immunofluorescence in the diagnosis of measles infections in children. J. Pediatr. 86:17–22.

Gardner, P.S. and McQuillin, J. 1968. Application of the immunofluorescent antibody technique in the rapid diagnosis of respiratory syncytial virus infection. Br. Med. J. 2:12–13.

Gardner, P.S. and McQuillin, J. 1980. Rapid Virus Diagnosis, Application of Immunofluorescence, ed. 2. London: Butterworths.

Gleaves, C.A., Lee, C.F., Kirsch, L., and Meyers, J.D. 1987. Evaluation of a direct fluorescein-conjugated monoclonal antibody for detection of cytomegalovirus in centrifugation culture. J. Clin. Microbiol. 25:1548–1550.

Gleaves, C.A., Smith, T.F., Shuster, E.A., and Pearson, G.R. 1985. Comparison of standard tube and shell vial cell culture techniques for detection of cytomegalovirus in clinical specimens. J. Clin. Microbiol. 21:217–222.

Liu, C. 1956. Rapid diagnosis of human influenza infection from nasal smears by means of fluorescein-labelled antibody. Proc. Soc. Exp. Biol. Med. 92:883–887.

Lucas, G., Seigneurin, J.M., Tamalet, J., Michelson, S., Baccard, M., Delagneau, J.F., and Deletoille, P. 1989. Rapid diagnosis of cytomegalovirus by indirect immunofluorescence assay with monoclonal antibody F6b in a commercially available kit. J. Clin. Microbiol. 27:367–369.

Marshall, W.F., Telenti, A., Proper, J., Akasamit, A.J., and Smith, T.F. 1990. Rapid detection of polyomavirus BK by shell vial cell culture assay. J. Clin. Microbiol. 28:1613–1615.

McQuillan, J., Madeley, C.R., and Kendal, A.P. 1985. Monoclonal antibodies for the rapid diagnosis of influenza A and B virus infections by immunofluorescence. Lancet 2:911–914.

Minnich, L.L. and Ray, C.G. 1980. Comparison of direct immunofluorescent staining of clinical specimens for respiratory virus antigens with conventional isolation techniques. J. Clin. Microbiol. 12:391–394.

Minnich, L.L., Goodenough, F., and Ray, C.G. 1991. Use of immunofluorescence to identify measles virus infection. J. Clin. Microbiol. 29:1148–1150.

Minnich, L.L., Shehab, Z.M., and Ray, C.G. 1987. Application of pooled monoclonal antibodies for one-hour detection of respiratory syncytial virus antigen in clinical specimens. Diagn. Microbiol. Infect. Dis. 7:137–141.

Minnich, L.L., Smith, T.F., and Ray, C.G. 1988. Cumitech 24, Rapid detection of viruses by immunofluorescence. S. Specter, (Coordinating ed.). Washington, D.C.: American Society for Microbiology.

Paya, C.V., Wold, A.D., and Smith, T.F. 1987. Detection of cytomegalovirus in specimens other than urine by the shell vial assay and conventional tube cell cultures. J. Clin. Microbiol. 25:755–757.

Rabella, N. and Drew, W.L. 1990. Comparison of conventional and shell vial cultures for detecting cytomegalovirus infection. J. Clin. Microbiol. 28:806–807.

Ray, C.G. and Minnich, L.L. 1987. Efficiency of immunofluorescence for rapid detection of common respiratory viruses. J. Clin. Microbiol. 24:355–357.

Stout, C., Murphy, M.D., Lawrence, S., and Julian, S. 1989. Evaluation of a monoclonal antibody pool for rapid diagnosis of respiratory viral infections. J. Clin. Microbiol. 27:448–452.

Waner, J.L., Whitehurst, N.J., Downs, T., and Graves, D.G. 1985. Production of monoclonal antibodies against parainfluenza 3 virus and their use in diagnosis by immunofluorescence. J. Clin. Microbiol. 22:535–538.

Radioimmunoassay

Isa K. Mushahwar and Thomas A. Brawner

8.1 INTRODUCTION

Ever since the introduction of radioimmunoassay (RIA) techniques (Yalow and Berson, 1960) for the determination of endogenous human plasma insulin, RIA has become the essential key analytical method in many sciences. Besides endocrinology, RIA has been a valuable tool in such fields as enzymology (Kolb and Grodsky, 1970), hematology (Rutland, 1984), toxicology (Shimada et al, 1983), pharmacology (Robinson and Smith, 1983), parasitology (Avraham et al, 1982), neurochemistry (Dowse et al, 1983), microbiology (Zollinger et al, 1976), plant pathology (Ghabrial and Shepherd, 1980), diagnostic medicine (Matsui et al, 1982), mycology (Poor and Cutler, 1979), and virology (Wiktor et al, 1972). Thus, RIA has contributed to the revolution of biology during the past 25 years in many other disciplines besides endocrinology. This trend will continue to have a positive impact on public health in the future through diagnostic and epidemiologic studies in the areas of infectious diseases and cancer.

8.2 COMPETITIVE BINDING RADIOIMMUNOASSAY

The original RIA described by Yalow and Berson (1960) and subsequently by other investigators (Faiman and Ryan, 1967; Odell et al, 1967; Goodfriend and Ball, 1969) was a competitive binding RIA, where the competition between an unlabeled antigen and a radiolabeled counterpart for a limited amount of antibody is monitored. This is illustrated as follows:

$$AG^* + AB + AG^0 \rightarrow AG^*AB + AG^0AB$$

where AG^* represents radiolabeled antigen; AG^0 is the unlabeled antigen; AB is the antibody; AG^*AB is a complex of radiolabeled antigen with antibody; and AG^0AB is a complex of unlabeled antigen with antibody. The higher the concentration of AG^0, the lower the concentration of radioactive AG^*AB and the higher the concentration of free AG^*. The reaction is designed to take place in a solution giving a mixture of bound and free radiolabeled antigen. Separation of antibody-bound antigen from the free antigen is achieved by electrophoretic or a variety of immunoprecipitation techniques (Morgan and Lazarow, 1962; Hales and Randle, 1963).

8.3 SOLID PHASE RADIOIMMUNOASSAYS

A simpler and widely used variation of the competitive binding RIA is a solid phase RIA, which is also based on competitive inhibition (Catt and Tregear, 1967) utilizing radiolabeled antigen and solid phase antibody for the separation of bound and free antigen as shown:

$$SP.AB + AG^* + AG^0$$

$$\rightarrow \begin{matrix} SP.AB.AG^0 \\ SP.AB.AG^* \end{matrix} + AG^* + AG^0$$

where the count rate of radioactive antigen (AG*), bound through antibody linkage to the surface of the solid phase (SP.AB) to form the radiolabeled antigen-antibody complex (SP.AB.AG*) is in direct proportion to the quantity of competitive unlabeled antigen (AG^0) in the reaction mixture. Removal of the reaction mixture and washing the solid phase (polystyrene tube) serves as an effective and highly reproducible method for separating bound and free antigen. Soon after the introduction of solid phase assays, it became apparent that RIA procedures based on principles other than competitive inhibition could be developed. These were the direct solid phase sandwich RIA (Wide et al, 1971). These methods were found applicable to all biological substances that have a minimum of two binding sites and, thus, can be utilized to assay for both antigen and antibodies. Soon, a direct solid phase sandwich RIA utilizing, for the first time, radiolabeled specific immunoglobulin as a probe for macromolecular viral antigens was introduced (Ling and Overby, 1972). The potential advantages of using radiolabeled antibody over radiolabeled antigen (e.g., viruses) is the greater availability, ease of labeling and immobilization of immunoglobulins.

The last decade has witnessed a rapid increase in the variety of solid phase RIA developed for the detection and quantification of a multitude of antigens and their corresponding antibodies (Overby and Mushahwar, 1979; Mushahwar and Overby, 1983). For the most part, these assays have been used extensively in highly specialized research environments. Most importantly, the vast majority of these tests never made the transition from a research procedure to a commercially available and standardized procedure (Hill and Matsen, 1983). Because the hepatitis viruses have been widely characterized with highly specific and commercially available RIA, the authors will use these models to describe the various direct and indirect (competitive) methods that employ solid phase sandwich RIA to detect the antigens and antibodies of viral hepatitis (Table 8.1).

8.4 HEPATITIS B SEROLOGIC MARKERS

Three distinct viral antigens and their respective antibodies are used for the diagnosis of hepatitis B virus (HBV) infections: hepatitis B surface antigen (HBsAg) and its antibody (anti-HBs), hepatitis B core antigen (HBcAg) and its antibody (anti-HBc); and hepatitis B e antigen (HBeAg) and its antibody (anti-HBe). These antigens and antibodies occur in serum sequentially during the course of disease and recovery (Mushahwar et al, 1981). Because HBsAg is a defective particle without nucleic acid and is produced in large quantities, it is an easily detectable marker of an ongoing HBV infection. Seroconversion to anti-HBs is an indication of recovery and eventual immunity. HBcAg is found in the internal core of the DNA-containing virus and has not been found to occur free in serum. Anti-HBc rises with the onset of

Table 8.1 Hepatitis

Hepatitis Marker	Reference
Hepatitis B virus (HBV)	
HBsAg	Hollinger et al, 1971
	Ling and Overby, 1972
	Purcell et al, 1973
Anti-HBc IgG	Neurath et al, 1978
	Overby and Ling, 1976
	Purcell et al, 1973
	Vyas and Roberts, 1977
Anti-HBs	Ginsberg et al, 1973
Anti-HBe	Blum et al, 1979
	Frosner et al, 1978
	Mushahwar et al, 1978
	Neurath et al, 1979
HBeAg	Blum et al, 1979
	Frosner et al, 1978
	Mushahwar et al, 1978
	Neurath et al, 1979
Anti-HBc IgM	Chau et al, 1983
HBcAg	Overby and Ling, 1976
	Purcell et al, 1973
	Vyas and Roberts, 1977
Hepatitis D virus (HDV)	
HDAg	Rizzetto et al, 1980
	Mushahwar and Decker, 1983
Anti-HD IgG	Mushahwar and Decker, 1983
	Rizzetto et al, 1980
Anti-HD IgM	Smedile et al, 1982
Hepatitis A virus (HAV)	
HAV	Decker et al, 1979
	Hollinger et al, 1975
	Hollinger and Maynard, 1976
	Purcell et al, 1976
Anti-HAV IgG	Decker et al, 1979
	Hollinger and Maynard, 1976
	Purcell et al, 1976
	Safford et al, 1980

Table 8.1 Hepatitis (*continued*)

Hepatitis Marker	Reference
Anti-HAV IgM	Bradley et al, 1977
	Decker et al, 1981
	Devine et al, 1979
	Flehmig et al, 1979
	Lemon et al, 1980
	Roggendorf et al, 1980
Anti-HAV IgA	Overby et al, 1981
Hepatitis C virus (HCV)	
Anti-HCV IgG	Kuo et al, 1989

viremia and persists through recovery and immunity. This antibody is sometimes the only detectable marker for exposure to the virus. HBeAg rises in serum at the same time as HBsAg, and its presence is associated with acute disease, chronic disease, and the presence of infectious virus. Seroconversion to anti-HBe signifies better clinical prognosis and a lower level of viremia.

8.5 DIRECT SOLID PHASE SANDWICH ASSAYS

8.5.1 Radiometric Assays for Hepatitis B Virus Antigens and Antibodies

The direct solid phase sandwich RIA procedure for antigen detection is divided into three main operational stages, as follows: adsorption of unlabeled antibody onto the solid phase, binding of the antigen by the adsorbed antibody, and detection of bound antigen by reaction with radiolabeled and highly specific antibody. The first successful direct solid phase RIA for the detection of viral antigens was the AUSRIA prototype (Ling and Overby, 1972) for the detection of HBsAg. In this system highly specific human anti-HBs was adsorbed to polystyrene tubes. The serum to be tested for HBsAg was incubated with the anti-HBs coated tube giving a complex of anti-HBs and HBsAg in the solid phase. The radioactive probe in the next reaction was

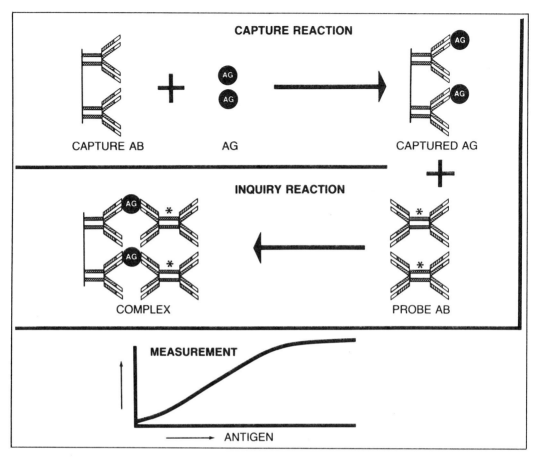

Figure 8.1 Schematic representation of a direct solid phase assay designed to detect HBeAg. The procedure involves capture of the antigen by antibody attached to the solid phase (*top*) and subsequent binding of labeled antibody to exposed sites on the captured antigen (*middle*). The amount of antigen measured is in direct proportion to the count rate of the washed solid phase (*bottom*).

affinity purified guinea pig anti-HBs labeled with radioactive iodine (^{125}I).

A direct solid phase RIA for HBeAg similar to AUSRIA was developed (Mushahwar et al, 1978) employing radiolabeled specific antibodies as illustrated in Figure 8.1. Briefly, 6-mm diameter polystyrene beads are coated with anti-HBe by incubating a dilution of antiserum with the beads at pH 9.0 for 24 hours at room temperature. These antibody-coated beads are then used as the solid phase antibody. Anti-HBe IgG preparations are radiolabeled with ^{125}I by the chloramine-T method (Greenwood et al, 1963) often resulting in a specific radio-

activity of 18–25 μCi/μg IgG. The ^{125}I-labeled anti-HBe solution is diluted to approximately 3 μCi/mL in a diluent containing 50% fetal calf serum. For HBeAg detection, serum samples of 0.2 mL are incubated overnight at room temperature with the antibody coated bead. These are then washed with distilled water, and the beads are further incubated in 0.2 mL of ^{125}I-labeled anti-HBe solution at 45 °C for four hours. The beads are then washed and counted for ^{125}I-labeled anti-HBe uptake. The resulting count rate of the multiple layered beads are directly proportional to HBeAg concentration in the sample. Hu-

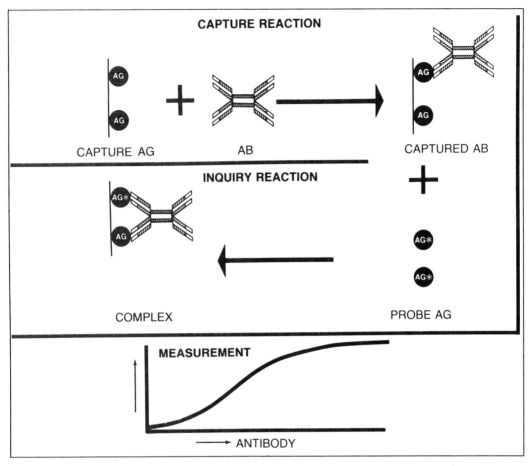

Figure 8.2 Schematic representation of direct, solid phase assay to detect anti-HBs. The two-step procedure involves capture of specific antibody by antigen bound to the solid phase (*top*) and binding of labeled antigen to specific, available sites on the captured antibody (*middle*). The amount of antibody measured is in direct proportion to the count rate of the washed solid phase (*bottom*).

man convalescent antisera containing anti-HBe are a source of reagents for the diagnostic RIA described, without recourse to animal immunizations with purified preparations of HBeAg. The manufacturers package insert for HBeAg determination describes the procedures in a stepwise manner.

The presence or absence of HBeAg is determined by comparing the counts per minute (cpm) of the unknown specimen to a predetermined cutoff value (Mushahwar et al, 1978). Specimens whose count rates are equal to or greater than this cutoff value are considered to be reactive for HBeAg.

For a test run to be valid the mean value for the positive control should be at least four times the negative control mean. It is recommended by the manufacturer that three negative and three positive HBeAg controls should be assayed with each run. A similar direct solid phase sandwich RIA for anti-HBs detection has been developed (Ginsberg et al, 1973). The principle is illustrated in Figure 8.2.

Highly purified HBsAg is bound to a polystyrene bead by adsorption to produce the solid phase antigen. The serum specimen to be assayed for anti-HBs is incubated with the solid phase antigen. If anti-HBs is present, it will complex with the antigen. ^{125}I-HBsAg is added in the next step, which will bind to the antibody already trapped on the bead, forming an antigen-antibody-^{125}I-HBsAg complex. The amount of radioactivity on the bead is in proportion to the concentration of anti-HBs in the serum.

The step-by-step procedure is described in the manufacturer's package insert. An example of this is the AUSAB test (Abbott Laboratories) for anti-HBs. It is recommended that seven negative and three positive controls should be assayed with each run of unknowns.

Again, as in the case of HBeAg detection, reactive and nonreactive specimens are determined by relating the net counts per minute of unknowns to a cutoff value calculated by multiplying the negative control mean by the factor 2.1. Specimens with values greater than the cutoff are reactive for anti-HBs.

8.6 INDIRECT (COMPETITIVE) RADIOMETRIC ASSAYS FOR HEPATITIS B VIRUS ANTIBODIES

In these assays, radiolabeled antibodies are utilized as probes for the detection of serum antibodies in either a one-step or a two-step procedure. The solid phase is an appropriate antigen-coated polystyrene tube or bead. The radiolabeled probe will have fewer antigen binding sites for the reaction when serum antibodies are present. Hence, the final count rate of the solid phase will be inversely proportional to the amount of antibody in the serum specimen being tested. These techniques have been applied to the detection of anti-HBc (Overby and Ling, 1976) and anti-HBe (Mushahwar et al, 1978).

8.6.1 One-Step Procedure

The anti-HBc assay is illustrated in Figure 8.3. The solid phase reagent is a polystyrene bead coated with purified HBcAg. The bead is first reacted with the serum specimen to be assayed for anti-HBc. The resulting HBcAg-anti-HBc complex is challenged with purified human ^{125}I-anti-HBc IgG in a second step. The level of radioactivity bound to the bead is inversely proportional to the amount of anti-HBc in the test specimen. A competitive binding RIA is available commercially (Table 8.2). The step-by-step procedure is described by the manufacturer's package insert.

Five positive and five negative controls are tested with each run of unknown specimens. The presence or absence of anti-HBc is determined by comparing the net counts per minute of the specimen to a cutoff value, calculated as the sum of the negative control and positive control means divided by 2. Specimens whose count rates are equal to or lower than the cutoff value are considered reactive for anti-HBc.

8.6.2 Two-Step Procedure

The anti-HBe assay has been described in detail (Mushahwar et al, 1978). This was the first successful indirect solid phase RIA for the detection of serum anti-HBe utilizing a two-step neutralization procedure. The principle of the assay is illustrated in Figure 8.4. In this assay, 0.1 mL of the patient's serum is mixed with an equal volume of standardized HBeAg-positive serum (neutralizing reagent). The mixture is incubated overnight at room temperature with the solid phase, a polystyrene bead coated with high-titered human anti-HBe serum. After washing, the bead is incubated with ^{125}I-labeled anti-HBe IgG for four hours at 45 °C, as described under the direct solid phase HBeAg assay. The quantity of neutralizing reagent in the initial incubation has been selected to give over

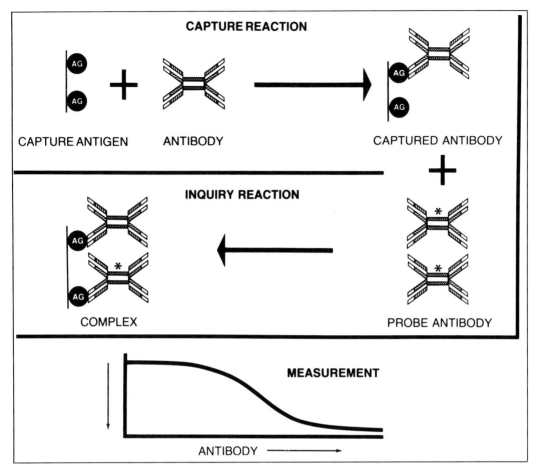

Figure 8.3 Schematic representation of an indirect, competitive solid phase assay for anti-HBc utilizing labeled antibodies. The two-step procedure involves capture of antibody by an antigen (HBcAg) coated solid phase (*top*) and binding of labeled antibody to available sites on the bound antigen (*middle*). The count rate of the washed solid phase will be inversely proportional to the amount of antibody in the assay specimen (*bottom*).

10,000 counts per minute on the bead in the presence of a serum that is negative for anti-HBe. A 50% or more reduction in the count rate indicates the presence of anti-HBe in the test sample. This assay is available commercially (Table 8.2). The step-by-step procedure is described by the manufacturer's package insert. Three negative and three positive anti-HBe controls are used each time the test is run for anti-HBe analysis.

The presence or absence of anti-HBe is determined by comparing the count rate of the specimen to a cutoff value calculated from the sum of the negative control and positive control divided by 2. Specimens whose count rates are equal to or less than the cutoff value are considered reactive for anti-HBe.

8.7 ASSAY FOR IGM CLASS ANTIBODIES

Virus-specific IgM antibody has been confirmed as a prominent early immune re-

Table 8.2 Commercial Hepatitis Radioimmunoassay Kits

Test	Serologic Marker	Product Name	Solid Phase Configuration	Manufacturer
Antigen	HBsAg detection	AUSRIA II	6-mm bead	Abbott Laboratories
		Heparia	6-mm bead	North American Biologicals, Inc.
		Rialyze	Gel column	Ames Co., Division Miles Laboratories
		AUK-3	6-mm bead	Sorin Biomedica
		Riasure	Microbeads	Electro-Nucleonics Laboratories, Inc.
		Clinical Assays	Plastic tube	Connought Laboratories (Travenol Laboratories)
		NML HBsAg RIA	6-mm bead	Nuclear-Medical Laboratories
	HBeAg	ABBOTT-HBe	6-mm bead	Abbott Laboratories
		EBK	6-mm bead	Sorin Biomedica
Antibody	Anti-HBs	AUSAB	6-mm bead	Abbott Laboratories
		AB-AUK	6-mm bead	Sorin Biomedica
		Hepab	6-mm bead	North American Biologicals, Inc.
		Clinical Assays	Plastic tube	Connought Laboratories (Travenol Laboratories)
	Anti-HBc (Total)	CORAB	6-mm bead	Abbott Laboratories
		AB-COREK	6-mm bead	Sorin Biomedica
	Anti-HBc (IgM)	CORAB-M	6-mm bead	Abbott Laboratories
		CORE-IGMK	6-mm bead	Sorin Biomedica
	Anti-HAV (IgM)	HAVAB-M	6-mm bead	Abbott Laboratories
	Anti-HBe	ABBOTT-HBe	6-mm bead	Abbott Laboratories
		EBK	6-mm bead	Sorin Biomedica
	Anti-HAV (Total)	HAVAB	6-mm bead	Abbott Laboratories
	Anti-Delta (Total)	ABBOTT-ANTI-DELTA	6-mm bead	Abbott Laboratories
	Anti-Rubella IgG	Gamma Coat	Plastic tube	Travenol-Genetech Diagnostics

sponse in many viral infections. Because it is relatively short lived, it is a good marker for acute disease (Chau et al, 1983). A reliable and reproducible RIA for the detection of hepatitis A virus IgM antibody (anti-HAV IgM) has been described (Decker et al, 1979). The anti-HAV IgM assay is based on the following reactions and is illustrated in Figure 8.5.

1. SP-Abμ + IgM → Sp-Abμ-IgM
2. SP-Abμ-IgM + HAV → SP-Abμ-IgM-HAV
3. SP-Abμ-IgM.HAV + *Ab-HAV → SP-Abμ-IgM HAV.*Ab-HAV

SP-Abμ is a polystyrene surface coated with μ chain-specific goat antihuman antibody. In step one, if anti-HAV

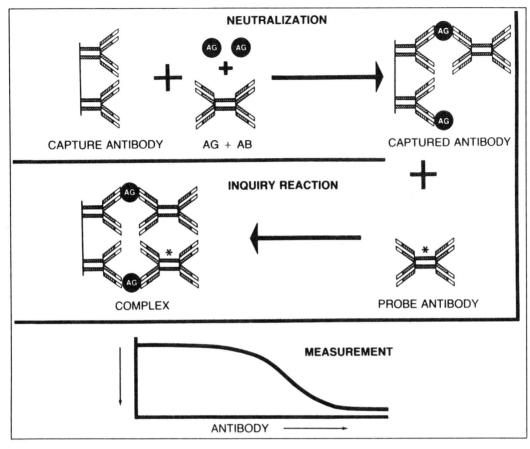

Figure 8.4 Schematic representation of a competitive solid phase sandwich assay for anti-HBe antibodies. The solid phase, coated with anti-HBe, is incubated with a standardized amount of HBeAg and a sample of the patient's serum. The anti-HBe in the patient's serum combines with the antigen and reduces the amount of antigen binding to the solid phase bound anti-HBe (*top*). The sandwich is formed by the addition of labeled anti-HBe (*middle*). The amount of labeled anti-HBe bound is inversely proportional to the amount of specific antibody present in the patient's serum (*bottom*).

IgM is present in a patient's serum, it will be bound by the μ chain-specific solid phase antibody. In step two, HAV will be attached to this complex to form the SP-Abμ-IgM-HAV. This complex is then detected in step three by incubation with the probe antibody, *Ab-HAV, an ^{125}I-labeled human anti-HAV IgG. The resulting count rate of the multiple layer product SP-Abμ-IgM-HAV.*Ab-HAV is in proportion to anti-HAV IgM concentration in the patient's serum. This test was shown to be highly specific. No crossreactions were observed in the presence of high-titered anti-HAV IgG, when hepatitis A convalescent serum samples were tested. This assay is commercially available. The step-by-step procedure is described by the manufacturer's package insert.

Two negative and three positive controls are used each time the test is run for anti-HAV IgM analysis. The presence or absence of anti-HAV IgM is determined by comparing the count rate of the specimen to a predetermined (Decker et al, 1979) cutoff value.

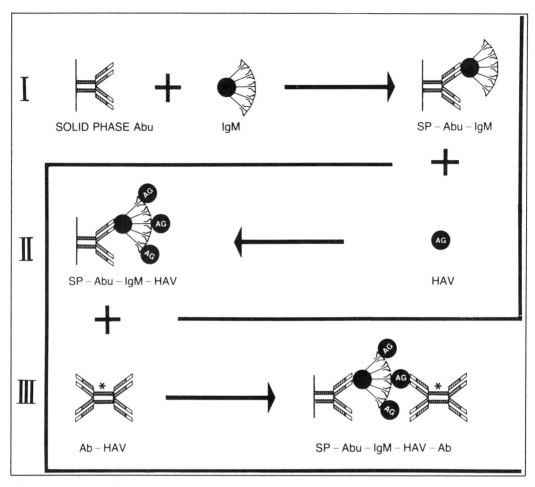

Figure 8.5 Schematic representation of a direct assay for IgM class anti-HAV antibody. The solid phase is coated with anti-human antibody specific for the μ chain. The three-step procedure involves capture of IgM antibodies by the solid phase μ chain specific antibody (*top*), subsequent binding of HAV to the captured antibody (*middle*) and binding of radiolabeled antibody to the solid-phase bound complex (*bottom*). The amount of antibody in the serum sample is directly proportional to the count rate of the washed solid phase.

8.8 OTHER VIRAL MARKERS

Several reviews (Overby and Mushahwar, 1979; Mushahwar and Overby, 1983) have surveyed the application of RIA for rapid diagnosis of viral, bacterial, and fungal diseases. The following review describes the use of RIA to identify viral antigens or antiviral antibodies appearing in the most recent literature. Tables 8.3 and 8.4 sum-marize the information presented in this review.

8.8.1 Enteroviruses

8.8.1.a Enterovirus Antibody Detection

A solid phase IgM capture RIA designed to detect antibody specifically directed against coxsackievirus B4 and B5 has been described (Morgan-Capner and McSorley,

Table 8.3 Antigen Detection

Organism	Radiolabeled Reagent	Sample	Separation Technique	Reference
Rotavirus	Ab	Stool	Polystyrene bead	Sarkkinen et al, 1979
	Ab	Stool	Polystyrene bead	Haikala et al, 1983
	Ab	Stool	Microtiter plate	Cukor et al, 1978
	Ab	Stool	Microtiter plate	Cukor et al, 1984
	Ab	Stool	Polystyrene tube	Middleton et al, 1977
	Ab	Stool	Polystyrene bead	Sarkkinen et al, 1980
Norwalk Agent	Ab	Stool	Microtiter plate	Greenberg et al, 1978
Adenovirus	Ab	Stool	Polystyrene bead	Halonen et al, 1980
	Ab	NP[a]	Polystyrene bead	Sarkkinen et al, 1981
	Ab	NP	Polystyrene bead	Vesikari et al, 1982
Herpes simplex	Protein A	Virus[b]	Microtiter plate	Marsden et al, 1984
	Protein A	Cell[c]	Filter paper disc	Cleveland et al, 1979
	Ab	Virus[b]	Polystyrene tube	Enlander et al, 1976
	Ab	Cell[c]	Glass vials	Forghani et al, 1974
Respiratory viruses				
(RSV)	Ab	NP	Polystyrene bead	Sarkkinen et al, 1981
	Ab	NP	Polystyrene bead	Meurman et al, 1984
	Ab	NP	Polystyrene bead	Vesikari et al, 1982
Paraflu type 2	Ab	NP	Polystyrene bead	Sarkkinen et al, 1981
Paraflu type 3	Ab	NP	Polystyrene bead	Vesikari et al, 1982
Influenza	AMP	NP	Polystyrene bead	Coonrod et al, 1984
	Ab	NP	Polystyrene bead	Sarkkinen et al, 1981
	Ab	NP	Polystyrene bead	Versikari et al, 1982

[a] Nasopharyngeal sample.
[b] Virus from cell culture.
[c] Antigens in cell culture.

1983). Positive results were obtained with sera from cases of heterologous enterovirus infection. These results may be due to anamnestic IgM response or shared, common antigenic determinants. The authors note that specificity may be improved by the development of monoclonal detector antibodies.

IgM class, as well as IgG class antibodies to coxsackievirus A7, A9, A16, B2, B4, B5 or echovirus 4, 17 or 25, could be detected using RIA procedures (Torfason et al, 1984). Immunoglobulin G titers were demonstrated to be higher in the convalescent phase sera. Modification of the indirect IgM test resulted in an antibody capture assay that demonstrated type specificity. The authors note that the antigen must be carefully standardized in order to maintain the type specificity.

8.8.2 Togavirus

8.8.2.a Anti-Alphaviruses Antibody

A previous review (Mushahwar and Overby, 1983) has detailed the sensitive and specific procedures used to identify antibody to Venezuelan, Western, and Eastern equine encephalitis viruses (Jarling et al, 1978).

Table 8.4 Antibody Detection

Organism	Ab Species Identified	Radiolabed Reagent	Separation Technique	Reference
Enteroviruses	IgM	Ab	Polystyrene bead	Morgan-Capner and McSorley, 1983
	IgG, IgM	Ab, Ag		Torfason et al, 1984
Togaviruses				
Rubella	IgM	Ab	Microtiter plate	Kangro et al, 1981
	IgM	Ab	Polystyrene bead	Mortimer et al, 1981
	IgG, IgM	Ab	Polystyrene bead	Meurman, 1978
	IgG, IgM	Ab	Polystyrene bead	Kalimo et al, 1976
	IgG, IgM	Ab	Polystyrene bead	Meurman and Granfors, 1977
	IgM	Ab	Polystyrene bead	Tedder et al, 1982
	IgM	Ab	Red cells	Sexton et al, 1982
St. Louis encephalitis virus	IgG, IgM	Ab	Polyvinyl microtiter plate	Wolff et al, 1981
	IgG	Ab	Bead	Trent et al, 1976
Rotavirus	IgG, A, M	Ab	Polystyrene bead	Sarkkinen et al, 1979
	IgG, A, M	Ab	Filter paper	Watanabe and Holmes, 1977
	IgA	Ab	Microtiter plate	Cukor et al, 1979
Herpesviruses				
Cytomegalovirus	IgG	Ab	Polyvinyl microtiter plate	Kimmel et al, 1980
	IgM	Ab	Microtiter plate, polystyrene well	Kangro et al, 1984
Herpes simplex	IgG	Ab	Microtiter plate	Dreesman et al, 1979
	IgG	Ab	Infected cells	Smith et al, 1974
	IgG, A, M	Ab	Plastic coated bead	Patterson et al, 1978
	IgG, IgM	Ab	Polystyrene bead	Kalimo et al, 1977
	IgG, IgM	Ab	Polystyrene bead	Kalimo et al, 1977
	IgG	Ab	Polyvinyl chloride plate	Matson et al, 1983
	IgG	Ab	Polyvinyl chloride plate	Adler-Storthz et al, 1983
Varicella-zoster	IgG	Ab	Polyvinyl microtiter plate	Friedman et al, 1979
	IgG, IgM	Ab	Polyvinyl microtiter plate	Arvin and Koropchak, 1980
	IgG	Ab	Microtiter plate	Benzie-Campbell et al, 1981
	IgM	Ab	Polystyrene tube and beads	Tedder et al, 1981
Adenovirus	IgA	Ab	Polystyrene bead	Halonen et al, 1979

8.8.2.b Anti-Flaviviruses Antibody

Trent et al, (1976) have described a solid phase assay for the detection of IgG and IgM antibodies to specific St. Louis encephalitis (SLE) virus structural protein. Other investigators (Wolff et al, 1981) have extended the solid phase system to use the crude antigens now commonly used to diagnose SLE infections. Their results indicated that the procedure was as sensitive as conventional serologic tests but not as specific. The procedure was capable of differentiating SLE from similar clinical infections with alphaviruses. However, infections caused by related flaviviruses could not be accurately differentiated.

8.8.2.c Anti-Rubella Virus Antibody

The ability to diagnose acute infections with serologic methods becomes more important when the clinical signs and symptoms are subtle. The principle of diagnosing rubella virus infections by the detection of virus-specific IgM antibody in a serum sample is now well accepted. Solid phase RIA has been shown to detect IgM, as well as IgG, antibodies to rubella virus. The adsorption of purified rubella virus to polystyrene beads resulted in a highly sensitive test capable of detecting IgM antibodies (Meurman et al, 1977; Meurman, 1978; Mortimer et al, 1981) or IgM or IgG antibodies (Kalimo et al, 1976). The development of a solid phase, IgM capture procedure (Mortimer et al, 1981; Tedder et al, 1982) resulted in a high degree of sensitivity and a decreased probability of interference by rheumatoid factor. The use of monoclonal antibody directed against the hemagglutinin protein of rubella virus resulted in a strong and specific reaction when used in the MACRIA assay (Tedder et al, 1982). Evaluation of clinical samples with this system may result in greater correlation with more traditional hemagglutination assays used to evaluate the immune status. Other solid phase systems, such as microtiter plates (Kangro et al, 1981) and the use of red cells (Sexton et al, 1982) have been successfully employed to detect IgM antibodies to rubella virus.

8.8.3 Rotavirus

8.8.3.a Antigen Detection

The use of solid phase detection systems for the identification of rotavirus antigens in stool specimens has become increasingly common. The assays have taken three general forms, as follows: capture of antigen on microtiter plates (Kalica et al, 1977; Cukor et al, 1978, 1984; Greenberg et al, 1978) on polystyrene beads (Sarkkinen et al., 1979a; Haikala et al, 1983), or using polystyrene tubes (Middleton et al, 1977).

Comparison of RIA procedures (Sarkkinen et al, 1979) to latex agglutination (Haikala et al, 1983) indicate that RIA is more sensitive and less subject to interference by particulate material. The key to a sensitive and specific test for antigen identification is the selection of immune reagents of the highest quality. Cukor et al (1978) noted that screening of antiserum by conventional procedures, such as complement fixation (CF), is not suitable or appropriate for selecting antibody capable of identifying rotavirus in stool samples with great sensitivity. The use of a monoclonal antibody directed against the common group-specific antigen has resulted in a sensitive and specific test (Cukor et al, 1984).

8.8.3.b Antibody Detection

Sarkkinen et al (1979b) described a solid phase RIA for the detection of human rotavirus-specific IgG, IgM, and IgA antibodies. This system, using viral antigens to capture specific antibodies, was found to be more sensitive than the reference CF test. Cukor et al (1979) have described a procedure for the detection of IgA in human milk. The RIA was capable of detecting antibody for longer periods of time than enzyme immunoassay (ELISA), immunofluorescence assay (IF), or neutralization assays.

8.8.4 Norwalk Virus

8.8.4.a Antigen Detection

A second viral agent has been linked to acute episodes of diarrhea or vomiting (Cukor and Blacklow, 1984). The Norwalk virus is a small, 27-nm nonenveloped particle. Although the particle was visualized in 1972, major advances were not made until an RIA capable of detecting the agent was reported (Greenberg et al, 1978). The use of a solid phase system employing microtiter plates allows a large number of patient specimens to be screened.

8.8.5 Adenovirus

8.8.5.a Antigen Detection

Adenovirus infections may result in virus shedding from two different sites. Virus shed in feces has been identified in a manner similar to that described for rotavirus. Halonen et al (1980) described a simple yet sensitive method for identifying adenovirus antigen in stool suspensions. Confirmatory tests have shown the procedure to be sensitive, while maintaining high specificity.

Identification of adenovirus antigens in nasopharyngeal secretions by RIA has been demonstrated (Sarkkinen et al, 1981). These investigators demonstrated complete correlation with reference to the IF assay.

8.8.6 Cytomegalovirus

8.8.6.a Antibody Detection

The presence or absence of IgG antibodies to cytomegalovirus (CMV) was determined (Kimmel et al, 1980) using an indirect RIA detection system. The procedure was capable of accurately detecting anti-CMV antibody in the presence of antibodies to other

members of the herpesvirus family. A modification of the original RIA incorporating a primary 1:100 dilution resulted in a rapid screening procedure with excellent correlation with serum titration experiments (Kimmel et al, 1980).

Kangro et al (1984) compared RIA procedures with ELISA and found comparable sensitivity with sera from adults. The RIA procedures were more sensitive, however, when cord serum was used. Results indicate that IFA procedures were less sensitive than RIA or ELISA. The identification of IgM antibodies with an indirect solid phase RIA has been described (Griffiths and Kangro, 1984). The procedure was shown to be highly specific for detecting congenital infection and sensitive for identifying primary CMV infection in pregnant women.

8.8.7 Herpes Simplex Virus

8.8.7.a Antibody Detection

Early application of RIA procedures for the detection of specific antibodies to herpes simplex virus (HSV) used virus infected, fixed monolayers to measure naturally occurring immunoglobulins (Smith et al, 1974). Extension of this procedure involved adsorption of viral antigens to the surface of a solid phase with results comparable to using monolayers (Smith et al, 1974). Refinement of serum dilution procedures and incubation times resulted in greater sensitivity (Patterson and Smith, 1973).

Improvements in the procedures, such as attachment of viral antigens to polystyrene beads (Kalimo et al, 1977c), use of specific viral antigens (Kalimo et al, 1977a) or removal of crossreacting antigens by sample adsorption (Patterson et al, 1978) resulted in lower background, greater sensitivity, and specificity.

Crossreactivity between members of the HSV group was observed to be a problem during the early phases of assay development (Kalimo et al, 1977b). Adsorption of patient serum with potential crossreacting organisms or using a solid phase coated with specific antigens resolve major problems of crossreactivity with other herpesviruses (Patterson et al, 1978; Smith and Kennell, 1981). The identification and use of specific HSV-1 and HSV-2 glycoproteins attached to the solid phase (Matson et al, 1983) allowed identification of an immune response to either HSV-1 or HSV-2 with high levels of sensitivity and specificity (Forghani et al, 1975; Adler-Storthz et al, 1983).

8.8.7.b Antigen Detection

Generally, the identification of HSV in clinical samples has required viral isolation in cell culture and subsequent identification by reaction with fluorescent or enzyme tagged antibody. RIA technology has been applied to the detection of HSV antigens in cell culture (Forghani et al, 1974; Cleveland et al, 1979) virus containing fluids (Enlander et al, 1976) or clinical specimens (Forghani et al, 1974). Identification of viral antigens was shown to be highly sensitive and specific when clinical specimens were examined (Forghani et al, 1974).

Commonly, labeled antibodies have been used to detect specific antigens. Staphylococcal protein A, labeled with ^{125}I, has been used to detect immune complexes (Cleveland et al, 1979). This generic approach lends itself to modification for the identification of other viral antigens.

8.8.8 Varicella-Zoster Virus

8.8.8.a Anti-Varicella-Zoster Virus Antibody

Several investigators have described methods for the detection of IgM (Tedder et al, 1981), IgG (Friedman et al, 1979; Benzie-Campbell et al, 1981; Richman et al, 1981), or IgG and IgM antibodies to varicella-

zoster virus (VZV) (Arvin and Koropchak, 1980). All of the assays were shown to be more specific and sensitive than the standard CF procedures.

8.8.9 Respiratory Viruses

8.8.9.a Influenza Antigen Detection

Coonrod et al (1984) observed that viral antigen detection was less sensitive during the first three days after onset of infection, and antigen could be detected longer than infectious virus. A comparison of two antigen detection systems (Sarkkinen et al, 1981) indicates that the sensitivity and specificity of RIA and IF are equivalent.

Solid phase assays for the detection of influenza A viral antigens from clinical samples have been described (Sarkkinen et al, 1981; Coonrod et al, 1984).

8.8.9.b Other Respiratory Viruses

Identification of viral antigens in the appropriate clinical specimen is the most direct method of identifying a causative agent. Sarkkinen et al (1981) reported the use of a solid phase antigen capture RIA procedure to identify the presence of respiratory syncytial virus (RSV), parainfluenza type 2, or adenovirus in nasopharyngeal secretions. Agreement between the RIA and IF assays was very good for all viruses. Correlation between IF and RIA was independent of the sample fraction, either mucus or cells, assayed for all viruses except RSV. Lower sensitivity was observed when only the mucus fraction (devoid of cells) was tested in the case of RSV.

The ability of RIA procedures to identify viral antigens in nasopharyngeal samples using RIA and a comparison with diagnosis by other serologic methods has been described (Meurman et al, 1984). These results indicate that the ability to identify viral antigen in the sample was dependent on the time after the onset of symptoms and the age of the patient. Antigen detection was more sensitive in specimens taken from children under six months of age when compared with older children.

8.8.10 Addendum

Since the publication of the first edition of this book in 1986, several recent RIAs were described in the literature for the detection of a variety of viral antigens and antibodies. The following summary describes these RIAs and covers the literature between 1984 and 1989.

8.8.10.a Adenovirus Antigen

Ruuskanen et al, (1984) have described a RIA for the detection of adenovirus antigen in the nasopharyngeal specimens of children with febrile exudative tonsillitis. The rapid detection of adenovirus antigen permitted withdrawal of unnecessary and ineffective antibiotic treatment in most of these children.

8.8.10.b Astrovirus Antibody

The development and evaluation of RIAs for the detection of IgM and IgG antibodies to astrovirus have been described (Wilson and Cubitt, 1988). The tests were shown to be sensitive and specific and suitable for screening large numbers of sera. The use of the assays has established that astrovirus type 1 is prevalent in the United Kingdom and that not only infants but also school-age children and elderly patients are affected by this virus.

8.8.10.c Coxsackie B Virus Antibody

IgM antibody capture RIAs (Pugh, 1984) were developed to detect coxsackie virus B1-B5 specific IgM. This specific antibody was detected in sera from all patients with coxsackie B virus infections that were con-firmed by virus isolation; however, sera from some patients with rising neutralizing antibody titers were negative in the assay. Frequent heterotypic responses were seen among the positive sera. When sera from patients with enterovirus infections other than coxsackie B virus were studied, they were also found to be reactive to the coxsackie B virus antigen used for this test.

8.8.10.d Dengue Virus

A monoclonal RIA was developed for detection of dengue virus (Monath et al, 1986a) in infected cell culture fluids and blood samples from dengue patients. The assay was found to be ten-fold more sensitive for dengue type 2 than for dengue types 1 and 3 viruses and 100-fold more sensitive than for dengue type 4 virus. Virus was more frequently detected in cases of primary infection than in cases of superinfection.

8.8.10.e Hepatitis C Virus Antibody

The isolation of a clone of nucleic acid from the genome of the non-A, non-B agent (Choo et al, 1989) allowed for the subsequent expression of an encoded protein (Choo et al, 1989; Kuo et al, 1989), and offered the first opportunity to develop specific tests for a new RNA virus subsequently named hepatitis C virus (HCV). A usable polypeptide (C-100) derived from about 11% of the viral genome has been found to contain at least two antigenic epitopes that are recognized by serum from many patients with chronic HCV infection. A very sensitive solid-phase RIA for anti-HCV has been developed (Kuo et al, 1989) utilizing this polypeptide.

8.8.10.f Herpes Simplex Virus Antibodies

A solid-phase RIA for the detection of IgG antibodies to herpes simplex virus, using a mouse monoclonal antibody specific for the Fc portion of human IgG as the radiolabeled detecting antibody has been developed (Berry et al, 1987). When compared to virus neutralization assays the RIA had a sensitivity of 100% and a specificity of 93%.

8.8.10.g Varicella-Zoster Virus (VZV) Antibody

Kangro et al (1988) described a simple and sensitive IgM antibody-capture RIA which utilizes VZV antigen and a single monoclonal anti-VZV antibody for the detection of anti-VZV IgM. This assay was also compared to the immunofluorescence test to detect IgM responses in patients with varicella (chicken pox) and patients with shingles (zoster). IgM antibodies were detected in all patients with varicella. The IgM antibodies appeared shortly after onset of the rash, reached peak levels one to four weeks later and then declined to low or undetectable levels in most patients. IgM antibodies were also detected in 98.2% of patients with zoster, but these IgM levels were significantly lower than the IgM produced in response to varicella. There was wider individual variation both in magnitude and duration of the IgM responses.

8.8.10.h Yellow Fever Virus

A solid-phase RIA was developed (Monath et al, 1986b) for the detection of yellow fever virus in infected cell culture supernatant fluid and clinical samples. The test employed a flavivirus group-reactive monoclonal antibody attached to a polystyrene bead as the solid support and a radiolabeled type-specific antibody probe in a simultaneous sandwich RIA format. The sensitivity of the assay was found to be 100 pg of gradient-purified virion protein per 100 mL. Specificity was approximately 99.4%.

8.9 Summary

Since its introduction, 25 years ago, as a highly sensitive, specific, and reproducible method for the determination of insulin levels in plasma (Yalow and Berson, 1960), RIA has been a valuable tool for the measurement of many molecules. Over the past 15 years, RIA has proven to be one of the key serologic procedures in diagnostic medicine and has been utilized in many clinical and epidemiologic investigations. The wide use of RIA in the field of infectious diseases has improved considerably our diagnostic accuracy and efficiency and enabled us to avoid diagnostic errors based on the use of less sensitive immunologic procedures.

REFERENCES

Adler-Storthz, K., Matson, D.O., Adam, E., and Dreesman, G.R. 1983. A micro solid-phase radioimmunoassay for detection of herpesvirus type-specific antibody: Specificity and sensitivity. J. Virol. Meth. 6:85–97.

Arvin, A.M. and Koropchak, C.M. 1980. Immunoglobulins M and G to Varicella-zoster virus measured by solid-phase radioimmunoassay: Antibody responses to varicella and Herpes zoster infection. J. Clin. Microbiol. 12:367–374.

Avraham, H., Golenser, J., Fazitt, Y., Spira, D.T., and Sulitzeanu, D. 1982. A highly sensitive solid-phase radioimmunoassay for the assay of Plasmodium falciparum antigens and antibodies. J. Immunol. Meth. 53:61–68.

Benzie-Campbell, A., Kangro, H.O., and Heath, R.B. 1981. The development and evaluation of a solid-phase radioimmunoassay (RIA) procedure for the determination of susceptibility to varicella. J. Virol. Meth. 2:149–158.

Berry, N.J., Grundy, J.E., and Griffiths, P.D.

1987. Radioimmunoassay for the detection of IgG antibodies to herpes simplex virus and its use as a prognostic indicator of HSV excretion in transplant recepients. J. Med. Virol. 21:147–154.

Blum, H.E., Dölken, G., and Gerok, W. 1979. Solid-phase radioimmunoassay for hepatitis B e-antigen. Klin. Wochenschr. 57:1129.

Bradley, D.W., Maynard, J.E., Hindman, S.H., Hornbeck, C.L., Fields, H.A., McCaustland, K.A., and Cook, E.H., Jr. 1977. Serodiagnosis of viral hepatitis A: Detection of acute-phase immunoglobulin M antihepatitis A virus by radioimmunoassay. J. Clin. Microbiol. 5:521–530.

Catt, K.J. and Tregear, G.W. 1967. Solid-phase radioimmunoassay in antibody coated tubes. Science 158:1570–1572.

Chau, K.H., Hargie, M.P., Decker, R.H., Mushahwar, I.K., and Overby, L.R. 1983. Serodiagnosis of recent hepatitis B infection by IgM Class Anti-HBc. Hepatology 3:142–149.

Choo, Q.L., Kuo, G., Weiner, A.J., Overby, L.R., Bradley, D.W., and Houghton, M. 1989. Isolation of a cDNA clone derived from a blood-borne non-A, non-B viral hepatitis genome. Science 244:359–361.

Cleveland, P.H., Richman, D.D., Oxman, M.N., Wickham, M.G., Binder, P.S., and Worthem, D.M. 1979. Immobilization of viral antigens on filter paper for a ^{125}I-staphylococcal protein A immunoassay: A rapid and sensitive technique for detection of herpes simplex virus antigens and antiviral antibodies. J. Immunol. Meth. 29:369–386.

Coonrod, J.D., Betts, R.F., Linnemann, C.C., Jr., and Hsu, L.C. 1984. Etiological diagnosis of influenza A virus by enzymatic radioimmunoassay. J. Clin. Microbiol. 19:361–365.

Cukor, G., Berry, M.K., and Blacklow, N.R. 1978. Simplified radioimmunoassay for detection of human rotavirus in stools. J. Infect. Dis. 138:906–910.

Cukor, G. and Blacklow, N.R. 1984. Human viral gastroenteritis. Microbiol. Rev. 48: 157–179.

Cukor, G., Perron, D.M., Hudson, R., and

Blacklow, N.R. 1984. Detection of rotavirus in human stools by using monoclonal antibody. J. Clin. Microbiol. 19:888–892.

Cukor, G., Blacklow, N.R., Capozza, F.E., Panjvani, Z.F.K., and Bednarek, F. 1979. Persistence of antibodies to rotavirus in human milk. J. Clin. Microbiol. 9:93–96.

Decker, R.H., Kosakowski, S.M., Vanderbilt, A.S., Ling, C.-M., and Overby, L.R. 1981. Diagnosis of acute hepatitis A by HAVAB-M, a direct radioimmunoassay for IgM anti-HAV. Am. J. Clin. Pathol. 76: 140–147.

Decker, R.H., Overby, L.R., Ling, C.-M., Frösner, G., Deinhardt, F., and Boggs, J. 1979. Serology of transmission of hepatitis A in humans. J. Infect. Dis. 139:74–82.

Devine, R.E., Sit, F., and Larke, R. 1979. Laboratory diagnosis of acute hepatitis A virus (HAV) infection by detection of HAV-specific IgM antibody using radioimmunoassay. Can. J. Pub. Health 70:58.

Dowse, C.A., Carnegie, P.R., Linthium, D.S., and Bernard, C.C.A. 1983. Solid-phase radioimmunoassay for human myelin basic protein using a monoclonal antibody. J. Neuroimmunol. 5:135–144.

Dreesman, G.R., Matson, D.O., Courtney, R.J., Adam, E., and Melnick, J.L. 1979. Detection of herpes virus type-specific antibody by a micro solid-phase radioimmunometric assay. Intervirology 12:115–119.

Enlander, D., Remedios, L.V.D., Weber, P.M., and Drew, L. 1976. Radioimmunoassay for herpes simplex virus. J. Immunol. Meth. 10:357–362.

Faiman, C. and Ryan, R.J. 1967. Radioimmunoassay for human follicle-stimulating hormone. J. Clin. Endocrinol. 27:444–447.

Flehmig, B., Ranke, M., Berthold, H., and Gerth, H.-J. 1979. A solid-phase radioimmunoassay for detection of IgM antibodies to hepatitis A virus. J. Infect. Dis. 140:169–175.

Forghani, B., Schmidt, N.J., and Lennette, E.H. 1974. Solid-phase radioimmunoassay for identification of Herpesvirus hominis types 1 and 2 from clinical materials. Appl. Microbiol. 28:661–667.

Forghani, B., Schmidt, N.J., and Lennette,

E.H. 1975. Solid-phase radioimmunoassay for typing herpes simplex viral antibodies in human sera. J. Clin. Microbiol. 2:410–418.

Friedman, M.G., Leventon-Kriss, S., and Sarov, I. 1979. Sensitive solid-phase radioimmunoassay for detection of human immunoglobulin G antibodies to varicella-zoster virus. J. Clin. Microbiol. 9:1–10.

Frösner, G.G., Brodersen, M., Papaevangelou, G., Sugg, Y., Hass, H., Mushahwar, I.K., Ling, C.-M., Overby, L.R., and Deinhardt, F. 1978. Detection of HBeAg and anti-HBe in acute hepatitis B by a sensitive radioimmunoassay. J. Med. Virol. 3:67–76.

Ghabrial, S.A. and Shepherd, R.J. 1980. A sensitive radioimmunosorbent assay for the detection of plant viruses. J. Gen. Virol. 48:311–317.

Ginsberg, A.L., Conrad, M.E., Bancroft, W.H., Ling, C.-M., and Overby, L.R. 1973. Antibody to Australia antigen: Detection with a simple radioimmune assay, incidence in military populations, and role in the prevention of hepatitis B with gammaglobulin. J. Lab. Clin. Med. 82:317–325.

Goodfriend, T.L. and Ball, D. 1969. Radioimmunoassay of bradykinin: Chemical modification to enable use of radioactive iodine. J. Lab. Clin. Med. 73:501–511.

Greenberg, H.B., Wyatt, R.G., Valdesco, J., Kalica, A.R., London, W.T., Chanock, R.M., and Kapikian, A.Z. 1978. Solid-phase microtiter radioimmunoassay for detection of the Norwalk strain of acute nonbacterial epidemic gastroenteritis virus and its antibodies. J. Med. Virol. 2:97–108.

Greenwood, F.C., Hunter, W.M., and Glover, J.S. 1963. The preparation of ^{131}I-labeled human growth hormone of high specific radioactivity. Biochem. J. 89:114–123.

Griffiths, P.O. and Kangro, H.O. 1984. A user's guide to the indirect solid-phase radioimmunoassay for the detection of cytomegalovirus-specific IgM antibodies. J. Virol. Meth. 8:271–282.

Haikala, O., Kokkonen, J.O., Leinonen, M.K., Nurmi, T., Mantyjarvi, R., and Sarkkinen, H.K. 1983. Rapid detection of rotavirus in stool by latex agglutination: Comparison with radioimmunoassay and electron microscopy and clinical evaluation of the test. J. Med. Virol. 11:91–97.

Hales, C.N. and Randle, P.J. 1963. Immunoassay of insulin with insulin-antibody precipitate. Biochem. J. 88:137–146.

Halonen, P., Bennich, H., Torfason, E., Karlsson, T., Ziola, B., Matikainen, M.-T., Hjertsson, E., and Wesslen, T. 1979a. Solid-phase radioimmunoassay of serum immunoglobulin A antibodies to respiratory syncytial virus and adenovirus. J. Clin. Microbiol. 10:192–197.

Halonen, P., Meurman, O., Matikainen, M.T., Torfason, E., and Bennick, H. 1979b. IgG antibody response in acute rubella determined by solid-phase radioimmunoassay. J. Hyg. (Camb.) 83:69–75.

Halonen, P., Sarkkinen H., Arstila, P., Hjertsson, E., and Torfason, E. 1980. Four-layer radioimmunoassay for detection of adenovirus in stool. J. Clin. Microbiol. 11:614–617.

Hill, H.R. and Matsen, J.M. 1983. Enzyme-linked immunosorbent assay and radioimmunoassay in the serologic diagnosis of infectious diseases. J. Infect. Dis. 147:258–263.

Hollinger, F.B., Vorndam, V., and Dreesman, G.R. 1971. Assay of Australia antigen and antibody employing double-antibody and solid-phase radioimmunoassay techniques and comparison with the passive hemagglutination methods. J. Immunol. 107:1099–1111.

Hollinger, F.B., Bradley, D.W., Maynard, J.E., Dreesman, G.R., and Melnick, J.L. 1975. Detection of hepatitis A viral antigen by radioimmunoassay. J. Immunol. 115:1464–1466.

Hollinger, F.B. and Maynard, J.E. 1976. Recent diagnostic techniques for detecting hepatitis A virus and antibody. Rush Presbyt. St. Luke's Med. Bull. 15:93–103.

Jahrling, P.B., Hesse, R.A., and Metzger, J.F. 1978. Radioimmunoassay for quantitation of antibodies to alphaviruses with staphylococcal protein A. J. Clin. Microbiol. 8:54–60.

Kalica, A.R., Purcell, R.H., Sereno, M.M., Wyatt, R.G., Kim, H.W., Chanock, R.M., and Kapikian, A.Z. 1977. A microtiter solid-phase radioimmunoassay for detection of the human reovirus-like agent in stools. J. Immunol. 118:1275–1279.

Kalimo, K.O.K., Meurman, O.H., Halonen, P.E., Ziola, B.R., Viljanen, M.K., Granfors, K., and Toivanen, P. 1976. Solid-phase radioimmunoassay of rubella virus immunoglobulin G and immunoglobulin M antibodies. J. Clin. Microbiol. 4:117–123.

Kalimo, K.O.K., Martilla, R.J., Granfors, K., and Viljanen, M.K. 1977a. Solid-phase radioimmunoassay of human immunoglobulin M and immunoglobulin G antibodies against herpes simplex virus type 1 capsid, envelope and excreted antigens. Infect. Immun. 15:883–889.

Kalimo, K.O.K., Martilla, R.J., Ziola, B.R., Matikainen, M.T., and Panelius, M. 1977b. Radioimmunoassay of herpes simplex and measles virus antibodies in serum and CSF of patients without infectious or demyelinating diseases of the central nervous system. J. Med. Virol. 10:431–438.

Kalimo, K.O.K., Ziola, B.R., Viljanen, M.K., Granfors, K., and Toivanen, P. 1977c. Solid-phase radioimmunoassay of herpes simplex virus IgG and IgM antibodies. J. Immunol. Meth. 14:183–195.

Kangro, H.O., Booth, J.C., Bakir, T.M.F., Tryhorn, Y., and Sutherland, S. 1984. Detection of IgM antibodies against cytomegalovirus: Comparison of two radioimmunoassays, enzyme-linked immunosorbent assay and immunofluorescent antibody test. J. Med. Virol. 14:73–80.

Kangro, H.O., Jackson, C., and Heath, R.B. 1981. Comparison of radioimmunoassay and the gel filteration technique for routine diagnosis of rubella during pregnancy. J. Hyg. (Camb.) 87:249–255.

Kangro, H.O., Ward, A., Argent, H., Heath, B.B., Cradock-Watson, J.E., and Ridehalgh, M.K. 1988. Detection of specific IgM in varicella and herpes zoster by antibody-capture radioimmunoassay. Epidemiol. Infect. 101:187–195.

Kimmel, N., Friedman, M.G., and Sarov, I. 1980. Detection of human cytomegalovirus-specific IgG antibodies by a sensitive solid-

phase radioimmunoassay and by the rapid screening test. J. Med. Virol. 5:195–203.

Kolb, H.J. and Grodsky, G.M. 1970. Biological and immunological activity of fructose 1,6-di-phosphatase. Application of a quantitative displacement radioimmunoassay. Biochemistry 9:4900–4906.

Kuo, G., Choo, Q.L., Alter, H.J., Gitnick, G.L., Redeker, A.G., Purcell, R.H., Miyamura, T., Dienstag, J.L., Alter, M.J., Stevens, C.E., Tegtmeier, G.E., Bonino, F., Colombo, M., Lee, W.S., Kuo, C., Berger, K., Shuster, J.R., Overby, L.R., Bradley, D.W., and Houghton, M. 1989. An assay for circulating antibodies to a major etiologic virus of human non-A, non-B hepatitis. Science 244:362–364.

Lemon, S.M., Brown, C.D., Brooks, D.S., Simms, T.E., and Bancroft, W.H. 1980. Specific immunoglobulin M response to hepatitis A virus determined by solid-phase radioimmunoassay. Infect. Immun. 28:927–936.

Ling, C.-M. and Overby, L.R. 1972. Prevalence of hepatitis B virus antigens as revealed by direct radioimmune assay with ^{125}I-antibody. J. Immunol. 109:834–841.

Marsden, H.S., Buckmaster, A., Palfreyman, J.W., Hope, R.G., and Minson, A.C. 1984. Characterization of the 92,000-dalton glycoprotein induced by herpes simplex type 2. J. Virol. 50:547–554.

Matson, D.O., Alder-Storthz, K., Adam, E., and Dreesman, G.R. 1983. A micro solid-phase radioimmunoassay for detection of herpesvirus type-specific antibody: Parameters involved in standardization. J. Virol. Meth. 6:71–83.

Matsui, A., Psacharopoulos, H.T., and Mowat, M.P. 1982. Radioimmunoassay of serum glycocholic acid, standard laboratory tests of liver function and liver biopsy findings: Comparative study of children with liver disease. J. Clin. Pathol. 35:1011–1017.

Meurman, O.H. 1978. Antibody in patients with rubella infection determined by passive hemagglutination, hemagglutination inhibiters, complement fixation, and solid-phase radioimmunoassay tests. Infect. Immunol. 19:369–372.

Meurman, O., and Granfors, K. 1977. Completion of hemagglutination inhibition test by solid-phase radioimmunoassay test in routine diagnostic rubella serology. Med. Biol. 55:241–244.

Meurman, O.H., Viljanen, M.K., and Granfors, K. 1977. Solid-phase radioimmunoassay of rubella virus immunoglobulin M antibodies: Comparison with sucrose density gradient centrifugation test. J. Clin. Microbiol. 5:257–262.

Meurman, O., Sarkkinen, H., Ruuskanen, O., Hanniners, P., and Halonen, P. 1984. Diagnosis of respiratory syncytial virus infection in children: Comparison of viral antigen detection and serology. J. Med. Virol. 14: 61–65.

Middleton, P.J., Holdway, M.D., Petric, M., Szymanski, M.T., and Tam, J.S. 1977. Solid-phase radioimmunoassay for the detection of rotavirus. Infect. Immun. 16:439–444.

Monath, T.P., Wands, J.R., Hill, L.J., Gentry, M.K., and Gubler, D.J. 1986a. Multisite monoclonal immunoassay for dengue viruses: Detection of viraemic human sera and interference by heterologous antibody. J. Gen. Virol. 67:639–650.

Monath, T.P., Hill, L.J., Brown, N.V., Cropp, C.B., Schlesinger, J.J., Saluzzo, J.F., and Wands, J.R. 1986b. Sensitive and specific monoclonal immunoassay for detecting yellow fever virus in laboratory and clinical specimens. J. Clin. Microbiol. 23:129–134.

Morgan, C.R. and Lazarow, A. 1962. Immunoassay of insulin using a two-antibody system. Proc. Soc. Exp. Biol. Med. 110:29–32.

Morgan-Capner, P. and McSorley, C. 1983. Antibody capture radioimmunoassay (MACRIA) for coxsackievirus B4 and B5-specific IgM. J. Hyg. (Camb.) 90:333–349.

Mortimer, P.P., Tedder, R.S., Hambling, M.H., Shafi, M.S., Burkhardt, F., and Schilt, U. 1981. Antibody capture radioimmunoassay for anti-rubella IgM. J. Hyg. (Camb.) 86: 139–153.

Mushahwar, I.K. and Decker, R.H. 1983. Prevalence of anti-delta in various HBsAg positive populations. In G. Verme, F. Bonino, and M. Rizzetto (eds.), Viral hepatitis and

delta infection. New York: Alan R. Liss, p. 269.

Mushahwar, I.K., Dienstag, J.L., Polesky, H.F., McGrath, L.C., Decker, R.H., and Overby, L.R. 1981. Interpretation of various serological profiles of hepatitis B virus infection. Am. J. Clin. Pathol. 76:773–777.

Mushahwar, I.K. and Overby, L.R. 1983. Radioimmune assays for diagnosis of infectious diseases. In F.S. Ashkar (ed.), Radiobioassays. Boca Raton, FL: CRC Press, pp. 167–194.

Mushahwar, I.K., Overby, L.R., Frösner, G., Deinhardt, F., and Ling, C.-M. 1978. Prevalence of hepatitis B e-antigen and its antibody as detected by radioimmunoassays. J. Med. Virol. 2:77–87.

Neurath, A.R., Szmuness, W., Stevens, C.E., Strick, N., and Harley, E.J. 1978. Radioimmunoassay and some properties of human antibodies to hepatitis B core antigen. J. Gen. Virol. 38:549–559.

Neurath, A.R., Strick, N., Szmuness, W., Stevens, C.E., and Harley, E.J. 1979. Radioimmunoassay of hepatitis B e-antigen (HBeAg); Identification for HBeAg not associated with immunoglobulins. J. Gen. Virol. 42:493.

Odell, W.D., Ross, G.T., and Rayford, P.L. 1967. Radioimmunoassay for luteinizing hormone in human plasma or serum: Physiological studies. J. Clin. Invest. 46:248–255.

Overby, L.R. and Ling, C.-M. 1976. Radioimmune assay for anti-core as evidence for exposure to hepatitis B virus. Rush Presbyt. St. Luke's Med. Bull. 15:83–92.

Overby, L.R., Ling, C.-M., Decker, R.H., Mushahwar, I.K., and Chau, K. 1981. Serodiagnostic profiles of viral hepatitis. W. Szmuness, H.J. Alter, and J.E. Maynard (eds.), Viral Hepatitis 1981 International Symposium. Philadelphia: Franklin Institute Press, pp. 169–182.

Overby, L.R. and Mushahwar, I.K. 1979. Radioimmune assays. In M.W. Rytel (ed.), Rapid Diagnosis in Infectious Disease. Boca Raton, FL: CRC Press, 39–69.

Patterson, W.R., Rawls, W.E., and Smith, K.O. 1978. Differentiation of serum antibodies to herpesvirus types 1 and 2 by radioimmuno-assay. Proc. Soc. Exp. Biol. Med. 157:273–277.

Patterson, W.R. and Smith, K.O. 1973. Improvement of radioimmunoassay for measurement of viral antibody in human sera. J. Clin. Microbiol. 2:130–133.

Poor, A.H. and Cutler, J.E. 1979. Partially purified antibodies used in a solid-phase radioimmunoassay for detecting candidal antigenemia. J. Clin. Microbiol. 9:362–368.

Pugh, S.F. 1984. Heterotypic reactions in a radioimmunoassay for coxsackie B virus specific IgM. J. Clin. Pathol. 37:433–439.

Purcell, R.H., Wond, D.C., Moritsugo, Y., Dienstag, J.L., Routenberg, J.A., and Boggs, J.D. 1976. A microtiter solid-phase radioimmunoassay for hepatitis A antigen and antibody. J. Immunol. 116:349–356.

Purcell, R.H., Wong, D.C., Alter, H.J., and Holland, P.V. 1973. Microtiter solid-phase radioimmunoassay for hepatitis B antigen. Appl. Microbiol. 26:478–484.

Richman, D.D., Cleveland, P.H., Oxman, M.N., and Zaia, J.A. 1981. A rapid radioimmunoassay using ^{125}I-labeled staphylococcal protein A for antibody to varicella-zoster virus. J. Infect. Dis. 143:693–699.

Rizzetto, M., Gocke, D.J., Verme, G., Shih, J.W.-K., Purcell, R.H., and Gerin, J.L. 1979. Incidence and significance of antibodies to delta antigen in hepatitis B virus infection. Lancet 2:986–990.

Rizzetto, M., Shih, J.W.-K., and Gerin, J.L. 1980. The hepatitis B virus-associated delta antigen: Isolation from liver, development of solid-phase radioimmunoassays for delta antigen and anti-delta and partial characterization of delta antigen. J. Immunol. 125:318–324.

Robinson, K. and Smith, R.N. 1983. Methadone radioimmunoassay: Two simple methods. J. Pharm. Pharmacol 35:566–569.

Roggendorf, M., Frösner, G.G., Deinhardt, F., and Scheidt, R. 1980. Comparison of solid-phase test systems for demonstrating antibodies against hepatitis A virus (anti-HAV) of the IgM-class. J. Med. Virol. 5:47–62.

Rutland, P.C. 1984. The development of a radio-

immunoassay for the measurement of fetal haemoglobin and its use in determining the distribution of HbF in the British population. Med. Lab. Sci. 41:84–85.

Ruuskanen, O., Sarkkinen, H., Meurman, O., Hurme, P., Rossi, T., Halonen, P., and Hanninen, P. 1984. Rapid diagnosis of adenoviral tonsillitis: A prospective clinical study. J. Pediatr. 104:725–728.

Safford, S.E.S., Needleman, S.B., and Decker, R.H. 1980. Radioimmunoassay for detection of antibody to hepatitis A virus. Am. J. Clin. Pathol. 74:25–31.

Sarkkinen, H.K., Halonen, P.E., and Arstila, P.P. 1979a. Comparison of four-layer radioimmunoassay and electron microscopy for detection of human rotavirus. J. Med. Virol 4:255–260.

Sarkkinen, H.K., Halonen, P.E., and Salmi, A.A. 1981. Detection of influenza A virus by radioimmunoassay and enzyme immunoassay from nasopharyngeal specimens. J. Med. Virol. 7:213–220.

Sarkkinen, H.K., Meurman, O.H., and Halonen, P.E. 1979b. Solid-phase radioimmunoassay of IgA and IgM antibodies to human rotavirus. J. Med. Virol. 3:281–289.

Sarkkinen, H.K., Tuokko, H., and Halonen, P.E. 1980. Comparison of enzyme-immunoassay and radioimmunoassay for detection of human rotaviruses and adenoviruses from stool specimens. J. Virol. Meth. 1:331–341.

Sexton, S.A., Hodgsen, J., and Morgan-Capner, P. 1982. The detection of rubella-specific IgM by an immunosorbent assay with solid-phase attachment of red cells (SPARC). J. Hyg. (Camb.) 88:453–461.

Shimada, N., Ushioda, K., Nagatsuka, S., Ueda, T., and Yokoshima, T. 1983. Comparison between radioreceptor assay and RIA for the determination of dihydroergotoxine in rabbit plasma samples. J. Immunol. Meth. 65:191–198.

Smedile, A., Lavarini, C., Crivelli, O., Raimondo, G., Fassone, M., and Rizzetto, M. 1982. Radioimmunoassay detection of IgM antibodies to the HBV-associated delta antigen: Clinical significance in delta infection. J. Med. Virol. 9:131–138.

Smith, K.O., Gehle, W.D., and McCracken,

A.W. 1974. Radioimmunoassay techniques for detecting naturally occurring viral antibody in human sera. J. Immunol. Meth. 5:337–344.

Smith, K.O. and Kennell, W. 1981. Differentiation of members of the human herpesviridae family by radioimmunoassay. Infect. Immun. 33:491–497.

Tedder, R.S., Mortimer, P.P., and Lord, R.B. 1981. Detection of antibody to varicella-zoster virus by competitive and IgM-antibody capture immunoassay. J. Med. Virol. 8:89–101.

Tedder, R.S., Yao, J.L., and Anderson, M.J. 1982. The production of monoclonal antibodies to rubella hemagglutinin and their use in antibody-capture assays for rubella-specific IgM. J. Hyg. (Camb.) 88:335–350.

Trent, D.W., Harvey, C.L., Quereshi, A., and LeStourgeon, D. 1976. Solid-phase radioimmunoassay for antibodies to flavivirus structural and nonstructural proteins. Infect. Immun. 13:1325–1333.

Torfason, E.G., Frisk, G., and Diderholm, H. 1984. Indirect and reverse radioimmunoassays and their apparent specificities in the detection of antibodies to enteroviruses in human sera. J. Med. Virol. 13:13–31.

Torfason, E.G., Källander, C., and Halonen, P. 1981. Solid-phase radioimmunoassay of serum IgG, IgM and IgA antibodies to cytomegalovirus. J. Med. Virol. 7:85–96.

Vesikari, T., Kuusela, A.-L., Sarkkinen, H.K., and Halonen, P.E. 1982. Clinical evaluation of radioimmunoassay of nasopharyngeal secretions and serology for diagnosis of viral infections in children hospitalized for respiratory infections. Ped. Infect. Dis. 1:391–394.

Vyas, G.N. and Roberts, I.M. 1977. Radioimmunoassay of hepatitis B core antigen and antibody with autologous reagents. Vox. Sang. 33:369.

Watanabe, H. and Holmes, I.H. 1977. Filter-paper solid-phase RIA for human rotavirus surface immunoglobulins. J. Clin. Microbiol. 6:319–324.

Wide, L., Kirkham, K.E., and Hunger, W.M., eds. 1971. Solid-phase antigen-antibody systems. In Radioimmunoassay Methods. Edinburgh: Churchill-Livingstone, p. 405.

Wilson, S.A. and Cubitt, W.D. 1988. The development and evaluation of radioimmune assays for the detection of immune globulins M and G against astrovirus. J. Virol. Methods. 19:151–159.

Wiktor, R.J., Koprowski, H., and Dixon, F. 1972. Radioimmunoassay procedure for rabies binding antibodies. J. Immunol. 109:464–470.

Wolff, K.L., Muth, D.J., Hudson, B.W., and Trent D.W. 1981. Evaluation of the solid-phase radioimmunoassay for diagnosis of St. Louis encephalitis infection in humans. J. Clin. Microbiol. 14:135–140.

Yalow, R.S. and Berson, S.A. 1960. Immunoassay of endogenous plasma insulin in man. J. Clin. Invest. 39:1157–1175.

Zollinger, W.D., Dalrymple, J.M., and Artenstein, M.S. 1976. Analysis of parameters affecting the solid-phase radioimmunoassay quantitation of antibody to meningococcal antigens. J. Immunol. 117:1788–1798.

9

Enzyme Immunoassay

Andrew J. O'Beirne and John L. Sever

9.1 INTRODUCTION

Over the past 20 to 25 years, heterogeneous enzyme immunoassays (EIA) have gained widespread acceptance as a diagnostic method of choice for a number of reasons: speed, sensitivity, specificity, accuracy, safety, low cost, potential for automation, long shelf-life, and broad adaptability to various user settings. First described by Avrameas and Guilbert (1971), Engvall and Perlmann (1971, 1972), and Van Weeman and Schuurs (1971), the enzyme-linked immunosorbent assay (ELISA or EIA) is related to two older techniques described elsewhere in this manual: fluorescence immunoassay (FIA) and radioimmunoassay (RIA). While the general principles of the three methods are quite similar (antigen or antibody bound to a solid phase is used to separate free antigen or antibody from a specimen and the resulting complex is measured by fluorescence, radioactivity, or color change in a substrate), ELISA provides the distinct advantages of long-shelf life and safety, i.e., no need for radiochemicals. ELISA is simple to perform, requiring only elementary dilution, incubation, and washing protocols. No secondary reactions, as with complement fixation (CF) or agglutination assays, are required, and quantitation is based upon the enzyme/substrate reaction producing a color change that can be evaluated visually, with a colorimeter or a spectrophotometer. ELISA is more sensitive than either CF or hemagglutination and is equal in sensitivity to RIA. A large number of reviews on the ELISA have been published (Schuurs and Van Weeman, 1977; Sever and Madden, 1977; Engvall and Pesce, 1978; Benjamin, 1979; Maggio, 1979; Malvano, 1979; O'Beirne and Cooper, 1979; Yolken, 1980; Voller et al, 1982; Wardley and Crowther, 1982; Avrameas, 1983) and have described both commercial and research uses of this methodology.

One of the most notable applications of ELISA has been the detection and quantitation of antiviral antibody. Table 9.1 summarizes the use of the indirect ELISA methodology for the detection and quantitation of specific antiviral and antibacterial antibodies. The test principle of the indirect ELISA methodology is shown in Figure 9.1. Theoretically, some modification of the ELISA methodology can be used to identify any viral agent for which a specific immune serum can be prepared. ELISA

153

Table 9.1 Applications of ELISA Methodology for the Detection and Quantitation of Specific Antibody

ADENOVIRUS	Sumaya, 1986b	Pearson, 1988	Park, 1979
Bidwell et al, 1977	Voller and Bidwell, 1976	Rasmussen et al, 1988	Ukkonen et al, 1977
McDermott et al, 1989	Yolken and Leister, 1981	Rhodes et al, 1985	Van de Perre et al, 1987
COXSACKIEVIRUSES	Zerbini et al, 1985	Rhodes et al, 1984	Vandervelde et al, 1977
Bidwell et al, 1977	*EPSTEIN-BARR VIRUS*	Roberts, 1989	Vandervelde, 1978
CYTOMEGALOVIRUS	Begovac et al, 1988	Robinson and Stevens, 1984	Wei et al, 1977
Adler et al, 1985	Costa et al, 1985	Shimakage et al, 1987	Wolters et al, 1975
Anderson and Anderson, 1977	Delisi et al, 1986	Steinitz et al, 1988	Wolters et al, 1976
Beckwith et al, 1985	Dillner et al, 1985	Sternas et al, 1986	Wolters et al, 1977
Begovac et al, 1988	Doerr et al, 1987	Sternas et al, 1983	*HERPES SIMPLEX VIRUSES*
Bidwell et al, 1977	Dolken et al, 1983	Sumaya, 1986a	Adler-Storthz et al, 1983
Boteler et al, 1985	Dolken et al, 1986	Sumaya, 1986b	Alexander et al, 1985
Buimovici-Klein et al, 1983	Dolken et al, 1987	Swanston et al, 1986	Anneren et al, 1986
Cappel et al, 1978	Dolken et al, 1984	Uen et al, 1988	Aurelian, 1982
Castellano et al, 1977	Effros et al, 1985	Venables et al, 1988	Baker et al, 1989
Cheeseman et al, 1984	Epstein, 1986	Voevodin and Pacsa, 1983	Bidwell et al, 1977
Chou et al, 1987	Freijd and Rosen, 1984	Vroman et al, 1985	Bos et al, 1987
Dannenmaier et al, 1985	Halprin et al, 1986	Wallen et al, 1977	Clayton et al, 1985
Delisi et al, 1986	Haskard and Archer, 1984	Webb et al, 1985	Clayton et al, 1986
Demmler et al, 1986	Ho et al, 1989	Wielaard et al, 1988	Coleman et al, 1983a
Doerr et al, 1987	Hopkins et al, 1982	Wolf et al, 1984	Coleman et al, 1983b
Dreikorn et al, 1985	Irving et al, 1985	*HEPATITIS VIRUSES*	Corey, 1986
Gerna, 1976	Kahan et al, 1987	Caldwell and Barrett, 1977	Dannenmaier et al, 1985
Gong et al, 1988	Kurstak et al, 1978	Dannenmaier et al, 1985	Dascal et al, 1989
Linde et al, 1988	Lennette and Henle, 1987	Duermeyer et al, 1977	Delisi et al, 1986
Miller et al, 1989	Linde et al, 1988	Duerymeyer and	Devillechabrolle et al, 1985
Morris et al, 1985	Luka et al, 1984	Van der Veen, 1978	Doerr et al, 1987
Najberg et al, 1985	Modrow and Wolf, 1986	Halbert and Anken, 1977	Dunkel et al, 1988
Nerurkar et al, 1987	Myrmel, 1988	Kacaki et al, 1977	Elitsur et al, 1983
Phipps et al, 1983	Nemerow et al, 1982	Lange et al, 1977	Evans et al, 1983
Schmitz et al, 1977	Nemerow et al, 1986	Locarnini et al, 1978	Felgenhauer et al, 1982
Stagno et al, 1985	Nerurkar et al, 1987	Mathiesen et al, 1978	Fortier et al, 1982

Table 9.1 (*continued*)

Fox et al, 1986	Nemerow et al, 1982	Cremer et al, 1985	Chao et al, 1978
Gilljam et al, 1985	Nerurkar et al, 1987	Deforest et al, 1989	Richardson et al, 1978
Gilman and Docherty, 1977	Njoo et al, 1988	Feldman et al, 1988	*ROTOVIRUS*
Gong et al, 1988	Pasquini et al, 1988	Kleiman et al, 1981	Ellens et al, 1978
Grauballe and Vestergaard, 1977	Rand et al, 1983	Melioli et al, 1985	Ghose et al, 1978
Grillner and Landqvist, 1983	Rapicetta et al, 1984	Murray and Lynch, 1988	Scherrer and Bernard, 1977
Hadar and Sarov, 1984	Ross et al, 1985	Neumann et al, 1985	Yolken et al, 1978
Hanada et al, 1988	al Samarai et al, 1989	Voller and Bidwell, 1976	*RUBELLA*
Hayashi et al, 1983	Shekarchi et al, 1987	Weigle et al, 1984	Arista et al, 1987
Hoffman et al, 1985	Shillitoe et al, 1983	*MUMPS VIRUS*	Bellamy et al, 1985
Jeansson et al, 1983	Skar et al, 1988	Deforest et al, 1989	Bellamy et al, 1986
Juto and Settergren, 1988	Smith et al, 1984	Meurman et al, 1982	Bidwell et al, 1977
Kahlon et al, 1986	Sugimoto et al, 1985	Sakata et al, 1984	Bonfanti et al, 1985
Kahlon and Whitley, 1988	Sullender et al, 1988	Ukkonen et al, 1981	Boteler et al, 1984
Katz et al, 1986	Sundqvist et al, 1984	*NEWCASTLE DISEASE VIRUS*	Braum et al, 1982
Kelley et al, 1986	Tellez et al, 1985	Bidwell et al, 1977	Buimovici-Klein et al, 1980
Keunen et al, 1987	Thongkrajai et al, 1986	*PARAINFLUENZA VIRUSES*	Chantler et al, 1985
Land et al, 1984	Vandvik et al, 1985	Bishai and Galli, 1978	Chantler et al, 1982
Lawrence et al, 1984	van Loon et al, 1985	*PNEUMONIA VIRUS OF MICE*	Chernesky et al, 1984
Lehtinen et al, 1985	van Loon et al, 1989	Payment and Descoteaux, 1978	Cleary et al, 1978
Liden, 1985	van Ulsen et al, 1987	*PLANT VIRUSES*	Cubie and Edmond, 1985
Linde et al, 1988	Warford et al, 1986	Clark and Adams, 1977	Cubie, 1983
Maltseva et al, 1987	Wu et al, 1989	Gera et al, 1978	Deforest et al, 1989
Mayer et al, 1986	Yoosook et al, 1987	Thresh et al, 1977	Echevarria et al, 1985a
McDermott et al, 1989	Ziaie et al, 1986	Voller et al, 1976	Echevarria et al, 1985b
Middeldorp et al, 1987	Ziegler and Halonen, 1985	*RABIES VIRUS*	Enders, 1985a
Mills et al, 1978	*INFLUENZA VIRUSES*	Antanasiu et al, 1977	Enders and Knotek, 1985
Miranda et al, 1977	Bishai and Galli, 1978	Antanasiu et al, 1978	Enders and Knotek, 1986
Morgan and Smith, 1984	Leinikki and Passila, 1977	Thraenhart and Kuwert, 1977	Enders and Knotek, 1989
Morris et al, 1985	*MEASLES VIRUSES*	*RESPIRATORY SYNCYTIAL*	Feldman et al, 1988
Najem et al, 1983	Arneborn et al, 1983	*VIRUS*	Field and Gong, 1984
Ndumbe and Levinsky, 1985	Boteler et al, 1983	Bidwell et al, 1977	Fitzgerald et al, 1988

Table 9.1 (*continued*)

Fogel et al, 1983
Gerna, 1976
Gerna et al, 1987
Gravell et al, 1977
Grillner et al, 1985
Hancock et al, 1986
Hedman et al, 1989
Hedman and Rousseau, 1989
Herrmann, 1985
Hodgson and Morgan-Capner, 1984
Hornstein et al, 1985
Horstmann et al, 1985
Hossain et al, 1988
Johnson et al, 1985
Just et al, 1985
Kasupski et al, 1984
Katow et al, 1989
Kawano and Minamishima, 1987
Kleeman et al, 1983
Lehtonen and Meurman, 1982
Leinikki and Passila, 1977
Leinikki et al, 1978a
Leinikki et al, 1978b
Linde, 1985
Maroto et al, 1986

Meegan et al, 1983
Morandi et al, 1986
Morris et al, 1985
Murray and Lynch, 1988
Paris-Hamelin and
  Fustec-Ibarbourne, 1985
Prevot and Guesdon, 1977
Revello et al, 1987
Rothe et al, 1986
Sander and Niehaus, 1985
Schiff et al, 1985
Serdula et al, 1984
Singh et al, 1986
Skendzel and Edson, 1985
Skurrie and Gilbert, 1983
Skvaril and Schilt, 1984
Steece et al, 1984
Steece et al, 1985
Thomas and
  Morgan-Capner, 1988a
Thomas and
  Morgan-Capner, 1988b
Traunt et al, 1983
Vejtorp, 1978
Voller and Bidwell, 1975

Voller and Bidwell, 1976
Wielaard et al, 1985
Wielaard et al, 1987
Wittenburg et al, 1984
SENDAI VIRUS
Parker et al, 1979
VARICELLA-ZOSTER VIRUS
Arneborn et al, 1983
Asano et al, 1988
Austgluen, 1985
Balfour et al, 1988
Bogger-Goren et al, 1984
Cox et al, 1984
Cremer et al, 1985
deOry et al, 1988
Doerr et al, 1987
Echevarria et al, 1989
Enders, 1982
Enders, 1985b
Florman et al, 1985
Forghani et al, 1978
Friedman et al, 1978
Gershon et al, 1981
Keller et al, 1986
Kovac et al, 1987

Landry et al, 1987
Larussa et al, 1987
Lipton and Brunell, 1989
Mayo, 1988
Morris et al, 1985
Moyner and Michaelsen, 1988
Murray and Lynch, 1988
Nassar and Touma, 1986
Ndumbe et al, 1985
Paryani et al, 1984
Schoub et al, 1985
Shanley et al, 1982
Shehab and Brunell, 1983
Sundqvist et al, 1982
Sundqvist et al, 1984
Taylor-Weideman et al, 1986
Taylor-Wiedeman et al, 1989
Tinker et al, 1983
Tomita et al, 1988
Trlifajova et al, 1986
Venkitaraman et al, 1984
Wreghitt et al, 1984
Zachar et al, 1988
Ziegler, 1984
Ziegler and Halonen, 1985

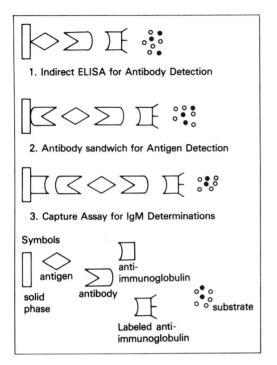

1. Indirect ELISA for Antibody Detection

2. Antibody sandwich for Antigen Detection

3. Capture Assay for IgM Determinations

Symbols

antigen

anti-immunoglobulin

solid phase

antibody

substrate

Labeled anti-immunoglobulin

Figure 9.1 Diagram of the bonding sequence of various ELISA assay methods used in virology. In all of these procedures an incubation step follows the addition of each reactant and unbound material is removed by careful washing after each incubation.

also can be used to detect and titrate antibody to any viral antigen.

Many protocol variations for performing ELISA's have been employed. Incubation times have ranged from a few minutes (Saunders and Wilder, 1974; Saunders et al, 1977) to several hours (Habermann and Heller, 1976; Voller et al, 1976; Bidwell et al, 1977). The enzyme employed has included alkaline phosphatase (Bidwell et al, 1977; Cleary et al, 1978; Forghani et al, 1978; Voller et al, 1978), horseradish peroxidase (Walls et al, 1977; Duerymeyer and van der Veen, 1978; Ruitenberg et al, 1978), and glucose oxidase (Prevot and Guesdan, 1977). The solid phase has included glass and polystyrene tubes (Engvall et al, 1971; Ruitenberg et al, 1976), isothiocyanate disks (Cleary et al, 1978), microtiter plates (Voller et al, 1977), and polystyrene beads

(Gilman and Docherty, 1977). Specific immunoglobulin conjugates such as anti-IgM and anti-IgE have been used (Goldblum et al, 1975); Duermeyer and van der Veen (1978) describe a reverse indirect method (RIMDA) for assaying IgM specific for antigens. An enhanced sensitivity comes from the use of an antibody to specific IgM and antigen enzyme conjugate, which avoids the problem of competing specific IgG and rheumatoid factors that may compete in the reaction system and which are normally encountered in indirect IgM systems.

In addition to variations in materials and test procedures, various systems for interpreting and reporting ELISA results have been developed. These include the use of a single sample dilution which is then measured and quantitated using a spectrophotometer and a program in which the absorbance related to the quantity of antibody present in comparison to known standard(s) (Voller, 1975; Vejtorp, 1978; Parker et al, 1979). Alternatively the product is quantitated by assaying several two-fold dilutions of the sample using a visual or by spectrophotometry (Bullock and Walls, 1977; Forghani et al, 1978; Parker et al, 1979). Although the variation in reporting methods may be of value to the individual investigator, there is a need for standardized reporting for comparison of diagnostic results from different clinical laboratories. Table 9.2 lists commercially available diagnostic test kits available for determination of viral antigen and antibody by EIA.

9.2 MATERIALS

9.2.1 Solid Phase

The solid phase employed in ELISA is the most critical component in determining variability and accuracy of an assay. Among the solid phases employed are polystyrene microtiter plates and tubes (Ruitenberg et al, 1974, 1976; Voller et al, 1974, 1976, 1980; Leinikki and Passila, 1977),

Table 9.2 Kits for Determination of Viral Antigen and Antiviral Antibody by EIA

Agent	Antibody	Antigen
Adenovirus		Analytab Products, Inc., Plainview, NY
		Cambridge BioTech, Worchester, MA
CMV	Abbott, Laboratories, Abbott Park, IL	
	Diamedix, Miami, FL	
	Incstar Corp., Stillwater, MN	
	Labsystems, Raleigh, NC	
	Pharmacia ENI, Columbia, MD	
	BioWhittaker, Walkersville, MD	
	Zeus, Raritan, NJ	
EBV	Incstar Corp.	
	Ortho Diagnostics, Raritan, NJ	
	Pharmacia ENI	
	BioWhittaker	
Hepatitis	Abbott	
	Dupont, Wilmington, DE	
	Labsystems	
	Organon Teknika, Durham, NC	
	Ortho	
HIV	Abbott	
	Coulter Diagnostics, Miami, FL	
	Dupont	
	Labsystems	
	Organon Teknika	
	Ortho Diagnostics	
	Syva, Palo Alto, CA	
HSV	Diamedix	
	Incstar (CSI)	Analytab Products, Inc.
	Pharmacia, ENI	Dupont
	BioWhittaker	Ortho
	Zeus	
HTLV-1	Abbott	
	Cambridge BioTech	
	Dupont	
	Organon Teknika	
	Ortho	
Mumps	Pharmacia ENI	
	BioWhittaker	
Rotavirus		Abbott
		Analytab Products, Inc.
		Cambridge BioTech
		Kallestad
RSV		Abbott
		Kallestad
		Ortho

Table 9.2 (*continued*)

Agent	Antibody	Antigen
Rubella	Abbott	
	Diamedix	
	Incstar	
	Kallestad, Chaska, MN	
	Labsystems	
	Pharmacia ENI	
	BioWhittaker	
VZV	Diamedix	
	Incstar	
	Pharmacia ENI	
	BioWhittaker	

polystyrene beads (Gilman and Docherty, 1977; Miranda, 1977), and specially designed sticks or cuvettes (Felgner, 1977, 1978; Glynn and Ison, 1978; Leinikki and Passila, 1977; Parker et al, 1979; Shekarchi et al, 1982). Reagents also have been passively adsorbed to other materials such as polyvinyl (Voller et al, 1976), polycarbonate (Smith and Gehle, 1977), amino-alkyl-silyl glass, (Hamaguchi et al, 1976; Kato et al, 1977), and silicone rubber (Hamaguchi et al, 1976). Other authors have described covalent coupling of antigen or antibody to the solid phase (Hendry and Herrmann, 1980; Neurath and Strick, 1981), employing cellulose (Van Weemen and Schuurs, 1974), isothicyanate disks (Halbert and Anken, 1977), and polyacrylamide (Van Weemen and Schuurs, 1974). In commercially available kits, polystyrene or polyvinyl chloride plates or tubes are most commonly employed today.

Early investigators reported variations in bonding between different types of plastic, between lots or batches of materials, and even between wells on the same plate (Shekarchi et al, 1984). With newer solid phases designed specifically for ELISA, this type of variation has been reduced, but since lot to lot differences still occur, bonding characteristics for each antigen must be checked with each new lot, a routine part of

procedure for quality control checked by the manufacturer (O'Beirne and Cooper, 1979).

Antigen or antibody diluted in "coating buffer" is attached to the solid phase by electrostatic bonding. Two commonly employed coating buffers are carbonate buffer, pH 9.6 (see 9.2.6.a) and phosphate buffered saline (PBS) pH 7.4 to 7.8. Binding occurs following overnight incubation at 2 °C to 4 °C. Usual coating concentrations are at 0.1 to 5 μg/mL, however, each lot of antigen (antibody) should be titrated to determine the optimal concentration.

9.2.2 Antigen

The type and degree of purity of ELISA antigens, including cell lysates, purified viruses, viral subunits, or single proteins, will influence the result obtained.

Cell lysates containing viral agents, although relatively crude reagents, are often satisfactory antigens if a large number of antigenic determinants are needed in a single preparation. The advantage of using cell lysates is that they are easy and inexpensive to produce. A typical procedure for preparing cell lysates is as follows:

1. Virus is grown in stationary cell cultures until CPE of 75% to 100% is observed.

2. Cells are washed and scraped into cold PBS and are pelleted at $700 \times g$ for 30 minutes.

3. The pellet is resuspended in 1/500 the original volume of PBS and cells are disrupted by sonication or repeated cycles of freezing and thawing.

4. Cell debris is removed from the preparation by low speed centrifugation. (1,000 rpm).

5. The supernatant fluid is centrifuged at 25,000 rpm for 30 minutes in a SW 27.1 rotor at $100,000 \times g$.

6. The pellet is resuspended in a small volume of PBS and the optimal dilution is determined by block titration.

7. Remaining antigen is aliquoted and stored at $-70\,°C$.

A control antigen consisting of similarly processed uninfected cell cultures must be prepared and used as a negative control in the assay.

Where greater specificity is required or for certain viral agents, e.g., rubella (Gravell et al, 1977; Forghani and Schmidt, 1979) or Epstein-Barr virus (Hopkins et al, 1982), cell lysate antigens may not be satisfactory. Purified antigens, prepared by differential centrifugation, sucrose, or other gradients, may be required for greater specificity. Because uninfected cell cultures processed by purification techniques may result in preparations with little or no protein to bond to the support matrix, control antigen preparations in test systems that used purified viral agents may not be available. Instead, positive and negative control sera are used in addition to these purified antigens.

Other antigens used for ELISA include antigens commercially prepared for hemagglutination and CF assays and virus infected cells grown in microtiter plates and fixed with acetone (Nerurkar et al, 1983) or dried to the solid phase (Saunders, 1977). Regardless of the method used to prepare the test antigen, the protocol must include removal of serum or other protein supplements from cell culture media as well as the removal of non-ionic detergents such as tween or Triton X, which could interfere with assay results by competing for binding sites on the solid phase.

Solubilization of the antigen is often beneficial, providing even coating of the antigen. Ionic detergents such as sodium dodecyl sulfate and deoxcholate are preferable over non-ionic detergents due to less competition for binding sites on the solid phase. Use of detergents must be carefully validated due to denaturation of some antigens.

The use of recombinant DNA procedures to prepare antigens may result in exquisite specificity in the ELISA assay, e.g., detecting antibody to herpes simplex virus (HSV) type 2 requires a very specific antigen preparation because HSV-1 and HSV-2 are very closely related viruses. Thus crude antigens share many common antigenic determinations making serological diagnosis of infection very difficult (Plummer, 1973). Recently, a recombinant HSV-2 glycoprotein G antigen has been developed and found to be useful in the detection of specific antibodies to HSV-2 using an ELISA assay (Ross et al, 1985). Recombinant antigens, however, are not without faults. The antigen may lack sensitivity due to lack of glycosylation or different tertiary structure from that of the native protein. In addition, contaminating (non-viral) antigens have to be removed to decrease the chance of false positives due to antibody reaction with the contaminating material. Provided that these potential pitfalls are considered, recombinant antigens should prove very useful in ELISA assays.

Because too little antigen reduces sensitivity and too much antigen may result in high background with negative samples, determination of proper antigen concentration is important in achieving desired assay results. See the Method section for titration of antigen to establish appropriate working concentrations.

9.2.3 Antibodies

ELISA assays for detection of viral antigens can be performed using antibodies to a specific viral agent. Such assays can be used to detect viral neuraminidase in nasopharyngeal washings (Yolken et al, 1980), HSV from vaginal swabs (Nerurkar et al, 1984), and rotavirus in stool samples (Yolken and Stopa, 1979). Antibody bound to a solid phase also is used to detect virus specific IgM (Naot and Remington, 1980; Forghani et al, 1983). Antibodies, including many monoclonal antibodies, to most classes of animal and human immunoglobulins and to specific infectious agents, are commercially available either as whole serum or purified immunoglobulin preparations. Although monoclonal or affinity purified antibodies (see section 9.2.4.a) are preferred, any high affinity antibody that is free of reactivity that would produce inaccurate results can be used. Optimal antibody concentration, as previously stated, is determined by block titration. Polyclonal antibody preparations in carbonate buffer (pH 9.6) bind easily to polystyrene and PVC microtiter plates. Monoclonal antibody coating may not be as straightforward. Immunoglobulins from certain clones may not attach by passive adsorption. Alternative attachment schemes, such as covalent binding, could be attempted or selection of antibody from a different clone should be considered for use as a solid phase capture antibody.

9.2.4 Conjugates and Substrates

To detect the immunologic reaction in ELISA, one of the reactants, usually the antibody, is labeled with an enzyme, which reacts with a substrate to produce a color change. The sensitivity of the ELISA is directly related to the properties of the enzyme label employed. To provide sensitivity, the enzyme must have a sufficiently high activity to permit adequate amplification of the serologic reaction within the desired time frame. The enzyme should have substrates which are degraded to stable, easily measured products and which are readily available in purified form at low cost. As noted above, enzymes that have been used in ELISA include alkaline phosphatase, horseradish peroxidase, and glucose oxidase, as well as beta-D-galactosidase. Methods for conjugating these enzymes have previously been described in the literature (Nakane and Pierce, 1967; Avrameas and Guilbert, 1971; Avrameas and Ternyek, 1971; Nakane and Kawaoi, 1974; Kato et al, 1976; Voller et al, 1976; Persijun and Jonker, 1978) and are detailed below in 9.2.4.b and c.

The most commonly used substrate for alkaline phosphatase is paranitrophenyl phosphate (pNPP) which is converted by hydrolysis to a stable, soluble yellow compound that can be readily quantitated spectrophotometrically. 5-Aminosalicyclic acid, diaminobenzidene, and orthophenylenediamine have successfully been used as substrates with peroxidase conjugates. 5-Aminosalicyclic acid and diaminobenzidene may result in partially insoluble enzymatic products, while diaminobenzidene and orthophenylenediamine are light sensitive and mutagenic. Methylumbelliferin galactoside derivatives are good substrates for beta-D-galactosidase, and because they are fluorogenic, may provide increased sensitivity but do not produce a qualitative colorimetric end point. They also require a more expensive spectrofluorometer.

9.2.4.a Purification of Antibodies

Antisera that are to be used as conjugates must be high titered and should be sufficiently purified to give the required specificity. Immunoglobulins are separated from other serum components by ammonium or sodium sulfate precipitation.

Ammonium Sulfate $(NH_4)_2SO_4$ Precipitation of Immunoglobulins

1. Prepare saturate $(NH_4)_2SO_4$ at room temperature (about 54 g in 70 mL H_2O); adjust to pH 7.8 with 2 M sodium hydroxide before use.
2. Add dropwise 5 mL of a saturated solution $(NH_4)_2SO_4$ to 10 mL of serum; after each drop shake to dissolve precipitate; continue until the precipitate remains.
3. Continue shaking for 2 hours at room temperature.
4. Centrifuge at room temperature for 30 minutes at 1400 × g.
5. Discard supernatant; dissolve precipitate in saline and bring volume up to 10 mL with saline.
6. Repeat steps 1 through 4.
7. Dissolve precipitate in 4 mL PBS.
8. Dialyse against several changes of PBS pH 7.4 at 4 °C to remove sulfate; check for sulfate ions by adding a few drops of 10% barium chloride to a small sample of dialysate.
9. When all ammonium sulfate has been removed, check for purity of antibody and for protein concentration.

Immunoglobulin Purification by Affinity Chromatography

Preparation of immunosorbent

1. Swell 2 g CNBr-activated Sepharose 4B (Pharmacia) in 200 mL of 0.001 M HCl in a beaker (about 15 minutes).
2. Pour into a small column (0.9 × 15 cm) and wash with carbonate buffer (C buffer, pH 8.2, 0.1M, which consists of $NaHCO_3$ 8.4 g, NaCl 29.2 g, plus distilled H_2O to make 1 L).

3. To the washed gel in a small beaker add 5 mg of protein in C buffer (e.g., purified human gamma-globulin 5 mg/mL).
4. Mix gently at room temperature for 2 hours (do not use a stirring bar).
5. Put in a small column (0.9 × 15 cm) and wash as follows:
   - PBS 20–50 mL
   - 8 M urea in PBS—add urea SLOWLY! at least 2 hours gradually increasing urea concentration to 8 M
   - 20 to 50 mL PBS—to wash out excess urea
   - Glycine HCl pH 2.5—until effluent pH is 2.5
   - PBS + 0.05 M 2 amino ethanol
   - PBS (+0.02% azide, if not to be used immediately)

Purification of antibody on immunosorbent column

1. Wash immunosorbent (described above) with PBS in a small (0.9 × 15 cm) column.
2. Add 1 mL of anti-globulin to the gel; let stand 15 minutes at room temperature
3. Wash column with PBS (~50 mL).
4. Elute column with glycine HCl using the smallest possible void volume; collect in 0.2 M Tris HCl—1 drop/tube.
5. Pool the tubes containing the eluted peak and neutralize immediately with 0.2 M Tris.
6. Concentrate to 5 mg/mL.

Affinity purified immunoglobulin is conjugated to one of several enzymes used in ELISA (Johnson et al, 1980; Nilsson et al, 1981; O'Sullivan and Marks, 1981). Conjugation methods for alkaline phosphatase and for horseradish peroxidase (HRP) are presented below. Also included are methods for preparation of substrates and reaction characteristics.

9.2.4.b Preparation of Alkaline Phosphatase Conjugates

One-step Glutaraldehyde Method
(Avrameas and Ternynck, 1969)

1. Centrifuge alkaline phosphase suspension (Sigma type VII) containing 5 mg of enzyme. Discard supernatant.

2. Add 2 mg protein (antibody) to be labeled in 1 mL PBS and mix.

3. Dialyse against several changes of PBS at 4 °C overnight.

4. Add 25% electron microscopy (EM) grade glutaraldehyde suspension to give a final concentration of 0.2%.

5. Hold at room temperature with occasional gentle mixing for 2 hours.

6. Dialyse against several changes of PBS at 4 °C overnight.

7. Dilute conjugate to 4 mL with Tris buffer (pH 8.0) containing 1% bovine serum albumin (BSA) and add 0.2% sodium azide. Store in the dark at 4 °C.

Alkaline Phosphatase Substrates
Paranitrophenyl phosphate (pNPP)

1. Prepare diethanolamine (DEA) buffer, pH 9.8 as follows:

   - Mix 97 mL diethanolamine in 800 mL H_2O.

   - Add 200 mg sodium azide (NaN_3) and 100 mg magnesium chloride ($MgCl_2$ $6H_2O$).

   - Adjust to pH 9.8 with 1 M HCl.

   - Make up volume to 1 L with H_2O; store at 4 °C in dark (for use, mix 5 mg pNPP (Sigma 104–105 phosphatase substrate) in 5 mL DEA buffer: Use within 30 minutes and protect from light; stop enzyme-substrate reaction with 3 M NaOH).

2. Positive reaction yields a yellow color.

3. Reading wavelength—405 nm.

4 Methylumbelliferylphosphate (4MUP)

1. Prepare DEA buffer as above.

2. Stock 4MUP (10 mg MUP in 40 mL DEA; store in the dark at 4 °C) (For use, mix 1 mL 4MUP stock with 9 mL DEA buffer; protect from light; read with fluorometer settings; 365 nm excitation; 450 nm emission).

9.2.4.c Preparation of Horse Radish Peroxidase (HRP) Conjugates

One-step Glutaraldehyde Method
(Avrameas, 1969)

1. Prepare 5 mg protein (antibody) in 1 mL of 0.1 M phosphate buffer pH 6.8

2. Add 12 mg HRP (Sigma Type IV RZ-3.0) and mix.

3. Add 0.05 mL of 1% aqueous glutaraldehyde dropwise while stirring.

4. Let stand at room temperature for 2 hours.

5. Dialyse at 4 °C against three changes of PBS over 12 to 24 hours.

6. Centrifuge at 4 °C to remove any precipitate present.

7. Remove free enzyme (see below) and store.

Two-step Glutaraldehyde Method
(Avrameas and Ternynck, 1969; 1971)

1. Dissolve 10 mg HRP (Type IV RZ-3.0) in 0.2 mL 0.1 M phosphate buffer pH 6.8 containing 1.25% EM glutaraldehyde.

2. Allow to stand overnight at room temperature.

3. Remove excess glutaraldehyde by dialysis against 0.15 M normal saline or on a Sephadex G25 column equilibrated with saline—the brown fraction is the activated peroxidase.

4. Concentrate to 1 mL and add 1 mL of 0.15 M saline containing 5 mg of the protein (antibody) to be labeled.

5. Add 0.1 mL of 0.2 M lysine and let stand 2 hours at room temperature.

6. Let stand at 4 °C overnight.

7. Add 0.1 mL of 0.2 M lysine and let stand 2 hours at room temperature.

8. Dialyse against several changes of PBS at 4 °C.

9. Remove free enzyme (see below) and store.

Periodate

1. To 4 mg of peroxidase in 1 mL of distilled water, add 0.2 mL of *freshly prepared* 0.1 M $NaIO_4$ and stir gently for 20 minutes. (On addition of the periodate, the solution should change color from gold to green; if it does not do so, new $NaIO_4$ is required.)

2. Dialyze the mixture against 1 mM sodium acetate buffer, pH 4.4, overnight at 4 °C.

3. Raise the pH by addition of 20 μL of 0.2 M sodium carbonate-bicarbonate buffer (pH 9.5).

4. *Immediately* add 8 mg of antibody in 1 mL of 0.01 M sodium carbonate buffer, pH 9.5. Stir for 2 hours at room temperature.

5. Add 0.1 mL of *freshly prepared* sodium borohydride (4 mg/mL in distilled water). Leave for 2 hours at 4 °C. Precipitate with saturated ammonium sulfate as described below.

Purification and Storage of HRP Conjugate

1. To remove free enzyme, precipitate the conjugate by adding dropwise an equal amount of cold, neutral, saturated ammonium sulfate, with stirring.

2. Centrifuge and wash the precipitate twice with cold half-saturated ammonium sulfate.

3. Redissolve the precipitate in PBS and dialyse against several changes of PBS to remove the ammonium sulfate.

4. Store at 4 °C in aliquots containing 50% glycerol and 1% BSA; they may also be frozen at −20 °C or lower.

Peroxidase Substrates

1. Add 50 mg 5-amino salicylic acid to 50 mL 0.2 M phosphate buffer pH 6.8.

2. Heat at 60 °C in a water bath until clear solution is obtained.

3. Add 2 mg charcoal and filter.

4. When cool, add 0.5 mL of 1% hydrogen peroxide solution.

5. Use within one hour and discard unused reagent (Stop reaction with 0.5% NaN_3 or 0.2 M NaOH; positive reaction yields a dark purple-brown color; reading wavelength—450 nm).

Ortho phenylenediamine (OPD)

1. Prepare phosphate-citrate buffer pH 5.0 (PC buffer, as follows: stock A, 0.1 M citric acid 19.2 g/L; stock B 0.2 M sodium phosphate NaHPO 28.4 g/L; mix 24.3 mL stock A with 25.7 mL stock B, add 50 mL H_2O).

2. For use, mix 40 mg OPD in 100 mL of PC buffer and add 40 μL of 30% H_2O_2; use immediately! Reagent is light-sensitive. (Stop reaction with 2.5 M H_2SO_4; positive reaction is orange to reddish-brown; reading wavelength—450 to 492 nm)

2.2′ amino-di[3 ethyl-benzthiazoline sulfonate (6)] (ABTS)

1. Prepare: stock 0.05 M citrate pH 4: 9.6 g citric acid in 1 L H_2O; adjust to pH 4 with 1 M NaOH.

2. Prepare: stock 40 mM ABTS: 548.7 mg ABTS in 25 mL H_2O: store at 4 °C.

3. Prepare: stock 0.5 M H_2O_2: add 0.5 mL 8 M solution to 7.5 mL H_2O: store at 4 °C (For use, mix 0.05 mL ABTS stock, 0.04 mL H_2O_2 stock, 10 mL citrate stock: stop reaction with 0.1 M HF pH 3.3: positive reaction is dark green: reading wavelength—405 to 414 nm).

9.2.5 Test Specimens

Acceptable specimens for ELISA assays include serum, cerebrospinal fluid, feces, milk, saliva, and washings from nasal, ear, or genital swabs. Pretreatment is usually not required for nonspecific inhibitors and nonreactive specimens can be removed from the reaction well by washing. Fecal suspensions occasionally cause capture antibody to be stripped from the solid phase, but this problem can be prevented by covalently bonding the antibody to the solid phase and adjusting the pH to neutral of the sample suspension.

Rheumatoid factor must be removed from samples being assayed for specific IgM. Absorption of IgG from serum with protein A or aggregated IgG (Leinikki et al, 1978b) will reduce false positive results due to rheumatoid factor. Absorption also is useful in detecting low levels of IgM masked by competition from high levels of specific IgG.

9.2.6 Buffers

9.2.6.a Coating Buffer pH 9.6

Sodium Carbonate Na_2CO_3 1.59 G

Sodium bicarbonate $NaHCO_3$ 2.93 g

Sodium Azide NaN_3 0.2 g

Make up to 1 L with distilled water; store at 4 °C

9.2.6.b PBS Tween pH 7.4–7.6

Sodium chloride NaCl 8 g

Potassium phosphate (monobasic) KH_2PO_4 0.2 g

Sodium Phosphate (dibasic) Na_2HPO_4 $12H_2O$ 2.9 g

Make up to 1 L with distilled water and add 0.5 mL Tween 20; store at 4 °C.

9.2.6.c Sample and Conjugate Diluent

PBS 90 mL

10% BSA stock 10 mL

EDTA 0.05%

9.3 ELISA TEST METHODS

Certain methods are common to all ELISA procedures, including washing and incubation. Washing is perhaps the most important step, since if properly performed, it removes all unbound reactants. Simple immersion in a wash solution is not adequate; however, each reactant vessel must be filled and emptied several times, either manually, with a mechanical washing device, or with automated equipment. Acceptable wash fluids include PBS with 0.05% tween 20, distilled water, PBS, or saline with or without added tween, 1% BSA, or serum. High salt concentration or pH below 7.2 may cause disruption of the antigen-antibody interaction. After washing, the empty reactant vessel should be vigorously tapped on absorbent paper to remove excess moisture.

Incubation time and temperature should be determined to optimize results, as well as to fit the laboratory's schedule, although the parameters within a laboratory must, once established, remain constant to assure reproducibility of results. The solid phase must be controlled to assure consistent temperature within a test run. Again, for commercially available kits, these requirements will be addressed in package inserts, usually with equipment recommendations to help assure consistency and reproducibility of results. Some automated ELISA systems provide for incubation, temperature, and timing.

9.3.1 Determination of Optimal Conjugate Concentration

Conjugates, whether commercial or laboratory prepared, should be titrated before use.

1. Dissolve human gamma globulin (HGG) in coating buffer to a concentration of 100 ng/mL (stock HGG 1 mg in

20 mL with buffer: mix 50 μL stock in 25 mL coating buffer).

2. Add 200 μL of diluted HGG to each of 48 wells: cover, incubate at 4 °C overnight.

3. Empty wells and fill with PBS Tween: let stand 10 minutes. Empty and fill two more times; empty and tap to remove all wash solution.

4. Dilute enzyme-anti-HGG conjugate 1:200, 1:400, 1:600, 1:800, 1:1000, 1:1200, 1:1600, and 1:2000 in conjugate diluent.

5. Test each conjugate dilution in four sensitized wells; to each well add 100 μL of the diluted conjugate.

6. Cover and incubate in a moist chamber at 37 °C for one hour.

7. Wash as above (step 3).

8. Prepare substrate.

9. Add 100 μL to each test well and to wells with HGG only.

10. Incubate in a moist chamber at 37 °C 30 minutes.

11. Stop reaction and read results, either visually or with a spectrophotometer. The dilution giving a reading close to an optical density of 1.0 indicates the range to be tested with the specific ELISA system being used.

9.3.2 Determination of Optimal Antigen Dilution

1. Dilute antigen and control antigen in coating buffer 1:100, 1:200, 1:400, and 1:800.

2. Across the plate (12 wells) add 200 μL per well of one antigen dilution; to the next row add the corresponding dilution of control antigen; repeat these two rows for the other antigen dilutions; cover.

3. Incubate at 4 °C overnight.

4. Empty wells and fill with PBS Tween; after 10 minutes empty and fill the wells two more times; empty the final wash and tap plate dry.

5. Dilute a positive and negative antigen specific antiserum 1:100, 1:200, 1:400, 1:800, and 1:1600. To one well of each antigen and control antigen dilution add 100 μL of the serum dilutions (eight vertical wells each contain same serum dilution); two rows of eight wells will serve as conjugate and substrate controls.

6. Cover and incubate in a moist chamber, at 37 °C for 90 minutes.

7. Wash as above (step 4).

8. Dilute conjugate (enzyme labeled antibody) in serum diluent and add 100 μL to all wells except eight for substrate control.

9. Cover and incubate in a moist chamber at 37 °C for 60 minutes.

10. Wash as above (step 4).

11. Prepare substrate for conjugate used and add 100 μL to each well.

12. Cover and incubate in a moist chamber at 37 °C for 30 minutes.

13. Stop with 50 μL of appropriate stop reagent.

14. Read results visually or spectrophotometrically; the highest dilution of antigen that gives an optical density reading of 1.0 with the selected serum dilution of positive serum and a reading ≤0.1 with the same dilution of negative serum is selected.

Block titrations of two or three dilutions of antigen against two or three dilutions of conjugate close to the optimal dilution determined by these methods with a single dilution of one positive and one negative sample will indicate optimal concentration of both for a specific test. The final dilutions to be used can be slightly varied without a significant effect on the sensitivity or specificity of the test. This may be useful, if it is necessary to conserve any of the reagents.

9.3.3 Determination of Optimal Antibody Dilution

1. Dilute antibody 1:500, 1:1000, 1:1500, and 1:2000 in coating buffer.
2. For each antibody dilution add 200 μL per well to 12 wells (across the plate).
3–4. As in 9.3.2.
5. Prepare antigen dilutions equivalent to 400, 200, 100, 50, and 25 plaque forming units of virus.
6. Add the antigen dilutions (100 μL per well) to the antibody sensitized wells so that all antibody dilutions are tested.
7–8. As in 9.3.2. steps 6 and 7.
9. Add labeled specific antibody (conjugate), 100 μL per well. If labeled specific antibody is not available, use a specific antibody from a different animal source and a conjugate suitable for that animal source.
10–15. As in 9.3.2, steps 9–14.
16. Select antibody dilution which offers the most sensitive detection of virus antigen.

9.3.4 ELISA for Detection of Antiviral Antibody

1. Sensitize the plates with appropriate dilution of antibody as determined above; then add antigen or control antigen (optimal dilution in coating buffer) to alternate rows, 200 μL per well.
2–3. As in 9.3.2. steps 3 and 4.
4. Add sample diluted in sample diluent. Include samples of known

positive and negative control specimens. Include conjugate and substrate control wells—diluent only at this step. 100 μL per well.
5. As in 9.3.2. steps 6–14.

9.3.5 ELISA for Detection of Viral Antigens

1. Sensitize the plates; add optimal dilution of antibody in coating buffer 200 μL per well.
2–3. As in 9.3.2. steps 3 and 4.
4. Add samples and known positive and negative controls diluted in sample diluent; test plate should also include wells for conjugate and substrate control; 100 μL per well.
5–6. As in 9.3.2. steps 6 and 7.
7. Add optimal dilution of enzyme labeled specific antiviral antibody. (Alternate: if labeled specific antiviral antibody is not available, add antibody to the virus prepared in a different species from the coating antibody to test wells and include a control well without test sample: 100 μL per well; cover and incubate at 37 °C for 60 minutes, then proceed as in 9.3.2. steps 7–14).
8–13. As in 9.3.2. steps 9–14. See 9.4 for quantitation.

9.3.6 Detection of Antiviral IgM

Two methods for detection of IgM by ELISA are currently used. One is referred to as indirect IgM ELISA and the other as IgM capture ELISA. The former is a modification of the assay described in section 9.3.4 in that serum is pretreated to remove IgG. The removal of IgG is necessary to prevent false positive results due to rheumatoid factor (RF) or false negative results in cases where specific IgG levels are sufficiently high to effectively block the binding of specific IgM. The pretreatment is done as a first step prior to performing the assay and can be accomplished in several ways. Initially, virus specific IgM determinations were done on IgM containing serum fractions obtained by sucrose density gradient centrifugation (Best et al, 1969) or gel filtration (Morgan-Capner et al, 1980). The "pretreatment step" has since been simplified by directly mixing the serum sample with substances such as protein A (Leinikki et al, 1978b), staphylococcus protein A and streptococci (Fuccillo et al, 1987), G-protein (Paxton et al, 1990) or anti-human IgG (Gerna et al, 1987). We prefer absorption with anti-human IgG for economic reasons and because the immune complexes formed during this treatment effectively remove RF. Following the pretreatment step, the immunoassay is performed as described using a class specific anti-IgM conjugate. The quality of the IgG removal step is determined by performing the assay on both absorbed and unabsorbed serum using an anti-IgG conjugate.

The IgM capture ELISA is similar to the assay described in 9.3.5 (Naot and Remington, 1980), see Figure 9.1. (Gerna et al, 1987). The solid phase is coated with anti-specific (e.g., human) IgM. The specimen is added and IgM is bound by the solid phase antibody. Unbound antibody is removed by a washing step. At this point, the assay design can vary depending on the detection system employed. Technically speaking the least difficult approach is to add and incubate with viral antigen which is followed by a wash step to remove unbound antigen. Enzyme labeled antiviral antibody (either poly or monoclonal) is then added to detect bound viral antigen and, therefore, antiviral IgM in the specimen. A second approach is to add enzyme labeled viral antigen which allows the direct detection of bound anti-viral IgM. The use of enzyme labeled antigen is very appealing because it saves a step in the assay but it can be technically more difficult to label viral antigens (particularly if multiple epitopes are involved) than antibody. IgM capture ELISA has been reported to elimi-

nate false negative and false positive results (Remington and Desmonts, 1983). However, other investigators have demonstrated that, in the absence of serum pretreatment (as described above), false positive or negative results can be observed due to high levels of specific IgG and/or RF when the IgM capture methodology is used (Forghani et al, 1983; Briantais et al, 1984).

9.4 QUANTITATION

ELISA results can be read and reported either qualitatively (positive or negative with the cutoff value equal to background color) or quantitatively. The end point titer also can be read visually or photometrically and is reported as the last sample dilution with color greater than the negative control or background.

In order to obtain a more subjective reading, the optical density of the sample can be determined and the titer can be interpolated from a single sample dilution. Several test standards (of known value) are run along with a single dilution of a test sample. The absorbance of the standards plotted against the known values (e.g., titer) defines the curve on which test samples can be evaluated (Parker et al, 1979). Another method of obtaining a qualitative result is to compare the absorbance of the test sample with the absorbance of a positive and negative control sample at the same dilution. The difference between positive and negative readings is divided into units, and results between the two are reported in relative units (Siegel and Remington, 1983). Results are also reported as optical density readings (Voller et al, 1980), effective dose (Leinikki and Passila, 1977), or area under the curve (Murphy et al, 1980).

Probably the most common means of reading test results, particularly in the clinical laboratory setting, is with a photometer. This enables use of single sample dilutions, automated reading, and a hard copy of the data. Most importantly, use of a single dilution lowers variability and increases reproducibility of test results (Siegel and Remington, 1983).

9.5 INSTRUMENTS

A wide variety of automated equipment is used in the performance of ELISAs. This includes pipettors, aspirating and washing systems, and readers with or without data reduction capabilities (Sever, 1983). The amount and type of automated equipment used depend, of course, on the volume of tests performed. Washing and pipetting devices of various degrees of sophistication are quite valuable, regardless of test volume. The accuracy and precision of these devices are critical to the reproducibility of test results. Pipettors should be calibrated regularly against a known standard. Fully automated systems which perform reagent addition, mixing, washing, and reading are available.

9.6 COMMENTS AND SOURCES OF ERROR

Although ELISA represents a relatively simple and reproducible method for viral serology and other applications, it must be remembered that for any test performance, other than using a commercially available kit and a totally automated system, technique is critical. The interaction of solid phase, antigen, antibody, conjugate, and substrate is dependent upon defined conditions of time, temperature, and counterstain. In commercially marketed products, variables have been standardized so the test performs within established parameters, provided manufacturer's instructions are followed. For nonstandardized assays, performance parameters must be carefully controlled within a run and standardized between runs to assure reproducible results.

As previously described, a carefully documented system of quality control within the laboratory is crucial. The solid phase must be tested for binding capacity and reagents must be prepared and stored to avoid contamination. The pH and osmolality of reagents must be checked and standardized and protocols for dilution and mixing should be developed and followed. Reagents generally are stored at refrigerated temperatures and brought to room temperature for use. Although serum containing reagents such as calibrators or controls may be frozen for storage, other reagents, such as alkaline phosphatase conjugates may not be stored frozen.

Because test volumes are small, consistent technique and accurate pipetting is essential. Equipment must be carefully selected, calibrated, used, and maintained. Thorough washing is important, but care must be taken not to remove bound material from the solid phase. Excess wash solution must be removed from the solid phase which is readily achieved by tapping the inverted plate on a surface containing absorbent paper. Test incubation temperature must be carefully controlled. For tests performed at room temperature, the temperature must be monitored and controlled within a narrow range that will assure interassay consistency. Conditions that may have a deleterious effect on test results include light (which may degrade the substrate) and exposure to sodium hypochlorite (household bleach) vapors which may emanate from waste disposal vessels and inactivate the enzyme.

ELISA technology has reached a high degree of technical sophistication and tests may be performed currently with greater ease than ten years ago. The range of tests available from commercial producers for use in settings from the physician's office to the large volume reference lab is great. With the advent of automatic or "walk away" instrumentation, ELISA has entered a new step in its developmental evolution and may finally live up to its expectations and become the ultimate system for detection of antibodies and/or antigens.

ACKNOWLEDGMENT

The authors gratefully acknowledge the editorial assistance of Patricia B. Shrader, Esq.

REFERENCES

Adler, S.P., McVoy, M., Biro, V.G., Britt, W.J., Hider, P., and Marshall, D. 1985. Detection of cytomegalovirus antibody with latex agglutination. J. Clin. Microbiol. 22: 68.

Adler-Storthz, K., Kendall, C., Kennedy, R.C., Henkel, R.D., and Dreesman, G.R. 1983. Biotin-avidin-amplified enzyme immunoassay for detection of herpes-simplex virus antigen in clinical specimens. J. Clin. Microbiol. 18:1329–1334.

Alexander, I., Ashley, C.R., Smith, K.J., Harbour, J., Roome, A.P., and Darville, J.M. 1985. Comparison of ELISA with virus isolation for the diagnosis of genital herpes. J. Clin. Pathol. 38:554–557.

Anderson, H.K. and Anderson, P. 1977. Detection of specific IgM antibodies in cytomegalovirus infection with the ELISA technique. Abstract from the 5th Scandinavian Virus Symposium Ouluu Finland, p. 15.

Anneren, G., Gronowitz, J.S., Kallander, C.F., and Sundqvist, V.A. 1986. Mothers of children with down syndrome have higher herpes simplex virus type 2 (HSV-2) antibody levels. Hum. Genet. 72:9–14.

Antanasiu, P., Savy, V., and Perrin, P.C. 1977. Rapid detection of rabies antibodies by immunoenzymatic assay. Ann. Microbiol. (Inst Pasteur) 128A:489.

Antanasiu, P., Savy, V., and Gibert, C. 1978. Rapid immunoenzymatic technique for titration of rabies antibodies IgG and IgM results. Med. Microbiol. Immunol. 166:201.

Arista, S., Pistoia, D., Titone, L., and Ammatuna, P. 1987. Comparison of serological methods for the detection of IgG and IgM antibodies to rubella virus. Microbiologica 10:151–160.

Arneborn, P., Biberfeld, G., Forsgren, M., and von Stedingk, L.V. 1983. Specific and nonspecific B cell activation in measles and varicella. Clin. Exp. Immunol. 51:165–172.

Asano, Y., Hiroishi, Y., Itakura, N., Hirose, S., Kajita, Y., Suga, S., and Yazaki, T. 1988. Placental transfer of IgG subclass-specific antibodies to varicella-zoster virus. J. Med. Virol. 26:1–6.

Aurelian, L. 1982. Herpes simplex virus diagnosis. Antigen detection by ELISA and flow microfluorometry. Diagn. Gynecol. Obstet. 4:375–388.

Austgluen, R. 1985. Immunization of children with malignant diseases with the Oka-strain varicella vaccine. Postgrad Med. J. 61:S93–95.

Avrameas, S. 1969. Coupling of enzymes to proteins with glutaraldehyde. Use of the conjugates for detection of antigens and antibodies. Immunochemistry 6:43–52.

Avrameas, S. 1983. Enzyme immunoassays and related techniques: Development and limitations. Curr. Topics in Microbiol. Immunol. 104:93–99.

Avrameas, S. and Guilbert, B. 1971. A method for quantitative determination of cellular immunoglobulins by enzyme-labelled antibodies. Eur. J. Immunol. 1:394–396.

Avrameas, S. and Ternynck, T. 1969. The crosslinking of proteins with glutaraldehyde and its use for the preparation of immunosorbents. Immunochemistry 6:53–66.

Avrameas, S. and Ternynck, T. 1971. Peroxidase labelled antibody and Fab conjugates with enhanced intracellular penetration. Immunochemistry 8:1175–1179.

Balfour, H.H. Jr., Edelman, C.K., Dirksen, C.L., Palermo, D.R., Suarez, C.S., Kelly,

J., Kentala, J.T., and Crane, D.D. 1988. Laboratory studies of acute varicella and varicella immune status. Diagn. Microbiol. Infect. Dis. 10:149–158.

Baker, D.A., Gonik, B., Milch, P.O., Berkowitz, A., Lipson, S., and Verma, U. 1989. Clinical evaluation of a new herpes simplex virus ELISA: A rapid diagnostic test for herpes simplex virus. Obstet. Gynecol. 73: 322–325.

Beckwith, D.G., Halstead, D.C., Alpaugh, K., Schweder, A., Blount-Pronefield, D.A., and Toth, K. 1985. Comparison of a latex agglutination test against CMV. J. Clin. Microbiol. 21:3.

Begovac, J., Soldo, I., and Presecki, V. 1988. Cytomegalovirus mononucleosis in children compared with the infection in adults and with Epstein-Barr virus mononucleosis. J. Infect. 17:121–125.

Bellamy, K., Hodgson, J., Gardner, P.S., and Morgan-Capner, P. 1985. Public health laboratory service IgM antibody capture enzyme-linked immunosorbent assay for detecting rubella specific IgM. J. Clin. Pathol. 38:1150–1154.

Bellamy, K., Rousseau, S.A., and Gardner, P.S. 1986. The development of an M antibody capture ELISA for rubella IgM. J. Virol. Methods 14:243–251.

Benjamin, D.R. 1979. Immunoenzymatic methods. In E.H. Lennette and N.J. Schmidt (eds.), Diagnostic procedures for viral rickettsial and chlamydial infections. Washington, D.C.: APHA Inc., pp. 153–170.

Best, J.M., Banatvala, J.E., and Watson, D. 1969. Serum IgM and IgG responses in postnatally acquired rubella. Lancet 2:65–68.

Bidwell, D.E., Bartlett, A., and Voller, A. 1977. Enzyme immunoassays for viral diseases. J. Infect. Dis. (Supplement) 136:S274–S278.

Bishai, F.R. and Galli, R. 1978. Enzyme-linked immunosorbent assay for detection of antibodies to influenza A and B and parainfluenza type 1 in sera of patients. J. Clin. Microbiol. 8:648.

Bogger-Goren, S., Bernstein, J.M., Gershon, A.A., and Ogra, P.L. 1984. Mucosal cell-mediated immunity to varicella-zoster virus: Role in protection against disease. J. Pediatr. 105:195–199.

Bonfanti, C., Meurman, O., and Halonen, P. 1985. Direct enzyme immunoassay for detection of specific IgG antibody to rubella virus by use of labelled antigen. J. Virol. Methods 11:161–170.

Bos, C.A., Olding-Stenkvist, E., Wilterdink, J.B., and Scheffer, A.J. 1987. Detection of viral antigens in cerebrospinal fluid of patients with herpes simplex virus encephalitis. J. Med. Virol. 21:169–178.

Boteler, W.L., Luipersbeck, P.M., Fuccillo, D.A., and O'Beirne, A.J. 1983. Enzyme-linked immunosorbent assay for detection of measles antibody. J. Clin. Microbiol. 17:814.

Boteler, W.L., Barnes, K.J., Buimovici-Klein, E., and O'Beirne, A.J. 1984. Multi-center evaluation of a 1-h enzyme-linked immunosorbent assay for rubella serology. J. Clin. Microbiol. 20:1140.

Boteler, W.L., Schroyer, K.C., Fuccillo, D.A., and O'Beirne, A. 1985. A one hour ELISA assay for cytomegalovirus serology. ASM 1985 Poster Session.

Braun, R., Hornig, C., Sann, G., and Doerr, H.W. 1982. Comparison of different methods for assessment of rubella infection and immunity. Zentralbl. Bakteriol. Mikrobiol. Hyg. 252:431–437.

Briantais, M.J., Keros, L.G., Pillot, J. 1984. Specificity and sensitivity of the IgM capture immunoassay: Studies of possible factors including false positive or false negative results. J. Virol. Methods 9:15–26.

Buimovici-Klein, E., Tinker, M.K., O'Beirne, A.J., Lange, M., and Cooper, Z. 1983. IgM detection by ELISA in the diagnosis of cytomegalovirus infections in homosexual and heterosexual immunosuppressed patients. Arch. Virol. 78:203.

Buimovici-Klein, E., O'Beirne, A.J., Millian, S.J., and Cooper, L.Z. 1980. Low level rubella immunity detected by ELISA and specific lymphocyte transformation. Arch. Virol. 66:321.

Bullock, S.L. and Walls, K.W. 1977. Evaluation of some of the parameters of the ELISA. J. Infect. Dis. (Supplement) 136:S279.

Caldwell, C.W. and Barrett, J.T. 1977. Enzyme immunoassay for hepatitis B and its comparison to other methods. Clin. Chim. Acta. 81:305.

Cappel, R., deCuyper, F., and deBrackeleer, J. 1978. Rapid detection of IgG and IgM antibodies for CMV by the enzyme-linked immunosorbent assay (ELISA). Arch. Virol. 58:253.

Castellano, G.A., Hazzard, G.T., Madden, D.L., and Sever, J.L. 1977. Comparison of the enzyme-linked immunosorbent assay and the indirect hemagglutination test for detection of antibody to cytomegalovirus. J. Infect. Dis. (Supplement) 136:S337–S340.

Chantler, J.K., Tingle, A.J., and Petty, R.E. 1985. Persistent rubella virus infection associated with chronic arthritis in children. N. Engl. J. Med. 313:1117–1123.

Chantler, S., Evans, C.J., Mortimer, P.P., Cradock-Watson, J.E., and Ridehalgh, M.K. 1982. A comparison of antibody capture radio- and enzyme immunoassays with immunofluorescence for detecting IgM antibody in infants with congenital rubella. J. Virol. Methods 4:305–313.

Chao, R.K., Fishaut, M., Schwartzman, J.D., and McIntosh, K. 1978. An enzyme linked immunosorbent assay (ELISA) for detection of respiratory syncytial virus in human nasal secretions. Abstract in Pan American Group for Rapid Viral Diagnosis, Vol. 1, No. 2.

Cheeseman, S.H., Doern, G.V., Ferriani, R.A., Keville, M.W., McGraw, B.R., and Stewart, J.A. 1984. Detection of cytomegalovirus antibody with two commercially available assays an indirect hemagglutination test and an enzyme immunosorbent assay. J. Clin. Microbiol. 20:1.

Chernesky, M.A., Wyman, L., Mahony, J.B., Castriciano, S., Unger, J.T., Safford, J.W., and Metzel, P.S. 1984. Clinical evaluation of the sensitivity and specificity of a commercially available enzyme immunoassay for detection of rubella virus-specific immunoglobulin M. J. Clin. Microbiol. 20:400–404.

Chou, S., Kim, D.Y., Scott, K.M., and Sewell, D.L. 1987. Immunoglobulin M to cytomegalovirus in primary and reactivation infections in renal transplant recipients. J. Clin. Microbiol. 25:1.

Clark, M.E. and Adams, A.N. 1977. Characteristics of the microplate method of enzyme linked immunosorbent assay for the detection of plant viruses. J. Gen. Virol. 34:475.

Clayton, A.L., Roberts, C., Godley, M., Best, J.M., and Chantler, S.M. 1986. Herpes simplex virus detection by ELISA: Effect of enzyme amplification, nature of lesion sampled and specimen treatment. J. Med. Virol. 20:89–97.

Clayton, A.L., Beckford, U., Roberts, C., Sutherland, S., Druce, A., Best, J., and Chantler, S. 1985. Factors influencing the sensitivity of herpes simplex virus detection in clinical specimens in a simultaneous enzyme-linked immunosorbent assay using monoclonal antibodies. J. Med. Virol. 17: 275–282.

Cleary, T.J., Cid, A., Ellis, B., Malkus, H., Noto, T., Halbert, S., and Castro, A. 1978. A direct enzyme-linked immunosorbent (ELISA) for detection of antibodies for rubella virus in human sera. Res. Commun. Chem. Pathol. Pharmacol. 19:281.

Coleman, R.M., Bailey, P.D., Whitley, R.J., Keyserling, H., and Nahmias, A.J. 1983a. ELISA for the detection of herpes simplex virus antigens in the cerebrospinal fluid of patients with encephalitis. J. Virol. Methods 7:117–125.

Coleman, R.M., Pereira, L., Bailey, P.D., Dondero, D., Wickliffe, C., and Nahmia, A.J. 1983b. Determination of herpes simplex virus type-specific antibodies by enzyme-linked immunosorbent assay. J. Clin. Microbiol. 18:287–291.

Corey, L. 1986. Laboratory diagnosis of herpes simplex virus infections. Principles guiding the development of rapid diagnostic tests. Diagn. Microbiol. Infect. Dis. 4:S111–119.

Costa, S., Barrasso, R., Terzano, P., Zerbini, M., Carpi, C., and Musiani, M. 1985. Detection of active Epstein-Barr infection in pregnant women. Eur. J. Clin. Microbiol. 4:335–336.

Cox, J.C., Moloney, M.B., Herrington, R.W., Hampson, A.W., and Hurrell, J.G. 1984. Enzyme immunoassay for antibodies to membrane associated antigen of varicella-zoster virus. J. Virol. Methods 8:137–145.

Cremer, N.E., Cossen, C.K., Shell, G., Diggs, J., Gallo, D., and Schmidt, N.J. 1985. Enzyme immunoassay versus plaque neutralization and other methods for determination of immune status to measles and varicella-zoster viruses and versus complement fixation for serodiagnosis of infections with those viruses. J. Clin. Microbiol. 21:869–874.

Cubie, H. and Edmond, E. 1985. Comparison of five different methods of rubella IgM antibody testing. J. Clin. Pathol. 38:203–207.

Cubie, H.A. 1983. Evaluation of a new ELISA method for estimating rubella-specific IgM antibody. J. Infect. 7:144–150.

Dannenmaier, B., Alle, W., Hoferer, E.W., Lorenz, D., Oertel, P.J., and Doerr, H.W. 1985. Incidences of antibodies to hepatitis B, herpes simplex and cytomegalovirus in prostitutes. Zentralbl. Bakteriol. Mikrobiol. Hyg. 259:275–283.

Dascal, A., Chan-Thim, J., Morahan, M., Portnoy, J., and Mendelson, J. 1989. Diagnosis of herpes simplex virus infection in a clinical setting by a direct antigen detection enzyme immunoassay kit. J. Clin. Microbiol. 27:700–704.

deOry, F., Echevarria, J.M., and Domingo, C.J. 1988. Screening of antibodies to varicella-zoster virus in leukemic patients: Comparison of commercial methods of enzyme immunoassay and fluoroimmunoassay. Diagn. Microbiol. Infect. Dis. 10:61–64.

Deforest, A., Gessner, A.R., Clark, J.L., and Zahradnik, J.M. 1989. Usefulness of enzyme-linked immunoassays (ELISA) in detecting seroconversion following measles-mumps-rubella (MMR) vaccines. Abstract from 1989 ASM.

Delisi, L.E., Smith, S.B., Hamovit, J.R., Maxwell, M.E., Goldin, L.R., Dingman, C.W., and Gershon, E.A. 1986. Herpes simplex virus, cytomegalovirus and Epstein-Barr virus antibody titers in sera from schizophrenic patients. Psychol. Med. 16:757–763.

Demmler, G.J., Six, H.R., Hurst, S.M., and Yow, M.D. 1986. Enzyme-linked immunosorbent assay for the detection of IgM-class antibodies to cytomegalovirus. J. Infect. Dis. 153:6.

Devillechabrolle, A., Hugues-Dorin, F., Fortier, B., Catalan, F., and Huraux, J.M. 1985. Prevalence of serum antibodies to herpes simplex virus types 1 and 2: Application of an ELISA technique to 100 cases of anogenital herpes. Sex. Transm. Dis. 12:40–43.

Dillner, J., Kallin, B., Klein, G., Jornvall, H., Alexander, H., and Lerner, R. 1985. Antibodies against synthetic peptides react with the second Epstein-Barr virus-associated nuclear antigen. EMBO J. 4:1813–1818.

Dillner, J., Szigeti, R., Henle, W., Henle, G., Lerner, R.A., and Klein, G. Cellular and humoral immune responses to synthetic peptides deduced from the amino-acid sequences of Epstein-Barr virus-encoded proteins in EBV-transformed cells. Int. J. Cancer 40:455–460.

Doerr, H.W., Rentschler, M., and Scheifler, G. 1987. Serologic detection of active infections with human herpes viruses (CMV, EBV, HSV, VZV): Diagnostic potential of IgA class and IgG subclass-specific antibodies. Infection 15:93–98.

Dolken, G., Boldt, C., Lange, W., Weitzmann, U., Lohr, G.W., and Rosen, A. 1983. Enzyme-linked immunosorbent assay for antibodies to Epstein-Barr virus associated early antigens and the viral capsid antigen. Cancer Detect. Prev. 6:287–292.

Dolken, G., Bross, K.J., Hecht, T., Brugger, W., Lohr, G.W., and Hirsch, F.W. 1986. Increased incidence of IgA antibodies to the Epstein-Barr virus-associated viral capsid antigen and early antigens in patients with chronic lymphocytic leukemia. Int. J. Cancer 38:55–59.

Dolken, G., Hecht, T., Weitzmann, U., Wagner, M., Lohr, G.W., and Hirsch, F.W. Purification of a polypeptide complex (p52) belonging to the D-subspecificities of Epstein-Barr virus-induced early antigens. Virology 148:58–73.

Dolken, G., Hecht, T., Rockel, D., and Hirsch, F.W. 1987. Characterization of the Epstein-Barr virus induced early polypeptide complex p50/58 EA-D using rabbit antisera, a monoclonal antibody, and human antibodies. Virology 157:460–471.

Dolken, G., Weitzmann, U., Boldt, C., Bitzer, M., Brugger, W., and Lohr, G.W. 1984. Enzyme-linked immunosorbent assay for IgG antibodies to Epstein-Barr virus associated early antigens and viral capsid antigen. J. Immunol. Methods 67:225–233.

Dreikorn, K., Doerr, H.W., and Geursen, R.G. 1985. Cytomegalovirus (CMV) infections in renal transplant recipients. Preliminary results of prophylaxis by an intramuscular human hyperimmune CMV IgG. Scand. J. Urol. Nephrol. Suppl. 92:15–21.

Duermeyer, W., Veen der Pric Jud, and Koster, B. 1977. Enzyme immunoassay for hepatitis A. Presented at the 16th Symposium of the European Association for Rapid Laboratory Viral Diagnosis. Amsterdam June 1977.

Duerymeyer, W. and Van der Veen. 1978. ELISA for specific IgM and applications for anti-hepatitis A IgM detection. Lancet 2:684.

Dunkel, E.C., Pavan-Langston, D., Fitzpatrick, K., and Cukor, G. 1988. Rapid detection of herpes simplex virus (HSV) antigen in human ocular infections. Curr. Eye Res. 7:661–666.

Echevarria, J.M., de Ory, F., Leon, P., and Tellez, A. 1989. Definition of high-proficiency serological markers for diagnosis of varicella-zoster virus infections by enzyme immunoassay. J. Med. Virol. (U.S.) 27:224–230.

Echevarria, J.M., Sainz, C., de Ory, F., and Najera, R. 1985a. Evaluation of commercial methods of enzyme immunoassay (EIA) for the measurement of rubella-specific IgM. J. Virol. Methods 11:177–187.

Echevarria, J.M., de Ory, F., and Najera, R. 1985b. Fluoroimmunoassay for detection of rubella-specific immunoglobulin M: Comparison with indirect enzyme immunoassay and Mu-Chain capture. J. Clin. Microbiol. 22:428–434.

Effros, R.B., Zeller, E., Dillard, L., and Walford, R.L. 1985. Detection of antibodies to cell surface antigens by a simplified cellular ELISA (CELISA). Tissue Antigens. 25:204–211.

Elitsur, Y., Carmi, R., and Sarov, I. 1983. HSV-specific serum/CSF antibody ratio in association with HSV serum IgM antibodies in diagnosis of herpes encephalitis in infants. Isr. J. Med. Sci. 19:943–945.

Ellens, D.J., deLeeuw, P.W., and Straver, P.J. 1978. The detection of rotavirus specific antibody in colostrum and milk by ELISA. Ann. Rech. Vet. 9:337.

Enders, G. 1982. Serodiagnosis of varicella-zoster virus infection in pregnancy and standardization of the ELISA IgG and IgM antibody tests. Dev. Biol. Stand. 52:221–236.

Enders, G. 1985a. Serologic test combinations for safe detection of rubella infections. Rev. Infect. Dis. 7:113–122.

Enders, G. 1985b. Management of varicella-zoster contact and infection in pregnancy using a standardized varicella-zoster ELISA test. Postgrad. Med. J. 61:S23–30.

Enders, G. and Knotek, F. 1985. Comparison of the performance and reproducibility of various serological methods and diagnostic kits or the detection of rubella antibodies. J. Virol. Methods 11:1–14.

Enders, G. and Knotek, F. 1986. Detection of IgM antibodies against rubella virus: Comparison of two indirect ELISAs and an anti-IgM capture immunoassay. J. Med. Virol. 19:377–386.

Enders, G. and Knotek, F. 1989. Rubella IgG total antibody avidity and IgG subclass-specific antibody avidity assay and their role in the differentiation between primary rubella and rubella reinfection. Infection 17:218–226.

Engvall, E., Jonsson, K., and Perlmann, P. 1971. ELISA II: Quantitative assay of protein antigen IgG by means of enzyme labelled antigen and antibody coated tubes. Biochem. Biophys. Acta. 251:427.

Engvall, E. and Perlman, P. 1971. Enzyme-linked immunosorbent assay (ELISA). Quantitative assay of immunoglobulin G. Immunochemistry 8:871–874.

Engvall, E. and Perlman, P. 1972. Enzyme-linked immunosorbent assay. III. Quantitation of specific antibodies by enzyme-labelled anti-immunoglobulin in antigen-coated tubes. J. Immunol. 109:120–136.

Engvall, E. and Pesce, A.J., eds. 1978. Quantitative Enzyme Immunoassay. Blackwell Scientific Publications. Scand. J. Immunol. 8:Suppl. 7.

Epstein, M.A. 1986. Vaccination against Epstein-Barr virus: Current progress and future strategies. Lancet 1:1425–1427.

Evans, L., Arsenakis, M., Sheppard, M., and May, J.T. 1983. An ELISA technique to detect IgG antibody to the early herpes simplex virus type 2 (HSV-2) antigen AG-4 in HSV-2 patients. J. Virol. Methods 6:245–254.

Feldman, S., Gigliotti, F., Bockhold, C., and Naegele, R. 1988. Measles and rubella antibody status in previously vaccinated children with cancer. Med. Pediatr. Oncol. (U.S.) 16:308–311.

Felgenhauer, K., Nekic, M., and Ackerman, R. 1982. The demonstration of locally synthesized herpes simplex IgG antibodies in CSF by a sepharose 4B linked enzyme immunoassay. J. Neuroimmunol. 3:149–158.

Felgner, P. 1977. Serologic diagnosis of extraintestinal amebiasis: A comparison of stick-ELISA and other immunological tests. Tropen Med. Parasitol. 28:491.

Felgner, P. 1978. A new technique of heterogeneous enzyme-linked immunosorbent assay, stick-ELISA I description of the technique. Zbl. Bakt. Hyg. I Abt. Org. A240:112.

Field, P.R. and Gong, C.M. 1984. Diagnosis of postnatally acquired rubella by use of three enzyme linked immunosorbent assays for specific immunoglobulins G and M and single radial hemolysis for specific immunoglobulin G. J. Clin. Microbiol. 20:951–958.

Fitzgerald, M.G., Pullen, G.R., and Hosking, C.S. 1988. Low affinity antibody to rubella antigen in patients after rubella infection in utero. Pediatrics 81:812–814.

Florman, A.L., Umland, E.T., Ballou, D., Cushing, A.H., McLaren, L.C., Gribble, T.J., and Duncan, M.H. 1985. Evaluation of a skin test for chicken pox. Infect. Control. 6:314–316.

Fogel, A., Barnea, B., and Gerichter, C.B. 1983. Susceptibility to rubella and borderline immunity. Isr. J. Med. Sci. 19:934–035.

Forghani, R., Schmidt, N.J., and Dennis, J.

1978. Antibody assays for varicella-zoster virus: Comparison of enzyme immunoassay with neutralization immune adherence hemagglutination and complement fixation. J. Clin. Microbiol. 8:545.

Forghani, B. and Schmidt, N.J. 1979. Antigen requirements, sensitivity and specificity of enzyme immunoassays for measles and rubella viral antibodies. J. Clin. Microbiol. 9:657–664.

Forghani, B., Myoraku, C., and Schmidt, N. 1983. Use of monoclonal antibodies to human immunoglobulin M in "capture" assays for measles and rubella immunoglobulin M. J. Clin. Microbiol. 18:652–657.

Fortier, B., Chabrolle, A.D., and Huraux, J.M. 1982. Comparison of an ELISA technique with quantal micro-neutralization test for serotyping of HSV-1 or HSV-2 infected patients. J. Virol. Meth. 5:11–20.

Fox, P.D., Khaw, P.T., McBride, B.W., McGill, J.I., and Ward, K.A. 1986. Tear and serum antibody levels in ocular herpetic infection: Diagnostic precision of secretory IgG. Br. J. Ophthalmol. 70:584–588.

Freijd, A. and Rosen, A. 1984. Epstein-Barr virus induced pneumococcal antibody production in man. A comparison of different lymphoid organs. Clin. Exp. Immunol. 55: 204–210.

Friedman, M.G., Haikin, H., Leventon-Kriss, S., Joffe, R., Goldstein, V., and Sarov, I. 1978. Detection of antibodies to varicella-zoster virus by radioimmunoassay and enzyme immunoassay techniques. Med. Microbiol. Immunol. 166:177.

Fuccillo, D.A., Madden, D.L., Tzan, N., Sever, J.L. 1987. Difficulties associated with serological diagnosis of *Toxoplasma gondii* infections. Diag. Clin. Immunol. 5:8–13.

Gera, A., Loebenstein, G., and Raccah, B. 1978. Detection of cucumber mosaic virus *vs* viruliferous aphids by enzyme-linked immunosorbent assay. Virology 86:542.

Gerna, G. 1976. Immunoperoxidase technique for rapid human cytomegalovirus and rubella virus identification and antibody assay. First International Symposium on Immunoenzymatic Techniques. INSERM Symposium No. 2. Amsterdam: North Holland Publishing Co., pp. 443–449.

Gerna, I., Zannino, M., Revello, M.G., Petruzzelli, E., and Dovis, M. 1987. Development and evaluation of a capture enzyme-linked immunosorbent assay for determination of rubella immunoglobulin M using monoclonal antibodies. J. Clin. Microbiol. 25:1033–1038.

Gershon, A.A., Frey, H.M., Steinburg, S.P., Seeman, M.D., Bidwell, D., and Voller, A. 1981. Determination of immunity to varicella using an enzyme-linked immunosorbent assay. Arch. Virol. 70:169.

Ghose, L.H., Schnagl, R.D., and Holmes, I.H. 1978. Comparison of an enzyme-linked immunosorbent assay for quantitation of rotavirus antibodies with complement fixation in an epidemiological survey. J. Clin. Microbiol. 8:268.

Gilljam, G., Sundqvist, V.A., Linde, A., Pihlstedt, P., Eklund, A.E., and Wahren, B. 1985. Sensitive analytic ELISAs for subclass herpes virus IgG. J. Virol. Methods 10:203–214.

Gilman, S.C. and Docherty, J.J. 1977. Detection of antibodies specific for herpes simplex virus in human sera by the enzyme-linked immunosorbent assay. J. Infect. Dis. (Supplement) 136:S286.

Glynn, A.A. and Ison, C. 1978. Serological diagnosis of gonorrhoea by an enzyme-linked immunosorbent assay (ELISA). Br. J. Vener. Dis. 54:97.

Goldblum, R.M., Ahlstedt, S., Carlsson, B., Hanson, L.A., Jodal, U., Lidin-Janson, G., and Sohlakerlund, A. 1975. Antibody forming cells in human colostrum after oral immunization. Nature 257:797.

Gong, A.K., Lipton, S.V., and Brunell, P.A. 1988. Establishment by enzyme-linked immunosorbent assay of seronegative range for herpes simplex virus and cytomegalovirus antibodies and evaluation of heterologous responses to live varicella vaccine. J. Clin. Microbiol. 26:781–783.

Grauballe, P.C. and Vestergaard, B.F. 1977. ELISA for herpes simplex virus type 2 antibodies. Lancet 2:1038.

Gravell, M., Dorsett, P.H., Gutenson, O., and Ley, A.C. 1977. Detection of antibody to rubella virus by enzyme-linked immunosorbent assay. J. Infect. Dis. (Supplement) 136:S300.

Grillner, L., Forsgren, M., and Nordenfelt, E. 1985. Comparison between a commercial ELISA, rubazyme, and hemolysis-in-gel test for determination of rubella antibodies. J. Virol. Methods 10:111–115.

Grillner, L. and Landqvist, M. 1983. Enzyme-linked immunosorbent assay for detection and typing of herpes simplex virus. Eur. J. Clin. Microbiol. 2:39–42.

Habermann, E. and Heller, I. 1976. Two enzyme immunoassays of tetanus antibodies using peroxidase coupled tetanus toxin as tracer. In H. Peeters (ed.), Protides of the Biological Fluids. Oxford: Pergamon Press, pp. 825–828.

Hadar, T. and Sarov, I. 1984. Specific IgG and IgA antibodies to herpes simplex virus (HSV)-induced surface antigen in patients with HSV infections and in healthy adults. J. Med. Virol. 14:201–207.

Halbert, S.P. and Anken, M. 1977. Detection of hepatitis B surface antigen (HBsAg) with use of alkaline phosphatase labelled antibody to HBsAg. J. Infect. Dis. (Supplement) 136:S318.

Hamaguchi, Y., Kato, K., Dukui, H., Ishikawa, I., Okawa, S., Ishikawa, E., Kobayashi, K., and Karubuma, N. 1976. Enzyme-linked sandwich immunoassay of macromolecular antigen using the rabbit antibody coupled glass rod as a solid phase. Eur. J. Biochem. 71:459.

Hamaguchi, Y., Kato, K., Ishikawa, E., Kobayzshi, K., and Kutunuma, N. 1976. Enzyme-linked sandwich immunoassay of macromolecular antigens using the rabbit antibody-loaded silicone piece as a solid phase. FEBS Lett. 69:11.

Hamaguchi, Y., Kato, K., Fukui, H., Shirakawa, I., Ishikawa, E., Kobayashi, K., and Katunuma, N. 1976. Enzyme-linked sandwich immunoassay of ornithine o-aminotransferase from rat liver using antibody coupled glass rods as solid phase. J. Biochem. 80:895.

Hanada, N., Kido, S., Terashima, M., Nishikawa, K., and Morishima, T. 1988. Noninvasive method for early diagnosis of herpes simplex encephalitis. Arch. Dis. Child. 63:1470–1473.

Hancock, E.J., Pot, K., Puterman, M.L., and Tingle, A.J. 1986. Lack of association between titers of HAI antibody and whole-virus ELISA values for patients with congenital rubella syndrome. J. Infect. Dis. 154:1031–1033.

Haskard, D.O. and Archer, J.R. 1984. The production of human monoclonal autoantibodies from patients with rheumatoid arthritis by the EBV-hybridoma technique. J. Immunol. Methods 74:361–367.

Hayashi, Y., Wada, T., and Mori, R. 1983. Protection of newborn mice against herpes simplex virus infection by prenatal and postnatal transmission of antibody. J. Gen. Virol. 64:1007–1012.

Hedman, K., Hietala, J., Tiilikainen, A., Hartikainen-Sorri, A.L., Raiha, K., Suni, J., Vaananen, P., and Pietilainen, M. 1989. Maturation of immunoglobulin G avidity after rubella vaccination studied by an enzyme linked immunosorbent assay (avidity-ELISA) and by haemolysis typing. J. Med. Virol. 27:293–298.

Hedman, K. and Rousseau, S.A. 1989. Measurement of avidity of specific IgG for verification of recent primary rubella. J. Med. Virol. 27:288–292.

Hendry, R.M. and Herrmann, J.E. 1980. Immobilization of antibodies on nylon for use in enzyme-linked immunoassay. J. Immunol. Methods 35:285–296.

Herrmann, K.L. 1985. Available rubella serologic tests. Rev. Infect. Dis. 7:S108–112.

Ho, D.W., Field, P.R., and Cunningham, A.L. 1989. Rapid diagnosis of acute Epstein-Barr virus infection by an indirect enzyme linked immunosorbent assay for specific immunoglobulin M (IgM) antibody without rheumatoid factor and specific IgG interference. J. Clin. Microbiol. 27:952–958.

Hoffman, B.E., Jungkind, D.L., Haller, G.J., Sharrar, R., Baker, R.A., and Weisberg, M. 1985. Evaluation of two rapid methods for the detection of herpes simplex virus antigen in patient specimens. Ann. Clin. Lab. Sci. 15:418–427.

Hodgson, J. and Morgan-Capner, P. 1984. Eval-uation of a commercial antibody capture enzyme immunoassay for the detection of rubella specific IgM. J. Clin. Pathol. 37:573–577.

Hopkins, R.F. III, Witmer, T.J., Neubauer, R.H., and Rabin, H. 1982. Detection of antibodies to Epstein-Barr virus antigens by enzyme-linked immunosorbent assay. J. Infect. Dis. 146:734–740.

Hornstein, L., Swartz, T.A., and Heimann, M. 1985. Rubella immunity measured by hemagglutination inhibition and enzyme linked immunosorbent assay. Isr. J. Med. Sci. 21:666–669.

Horstmann, D.M., Schluederberg, A., Emmons, J.E., Evans, B.K., Randolph, M.F., and Andiman, W.A. 1985. Persistence of vaccine-induced immune responses to rubella: Comparison with natural infection. Rev. Infect. Dis. 7:S80–85.

Hossain, A., Ramia, S., and Bakir, T.M. 1988. Comparison of haemagglutination test, enzyme linked immunosorbent assay and indirect immunofluorescence antibody test for determination of rubella immune status. J. Trop. Med. Hyg. 91:216–221.

Irving, W.L., Walker, P.R., and Lydyard, P.M. 1985. Abnormal responses of rheumatoid arthritis lymphocytes to Epstein-Barr virus infection in vitro: Evidence for multiple defects. Ann. Rheum. Dis. 44:462–468.

Jeansson, S., Forsgren, M., and Svennerholm, B. 1983. Evaluation of solubilized herpes simplex virus membrane antigen by enzyme-linked immunosorbent assay. J. Clin. Microbiol. 18:1160.

Johnson, R.B., Libby, R.M., and Nakamura, R.M. 1980. Comparison of glucose oxidase and peroxidase as labels for antibody in enzyme-linked immunosorbent assay. J. Immunoassay 1:27–37.

Johnson, S., McAnerney, J.M., Schoub, B.D., and Kidd, A.H. 1985. Laboratory monitoring of rubella. S. Afr. Med. J. 67:721–723.

Just, M., Just, V., Berger, R., Burkhardt, F., and Schilt, U. 1985. Duration of immunity after rubella vaccination: A long-term study in Switzerland. Rev. Infect. Dis. 7:S91–94.

Juto, P. and Settergren, B. 1988. Specific serum IgA, IgG and IgM antibody determination by a modified indirect ELISA technique in primary and recurrent herpes simplex virus infection. J. Virol. Methods 20:45–55.

Kacaki, J., Wolters, G., Kuijpers, L., and Schuurs, A. 1977. A specificity control in solid phase enzyme immunoassay for HBsAg by one step *in situ* blocking with human anti-HBs. J. Clin. Pathol. 30:894.

Kahan, A., Lefloch, J.P., and Charreire, J. 1987. Normal suppressive T cell function of Epstein-Barr virus-induced B cell activation in Graves' disease. J. Clin. Endocrinol. Metab. 65:555–560.

Kahlon, J., Lakeman, F.D., Ackermann, M., and Whitley, R.J. 1986. Human antibody response to herpes simplex virus-specific polypeptides after primary and recurrent infection. J. Clin. Microbiol. 23:725–730.

Kahlon, J. and Whitley, R.J. 1988. Antibody response of the newborn after herpes simplex virus infection. J. Infect. Dis. 158:925–933.

Kasupski, G.J., Lo, P.L., Gobin, G., and Leers, W.D. 1984. Clinical evaluation of the Ortho rubella ELISA test system. Am. J. Clin. Pathol. 81:230–232.

Kato, K., Fukui, H., Hamaguchi, Y., and Ishikawa, E. 1976. Enzyme-linked immunoassay. Conjugation of Fab fragment of rabbit IgG with B-D-galactosidase from *E. coli* and its use for immunoassay. J. Immunol. 116:1554.

Kato, K., Hamaguchi, Y., Fukui, H., and Ishikawa, E. 1976. Enzyme-linked immunosorbent conjugation of rabbit antibody with B-D-galactosidase. Eur. J. Biochem. 62:285.

Kato, K., Hamaguchi, Y., Okawa, S., Ishikawa, E., Kobayashi, K., and Katunuma, N. 1977. Use of rabbit antibody IgG bound onto plain and aminoalkylsilyl glass surface for the enzyme-linked sandwich assay. J. Biochem. 82:261.

Katow, S., Sugiura, A., and Janejai, N. 1989. Single-serum diagnosis of recent rubella infection with the use of hemagglutination inhibition test and enzyme-linked immunosorbent assays. Microbiol. Immunol. (Japan) 33:141–145.

Katz, D., Hilliard, J.K., Mirkovic, R.R., and

Word, R.A. 1986. Elisa for detection of IgG and IgM antibodies to HSV-1 and HSV-2 in human sera. J. Virol. Methods 14:43–55.

Kawano, K. and Minamishima, Y. 1987. Removal of nonspecific hemagglutination inhibitors, immunoglobulin G, and immunoglobulin A with streptococcal cells and its application to the rubella hemagglutination inhibition test. Arch. Virol. 95:41–52.

Keller, P.M., Lonergan, K., Neff, B.J., Morton, D.A., and Ellis, R.W. 1986. Purification of individual varicella-zoster virus (VZ) glycoproteins gpI, gpII, and gpIII and their use in ELISA for detection of VZV glycoprotein-specific antibodies. J. Virol. Methods 14:177–188.

Kelley, G.R., Ashizawa, T., and Gyorkey, F. 1986. Herpes simplex virus encephalitis. A case with dysplastic plasma cell infiltration. Arch. Pathol. Lab. Med. 110:82–85.

Keunen, R.W., ter Bruggen, J.P., Kuijper, E.J., Peeters, M.F., Op de Coul, A.A., and van Loon, A.M. 1987. Rapid diagnosis of herpes encephalitis by enzyme immuno-assay. Clin. Neurol. Neurosurg. 89:97–101.

Kleeman, K.T., Kiefer, D.J., and Halbert, S.P. 1983. Rubella antibodies detected by several commercial immunoassays in hemagglutination inhibition-negative sera. J. Clin. Microbiol. 18:1131–1137.

Kleiman, M.B., Blackburn, C.K.L., Zimmerman, S.E., and French, M.L.V. 1981. Comparison of enzyme-linked immunosorbent assay for acute measles with hemagglutination inhibition, complement fixation, and fluorescent-antibody methods. J. Clin. Microbiol. 14:147.

Kovac, J., Zachar, V., Reichel, M., and Necas, S. 1987. Varicella-zoster virus IgM antibody fraction separated by minicolumn gel filtration in serodiagnosis of acute infection. Acta. Virol. 31:19–24.

Kurstak, E., de The G., Van den Hurk, J., Charpentier, G., Kurstak, C., Tijssen, P., and Morisset, R. 1978. Detection of Epstein-Barr virus antigens and antibodies by peroxidase-labeled specific immunoglobulins. J. Med. Virol. 2:189.

Land, S.A., Skurrie, I.J., and Gilbert, G.L. 1984. Rapid diagnosis of herpes simplex virus infections by enzyme-linked immunosorbent assay. J. Clin. Microbiol. 19:865–869.

Landry, M.L., Cohen, S.D., Mayo, D.R., Fong, C.K.Y., and Andiman, W.A. 1987. Comparison of fluorescent-antibody-to-membrane-antigen test, indirect immunofluorescence assay, and a commercial enzyme-linked immunosorbent assay for determination of antibody to varicella-zoster virus. J. Clin. Microbiol. 25:5.

Lange, W., Kohler, H., Aprodaca, J., Ott, A., and Sucker, U. 1977. Demonstration of HBs antigen with the Hepanostika test on enzyme-immunoassay. Dtsch. Med. Wochenschr. 102:1581.

Larussa, P., Steinberg, S., Waithe, E., Hanna, R., and Holzman, R. 1987. Comparison of five assays for antibody to varicella-zoster virus and the fluorescent-antibody-to-membrane-antigen test. J. Clin. Microbiol. 25:11.

Lawrence, T.G., Budzko, D.B., and Wilcke, B.W. Jr. 1984. Detection of herpes simplex virus in clinical specimens by an enzyme-linked immunosorbent assay. Am. J. Clin. Pathol. 81:339–341.

Lehtinen, M., Lehtinen, T., Koivisto, V., Paavonen, J., and Leinikki, P. 1985. Serum antibodies to the major HSV-2 specified DNA binding protein in patients with an acute HSV infection or cervical neoplasia. J. Med. Virol. 16:245–256.

Lehtonen, O.P. and Meurman, O.H. 1982. An ELISA for the estimation of high-avidity and total specific IgG and IgM antibodies to rubella virus. J. Virol. Methods 5:1–10.

Leinikki, P. and Passila, S. 1977. Quantitative semi-automated enzyme-linked immunosorbent assay for viral antibodies. J. Infect. Dis. (Supplement) 136:S294–S299.

Leinikki, P., Shekarchi, I., Dorsett, P., and Sever, J. 1978a. Enzyme-linked immunosorbent assay determination of specific rubella antibody levels in microgram of immunoglobulin G per milliliter of serum in clinical samples. J. Clin. Microbiol. 8:419.

Leinikki, P., Shekarchi, I., Dorsett, P., and Sever, J. 1978b. A determination of virus specific IgM antibodies by using ELISA: Elimination of false-positive results with protein A sepharose absorption and subsequent IgM antibody assay. J. Lab. Clin. Med. 92:849–857.

Leinikki, P.O., Shekarchi, I., Tzan, N., Madden, D.L., and Sever, J.L. 1979. Evaluation of enzyme-linked immunosorbent assay (ELISA) for mumps virus antibodies. Proc. Soc. Exp. Biol. Med. 160:363–367.

Lennette, E.T. and Henle, W. 1987. Epstein-Barr virus infections: Clinical and serologic features. Lab. Mgt. June.

Liden, S. 1985. Herpes simplex. Clinical and pathogenetic aspects. Surv. Immunol. Res. 4:24–29.

Linde, G.A. 1985. Subclass distribution of rubella virus-specific immunoglobulin G. J. Clin. Microbiol. 21:117–121.

Linde, A., Dahl, H., Wahren, B., Fridell, E., Salahuddin, Z., and Biberfeld, P. 1988. IgG antibodies to human herpesvirus-6 in children and adults and in primary Epstein-Barr virus infections and cytomegalovirus infections (corrected) (published erratum appears in J. Virol. Methods 1989 Feb;23(2):231). J. Virol. Methods 21:117–123.

Lipton, S.V. and Brunell, P.A. 1989. Management of varicella exposure in a neonatal intensive care unit. JAMA 261:1782–1784.

Locarnini, S.A., Garland, S.M., Lehmann, N.I., Pringle, R.C., and Gust, I.D. 1978. Solid phase enzyme-linked immunosorbent assay for detection of hepatitis A virus. J. Clin. Microbiol. 8:277.

Luka, J., Chase, R.C., and Pearson, G.R. 1984. A sensitive enzyme-linked immunosorbent assay (ELISA) against the major EBV-associated antigens. I. Correlation between ELISA and immunofluorescence titers using purified antigens. J. Immunol. Methods 67:145–156.

Maggio, E.T. (ed.). 1979. Enzyme-Immunoassay. Boca Raton, FL: CRC Press.

Maltseva, N.N., Zvonarev, A. Yu., Matsevich, G.R., and Ebralidze, L.K. 1987. Enzyme immunoassay for detection of antibodies against herpes simplex virus with the use of different viral antigens. Acta. Virol. 31:226–233.

Malvano, R. (ed.). 1979. Immunoenzymatic Assay Techniques (Development in Clinical

Biochemistry, Vol. 1). Hague: Martinus Nijhoff Publishers.

Maroto, Vela, M.C., Bernal, Zamora, M.C., Levya, Garcia, A., and Piedrola, G. 1986. Detection of specific IgG and IgM antibodies in the hemagglutination inhibition test and the enzyme-linked immunoassay for the diagnosis of rubella infection. Infection. 14:159–162.

Mathiesen, L.R., Feinstone, S.M., Wong, D.C., Skinhoej, P., and Purcell, R.H. 1978. Enzyme-linked immunosorbent assay for detection of hepatitis A antigen in stool and antibody to hepatitis A antigen in sera. Comparison with solid phase radioimmunoassay immune electron microscopy and immune adherence hemagglutination assay. J. Clin. Microbiol. 7:184.

Mayer, V., Zachar, V., Schmidtmayerova, H., Hruzik, J., Rajcani, J., Mitrova, E., Lackovic, V., Raucina, J., Kotuliak, J., and Faybik, M. 1986. Early intrathecal production of specific IgM and IgG antibodies and alpha-interferon in herpes simplex virus encephalitis. Acta. Virol. (Praha) 30:149–157.

Mayo, D.R. 1988. Single high ELISA values suggest recent varicella-zoster virus infection. Serodiagnosis and Immunotherapy. Vol 2.

McDermott, M.R., Graham, F.L., Hanke, T., and Johnson, D.C. 1989. Protection of mice against lethal challenge with herpes simplex virus by vaccination with an adenovirus vector expressing HSV glycoprotein B. Virology 169:244–247.

Meegan, J.M., Evans, B.K., and Horstmann, D.M. 1983. Use of enzyme immunoassays and the latex agglutination test to measure the temporal appearance of immunoglobulin M and G antibodies after natural infection or immunization with rubella virus. J. Clin. Microbiol. 18:745–748.

Melioli, G., Pedulla, D., Merli, A.L., Arata, L., and Leonardi, A. 1985. A simple method to detect intrathecal production of specific antimeasles antibodies in cerebrospinal fluid during subacute sclerosing panencephalitis. Diagn. Microbiol. Infect. Dis. 3:411–417.

Meurman, O., Hanninen, P., Krishna, R.V., and Ziegler, T. 1982. Determination of IgG and IgM class antibodies to mumps virus by

solid-phase enzyme-immunoassay. J. Virol. Meth. 4:249.

Middeldorp, J.M., Hooymans, A.M., Kocken, A.J., van Loon, A.M., Emsbroek, J.A., and Coutinho, R.A. 1987. A sensitive enzyme-linked immunosorbent assay for the detection of herpes simplex virus antigens. J. Virol. Methods 17:159–174.

Miller, H., McCulloch, B., Landini, M.P., and Rossier, E. 1989. Comparison of immunoblotting with other serological methods and virus isolation for the early detection of primary cytomegalovirus infection in allograft recipients. J. Clin. Microbiol. 27:12.

Mills, K.W., Gerlach, E.H., Bell, J.W., Farkas, M.E., and Taylor, R.J. 1978. Serotyping herpes simplex virus isolates by enzyme-linked immunosorbent assays. J. Clin. Microbiol. 7:73.

Miranda, Q.R., Bailey, G.D., Fraser, A.S., and Tenoso, H.J. 1977. Solid phase enzyme immunoassay for herpes simplex virus. J. Infect. Dis. (Supplement) 136:S304.

Modrow, S. and Wolf, H. 1986. Characterization of two related Epstein-Barr virus-encoded membrane proteins that are differentially expressed in Burkitt lymphoma and in vitro-transformed cell lines. Proc. Natl. Acad. Sci. 83:5703–5707.

Morandi, N., Gemme, G., Miletich, F., Ahmed Farah, A., and Terragna, A. 1986. The prevalence of rubella infection in Somilia. Trop. Med. Parasitol. 37:298–299.

Morgan, M.A. and Smith, T.F. 1984. Evaluation of an enzyme-linked immunosorbent assay for the detection of herpes simplex virus antigen. J. Clin. Microbiol. 19:730–732.

Morgan-Capner, P., Davies, E., and Pattison, J.R. 1980. Rubella-specific IgM detection using Sephacryl S-300 gel filtration. J. Clin. Pathol. 32:1082–1085.

Morris, G.E., Coleman, R.M., Best, J.M., Benetato, B.B., and Nahmias, A.J. 1985. Persistence of serum IgA antibodies to herpes simplex, varicella-zoster, cytomegalovirus, and rubella virus detected by enzyme-linked immunosorbent assays. J. Med. Virol. 16:343–349.

Moyner, K. and Michaelsen, T.E. 1988. IgG subclass distribution among antibodies to varicella-zoster virus in human varicella-zoster immunoglobulin preparation and the corresponding donor plasma. J. Biol. Stand. 16:157–164.

Murphy, B.R., Tierney, E.L., Barbour, B.A., Yolken, R.H., Alling, D.W., Holley, H.P., Mayner, R.E., and Chanock, R.M. 1980. Use of enzyme-linked immunosorbent assay to detect serum antibody responses of volunteers who received attenuated influenza A virus vaccines. Infect. Immun. 29:342–347.

Murray, D.L. and Lynch, M.A. 1988. Determination of immune status to measles, rubella, and varicella-zoster viruses among medical students: Assessment of historical information. Am. J. Public Health 78:836–838.

Myrmel, H. 1988. Comparison on tests for heterophile antibodies with a test for specific IgM-antibodies to Epstein-Barr virus. Brief Report. APMIS (Denmark). 96:280–281.

Najberg, G., Jankowski, M., Krotochwil-Skrzypkowa, M., and Gut, W. 1985. Cytomegalovirus in the mononucleosis syndrome in children. Acta. Microbiol. Pol. 34:137–144.

Najem, S.N., Vestergaard, B.F., and Potter, C.W. 1983. Herpes simplex virus type-specific antibodies detected by indirect and competition ELISA. Comparison of sera from patients with carcinoma of the uterine cervix, age matched controls and patients with recurrent genital herpes. Acta. Pathol. Microbiol. Immunol. Scand. 91:205–207.

Nakane, P.K. and Pierce, G.B. 1967. Enzyme-labeled antibodies: Preparation and application for the localization of antigens. J. Histochem. Cytochem. 14:929.

Nakane, P.K. and Kawaoi, A. 1974. Peroxidase-labeled antibody. A new method of conjugation. J. Histochem. Cytochem. 22:1084–1091.

Naot, Y. and Remington, J.S. 1980. An enzyme-linked immunosorbent assay for detection of IgM antibodies to *Toxoplasma gondii*: Use for diagnosis of acute acquired toxoplasmosis. J. Infect. Dis. 142:757–766.

Nassar, N.T. and Touma, H.C. 1986. Brief report: Susceptibility of Filipino nurses to the varicella-zoster virus. Infect. Control. 7:71–72.

Ndumbe, P.M. and Levinsky, R.J. 1985. Immunological cross-reactivities among three herpesviruses. J. Immunol. Methods 83:337–342.

Ndumbe, P.M., Macqueen, S., Holzel, H., and Levinsky, R.J. 1985. Immunity to varicella-zoster virus in a normal adult population. J. Med. Microbiol. 20:105–111.

Nemerow, G.R., Jensen, F.C., and Cooper, N.R. 1982. Neutralization of Epstein-Barr virus by nonimmune human serum. Role of cross-reacting antibody to herpes simplex virus and complement. J. Clin. Invest. 70:1081–1091.

Nemerow, G.R., Siaw, M.F., and Cooper, N.R. 1986. Purification of the Epstein-Barr virus/C3d complement receptor of human B lymphocytes: Antigenic and functional properties of the purified protein. J. Virol. 58:709–712.

Nerurkar, L.S., Jacob, A.J., Madden, D., and Sever, J.L. 1983. Detection of genital herpes simplex infection by a tissue culture fluorescent-antibody technique with biotinavidin. J. Clin. Microbiol. 17:149–154.

Nerurkar, L.S., Namba, M., Brashears, G., Jacob, A.J., Lee, Y.J., and Sever, J.L. 1984. Rapid detection of herpes simplex virus in clinical specimens by use of a capture biotin-streptavidin enzyme-linked immunosorbent assay. J. Clin. Microbiol. 20:109–114.

Nerurkar, L.S., Biggar, R.J., Goedert, J.J., Wallen, W., Becker, P., West, F., Tzan, N., Traub, R., Lee, Y.J., and Botelar, W. 1987. Antiviral antibodies in the sera of homosexual men: Correlation with their lifestyle and drug usage. J. Med. Virol. 21:123–135.

Neumann, P.W., Weber, J.M., Jessamine, A.G., and O'Shaughnessy, M.V. 1985. Comparison of measles antihemolysin test, enzyme-linked immunosorbent assay, and hemagglutination inhibition test with neutralization test for determination of immune status. J. Clin. Microbiol. 22:2.

Neurath, A.R. and Strick, N. 1981. Enzyme-linked fluorescence immunoassays using B-galactosidase and antibodies covalently bound to polystyrene plates. J. Virol. Meth. 3:155–165.

Nilsson, P., Bergquist, N.R., and Grundy, M.S. 1981. A technique for preparing defined conjugates of horseradish peroxidase and immunoglobulin. J. Immunol. Meth. 41:81–93.

Njoo, F.L., Wertheim-van Dillen, P., and Devriese, P.P. 1988. Serology in facial paralysis caused by clinically presumed herpes zoster infection. Arch. Otorhinolaryngol. 245:230–233.

O'Beirne, A.J. and Cooper, H. 1979. Heterogeneous enzyme immunoassays. J. Histochem. Cytochem. 27:1148–1162.

O'Sullivan, M.J. and Marks, V. 1981. Methods for the preparation of enzyme-antibody conjugates for use in enzyme immunoassay. Enzymology 73:147–166.

Paris-Hamelin, A. and Fustec-Ibarbourne, S. 1985. An evaluation of ELISA kits for rubella IgG and IgM antibodies. J. Virol. Methods 10:355–361.

Park, H. 1979. New technique for solid phase immunoassay: Application to hepatitis B surface antigen. Clin. Chem. 25:178.

Parker, J.C., O'Beirne, A.J., and Collins, M.J. 1979. Sensitivity of enzyme-linked immunosorbent assay complement fixation hemagglutination-inhibition serological tests for detection of sendai virus antibody in laboratory mice. J. Clin. Microbiol. 9:444–447.

Paryani, S.G., Arvin, A.M., Koropchak, C.M., Dobkin, M.B., Wittek, A.E., Amylon, M.D., and Budinger, M.D. 1984. Comparison of varicella-zoster antibody titers in patients given intravenous immune serum globulin or varicella-zoster immune globulin. J. Pediatr. 105:200–205.

Pasquini, P., Mele, A., Franco, E., Ippolito, G., and Svennerholm, B. 1988. Prevalence of herpes simplex virus type 2 antibodies in selected population groups in Italy. Eur. J. Clin. Microbiol. Infect. Dis. 7:54–56.

Paxton, H., Meyer, W., Malloy, D., and Firestone, C. 1990. Use of a recombinant G-protein device in infectious disease testing. Am. Clin. Lab. Dec. Jan. pp. 24–25.

Payment, P. and Descoteaux, J-P. 1978. Enzyme-linked immunosorbent assay for the detection of antibodies to pneumonia virus of mice in rat sera. Lab. Anim. Sci. 28:676.

Pearson, G.R. 1988. ELISA tests and monoclonal antibodies for EBV. J. Virol. Methods (Netherlands). 21:97–104.

Persijun, J.P. and Jonker, K.M. 1978. A terminating reagent for the peroxidase-labelled enzyme immunoassay. J. Clin. Chem. Clin. Biochem. 16:531.

Phipps, P.H., Gregoire, L., Rossier, E., and Perry, E. 1983. Comparison of five methods of cytomegalovirus antibody screening of blood donors. J. Clin. Microbiol. 18:6.

Plummer, G. 1973. A review of the identification and titration of antibodies to herpes simplex viruses type 1 and type 2 in human sera. Can. Res. 33:1469–1476.

Prevot, J. and Guesdon, J-L. 1977. Titrage immunoenzymatique des anti-corps IgG et IgM specifiques de la rubeole. Ann. Microbiol. (Inst. Pasteur) 128B:531.

Rand, K.H., Jacobson, D.G., Cottrell, C.R., Koch, K.L., Guild, R.T., and McGuigan, J.E. 1983. Antibodies to herpes simplex type 1 in patients with active duodenal ulcer. Arch. Intern. Med. 143:1917–1920.

Rapicetta, M., Morace, G., Alema, G., and Mancini, G. 1984. Detection of antibodies of the IgG and IgM classes to herpes simplex virus type 1 in cerebrospinal fluid and serum from four patients with herpes encephalitis. Microbiologica. 7:273–277.

Rasmussen, J.M., Marquart, H.V., Rask, R., Jepsen, H.H., and Svehag, S.E. 1988. Quantification of C3dg/Epstein-Barr virus receptors on human B cells and B cell lines. Complement. 5:98–107.

Remington, J.S. and Desmonts, G. 1983. Toxoplasmosin. In Remington, J.S. and Klein, J.E., (eds.), Infectious Disease of the Fetus and Newborn Infant. Philadelphia: W.B. Saunders Co., pp. 143–263.

Revello, M.G., Percivalle, E., Zavattoni, M., and Gerna, G. 1987. Rubella IgM antibody determination: Comparison of two indirect and two capture commercial enzyme immunoassays. Microbiologica 10:393–401.

Rhodes, G., Carson, D.A., Valbracht, J., Houghten, R., and Vaughan, J.H. 1985. Human immune responses to synthetic peptides from the Epstein-Barr nuclear antigen. J. Immunol. 134:211–216.

Rhodes, G., Houghten, R., Taulane, J.P., Carson, D., and Vaughan, J. 1984. The immune response to Epstein-Barr nuclear antigen: Conformational and structural features of antibody binding to synthetic peptides. Mol. Immunol. 21:1047–1054.

Richardson, L.S., Yolken, R.H., Belshe, R.B., Camargo, E., Kim, H.W., and Chanock, R.M. 1978. Enzyme-linked immunosorbent assay for the measurement of serological response to respiratory syncytial virus infection. Infect. Immun. 20:660.

Roberts, G.H. 1989. The many faces of Epstein-Barr virus. Diag. Clin. Test. 27:7.

Robinson, J.E. and Stevens, K.C. 1984. Production of autoantibodies to cellular antigens by human B cells transformed by Epstein-Barr virus. Clin. Immunol. Immunopathol. 33:339–350.

Ross, C., Glorioso, J., Sacks, S., Lavery, C., and Rawls, W.E. 1985. Competitive inhibition by human sera of mouse monoclonal antibody binding to glycoproteins C and D of herpes simplex virus types 1 and 2. J. Virol. 54:851–855.

Rothe, R., Lasch, J., Romer, I., Sandow, D., Hubner, G., Rosmus, K., Kiessig, S.T., and Porstmann, T. 1986. Determination of antibodies to rubella virus with the disperse dye immunoassay (DIA) in comparison with an enzyme-linked immunosorbent assay (ELISA). Biomed. Biochim. Acta. 45:1325–1352.

Ruitenberg, E.J., Steerenbert, P.A., Brosi, B.J.M., and Buys, J. 1974. Serodiagnosis of *Trichinella spiralis* infections in pigs by enzyme-linked immunosorbent assays. Bull. WHO 51:108–109.

Ruitenberg, E.J., Steerenberg, P.A., Brosi, B.J.M., and Buys, J. 1976. Reliability of ELISA as control method for the detection of *Trichinella spiralis* infections in conventionally raised pigs. J. Immunol. Methods 10:67.

Ruitenberg, E.J., Capron, A., Bout, D., and Van Knapen, F. 1978. Enzyme immunoassay for the serodiagnosis of parasitic infections. Biomedicine 26:311.

Sakata, H., Hishiyama, M., and Sugivra, A. 1984. Enzyme-linked immunosorbent assay compared with neutralization tests for eval-uation of live mumps vaccines. J. Clin. Microbiol. 19:21.

Samarai, A.M., Shareef, A.A., Kinghorn, G.R., and Potter, C.W. 1989. Sequential genital infections with herpes simplex virus types 1 and 2. Genitourin Med. 65:39–41.

Sander, J. and Niehaus, C. 1985. Screening for rubella IgG and IgM using an ELISA test applied to dried blood on filter paper. J. Pediatr. 106:457–461.

Saunders, G.C. and Wilder, M.E. 1974. Disease screening with enzyme labelled antibodies. J. Infect. Dis. 129:362.

Saunders, G.C., Clinard, E.H., Bartlett, M.L., and Saunders, W.M. 1977. Application of the indirect enzyme labelled antibody microtest to the detection and surveillance of animal disease. J. Infect. Dis. (Supplement) 136:S258.

Saunders, G.C. 1977. Development and evaluation of an enzyme-labeled antibody (ELA) test for the rapid detection of hog cholera antibodies. Am. J. Vet. Res. 38:21–25.

Scherrer, R. and Bernard, S. 1977. Application of enzyme-linked immunosorbent assay (ELISA) to the detection of calf rotavirus antibodies. Ann. Microbiol. 128A.

Schiff, G.M., Young, B.C., Stefanovic', G.M., Stamler, E.F., Knowlton, D.R., Grundy, B.J., and Dorsett, P.H. 1985. Challenge with rubella virus after loss of detectable vaccine-induced antibody. Rev. Infect. Dis. 7:S157–163.

Schmitz, H., Doerr, H.W., Kampa, D., and Vogt, A. 1977. Enzyme immunoassay for cytomegalovirus antibody. J. Clin. Microbiol. 5:629.

Schoub, B.D., Johnson, S., and McAnerney, J.M. 1985. Prevalence of antibodies to varicella-zoster virus in healthy adults. S. Afr. Med. J. 67:929–931.

Schuurs, A.H.W.M. and Van Weeman, B.K. 1977. Enzyme-immunoassay. Clin. Chim. Acta 81:1–40.

Serdula, M.K., Halstead, S.B., Wiebenga, N.H., and Herrmann, K.L. 1984. Serological response to rubella revaccination. JAMA 251:1974–1977.

Sever, J.L. 1983. Automated systems in viral diagnosis. In M. Cooper, P.H. Hofschneider, H. Koprowski, F. Melchers, R. Rott, H.G. Schweiger, P.K. Vogt, and R. Zinkernagel (eds.), Current Topics in Microbiology, Berlin: Springer-Verlag, pp. 57–75.

Sever, J.L. and Madden, D.L. (eds.). 1977. Enzyme-linked Immunosorbent Assay (ELISA) for infectious agents. J. Infect. Dis. 136:S257–S340.

Shanley, J., Myers, M., Edmond, B., and Steele, R. 1982. Enzyme-linked immunosorbent assay for detection of antibody to varicella-zoster virus. J. Clin. Microbiol. 15:208.

Shehab, Z. and Brunell, P.A. 1983. Enzyme-linked immunosorbent assay for susceptibility to varicella. J. Infect. Dis. 148:472–476.

Shekarchi, I.C., Fuccillo, D.A., Strouse, R., and Sever, J.L. 1987. Capillary enzyme immunoassay for rapid detection of herpes simplex virus in clinical specimens. J. Clin. Microbiol. 25:320–322.

Shekarchi, I.C., Sever, J.L., Ward, L.A., and Madden, D.L. 1982. Microsticks as solid-phase carriers for enzyme-linked immunosorbent assays. J. Clin. Microbiol. 16: 1012–1018.

Shekarchi, I.C., Sever, J.L., Lee, Y.J., Castellano, G., and Madden, D.L. 1984. Evaluation of various plastic microtiter plates with measles, toxoplasma and gamma globulin antigens in enzyme-linked immunosorbent assays. J. Clin. Microbiol. 19:89–96.

Shillitoe, E.J., Greenspan, D., Greenspan, J.S., and Silverman, S. Jr. 1983. Immunoglobulin class of antibody to herpes simplex virus in patients with oral cancer. Cancer. 51:65–71.

Shimakage, M., Ikegami, N., Chatani, M., Yoshino, K., and Sato, T. 1987. Serological follow-up study on the antibody levels to Epstein-Barr virus-determined nuclear antigen (EBNA) patients with nasopharyngeal carcinoma (NPC) after radiation therapy. Biken. J. 30:45–51.

Siegel, J.P. and Remington, J.S. 1983. Comparison of methods for quantitating antigen specific immunoglobulin M antibody with a reverse enzyme-linked immunosorbent assay. J. Clin. Microbiol. 18:63–70.

Singh, V.K., Tingle, A.J., and Schulzer, M. 1986. Rubella-associated arthritis. II. Relationship between circulating immune complex levels and joint manifestations. Ann. Rheum. Dis. 45:115–119.

Skar, A.G., Middeldorp, J., Gundersen, T., Rollag, H., and Degre, M. 1988. Rapid diagnosis of genital herpes simplex infection by an indirect ELISA method. NIPH Ann. 11:59–65.

Skendzel, L.P. and Edson, D.C. 1985. Evaluation of enzyme immunosorbent rubella assays. Arch. Pathol. Lab. Med. 109:391–393.

Skurrie, I.J. and Gilbert, G.L. 1983. Enzyme-linked immunosorbent assay for rubella immunoglobulin G: New method for attachment of antigens to microtiter plates. J. Clin. Microbiol. 17:738–743.

Skvaril, F. and Schilt, U. 1984. Characterization of the subclasses and light chain types of IgG antibodies to rubella. Clin. Exp. Immunol. 55:671–676.

Smith, K.J., Ashley, C.R., Darville, J.M., Harbour, J., and Roome, A. 1984. Comparison of a commercial ELISA system with restriction endonuclease analysis for typing herpes simplex virus. J. Clin. Pathol. 37:937–941.

Smith, K.O. and Gehle, W.D. 1977. Magnetic transfer devices for use in solid-phase radioimmunoassays and enzyme-linked immunosorbent assays. J. Infect. Dis. 136: S329–S336.

Stagno, S., Tinker, M.K., Elrod, C., Fuccillo, D.A., Cloud, G., and O'Beirne, A.J. 1985. Immunoglobulin M antibodies detected by enzyme-linked immunosorbent assay and radioimmunoassay in the diagnosis of cytomegalovirus infections in pregnant women and newborn infants. J. Clin. Microbiol. 21:930.

Steece, R.S., Talley, M.S., Skeels, M.R., and Lanier, G.A. 1984. Problems in determining immune status in borderline specimens in an enzyme immunoassay for rubella immunoglobulin G antibody. J. Clin. Microbiol. 19:923–925.

Steece, R.S., Talley, M.S., Skeels, M.R., and Lanier, G.A. 1985. Comparison of enzyme-linked immunosorbent assay, hemagglutination inhibition, and passive latex agglutination for determination of rubella immune status. J. Clin. Microbiol. 21:140–142.

Steinitz, M., Tamir, S., Sela, S.B., and Rosenmann, E. 1988. The presence of non-isotype-specific antibodies in polyclonal anti-IgE reagents: Demonstration of their binding to specifically selected Epstein-Barr virus-transformed cell lines. Cell Immunol. 113:10–19.

Sternas, L., Eliasson, L., Lerner, R., and Klein, G. 1986. Quantitation of Epstein-Barr virus (EBV)-determined nuclear antigen (EBNA) by a two-site enzyme immunoassay, in parallel with EBV-DNA. J. Immunol. Methods 89:151–158.

Sternas, L., Luka, J., Kallin, B., Rosen, A., Henle, W., Henle, G., and Klein, G. 1983. Enzyme-linked immunosorbent assay for the detection of Epstein-Barr virus induced antigens and antibodies. J. Immunol. Methods 63:171–185.

Sugimoto, T., Sakane, Y., and Kobayashi, Y. 1985. Early diagnosis of herpes simplex virus encephalitis in childhood by the enzyme-linked immunosorbent assay. Acta. Neurol. Scand. 71:359–363.

Sullender, W.M., Yasukawa, L.L., Schwartz, M., Pereira, L., Hensleigh, P.A., Prober, C.G., and Arvin, A.M. 1988. Type-specific antibodies to herpes simplex virus type 2 (HSV-2) glycoprotein G in pregnant women, infants exposed to maternal HSV-2 infection at delivery, and infants with neonatal herpes. J. Infect. Dis. 157:164–171.

Sumaya, C.V. 1986a. Epstein-Barr virus serologic testing: Diagnostic indications and interpretations. Pediatr. Infect. Dis. 5:337–342.

Sumaya, C.V. 1986b. Infectious mononucleosis and other EBV infections: Diagnostic factors. Lab. Mgt. Oct.

Sundqvist, V.A. 1982. Frequency and specificity of varicella-zoster virus IgM response. J. Virol. Methods 5:219–227.

Sundqvist, V.A., Linde, A., and Wahren, B. 1984. Virus-specific immunoglobulin G subclasses in herpes simplex and varicella-zoster virus infections. J. Clin. Microbiol. 20:94–98.

Swanston, W., Mahony, J., McLaughlin, B.,

and Chernesky, M. 1986. Assessment of serologic markers for Epstein-Barr virus. Diag. Microbiol. Dis. Vol. 5.

Taylor-Weideman, J., Brunell, P.A., Geiser, C., Shehab, Z.M., and Frierson, L.S. 1986. Effect of transfusions on serologic testing for antibody to varicella. Med. Pediatr. Oncol. 14:316–318.

Taylor-Wiedeman, J., Yamashita, K., Miyamura, K., and Yamazaki, S. 1989. Varicella-zoster virus prevalence in Japan: No significant change in a decade. Jpn. J. Med. Sci. Biol. 42:1–11.

Tellez, P.A., Odom, L., and Hayward, A.R. 1985. Immunity to herpes simplex virus in children receiving treatment for acute lymphoblastic leukemia (ALL). Clin. Exp. Immunol. 62:525–529.

Thomas, H.I. and Morgan-Capner, P. 1988a. Specific IgG subclass antibody in rubella virus infections. Epidemiol. Infect. 100:443–454.

Thomas, H.I. and Morgan-Capner, P. 1988b. Rubella-specific IgG subclass avidity ELISA and its role in the differentiation between primary rubella and rubella reinfection. Epidemiol. Infect. 101:591–598.

Thongkrajai, P., Pothinam, S., and Hahnvajanawong, C. 1986. A rapid detection of herpes simplex virus by ELISA in asymptomatic pregnant women and neonates. Southeast Asian J. Trop. Med. Public Health 17:91–95.

Thraenhart, O. and Kuwert, E.K. 1977. Enzyme immunoassay for demonstration of rabies-virus antibodies after immunization. Lancet 2:399.

Thresh, J.M., Adams, A.N., Barbara, D.J., and Clark, M.F. 1977. The detection of three viruses of hop (*humulus lupulus*) by enzyme-linked immunosorbent assay (ELISA). Ann. Appl. Biol. 87:57.

Tinker, M.K., Progar, N.A., Fuccillo, D.A., and O'Beirne, A.J. 1983. Enzyme-linked immunosorbent assay (ELISA) to determine immune status for varicella. ASM. Poster Session.

Tomita, H., Tanaka, M., Kukimoto, N., and Ikeda, M. 1988. An ELISA study on varicella-zoster virus infection in acute peripheral facial palsy. Acta. Otolaryngol. Suppl. (Stockh) 446:10–16.

Trlifajova, J., Kyselova, M., Bruckova, M., and Svandova, E. 1986. Rapid assay for antibodies to varicella-zoster virus (VZ) using a VZV-HEM preparation. J. Hyg. Epidemiol. Microbiol. Immunol. 30:171–176.

Truant, A.L., Barksdale, B.L., Huber, T.W., and Elliott, L.B. 1983. Comparison of an enzyme-linked immunosorbent assay with indirect hemagglutination and hemagglutination inhibition for determination of rubella virus antibody: Evaluation of immune status with commercial reagents in a clinical laboratory. J. Clin. Microbiol. 17:106–108.

Uen, W.C., Luka, J., and Pearson, G.R. 1988. Development of an enzyme-linked immunosorbent assay (ELISA) for detecting IgA antibodies to the Epstein-Barr virus. Int. J. Cancer. 41:479–482.

Ukkonen, P., Koistinen, V., and Penttinen, K. 1977. Enzyme immunoassay in the detection of hepatitis B surface antigen. J. Immun. Methods 15:343.

Ukkonen, P., Granstrom, M.L., and Penttinen, K. 1981. Mumps specific immunoglobulin M and G antibodies in natural mumps infection as measured by enzyme-linked immunosorbent assay. J. Med. Virol. 8:131.

Van de Perre, P., Clumeck, N., Steens, M., Zissis, G., Carael, M., Lagasse, R., De Wit, S., Lafontaine, T., De Mol, P., and Butzler, J.P. 1987. Seroepidemiological study on sexually transmitted diseases and hepatitis B in African promiscuous heterosexuals in relation to HTLV-III infection. Eur. J. Epidemiol. 3:14–18.

Vandervelde, E.M., Cohen, B.J., and Cossart, Y.E. 1977. An enzyme-linked immunosorbent assay test for hepatitis B surface antigen. J. Clin. Pathol. 30:714.

Vandervelde, E.M. 1978. Enzyme-linked immunoassay (ELISA). Its practical application to the diagnosis of hepatitis B. J. Med. Virol. 3:17.

Vandvik, B., Skoldenberg, B., Forsgren, M., Stiernstedt, G., Jeansson, S., and Norrby, E. 1985. Long-term persistence of intrathecal virus-specific antibody responses after herpes simplex virus encephalitis. J. Neurol. 231:307–312.

van Loon, A.M., Van der Logt, J.T., Heessen, F.W., and Van der Veen, J. 1985. Use of enzyme labeled antigen for the detection of immunoglobulin M and A antibody to herpes simplex virus in serum and cerebrospinal fluid. J. Med. Virol. 15:183–195.

van Loon, A.M., van der Logt, J.T., Heessen, F.W., Postma, B., and Peeters, M.F. 1989. Diagnosis of herpes simplex virus encephalitis by detection of virus-specific immunoglobulins A and G in serum and cerebrospinal fluid by using an antibody-capture enzyme-linked immunosorbent assay. J. Clin. Microbiol. 27:1983–1987.

van Ulsen, J., Dumas, A.M., Wagenvoort, J.H., van Zuuren, A., van Joost, T., and Stolz, E. 1987. Evaluation of an enzyme immunoassay for detection of herpes simplex virus antigen in genital lesions. Eur. J. Clin. Microbiol. 6:410–413.

van Weeman, B.K. and Schuurs, A.H.W.M. 1971. Immunoassay using haptoenzyme conjugates. FEBS Lett. 24:77–81.

van Weemen, B.K. and Schuurs, A.H.W.M. 1974. Immunoassay using antibody-enzyme conjugates. FEBS Lett. 43:215.

Vejtorp, M. 1978. Enzyme-linked immunosorbent assay for determination of rubella IgG antibodies. Acta. Pathol. Microb. Scand., Sect. B, 86:387.

Venables, P.J., Pawlowski, T., Mumford, P.A., Brown, C., Crawford, D.H., and Maini, R.N. 1988. Reaction of antibodies to rheumatoid arthritis nuclear antigen with a synthetic peptide corresponding to part of Epstein-Barr nuclear antigen 1. Ann. Rheum. Dis. 47:270–279.

Venkitaraman, A.R., Seigneurin, J.M., Baccard, M., Lenoir, G.M., and John, T.J. 1984. Measurement of antibodies to varicella-zoster virus in a tropical population by enzyme linked immunosorbent assay. J. Clin. Microbiol. 20:582–583.

Voevodin, A.F. and Pacsa, A.S. 1983. Enzyme-linked immunosorbent assay for detection of antibodies to Epstein-Barr virus antigens. Acta. Microbiol. Hung. 30:125–129.

Voller, A., Bidwell, D.E., Huldt, G., and Engvall, E. 1974. A microplate method of enzyme-linked immunosorbent assay and its application to malaria. Bull. WHO 51: 209.

Voller, A. and Bidwell, D.E. 1975. A simple method for detecting antibodies to rubella. Br. J. Exp. Pathol. 56:338.

Voller, A. and Bidwell, D.E. 1976. Enzyme immunoassays for antibodies in measles, cytomegalovirus infections and after rubella vaccination. Br. J. Exp. Pathol. 57:243.

Voller, A., Bartlett, A., Bidwell, D.E., Clark, M.F., and Adams, A.W. 1976. The detection of plant viruses by enzyme-linked immunosorbent assay (ELISA). J. Gen. Virol. 33:165.

Voller, A., Bidwell, D.E., and Bartlett, A. 1977a. The Enzyme-Linked Immunosorbent Assay (ELISA). Guernsey: Flow-line Publications, pp. 1–48.

Voller, A., Bidwell, D.E., Bartlett, A., and Edwards, R. 1977b. A comparison of isotopic and enzyme immunoassays for tropical parasitic diseases. Trans. R. Soc. Med. Hyg. 71:431.

Voller, A., Bartlett, A., and Bidwell, E. 1978. Enzyme immunoassays special reference to ELISA techniques. J. Clin. Pathol. 31:507.

Voller, A., Bidwell, D.E., and Bartlett, A. 1980. Microplate enzyme immunoassays for the immunodiagnosis of virus infections. In N.R. Rose and H. Friedman (eds.), Manual of Clinical Immunology. Washington, D.C.: American Society of Microbiologists, pp. 359–371.

Voller, A., Bidwell, D.E., and Bartlett, A. 1982. ELISA techniques in virology. In C.R. Howard (ed.), New Developments in Practical Virology. New York: Alan R. Liss. pp. 59–81.

Vroman, B., Luka, J., Rodriguez, M., and Pearson, G.R. 1985. Characterization of a major protein with a molecular weight of 160,000 associated with the viral capsid of Epstein-Barr virus. J. Virol. 53:107–113.

Wallen, W.C., Mattison, J.M., and Levine, P.H. 1977. Detection of soluble antigen of Epstein-Barr virus by the enzyme-linked immunosorbent assay. J. Infect. Dis. (Supplement) 136:S324.

Walls, K.W., Bullock, S.L., and English, D.K. 1977. Use of the enzyme-linked immunosorbent assay (ELISA) and its micro-adaptation for the serodiagnosis of toxoplasmosis. J. Clin. Microbiol. 5:273.

Wardley, R.C. and Crowther, J.R. (eds.). 1982. The ELISA: Enzyme-linked immunosorbent assay in veterinary research and diagnosis. (Current Topics in Veterinary Medicine and Animal Science, Vol 22). Hague: Martinus Nijhoff Publishers.

Warford, A.L., Levy, R.A., Rekrut, K.A., and Steinberg, E. 1986. Herpes simplex virus testing of an obstetric population with an antigen enzyme linked immunosorbent assay. Am. J. Obstet. Gynecol. 154:21–28.

Webb, C.F., Cooper, M.D., Burrows, P.D., and Griffin, J.A. 1985. Immunoglobulin gene rearrangements and deletions in human Epstein-Barr virus-transformed cell lines producing different IgG and IgA subclasses. Proc. Natl. Acad. Sci. USA 82:5495–5499.

Wei, R., Knight, G.J., Zimmerman, D.H., and Bond, H.E. 1977. Solid-phase enzyme immunoassay for hepatitis B surface antigen. Clin. Chem. 23:813.

Weigle, K.A., Murphy, M.D., and Brunell, P.A. 1984. Enzyme-linked immunosorbent assay for evaluation of immunity to measles virus. J. Clin. Microbiol. 19:376.

Wielaard, F., Denissen, A., Van Elleswijk-V.D., Berg, J., and Van Gemert, G. 1985. Clinical validation of an antibody-capture anti-rubella IgM-ELISA. J. Virol. Methods. 10:349–354.

Wielaard, F., Denissen, A., van der Veen, L., and Rutjes, I. 1987. A sol-particle immunoassay for determination of anti-rubella antibodies: Development and clinical validation. J. Virol. Methods 17:149–158.

Wielaard, F., Scherders, J., Dagelinckx, C., Middeldorp, J.M., Sabbe, L.J., and Van Belzen, C. 1988. Development of an antibody-capture IgM enzyme-linked immunosorbent assay for diagnosis of acute Epstein-Barr virus infections. J. Virol. Methods 21:105–115.

Wilson, M.B. and Nakane, P.K. 1978. Recent developments in the periodate method of conjugating horseradish peroxidase. In W. Knapp, K. Holubar, and G. Wick (eds.), Immunofluorescence and Related Staining

Techniques. Elsevier/North Holland: Biomedical Press. pp. 215–224.

Wittenburg, R.A., Roberts, M.A., Elliott, L.B., and Little, L.M. 1984. Comparative evaluation of commercial rubella virus antibody kits. Am. Soc. Microbiol. 21:161.

Wolf, H., Motz, M., Kuhbeck, R., Seibl, R., Jilg, W., Bayliss, G.J., Barrell, B., Golub, E., Zeng, Y., and Gu, S.Y. 1984. Strategies for the economic preparation of Epstein-Barr virus proteins of diagnostic and protective value by genetic engineering: A new approach based on segments of virus-encoded gene products. IARC Sci. Publ. 63: 525–539.

Wolters, G., Kuipers, L.P.C., and Schuurs, A.H.W.M. 1975. Enzyme immunoassay (EIA) for HBsAg in microtiter plates. Hepatitis Scientific Memoranda H. 908:1.

Wolters, G., Juijpers, L., Kacaki, J., and Schuurs, A.H.W.M. 1976. Solid phase enzyme immunoassay for detection of hepatitis B surface antigen. J. Clin. Pathol. 29: 873.

Wolters, G., Kuipers, L.P.C., Kacaki, J., and Schuurs, A.H.W.M. 1977. Enzyme-linked immunosorbent assay for hepatitis B surface antigen. J. Infect. Dis. (Supplement) 136:S311.

Wreghitt, T.G., Tedder, R.S., Nagington, J., and Ferns, R.B. 1984. Antibody assays for varicella-zoster virus: Comparison of competitive enzyme-linked immunosorbent assay (ELISA), competitive radioimmunoassay (RIA), complement fixation, and indirect immunofluorescence assays. J. Med. Virol. 13:361–370.

Wu, T.C., Zaza, S., and Callaway, J. 1989. Evaluation of the DuPont HERPCHEK herpes simplex virus antigen test with clinical specimens. J. Clin. Microbiol. 27:1903–1905.

Yolken, R.H., Wyatt, R.G., Barbour, B.A., Kim, H.W., Kapikian, A.Z., and Chanock, R.M. 1978. Measurement of rotavirus antibody by an enzyme-linked immunosorbent assay blocking assay. J. Clin. Microbiol. 8:283.

Yolken, R. and Stopa, P. 1979. Enzyme-linked fluorescence assay: Ultrasensitive solid phase assay for detection of human rotavirus. J. Clin. Microbiol. 10:317–321.

Yolken, R.H. 1980. Enzyme-linked immunosorbent assay (ELISA): A practical tool for rapid diagnosis of viruses and other infectious agents. Yale J. Biol. Med. 53:85–92.

Yolken, R., Torch, V., Berg, R., Murphy, B., and Lee, Y.C. 1980. Fluorometric assay for measurement of viral neuraminidase-application to the rapid detection of influenza virus in nasal wash specimens. J. Infect. Dis. 142:516–523.

Yolken, R.H. and Leister, F.J. 1981. Enzyme immunoassays for measurement of cytomegalovirus immunoglobulin M antibody. J. Clin. Microbiol. 14:427.

Yoosook, C., Rimdusit, P., Chantratita, W., Leechanachai, P., and Bhattarakosol, P. 1987. Evaluation of biotin-streptavidin enzyme-linked immunosorbent assay for detection of genital herpes simplex virus infection. Asian Pac. J. Allergy Immunol. 5:143–148.

Zachar, V., Mayer, V., and Schmidtmayerova, H. 1988. Varicella-zoster virus IgG antibodies during primoinfection in competent and transfer factor modulated immunocompromised host: Comparison of three indirect assays. Acta. Virol. 32:243–251.

Zerbini, M., Plazzi, M., and Musiani, M. 1985. Rapid neutralization assay for human cytomegalovirus antibody. J. Clin. Microbiol. 21:969.

Ziaie, Z., Friedman, H.M., and Kefalides, N.A. 1986. Suppression of matrix protein synthesis by herpes simplex virus type 1 in human endothelial cells. 1986. Coll. Relat. Res. 6:333–349.

Ziegler, T. 1984. Detection of varicella-zoster viral antigens in clinical specimens by solid phase enzyme immunoassay. J. Infect. Dis. 150:149–154.

Ziegler, T. and Halonen, P.E. 1985. Rapid detection of herpes simplex and varicella-zoster virus antigens from clinical specimens by enzyme immunoassay. Antiviral Res. Suppl 1:107–110.

10

Immunoperoxidase Detection of Viral Antigens in Cells

Helen Gay, John J. Docherty, and
Joseph E. Kall, Jr.

10.1 INTRODUCTION

Immunocytochemical staining, a sensitive and specific method to detect localized antigens with labeled antibodies, has been used extensively to detect viral antigens. Coons introduced immunocytochemistry in 1942 by using fluorescein labeled anti-pneumococcal III antibody to detect pneumococcal antigens in the livers and the spleens of experimentally infected mice (Coons et al, 1941, 1942). In the following years, immunofluorescence (IF) was used to detect many different bacterial and viral antigens in vivo and in vitro (Coons et al, 1950; Kaplan et al, 1950; Kurstak, 1971). Information gathered from such studies advanced our knowledge of the structure and function of viral proteins, transforming mechanisms of viruses, and viral replication cycles. Indeed, IF continues to be used successfully in a variety of ways, in both the research and the clinical laboratory.

As with any procedure there has been a continuing quest to improve on the original procedures of Coons. By the mid 1960s a procedure began to emerge that used the basic methodology of previously described fluorescent methods with the fluor being replaced by an enzyme. When used to detect virus antigens, a precipitate formed marking the location of the antibody-antigen reaction. This immunoenzymatic procedure, which was reported by Ram et al (1966), used acid phosphatase conjugated antibodies to localize tissue antigens. Because of rapid loss of enzymatic activity of that antibody-enzyme conjugate (Nakane and Pierce, 1967), acid phosphatase was replaced with the more stable enzyme horseradish peroxidase (HRP). Introduced by Nakane and Pierce, immunoperoxidase staining involves reacting the antigen with HRP-conjugated antibody, then visualizing the complex by exposing the tissue to a solution containing appropriate electron transfer substrates, such as 3,3'-diaminobenzidine tetrahydrochloride (DAB) and hydrogen peroxide (Nakane and Pierce, 1967). The bound enzyme first reduces the hydrogen peroxide to water and then oxidizes DAB. The resulting oxidized DAB polymerizes to form a nondiffusable, insoluble dark brown precipitate that localizes at the site of the antigen (Pearse, 1972). This immunoenzymatic staining method overcame many of the weaknesses of IF. The peroxidase stained preparations, once mounted, are essentially permanent, can be viewed with an ordinary light microscope, and avoid cellular autofluorescence (Spendlove, 1967). The peroxidase enzyme is available in pure form, is stable after chemical conjugation, and because it is small it readily penetrates tissue (Sternberger, 1974). The practical advantages offered by the immunoperoxidase staining method have made it an attractive procedure for the detection of a wide variety of viral antigens. The use of peroxidase does have one major drawback. This enzyme is endogenous to some mammalian tissues (Straus, 1971; Streefkerk, 1972; Wier et al, 1974; Blain et al, 1975), which results in nonspecific staining that can interfere with the interpretation of the immunoperoxidase staining test results. This problem can be solved by eliminating endogenous peroxidase by pretreatment of tissues with peroxidase inactivating reagents (Straus, 1971, 1979; Streefkerk, 1972; Weir et al, 1974). Most of the endogenous enzyme activity is eliminated by this treatment, but some investigators have reported that some of these treatments also damage antigens (Fink et al, 1979; Straus, 1979).

During the years of the refinement of both the IF and the immunoperoxidase staining techniques, improvements were introduced that greatly increased the sensitivity of these staining methods. Coons had first introduced IF as a direct stain in which the fluorochrome was directly attached to the specific- or primary-antibody (Coons et al, 1941, 1942). Twelve years later, Weller and Coons showed that IF staining sensitivity could be increased tenfold if the tissue was first treated with unlabeled specific antibody and then treated with fluorescein labeled anti-immunoglobulin (e.g., anti-IgG) antibody (Weller and Coons , 1954). Through this indirect staining, several anti-IgG molecules could attach to a single bound primary antibody, thus greatly amplifying the fluorescent signal and thereby detection of the antigen (Sternberger,

1974). In addition, indirect staining made the technique more general because a single preparation of fluorochrome labeled anti-IgG could be used to detect antibodies raised in the same species to a variety of different antigens, avoiding the necessity of labeling each specific antibody.

The indirect staining method has been applied to immunoperoxidase staining to detect a wide variety of viruses including: canine distemper virus (Higgins et al, 1982), enteric adenovirus (Cevenini et al, 1984), enteroviruses (Herrmann et al, 1974), hepatitis B virus (HBV) associated delta antigen (Recchia et al, 1981), hepatitis B surface antigen (Hsu et al, 1983), human cytomegalovirus (CMV; Gerna et al, 1976), mouse mammary tumor virus (Keydar et al, 1978), papillomavirus antigen (Kurman et al, 1981; Syrjanen et al, 1982), rotavirus (Chasey, 1980; Cervenini et al, 1984; Grom et al, 1985), yellow fever virus (de la Monte et al, 1983), Epstein-Barr virus (Musiani et al, 1986), influenza virus type A (Gardner et al, 1978), and respiratory syncytial virus (RSV; Gardner et al, 1978). Two technical problems have come to light regarding this method. First, the chemical conjugation process inactivates some of the HRP that is coupled to the antibody; second, some of the antibody remains unconjugated. Both the unconjugated antibody and the antibody conjugated to inactive enzyme compete with antibody molecules labeled with active enzyme for the antigenic sites and, thus, decrease the sensitivity of the assay.

In 1970, Sternberger et al (1970) introduced an immunoperoxidase procedure that required no chemical conjugation of enzyme to antibody. Peroxidase was bound to an anti-peroxidase antibody via specific antibody-antigen interactions. Viral antigens were localized by this unlabeled antibody-enzyme technique, schematically presented in Figure 10.1, when the specimen was treated sequentially with the following reagents:

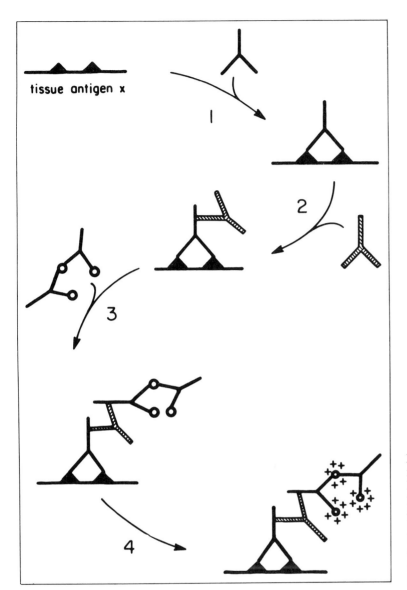

tissue antigen x

Figure 10.1 Schematic representation of the un-labeled peroxidase anti-peroxidase method of antigen localization. (1) Primary antibody; (2) bridge antibody; (3) peroxidase anti-peroxidase complexes; (4) DAB and peroxide.

1. Antibody to a specific antigen raised in species A (primary antibody)

2. Anti-species-A serum raised in species B (bridge antibody) applied in sufficient excess that one binding site of the antibody attaches to the primary antibody and the other binding site remains free

3. A purified antibody-enzyme complex made up of anti-HRP raised in species A that had been combined with HRP

4. DAB and peroxide

This method, known as the unlabeled peroxidase anti-peroxidase (PXAPX) technique, is 20 times more sensitive than the indirect immunoperoxidase staining method and 100 to 1000 times more sensitive than IF (Pearson et al, 1979). Sensitivity is greatly increased because each viral antigen is attached to at least one, if not more, active enzyme molecules.

The PXAPX technique has been used to detect a wide variety of viruses including

Table 10.1 Applications of Unlabeled Peroxidase Anti-Peroxidase Technique to Detect Viral Antigens

Virus	References
Canine Distemper Virus	Ducatelle et al, 1980
Cytomegalovirus	Jiwa et al, 1990
Dengue-2 Virus	Churdboonchart et al, 1984
Hepatitis B Virus	Machado-Vieira et al, 1982
Surface antigen	
Core antigen	
Herpes Simplex Virus	Adams et al, 1984
Type 1 or 2	Flaitz et al, 1988
	Kurchak et al, 1977
	Pearson et al, 1979
Paramyxovirus	Schwendemann et al, 1982
Transmissable Gastroenteritis Virus	Chu et al, 1982

canine distemper virus (Ducatelle et al, 1980), cytomegalovirus (Jiwa et al, 1990), dengue-2 virus (Churdboonchart et al, 1984), hepatitis B virus (Machado-Vieira et al, 1982), herpes simplex virus (HSV; both HSV type 1 and HSV type 2) (Kurchak et al, 1977; Pearson et al, 1979; Adams et al, 1984; Flaitz et al, 1988), paramyxovirus (Schwendemann et al, 1982), and transmissable gastroenteritis virus (Chu et al, 1982). (See Table 10.1).

We have used the PXAPX technique routinely to detect HSV in either tissue culture or tissue scrapings from human genitalia. As proof of this method's increased sensitivity, we have successfully been able to detect a single HSV polypeptide in infected cells using polyclonal primary antibody prepared against that single viral peptide. In this chapter, we will provide the reader with our insights and experience using immunoenzymatic PXAPX staining to detect viral antigens. It is not our intent to belabor the pros and cons of one procedure over another, but rather to describe the PXAPX method that is routinely used in our laboratory. The procedure outlined can serve as a foundation and may be adapted and modified to the specifications of any laboratory or virus, as conditions dictate.

10.2 STAINING CONSIDERATIONS

Successful staining by the PXAPX method requires consideration of several critical factors. One of the most important considerations is the choice of tissue fixative. Detection of the antigens cannot occur by any immunocytochemical method if they have been destroyed or significantly altered by the fixative. An appropriate yet effective fixative for the antigen(s) to be examined must first be determined and may vary for different viruses. Secondly, optimum concentrations of the three different antisera used must be established, regardless of whether the sera were prepared in the research laboratory or obtained from a commercial source. Establishment of appropriate dilutions is carried out by block titration of the three antisera and recording those combinations that produce optimum results. Once approximate values are obtained we frequently adjust to final concentrations by titrating one of the three (frequently the primary antibody) while holding the other two constant and subjectively evaluating the quality of the stain for optimum results. Optimum results are defined as maximum staining intensity of the antigen with minimal staining of tissue that does not contain the antigen. This may be repeated for the bridging antibody and the anti-peroxidase antibody. The dilutions and incubations used in the procedure detailed below were optimized for the detection of HSV-2 in acetone (Pearson et al, 1979) or 70% ethanol (Steinkamp and Crissman, 1974) fixed tissue culture cells or specimens obtained from patients. The staining procedure is easily adaptable to other viral systems, and the specific dilutions and incubation times described here can be used as a starting point toward optimizing the procedure for other viruses.

10.3 MATERIALS FOR IMMUNOPEROXIDASE STAINING

The materials listed below are for the PXAPX staining system in which the antiviral serum and anti-peroxidase serum was produced in rabbits and the bridge antirabbit antibody was produced in a goat. Other combinations of hosts can be used but it is essential that consistency be maintained, i.e., that the two hosts are not too closely related, and that the anti-viral and anti-peroxidase antibodies have been raised in the same host. Normal, nonimmune serum of the host in which the bridge antibody was raised (in this case, goat) is used to block nonspecific binding of antibodies and is layered on the samples before addition of each of the three different antibodies. The nonimmune serum contains an insignificant amount of crossreacting antibody that does not interfere with the specific binding of any of the antibodies.

1. Sera
   - Nonimmune goat serum (NGS): prepared in our laboratory but can be commercially obtained
   - Anti-viral serum raised in rabbit: pre-

pared in our laboratory but can be commercially obtained
   - Preimmune serum from rabbit: prepared in our laboratory but can be commercially obtained
   - Anti-rabbit serum raised in goat: commercially obtained, antibody protein concentration is 15 mg/mL
   - Rabbit peroxidase/anti-peroxidase (PXAPX): commercially obtained, antibody protein concentration is 3 mg/mL
2. Phosphate buffered saline (PBS) (0.01M sodium phosphate pH 7.2; 0.15M NaCl)
3. Absolute methanol
4. Hydrogen peroxide (fresh)
5. 3,3'-diaminobenzidine tetrahydrochloride (DAB)
6. Absolute ethanol, 90% ethanol and 70% ethanol
7. Xylene
8. Permount

10.4 METHODS

Phosphate buffered saline is used throughout for all dilutions and washes. All procedures are performed at room temperature (~22 °C).

1. If samples are immersed in fixative, rinse with PBS (5 minutes) before beginning to stain; this is not necessary if samples have been previously fixed and air-dried. (We routinely fix in acetone at 4 °C for 5 to 10 minutes, air dry, and store desiccated at 4 °C for up to 21 days before staining.)
2. Cover samples with a few drops of a 1:30 dilution of NGS for 5 minutes; if the samples are dry at the beginning of the staining procedure prewet by dipping into PBS before adding NGS.
3. Drain NGS by tipping the sample onto a paper towel at about a 90° angle; this

removes the liquid that has accumulated at the edges of the coverslip or slide.

4. Overlay samples with a few drops of an appropriate dilution of anti-viral serum and incubate for one hour in a humid chamber; the optimum dilution for anti-HSV-2 serum raised in rabbits in this laboratory has ranged from 1:200 to 1:2000. (Controls include the following: positive control—virus infected cells plus immune serum; negative control—virus infected cells plus preimmune serum, virus infected cells treated with PBS in place of the primary antibody, uninfected cells treated with immune serum.)

5. Wash samples twice in PBS (5 minutes each).

6. Cover samples with a few drops of diluted NGS for 5 minutes, at room temperature, then drain.

7. Overlay each sample with a few drops of a 1:10 dilution of goat anti-rabbit serum; incubate for 20 minutes and drain.

8. Wash samples twice in PBS (5 minutes each).

9. Immerse samples in absolute methanol for 15 minutes.

10. Wash samples three times in PBS (5 minutes each).

11. Cover samples with a few drops of diluted NGS for 5 minutes, then drain.

12. Apply a few drops of rabbit PXAPX diluted 1:100 in 1% NGS-PBS (1 mL NGS + 99 mL PBS) for 20 minutes, then drain.

13. Wash samples twice in PBS (5 minutes each).

14. Stain samples by immersing them in the disclosing reaction for 7 minutes while protected from light; the disclosing reaction consists of 0.025% DAB in PBS, freshly prepared and filtered (0.45 micron filter) to which hydrogen peroxide is added to a final concentration of 0.005%; allow the DAB to be stirred rapidly for at least 30 minutes (in the dark) before filtering and adding hydrogen peroxide.

15. Wash stained samples twice in PBS (5 minutes each).

16. Mount the samples. (Dehydrate the stained samples by immersing them sequentially for 1 minute in 70% ethanol, 90% ethanol, absolute ethanol, and then xylene; add a drop of permount and place a glass coverslip on top.)

10.5 RESULTS

Immunoperoxidase stained samples contain dark brown deposits at the site of the antigen (if DAB has been used as a substrate) and very little staining elsewhere [Figure 10.2(A) and (C)]. Control specimens, in which preimmune serum or PBS was used in place of the primary serum, will often show some diffuse light-brown nonspecific staining throughout the cell or tissue specimen [Figure 10.2(B) and (D)]. As demonstrated in Figure 10.2, the immunoperoxidase stain worked equally well in detecting HSV-2 antigens in either HEp-2 or VERO cells. The intensity of the stain can be regulated to varying degrees by the length of time the sample is in the disclosing reaction. As the staining time is increased, however, DAB can be oxidized by atmospheric oxygen, resulting in progressive deposition of product over the entire sample. This reduces the contrast between stained and unstained areas in the sample and may make interpretation of the results difficult. Manipulation of the DAB and H_2O_2 concentrations can result in faster or slower substrate conversion and is frequently adjusted in a subjective manner to meet the demands or constraints of a particular system. It is possible, however, to overstain the sample so that normal tissue is intensely stained and cannot be distinguished from infected tissue.

An added attraction of the PXAPX method is that it can easily be interfaced with other staining procedures that may assist in viewing specific virus-induced cellular changes. In this regard, we have successfully interfaced the procedure with the Papanicolaou (PAP) stain by merely taking the sample through a standard PAP procedure after it was stained by PXAPX (Pearson et al, 1979). In Figure 10.3(A) and (B) the combination PXAPX-PAP procedure was used to stain HSV-2 infected VERO cells [Figure 10.3(A)] and the genital specimen from a female patient with a confirmed HSV-2 infection [Figure 10.3(B)]. Although the black and white photograph does not readily reveal color differences, the HSV-2 PXAPX-positive cells are stained dark brown [Figure 10.3(A), (B), arrows]. The uninfected cells in the human specimen, PAP stained, are light orange, blue, or pink with dark blue-purple nuclei. In the center of Figure 10.3(A) is an HSV syncytium that is characterized by multiple dark purple nuclei surrounded by a brown cytoplasm that is antigen-positive. For comparative purposes Figure 10.3(C) is a sample from the same patient as in Figure 10.3(B), but Figure 10.3(C) is only PXAPX stained revealing two PXAPX-positive syncytia (arrows). Uninfected cells from this specimen were light tan and the nuclei were not visible [Figure 10.3(C)]. Although we chose to use a common laboratory staining procedure (i.e., PAP), many other staining procedures could be used in conjunction with the PXAPX method should the staining of specific cellular elements, viral inclusions, or pathologic changes be required.

10.6 DISCUSSION

It has been our experience that immunoperoxidase staining produces consistent, sensitive, and reliable results. Some of the advantages of the procedure are that it is straightforward and uncomplicated, the reagents needed can be easily prepared in the laboratory or are commercially available, no extraordinary equipment is required to carry out the procedure or view the results, and the stained preparation is permanent. We have found this latter characteristic to be particularly valuable for reviewing samples months after staining.

As with most procedures, the PXAPX method described has inherent weaknesses. The disadvantages include the rather long period of time required to complete the PXAPX procedure, which precludes its use when rapid results are required. Indirect immunoperoxidase is more rapid than the PXAPX procedure described and it should be considered when speed of obtaining results overrides test sensitivity (Benjamin, 1975). Additionally, DAB is a suspected carcinogen and care must be taken when handling and disposing of this material. Because of the carcinogenic properties of DAB, it may be wiser to use 4-chloro-1-napthol (Hawkes et al, 1982) or p-phenylenediamine dihydrochloride/pyrocatechol (Hanker et al, 1977).

Perhaps the biggest drawback to immunoperoxidase staining that has occasionally plagued its use is the background staining caused by endogenous peroxidase in mammalian mucosecretions (Blain et al, 1975) and inflammatory cells (Straus, 1971). This was particularly troublesome because most of our studies have been of HSV infections of the female genital tract. This anatomical site is rich in mucosecretions and, during an active herpetic infection, inflammatory cells. Because of these two sources of endogenous peroxidase, nonspecific staining can make the detection of a positive reaction extremely difficult or impossible. Consequently, the endogenous enzyme must be inactivated in order to obtain a satisfactory immunoperoxidase stain. Fortunately there are several methods to inactivate endogenous peroxide such as pretreatment with methanol (Straus, 1971, 1979; Streefkerk, 1972; Wier et al, 1974). However, some reports suggest that

Figure 10.2 Immunoperoxidase stained HSV-2 infected VERO and HEp-2 cells. (A) HSV-2 infected VERO cells + anti-HSV-2 serum; (B) HSV-2 infected VERO cells + preimmune serum; (C) HSV-2 infected HEp-2 cells + anti-HSV-2 serum; (D) HSV-2 infected HEp-2 cells + preimmune serum.

such treatment may alter antigenic structure (Fink et al, 1979; Straus, 1979) and we have found this to be true for HSV antigens. Indeed, if an HSV positive sample is treated with methanol prior to the PXAPX procedure, the HSV antigens are altered to such an extent that they are not recognized by the primary antibody. This may result in a false-negative, a highly undesirable result in either the clinical or research laboratory.

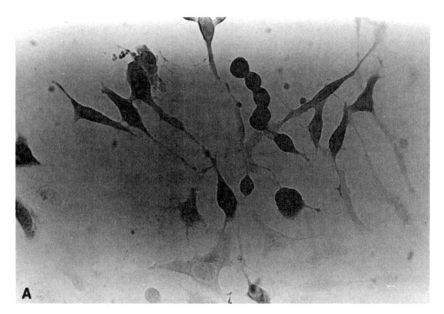

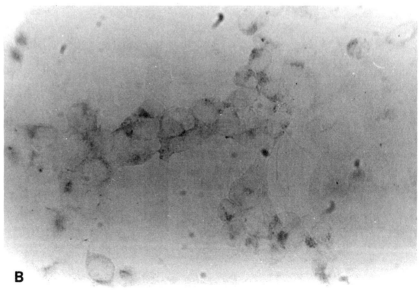

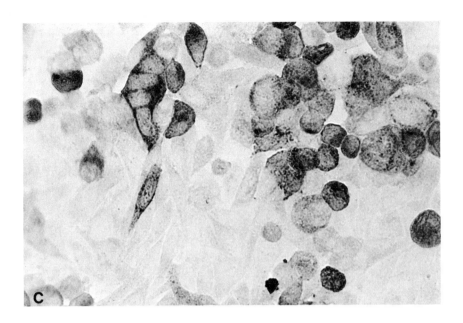

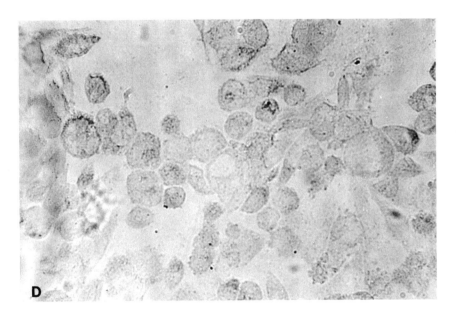

When we were confronted with this apparent impasse we reasoned that while the methanol adversely affected the viral antigens, it may not affect the antibodies after they have reacted with HSV antigens. Therefore, we ran a series of experiments in which methanol treatment was attempted at each stage of the procedure after the reaction with primary antibody (Figure 10.1). These studies eventually demonstrated that methanol treatment after goat anti-rabbit IgG, but before rabbit PXAPX, preserved the viral antigen primary antibody reaction, inactivated endogenous en-

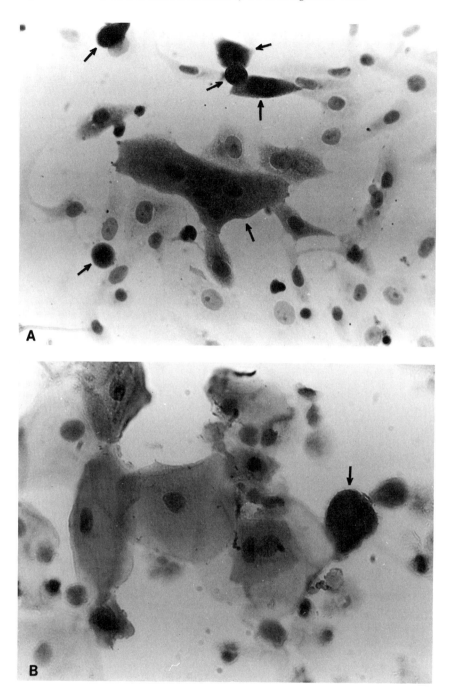

Figure 10.3 Immunoperoxidase-Papanicolaou stained VERO cells and human cells. Primary antibody in (A), (B), and (C) was rabbit anti-HSV-2. (A) HSV-2 infected VERO cells; (B) human cells with confirmed HSV-2 infections; (C) human cells with confirmed HSV-2 infection stained only by the immunoperoxidase method.

C

zyme, and did not interfere with the detection of viral antigen (Pearson et al, 1979).

Nonetheless, because the possibility for antigenic alteration exists by such pretreatment, we have used an immunoenzymatic staining method that uses a non-mammalian enzyme, glucose oxidase (Gay et al, 1984). Essentially, the procedure is the same as the PXAPX method, but glucose oxidase is used in place of peroxidase. Because glucose oxidase is not present in mammalian cells, this modification has allowed us to eliminate methanol, yet preserve the sensi-

tivity and permanency of the PXAPX method (Campbell and Bhatnager, 1976; Suffin et al, 1979; Clark et al, 1982; Gay et al, 1984).

The PXAPX method provides a permanent preparation and is an extremely sensitive method for detecting viral antigens. It has been used for several different viruses. Currently, it is more commonly used in the research laboratory, but is gaining wider acceptance as evidenced by the appearance of commercially prepared peroxidase kits for use by the clinical laboratory for the diagnosis of viral diseases.

REFERENCES

Adams, R.L., Springall, D.R., and Levene, M.M. 1984. The immunocytochemical detection of herpes simplex virus in cervical smears—a valuable technique for routine use. J. Pathol. 143:241–247.

Benjamin, D.R. 1975. Use of immunoperoxidase for rapid viral diagnosis. In D. Schlessinger (ed.), Microbiology—1975. Washington, D.C.: American Society for Microbiology, pp. 89–96.

Blain, J.A., Heald, P.J., Mack, A.E., and Shaw, C.E. 1975. Peroxidase in human cervical mucus during the menstrual cycle. Contraception 11:677–680.

Campbell, G.T. and Bhatnager, A.S. 1976. Simultaneous visualization by light microscopy of two pituitary hormones in a single tissue section using a combination of indirect immunohistochemical methods. J. Histochem. Cytochem. 24:448–452.

Cevenini, R., Rumpianesi, F., Mazzaracchio, R., Donati, M., Falcieri, E., and Sarov, I. 1984. A simple immunoperoxidase method for detecting enteric adenovirus and rotavirus in cell culture. J. Infect. 8:22–27.

Chasey, D. 1980. Investigation of immunoperoxidase—labelled rotavirus in tissue culture by light and electron microscopy. J. Gen. Virol. 50:195–200.

Chu, R.M., Li, N., Glock, R.D., and Ross, R.F. 1982. Applications of peroxidase-antiperoxidase staining technique for detection of transmissible gastroenteritis virus in pigs. Am. J. Vet. Res. 43:77–81.

Churdboonchart, V., Kamsattaya, K., Yoksan, S., Sinarachatanant, P., and Bhamarapravati, N. 1984. Application of peroxidase-antiperoxidase (PAP) staining for detection and localization of dengue-2 antigen. I. In an endogenous peroxidase containing cell systems. Southeast Asian J. Trop. Med. Public Health. 15:547–553.

Clark, C.A., Downs, E.C., and Primus, F.J. 1982. An unlabeled antibody method using glucose oxidase-antiglucose oxidase complexes (GAG): A sensitive alternative to immunoperoxidase for the detection of tissue antigens. J. Histochem. Cytochem. 30: 27–34.

Coons, A.H., Creech, H.J., and Jones, R.N. 1941. Immunological properties of an antibody containing a fluorescent group. Proc. Soc. Exp. Biol. Med. 47:200–202.

Coons, A.H., Creech, H.J., Jones, R.N., and Berliner, E. 1942. The demonstration of pneumococcal antigens in tissues by the use of the fluorescent antibody. J. Immunol. 45:159–170.

Coons, A.H., Snyder, J.C., Cheever, F.S., and Murray, E.S. 1950. Localization of antigen in tissue cells. IV. Antigens of rickettsiae and mumps virus. J. Exp. Med. 91:31–38.

de la Monte, S.M., Linhares, A.L., Travassos

Da Rosa, A.P.A., and Pinheiro, F.P. 1983. Immunoperoxidase detection of yellow fever virus after natural and experimental infections. Trop. Geogr. Med. 35:235–241.

Ducatelle, R., Coussement, W., and Hoorens, J. 1980. Demonstration of canine distemper viral antigen in paraffin sections, using unlabeled antibody—enzyme method. Am. J. Vet. Res. 41:1860–1862.

Fink, B., Loepfe, E., and Wyler, R. 1979. Demonstration of viral antigens in cyrostat sections by a new immunoperoxidase procedure eliminating endogenous peroxidase activity. J. Histochem. Cytochem. 27: 686–688.

Flaitz, C.M. and Hammond, H.L. 1988. The immunoperoxidase method for the rapid diagnosis of intraoral herpes simplex virus infection in patients receiving bone marrow transplants. Spec. Care Dentist. 8:82–85.

Gardner, P.S., Grandien, M., and McQuillin, J. 1978. Comparison of immunofluorescence and immunoperoxidase methods for viral diagnosis at a distance: A WHO collaborative study. Bull. WHO 56(1):105–110.

Gay, H., Clark, W.R., and Docherty, J.J. 1984. Detection of herpes simplex virus infection using glucose oxidase-antiglucose oxidase immunoenzymatic stain. J. Histochem. Cytochem. 32:447–451.

Gerna, G., Vasquez, A., McCloud, C.J., and Chambers, R.W. 1976. The immunoperoxidase technique for rapid human cytomegalovirus identification. Arch. Virol. 50: 311–321.

Grom, J. and Bernard, S. 1985. Virus enzyme—linked cell immunoassay (VELCIA): Detection and titration of rotavirus antigen and demonstration of rotavirus neutralizing and total antibodies. J. Virol. Meth. 10:135–144.

Hanker, J.S., Yates, P.E., Metz, C.B., and Rustioni, A. 1977. A new specific, sensitive and non-carcinogenic reagent for the demonstration of horseradish peroxidase. Histochem. J. 9:789–792.

Hawkes, R., Niday, E., and Gordon, J. 1982. A dot-immunoblotting assay for monoclonal and other antibodies. Analyt. Biochem. 119:142–147.

Herrmann, J.E., Morse, S.A., and Collins, F.

1974. Comparison of techniques and immunoreagents used for indirect immunofluorescence and immunoperoxidase identification of enteroviruses. Infect. Immun. 10:220–226.

Higgins, R.J., Krakowka, S., Metzler, A.E., and Koestner, A. 1982. Immunoperoxidase labeling of canine distemper virus replication cycle in VERO cells. Am. J. Vet. Res. 43:1820–1824.

Hsu, H., Kao, M., Lin, Y., Chen, D., and Lee, C. 1983. Detection of hepatitis B surface antigen in liver tissue: A comparative study on the sensitivity of histochemical and immunoperoxidase techniques. J. Formosan Med. Assoc. 82:657–666.

Jiwa, M., Steebergen, R.D.M., Zwaah, F.E., Kluin, P.M., Raap, A.K., and van der Ploeg, M. 1990. Three sensitive methods for the detection of cytomegalovirus in lung tissue of patients with interstitial pneumonitis. Am. J. Clin. Pathol. 93:491–494.

Kaplan, M.H., Coons, A.H., and Deane, H.W. 1950. Localization of antigen in tissue cells. III. Cellular distribution of pneumococcal polysaccharides types II and III in the mouse. J. Exp. Med. 91:15–30.

Keydar, I., Mesa-Tejada, R., Ramanarayanan, M., Ohno, T., Fenoglio, C., Hu, R., and Spiegelman, S. 1978. Detection of viral proteins in mouse mammary tumors by immunoperoxidase staining of paraffin sections. Proc. Natl. Acad. Sci. USA 75:1524–1528.

Kurchak, M., Dubbs, D.R., and Kit, S. 1977. Detection of herpes simplex virus-related antigens in the nuclei and cytoplasm of biochemically transformed cells with peroxidase/anti-peroxidase immunological staining and direct immunofluorescence. Intl. J. Cancer 20:371–380.

Kurman, R.J., Shah, K.H., Lancaster, W.D., and Jensen, A.B. 1981. Immunoperoxidase localization of papillomavirus antigens in cervical dysplasia and vulvar condylomas. Gynecology 140:931–935.

Kurstak, E. 1971. The immunoperoxidase technique: Localization of viral antigens in cells. In K. Maramorosch and H. Koprowski (eds.), Methods in Virology, Vol. 5. New York: Academic Press, pp. 423–444.

Machado-Vieira, L.M. and Sarno, E.N. 1982. The efficacy of the immunoperoxidase technique for the detection of several antigens after destaining of stored stained sections. Brazilian J. Med. Biol. Res. 15:265–268.

Musiani, M., Zerbini, M., Plazzi, M., and La Placa, M. 1986. Double immunoenzymatic staining for the simultaneous detection of Epstein-Barr virus induced antigens. Histochemistry 84:15–17.

Nakane, P.K. and Pierce, G.B. 1967. Enzyme-labeled antibodies: Preparation and application for the localization of antigens. J. Histochem. Cytochem. 14:929–931.

Pearse, A.G.E. 1972. Histochemistry: Theoretical and Applied, Vol. 2. Edinburgh: Churchill Livingston.

Pearson, N.S., Fleagle, G., and Docherty, J.J. 1979. Detection of herpes simplex virus infection of female genitalia by the peroxidase-antiperoxidase method alone or in conjunction with the Papanicolaou stain. J. Clin. Microbiol. 10:737–746.

Ram, J.S., Nakane, P.K., Rawlinson, D.G., and Pierce, G.B. 1966. Enzyme-labeled antibodies for ultrastructural studies. Fed. Proc. 25:732.

Recchia, S., Rizzi, R., Acquauiva, F., Rizzetto, M., Tison, V., Bonino, F., and Verme, G. 1981. Immunoperoxidase staining of the HBV-associated delta antigen in paraffinated liver specimens. Pathologica 73:773–777.

Schwendemann, G., Wolinsky, J.S., Hatzidimitrou, G., Merz, D.C., and Waxham, M.N. 1982. Postembedding immunocytochemical localization of paramyxovirus antigens by light and electron microscopy. J. Histochem. Cytochem. 30:1313–1319.

Spendlove, R.S. 1967. Microscopic techniques. In K. Maramorosch and H. Koprowski (eds.), Methods in Virology, Vol. 3. New York: Academic Press, pp. 482–520.

Steinkamp, J.A. and Crissman, H.A. 1974. Automated analysis of deoxyribonucleic acid, protein and nuclear to cytoplasmic relationships in tumor cells and gynecologic specimens. J. Histochem. Cytochem. 22:616–621.

Sternberger, L.A. 1974. Immunocytochemistry. Englewood Cliffs, N.J.: Prentice-Hall.

Sternberger, L.A., Hardy, P.H., Cuculis, J.J., and Meyer, H.G. 1970. The unlabeled antibody enzyme method of immunohistochemistry. Preparation and properties of soluble antigen-antibody complex (horseradish peroxidase-antihorseradish peroxidase) and its use in identification of spirochetes. J. Histochem. Cytochem 18:315–333.

Straus, W. 1971. Inhibition of peroxidase by methanol and by methanol-nitroferricyanide for use in immunoperoxidase procedures. J. Histochem. Cytochem. 19:682–688.

Straus, W. 1979. Peroxidase procedures. Technical problems encountered during their application. J. Histochem. Cytochem. 27:1349–1351.

Streefkerk, J.G. 1972. Inhibition of erythrocyte pseudoperoxidase activity by treatment with hydrogen peroxide following methanol. J. Histochem. Cytochem. 20:829–831.

Suffin, S.C., Much, K.B., Young, J.C., Lewin, K., and Porter, D.D. 1979. Improvement of the glucose oxidase immunoenzymatic technic. Am. J. Clin. Pathol. 71:492–496.

Syrajanen, K.J. and Pyrhonen, S. 1982. Demonstration of human papilloma virus antigen in the condylomatous lesions of the uterine cervix by immunoperoxidase technique. Gynecol. Obstet. Invest. 14:90–96.

Weir, E.E., Pretlow, T.G., Pitts, A., and Williams, E.E. 1974. Destruction of endogenous peroxidase activity in order to locate cellular antigens by peroxidase-labeled antibodies. J. Histochem. Cytochem. 22:51–54.

Weller, T.H. and Coons, A.H. 1954. Fluorescent antibody studies with agents of varicella and herpes zoster propagated in vitro. Proc. Soc. Exp. Biol. Med. 86:789–794.

11

Complement Fixation Test

Lillian M. Stark and Arthur L. Lewis

11.1 INTRODUCTION

The introduction of new or improved techniques often results in the rejection of older methodologies as inadequate. In the early years of clinical virology the neutralization test served as the standard against which the specificity and sensitivity of alternative methods were evaluated. However, the refinement of the complement fixation (CF) test and the purification of its reagents, subsequently made this test the standard. Presently, enzyme-linked immunosorbent assay (ELISA) and radio-immunoassay (RIA) tests predominate because of their increased sensitivity, speed, and ability to be automated. Nevertheless, the CF test is still valuable for antibody studies, as well as virus identification and typing. Automation of dilution and pipetting steps are possible, but not necessary; the test does not require expensive equipment, such as a microplate spectrophotometer. The same protocol is used for all CF tests, regardless of which antigen or antibody is assayed, allowing convenient screening against a large battery of agents simultaneously.

11.1.1 Test Principles

The complement fixation test is comprised of two antigen-antibody reactions. The first is the specific reaction between either a known antigen and unknown (patient) serum or an unknown antigen and specific antiserum. This occurs in the presence of a predetermined amount of complement. The specific antigen-antibody reaction binds (fixes) the complement, preventing it from reacting in the second stage of the test. If the antigen and antibody used in the test do not react with each other, the complement is not fixed and thus is free to react in the next step.

A hemolytic system is used to detect a lack of complement fixation. The second antigen-antibody reaction occurs when sheep red blood cells (SRBC) are reacted with rabbit antibody to SRBC (hemolysin). This sensitizes the cells, causing them to lyse in the presence of free complement, but not when the complement has been previously fixed by a specific antigen-antibody reaction. Thus, hemolysis occurs when the test antigen and serum lacking antibody do not react specifically with each

Step 1. Primary test reaction

ANTIGEN-X + ANTISERUM-A + COMPLEMENT → ANTIGEN-X + ANTISERUM-A
+ COMPLEMENT

ANTIGEN-X + ANTISERUM-X + COMPLEMENT → (ANTIGEN-X − ANTIBODY-X
− COMPLEMENT

Step 2. Detection of primary test reaction

[S-RBC = sheep red blood cells sensitized with hemolysin]

ANTIGEN-X + ANTISERUM-A + COMPLEMENT + S-RBC → HEMOLYSIS OF RBC

(ANTIGEN-X − ANTIBODY-X − COMPLEMENT) + S-RBC → NO HEMOLYSIS
(RBC BUTTON)

Figure 11.1 Test principles.

other (negative test), and conversely, a lack of hemolysis indicates a positive test antigen-antibody reaction. The reactions are diagrammed in Figure 11.1.

11.1.2 Background

The CF test was introduced in 1909 by Wasserman et al for syphilis serology. Thereafter the test was applied in veterinary studies: vaccinia in calves, foot and mouth disease, psittacosis, and the arthropod-borne encephalitides (Ciuca, 1929; Rice, 1948a, 1948b; Bankowski et al, 1953). There were, however, difficulties in adapting the test to routine diagnosis of human diseases of viral etiology.

Early work used crude antigens of low titer, containing extraneous material, which were nonspecific and anticomplementary (Howitt, 1937; Sosa-Martinez and Lennette, 1955). Antigens prepared from infected mouse brain contained interfering substances which could be removed by high speed centrifugation (Kidd and Friedewald, 1942; Havens et al, 1943; Lennette et al, 1956a). Seitz filtration removed these substances, but lowered the viral titer (Casals and Palacios, 1941) whereas repeated freeze-thaw cycles prior to centrifugation increased specificity without titer loss. A cephalin-like substance in tissue extracts was responsible for some nonspe-

cific reactions with normal serum (Maltaner, 1946).

Lipid extraction was used for the preparation of CF antigens for the equine arthropod-borne encephalitides, lymphocytic choriomeningitis (LCM), and St. Louis encephalitis (SLE) viruses (Howitt, 1937; DeBoer and Cox, 1947; Espana and Hammon, 1948; Casals and Palacios, 1949). Virus antigens were prepared from the brains of week old mice, which had been inoculated intra-cerebrally when less than three days old, using a sucrose-acetone extraction method (Clark and Casals, 1958). This procedure provided an antigen suitable for both CF and hemagglutination (HA) testing (Hammon and Sather, 1969). An LCM antigen was prepared from guinea pig spleen (Smadel et al, 1939). The CF test has been used extensively in arbovirology; studies indicate the California-encephalitis group of arthropod-borne viruses to be large and complex. Cross antigenic comparisons among isolates by complement fixation provided data indicating differences which were not apparent by other examinations, such as hemagglutination inhibition (HI) and neutralization tests. CF findings indicated the California group was composed of thirteen serotypes, eleven present in the Western and two in the Eastern hemisphere (Sather and Hammon, 1967).

Virus propagation in embryonated chicken eggs (Woodruff and Goodpasture, 1931; Goodpasture et al, 1931, 1933; Stevenson and Butler, 1933) allowed for the production of high-titer antigens for various orthomyxoviruses and paramyxoviruses (Beveridge and Burnet, 1946; French, 1952; Whitney et al, 1953; Sosa-Martinez and Lennette, 1955; Lennette et al, 1956a, 1956b).

Virus propagation in cell cultures (Enders et al, 1949; Robbins et al, 1950, 1951; Dulbecco, 1952; Rowe et al, 1953; Enders and Peebles, 1954; Hilleman and Werner, 1954; Huebner et al, 1954; Henle et al, 1955) allowed the utilization of cell culture fluids as a source of CF antigens. Problems of low antigen titer (Ruckle and Rogers, 1957; Schmidt, 1957; Giardi et al, 1958; Weller and Witton, 1958; Taylor-Robinson and Downie, 1959) and nonspecific and anticomplementary reactions occurred (Svedmyr et al, 1952; Black and Melnick, 1954). Attempts to concentrate the antigens or remove nonspecific inhibitors by heat were unsuccessful (Svedmyr et al, 1953; LeBouvier et al, 1954; Black and Melnick, 1955; Schmidt and Lennette, 1956). Other attempts to improve antigen preparations included: reduced volume of culture medium (Schmidt et al, 1957); use of a roller bottle (Churcher et al, 1959) or suspension cultures (Westwood et al, 1960; Suggs et al, 1961; Halonen et al, 1967); antigen purification on sucrose density gradients (Julkunen et al, 1984); varying the composition of the culture medium and the multiplicity of infection (Schmidt, 1969).

In addition to identification of unknown agents in infected host systems such as cell cultures, or mouse brain, the CF test has been used to detect rotavirus antigens directly in stool specimens. Anti-complementary activity present in some specimens was removed by absorption of the clarified stool suspension supernatent with calf serum (Pauri et al, 1981).

Some nonspecific reactions in the CF test were also attributed to the presence of natural antibody in normal serum, which could be removed by heating the serum for 30 minutes at 60 °C to 65 °C (Kidd and Friedewald, 1942). Non-specific complement fixation also occurred with sera that were positive in the Wasserman test (Lichter, 1953), as well as when auto-antibodies were present (Thorn et al, 1988).

The early CF tests were performed in test tubes, required large quantities of reagents, and used various protocols. The Communicable Disease Center (presently, the Centers for Disease Control, U.S. Department of Health and Human Services, Atlanta, GA) developed a standardized CF test in the late 1950s, which, with the development of microtechniques (Takatsy, 1950; Sever, 1962), became the Laboratory Branch Complement Fixation Test (LBCF) (Casey, 1965). Revisions to the protocol were made in 1974 and 1981. Because it requires minimal quantities of reagents, microtiter CF has become the standard methodology for CF.

New variations on the basic complement fixation test protocol have included an ELISA-CF, where peroxidase-labeled complement was used to detect CF antibodies, rather than the hemolytic system (Taguchi, 1988). Rapid rate–kinetic turbidometric assays (Fulton and Dininno, 1985) and complement mediated neutralization tests (Hishiyama et al, 1988), as well as a single radial complement fixation test in agarose plates (Sato et al, 1983, 1988; Ochiai et al, 1987) have also been developed. A monoclonal antibody to substitute for the polyclonal hemolysin has been investigated (Ossewaarde and deBooij, 1989).

11.1.3 Comparison With Other Methods

CF has been compared with various immunoassay procedures to detect antibodies to cytomegalovirus (CMV). The specific clinical population tested influenced the correlation of positive results between test types (Ravaoarino et al, 1984). Although the CF test generally gave lower titers than the ELISA test, they detected similar antibody prevalence rates in the group of healthy blood donors tested (Dzierzanowska et al, 1986). Miller et al (1989) compared six serological techniques, including CF, with virus isolation in transplant patients and concluded no single serological test could reliably and rapidly detect primary CMV infection. They suggest that a combination of different methods be used; e.g., CF and ELISA procedures do not necessarily test for the same category of specific antibodies. Thorn et al (1988) has indicated that preferential elevation of CF antibodies against CMV relative to the CMV immunoglobins detected with ELISA may be reflective of the pathogenic process in Sjogren's syndrome.

CF, ELISA, and HI tests were compared for measles antibody determination (Ferrante et al, 1987). The range of titers (sensitivity) was greater with ELISA than CF; both were considered superior to HI. For mycoplasma infections, good correlations were seen between CF and ELISA tests (Fischer et al, 1986; Wreghitt and Sillis, 1987). Inouye et al (1981) compared CF with immune adherence hemagglutination (IAHA) and felt the ratio of antibody titer detected in each test was of value in diagnosing recent infection for rotavirus and Japanese encephalitis. In a study monitoring seroconversion after administration of a rotavirus vaccine, CF was not as sensitive as the immunoassay used when the entire population was evaluated, but was better in the older than in the younger recipients (Midthun et al, 1989).

Antigens produced from influenza virus may be either the S, soluble (internal nucleoprotein), or V, virion (hemagglutinin, neuraminidase), type. The S-antigen is type specific for influenza A, B, or C; the V-antigen is strain specific and stimulates the formation of protective antibody.

(Hoyle, 1952; Lief and Henle, 1956a, 1956b) The antibody response to the S-antigen is a rapid development of IgM, which is more readily measured with the CF than with the HI test (Duca et al, 1979). V antigen neuraminidases have also been distinguished using monoclonal antibodies in a CF test (Holmes et al, 1982). ELISA was shown to be more sensitive than CF (Julkunen et al, 1984), but these authors also state that during nonepidemic periods the CF might be preferable because it allows routine simultaneous testing against several viral antigens.

The ability to screen against a battery of antigens (or antisera) with a single test is one of the major advantages of the complement fixation procedure. It has been used that way in studies to test for possible etiologic agents of a specific syndrome (Hudson et al, 1981) as well as in retrospect to determine clinical conditions associated with infection (Puolakkainen et al, 1987). In the author's laboratory, sera are tested routinely against a battery of respiratory or central nervous system (CNS) agents using CF to evaluate the association between potential etiologic agents and the clinical picture.

11.2 GENERAL CONSIDERATIONS

11.2.1 Technique

The CF test requires a capable, well-trained technologist for accurate results. Since it is performed with a microtiter procedure, minute amounts of reagents are used and precision in delivery is crucial. Calibrated micro-pipets with disposable tips are beneficial. When dropper pipets and loops are used for delivery and dilution they must be properly cleaned, calibrated, and used carefully to assure reliability. If large numbers of specimens are to be examined, automatic microdilution and reagent delivery systems are of value.

All reagents must be very carefully

prepared according to established laboratory protocol; the test is very sensitive to errors in reagent composition and concentration. All reagents should be properly stored according to the supplier's directions. Multiple freeze-thaw cycles are to be avoided, thus, it is best to aliquot sensitive reagents in small volumes for storage. New lots of hemolysin, antisera, and antigens must be titrated before use in routine testing. The SRBC should be handled gently and not used if there is any hemolysis during washing or storage.

11.2.2 Specimens

Good quality specimens are essential for a meaningful test. Serum should be collected in a sterile tube without preservatives or anticoagulants. Hemolysis must be avoided. Sufficient blood must be drawn to provide at least 1 to 2 mL of serum. If any red blood cells remain after the serum is decanted from the clot, they should be removed by centrifugation prior to freezing the specimen. As bacterial contamination of the serum may produce erroneous test results, the specimen should be handled aseptically.

Sera may sometimes react with complement in the test. This "anticomplementary" behavior may be present in the serum when collected, or may develop during storage. Heat inactivation of the serum (incubate serum aliquot in a 56 °C waterbath for 30 minutes) may alleviate this, otherwise the serum should be treated with complement to remove "anticomplement" components. If this is not effective, another serum specimen should be collected. Unfortunately, some individuals may persistently contain anticomplementary factors in their serum. These factors are usually present in the lower serum dilutions so that a high specific titer may still be recognizable in the specimen.

The CF test may be used to identify an unknown antigen, such as a virus isolate made in cell culture. The culture should be rapidly frozen (dry ice-alcohol bath) and thawed three times, and clarified by centrifugation. An uninoculated culture should be similarly processed for use as the tissue control in the test. If nonspecific reactions occur, the antigen may be clarified by extraction with an equal volume of Freon. The test antiserum may be absorbed with noninfected (control) tissue to eliminate nonspecific reactions.

11.2.3 Reagents

Complement from different animal sources is commercially available, however, they differ in sensitivity in the CF test (Sethi et al, 1981). Most frequently, guinea pig complement is used at five 50% hemolytic units (CH50) (LBCF). Some test protocols call for complement to be used at two full hemolytic units, in which case it should be titrated in the presence of each test antigen, as antigens vary in their ability to bind complement (Lennette, 1969).

Antigens may be whole virion, V-antigens, or produced by the fragmentation of the virus, S-antigens. The latter are a mixture including nucleoprotein (Craigie and Wishart, 1936a, 1936b; Hoyle, 1952; Ada and Perry, 1954; Schafer and Zillig, 1954; Panucker et al, 1956; Westwood et al, 1965; Cohen and Wilcox, 1966). S-antigens have been demonstrated for a number of viruses and vary in stability and specificity (Chambers et al, 1950; Black and Melnick, 1955; Ende et al, 1957). In mumps (Henle et al, 1948) and influenza (Duca et al, 1979) infections, anti-S antibodies rise more rapidly than anti-V antibodies, and can thus be used to distinguish recent from previous infections.

Reagents for the CF test can be prepared in-house, but this may lead to difficulties in quality control and in test comparisons. Commercial suppliers known to the authors at the time of this chapter revision are listed in Table 11.1. No assur-

Table 11.1 Commercial Suppliers of Reagents for the CF Test

	Antigens	Antisera	Complement	Hemolysin	RBC
Accurate Chemical & Scientific Corp. 300 Shames Drive Westbury, NY 11590 (516) 333-2221 (800) 645-6264	X	X	X	X	
Diamedix 2140 N. Miami Avenue Miami, FL 33217 (305) 324-2300 (800) 327-4565			X	X	X
Hillcrest Biologicals 10703 Progress Way Cypress, CA 90630-4714 (213) 420-2657 (800) 445-0185	X				
Microbix 341 Bering Avenue Toronto, Ontario M8Z 3A8 (416) 234-1624	X	X	X		
Rockland, Inc. Box 316 Gilbertsville, PA 19525 (215) 369-1008			X	X	X
Virion (U.S.), Inc. 4 Upperfield Road Morristown, NJ 07960 (201) 993-8219 (800) 524-2689	X	X	X	X	
BioWhittaker Inc. 8830 Biggs Ford Road Walkersville, MD 21793-0127 (301) 898-7025 (800) 638-3976	X	X	X	X	X

ances are made as to the quality or availability of products from these sources.

11.2.4 Test Interpretation

For a test to be considered valid, all controls must be within the limits specified in the test procedure. These include: sensitivity of the red blood cells to lysis, concentration of complement used and its reaction with the RBCs, sensitivity of the antiserum-antigen reaction, and no reaction with the tissue (antigen) or serum controls. If the controls fail, corrective action and a repeat test must be performed before results are reported.

CF serology can only identify an agent responsible for a recent infection when both acute and convalescent sera are tested simultaneously. The acute specimen should be collected as close to onset of illness as possible. The convalescent specimen should be collected two to three weeks later. In immunologically impaired patients or in the presence of an agent sensitive to the presence of administered antibiotics, where antibody production may be delayed, a third serum collected four to six weeks after onset may be of value. It is generally accepted that a fourfold rise in titer between the sera to a specific agent is indicative of recent infection. Nevertheless, the entire patient history must be considered, since heterologous anamnestic reactions may occur. In such cases the individual had experienced a previous infection, or immunization, and a later infection with another virus spurred a sharp, rapid rise in IgG antibodies to the first agent. This is not uncommon with the Group A and B arboviruses or paramyxoviruses. In addition, stable elevated titers to multiple agents may indicate a generalized stimulation of immunological activity unrelated to those agents.

11.3 TEST SETUP

11.3.1 Equipment

Freezer $-20\,°C$ ($-70\,°C$ freezer preferable)
Refrigerator $4\,°C$
Incubator $37\,°C$
Water baths $37\,°C$ and $56\,°C$
Calibrated centrifuge
Centrifuge microtiter plate carriers
pH meter
Balance
Liquid aspiration device
Microtiter plate vibrator
Timer
Test tube racks
Test tubes: 13×100 mm, 15×125 mm
Microtiter plates, U style, 96 well format

Diluter (automatic, calibrated loops, or pipettor type)
Microtiter dropper pipettes (0.025 and 0.05 mL)
Serological pipettes (1, 2, 5, and 10 mL)
Pipettors and disposable tips (0.025 and 0.05 mL)
Glassware (Erlenmeyer and volumetric flasks; graduated cylinders; beakers; centrifuge tubes)
Mirror reading device (optional)

11.3.2 Reagents

pH buffer standards
Hydrochloric acid, 1 N
Stock 1M $MgCl_2$—0.3M $CaCl_2$ solution
Veronal buffer diluent, stock
Gelatin
Sheep red blood cells—2.8% suspension
Hemolysin
Complement
Antigens
Antigen controls (tissue controls)
Sera, known positive antibody status
Sera, known negative antibody status
Sera, test (unknown)

11.3.3 Reagent Preparation

Label all reagents with their name, concentration, date prepared, technician's initials.

11.3.3.a 1M $MgCl_2$—0.3M $CaCl_2$ Solution

Add 20.3 g $MgCl_2 \cdot 6H_2O$ and 4.4 g $CaCl_2 \cdot 2H_2O$ to 70 mL reagent grade water in a 100 mL volumetric flask.

Mix by swirling.

Fill to the 100 mL mark with reagent grade water.

This reagent may be filter sterilized for extended storage.

Store in the refrigerator.

11.3.3.b Veronal Buffered Diluent, Stock Solution (5× VBD)

Add 83.0 g NaCl to 1500 mL reagent grade water in a 2 liter volumetric flask.

Add 10.19 g Na-5,5-diethyl barbiturate (Sodium barbital, veronal).

Mix by swirling until the chemicals are completely dissolved.

Add 34.0 mL 1N HCl and mix by swirling.

Add 5.0 mL stock $MgCl_2 \cdot CaCl_2$.

Bring the total solution volume up to 2 liters with reagent grade water.

Check the pH of a 1:5 dilution of the buffer (1 mL of 5× VBD + 4 mL reagent grade water); if the pH is below 7.3 or above 7.4, discard the buffer and prepare fresh.

This reagent may be filter sterilized for extended storage.

Store in the refrigerator.

11.3.3.c Gelatin Water

Add 1 g gelatin to 200 mL reagent grade water in a 2 liter flask.

Bring the solution to a boil, swirling to dissolve the gelatin.

Remove from heat and allow to cool.

Add 600 mL sterile reagent grade water.

Cover tightly.

Store in the refrigerator for no longer than one week.

11.3.3.d Veronal Buffered Diluent, Working Solution (VBD)

This should be prepared on the day of use.

Add 200 mL stock 5× VBD to 800 mL gelatin water.

Check the pH; if the pH is below 7.3 or above 7.4, discard the buffer and prepare fresh stock 5× VBD.

11.3.3.e Stock Hemolysin (1:100)

Dissolve 0.85 g NaCl in 100 mL reagent grade water; add 5.0 g phenol and swirl to dissolve.

Aliquot 4.0 mL of 5% phenol saline into a 125 mL Erlenmeyer flask.

Add 94.0 mL cold VBD to the flask and mix by swirling.

Add 2.0 mL glycerinized hemolysin to the flask and mix by swirling.

Store at 4 °C.

11.3.4 Test Outline

The steps required for the performance of a complement fixation test, according to the protocols used in the author's laboratory, are as follows:

1. Prepare reagents. The gelatin water, veronal buffered diluent (VBD), color standards, sensitized SRBC, diluted complement, antisera, and antigen must be prepared fresh for each test.
2. Wash and standardize SRBC (11.4.1).
3. If a new lot of hemolysin is to be used, perform a hemolysin titration (11.4.2).
4. Prepare color standards (11.4.3).
5. Perform a complement titration (11.4.4). Complement must be titrated each time an antigen titration or diagnostic test is performed; the minimal acceptable hemolytic titer is 250 CH50/mL.
6. If a new lot of antigen is to be used, perform an antigen titration (11.4.5). This procedure may also be used for the identification of an unknown antigen by testing versus antisera to suspect agents.
7. Perform the diagnostic serology test (11.4.6).

11.4 TEST PERFORMANCE

11.4.1 Preparation of Sheep Erythrocytes

11.4.1.a Cell Washing

1. Place 10 mL of the SRBC suspension in a 250 mL centrifuge bottle.
2. Add approximately 100 mL cold VBD (11.3.3.d) and mix gently.
3. Centrifuge at 600 × g at 4 °C for 10 minutes.
4. Aspirate the supernatant.
5. Gently resuspend the cells in 100 mL fresh cold VBD, centrifuge and aspirate as above.
6. Repeat this procedure again, for a total of three washes.
7. Resuspend the cells in approximately 20 mL cold VBD, transfer to a 40 mL conical, graduated centrifuge tube and fill to the 40 mL mark with VBD.
8. Centrifuge at 600 × g at 4 °C for 10 minutes.
9. Note the volume of the packed cells, and aspirate the supernatant without disturbing the cells. If the supernatant is colorless, proceed with cell standardization. If hemolysis is observed, the cells are not suitable for use, and a new lot should be obtained.

11.4.1.b Cell Standardization—Centrifugation Method

1. To calculate the volume of VBD required for a 2.8% cell suspension, multiply the volume of packed cells obtained after the final wash centrifugation (11.4.1.a) by 34.7.
2. Suspend the cells in the calculated volume of VBD in an appropriate size Erlenmeyer flask by gentle swirling.
3. To check the accuracy of the dilution, pipette 7.0 mL of the suspension into a 10 mL graduated centrifuge tube having an accuracy of ±0.025 mL in the 0 to 1 mL range.

4. Centrifuge at 600 × g at 4 °C for 10 minutes.
5. If the volume of the packed cells is 0.2 mL, the suspension is accurate, and the 7 mL sample may be resuspended and returned to the original flask for use in the test.
6. If the volume of the packed cells is not 0.2 mL, calculate the correction factor, which is the actual volume of packed cells read from the centrifuge tube divided by 0.2. This factor is multiplied by the original volume of cell suspension, minus 7.0 mL, to determine the corrected cell suspension volume. Compute the difference between the corrected volume and the existing volume; this is the volume of VBD that is used to adjust the cell suspension (steps 7 and 8).
7. If the corrected volume is less than the existing volume, aliquot a portion of cell suspension from the flask sufficient to include the volume of VBD to be removed into a centrifuge tube. Centrifuge at 600 × g at 4 °C for 5 minutes. Pipette and discard the calculated excess VBD from the supernatant. Resuspend the cells, returning them to the flask. Recheck the dilution accuracy (step 3).
8. If the corrected volume is greater than the existing volume, add the required VBD to the flask, resuspend the cells, and recheck the dilution accuracy (step 3).

11.4.2 Hemolysin Titration

11.4.2.a Preparation of a 1:1000 Hemolysin Dilution

1. Place 9.0 mL of cold VBD in a 15 × 125 mm test tube labeled 1:1000 hemolysin.
2. Add 1.0 mL of the 1:100 stock hemolysin and mix well with the pipette.

11.4.2.b Preparation of Additional Hemolysin Dilutions

1. The dilution protocol is outlined in Table 11.2.

Table 11.2 Preparation of Additional
Hemolysin Dilutions

Final Dilution	VBD (mL)	1:1000 Hemolysin (mL)
1:1500	0.5	1.0
1:2000	1.0	1.0
1:2500	1.5	1.0
1:3000	2.0	1.0
1:4000	3.0	1.0
1:8000	7.0	1.0

2. Label six 15 × 125 mm test tubes with the dilutions listed.
3. Using a 5.0 mL pipette, place the designated volumes of VBD into the appropriately labeled tubes.
4. Using a 5.0 mL pipette, add 1.0 mL of the 1:1000 hemolysin dilution to each tube.

11.4.2.c Preparation of a 1:400 Dilution of Complement

1. Place undiluted complement in an ice bath.
2. Measure 100 mL cold VBD in a 100 mL graduated cylinder; use a 1 mL pipette to remove 0.25 mL; transfer the remainder to a 125 mL Erlenmeyer flask.
3. Draw up the undiluted complement in a 1.0 mL pipette to the 0.6 mL mark; wipe the pipette tip. Holding the pipette vertically, deliver 0.25 mL of complement dropwise into the VBD in the flask.
4. Mix by swirling gently.
5. Place the 1:400 dilution of complement at 4 °C for at least 20 minutes before using; do not use after 2 hours.

11.4.2.d Preparation of Hemolysin Sensitized Cells

1. Place seven 13 × 100 mm test tubes in a rack.

2. Label the first 1:1000; label the remaining six tubes with the dilutions listed in Table 11.2.
3. Add 1.0 mL of the standardized 2.8% SRBC to each of the seven tubes.
4. Thoroughly mix the 1:1000 hemolysin dilution (11.4.2.a) and add 1.0 mL slowly, with constant swirling, to the tube labeled 1:1000.
5. To each of the labeled tubes add 1.0 mL of the appropriate hemolysin dilution (11.4.2.b) in a like manner.
6. Shake the rack and place it in a 37 °C water bath for 15 minutes.

11.4.2.e Hemolysin Titration

1. Place seven 12 × 75 mm test tubes in a rack. Label the first 1:1000; label the remaining six tubes with the dilutions listed in Table 11.2.
2. Add 0.4 mL VBD to each tube.
3. Add 0.4 mL of the 1:400 complement dilution (11.4.2.c) to each tube; shake the rack.
4. Add 0.2 mL of the SRBC sensitized with each of the hemolysin dilutions (11.4.2.d) to the comparably labeled tube containing VBD and complement; mix each tube by shaking.
5. Incubate the tubes in a 37 °C water bath for 1 hour; shake the rack after 30 minutes.
6. Prepare color standards (11.4.3).
7. Centrifuge the tubes at 600 × g for 5 minutes.
8. Compare each tube with the prepared color standards. If the tube matches a color standard, record the percent hemolysis; if not, interpolate to the nearest 5% and record.
9. Plot on arithmetic (linear) graph paper the percent hemolysis obtained at each hemolysin dilution (Figure 11.2). Draw a line through the plotted points and determine the plateau region. This is the area where additional hemolysin

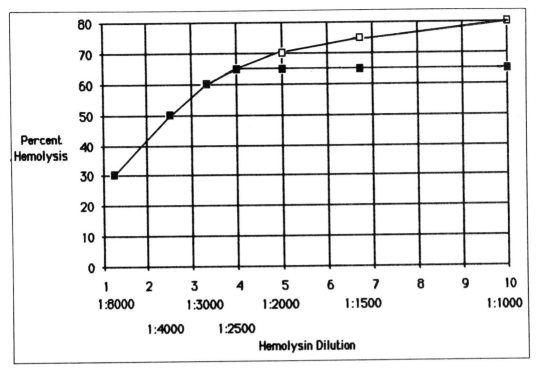

Figure 11.2 Determination of the optimal hemolysin dilution.
The location for each of the hemolysin dilutions on the X axis is determined by dividing the lowest dilution by the next higher dilution and multiplying by the number of blocks allotted on the X axis.

> Example:
> 1000/1500 × 10 blocks = 6.7 (hemolysin dilution 1:1500)
> 1000/2000 × 10 blocks = 5.0 (hemolysin dilution 1:2000)
> 1000/2500 × 10 blocks = 4.0 (hemolysin dilution 1:2500)

The closed squares show an acceptable titration and the open squares, an unacceptable titration. The optimal dilution in this titration is 1:2000 (i.e., the second dilution on the plateau).

produces no marked increase in hemolysis.

10. The second dilution on the plateau is the optimal dilution for the tested lot of hemolysin to be used for cell sensitization. (In Figure 11.2, this is a 1:2000 dilution). The minimal acceptable titer (dilution) is ≥1:2000.

11. Note: commercial complement diluted 1:400 generally will yield 30% to 80% hemolysis for cells optimally sensitized as above. To obtain the correct percent hemolysis with a less active complement sample, it may be necessary to use a 1:300 dilution. With a very potent complement preparation a 1:500 dilution may be needed.

11.4.3 Color Standards

11.4.3.a Preparation of Hemoglobin Solution

1. Pipette 1.0 mL of the thoroughly mixed 2.8% SRBC suspension into a 15 × 125 mm test tube.

2. Add 7.0 mL distilled water and shake the tube until all the cells are lysed.

3. Add 2.0 mL stock buffer (5× VBD, 11.3.3.b) to the tube and mix well.

Table 11.3 Preparation of Color Standards

Reagents	Percent Hemolysis										
	0	10	20	30	40	50	60	70	80	90	100
Hemoglobin Solution	0.0	0.1	0.2	0.3	0.4	0.5	0.6	0.7	0.8	0.9	1.0
0.28% Cell Suspension	1.0	0.9	0.8	0.7	0.6	0.5	0.4	0.3	0.2	0.1	0.0

11.4.3.b Preparation of 0.28% Red Blood Cell Suspension

1. Pipette 1.0 mL of the thoroughly mixed 2.8% SRBC suspension into a 15 × 125 mm test tube.
2. Add 9.0 mL cold VBD and mix well.

11.4.3.c Preparation of Color Standards

1. Label eleven 12 × 75 mm test tubes with the percent hemolysis given in Table 11.3 (0, 10, . . ., 100) and place in a rack in ascending order.
2. Using a 2.0 mL pipette, deliver the volume of hemoglobin solution designated in Table 11.3 appropriately to each tube.
3. Using a 2.0 mL pipette, deliver the volume of 0.28% SRBC suspension designated in the Table appropriately to each tube.
4. Shake the rack vigorously to mix well.
5. Centrifuge the tubes at 600 × g for 5 minutes at ambient temperature.
6. Remove the tubes from the centrifuge without agitation and store them in a rack in the refrigerator until needed.

11.4.4 Complement Titration

11.4.4.a Preparation of Sensitized Cells

1. Add 8.0 mL of the standardized 2.8% SRBC suspension to a 125 mL Erlenmeyer flask.
2. Prepare 10 mL of the optimal hemolysin dilution from the 1:100 stock hemolysin solution and add 8.0 mL of this dilution to the SRBC with rapid stirring.

3. Incubate 15 minutes in a 37 °C water bath, mixing at 7 minutes.

11.4.4.b Preparation of a 1:400 dilution of complement

1. See 11.4.2.c.
2. A less potent complement may be tested at a 1:300 dilution. Use exactly 99.67 mL VBD, adding 0.33 mL undiluted complement.

11.4.4.c Complement Titration

1. Label two sets of five 12 × 75 test tubes with numbers 0 through 4 and place in a rack; the titration is performed in duplicate.
2. Add VBD to each tube in the amounts given in Table 11.4.
3. Add the 1:400 diluted complement to each tube in the amounts given in Table 11.4.
4. Shake the rack to mix the reagents.
5. Add 0.2 mL of sensitized SRBC to each tube.
6. Shake the rack and place in a 37 °C water bath; after 15 minutes incubation, shake the rack; continue incubation an additional 15 minutes.
7. Prepare color standards (11.4.3) while the titration tubes are incubating.
8. Remove rack from the water bath after the total 30 minutes incubation; centrifuge the titration tubes and color standard tubes at 600 × g for 5 minutes.
9. Compare each tube with the color standards; record the percent hemolysis,

Table 11.4 Complement Titration

| Tube Number | Reagent (mL) | | |
	VBD	Complement (1:400)	Sensitized RBC
0 (cell control)	0.80	0	0.20
1	0.60	0.20	0.20
2	0.55	0.25	0.20
3	0.50	0.30	0.20
4	0.40	0.40	0.20

interpolating to the nearest 5% if an exact match does not occur.

10. Calculate the average percent hemolysis for each duplicate pair of tubes.

11. If lysis is greater than 90%, or less than 10%, repeat the procedure using complement that has been diluted more, or less, than 1:400.

11.4.4.d Computation of the Complement Volume producing 50% Hemolysis (CH_{50})

1. Use Table 11.5 to determine the ratio (X/100-X) for the average percent hemolysis calculated for each pair of tubes in 11.4.4.d, step 10.

2. For each of the four tubes containing complement, plot on log graph paper the ratio value (X/100-X) on the X axis versus the volume of 1:400 complement in mL on the Y axis (Figure 11.3).

3. Examine the graph to see whether two of the plotted points are on the left and two on the right of the vertical 1.0 line; if so, proceed with the next step, other-

wise the complement titration must be repeated. If more than two points are right of the line, too much complement was used; repeat using a more dilute complement (1:500). If more than two points occur left of the line, there was insufficient complement in the test; repeat using a less dilute complement (1:300).

4. Draw a line between the two plotted points for tubes 1 and 2, and find its midpoint.

5. Draw a line between the two plotted points for tubes 3 and 4, and find its midpoint.

6. Draw a line between the two midpoints, and determine the slope: From any point near the left end of the line, measure horizontally to a point 10 cm to the right; measure the vertical distance, in centimeters, from that point upward to the midpoint line; divide the vertical distance by 10 cm to obtain the slope.

7. Test reproducibility requires that the slope be between 0.18 to 0.22; if the

Table 11.5 Percent Ratio

X	X/100-X	X	X/100-X	X	X/100-X
10	0.111	40	0.67	70	2.33
15	0.176	45	0.82	75	3.00
20	0.250	50	1.00	80	4.00
25	0.330	55	1.22	85	5.70
30	0.430	60	1.50	90	9.00
35	0.540	65	1.86		

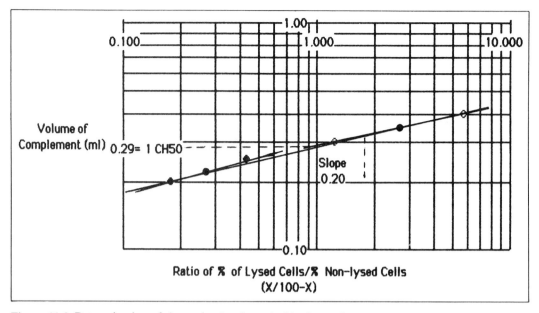

Figure 11.3 Determination of the optimal volume (mL) of complement.

computed slope is not within this range, the complement titration must be repeated.

8. From the intersection of the vertical 1.0 line with the line joining the two midpoints, draw a dotted horizontal line to the Y axis. Read the volume in mL of the 1:400 complement dilution at the point where the dotted line intersects the Y axis. This volume contains one CH_{50}; this volume is multiplied by 5 to obtain the volume for 5 CH_{50}.

9. The test requires 5 CH_{50} in 0.4 mL. The dilution of complement which will provide this (X∗) is calculated by:

$$\frac{\text{Dilution of complement used in titration}}{\text{Volume containing 5 } CH_{50}} = \frac{X*}{0.4}$$

For example: if the volume of 1:400 complement dilution containing 5 CH_{50} is 1.3 ml:

$$\frac{400}{1.3} = \frac{X*}{0.4} \text{ and } X* = 123$$

Thus, in this example, complement should be used in the test at a dilution of 1:123.

11.4.5 Antigen Titration

11.4.5.a Setup

1. Label a 96 well microtiter plate according to Figure 11.4; each square represents a well and the diagram indicates which reactants will be added to each well.

2. Prepare the 1:8 starting antiserum dilution by adding 0.2 mL of specific antiserum to a 12 × 75 mm test tube; add 1.4 mL VBD and mix; inactivate in a 56 °C water bath for 30 minutes; cool to room temperature.

3. Label tubes for the known positive antigen at its optimal dilution and for the antigen tissue control.

4. Label 13 × 100 mm test tubes for serial twofold dilutions of the test antigen; add 1.0 mL cold VBD to each of the antigen dilution tubes; if the test is to identify an unknown antigen, use the antigen undiluted, and at 1:2, 1:4, and 1:8 dilutions. Include a tube of undiluted tissue control for this antigen. If the test is to determine the optimal dilution of a new

ANTISERUM DILUTION

	8	16	32	64	128	256	512	1024	
A-1	S								antigen dilutions
A-2	S								
A-3	S								
A-4	S								
A-5	S								
+A	S								reference antigen
+TAC	S								tissue control
SC	S								serum control
A-1	5u	2.5u	1.25u	V 5u		5u	2.5u	1.25u	A-5 — complement controls
A-2	5u	2.5u	1.25u	V 2.5u		5u	2.5u	1.25u	+A
A-3	5u	2.5u	1.25u	V 1.25u		5u	2.5u	1.25u	+TAC
A-4	5u	2.5u	1.25u	CC	CC	S 5u	S 2.5u	S 1.25u	SC
	5	2.5	1.25			5	2.5	1.25	

UNITS OF COMPLEMENT

Figure 11.4 Antigen Titration Microtiter Plate

lot of a previously used known CF antigen, dilute the antigen so as to include two dilutions above and two below the previous lot optimal dilution in the titration. The same tissue control used for the known positive antigen (step 3) may be used.

11.4.5.b Addition of Reactants to the Microtiter Plate

1. Deliver 0.05 mL of "heat" inactivated antiserum diluted 1:8 to each of the wells indicated with an "S" in Figure 11.4.

2. Using a calibrated dropper pipet, deliver 0.025 mL VBD to each of the remaining serum dilution wells.

3. Prepare serial twofold serum dilutions; (an automatic programmable dilutor, microtiter pipetting devices with disposable tips, or pre-tested metal loops may be used); transfer 0.025 mL of serum from the well labeled 8 into the well labeled 16, mix with the VBD, and continue to transfer and mix across the plate.

4. All reagents should be kept cold, preferably in a ice bath, prior to delivery to the microtiter plate.

5. Prepare the optimal dilution of the known positive antigen with VBD; a tissue antigen control (TAC) is prepared from an uninfected host (e.g., cell culture, mouse brain) of the same type in which the antigen was produced; a tissue control must be used for each type of antigen tested.

6. Prepare serial twofold dilutions of test antigen in the labeled test tubes: Add 1.0 mL of antigen to the tube labeled 1:2; mix well with the pipette; prepare the 1:4 dilution by adding 1.0 mL of the 1:2 dilution to the VBD in the tube labeled 1:4; mix well with the pipette and continue serial transfer through the last required dilution.

7. Using a 0.025 mL dropper pipette, add 0.025 mL of antigen dilution 1 to the eight wells labeled A-1, and to the three wells of the complement controls labeled A-1; repeat for each of the remaining antigen dilutions; use row A-5 for the test TAC if it differs from the tissue of the known positive antigen.

8. Using a 0.025 mL dropper pipette, add 0.025 mL of the optimal dilution of the known positive antigen to the eight +A wells and to the three wells of the complement control labeled +A; similarly add the positive TAC to the 8 +TAC wells and the three +TAC complement control wells.

9. Using a 0.025 mL dropper pipette, add 0.025 mL of VBD to the eight serum control wells (SC), and to each of the complement control wells; add 0.05 mL of VBD to the three VBD control wells (V); add 0.1 mL VBD to each cell control well (CC).

10. Shake the plates on an orbital shaker for 20 to 30 seconds; cover and hold the plates in the refrigerator while preparing the diluted complement.

11.4.5.c Preparation of Complement

1. Prepare 5 mL of complement per plate, plus an additional 5 mL for the complement controls, at the dilution previously determined by complement titration; to calculate the volume of undiluted complement required, multiply the total volume needed times the reciprocal of the dilution determined in the complement titration. For example: 40 mL of a 1:133 dilution of complement are needed; the volume of undiluted complement necessary = 40/133 = 0.3 mL; therefore, use 0.3 mL undiluted complement + 39.7 mL VBD.

2. Add the calculated volume of cold VBD to a flask; add the calculated volume of complement dropwise to the VBD; mix by swirling and allow to stand 20 minutes on ice or in the refrigerator.

3. Label a 15 × 125 mm test tube "2.5"; add 2.0 mL VBD and 2.0 mL of the 5 CH_{50} complement dilution; mix gently. Label a second tube "1.25"; add 3.0 mL VBD and 1.0 mL of the 5 CH_{50} complement dilution; mix gently.

4. Using a 0.05 mL dropper pipette add 0.05 mL of the 5 CH_{50} complement dilution to all wells in the antigen titration, and to the wells labeled 5u (u = units) in the complement control.

Table 11.6 Acceptable Percent Hemolysis in Control Test Reactions

Type of Control	Number of CH_{50} units		
	5	2.5	1.25
Antigen	100	85–100	0–75
VBD	100	90–100	40–75
Serum	100	90–100	0–75
Tissue	100	85–100	0–75

5. Using a 0.05 mL dropper pipette add 0.05 mL of the 2.5 CH_{50} complement dilution to all wells labeled 2.5 in the complement control; proceed likewise with the 1.25 CH_{50} dilution.

6. Shake the plates on an orbital mixer for 20 to 30 seconds; cover and incubate at 4 °C for 15 to 18 hours (refrigerate overnight).

11.4.5.d Preparation of Hemolysin Sensitized SRBC

1. Determine the volume of sensitized SRBC needed for the test by multiplying the total number of plates plus one times 3 mL per plate.

2. Prepare a volume of optimally diluted hemolysin equal to one half of the total volume calculated in step 1 plus about 10%.

3. Swirl the 2.8% standardized SRBC suspension to achieve a homogeneous suspension; into a small flask, aliquot a volume of suspended SRBC equal to one half the total volume calculated in step 1.

4. Place the flask of SRBC on a rotating platform, and while the flask is rotating, use a 10 mL pipette to add a volume of the diluted hemolysin equal to the volume of SRBC suspension; mix well.

5. Incubate the cells in a 37 °C water bath for 15 minutes, swirling to mix after 7 minutes.

6. Meanwhile, remove the prepared test plates from the refrigerator and allow to warm to room temperature.

7. Using a 0.025 mL dropper pipette add 0.025 mL of sensitized SRBC to each well on the plate; tap the plate edge to mix; seal the plate with transparent sealing tape; vibrate the plate for 20 seconds on an orbital shaker.

8. Place the plates on a level shelf in a 37 °C incubator for 30 minutes; do not stack the plates; tap plate edge at 15 minutes to resuspend the SRBC.

11.4.5.e Reading the Test

1. If color standards less than 24 hours old are available, they may be used, otherwise prepare fresh color standards (11.4.3).

2. Centrifuge the plates at 300 × g for 5 minutes.

3. Read and record the percentage hemolysis of the complement controls using the color standards; interpolate to the nearest 5%.

4. Compare the readings with those in Table 11.6 to determine if they are acceptable: any values less than those shown are considered to be anticomplementary, therefore the test is not acceptable; the test should then be repeated.

5. Read and record the percentage hemolysis for each titration well; an example is given in Table 11.7.

6. Draw a line through the approximate 30% hemolysis endpoints of each anti-

Table 11.7 Determination of Optimal Antigen Dilution

	Antigen Dilution	Reference Antiserum Dilutions								Complement Controls		
		1:8	1:16	1:32	1:64	1:128	1:256	1:512	1:1024	5u	2.5u	1.25u
Test Antigen	1:2	0*	0	0	10	40	70	100	100	70	50	0
	1:4	0	0	0	30	45	80	100	100	100	85	0
	1:8	0	0	0	0	20	60	90	100	100	95	10
	1:16	0	0	0	0	20	70	100	100	100	100	40
	1:32	0	0	0	20	50	100	100	100	100	100	40
Reference Antigen	optimal	0	0	0	0	10	50	100	100	100	100	35
Tissue Control	optimal	100	100	100	100	100	100	100	100	100	100	40
Serum Control	none	100	100	100	100	100	100	100	100	100	100	40
VBD Control	none	—	—	—	—	—	—	—	—	100	90	60

* = Values indicate percent hemolysis.
The 1:2 dilution of the test antigen is anticomplementary. The optimal dilution is 1:8, based on the criteria described in the text.

gen dilution, interpolating where necessary; do not include dilutions showing anticomplementary activity.

7. The optimal antigen dilution is the one at which the greatest amount of complement is fixed, based on the following criteria:

All dilutions showing anticomplementary activity plus the next highest dilution must be excluded from the curve.

If two antigen dilutions give identical reactions within the optimal dilution curve, select the one which gives greater fixation to the right of the curve.

If two antigen dilutions give identical fixation reactions, select the lower dilution.

The titer of the reference antiserum with the test antigen must be within one twofold dilution of the titer obtained with the reference antigen.

8. The antigen titration procedure may be used to identify an unknown antigen by testing against known antisera; the degree of fixation by the unknown antigen and specific antiserum is compared with that of the reference antigen with its specific antiserum.

11.4.6 Diagnostic Serology Test

11.4.6.a Setup

1. Each serum specimen may be tested against any number of antigens; however, for each antigen both known positive and known negative antisera must be run concurrently in the test.

2. Label one 12 × 75 mm test tube for each unknown serum, known positive and known negative serum to be tested.

3. Aliquot 0.2 mL of each serum into the appropriate tube; to each tube add 1.4 mL cold VBD, resulting in a 1:8 starting serum dilution.

4. Inactivate the serum(a) by placing the tubes in a 56 °C water bath for 30 minutes.

5. Label microtiter plates; an example is given in Figure 11.5; if numerous sera are to be run, it is advantageous to arrange the plates so the unknown sera are batched for each antigen tested, all the test serum controls are on one plate, the antigen controls (known positives and negative sera) are on another plate, and the complement controls are on a third plate; the exact arrangement will be dictated in part by the equipment

available to dilute the reagents and samples.

6. Prepare the test antigen at its previously determined optimal dilution; to determine the total volume required, multiply the number of wells receiving the test antigen by 0.025 mL, add 0.5 to 1 mL to allow excess for some loss of sample during pipetting. Prepare the tissue control for each antigen at the same dilution as the test antigen; compute the total volume needed in the same manner; store all antigens at 4 °C until needed.

11.4.6.b Addition of Reactants to the Microtiter Plate

1. Deliver 0.05 mL of each cooled, inactivated serum to the appropriately labeled wells (1:8) on the test and control plates.

2. Using a calibrated dropper pipette, deliver 0.025 mL VBD to each of the remaining serum dilution wells.

3. Prepare serial twofold serum dilutions; an automatic programmable dilutor, microtiter pipetting devices with disposable tips, or pretested metal loops may be used. Transfer 0.025 mL of serum from the well labeled 8 into the well labeled 16, mix with the VBD, and continue to transfer and mix across the plate.

4. Using a calibrated dropper pipette, deliver 0.025 mL of the optimal dilution of test antigen to each of the wells in the antigen portion of the test; add antigen to the positive and negative control serum, and the complement control wells.

5. Using a calibrated dropper pipette, deliver 0.025 mL of the diluted tissue control to each of the wells in the tissue control portion of the test plate; add this volume also to the positive and negative control serum, and the complement control wells.

6. Using a calibrated dropper pipette, deliver 0.025 mL of VBD to the complement control wells, 0.05 mL to the VBD control wells, and 0.1 mL to the cell control wells.

7. Shake the plates on an orbital shaker for 20 to 30 seconds; cover and hold the plates in the refrigerator while preparing the diluted complement.

11.4.6.c Preparation of Complement (see 11.4.5.c)

11.4.6.d Preparation of Hemolysin Sensitized SRBC (see 11.4.5.d)

11.4.6.e Reading the Test

1. If color standards less than 24 hours old are available, they may be used, otherwise prepare fresh color standards (11.4.3).

2. Centrifuge the plates at $300 \times$ g for 5 minutes.

3. Read and record the percentage hemolysis of the complement controls using the color standards; interpolate to the nearest 5%.

4. Compare the readings with those in Table 11.6 to determine if they are acceptable: any values less than shown are considered to be anticomplementary, thus not acceptable; the test should then be repeated.

5. Read and record the percent hemolysis in each well of the test; the serum titer is the highest serum dilution resulting in less than 30% hemolysis.

6. Diagnostic sera that do not react with the tissue control and are not anticomplementary (i.e., do not have titer in the serum control plate) may have their results reported; if the serum is anticomplementary or reacts with tissue at the 1:8 dilution, but not at higher dilutions, reactions observed at the higher dilutions are considered valid.

		Components (ml)	Serum Dilution					
			1:8	1:16	1:32	1:64	1:128	1:256
Unknown Serum Assay	Test	Unknown Serum Test Antigen 5 CH$_{50}$	0.025 0.025 → 0.050					
	Serum Controls	Unknown Serum VBD 5 CH$_{50}$	0.025 0.025 → 0.050					
	Tissue Controls	Unknown Serum Normal Tissue Antigen* 5 CH$_{50}$	0.025 0.025 → 0.050					
Known Positive Serum Assay	Test	Known Positive Serum Test Antigen 5 CH$_{50}$	0.025 0.025 → 0.050					
	Serum Controls	Known Positive Serum VBD 5 CH$_{50}$	0.025 0.025 → 0.050					
	Tissue Controls	Known Positive Serum Normal Tissue Antigen[a] 5 CH$_{50}$	0.025 0.025 → 0.050					
Known Negative Serum Assay	Test	Known Negative Serum Test Antigen 5 CH$_{50}$	0.025 0.025 → 0.050					
	Serum Controls	Known Negative Serum VBD 5 CH$_{50}$	0.025 0.025 → 0.050					
	Tissue Controls	Known Negative Serum Normal Tissue Antigen[a] 5 CH$_{50}$	0.025 0.025 → 0.050					
	Cell Control	VBD	0.1					

[a] Normal tissue antigen used at same dilution as test antigen.
[b] The 2.5-unit and 1:25-unit are a 1:2 and a 1:4 dilution of the 5 CH$_{50}$ in 0.05 ml of complement, respectively.

Figure 11.5 Diagnostic Serology Microtiter Test

7. If the serum reacts with the tissue control, it must be tissue treated and retested. Mix 0.2 mL of undiluted serum with 0.2 mL of the tissue control in a 12 × 75 mm test tube, making a 1:2 dilution of the serum; incubate in a 37 °C water bath for one hour. Add 1.2 mL VBD, making a 1:8 dilution of the serum. Inactivate in a 56 °C water bath for 30 minutes. The serum is now ready for retesting.

8. If the serum is anticomplementary, it is

Complement Controls for Test

Complement Controls	Components (ml)	5-Unit CH_{50}	2.5-Unit[b] CH_{50}	1.25-Unit[b] CH_{50}
Antigen	Test Antigen	0.025	0.025	0.025
	VBD	0.025	0.025	0.025
	Complement	0.050	0.050	0.050
Tissue[a]	Normal Tissue Antigen	0.025	0.025	0.025
	VBD	0.025	0.025	0.025
	Complement	0.050	0.050	0.050
VBD	VBD	0.050	0.050	0.050
	Complement	0.050	0.050	0.050

Figure 11.5 (*continued*)

treated in the same manner, substituting complement for the tissue control.

9. If the serum is both anticomplementary and tissue reactive, treat 0.2 mL serum with 0.2 mL tissue and 0.4 mL complement; after the one hour incubation at 37 °C add 0.8 mL VBD and inactivate in a 56 °C water bath for 30 minutes.

10. Serum lipids, if excessive, may interfere with the test; treat 0.5 mL serum with 0.5 mL Freon; mix well on a vortex mixer; centrifuge at 600 × g for 15 minutes; carefully aspirate the supernatant fluid containing the serum; proceed with 11.4.6.a.3.

REFERENCES

Ada, G.L. and Perry, B.T. 1954. Studies on the soluble complement-fixing antigens of influenza virus III: The nature of the antigens. Austral. J. Exp. Biol. Med. Sci. 32:177–186.

Bankowski, R.A., Wichmann, R.W., and Kummer, M. 1953. A complement-fixation test for identification of immunological types of the virus of vesicular exanthema of swine. Am. J. Vet. Res. 14:145–149.

Beveridge, W.I.B. and Burnet, F.M. 1946. The cultivation of viruses and rickettsiae in the chick embryo. Medical Research Council, Special Report Series No. 256. London: His Majesty's Statistics Office, pp. 1–92.

Black, F.L. and Melnick, J.L. 1954. The specificity of the complement fixation test in poliomyelitis. Yale J. Biol. Med. 26:385–393.

Black, F.L. and Melnick, J.L. 1955. Appearance of soluble and cross-reactive complement-fixing antigens on treatment of poliovirus with formalin. Proc. Soc. Exp. Biol. Med. 89:353–355.

Casals, J. and Palacios, R. 1941. The complement fixation test in the diagnosis of virus infections of the central nervous system. J. Exp. Med. 99:429–449.

Casals, J. and Palacios, R. 1949. Acetone-ether extracted antigens for complement fixation with certain neurotropic viruses. Proc. Soc. Exp. Biol. Med. 70:339–343.

Casey, H.L. 1965. Adaptation of LBCF method of microtechnique. In Standard Diagnostic Complement-Fixation Method and Adaptation to Microtest. Public Health Monograph

No. 74, Public Health Service Publication No. 1228. Washington, D.C.: U.S. Government Printing Office.

Chambers, L.A., Cohen, S.S., and Clawson, J.R. 1950. Studies on commercial typhus vaccines II. The antigenic fractions of disrupted epidemic typhus rickettsiae. J. Immunol. 65:459–463.

Churcher, G.M., Sheffield, F.W., and Smith, W. 1959. Poliomyelitis virus flocculation: The reactivity of unconcentrated cell-culture fluids. Br. J. Exp. Pathol. 40:87–95.

Ciuca, A. 1929. The reaction of complement-fixation in foot-and-mouth disease as a means of identifying the different types of virus. J. Hyg. 28:325–339.

Clark, D.H. and Casals, J. 1958. Techniques for hemagglutination and hemagglutination-inhibition with arthropod-borne viruses. Am. J. Trop. Med. Hyg. 7:561–573.

Cohen, G.H. and Wilcox, W.C. 1966. Soluble antigens of vaccinia infected mammalian cells. I. Separation of virus-induced soluble antigens into two classes on the basis of physical characteristics. J. Bacteriol. 92: 676–686.

Craigie, J. and Wishart, F.O. 1936a. The complement fixation reaction in variola. Can. Pub. Health J. 27:371–379.

Craigie, J. and Wishart, F.O. 1936b. Studies on the soluble precipitable substances of vaccinia. II. The precipitable substances of dermal vaccine. J. Exp. Med. 64:819–830.

DeBoer, C.J. and Cox, H.R. 1947. Specific complement-fixing diagnostic antigens for neurotropic virus diseases. J. Immunol. 55: 193–204.

Duca, M., Moroisanu, V., Handrache, L., Ciubitaru, V., Ionescu, L., and Liscia, M. 1979. Efficiency of complement fixation (CF) test with internal nucleoprotein (NP) antigen and hemagglutination-inhibition (HI) test in the serodiagnosis of A and B influenza infections. Arch. Roum. Pathol. Exp. Microbiol. 38:331–337.

Dulbecco, R. 1952. Production of plaques in monolayer tissue cultures by single particles of an animal virus. Proc. Natl. Acad. Sci. 38:747–752.

Dzierzanowska, D., Kolodziejczyk, J., and

Lourie, I. 1986. Determination of cytomegalovirus antibody titers in sera of volunteer blood donors by use of complement fixation and two immunoenzymatic tests. Arch. Immunol. Ther. Exp. (Warsz) 34:397–402.

Ende, M., Van den Polson, A., and Turner, G.S. 1957. Experiments with the soluble antigen of rabies in suckling mouse brain. J. Hyg. 55:361–373.

Enders, J.F., Weller, T.H., and Robbins, F.C. 1949. Cultivation of the Lansing strain of poliomyelitis virus in cultures of various human embryonic tissues. Science 109:85–87.

Enders, J.F. and Peebles, T.C. 1954. Propagation in tissue cultures of cytopathogenic agents from patients with measles. Proc. Soc. Exp. Biol. Med. 80:277–286.

Espana, C. and Hammon, W.McD. 1948. An improved benzene-extracted complement-fixing antigen applied to the diagnosis of the arthropod-borne virus encephalitides. J. Immunol. 59:31–44.

Ferrante, P., Achilli, G., Gerna, G., and Bergamini, F. 1987. Subacute sclerosing panencephalitis: Detection of measles antibody in serum and cerebrospinal fluid by enzyme-linked immunosorbent assay, complement fixation, and hemagglutination inhibition. Microbiologica 10(1):111–118.

Fischer, G.S., Sweimler, W.I., and Kleger, B. 1986. Comparison of Mycoplasmelisa with complement fixation test for measurement of antibodies to *Mycoplasma pneumoniae*. Diagn. Micro. Infect. Dis. 4:139–145.

French, E.L. 1952. Murray Valley encephalitis. Isolation and characterization of the aetiological agent. Med. J. Australia 1:100–103.

Fulton, R.E. and Dininno, V.L. 1985. Rapid rate-kinetic turbidometric assay for quantitation of viral complement fixing antibodies. J. Virol. Methods 12:13–24.

Girardi, A.J., Warren, J., Goldman, C., and Jeffries, B. 1958. Growth and CF antigenicity of measles virus in cells deriving from human heart. Proc. Soc. Exp. Biol. Med. 98:18–22.

Goodpasture, E.W., Woodruff, A.M., and Buddingh, G.J. 1931. The cultivation of vaccine and other viruses in the chorio-allantoic membrane of chick embryos. Science 74: 371–372.

Goodpasture, E.W., Woodruff, A.M., and Buddingh, G.J. 1933. Use of embryo chick in investigation of certain pathological problems. South. Med. J. 26:418–420.

Halonen, P.E., Casey, H.L., Stewart, J.A., and Hall, A.D. 1967. Rubella complement fixing antigen prepared by alkaline extraction of virus grown in suspension culture of BHK-21 cells. Proc. Soc. Exp. Biol. Med. 125:167–172.

Hammon, W.McD. and Sather, G.E. 1969. Arboviruses. In E.H. Lennette and N.J. Schmidt (eds.), Diagnostic Procedures for Viral and Rickettsial Infections, 4th ed. American Public Health Association, pp. 227–280.

Havens, W.P. Jr., Watson, D.W., Green, R.H., Lavin, G.I., and Smadel, J.E. 1943. Complement fixation with neurotropic viruses. J. Exp. Med. 77:139–153.

Henle, G., Harris, S., and Henle, W. 1948. The reactivity of various human sera with mumps complement fixation antigens. J. Exp. Med. 88:133–147.

Henle, G., Harris, S., Henle, W., and Deinhardt, F. 1955. Propagation and primary isolation of mumps virus in tissue culture. Proc. Soc. Exp. Biol. Med. 89:556–560.

Hilleman, M.R. and Werner, J.H. 1954. Recovery of new agent from patients with acute respiratory illness. Proc. Soc. Exp. Biol. Med. 85:183–188.

Hishiyama, M., Tsurudome, M., Ito, Y., Yamada, A., and Sugiura, A. 1988. Complement mediated neutralization test for determination of mumps vaccine induced antibody. Vaccine 6:423–427.

Holmes, K.T., Hampson, A.W., Raison, R.L., Webster, R.G., O'Sullivan, W.J., and Mountford, C.E. 1982. A comparison of two antineuraminidase monoclonal antibodies by complement activation. Eur. J. Immunol. 12:523–526.

Howitt, B.F. 1937. The complement fixation reaction in experimental equine encephalomyelitis, lymphocytic choriomeningitis and St. Louis type of encephalitis. J. Immunol. 33:235–250.

Hoyle, L. 1952. Structure of the influenza virus. The relation between biological activity and chemical structure of virus fractions. J. Hyg. 50:229–245.

Hudson, L.D., Adelman, S., and Lewis, C.W. 1981. Pityriasis rosea. Viral complement fixation studies. J. Am. Acad. Dermatol. 4:544–546.

Huebner, R.J., Rowe, W.P., Ward, T.G., Parrott, R.H., and Bell, J.A. 1954. Adenoidal-pharyngeal conjunctival agents. A newly recognized group of common viruses of the respiratory system. N. Engl. J. Med. 251:1077–1086.

Inouye, S., Matsuno, S., and Kono, R. 1981. Difference in antibody reactivity between complement fixation and immune adherence hemagglutination tests with virus antigens. J. Clin. Micro. 14:241–246.

Julkunen, I., Kleemola, M., and Hovi, T. 1984. Serological diagnosis of influenza A and B infections by enzyme imunoassay. Comparison with the complement fixation test. J. Virol. Methods 9:7–14.

Kidd, J.G. and Friedewald, W.F. 1942. A natural antibody that reacts in vitro with a sedimentable constituent of normal tissue cells. I. Demonstration of the phenomenon. J. Exp. Med. 76:543–556.

LeBouvier, G.L., Laurence, G.D., Parfitt, E.M., Jennens, M.G., and Goffe, A. 1954. Typing of poliomyelitis virus by complement fixation. Lancet 2:531–532.

Lennette, E.H. 1969. General principles underlying laboratory diagnosis of viral and rickettsial infections. In E.H. Lennette and N.J. Schmidt (eds.), Diagnostic Procedures for Viral and Rickettsial Diseases, 4th ed. New York: American Public Health Association, pp. 1–65.

Lennette, E.H., Wiener, A., Neff, B.J., and Hoffman, M.N. 1956a. A chick embryo-derived complement-fixing antigen for western equine encephalomyelitis. Proc. Soc. Exp. Biol. Med. 92:575–577.

Lennette, E.H., Wiener, A., Ota, M.I., Fujimoto, F.Y., and Hoffman, M.N. 1956b. Rapid identification of isolates of western equine encephalomyelitis virus by the complement fixation technique. Am. J. Hyg. 64:270–275.

Lichter, A.G. 1953. Observation on anticomplementary reactions. A.M.A. Arch. Dermat. Syph. 67:362–368.

Lief, F.S. and Henle, W. 1956a. Studies on the soluble antigen of influenza virus. I. The release of S antigen from elementary bodies by treatment with ether. Virology 2:753–771.

Lief, F.S. and Henle, W. 1956b. Studies on the soluble antigen of influenza virus. III. The decreased incorporation of S antigen into elementary bodies of increasing incompleteness. Virology 2:772–797.

Maltaner, F. 1946. Significance of thromboplastic activity of antigens used in complement fixation tests. Proc. Soc. Exp. Biol. Med. 62:302–204.

Midthun, K., Pang, L.I., Flores, J., Kapikian, A.Z. 1989. Comparison of immunoglobin A (IgA), IgG, and IgM enzyme-linked immunosorbent assays, plaque reduction neutralization assay, and complement fixation in detecting seroresponses to rotavirus vaccine candidates. J. Clin. Micro. 27:2799–2804.

Miller, H., McCulloch, B., Landini, M.P., and Rossier, E. 1989. Comparison of immunoblotting with other serological methods and virus isolation for the early detection of primary cytomegalovirus infection in allograft recipients. J. Clin. Micro. 27:2672–2677.

Ochiai, H., Shibata, M., Sato, S., and Niwayama, S. 1987. Single radial complement fixation test using NP containing plates: A simple and sensitive method for the detection of influenza infection. J. Virol. Methods 15:151–158.

Ossewaarde, J.M. and deBooij, A.D. 1989. Development of a monoclonal antibody for use as an amboceptor in complement fixation tests. J. Virol. Methods 25:13–20.

Panucker, K., Lief, F.S., and Henle, W. 1956. Studies on the soluble antigen of influenza virus, IV. Fractionation of elementary bodies labeled with radio-active phosphorus. Virology 2:798–810.

Pauri, P., Bagnarelli, P., and Clementi. M. 1981. Complement fixation test for rotavirus detection: Comparison and analysis of different methods to reduce anticomplementary activity of some specimens. J. Virol. Methods 3:329–335.

Puolakkainen, M., Kousa, M., and Saikku, P. 1987. Clinical conditions associated with positive complement fixation serology for *Chlamydiae*. Epidem. Infect. 98:101–108.

Ravaoarinoro, M., Reginster, M., Doraal, A., and Sondag-Thull, D. 1984. Comparison between an enzyme-linked immunosorbent assay and a complement fixation test in assessing the age related acquisition of cytomegalovirus antibodies in groups of the population living in Belgium. Acta Clin. Belg. 39:6–12.

Rice, C.E. 1948a. Inhibitory effects of certain avian and mammalian antisera in specific complement fixation systems. J. Immunol. 59:365–378.

Rice, C.E. 1948b. Some factors influencing the selection of a complement-fixation method. II. Parallel use of the direct and indirect techniques. J. Immunol. 60:11–21.

Robbins, F.C., Enders, J.F., and Weller, T.H. 1950. Cytopathogenic effect of poliomyelitis viruses in vitro on human embryonic tissues. Proc. Soc. Exp. Biol. Med. 75:370–374.

Robbins, F.C., Enders, J.F., Weller, T.H., and Florentino, G.L. 1951. Studies on the cultivation of poliomyelitis viruses in tissue culture. V. The direct isolation and serologic identification of virus strains in tissue culture from patients with nonparalytic and paralytic poliomyelitis. Am. J. Hyg. 54: 286–293.

Rowe, W.P., Huebner, R.J., Gilmore, L.K., Parrott, R.H., and Ward, T.G. 1953. Isolation of a cytopathogenic agent from human adenoids undergoing spontaneous degeneration in tissue culture. Proc. Soc. Exp. Biol. Med. 84:570–573.

Ruckle, G. and Rogers, K.D. 1957. Studies with measles virus. II. Isolation of virus and immunological studies in persons who have had the natural disease. J. Immunol. 78: 341–355.

Sather, G.E. and Hammon, W.McD. 1967. Antigenic patterns within the California-Encephalitis-Virus group. Am. J. Trop. Med. Hyg. 16:548–557.

Sato, S., Motoda, S., Iwase, I., and Jo, K. 1983.

Single radial complement fixation test using complement film. Assay of the antibody response to strain and type specific antigens of influenza. J. Virol. Methods 7:57–64.

Sato, S., Ochiai, H., and Niwayama, S. 1988. Application of the single radial complement fixation test for serodiagnosis of influenza, respiratory synctial, mumps, adeno type 3 and herpes simplex type 1 virus infections. J. Med. Virol. 24:395–404.

Schafer, W. and Zillig, W. 1954. Über den Aufbau des Virus-Elementarteilchens der klassischen Geflügelpest. I. Gewinnung, physikalisch-chemische und biologische Eigenschaften einiger Spaltprodukte. Ztschr. Naturforsch 9b:779–788.

Schmidt, N.J. 1957. An inquiry into the use of the complement fixation test for the typing of poliomyelitis viruses. Am. J. Hyg. 66:119–130.

Schmidt, N.J. 1969. Tissue culture methods and procedures for diagnostic virology. In E.H. Lennette and N.J. Schmidt (eds.), Diagnostic Procedures for Viral and Rickettsial Disease, 4th Ed. New York: American Public Health Association, pp. 79–178.

Schmidt, N.J. and Lennette, E.H. 1956. Modification of the homotypic specificity of poliomyelitis complement fixing antigens by heat. J. Exp. Med. 104:99–102.

Schmidt, N.J., Lennette, E.H., Doleman, J.H., and Hagens, S.J. 1957. Factors influencing the potency of poliomyelitis complement-fixing antigens produced in tissue culture systems. Am. J. Hyg. 66:1–9.

Sethi, J., Pei, D., and Hirshaut, Y. 1981. Choice and specificity of complement in complement fixation assay. J. Clin. Micro. 13:888–890.

Sever, J.L. 1962. Application of a microtechnique to viral serological investigations. J. Immunol. 88:320–329.

Smadel, J.E., Baird, R.D., and Wall, M.J. 1939. Complement fixation in infections with the virus of lymphocytic choriomeningitis. Proc. Soc. Exp. Biol. Med. 40:71–73.

Sosa-Martinez, J. and Lennette, E.H. 1955. Studies on a complement fixation test for herpes simplex. J. Bacteriol. 70:205–215.

Stevenson, W.D.H. and Butler, M.B.E. 1933. Dermal strains of vaccinia virus grown on the chorio-allantoic membrane of chick embryos. Lancet 1(225):228–230.

Suggs, M.R., Jr., Casey, H.L., Sligh, D.D., Fodor, A.R., and McLimmans, W.F. 1961. A batch-type concentration and purification procedure for poliovirus complement fixing antigen. J. Bacteriol. 82:789–791.

Svedmyr, A., Enders, J.R., and Holloway, A. 1952. Complement fixation with Brunhilde and Lansing poliomyelitis viruses propagated in tissue culture. Proc. Soc. Exp. Biol. Med. 79:296–309.

Svedmyr, A., Enders, J.R., and Halloway, A. 1953. Complement fixation with the three types of poliomyelitis viruses propagated in tissue culture. Am. J. Hyg. 57:60–70.

Taguchi, F. 1988. New complement fixation test with peroxidase labeled complement C1q for direct and quantative determination of antibodies to herpes simplex virus. Microbiol. Immunol. 32:1167–1173.

Takatsy, G. 1950. A new method for the preparation of serial dilutions in a quick and accurate way. Kiserletes Orvostudomany 2:293–296.

Taylor-Robinson, D. and Downie, A.W. 1959. Chickenpox and herpes zoster. I. Complement fixation studies. Br. J. Exp. Pathol. 40:398–409.

Thorn, J.J., Oxholm, P., and Andersen, H.K. 1988. High levels of complement fixing antibodies against cytomegalovirus in patients with primary Sjogren's syndrome. Clin. Exp. Rheumatol. 6:71–74.

Weller, T.H. and Witton, H.M. 1958. The etiologic agents of varicella and herpes zoster. Serologic studies with the viruses as propagated in vitro. J. Exp. Med. 108:869–890.

Westwood, J.C.N., Appleyard, G., Taylor-Robinson, D., and Zwartouw, H.T. 1960. The production of high titre poliovirus in concentrated suspensions of tissue culture cells. Br. J. Exp. Pathol. 41:105–111.

Westwood, J.C.N., Zwartouw, H.T., Appleyard, G., and Titmuss, D.H.J. 1965. Comparison of the soluble antigens and virus particle antigens of vaccinia virus. J. Gen. Microbiol. 38:47–53.

Whitney, E., Kraft, L.M., Lawson, W.B., and Gordon, I. 1953. Noninfectious complement-fixing antigen from embryonated hens' eggs infected with lymphocytic choriomeningitis virus. Proc. Soc. Am. Bact. p. 50.

Woodruff, A.M. and Goodpasture, E.W. 1931. The susceptibility of the chorioallantoic membrane of chick embryos to infection with the fowl-pox virus. Am. J. Pathol. 7:209–222.

Wreghitt, T.G. and Sillis, M. 1987. An investigation of the *Mycoplasma pneumoniae* infections in Cambridge in 1983 using mu-capture enzyme-linked immunosorbent assay (ELISA), indirect immunofluorescence (IF) and complement fixation (CF) tests. Isr. J. Med. Sci. 23:704–708.

12

Neutralization

Harold C. Ballew

12.1 INTRODUCTION

Neutralization of a virus is defined as the loss of infectivity through reaction of the virus with specific antibody. To perform the virus neutralization test, virus and serum are mixed under appropriate conditions, incubated, and then used to inoculate a susceptible living host to detect unneutralized virus. The presence of unneutralized virus may be detected by reactions such as cytopathic effect (CPE), plaque formation, and metabolic inhibition in cell cultures; death of or paralysis in animals; and pock formation on the chorioallantoic membrane (CAM) of embryonated hens' eggs.

The neutralization test has been used in virology longer than any other serologic procedure. Despite its relative antiquity, neutralization is one of the most specific and widely used serologic procedures in diagnostic virology. Neutralization techniques can be used in virology to identify a virus isolate or to measure the antibody response of an individual to a virus. Because of its high immunologic specificity, the neutralization test is often the standard against which the specificity of other serologic procedures is evaluated.

One of the first principles formulated in virus serologic testing is the so-called "percentage law" (Andrewes and Elford, 1933). This law states that when virus is added to excess antibody, the percentage of virus not neutralized is the same regardless of the amount of virus added. With the advent of plaquing procedures for viruses, the neutralization reaction could be accurately evaluated (Dulbecco et al, 1956). In their studies on the kinetics of the reaction, they found that there is a linear interdependence between the rate of neutralization and the concentration of antibody, and there is always a fraction of virus that is not inactivated.

Other investigators who have helped establish the basic properties of the neutralization test are Tyrrell and Horsfall (1953), Salk et al (1954), and Mandel (1960). The contributions of these and other investigators to the neutralization test are not described in this chapter but are adequately reviewed in Horsfall and Tamm (1960) and Maramorosch and Koprowski (1967).

229

12.2 STANDARDIZATION OF TEST MATERIALS

Before the neutralization test is performed, the known components that are to be used must be standardized. To identify a virus isolate, a known pretitered antiserum or standardized serum pool is used. Conversely, to measure the antibody response of an individual to a virus, a known pretitered virus is employed.

12.2.1 Virus

The known virus consists of extracts from infected tissues, or fluids from cell cultures and embryonated hens' eggs. This virus can be purchased commercially or prepared by inoculating a susceptible host system with a stock virus and harvesting it at the optimal time. This known virus should always be titrated in the host system to be used for the neutralization test, whether purchased commercially or prepared by host system inoculation.

To titrate a known virus or virus isolate, prepare serial tenfold dilutions in a maintenance medium and inoculate a susceptible host system with fixed volumes of each dilution. Observe the host for signs of infection, which indicates virus that is not neutralized. The virus endpoint titer is the reciprocal of the highest dilution of virus that infects 50% of the host systems. This endpoint dilution contains one 50% tissue culture infecting dose ($TCID_{50}$) of virus per unit volume, or the amount of virus that will infect 50% of the cell cultures inoculated. In animals, if death is the criterion employed, then this endpoint dilution is called the 50% lethal dose (LD_{50}). The concentration of virus generally used in the neutralization test is 100 TCD_{50} or 100 LD_{50} per unit volume (Figure 12.1).

12.2.2 Serum

A specific immune serum can be purchased commercially or prepared by immunizing susceptible animals and harvesting the serum at the optimal time. The antiserum should always be titrated in the neutralization test against its homologous virus, whether it is purchased commercially or prepared by immunization.

To titrate a specific immune serum or test serum, prepare serial twofold dilutions of the serum and mix each dilution with an equal volume of standardized virus (usually 100 TCD_{50}). The virus and serum mixtures are usually incubated for one hour at room temperature or at 37 °C. The time and temperature for incubating the virus and serum mixtures varies with different viruses. Inoculate a susceptible host system with each virus–serum mixture. The serum antibody titer is the reciprocal of the highest dilution of the antiserum protecting against the virus. The endpoint dilution contains one antibody (Ab) unit per unit volume. The standardized concentration of antiserum generally used in the neutralization test is 20 antibody units per unit volume (Figure 12.2). In Figure 12.2, 1 Ab unit/0.1 mL is used as an example to demonstrate how antibody units are calculated.

Earlier studies have shown that the neutralization titer of a serum is reduced in certain circumstances by heat, dilution, repeated freezing–thawing, and prolonged storage at 4 °C. Part of the neutralization titer may be restored by adding "accessory factors," undefined substances that often are present in freshly collected serum or serum that has been maintained in a frozen state, to the virus–serum mixtures when they are prepared. However, it is usually possible to detect seroconversion in the absence of accessory factors. In addition, these factors may be eliminated by heating the sera to be tested at 56 °C for 30 minutes. Many laboratories do nothing about these factors. The sera may be used without inactivation or without the addition of fresh serum. There is no general agreement on the importance of accessory factors, but accessory factors and nonspecific neutral-

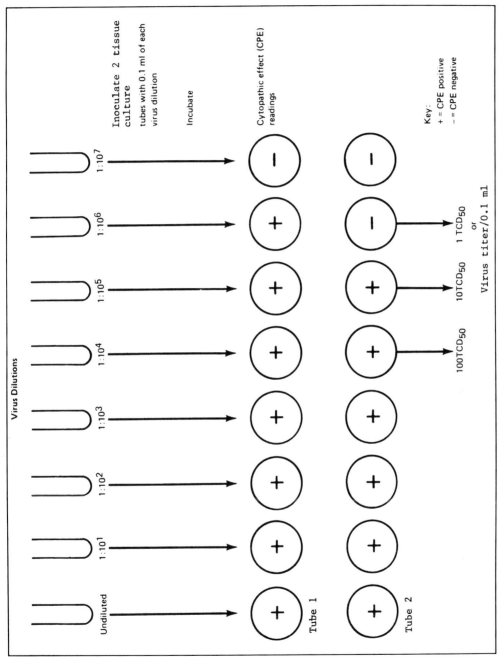

Figure 12.1 This virus titration demonstrates that the endpoint titer is the highest dilution of the virus that infects 50% of the cell cultures tubes inoculated.

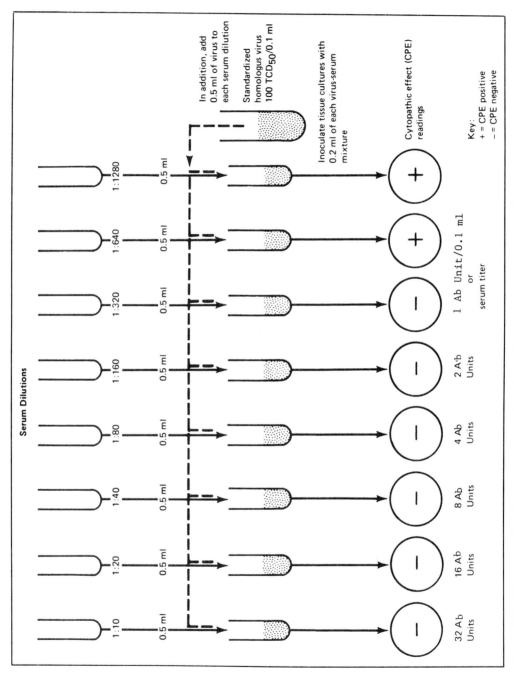

Figure 12.2 This serum titration demonstrates that the antibody titer is the highest dilution of the antiserum protecting against the standardized virus.

izing substances have been discussed in detail by Ginsberg and Horsfall (1949), Sabin (1950), Tyrrell and Horsfall (1953), and Mandel (1960).

12.2.3 Host System

One of the basic requirements for all neutralization tests is a living host system to demonstrate unneutralized virus. In the neutralization procedure, virus and serum are mixed, incubated, and then injected into or exposed to a susceptible host system in which the presence of surviving virus may be detected. The host used is primarily determined by the infectious and lethal capacity of the virus, as well as by host availability. The three host systems commonly used for the neutralization test are cell cultures, embryonated hens' eggs, and mice.

12.2.3.a Cell Cultures

Where feasible cell cultures are the preferred host. Cell cultures are the most important hosts for performing the neutralization test because they are susceptible to a wide range of viruses, are readily available, and have no immune system to influence the test. The two types of cell cultures used in the neutralization test are suspension and monolayer cultures. In the suspension culture, the virus-infected cells grow floating in the medium. After the virus replicates in the cells, progeny are released into the medium and the unneutralized virus may be detected by reactions such as CPE, change in pH of the medium, or by hemadsorption. In monolayer cell cultures the virus replicates in cell cultures that have been overlaid with agar (Dulbecco et al, 1956). This "agar cover," prevents the virus from establishing secondary foci of infection and keeps the initial infection localized, giving rise to plaques. One plaque is produced by each infectious or plaque

forming unit (PFU) in the original virus suspension. The counting of plaques is a very accurate method for quantitating virus. The prevention of PFU with specific antiserum in the neutralization test is called "plaque reduction."

12.2.3.b Embryonated Hens' Eggs

The inoculation of various tissues or cavities of the developing embryo for attempted virus isolation and identification is standard procedure in many laboratories. The route of inoculation depends on the virus to be isolated. For example, the amniotic cavity is used for isolating influenza viruses and mumps virus, and the CAM is often used for isolating variola virus, vaccinia virus, and herpes simplex virus (HSV). Unneutralized virus will replicate in the embryonated hens' egg and can be detected by the death of the embryo, by pock formation on the CAM, or by the agglutination of erythrocytes using amniotic or allantoic fluids containing hemagglutinating virus.

12.2.3.c Mice

Adult and suckling mice are the animals most frequently used for virus isolation and identification. The white Swiss mouse is used extensively for isolating and determining unneutralized arboviruses, certain enteroviruses, and rabiesvirus. The age of the mouse has a great influence on its susceptibility to disease. Thus, suckling mice may be more susceptible to viruses than adult mice. Unneutralized virus is usually determined by death of or paralysis in infected mice.

12.3 TEST PROCEDURES

Cell culture is the most frequently used host system for performing the neutralization test. In the following types of neutralization tests, suspension cell cultures primarily will be used as the host system.

12.3.1 Constant Virus, Varying Serum

Mix a selected dilution of virus (usually 100 TCD_{50}, as determined from previous titration) with varying dilutions of acute- and convalescent-phase sera. Incubate the virus–serum mixture and then inoculate a susceptible host system with the mixture. The reciprocal of the highest dilution of acute- and convalescent-phase sera protecting the host against the virus is the serum titer.

12.3.1.a Materials

1. Susceptible host system such as cell cultures, embryonated hens' eggs, or mice
2. Maintenance medium
3. Known positive virus (pretitered and standardized to contain 100 TCD_{50} or $LD_{50}/0.1$ mL)
4. Acute- and convalescent-phase sera (usually inactivated at 56 °C for 30 minutes)

12.3.1.b Procedures

1. Prepare serial twofold dilutions of the acute- and convalescent-phase sera (1:10 through 1:5120) in 0.5-mL volumes
2. Mix 0.5 mL of standardized known positive virus with each serum dilution
3. Dilute the known standardized virus (1:10 through 1:1000)
4. Incubate the virus-serum mixtures at about 37 °C for one hour
5. After incubation, inoculate each of three cell culture tubes with 0.2 mL of each virus-serum mixture (when inoculating mice, use 0.02–0.04 mL of mixture)

6. Inoculate three cell cultures with 0.1 mL of known positive virus dilutions (undiluted through 1:1000); these dilutions are used as a back titration to confirm the test potency of 100 TCD_{50} (Figure 12.1)
7. Include three uninoculated cell cultures for controls
8. Incubate the cell cultures at 33 °C to 35 °C in a slanted position
9. Observe the cell cultures daily for CPE

12.3.1.c Interpretation

A fourfold rise in antibody titer between the acute- and convalescent-phase sera is considered to be diagnostically significant. In this test, CPE in cell cultures was used to detect unneutralized virus. This unneutralized virus may be detected in embryonated hens' eggs by pocks on the CAM and in mice by death or paralysis.

12.3.2 Constant Antiserum, Varying Virus

For virus identification varying dilutions of the virus are mixed with a constant antiserum dilution. These dilutions are incubated to allow the virus and antiserum to react. Each virus and serum mixture is then inoculated into a susceptible host system. The dilution of virus that infects 50% of the host systems is considered the endpoint dilution.

12.3.2.a Materials

1. Susceptible host system such as cell cultures, embryonated hens' eggs, or mice
2. Maintenance medium, such as Eagle's modified essential medium (EMEM)
3. Known positive antiserum (inactivated at 56 °C for 30 minutes and standardized to contain 20 Ab units/0.1 mL)

4. Known negative serum (inactivated at 56 °C for 30 minutes)

5. Virus isolate

12.3.2.b Procedures

1. Prepare serial tenfold dilutions (1:10 through 1:10^8) of the virus isolate in maintenance medium

2. Mix 0.5 mL of each isolate dilution (undiluted through 1:10^8) with 0.5 mL of known standardized antiserum

3. Mix 0.5 mL of each isolate dilution (undiluted through 1:10^8) of the virus with 0.5 mL of known negative serum

4. Incubate the virus–serum mixtures at 37 °C for one hour

5. Inoculate each of three cell culture tubes with 0.2 mL of each virus–serum mixture (when inoculating mice, use 0.02 to 0.04 mL of the mixture)

6. Include three uninoculated cell cultures for controls

7. Incubate the cell cultures at 33 °C to 35 °C in a slanted position

8. Observe the cell cultures daily for CPE

12.3.2.c Interpretation

The dilution of virus that infects 50% of the host system is considered the endpoint dilution. A difference of at least two logs or two tubes must be demonstrated between the normal and the immune antiserum to show significant neutralization. Virus dilutions are generally made using tenfold dilutions. The standardized antiserum dilution in this test is selected based on its ability to neutralize 100 TCD$_{50}$ or 100 LD$_{50}$ of virus. The neutralization of 100 TCD$_{50}$ (LD$_{50}$) of virus gives the greatest amount of sensitivity and specificity for the test.

12.3.3 Constant Virus, Constant Antiserum

A selected dilution of virus (usually 100 TCD$_{50}$ as determined from prior titration) is mixed with a selected dilution of known antiserum (usually 20 Ab units/0.1 mL). The mixture is incubated and then injected into a susceptible host system for observation of unneutralized virus. The virus is identified if the antiserum neutralizes the infectivity of the virus.

12.3.3.a Material

1. Susceptible host system such as cell cultures, embryonated hens' egg, or mice

2. Maintenance medium

3. Known positive antiserum (inactivated at 56 °C for 30 minutes and standardized to contain 20 Ab units/0.1 mL)

4. Unknown virus isolated (pretitered and standardized to contain 100 TCD$_{50}$/0.1 mL)

12.3.3.b Procedures

1. Mix 0.5 mL of the standardized unknown virus isolate with 0.5 mL of the known standardized positive antiserum

2. Dilute the standardized unknown virus isolate (1:10 through 1:1000)

3. Incubate the virus-serum mixture at about 37 °C for one hour

4. Inoculate each of three susceptible cell culture tubes with 0.2 mL of the virus-serum mixture (when inoculating mice, use 0.02 to 0.04 mL of the mixture)

5. Inoculate three susceptible cell culture tubes with 0.1 mL of the standardized unknown virus dilutions (undiluted through 1:1000). These dilutions are used as a back titration to confirm the test potency of 100 TCD$_{50}$

6. Include three uninoculated cell culture tubes for controls

7. Incubate the cell cultures at 33 °C to 35 °C in a slanted position

8. Observe the cell cultures daily for CPE

12.3.3.c Interpretation

The absence of CPE in the cell cultures inoculated with the virus-serum mixtures identifies the virus isolate. Antibody in serum can be determined also by running a single dilution of a test serum against a standard dose of known virus. Neutralization indicates the presence of specific antibody.

12.3.4 Varying Virus, Varying Antiserum

This type of neutralization test should be used with caution. The titers of the unknown virus and the known antiserum have not been predetermined. Varying dilutions of the virus and antiserum are made, combined, incubated, and then inoculated into a susceptible host system in which the presence of unneutralized virus may be detected. The test can actually give maximum information about both the virus and the antiserum in relation to one another. Although this test design can be used for neutralization results, it is mainly used as a routine procedure in the block titration of antigen in the complement fixation test.

12.4 CALCULATIONS OF 50 PERCENT ENDPOINTS

The 50% endpoint can be used in several different reactions. A 50% mortality ratio is expressed as LD_{50} and a 50% infective dose, as ID_{50}; in cell cultures, a CPE in 50% of the cultures is expressed as TCD_{50}.

12.4.1 Reed-Muench Method

See Tables 12.1 and 12.2.

12.4.2 Karber Method for Calculating 50 Percent Mortality

Karber Formula: Negative logarithm of the 50% end-point titer = negative logarithm of the highest virus concentration used

$$\left[\frac{(\text{Sum of \% Mortality at each dilution} - 0.5)}{100}\right.$$

$$\left. \times (\text{logarithm of dilution})\right]$$

Example:

Virus Dilution	Mortality Ratio	% Mortality
10^{-1} (1:10^1)	8/8	100
10^{-2} (1:10^2)	8/8	100
10^{-3} (1:10^3)	8/8	100
10^{-4} (1:10^4)	4/8	50
10^{-5} (1:10^5)	0/8	0

$= -1.0 -$

$$\left[\left(\frac{(100 + 100 + -100 + 50)}{100} - 0.5\right)\right.$$

$$\left. \times (\log_{10})\right]$$

$$= -1.0 - \left[\left(\frac{350}{100} - 0.5\right) \times (\log_{10})\right]$$

$$= 1.0 \, [(3.5 - 0.5) \times 1]$$

$$= -1.0 - 3.0$$

$$= 4.0$$

The 50% endpoint dilution = $10^{-4.0}$.
The 50% endpoint titer = $10^{4.0}$.

12.5 PREPARATION OF SERUM POOLS

To avoid the time and expense required for typing each enterovirus isolate using individual antiserum, the use of serum pools is recommended (Lim and Benyesh-Melnick, 1960; Schmidt et al, 1961). In one type of serum pool, several antisera that react against different enterovirus serotypes are combined into a pool. A number of pools

Table 12.1 Calculation of 50% Mortality (Virus Titration)

Virus Dilution	Deaths per Number Inoculated	Cumulative Deaths	Cumulative Survivors	Mortality Ratio	Percent Mortality
10^{-4} (1:10^4)	5/5	10	0	10/10	100%
10^{-5} (1:10^5)	4/5	5	1	5/6	83%
10^{-6} (1:10^6)	1/5	1	5	1/6	17%
10^{-7} (1:10^7)	0/5	0	10	0/10	0%

Interpolation Formula: $\dfrac{\% \text{ Mortality Greater than } 50\%-50\%}{\% \text{ Mortality Greater than } 50\%-\% \text{ Mortality less than } 50\%}$

Substituting: $\dfrac{83-50}{83-17} = \dfrac{33}{66} = 0.5$

a. Multiply the interpolative value times the negative $\log_{10}$ of the dilution ratio.
   Negative $\log_{10}$ of the dilution ration 10 $= -1$
   Interpolative value $= \times 0.5$
   Corrected interpolative value $= -0.5$
b. The endpoint dilution associated with 50% mortality is located between the 10^{-5} and 10^{-6} dilution.
c. The $\log_{10}$ of the 50% endpoint dilution is estimated by adding the corrected interpolative value to the $\log_{10}$ of the dilution above 50%

$$-5 + (-0.5) = -5.5$$

d. The 50% endpoint dilution is estimated at $10^{-5.5}$.
e. The 50% $-$ titer is estimated at $10^{5.5}$.

containing different antisera are set up. Neutralization of the isolate by one of the serum pools indicates that the isolate is one of the enteroviruses whose antiserum is included in the pool. For final identification, the isolate is run in a neutralization test against each antiserum in the pool. In the second type of serum pool, each antiserum is present in two different intersecting pools. For example, antibody to poliovirus type 1 (PV-1) is the only antiserum that is included in both pools C and F, and PV-1 is the only virus that will be neutralized by both of these pools. The method of incorporating antisera into intersecting pool schemes is not included in the chapter but is adequately reviewed by Lennette and Schmidt (1979) and Melnick et al (1977).

Serum pools should be prepared with only known pretitered antisera. The procedure for preparing enterovirus serum pools follows.

High-titered antiserum must be used. The titers of each serum must be high enough so that their combination in a pool will result in a total serum concentration not greater than 10%. Titer each serum individually against 100 TCD_{50}/0.1 mL of its prototype virus (Figure 12.2).

1. Prepare serial twofold dilutions (1:10 through 1:1280) of the serum and mix each dilution with an equal volume containing 100 TCD_{50} of the virus

2. After incubation at 37°C for one hour, inoculate three cell culture tubes (animals may be used) with 0.2 mL of each virus-serum mixture

Table 12.2 Calculation of 50% Endpoint Dilution of the Virus Plus Constant Immune Serum

Virus Dilution	+	Constant Serum	Deaths per Number Inoculated	Cumulative Deaths	Cumulative Survivors	Mortality Ratio	Percent Mortality
10^{-2} $(1:10^2)$	+	IS	5/5	10	0	10/10	100
10^{-3} $(1:10^3)$	+	IS	3/5	5	2	5/7	71
10^{-4} $(1:10^4)$	+	IS	2/5	2	5	2/7	29
10^{-5} $(1:10^5)$	+	IS	0/5	0	10	0/10	0

Interpolation Formula: $\dfrac{\text{\% Mortality Greater than 50\%}-50\%}{\text{\% Mortality Greater than 50\%}-\text{\% Mortality less than 50\%}}$

Substituting: $\dfrac{71-50}{71-29}=\dfrac{21}{42}=0.5$

a. Multiply the interpolative value times the negative $\log_{10}$ of the dilution ratio.
   Negative $\log_{10}$ of the dilution ration 10 $\quad = -1$
   Interpolative value $\qquad\qquad\qquad\qquad\quad = \times\ 0.5$
   Corrected interpolative value $\qquad\qquad = -0.5$

b. The endpoint dilution associated with 50% mortality is located between the 10^{-3} and 10^{-4} dilutions.

c. The $\log_{10}$ of the 50% endpoint dilution is estimated by adding the corrected interpolative value to the $\log_{10}$ of the dilution above 50%.

$$-3 + (-0.5) = -3.5$$

d. The 50% endpoint dilution is estimated at $10^{-3.5}$.

e. The 50% − titer is estimated at $10^{3.5}$.

The logarithmic difference between the 50% titers of the virus titration in procedure 1 and the virus plus immune serum titration in procedure 2 is 2.0 [5.5 − 3.5]. Generally, a reduction of at least 2 in $\log_{10}$ (titer) of the virus must be demonstrated by a test serum to show significant neutralization.

3. Determine the titer of each serum by examining the cell cultures for CPE; this endpoint dilution or antibody titer is the last dilution of serum demonstrating complete neutralization (no CPE); this endpoint dilution contains 1 Ab unit/0.1 mL

4. 20 Ab units/0.1 mL are required in the pool; to calculate 20 Ab units/0.1 mL, divide 20 into the denominator of the dilution containing 1 Ab unit/0.1 mL

Example: $\dfrac{320}{20}=16$ (thus, a 1/16 dilution contains 20 Ab units/0.1 mL)

With the sera titered and the dilution containing 20 Ab units of serum/0.1 mL calculated, incorporate the sera into the pool at the dilution that will give 20 Ab units/0.1 mL

Example: To make a pool of polio 1, 2, and 3, immune serum (IS) have the following titers:
Polio 1 IS, 1:5120
Polio 2 IS, 1:20,480
Polio 3 IS, 1:10,240

Divide the titers by 20 to get 20 Ab units/0.1 mL.

Dilution containing 20 Ab units/0.1 mL:

$$\text{Polio 1 IS } \frac{5120}{20} = 256$$

$$\text{Polio 2 IS } \frac{20,480}{20} = 1024$$

$$\text{Polio 3 IS } \frac{10,240}{20} = 512$$

Decide on the volume of the pool you want to make. To make a 1000 mL polio pool using the figures in the example above, divide the denominator of the dilution containing 20 Ab units/0.1 mL into the volume of the pool desired to obtain the amount of each undilute serum to be added to the pool.

$$\text{\textit{Example:} Polio 1 IS } \frac{1000}{256} = 3.90 \text{ mL}$$

$$\text{Polio 2 IS } \frac{1000}{1024} = 0.98 \text{ mL}$$

$$\text{Polio 3 IS } \frac{1000}{512} = 1.95 \text{ mL}$$

Total the volumes of sera to be added:

$$
\begin{array}{r}
3.90 \text{ mL} \\
0.98 \text{ mL} \\
+ \ 1.95 \text{ mL} \\
\hline
6.83 \text{ mL}
\end{array}
$$

Subtract this from the total volume of the pool to determine the amount of diluent to be used:

$$
\begin{array}{r}
1000.00 \text{ mL} \\
- \ 6.83 \text{ mL} \\
\hline
993.17 \text{ mL}
\end{array}
$$

Combine the diluent (maintenance medium) with the immune sera in the following volumes:

Maintenance medium	993.17 mL
Polio 1 IS	3.90 mL
Polio 2 IS	0.98 mL
Polio 3 IS	1.95 mL

Aliquot the pools and store frozen at $-20\,°C$ or lower. This results in a pool containing 20 Ab units/0.1 mL of each serum.

12.6 INTERPRETATION OF SEROLOGIC RESULTS

The neutralization test is frequently used to identify a virus isolate and is also used to indicate a recent virus infection by demonstrating a significant rise in antibody titer to a specific virus using paired sera. Usually, a fourfold rise in antibody titer between an acute- and a convalescent-phase serum run in the same test is considered to be diagnostically significant. Recent infections cannot be distinguished from past infections by examining a single serum specimen. A high antibody titer with a single serum specimen is no guarantee of recent infection.

However, a single serum specimen may be of value in determining if an individual has been exposed to a viral agent at some time in the past. In some diseases, the presence of antibody against the causative agent indicates immunity to the disease. Information gained from a single serum specimen may also be of epidemiologic value in determining the number of individuals in a population exposed to certain agents and, where applicable, in assessing the herd immunity.

The serum collection time is extremely important in all serologic tests. The acute serum should be collected as soon as possible after the onset of disease. The convalescent serum is usually collected two to three weeks later. If the acute serum is drawn too late, antibodies may be approaching maximal levels; therefore, a significant rise in antibody titer may not be demonstrated.

12.6.1 Interpretation of Tests for Viral Antibodies

The following test examples demonstrate the basic principles for interpreting sero-

Test 1

	Undiluted	1:10	1:20	1:40	1:80	1:160	1:320	1:640	1:1280	Serum Controls
S-1 Acute-phase serum	+	+	+	+	+	+	+	+	+	−
S-2 Convalescent-phase serum	+	+	+	+	+	+	+	+	+	−

The results in this test indicate the individual has not been exposed to the virus tested.

Test 2

	Undiluted	1:10	1:20	1:40	1:80	1:160	1:320	1:640	1:1280	Serum Controls
S-1 Acute-phase serum	−	−	+	+	+	+	+	+	+	−
S-2 Convalescent-phase serum	−	−	−	−	−	+	+	+	+	−

The results demonstrate an eightfold rise in antibody titer between the acute and convalescent serum. The results suggest recent infection with the virus tested.

Test 3

	Undiluted	1:10	1:20	1:40	1:80	1:160	1:320	1:640	1:1280	Serum Control
Single Serum	−	−	−	−	−	−	+	+	+	−

The 160 antibody titer on a single serum indicates past infection with the antigen tested at some time in the past. A high antibody titer is no guarantee of a recent infection.

Test 4

	Undiluted	1:10	1:20	1:40	1:80	1:160	1:320	1:640	1:1280	Serum Controls
S-1 Acute-phase serum	−	−	−	−	−	+	+	+	+	−
S-2 Convalescent-phase serum	−	−	−	−	−	+	+	+	+	−

The 80 antibody titers of the paired sera indicate infection with the antigen tested at some time in the past. It is very important to collect the sera at the proper time. If the acute serum is drawn too late, the antibody titer may have already risen to such a point that, when it is compared with the convalescent serum titer, an antibody rise may not be present.

logic results and do not represent tests for any particular virus.

Key: + = Virus infectivity, − = No virus infectivity.

REFERENCES

Andrewes, C.H. and Elford, W.J. 1933. Observation on anti page 1: The percentage law. Br. J. Exp. Pathol. 14:367–374.

Dulbecco, R., Voit, M., and Strickland, A.G.R. 1956. A study on the basic aspects of neutralization of two animal viruses, western equine encephalitis virus and poliomyelitis virus. Virology 2:162–205.

Ginsberg, H.S. and Horsfall, F.L. Jr. 1949. A labile component of normal serum which combines with various viruses. Neutralization of infectivity and inhibition of hemagglutination by the component. J. Exp. Med. 90:475–495.

Horsfall, F.L. Jr. and Tamm, I. (eds.). 1960. Viral and Rickettsial Infections of Man, 4th ed. Philadelphia: JB Lippincott.

Lennette, E.H. and Schmidt, N.J., eds. 1979. Diagnostic Procedures for Viral, Rickettsial and Chlamydia Infections, 5th ed. Washington, D.C.: American Public Health Association.

Lim, K.A. and Benyesh-Melnick, M. 1960. Typing of viruses by combination of antiserum pools. Application of typing of enteroviruses (coxsackie and ECHO). J. Immunol. 84:309–317.

Mandel, B. 1960. Neutralization of viral infectivity. Characterization of virus-antibody complex, including association, disassociation and host cell interaction. Ann. N.Y. Acad. Sci. 83:515–527.

Maramorosch, K. and Koprowski, H. (eds.). 1967. Methods in Virology, Vol. III. New York: Academic Press.

Melnick, J.L., Schmidt, N.J., Hampil, B., and Ho, H.H. 1977. Lyophilized combination pools of enterovirus equine antisera: Preparation and test procedures for the identification of field strains of 19 group A coxsackie serotypes. Intervirology 8:172–181.

Sabin, A.B. 1950. The dengue group of viruses and its family relationship. Bact. Rev. 14:225–232.

Salk, J.E., Younger, J.S., and Ward, E.N. 1954. Use of color change of phenol red as the indicator in titrating poliomyelitis virus or its antibody in a tissue-culture system. Am. J. Hyg. 60:214–230.

Schmidt, N.J., Guenther, R.W., and Lennette, E.H. 1961. Typing of ECHO virus isolates of immune serum pools, the intersecting serum scheme. J. Immunol. 87:623–626.

Tyrrell, D.A.J. and Horsfall, F.L. Jr. 1953. Neutralization properties. J. Exp. Med. 97:845–862.

BIBLIOGRAPHY

Ballew, H.C., Forrester, F.T., Lyerla, H.C., Velleca, W.M., and Bird, B.R. 1977. Laboratory Diagnosis of Viral Diseases. Atlanta: Centers for Disease Control.

Ballew, H.C., Forrester, F.T., Lyerla, H.C., Velleca, W.M., Bird, B.R., and Roberts, J.D. 1979a. Basic Laboratory Methods in Virology. Atlanta: Centers for Disease Control.

Ballew, H.C., Forrester, F.T., and Lyerla, H.C. 1983. Laboratory Methods for Diagnosing Respiratory Virus Infections. Atlanta: Centers for Disease Control.

Ballew, H.C., Lyerla, H.C., and Forrester, F.T. 1979b. Laboratory Methods for Diagnosing Herpesvirus Infections. Atlanta: Centers for Disease Control.

Habel, K. and Salzman, N.P. 1969. Fundamental Techniques in Virology. New York: Academic Press.

Hatch, M.H. and Marchetti, G.E. 1971. Isolation of echoviruses with human embryonic lung fibroblast cells. Appl. Microbiol. 22:736–737.

Hsiung, G.D. 1982. Diagnostic Virology, 3rd ed. New Haven: Yale University Press.

Lennette, E.H., Balows, A., Hausler, W.J. Jr., and Truant, J.P. (eds.). 1980. Manual of Clinical Microbiology, 3rd ed. Washington, D.C.: American Society of Microbiology.

Lennette, D.A., Specter, S., and Thompson, K.D. (eds.). 1979. Diagnosis of Viral Infections. Baltimore: University Park Press.

Melnick, J.L. and Hampil, B. 1965. WHO collaborative studies on enterovirus reference antisera. (Bulletin WHO 33) Geneva: World Health Organization.

Wallis, C. and Melnick, J.F. 1967. Virus aggregation as the cause of non-neutralizable persistent fraction. J. Virol. 1:478–488.

13

Hemagglutination Inhibition and Hemadsorption

Leroy C. McLaren

13.1 HEMAGGLUTINATION INHIBITION TEST

A wide variety of viruses possess the property of hemagglutination [i.e., the ability to agglutinate the erythrocytes (RBC) of one or more animal species under certain conditions]. Influenza, parainfluenza, arboviruses, adenoviruses, rubella, and some strains of picornaviruses have this property. Antibodies to these viruses have the ability to react with the virus and prevent viral hemagglutination. This is the principle of the hemagglutination inhibition test (HAI), which is extremely valuable in the serologic diagnosis of infections. The test is most frequently used in the routine clinical virology laboratory for influenza and parainfluenza viruses, but laboratories with special interests also use this test for the serologic diagnosis, or viral identification, of other hemagglutinating viruses of medical importance.

The HAI test is simple to perform and requires relatively inexpensive equipment and reagents. Serial dilutions of patients' sera are allowed to react with a fixed dose of viral hemagglutinin (HA), followed by the addition of agglutinable erythrocytes. In the presence of antibody, the ability of the virus to agglutinate the erythrocytes is inhibited. In some viral systems, the test can be complicated by the presence of nonspecific viral inhibitors (nonantibody) in sera that must, therefore, be treated prior to being used. The presence of such inhibitors can give rise to false-positive results in the HAI test.

The specificity of the HAI test varies with the virus. The reaction can be highly specific for certain viruses (influenza–parainfluenza groups) and less specific for other viruses (arboviruses).

The method presented in this section will concern the influenza virus system with which most routine clinical virology laboratories are concerned. Conditions for conducting the HAI test for other viruses are presented in the Appendix. Specific methods for these viruses can be found in a number of excellent references (Howard, 1982; Hsiung, 1982; Washington, 1985; Herrmann, 1988; Schmidt and Emmons, 1989).

243

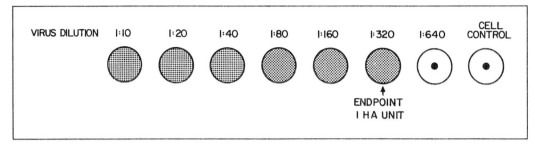

Figure 13.1 Determination of viral hemagglutinin titer.

13.1.1 Materials

1. Blood from a guinea pig or other mammalian or avian species
2. Amniotic, allantoic, or tissue culture fluid containing influenza virus
3. Alsever's solution
4. Phosphate buffered saline (PBS), 0.01 M, pH 7.2
5. Disposable microtiter plates, "U" type
6. Calibrated diluting loops, 0.025 and 0.050 mL
7. Dropping pipettes, 0.025 and 0.050 mL per drop
8. Microtiter plate reading mirror
9. 37 °C and 56 °C water baths
10. Receptor destroying enzyme (RDE), 100 units/mL

See Appendix for sources.

13.1.2 Preparation of Red Blood Cells Suspension

1. Guinea pig, human, or chicken blood is collected in either Alsever's solution or in heparin to prevent clotting; cells are washed three times in PBS by centrifugation at 1800 × g for 5 minutes and suspended in PBS as a 10% stock suspension, which should be stored for no more than one week at 4 °C.
2. For use in the following tests the red blood cell (RBC) stock suspension should be diluted to 0.5% in PBS prior to use (1 mL of 10% RBC added to 19 mL of PBS).

13.1.3 Titration of Hemagglutinin

1. Influenza stock suspension of a contemporary strain of each influenza type A and type B strain are diluted twofold, from 1:10 to 1:2560, employing a 0.05-mL volume to each well.
2. Include an RBC control well by adding 0.05 mL PBS.
3. Add 0.05 mL of the diluted RBC suspension to each mitrotiter well, gently mix, and incubate at room temperature until the RBC settle to the bottom of each well (one to two hours).
4. Read the HA titer by determining the highest dilution of virus capable of causing agglutination; agglutinated RBC will form a confluent pattern of RBC on the bottom of the well; nonagglutinated RBC will form a discrete button at the bottom of the well (Figure 13.1).
5. The HA titer is the reciprocal of the highest dilution of virus showing agglutination and represents 1 HA unit/0.05 mL of virus.
6. For the HAI test, the virus suspension will be diluted to contain 4 HA units/0.025 mL (or 8 HA units/0.05 mL); thus, if the HA titer is 160 then, for the HAI test, the original virus stock will be diluted 1:20.

13.1.4 Serum Treatment

Many influenza and parainfluenza virus strains are sensitive to serum factors that may inhibit hemagglutination nonspecifically. In most instances, these inhibitors can be removed from human sera by treatment with RDE and by heat inactivation, potassium periodate (KIO) treatment, or with kaolin.

13.1.5 RDE Treatment

1. Add 0.4 mL of RDE (100 units/mL), which is commercially available, to 0.1 mL of serum and incubate overnight in a water bath at 37 °C.
2. Add 0.3 mL of 2.5% sodium citrate to the RDE–serum mixture and incubate in a 56 °C water bath for 30 minutes.
3. Add 0.2 mL of PBS to each inactivated serum mixture to give a final serum dilution of 1:10.

13.1.6 KIO Treatment

1. Add 0.3 mL of 0.01 M KIO_4 to 0.1 mL of serum and incubate the serum–KIO_4 mixture for 15 minutes at room temperature.
2. Add 0.3 mL of 1% glycerol (in PBS) to neutralize excess KIO_4.
3. Add 0.3 mL PBS to each serum mixture to give a final serum dilution of 1:10.

13.1.7 Kaolin Treatment

1. Prepare a 1:5 dilution of serum.
2. Add an equal volume of acid washed kaolin (25% suspension in PBS) to the diluted serum; mix thoroughly and allow to stand at room temperature for 30 minutes with intermittent shaking.
3. Centrifuge at 3000 rpm for 30 minutes.
4. Carefully remove the supernatant serum, which has a final dilution of 1:10.

13.1.8 Removal of Naturally Occurring Agglutinins for Red Blood Cells

Some sera may contain agglutinins for human O or chicken RBC. In the HAI test that follows, if the serum control wells (serum and RBC, but no virus) show agglutination, the agglutinins can be adsorbed with erythrocytes and the HAI test repeated.

1. Add 0.1 mL of a 50% suspension of the RBC being agglutinated to 1.0 mL of the 1:10 dilution of heat-inactivated (56 °C for 30 minutes) serum.
2. Incubate for one hour at 4 °C and remove the RBC by centrifugation at 2000 rpm for 10 minutes.

13.1.9 Hemagglutination Inhibition Test for Influenza

Paired sera (acute and convalescent) treated to remove nonspecific inhibitors should be tested for antibodies to one or more contemporary strains of virus. The hemagglutinin titer for each virus employed should be previously titrated for preparation of 4 HA units/0.025 mL in PBS. Reference antiserum for each virus should be included to confirm the identity of viral strains used in the test.

1. Prepare twofold dilutions of each treated serum, from 1:10 to 1:2560.
2. Add 0.025 mL serum dilution to wells for each virus employed.
3. Add 0.025 mL of virus suspension (containing 4 HA units/0.025 mL) to each serum dilution.
4. Include a hemagglutinin antigen control for each virus used (0.025 mL PBS and 0.025 mL hemagglutinin).
5. Include a serum control for each serum being titrated (0.025 mL of 1:10 dilution of serum and 0.025 mL PBS).

A back titration of the test hemagglutinin antigen dilution is necessary to insure that the amount of hemagglutinin antigen

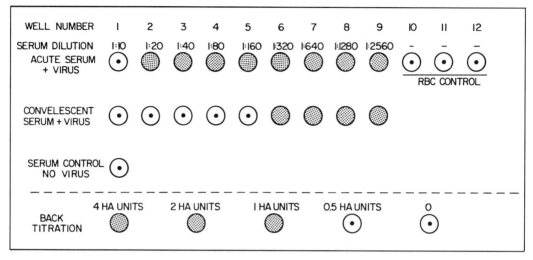

Figure 13.2 Hemagglutination inhibition.

used in the HAI test contains 4 HA units/ 0.025 mL.

- Add 0.05 mL PBS to each of five wells.
- Add 0.05 mL diluted test antigen to the first well.
- Make serial twofold dilutions with a 0.05-mL loop.
- Add 0.05 mL RBC to each well.

6. Include an RBC control (0.05 mL PBS).

7. Gently shake the microtiter plate after serum dilutions and hemagglutinin have been added and incubate at room temperature for 30 minutes.

8. Add 0.05 mL RBC suspension to each well.

9. Gently shake the microtiter plate and incubate the plate at room temperature until the RBC control shows a button at the bottom of the well.

10. The serum controls should show the absence of RBC agglutination.

11. The hemagglutinin control should show positive agglutination.

12. The HAI titer of each serum is defined as the highest dilution of serum that completely inhibits hemagglutination.

13. The HAI titer of the acute phase serum is 10, and for the convalescent serum the titer is 160; fourfold or greater rise in HAI titers is interpreted as significant and is indicative of recent influenza virus infection or vaccination (Figure 13.2 illustrates an example of an acute and convalescent serum titrated against a single influenza strain).

13.1.10 Other Viruses for Which the Hemagglutination Inhibition Test Is Applicable

The procedure described above is applicable with some modifications for the parainfluenza, rubella, measles, reovirus, adenovirus, enterovirus, and togavirus groups. The appropriate chapter in this volume should be consulted for details. However, examples of modifications that are necessary for these viruses are summarized below.

1. Members of the parainfluenza virus group agglutinate human O, guinea pig, and chicken RBC; however, in addition to RDE heat treatment of sera, adsorption of sera with guinea pig RBC is usually necessary if guinea pig RBC are used in the HAI test.

2. For rubella virus HAI tests, one-day-old chick or goose RBC are used. The diluent employed is not PBS. HEPES-Saline-albumen-gelatin (HSAG) diluent is employed. In addition, sera must be treated with $MnCl_2$-heparin and adsorbed with chick RBC to remove non-specific inhibitors and agglutinins (see Appendix).

3. For group I adenovirus serotypes, rhesus monkey RBC are usually used, and for adenoviruses in group II, rat RBC. Sera must be heat-inactivated at 56 °C for 30 minutes and adsorbed with the type of RBC being used in the HAI test before testing.

4. For reoviruses, human O RBC are used. Sera must be heat inactivated and adsorbed with kaolin.

5. For members of the togavirus group, hemagglutinin occurs optimally at different pH values ranging from pH 6.0 to 7.4. Heat inactivation and kaolin adsorption of sera are usually employed.

6. Some serotypes of enteroviruses agglutinate human O RBC but the incubation temperature of the hemagglutinin and HAI tests is critical. Coxsackie A-20, A-21, A-24, and echovirus types 3, 11, 13, and 19 hemagglutinate at 4 °C. Coxsackie B-1, B-3, B-5, and echovirus 6, 7, 12, 20, and 21 hemagglutinate at 37 °C. Heat inactivation and kaolin adsorption of sera are employed for the HAI test.

13.2 HEMADSORPTION

Influenza and parainfluenza viruses, as well as mumps and Newcastle's disease virus (NDV), can replicate in a variety of cell culture systems. Commonly employed cell cultures in clinical virology laboratories include primary monkey kidney cells, and continuous cultures of Madin-Darby canine kidney (MDCK) cells and the LLC-MK$_2$ cell line derived from rhesus monkey kidney. These viruses frequently do not produce extensive CPE; however, infected cells can be detected by the hemadsorption (HAd) technique. These viruses mature by budding from the plasma membrane of infected cells and, because they can react with human O, chicken, or guinea pig RBC, the addition of a RBC suspension to the cell monolayers results in the RBC adsorbing onto the surface of infected cells (Figure 13.3).

13.2.1 Materials

1. Guinea pig blood washed three times in PBS by centrifugation and suspended in Alsever's solution at a concentration of 20%; RBC should not be kept for more than one week at 4 °C

2. Uninoculated control cell cultures

3. Cell cultures inoculated with clinical specimens

4. Cell cultures inoculated with hemadsorbing strain of virus

5. Cold PBS

See Appendix for sources.

13.2.2 Methods

1. Remove fluids from control and inoculated cell cultures to sterile tubes.

2. Add 0.25 mL of 0.5% guinea pig RBC suspension to each cell culture.

3. Incubate the cell cultures at either 4 °C or room temperature for 30 minutes.

4. Carefully remove the unadsorbed RBC by rinsing the cell monolayers with cold PBS.

5. Observe the cells microscopically for HAd; the RBC should not adsorb to uninoculated cells (RBC suspensions stored for more than one week frequently adsorb to uninoculated cell cultures of kidney origin).

6. Culture fluids from cells showing positive HAd can be subcultured to newly prepared cell cultures for subsequent identification by HAI, or HAd-inhibition

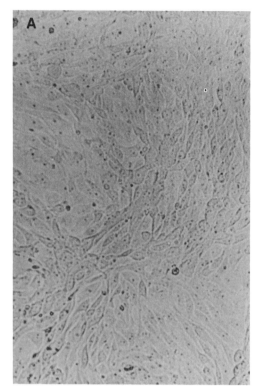

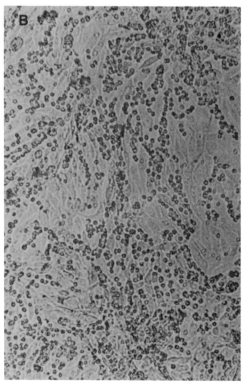

employing reference antisera to influenza and parainfluenza viruses.

13.3 HEMADSORPTION INHIBITION TEST

Inhibition of HAd by means of reference antisera can be used for identification of viral isolates. Second passage to replicate cell cultures of viral isolates produces HAd. At least four cell cultures need to be inoculated for identification of influenza type A and B isolates, and if parainfluenza types 1, 2, and 3 are to be included, six additional cultures.

13.3.1 Methods

1. Infected and uninoculated control cell cultures are rinsed with Hanks' balanced salt solution (HBSS).
2. Specific viral antisera that have been RDE heat treated and absorbed with guinea pig RBC are added to one set of inoculated and control cultures with 0.8 mL of each antiserum, which have been diluted to 2.5% concentration in HBSS.
3. The antisera are allowed to cover the entire cell monolayer at room temperature for 30 minutes.
4. Add 0.2 mL of 0.4% guinea pig RBC to each cell culture and incubate at room temperature for 30 minutes.
5. Examine the cell cultures microscopically for inhibition of HAd; isolates are identified by the serum inhibition of HAd.

APPENDIX

Reagents

1. *Phosphate buffered saline* (PBS) pH 7.2 is prepared as follows:
   Disodium phosphate (Na_2HPO_4), 1.096 g
   Monosodium phosphate (NaH_2PO_4), 0.315 g

Figure 13.3 Hemadsorption. (A) Uninoculated primary monkey kidney cells. (B) Primary monkey kidney cells infected with influenza A.

Sodium chloride, 8.500 g
Distilled water, q.s.ad, 1.000 L

2. *Alsever's solution* is available from commercial sources or can be prepared as follows:
Dextrose, 20.500 g
Sodium citrate ($Na_3C_6H_5O_7 \cdot 2H_2O$), 8.000 g
Citric acid ($C_6H_8O_7 \cdot H_2O$), 0.550 g
Sodium chloride, 4.200 g
Distilled water, q.s.ad, 1.000 L

3. *Erythrocytes* from human and a variety of animal species in Alsever's solution are available commercially from Flow Laboratories, M.A. Bioproducts, and numerous other suppliers of virologic products.

4. *Hemagglutinating viral antigens* are prepared by inoculation of ten-day-old embryonated chicken eggs or sensitive cell cultures; for most clinical virology laboratories it is more convenient to obtain contemporary strains of influenza A and B or parainfluenza virus strains from commercial sources (Flow Laboratories or M.A. Bioproducts); other hemagglutinating viral antigens are also available from these sources and include measles, mumps, and rubella viruses.

5. *Reference antisera* for most hemagglutinating viruses are available from Flow Laboratories or M.A. Bioproducts.

6. *Receptor destroying enzyme* (RDE) is available from Flow Laboratories and M.A. Bioproducts and should be shipped on dry ice; upon receipt, RDE should be stored at $-70\ °C$.

7. *Kaolin* for serum treatment is available from Flow Laboratories or prepared as a 25% suspension of acid-washed kaolin in PBS.

8. *HEPES-saline-albumen-gelatin* (HSAG) diluent is available from Flow Laboratories and M.A. Bioproducts.

9. *Heparin–MnCl₂* reagent is available from Flow Laboratories or can be prepared by adding equal parts of heparin (5000 units/mL) and 1 M $MnCl_2$ (39.6 g $MnCl_2 \cdot 4H_2O$ in 200 mL distilled water); the heparin–$MnCl_2$ solution should be stored at 4 °C and prepared every one to two weeks.

REFERENCES

Herrmann, K.L. 1988. Antibody Detection. In E.H. Lennette, P. Halonen, and F.A. Murphy (eds.). Laboratory Diagnosis of Infectious Diseases—Principles and Practice, Vol. II. New York: Springer-Verlag.

Howard, C.R. (ed.). 1982. New Developments in Practical Virology. New York: Alan R. Liss.

Hsiung, G.D. 1982. Diagnostic Virology. New Haven: Yale University Press.

Schmidt, N.J. and Emmons, R.W. 1989. Diagnostic Procedures for Viral, Rickettsial and Chlamydial Infections, 6th edition. Washington, D.C.: American Public Health Association.

Washington, J.A. II (ed.). 1985. Laboratory Procedures in Clinical Microbiology, 2nd edition. New York: Springer-Verlag.

Immune Adherence
Hemagglutination

Evelyne T. Lennette and David A. Lennette

14.1 INTRODUCTION

Each of the various serologic test procedures described in this volume has its advantages and disadvantages. Their relative utilities depend on the specific requirements of the laboratorian, and selection of an appropriate assay requires careful analysis, as no assay is ideal for all situations. The following discussion is oriented to the needs of a clinical diagnostic laboratory. Clinical laboratories need serologic test procedures of high specificity and with sensitivity suitable both for determination of immunity and for diagnosis of current infections. Procedures should be convenient and rapid to perform, on a single sample or on a hundred serum samples. The demand on equipment and reagents should be within the reach of smaller laboratories with limited resources. Test antigens should be varied and readily available. Of significance to a reference laboratory is the ease with which assays may be adapted for use with new or additional test antigens.

Some years ago, when we reviewed many serologic procedures applicable for use in virology, two procedures appeared to meet most of the requirements just outlined. One was the complement fixation (CF) test, and the other was immune adherence hemagglutination (IAHA), a test virtually unknown in the United States, although well-developed in Japan. Despite much progress in the development of potentially useful serologic procedures during recent years, the situation has not changed greatly. CF tests remain in use in many reference virology laboratories, despite their low sensitivity compared to other serologic techniques: e.g., radioimmunoassay (RIA), enzyme-linked immunosorbant assay (ELISA), hemagglutination inhibition, and immunofluorescence assay (IFA). Continued use of the CF test is probably due to the wide range of antigens commercially available for use in this assay, as well as a large amount of experience with the CF procedure for clinical diagnosis, as suggested by publications in the medical literature. In contrast, the limited inventory of commercially available ELISA tests, users' reluctance to deal with hazards and waste disposal problems of RIA procedures, and the unsuitability of IFA tests for examining large numbers of specimens, have all acted to limit the adoption of these assays for more widespread use.

When we first evaluated the IAHA

assay in 1976, immune adherence assays were known in the United States only to a few researchers studying hepatitis A virus (HAV) and varicella-zoster virus (VZV) and to a number of Japanese workers who were instrumental in developing the procedure for use in clinical laboratories. Our experience with the IAHA assay, and our continued comparisons of the IAHA assay to other serologic methods, confirms our belief that the IAHA test could replace the CF test as a general serodiagnostic method. As indicated in section 14.7, IAHA is as specific as CF, but is four to eight times as sensitive and is easier to perform. Additional uses for the IAHA test have developed since publication of the first edition of this book and are referenced in Table 14.1.

14.2 HISTORIC BACKGROUND

The IAHA assay is based on a phenomenon known as serological adhesion, observed by Levaditi (1901). After injecting antibody-coated cholera vibrios into guinea-pigs, he noted platelet aggregation (or adhesion) to the vibrios. In the same year, similar observations were reported by Laveran and Mesnil (1901) with *Trypanosoma lewisi* using immune rats. Kritchewski later showed that this adhesion reaction required complement. Other French and Russian workers then improved and applied the adhesion test for the in vitro assay of antibodies to trypanosomes, leishmania, leptospira, and spirochetes (Lamana, 1957). In 1952, Nelson showed that similar adhesion occurs if human erythrocytes are substituted for platelets. The agglutination was examined under a microscope after mixing antigen, serum, complement, and erythrocytes on a slide. Nelson coined the name immune adherence for the reaction (1963). The simple procedure was improved when Ito Tagaya adapted it for use with microtest plates (1966). Although the potential utility of the IAHA procedure was

recognized early, it failed to come into widespread use due to a shortcoming: the agglutination reaction was reversible, and was not stable enough to be able to reliably read the reaction. This problem was finally overcome when Mayumi and coworkers (1971) introduced the use of dithiothreitol (DTT) as a stablilizer. With that single improvement, the repertoire of antigens found suitable for use in the IAHA assay grew quickly to include enteroviruses, adenoviruses, and hepatitis B virus (HBV). Japanese investigators also found IAHA was sensitive and could detect HB surface antigen in serum. In the United States, the IAHA test was first used in HAV seroepidemiology studies. Table 14.1 provides a list of many of the successful application of IAHA assays, showing that it has been used in conjunction with many viral, bacterial, fungal, and mycoplasmal antigens.

14.3 IAHA MICROTITER PROCEDURE

In the IAHA test, antibodies (Ab) and antigens (Ag) are allowed to form complexes in the first incubation. This is followed by the addition of complement (C). The resultant Ab-Ag-C complexes can then react with C3b receptors on human erythrocytes to cause hemagglutination.

The equipment, reagents, and manipulations required for the performance of the IAHA test are similar to those used for the CF test. The following is a step-by-step description of the procedure. The composition of solutions used for IAHA is found in the appendix at the end of this chapter.

1. Dilute test sera in diluting buffer, veronal non-buffered saline (VBS, pH 3), at 1:4 and then inactivate the sera by heating at 56 °C for 30 minutes.

2. Rinse V-well microtiter plates prior to their use, with VBS containing gelatin (GVB); discard the rinse solution by

Table 14.1 Applications of IAHA

Organism	Antigens	References
Bacterial	*Legionella pneumophila*	Lennette et al, 1978
Chlamydial	*C. psittaci*	Lennette and Lennette, 1978
	C. trachomatis	
Fungal	*Blastomyces*	Lennette E.T., unpublished
	Histoplasma capsulatum (mycelial phase)	
Rickettsial	Q-Fever	Lennette E.T., unpublished
Viral	Adenoviruses	Ito and Tagaya, 1966
		Lennette and Shah,
	BK papovavirus	unpublished
	Cytomegalovirus (CMV)	Dienstag et al, 1976
	Dengue virus	Inouye et al, 1980
	Enteroviruses	Ito and Tagaya, 1966
	Epstein-Barr virus (EBV)	Lennette et al, 1982
	Herpes simplex virus	Ito and Tagaya, 1966
	Hemorrhagic fever with renal syndrome	Sugiyama et al, 1987
	Hepatitis A virus	Miller et al, 1975
	Hepatitis B surface	Okochi et al, 1970
	Hepatitis B core	Tsuda et al, 1975
	Japanese encephalitis	Inouye et al, 1981
	Mammary tumor virus	Nagayoshi et al, 1981
	Measles (rubeola) virus	Lennette and Lennette, 1978
	Norwalk virus	Kapikian et al, 1978
	Rabies virus	Budzko et al, 1983
		Bota et al, 1987
	Rotavirus	Kapikian et al, 1981
		Nagayoshi et al, 1978
	SV40	Ichikawa et al, 1987
	Varicella-zoster (VZV)	Gershon et al, 1976
Soluble	Paul-Bunnel heterophil	Lennette et al, 1982

inverting the plates over a sink, then rap the inverted plates against a hard surface to remove residual buffer.

3. Add one drop (25 mL) of VBS containing bovine serum albumin (BVB) to every well that will be used.

4. Add one drop of each inactivated serum to the first and eighth well of one row on the test plate; prepare serial twofold dilutions of the added serum (two sets of 7 and 5 wells), using microdilutors.

5. Add one drop of positive test antigen (previously titrated to contain 1 to 2 units of reactivity) to the first seven wells of each row. Add identically diluted negative (control) antigen to wells 8 through 12; the correct antigen concentrations are determined by block titrations with control reference sera.

6. Shake the plates for 10 seconds with a vibrating mixer, then incubate them at 37 °C for 30 minutes; it is also acceptable

to incubate the plates overnight in a refrigerator (useful with heat labile antigens); all plates should be covered during the incubations to minimize losses due to evaporation.

7. Add one drop of diluted guinea-pig complement (1:100 in BVB) to all wells on the test plates; again mix and incubate the plates at 37 °C for 40 minutes; determine the exact dilution of complement by a prior block titration; ordinarily no excess complement is used.

8. At the end of the incubation, stop the reaction by adding one drop of DTT-EDTA to each well, then add one drop of a 0.8% suspension of human erythrocytes, type 0, to each well.

9. Hemagglutination should be complete and readable within one hour; positive reactions are those showing >50% agglutination; the agglutination pattern is usually stable at room temperature; for best results, we chill the test plates (overnight), if they are not read within an hour or so.

Either U- or V-bottom microtest plates can be used for IAHA tests, although the hemagglutination patterns are easier to read on the V plates. We find that polystyrene plastic plates result in non-specific binding of erythrocytes which can be eliminated by pre-wetting the plates with a buffer containing carrier protein. If gelatin is used in the pre-wetting buffer, each lot of gelatin should be screened for suitability, as some lots contain heat-labile substances that interfere in the IAHA test. The interference can be destroyed by autoclaving the gelatin stock solution.

Many acute-phase sera appear to contain immune-complexes, which persist even after the diluted serum has been inactivated at 56 °C. These complexes give positive IAHA reactions, even in the absence of added test antigens. Sera containing these immune-complexes are not found as

frequently as are sera that are reported as "anticomplementary" in the CF test, nevertheless, the immune-complexes are an annoyance. We have been able to eliminate the non-specific reactions due to these immune complexes by diluting test sera in barbital-saline adjusted to pH 3 (essentially unbuffered) and then heating the diluted sera at 56 °C for 30 minutes. It appears that this modified inactivation procedure removes interference from complexes. If the serum is not diluted more than 1:5, no subsequent adjustment of pH is necessary.

Commercially available antigens for use with the CF test have been satisfactory for use in the IAHA test, provided the antigen CF test titer was at least 1:8. Antigen concentrations needed for the IAHA test are generally only one fourth or one eighth those required for the CF test, which may result in significant savings of cost and material. We have prepared many different viral antigens and compared them to commercially available materials. In nearly all cases, antigens prepared from tissue culture material "in-house" has been of better quality and higher titer than the commercially prepared antigens. The following procedure is satisfactory for the preparation of many viral antigens for use in the IAHA test.

Infected and noninfected cell cultures are separately suspended in saline equal in volume to one fifth their original growth medium. The cell suspensions are then disrupted by sonication and clarified by centrifugation at 10,000 × g for 10 minutes. Further purification or concentration of their antigen preparation is usually unnecessary. The preparation can be dispensed in small volumes and stored frozen at −70 °C for indefinite periods. When proper conditions for viral replication are used, antigen titers of 32 or greater should be expected. Photochemical inactivation of virus infectivity in the antigen preparation using psoralens and long-wave UV irradiation has given us excellent results, using the methods described by Hanson et al (1978).

We have prepared bacterial antigens using the confluent growth obtained on appropriate agar media, which can be scraped, suspended in saline, and inactivated by heating or the addition of formalin (at 0.1%). The suspension is then washed twice by centrifugation at 10,000 × g for 20 minutes. The washed sediment is then adjusted to a 10% suspension and sonicated to disrupt the cells. For disruption of 1 to 2 mL of suspension, we use three cycles of 15 seconds each at 10 watts output of a "micro-probe" type sonicator. The preparation is clarified at 10,000 × g and stored at −70 °C until use. The antigens can be prepared very efficiently in this manner, and the antigen titer achievable is usually between 500 and 2000.

Egg-derived antigens have not been satisfactory in the IAHA test. Such antigens include rickettsiae and chlamydiae grown in yolk sacs. Extraction of lipids from these antigens with Freon 113 improves the performance of the testing using these antigens, but at the cost of decreased antigen titers. Freon 113 extraction is also needed to obtain a satisfactory rotavirus antigen from stools and for marmoset-liver derived HAV purified by density gradient centrifugation. The use of Freon 113 and other chlorofluorocarbons is expected to be reduced or eliminated in most laboratories during the next few years as environmentally undesirable.

The stability of viral and bacterial antigens for the IAHA test is very good, provided they do not become contaminated with microbes. Addition of 0.01% thimerosal as a preservative reduces, but will not eliminate, such contamination. A minor problem encountered with frozen antigen preparations is the precipitation of antigen that has been stored and thawed. This can usually be counteracted by brief sonication of the freshly thawed antigen preparation.

14.4 RED BLOOD CELLS

Primate erythrocytes and non-primate platelets are suitable indicator cells for the IAHA test (Nelson, 1963). The only widely used cell type is human type O erythrocytes. The nature of the reactive site on the cells is unknown—the presence of C3b receptors on the cells is necessary, but not sufficient to provide IAHA reactivity, as only about one out of three O positive donors have suitable erythrocytes (Klopstock, et al, 1963). The receptors are sensitive to proteolytic enzymes and to neuraminidase, consistent with a glycoprotein composition (E. T. Lennette, unpublished observations). The ability of cells to react in the IAHA procedure, in the presence of immune complexes, is apparently a permanent property of the donor: erythrocytes obtained from one of the authors are known to have been reactive over a span of more than ten years. Hence the suitability of any particular donor is permanent. Obtaining a supply of reactive erythrocytes is the main obstacle preventing some laboratories from adopting IAHA procedures. There is no commercial supply of pretested blood, so that each laboratory needs to find its own supply. Our laboratory uses two sources of blood: laboratory personnel and local blood banks. Type O blood from a blood bank can be used for two to three weeks; the sample segments (which are often discarded) from blood bags may supply enough cells for a useful number of IAHA tests—being used at a rate of one segment per day by our laboratory. Often there are as many as eight segments from each blood bag that are unused by the blood bank and that could be made available to the clinical laboratory. Blood collected in EDTA is not suitable for use in the IAHA test.

In our laboratory, a reference antigen and a set of reference sera are used to screen blood samples for suitable erythrocytes. Only erythrocytes that give acceptable serum titers are selected for use. Donated blood segments can be stored for up to two weeks at 4 °C before use. Although this means of obtaining erythrocytes is less convenient than having access to a panel of pre-selected donors, we find it presents the fewest problems. We do not recommend pooling cells from randomly selected donors; using pooled erythrocytes with the IAHA procedure gives substantially lower sensitivity. Limited efforts have been made towards increasing the shelf life of erythrocytes for use in the IAHA test. Cells fixed with glutaraldehyde or formaldehyde can be used in IAHA procedures, but fixed cells are less satisfactory than fresh cells. A method for freezing human erythrocytes for later use in the IAHA test has been reported (Lawrence and Wentworth, 1985) which eliminates the need to screen blood donors.

14.5 HEMAGGLUTINATION PATTERNS

The agglutination pattern seen in the IAHA test is uniform and does not vary appreciably with different test antigens. However, one should become familiar with the agglutination patterns produced by individual sera. The patterns in microtest plates are best examined with the aid of a magnifying mirror. A specific positive reaction should appear as uniformly granular agglutination and a negative reaction should appear as a cell button with a smooth outline. While reading test results, one should compare the patterns obtained with test sera to those obtained with the control or reference sera. With practice, it is usually possible to differentiate specific agglutination from nonspecific reactions. In addition to the agglutination sometimes seen with "negative control antigens," some sera give an uncharacteristic agglutination with the appropriate test antigen. Specific reactions

give a slightly granular or coarse agglutination; non-specific reactions often produce a fine textured or matte agglutination. Even when freshly collected sera are inactivated in low-pH diluent, about 1% of the specimens tested have equal titers with positive and control antigens. It is usually necessary to resort to another test system to evaluate such sera.

14.6 DETECTION OF IGM ANTIBODIES

IgM antibodies are not efficiently detected by the CF assay. As both CF and IAHA tests detect complement-activating immune complexes, it was somewhat surprising to discover that IgM antibodies are detectable by IAHA assay. The first evidence that IgM antibodies react well in IAHA tests came from the application of IAHA testing to the detection of the Paul-Bunnell antibodies associated with Epstein-Barr virus (EBV) mediated infectious mononucleosis. The IAHA test was both more sensitive and more specific than the standard differential heterophil agglutination test and ox-cell hemolysis test (Lennette et al, 1978). Paul-Bunnell antibodies are exclusively of the IgM class. Other evidence that IgM antibodies are reactive in IAHA came from work employing sera that were fractionated by sucrose gradient density centrifugation in studies using sera obtained from varicella and zoster patients (Gershon et al, 1981). Fractions containing either VZV specific IgM or IgG were both reactive in the IAHA test, provided that the IAHA test was performed promptly after fractionation of the sera. In our experience, sucrose has a detrimental effect on IAHA reactive IgM even after a brief storage; thus, we recommend other methods for physical separation of IgM, e.g., Quick-Sep columns (Isolab, Inc., Akron, OH).

We have observed that the detection of IgM antibodies by IAHA depends on the quality of the test antigens used. Using sera containing Paul-Bunnell heterophil antibody, it was possible to show that antigen lots varied greatly in reactivity. Results obtained by box-titration of different lots of antigen can differ by as much as tenfold. This observation has been extended to include other commercially available viral antigens for use in CF tests, e.g., cytomegalovirus (CMV) and VZV antigens. Using IgM containing reference sera, the lot-to-lot variation of an antigen preparation provided by any one supplier was seen as a shift in the optimal titer of each lot of antigen. Although the optimal titer of antigen varies from lot to lot, the reference sera titers do not vary. In contrast, the variations in antigen preparations from supplier to supplier affect the titers of the test sera. That is, the serum titer obtained can vary greatly depending upon the source of antigen. This variation is seen mainly with IgM-containing sera, and not appreciably with sera containing IgG principally, i.e., from immune donors, suggesting different reactivities of these Ig classes to viral antigens. At present, there is no standardization among suppliers as to the composition of viral test antigens used in serologic reactions to contain a particular specificity. The differences are noticeable in the CF test, but are more pronounced in the IAHA test, especially when sera containing IgM antibodies are tested. We recommend that serum panels for antigen titration and evaluation should include known IgM-positive sera when possible.

14.7 SENSITIVITY AND SPECIFICITY

IAHA is most often compared to CF, probably due to their similarities in manipulations and reagent requirements. Also, both procedures measure complement activation by immune complexes, the IAHA directly, and the CF by complement depletion.

There are, however, a few differences between the two tests that explain the advantages found with IAHA.

First, the tests differ in the nature of the indicator system used to measure the immune complexes formed in vitro. In the CF test, a measured amount of complement is added to the test wells; any complement not bound in the test reactions is then indicated by the addition of sheep red blood cell(s) (SRBC)–anti-sheep RBC complex. Remaining complement reacts to lyse the SRBC. This procedure is inherently insensitive to small differences in depletion of the complement added at the beginning of the test. In the IAHA procedure, only the complement reacted in the test is measured; the test is insensitive to excess unreacted complement. Any agglutination above a very low background is significant. This difference in mechanism between the two tests accounts for the four to eightfold increase in sensitivity shown by the IAHA method.

The increased sensitivity of the IAHA test has been shown to be adequate for the reliable use of the IAHA test for determination of immune status, even of immunosuppressed patients (Gershon et al, 1976). Agents for which this has been shown include VZV, CMV, EBV, and others (see Table 14.1). This increased sensitivity, together with the ability to detect IgM antibodies allows the IAHA test to demonstrate a more rapid and pronounced titer increase during an infection than can be seen with the CF test. Using the CF test, a minimum of two weeks is advocated for the reliable detection of antibody titer change between acute-phase and convalescent-phase sera. With the IAHA test, 8- to 16-fold titer changes can frequently be found with sera collected three to five days apart, during the acute or early convalescent phase of illness.

IAHA is comparable to IFA, rather than CF, in its sensitivity, as shown by parallel serologic testing of numerous sera using varied antigens. Although IFA titers are often slightly higher than those obtained with IAHA, the ability of the IFA test to differentiate positive from negative sera is the same as that of the IAHA test. In our laboratory, IFA and IAHA tests can be used almost interchangeably, providing useful and complementary test systems for a wide range of antigens.

IAHA is comparable in reactivity to both FAMA and neutralization tests (Gershon et al, 1976) for determination of varicella-zoster immune status. Data in Baba et al, (1987) showed that the IAHA test clearly differentiates infections with VZV and herpes simplex virus. For determination of post-immunization immunity to rabies, IAHA was found to be as suitable as the "rapid" fluorescent-focus inhibition (RFFIT) assay which is commonly used (Budzko et al, 1983; Bota et al, 1987). The RFFIT assay takes several days to complete and requires the use of live rabies virus.

Limited comparisons of IAHA against RIA and EIA tests indicate that IAHA is slightly less sensitive than these methods. The sensitivity difference does not appear to be significant in routine clinical applications.

IAHA tests are useful to detect antigenic differences among related viruses and offer very good type-specificity. For example, IAHA has been used to sub-group rotaviruses in a different system than that developed by neutralization tests (Kapikian et al, 1981). IAHA is also reported to be useful for serotyping hemorrhagic fever with renal syndrome viruses into four serotypes (Sugiyama et al, 1987), where it proved superior to IFA testing for classification. In our own laboratory, IAHA has been used to subtype *Legionella pneumophila*, using hyperimmune rabbit sera. The ability to detect antigenic differences, against a background of crossreacting specificities, allows the IAHA to be used in monitoring antigen purification, e.g., purification of rotavirus extracted from stools. In the purification of EBV antigens, IAHA was found to detect specific EBV antigens and was unaffected by the presence of other antigens (impurities) in the antigen preparations of the virus.

14.8 ADVANTAGES AND DISADVANTAGES

The principal limitation of the IAHA lies in its inability to differentiate antibodies of different immunoglobulin subclasses. Although the IAHA assay detects IgM antibodies as well as IgG antibodies, the two classes cannot be measured separately, unless they are physically separated, such as by column chromatography or density gradient centrifugation. In addition, IAHA will not react with antibodies that do not fix complement, e.g., IgA. Another disadvantage of the IAHA is that it is difficult to use with viruses that agglutinate RBC. In theory, it is possible to obtain antigens free of hemagglutinins. In practice, commercial CF antigens for viruses such as influenzas and mumps often do contain hemagglutinin activity and are not suitable for use in the IAHA test.

We find the IAHA test to be satisfactory for routine use for reasons other than sensitivity and specificity. The procedure is quite economical relative to costs of reagents and supplies. Due to its increased sensitivity, compared to the CF test, the amounts of antigen and complement used may be reduced about fourfold, on the average. Although no special equipment is required to perform IAHA tests, equipment is sold that would permit partial to full automation of the test procedure. The end points are usually very sharp, with 4+ agglutination in one well and no agglutination in the next well. Thus, there is little uncertainty or subjectivity in obtaining an accurate titer—which is often a problem with reading IFA tests, for example. Titers are reproducible, both within test runs, and between runs, which makes it easy to detect significant titer changes.

Another advantage of the IAHA assays is the simplicity of pretest preparations. Every test component of the CF procedure has to be monitored carefully and titered for each test run. With the IAHA procedure, every component is added in excess and needs to be titered only once for each lot of reagent, as long as the reagent is stable during storage. Also, the CF test usually requires overnight incubation, whereas the IAHA test is completed within three to four hours.

14.9 APPENDIX

Veronal Buffer (VB) 5× stock: Dissolve 43.0 g of NaCl and 4.6 g of diethylbarbituric acid in 950 mL of warmed (lower than 65 °C) deionized water. Adjust the pH with NaOH to 7.4. Add 2.5 mL of $MgCl_2/CaCl_2$ solution and adjust final volume to 1 L. The $MgCl_2/CaCl_2$ solution should contain 20.33 g $MgCl_2 \cdot 6H_2O$ and 4.4 g of $CaCl_2$ in 100 mL of water.

Serum dilution buffer: VB 1× is prepared by diluting above stock solution fivefold with water. The pH is adjusted with 2N HCl to a pH of 3.0

GVB: Bovine serum albumin, fraction V, is added to 1× VB to a final concentration of 1 mg/mL.

EDTA-DTT-VB buffer: Two parts of 0.1 M EDTA pH 7.5 (disodium ethylenediamine-tetraacetic acid) is added to three parts of 1× VB. DTT is added to a final concentration of 3 mg/mL. EDTA-VB mixture can be stored indefinitely at 5 °C. Once DTT is added, the solution can be used for up to four weeks, if stored at 5 °C.

REFERENCES

Baba, K., Shiraki, K., Kanesakii, T., Yamanishi, K., Ogra, P., Yabuuchi, H., and Takahasi, M. 1987. Specificity of skin test with varicella-zoster virus antigen in varicella-zoster and herpes simplex virus infections. J. Clin. Microbiol. 25:2193–2196.

Bota, C., Anderson, R., Goyal, S., Charamella, L., Howard, D., and Briggs, D. 1987. Comparative prevalence of rabies antibodies among household and unclaimed/stray dogs as determined by the immune adherence hemagglutination assay. Int. J. Epidemiol. 16:472–476.

Budzko, D.B., Charamella, L.J., Jelinek, D., and Anderson, G.R. 1983. Rapid test for detection of rabies antibodies in human serum. J. Clin. Microbiol. 17:481–484.

Dienstag, J.L., Cline, W.L., and Purcell, R.H. 1976. Detection of cytomegalovirus antibody by immune adherence hemagglutination. Proc. Soc. Exp. Biol. Med. 153:543–548.

Gershon, A.A., Kalter, Z.G., and Steinberg, S. 1976. Detection of antibody to varicella-zoster virus by immune adherence hemagglutination. Proc. Soc. Exp. Biol. Med. 151:762–765.

Gershon, A.A., Steinberg, S.P., Borkowsky, W., Lennette, D., and Lennette, E. 1981. IgM to varicella-zoster virus: Demonstration in patients with and without clinical zoster. Ped. Inf. Dis. 1:164–166.

Hanson, C.V., Riggs, J.L., and Lennette, E.H. 1978. Photochemical inactivation of DNA and RNA viruses by psoralen derivatives. J. Gen. Virol. 40:345–358.

Ichikawa, T., Minamoto, N., Kinjo, T., Matsubayashi, N., Matsubayashi, K., and Narama, I. 1987. A serological survey of simian virus 40 in monkeys. Microbiol. Immunol. (Japan) 31:1001–1008.

Inouye, S., Matsuno, S., Hasegawa, A., Miyamura, K., Kono, R., and Rosen, L. 1980. Serotyping of dengue viruses by an immune adherence hemagglutination test. Am. J. Trop. Med. Hyg. 29:1389–1393.

Inouye, S., Matsuno, S., and Kono, R. 1981. Difference in antibody reactivity between complement fixation and immune adherence hemagglutination tests with virus antigens. J. Clin. Microbiol. 14:241–246.

Ito, M. and Tagaya, I. 1966. Immune adherence hemagglutination test as a new sensitive method for titration of animal virus antigens and antibodies. Japan. J. Med. Sci. Biol. 19:109–126.

Kapikian, A.Z., Greenberg, H.B., Cline, W.L., Kalica, A.R., Wyatt, R.G., James, H.D., Jr., Lloyd, N.L., Chanock, R.M., Ryder, R.W., and Kim, H.W. 1978. Prevalence of antibody to the Norwalk agent by a newly developed immune adherence hemagglutination assay. J. Med. Virol. 2:281–294.

Kapikian, A.Z., Cline, W.L., Greenberg, H.B., Wyatt, R.G., Kalica, A.R., Banks, C.E., James, H.D., Jr., Flores, J., and Chanock, R.M. 1981. Antigenic characterization of human and animal rotaviruses by immune adherence hemagglutination assay (IAHA): Evidence for distinctness of IAHA and neutralization antigens. Infect. Immun. 33:415–425.

Klopstock, A., Schwartz, J., and Zipkis, N. 1963. Individual differences of the reactivity of human erythrocytes in the immune adherence hemagglutination test. Vox Sang. 8:382–383.

Lamanna, C. 1957. Adhesion of foreign particles to particulate antigens in the presence of antibody and complement (serological adhesion). Bact. Rev. 21:30–45.

Laveran, A. and Mesnil, F. 1901. Recherches morphologiques et experimentales sur le trypanosome des rats (Tr. lewisi Kent.). Ann. Inst. Pasteur 15:673–714.

Lawrence, T. and Wentworth, B. 1985. Freezing and rejuvenation of human 0 erythrocytes for use in the immune adherence hemagglutination test. J. Clin. Microbiol. 22:654–655.

Lennette, E.T., Henle, G., Henle, W., and Horwitz, C.A. 1978. Heterophil antigen in bovine sera detectable by immune adherence hemagglutination with infectious mononucleosis sera. Infect. Immun. 19:923–927.

Lennette, E.T. and Lennette, D.A. 1978. Immune adherence hemagglutination: Alternative to complement-fixation serology. J. Clin. Microbiol. 7:282–285.

Lennette, D.A., Lennette, E.T., Wentworth, B.B., French, M.L.V., and Lattimer, G.L. 1979. Serology of Legionnaires' Disease: Comparison of indirect immunofluorescent antibody, immune adherence hemagglutination and indirect hemagglutination tests. J. Clin. Microbiol. 10:876–879.

Lennette, E.T., Ward, E., Henle, G., and Henle, W. 1982. Detection of antibodies to Epstein-Barr virus capsid antigen by immune adherence hemagglutination. J. Clin. Microbiol. 15:69–73.

Levaditi, C. 1901. Sir l'etat de la cytose dans la plasma des animaux normaux et des organismes vaccine contre le vibrion chloerique. Ann Inst. Pasteur 15:894–927.

Mayumi, M.K., Okochi, K., and Nishioka, K. 1971. Detection of Australian antigen by means of immune adherence hemagglutination test. Vox Sang. 20:178–181.

Miller, W.J., Provost, P.J., McAller, W.J., Ittensohn, O.L., Villarejos, V.M., and Hilleman, M.R. 1975. Specific immune adherence assay for human Hepatitis A antibody application to diagnostic and epidemiologic investigations. Proc. Soc. Exp. Biol. Med. 149:254–261.

Nagayoshi, S., Yamaguchi, H., Ichikawa, T., Miyazu, M., Morishima, T., Ozaki, T., Isomura, S., Suzuki, S., and Hoshino, M. 1980. Changes of the rotavirus concentration in faeces during the course of acute gastroenteritis as determined by the immune adherence hemagglutination test. Eur. J. Pediatr. 134:99–102.

Nagayoshi, S., Imai, M., Tsutsui, Y., Saga, S., Takahashi, M., and Hoshino, M. 1981. Use of the immune adherence hemagglutination test for titration of breast cancer patients' sera cross-reacting with purified mouse mammary tumor virus. GANN 72:98–103.

Nelson, D.S. 1963. Immune adherence. Adv. Immunol. 3:131–180.

Okochi, K., Mayumi, M., Haguino, Y., and

Saito, N. 1970. Evaluation of frequency of Australia antigen in blood donors of Tokyo by means of immune adherence hemagglutination technique. Vox Sang. 19:332–337.

Sugiyama, K., Morikawa, S., Matsuura, Y., Tkachenko, E., Morita, C., Komatsu, T., Adao, Y., and Kitamura, T. 1987. Four serotypes of hemorrhagic fever and renal syndrome viruses identified by polyclonal and monoclonal antibodies. J. Gen. Virol. 68:979–987.

Suntharee, R., Charnchudhi, C., Sompop, A.,

Kanai, C., Igarashi, A., and Inouyes, S. 1981. Isolation and identification of dengue viruses combined use of C6/36 cells and the immune adherence hemagglutination test. Jpn. J. Med. Sci. Biol. 34:375–379.

Tsuda, F., Takahashi, T., Takahashi, K., Miyakawa, Y., and Mayumi, M. 1975. Determination of antibody to hepatitis B core antigen by means of immune adherence hemagglutination. J. Immunol. 115:834–838.

IgM Determinations

Kenneth L. Herrmann and Dean D. Erdman

15.1 INTRODUCTION

The presence of specific antibody activity due to immunoglobulins (Ig) in serum was reported as early as the 1930s (Heidelberger and Pederson, 1937). Subsequent studies demonstrated that the first immunoglobulins to appear after a primary antigenic stimulus were of the IgM class. These IgM antibodies reportedly disappeared rapidly, usually within a few weeks, and were replaced by IgG antibodies that persisted for a longer period.

The transient nature of the IgM antibody response appears to hold true for most primary viral infections, and the determination of specific antiviral IgM antibodies is now well established as a potentially valuable method for the rapid diagnosis of recent or current viral infections. Such an approach provides a considerable advantage over classic serologic testing, which required the demonstration of a significant rise in antibody titer between paired acute- and convalescent-phase serum specimens. For this approach to be successful, the IgM antibody response must be specific, must be measurable with adequate reliability and sensitivity,

and must be transient (i.e., present only with recent active infection by the particular virus).

15.2 METHODS USED FOR IgM ANTIBODY DETERMINATION

Since the introduction of the first diagnostic applications of IgM determination, a variety of methods have been developed and applied for this purpose (Table 15.1). These methods can generally be separated into four groups: a) those based on comparing titers before and after chemical inactivation of serum IgM proteins, b) those based on physicochemical separation of IgM from the other serum immunoglobulin classes, c) those based on solid phase indirect immunoassays using labeled anti-human IgM, and d) reverse "capture" solid phase IgM assays. This chapter will discuss each of these approaches.

15.2.1 Methods Based on Chemical Inactivation of IgM

Mercaptans have the capacity to selectively split IgM molecules into immunologically

Table 15.1 Assay Methods for IgM Antibody

Methods based on chemical inactivation of IgM
  Alkylation-reduction by mercaptans
Methods based on physicochemical separation of IgM
  Sucrose density gradient ultracentrifugation
  Gel filtration
  Ion exchange chromatography
  Affinity chromatography
  Immune precipitation
  Protein A absorption
  Protein G absorption
Solid phase indirect immunoassays using labeled anti-human IgM for detector antibody
  Immunofluorescence assay (IFA)
  Radioimmunoassay (RIA)
  Enzyme immunoassay (EIA)
Solid phase "capture" immunoassays using unlabeled anti-human IgM for coating antibody
  Immunofluorescence assay (IFA)
  IgM antibody capture radioimmunoassay (MACRIA)
  Solid phase immunosorbent technique (SPIT)
  Hemadsorption immunosorbent technique (HIT)
  Enzyme-labeled antigen (ELA)
Other assays
  Radioimmunodiffusion
  Radioimmunoprecipitation
  Counterimmunoelectrophoresis
  Anti-IgM hemagglutination
  Latex-IgM agglutination

---

compromised monomers by breaking the disulfide bonds between the polypeptide chains. Some laboratories have attempted to use this approach to detect virus-specific IgM antibodies (Banatvala et al, 1967). Briefly, serum antibody titers are measured by standard serologic methods before and after mercaptan treatment. The presence of specific IgM antibody is indicated by a significant (fourfold or greater) decrease in titer of the treated serum sample. The method is simple, but relatively insensitive. For this test to be positive (i.e., demonstration of a fourfold or greater decrease in titer between treated and untreated serum), at least 75% of the total virus-specific antibody must be of the IgM class. This would be true only during the very early stages of most viral infections, and therefore, the diagnostic value of this approach is quite limited. Furthermore, because mercaptans are volatile, IgM monomers can reassociate and regain immune reactivity, resulting in false-negative reactions. Alternatively, too rigorous mercaptan treatment will reduce IgG molecules producing a false-positive test result. For these reasons, this method is not considered acceptable for detection of virus-specific IgM. However, serum treatment with 2-mercaptoethanol (2-ME) or dithiothreitol (DTT) may be an effective control step when used in conjunction with various physicochemical immunoglobulin separation methods described later in this chapter (Caul et al, 1974; Pattison, 1982).

15.2.2 Methods Based on Physicochemical Separation of IgM

15.2.2.a Gel Filtration

Column chromatographic methods have been used for many years to separate and isolate serum IgM antibodies (Frisch-Niggemeyer, 1982). Sephacryl S-300 or Sephadex G-200 (Pharmacia, Inc.) are two commercially available products used to fractionate serum immunoglobulins. Sephacryl S-300, a mixture of sepharose and acrylamide, offers several advantages over Sephadex G-200, because it does not need to be rehydrated and will allow high flow-rate filtration of serum under pressure without problems of overpacking or deforming the column (Morgan-Capner et al, 1980). Chromatography columns are prepared and used as directed by the manufacturer. Since serum lipoproteins and nonspecific cell agglutinins elute from these columns along with the IgM, these substances must be removed prior to serum fractionation if they interfere with the reliability of the assay such as interfering with the activity of the IgM antibody. Serum pretreated with heparin and $MnCl_2$ to remove lipoproteins and cell adsorbed to remove nonspecific agglutinins is layered on the top of the gel column and eluted through the column with Tris-buffered saline (0.02 M Tris in 0.15 M NaCl). Discrete fractions are collected for titration of antibody activity. Each new column should be standardized with known specific IgM-positive and IgM-negative sera. With both Sephacryl S-300 and Sephadex G-200 columns, IgM is eluted in the first protein peak and IgG in the second. IgA may also be present in the first peak eluted from the Sephadex G-200 column but not from the Sephacryl S-300 column. Both columns are equally useful for separating IgM from IgG molecules.

The specificity of the gel filtration IgM test for diagnosis of viral infections is very high, provided a number of factors that can cause false-positive results are taken into consideration (Pattison et al, 1976). Prolonged storage of sera at $-20\,°C$ or bacterial contamination of the sera may make the serum pretreatment ineffective resulting in false positive results. Also, if the serum has been preheated at 56 °C or higher, IgG may aggregate and therefore could elute in the IgM fractions after gel filtration. To minimize misinterpretation of the gel fractionation test, any presumptive IgM antibody activity in the first peak should be shown to be 2-ME sensitive.

15.2.2.b Sucrose Density Gradient Ultracentrifugation

The most commonly used method of fractionating sera for detecting virus-specific IgM antibodies employs high-speed ultracentrifugation of serum on sucrose gradients. Because IgM proteins have a higher sedimentation coefficient (19S) relative to the other immunoglobulins (7-11S), IgM antibodies can be physically separated from other antibodies by rate-zonal centrifugation on sucrose gradients. Lipoprotein molecules, including most of the nonspecific inhibitors of rubella hemagglutination, have a low density and therefore remain near the top of the gradient following centrifugation. The technique was introduced in 1968 for the rapid diagnosis of recent rubella virus infection by demonstrating the presence of IgM antibodies (Vesikari and Vaheri, 1968). Since that time, various modifications of the method have been published (Forghani et al, 1973; Caul et al, 1976), and the test has been applied to the diagnosis of other viral infections as well (Al-Nakib, 1980; Hawkes et al, 1980). Provided the necessary equipment is available, this separation procedure is relatively simple to perform.

The method as performed in our laboratory is described as follows. A density gradient is prepared by layering 1.4 mL amounts of 37%, 23%, and 10% (w/v) solutions of sucrose in phosphate-buffered sa-

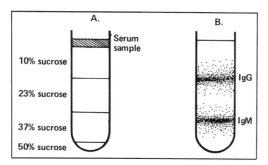

Figure 15.1 Diagram of sucrose density gradient before (A) and after (B) ultracentrifugation.

line (PBS), pH 7.2, on a 0.2 mL cushion of 50% sucrose in a 5 mL ultracentrifuge tube (Figure 15.1(A)). The sucrose layers are allowed to equilibrate for four to six hours at 4 °C. The test serum is diluted 1:2 in PBS (0.15 mL of serum mixed with 0.15 mL PBS), pretreated (if necessary) to remove nonspecific serum components that would interfere with the assay of IgM antibody, and 0.2 mL of the diluted serum is carefully layered on top of the gradient. The gradient is then centrifuged at 157,000 × g for 16 hours in a swinging-bucket rotor. Ten to 12 fractions (about 0.4 mL each) are collected by puncturing the bottom of the tube. The IgM antibodies concentrate in the bottom three or four fractions, IgG antibodies separate primarily in fractions six to eight, and the lipoproteins (nonspecific inhibitors) remain near the top of the gradient (Figure 15.1(B)).

The isolated IgM fractions can then be tested for antiviral activity using any suitable serologic test. The first four fractions (i.e., those presumed to contain IgM) must be checked for the presence of contaminating human IgG by radial immunodiffusion using agar gel immunodiffusion plates that will detect low levels of IgG. Suitable IgG assay plates are available commercially. In addition, the specificity of the IgM antibody activity in fractions one to four may be confirmed by demonstrating its sensitivity to 2-ME treatment.

Studies have shown that sucrose gra-

dient ultracentrifugation may not be as sensitive as some of the more recently developed indirect immunoassays or IgM capture immunoassays. However, because of its high degree of specificity and overall reliability, this method is generally considered to be the standard for comparison when new IgM antibody tests are developed. Sucrose gradient ultracentrifugation is a rather laborious procedure, and the high cost of the equipment places it out of the financial reach of most clinical laboratories. The more recent introduction of vertical rotors with reorienting gradients, however, now makes it possible to reduce the centrifugation time from 16 hours to only two hours, making it feasible to process a greater number of specimens in a given time period.

15.2.2.c Other Physicochemical Separation Methods

Other less frequently utilized methods for physically separating IgM from the other serum immunoglobulins include ion-exchange chromatography (Johnson and Libby, 1980), affinity chromatography (Barros and Lebon, 1975), immune precipitation, staphylococcal protein A absorption (SPA-Abs) (Ankerst et al, 1974), and streptococcal protein G adsorption (SPG-Abs) (Weiblen et al, 1990). Ion exchange chromatography is based on the differential binding of IgM and IgG to anion exchange resins. The commercially available Quik-Sep IgM# Isolation System (Isolab, Inc., Akron, OH) is an example of such a filtration system. Affinity chromatography employs columns of anti-human IgM covalently bound to sepharose beads to isolate serum IgM for subsequent assay for specific viral antibodies. Immune precipitation uses unbound anti-human IgG to remove interfering IgG antibodies prior to IgM assay as exemplified by the commercially prepared IgG Blocking Solution (Clinical Sciences, Inc., Whippany, NJ).

These methods have not received much attention for IgM antibody assay. On the other hand, SPA-Abs has attracted considerable interest as a simple and rapid screening method for IgM antibody. SPA, a cell wall protein present in some *Staphylococcus aureus* strains, binds to the Fc receptor of the IgG molecule, and can be used to absorb and remove the majority of the IgG component of serum. SPA-Abs does not remove all IgG subclasses, however, and up to 5% of the original IgG antibody activity may still remain following absorption. Care must be taken in interpreting test results as this residual antibody activity must not be mistakenly interpreted as representing IgM antibody. Absorbed and unabsorbed serum samples are run in parallel in a standard doubling dilution assay, so the percentage of residual antibody activity can be approximated. A decrease in titer of less than four doubling dilutions (i.e., residual of 12.5% or more of original titer) is presumptive evidence of specific IgM. However, results should be confirmed using a more definitive IgM assay. For example, a serum with an unabsorbed titer of 256 and a titer of 64 after SPA absorption (i.e., a two dilution decrease or 25% of original titer) would be considered presumptive positive for specific IgM, whereas a serum with titers of 256 and 16, before and after SPA-Abs (i.e., a four dilution decrease or about 6% residual), would be interpreted as negative for specific IgM.

Another frequent use of the SPA absorbent reagent is for pretreating serum to remove excess IgG and possible IgG-IgM immune complexes before testing by one of the solid phase indirect immunoassays, using labeled anti-human IgM (see below). This potentially increases the sensitivity and specificity of the assays. More recently introduced, SPG, a cell wall protein of some strains of group G streptococci, offers the advantage of removing all IgG subclasses with minimal loss of IgM antibodies (Bjork and Kronvall, 1984). Recombinant SPA and SPG are now available from several commercial sources in the United States.

15.2.3 Solid Phase Indirect Immunoassays

The availability of class-specific antiglobulins has led to the adaptation of several other serologic techniques, including indirect immunofluorescence assay (IFA), enzyme immunoassay (EIA), and radioimmunoassay (RIA) for detecting virus-specific IgM antibodies. The general principle for these methods is that test sera are incubated with viral antigens that have been previously bound to a solid phase surface (typically plastic microtiter plates, beads or tubes). Specific IgM antibodies bound to the antigen are subsequently detected with anti-human IgM antibody labeled with a suitable marker (Figure 15.2).

Figure 15.2 Schema of solid phase indirect immunoassay for IgM antibody.

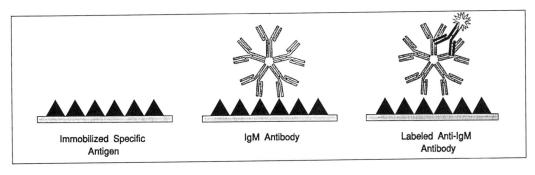

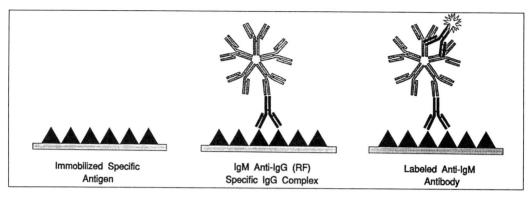

| Immobilized Specific | IgM Anti-IgG (RF) | Labeled Anti-IgM |
| Antigen | Specific IgG Complex | Antibody |

Figure 15.3 Schema of false-positive IgM due to RF interference in the solid phase indirect immunoassay.

Because of the technical simplicity of these methods, commercially prepared indirect IgM immunoassay kits for several viruses, including rubella virus, herpes simplex virus (HSV), cytomegalovirus (CMV), Epstein-Barr virus (EBV), rotavirus, and hepatitis A virus and hepatitis B surface antigen (HBsAg) are available in the United States and Europe. However, there are concerns over a number of pitfalls in such methods that may limit the sensitivity and specificity of these assays. These pitfalls can be grouped into three categories: a) the quality of available reagents, b) interference by IgM-class rheumatoid factor (RF), and c) competition between specific IgG and IgM antibodies in patient serum specimens for available antibody-binding sites on the antigen bound to the solid phase. Each of these factors can play an important role in the reliability of an indirect IgM antibody assay system.

Substantial improvements in the quality of reagents used in these assays have occurred during the past few years. More highly purified antigens and highly specific anti-IgM globulin conjugates are now available, although no standards for the specificity and potency of these reagents have as yet been established. False-positive results can occur in these IgM antibody assays if an IgM that has an anti-IgG activity but has no antiviral specificity is present in the

serum (i.e., RF). It would attach to complexes of IgG antibody bound to the solid phase antigen (Figure 15.3). On the other hand, failure to detect specific antiviral IgM may occur because a serum has a high level of specific antiviral IgG that competes better than IgM for antigen bound to the solid phase. These problems can be minimized by preadsorbing sera with SPA or protein A-Sepharose. This preadsorption has been shown to effectively eliminate nonspecific IgM activity from sera with known RF and to significantly increase the sensitivity of the specific IgM assay by removing most competing IgG (Kronvall and Williams, 1969).

15.2.3.a Immunofluorescence Assay

The first of the indirect assays to be applied for the detection of antiviral IgM antibodies was the IFA (Baublis and Brown, 1968). The antigen most often used is infected cells fixed on microscope slides. The method is essentially identical to IFA for IgG antibodies, except that fluorescein-labeled antihuman IgM is used. Polyclonal and monoclonal antihuman IgM conjugates are now commercially available.

The reading of IFA-IgM tests requires considerable skill and experience. Nonspecific staining may cause false-positive readings, but an experienced FA microscopist can minimize false-positive results by differentiating patterns of specific and nonspecific fluorescence, a possibility that does not exist in RIA and EIA tests. In experienced hands and with the use of high quality reagents, the IFA–IgM test can be both a sensitive and reliable method. This method should be avoided by those laboratories lacking substantial skill and past experience with IFA.

15.2.3.b Radioimmunoassay

Solid phase RIA has been used to detect viral antibodies since the early 1970s. The use of purified viral antigens bound to a suitable solid phase surface (polyvinylchloride or polystyrene) has eliminated much of the nonspecific background reactivity caused by anticellular or antinuclear antibodies found to be problematic in earlier RIA tests. When using highly purified antigens, the preparation of comparable control antigens is often impossible, and the specificity of the assay must be evaluated by comparison with a reference method and with results obtained after blocking by specific hyperimmune serum.

The major advantages of RIA for specific IgM antibody are the high sensitivity of the method and the potential for automation. Blocking the activity of specific IgM antibodies by IgG antibodies, a common drawback in IFA–IgM assays, appears not to be a problem in RIA–IgM tests (Knez et al, 1976). The major disadvantage of RIA has been the relatively short shelf-life of the ^{125}I-labeled antihuman immunoglobulin conjugates compared with IFA or EIA reagents.

15.2.3.c Enzyme Immunoassay

Enzyme immunoassay was first reported for the detection of specific antiviral IgM antibodies by Voller and Bidwell (1976). In principle, this method is identical to the RIA. The EIA uses antihuman IgM labeled with enzyme, usually alkaline phosphatase or horseradish peroxidase. The antigen preparations and solid phase supports used in EIA are identical to those used in RIA. The sensitivity and specificity of the two assays are comparable. Competition between specific IgM and IgG antibodies has been reported with EIA (Heinz et al, 1981), necessitating the separation of IgM from serum proteins. The enzyme conjugates for EIA tests have a long shelf-life, compared with the iodinated conjugates for RIA, and thus are more practical for commercial development. Consequently, commercial EIA–IgM test kits are now widely available. These tests generally may be read and results obtained using automated multichannel spectrophotometers linked to microcomputers, which may also be used to aid data interpretation.

15.2.4 Solid Phase "Capture" Immunoassays

Another approach for avoiding the problems of competitive interference and nonspecific reactivity seen with the traditional indirect immunoassays previously described is the reverse or "capture" IgM method (Figure 15.4). This approach, first reported for detecting antiviral IgM antibodies by Flehmig (1978) and more fully described by Duermeyer et al (1979), employs a solid phase surface sensitized with an anti-human-IgM antibody to "capture" and bind the IgM antibodies in a serum specimen. The washing process removes IgG and any immune complexes in the specimen. The addition of specific viral antigen, followed by a second labeled antiviral antibody, completes the test. Alternatively, antigen can be directly labeled (Schmitz et al, 1980), obviating detector antibody, or when antigen is a hemagglutinating virus, red blood cells can be used as indicators to produce hemagglutination (Krech and Wilhelm, 1979) or hemadsorption (Van der Logt et al, 1981; 1985). This

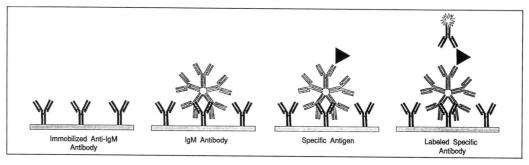

| Immobilized Anti-IgM Antibody | IgM Antibody | Specific Antigen | Labeled Specific Antibody |

Figure 15.4 Schema of reverse or "capture" solid phase immunoassay for IgM antibody.

approach has attracted considerable support and has become a standard method for detecting IgM antibodies to many viruses (Table 15.2).

Capture IgM assays have proven to be very sensitive and specific, and where comparisons have been made, superior to other assay designs (Gerna et al, 1987; Besselaar et al, 1989). Because the first step in the IgM capture assay leads to separation of IgM antibodies from other serum components, competition between IgG and IgM does not occur. However, because specific to nonspecific IgM antibodies must compete for available sites on the capture phase, assay sensitivity can vary with the relative proportion of specific to nonspecific IgM antibodies in the specimen. RF interference potentially exists with IgM capture assays as it does with indirect IgM immunoassays. For example, IgM–RF may be captured and bound on the solid phase and then, in turn, bind labeled antiviral antibodies. RF interference by this mechanism can be minimized, however, by 1) the substitution of monoclonal detector antibodies for human and animal polyclonal reagents which appears to reduce unwanted RF activity (Wielaard et al, 1985; Chantler and Evans, 1986); 2) using labeled F(ab)'2 fragments as detector antibody and eliminating the Fc portion of the IgG molecule which binds RF; 3) eliminating detector antibody entirely through direct labeling of antigen with enzyme or biotinylation of antigen followed by avidin-peroxidase

(Tuokko, 1988); 4) blocking the binding sites of RF by the addition of aggregated IgG to the serum diluent to which RF binds with greater affinity than native IgG (Forghani et al, 1983); and 5) assaying each specimen against negative cell controls to identify potential false positives. A more significant, but fortunately less common, problem is the binding of RF-antiviral IgG complexes by the capture phase which in turn can bind viral antigen (Figure 15.5). These specimens generally have very high levels of both RF and antiviral IgG antibodies, and are consequently less commonly encountered. In any case, capture IgM assays appear less susceptible to interference by RF than traditional indirect IgM assays and have become the assay of choice for the simple, sensitive, and specific detection of anti-viral IgM antibodies.

15.3 INTERPRETATION OF ASSAY TO DETECT IgM ANTIBODIES

The demonstration of specific antiviral IgM antibodies may be interpreted as indicating a recent or current infection with the virus in question only if the IgM response is specific (i.e., these IgM antibodies are not produced by any other infection or condition). In contrast, the absence of specific IgM antibodies rarely can be used as evidence to exclude a recent infection with a given virus. Variations in the temporal appearance of IgM antibodies, including the

Table 15.2 Capture Solid Phase Immunoassays for IgM Antibody

Virus	Method	Capture	Detector	Reference
Enteroviruses				
Coxsackie B	EIA	Goat	Mouse	Bell et al, 1986
	EIA	Sheep	Mouse	McCartney et al, 1986
Enterovirus 70	EIA	Goat	MAb	Wulffet et al, 1987
Hepatitis A	EIA	Rabbit	Human	Moller & Mathiesen, 1979
	RIA	Rabbit	Human	Flehmig et al, 1979
	EIA	Goat	Rabbit	Roggerndorf, 1980
	RIA	Goat	Human	Parry, 1984
Paramyxoviruses				
Measles	EIA	Rabbit	Rabbit	Pederson et al, 1982
	EIFA	MAb	Guinea pig	Forghani et al, 1983
	ELA	Swine	Labeled-Ag	Tuokko, 1988
Mumps	ELA	Rabbit	Labeled-Ag	Gut et al, 1985
Respiratory Syncytial Virus	EIA	Rabbit	Bovine	Cevenini et al, 1986
Parainfluenza	HIT	Goat	RBCs	Van der Logt et al, 1985
Influenza A	HIT	Goat	RBCs	Vikerfors et al, 1989
Hepatitis B	EIA	Rabbit	Human	Roggendorf et al, 1981
	EIA	Rabbit	Human	Briantais et al, 1984
	EIA	Rabbit	Labeled-Ag	Vilja et al, 1985
Herpesviruses				
Herpes simplex	ELA	Goat	Labeled-Ag	Van Loon et al, 1985
Varicella-zoster	EIA	Goat	MAb	Forghani et al, 1984
Cytomegalovirus	EIA	Goat	Goat	Yolken & Leister, 1981
	ELA	Rabbit	Labeled-Ag	Schmitz et al, 1980
	EIA	Rabbit	?	Re et al, 1989
	ELA	Rabbit	Labeled-Ag	Nielsen et al, 1987
Epstein Barr virus	EIA	Sheep	Sheep	Wielaard et al, 1988
	ELA	Rabbit	Labeled-Ag	Schmitz, 1982
Togaviruses & Flaviviruses				
Rubella	SPIT	Rabbit	RBCs	Krech & Wilhelm, 1979
	EIA	Rabbit	Rabbit	Kurtz & Malic, 1981
	EIA	Rabbit	Human	Vejtorp, 1981
	HIT	Sheep	RBCs	Briantais et al, 1984
	EIA	Sheep	MAb	Wielaard et al, 1985
	ELA	Rabbit	Labeled-Ag	Bonfanti et al, 1985
	EIA	MAb	MAb	Chantler & Evans, 1986
	EIA	Goat	MAb	Gerna et al, 1987
	RIA	Rabbit	Human	Mortimer et al, 1981
Japanese encephalitis	RIA	Goat	Monkey	Burke & Nisalak, 1982
	EIA	Goat	Human	Bundo & Igarashi, 1985
Tick-borne encephalitis	EIA	Rabbit	Rabbit	Heinz et al, 1981
Saint Louis encephalitis	EIA	Goat	MAb	Monath et al, 1984
Western & Eastern equine encephalitis	EIA	Goat	Mouse	Calisher et al, 1986
Colorado tick fever virus	EIA	Goat	Mouse	Calisher et al, 1985
Dengue	EIA	Goat	Human	Bundo & Igarashi, 1985
West Nile & Sindbis virus	EIA	Goat	MAb	Besselaar et al, 1989
Parvovirus B19	EIA	Goat	MAb	Anderson et al, 1986
	EIA	Goat	MAb	Brown et al, 1989
	EIA	Goat	MAb	Yaegashi et al, 1989
Norwalk virus	EIA	Goat	Human	Erdman et al, 1988
Rotavirus	EIA	Goat	MAb	Coulson, 1989
Rabies virus	EIA	Rabbit	Rabbit	Tingpalapong et al, 1986
Delta Agent	EIA	Rabbit	Human	Shattock et al, 1989

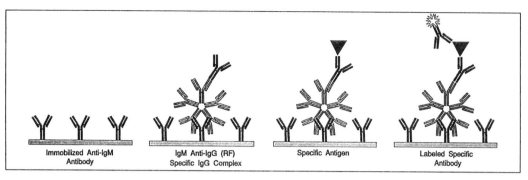

Immobilized Anti-IgM Antibody   IgM Anti-IgG (RF) Specific IgG Complex   Specific Antigen   Labeled Specific Antibody

Figure 15.5 Schema of false-positive IgM due to RF interference in the reverse capture solid phase immunoassay.

occurrence of prolonged IgM antibody responses, can result in difficulties in interpreting the significance of the test results in relation to the clinical illness in question.

False-positive IgM antibody results may occur due to cross-reactions between closely related viruses. Such cross-reactions have been reported for togaviruses (Wolff et al, 1981) and in coxsackie B virus infections (Schmidt et al, 1968). In general, the heterologous IgM antibody responses are low compared with homologous titers.

Evidence suggesting the occurrence of true polyclonal IgM production in cases of acute infectious mononucleosis has been reported by Morgan-Capner et al (1983). Their report suggests that production of various IgM antibodies may result from EBV-induced stimulation of B lymphocytes already committed by prior antigenic stimulation. These results emphasize the importance for careful interpretation of positive virus-specific IgM together with the complete clinical picture.

In infections with viruses belonging to groups of closely related strains or serotypes (adenoviruses, enteroviruses, parainfluenza, or togaviruses), serodiagnosis using specific IgM testing may be complicated by the possible absence of a specific IgM response, as well as by possible false-positive reactions to related viruses. Specific IgM antibody responses generally are absent in reinfections or reactivations of

latent virus infections and may be very weak or absent in certain immunocompromised patients.

Finally, the expected duration of the specific IgM response must be considered when interpreting the significance of observed specific IgM antibody. Generally, the IgM antibody response following an acute viral infection is of limited duration, usually one to two months. However, prolonged IgM antibody responses have been observed in complicated infections, chronic infections, congenital infections, and in some immunosuppressed patients. The persistence of specific IgM in these cases appears to be related to the persistence of viral antigen (or even replicating virus) in the patient. Occasionally, prolonged IgM antibody responses have been observed without any apparent reason. Also, as more sensitive methods are developed for the detection of specific antiviral IgM antibodies, the time following an acute infection during which specific IgM is detectable will be extended. For the diagnosis of an acute infection, the ideal maximum duration of specific IgM antibodies should be two to three months. It may therefore be necessary to limit the sensitivity of some assays to retain the optimal diagnostic usefulness of the methods.

The diagnostic value of specific IgM antibody assay is variable and is dependent on the virus and the infection in question.

Generally transient IgM responses are characteristic of acute viral infections caused by viruses that elicit long-lasting immunity. Such responses are seen with rubella, measles, mumps, parvovirus B19 and hepatitis A viruses. In these infections, a reliable diagnosis can usually be made by specific IgM antibody testing of a single serum specimen taken early in the illness. For other virus infections, such as HSV or CMV, the diagnostic usefulness of such tests is much more limited.

Several sensitive and reliable methods for the determination of specific antiviral IgM antibodies have now been developed. Reagents and kits for the performance of these tests are now available commercially for some virus infections. These methods, when adopted for routine use in clinical laboratories, should bring considerable improvement to viral diagnostic services.[1]

---

[1] Use of trade names is for identification only and does not imply endorsement by the Public Health Service or by the U.S. Department of Health and Human Services.

REFERENCES

Al-Nakib, W. 1980. A modified passive-haemagglutination technique for the detection of cytomegalovirus and herpes simplex virus antibodies: Application in virus-specific IgM diagnosis. J. Med. Virol. 5:287–293.

Anderson, L.J., Tsou, C., Parker, R.A., Chorba, T.L., Wulff, H., Tattersall, P., and Mortimer, P.P. 1986. Detection of antibodies and antigens of human parvovirus B19 by enzyme-linked immunosorbent assay. J. Clin. Microbiol. 24:255–256.

Ankerst, J., Christensen, P., Kjellen, L., and Kronvall, G. 1974. A routine diagnostic test for IgA and IgM antibodies to rubella virus: Absorption of IgG with *Staphylococcus aureus*. J. Infect. Dis. 130:268–273.

Banatvala, J.E., Best, J.M., Kennedy, E.A., Smith, E.E., and Spencer, M.E. 1967. A serological method for demonstrating recent infection by rubella virus. Br. Med. J. 3:285–286.

Barros, M.F. and Lebon, P. 1975. Separation des anticorps IgM anti-rubeole par chromatographie d'affinite. Biomedicine (Express) 23:184–188.

Baublis, J.V. and Brown, G.C. 1968. Specific responses of the immunoglobulins to rubella infection. Proc. Soc. Exp. Biol. Med. 128:206–210.

Bell, E.J., McCartney, R.A., Basquill, D., and Chaudhuri, A.K.R. 1986. μ-Antibody capture ELISA for the rapid diagnosis of enterovirus infection in patients with aseptic meningitis. J. Med. Virol. 19:213–217.

Besselaar, T.G., Blackburn, N.K., and Aldridge, N. 1989. Comparison of an antibody-capture IgM enzyme-linked immunosorbent assay with IgM-indirect immunofluorescence for the diagnosis of acute Sindbis and West Nile infections. J. Virol. Meth. 25:337–346.

Bjork, L. and Kronvall, B. 1984. Purification and some properties of streptococcal protein G, a novel IgG-binding reagent. J. Immunol. 133:969–974.

Bonfanti, C., Meurman, O., and Halonen, P. 1985. Detection of specific immunoglobulin M antibody to rubella virus by use of enzyme-labelled antigen. J. Clin. Microbiol. 21:963–968.

Briantais, M-J., Grangeot-Keros, L., and Pillot, J. 1984. Specificity and sensitivity of the IgM capture immunoassay: Studies of possible factors inducing false positive or false negative results. J. Virol. Meth. 9:15–26.

Brown, K.E., Buckley, M.M., Cohen, B.J., and Samuel, D. 1989. An amplified ELISA for the detection of parvovirus B19 IgM using monoclonal antibody to FITC. J. Virol. Meth. 26:189–198.

Bundo, K. and Igarashi, A. 1985. Antibody-capture ELISA for detection of immunoglobulin M antibodies in sera from Japanese encephalitis and dengue hemorrhagic fever patients. J. Virol. Meth. 11:15–22.

Burke, D.S. and Nisalak, A. 1982. Detection of Japanese encephalitis virus immunoglobulin M antibodies in serum by antibody capture radioimmunoassay. J. Clin. Microbiol. 15: 353–361.

Calisher, C.H., Poland, J.D., Calisher, S.B., and Warmoth, L.A. 1985. Diagnosis of Colorado tick fever virus infection by enzyme immunoassays for immunoglobulin M and G antibodies. J. Clin. Microbiol. 22:84–88.

Calisher, C.H., Berardi, V.P., Muth, D.J., and Buff, E.E. 1986. Specificity of immunoglobulin M and G antibody responses in humans infected with Eastern and Western equine encephalitis viruses: Application to rapid serodiagnosis. J. Clin. Microbiol. 23:369–372.

Caul, E.O., Hobbs, S.J., Roberts, P.C., and Clarke, S.K.R. 1976. Evaluation of simplified sucrose gradient method for the detection of rubella-specific IgM in routine diagnostic practice. J. Med. Virol. 2:153–163.

Caul, E.O., Smyth, G.W., and Clarke, S.K.R. 1974. A simplified method for the detection of rubella-specific IgM employing sucrose density fractionation and 2-mercaptoethanol. J. Hyg. (Camb.) 73:329–340.

Cevenini, R., Donati, M., Bertini, S., Moroni, A., and Sambri, V. 1986. Capture-ELISA for serum IgM antibody to respiratory syncytial virus. J. Hyg. (Camb.) 97:511–517.

Chantler, S. and Evans, C.J. 1986. Selection and performance of monoclonal and polyclonal antibodies in an IgM antibody capture enzyme immunoassay for rubella. J. Immunol. Meth. 87:109–117.

Coulson, B.S. 1989. Evaluation of end-point titration, single dilution and capture enzyme immunoassays for measurement of antirotaviral IgA and IgM in infantile secretions and serum. J. Virol. Meth. 53:53–66.

Duermeyer, W., Wielaard, F., and van der Veen, J. 1979. A new principle for the detection of specific IgM antibodies applied in an ELISA for hepatitis A. J. Med. Virol. 4:25–32.

Erdman, D.D., Gary, G.W., and Anderson, L.J. 1988. Development and evaluation of an IgM capture enzyme immunoassay for diagnosis of recent Norwalk virus infection. J. Clin. Microbiol. 24:57–66.

Flehmig, B. 1978. Laboratoriumdiagnose der Hepatitis A-Infektion. Bundesgesundheitsblatt 21:277–283.

Flehmig, B., Ranke, M., Berthold, H., and Gerth, H-J. 1979. A solid-phase radioimmunoassay for detection of IgM antibodies to hepatitis A virus. J. Infect. Dis. 140:169–175.

Forghani, B., Myoraku, C.K., and Schmidt, N.J. 1983. Use of monoclonal antibodies to human immunoglobulin M in "capture" assays for measles and rubella immunoglobulin M. J. Clin. Microbiol. 18:652–657.

Forghani, B., Myoraku, C.K., Dupuis, K.W., and Schmidt, N.J. 1984. Antibody class capture assays for varicella-zoster virus. J. Clin. Microbiol. 19:606–609.

Forghani, B., Schmidt, N.J., and Lennette, E.H. 1973. Demonstration of rubella IgM antibody by indirect fluorescent antibody staining, sucrose density gradient centrifugation and mercaptoethanol reduction. Intervirology 1:48–59.

Frisch-Niggemeyer, W. 1982. Simple and rapid chromatographic separation of IgM using microcolumns and stained sera. J. Virol. Methods 5:135–142.

Gerna, G., Aannino, G.M., Revello, M.G., Petruzelli, E., and Dovis, M. 1987. Development and evaluation of a capture enzyme-linked immunosorbent assay for determination of rubella immunoglobulin M using monoclonal antibodies. J. Clin. Microbiol. 25:1033–1038.

Gut, J-P., Spiess, C., Schmitt, S., and Kirn, A. 1985. Rapid diagnosis of acute mumps infection by a direct immunoglobulin M antibody capture enzyme immunoassay with labeled antigen. J. Clin. Microbiol. 21:346–352.

Hawkes, R.A., Boughton, C.R., Ferguson, F., and Lehmann, N.I. 1980. Use of immunoglobulin M antibody to hepatitis B core antigen in diagnosis of viral hepatitis. J. Clin. Microbiol. 11:581–583.

Heidelberger, M. and Pedersen, K.O. 1937. The molecular weight of antibodies. J. Exp. Med. 65:393–414.

Heinz, F.X., Roggendorf, M., Hofmann, H., Kunz, C., and Dienhardt, F. 1981. Comparison of two different enzyme immunoassays for detection of immunoglobulin M antibodies against tick-borne encephalitis virus in serum and cerebrospinal fluid. J. Clin. Microbiol. 14:141–146.

Johnson, R.B. Jr. and Libby, R. 1980. Separation of immunoglobulin M (IgM) essentially free of IgG from serum for use in systems requiring assay of IgM type antibodies without interference from rheumatoid factor. J. Clin. Microbiol. 12:451–454.

Knez, V., Stewart, J.A., and Zeigler, D.W. 1976. Cytomegalovirus-specific IgM and IgG response in humans studied by radioimmunoassay. J. Immunol. 117:2006–2013.

Krech, U. and Wilhelm, J.A. 1979. A solid-phase immunosorbent technique for the rapid detection of rubella IgM by hemagglutination inhibition. J. Gen. Virol. 44:181–286.

Kronvall, G. and Williams, R.C. Jr. 1969. Differences in anti-protein A activity among IgG subgroups. J. Immunol. 103:828–833.

Kurtz, J.B. and Malic, A. 1981. Rubella-specific IgM detected by an antibody capture assay/ELISA technique. J. Clin. Pathol. 34:1392–1395.

McCartney, R.A., Banatvala, J.E., and Bell, E.J. 1986. Routine use of μ-antibody-capture ELISA for the serological diagnosis of coxsackie B virus infections. J. Med. Virol. 19:205–215.

Moller, A. and Mathiesen, L.R. 1979. Detection of immunoglobulin M antibodies to hepatitis A virus by enzyme-linked immunosorbent assay. J. Clin. Microbiol. 10:628–632.

Monath, T.P., Nystrom, R.R., Bailey, R.E., Calisher, C.H., and Muth, D.J. 1984. Immunoglobulin M antibody capture enzyme-linked immunosorbent assay for diagnosis of St. Louis encephalitis. J. Clin. Microbiol. 20:784–790.

Morgan-Capner, P., Davies, E., and Pattison, J.R. 1980. Rubella-specific IgM detection using Sephacryl S-300 gel filtration. J. Clin. Pathol. 33:1082–1085.

Morgan-Capner, P., Tedder, R.S., and Mace, J.E. 1983. Rubella-specific IgM reactivity in sera from cases of infectious mononucleosis. J. Hyg. (Camb.) 90:407–413.

Mortimer, P.P., Tedder, R.S., Hambling, M.H., Shafi, M.S., Burkhardt, F., and Schilt, U. 1981. Antibody capture radioimmunoassay for anti-rubella IgM. J. Hyg. (London) 86:139–153.

Nielsen, C.M., Hansen, J., Andersen, H.M.K., Gerstoft, J., and Vestergaard, B.F. 1987. An enzyme labelled nuclear antigen immunoassay for detection of cytomegalovirus IgM antibodies in human serum: Specific and nonspecific reactions. J. Med. Virol. 22:67–76.

Parry, J.V. 1984. Diagnosis of hepatitis A infection: Comparative specificity of IgM capture assays using antigens derived from tissue cultures and marmoset faeces. J. Virol. Meth. 9:35–44.

Pattison, J.R. 1982. Laboratory Investigation of Rubella. Public Health Laboratory Service Monograph Series No. 16. London: Her Majesty's Stationery Office.

Pattison, J.R., Mace, J.E., and Dane, D.S. 1976. The detection and avoidance of false-positive reactions in tests for rubella-specific IgM. J. Med. Microbiol. 9:355–357.

Pedersen, I.R., Antonsdottir, A., Evald, T., and Mordhorst, C.H. 1982. Detection of measles IgM antibodies by enzyme-linked immunosorbent assay (ELISA). Acta. Path. Microbiol. Immunol. Scand. 90:153–160.

Re, M.C. and Landini, M.P. 1989. IgM to human cytomegalovirus: Comparison to two enzyme immunoassays and IgM reactivity to viral polypeptides detected by immunoblotting. J. Clin. Lab. Anal. 3:169–173.

Roggendorf, M., Deinhardt, F., Frosner, G.G., Scheid, R., Bayerl, B., and Zachoval, R. 1981. Immunoglobulin M antibodies to hepatitis B core antigen: Evaluation of enzyme immunoassay for diagnosis of hepatitis B virus infection. J. Clin. Microbiol. 13:618–626.

Roggendorf, M., Frosner, G.G., Deinhardt, F., and Scheid, R. 1980. Comparison of solid phase test systems for demonstrating antibodies against hepatitis A virus (anti-HAV) of the IgM-class. J. Med. Virol. 5:47–62.

Schmidt, N.J., Lennette, E.H., and Dennis, J. 1968. Characterization of antibodies produced in natural and experimental coxsackievirus infections. J. Immunol. 100:99–106.

Schmitz, H. 1982. Detection of immunoglobulin M antibody to Epstein-Barr virus by use of an enzyme-labeled antigen. J. Clin. Microbiol. 16:361–366.

Schmitz, H., von Deimling, U., and Flehmig, B. 1980. Detection of IgM antibodies to cytomegalovirus (CMV) using an enzyme-labelled antigen (ELA). J. Gen. Virol. 50: 59–68.

Shattock, A.G., Morris, M., Kinane, K., and Fagan, C. 1989. The serology of delta hepatitis and the detection of IgM anti-HD by EIA using serum derived delta antigen. J. Virol. Meth. 23:233–240.

Tingpalapong, M., Hoke, C.H., Ward, G.S., Burke, D.S., Elwell, M.R., Lohytyothin, S., and Saisombat, S. 1986. Anti-rabies virus IgM in serum and cerebrospinal fluid from rabid dogs. Southeast Asian J. Trop. Med. Public Health 17:550–557.

Tuokko, H. 1988. The detection of measles specific immunoglobulin M antibodies using biotylated antigens. Acta Path. Microbiol. Immunol. Scand. 96:491–496.

van Loon, A.M., Van der Logt, J.T.M., Heessen, F.W.A., and Van der Veen, J. 1985. Use of enzyme-labeled antigen for the detection of immunoglobulin M and A antibody to herpes simplex virus in serum and cerebrospinal fluid. J. Med. Virol. 15:183–195.

Van der Logt, J.T.M., van Loon, A.M., and Van der Veen, J. 1981. Hemadsorption immunosorbent technique for determination of rubella immunoglobulin M antibody. J. Clin. Microbiol. 13:410–415.

Van der Logt, J.T.M., van Loon, A.M., and Van der Veen, J. 1985. Detection of parainfluenza IgM antibody by hemadsorption immunosorbent technique. J. Med. Virol. 10: 213–221.

Vejtorp, M. 1981. Solid phase anti-IgM ELISA for detection of rubella specific IgM antibodies. Acta Path. Microbiol. Immunol. Scand. Section B. 89:123–128.

Vesikari, T. and Vaheri, A. 1968. Rubella: A method for rapid diagnosis of a recent infection by demonstration of the IgM antibodies. Br. Med. J. 1:221–223.

Vikerfors, T., Lindegren, G., Grandien, M., and Van der Logt, J. 1989. Diagnosis of influenza A virus infections by detection of specific immunoglobulins M, A, and G in serum. J. Clin. Microbiol. 27:453–458.

Vilja, P., Turunen, H.J., and Leinikki, P.O. 1985. Determination of immunoglobulin M antibodies for hepatitis B core antigen with a capture enzyme immunoassay and biotin-labeled core antigen produced in *Escherichia coli*. J. Clin. Microbiol. 22:637–640.

Voller, A. and Bidwell, D.E. 1976. Enzyme-immunoassays for antibodies in measles, cytomegalovirus infections and after rubella vaccination. Br. J. Exp. Pathol. 57:243–247.

Weiblen, B.J., Schumacher, R.T., and Hoff, R. 1990. Detection of IgM and IgA HIV antibodies after removal of IgG with recombinant protein G. J. Immunol. Meth. 126: 199–204.

Wielaard, F., Denissen, A., van Elleswijk, V.D., Berg, J., and van Gemert, G. 1985. Clinical validation of an antibody-capture anti-rubella IgM–ELISA. J. Virol. Meth. 10:349–354.

Wielaard, F., Scherders, J., Dagelinckx, C., Middeldorp, J.M., Sabbe, L.J.M., and Belzen, C.V. 1988. Development of an antibody-capture IgM-enzyme-linked immunosorbent assay for diagnosis of acute Epstein-Barr virus infections. J. Virol. Meth. 21:105–115.

Wolff, K.L., Muth, D.J., Hudson, B.W., and Trent, D.W. 1981. Evaluation of the solid-phase radioimmunoassay for diagnosis of St. Louis encephalitis infection in humans. J. Clin. Microbiol. 14:135–140.

Wulff, H., Anderson, L.J., Pallansch, M.A., and de Souza Carvalho, R.P. 1987. Diagnosis of enterovirus 70 infection by demonstration of IgM antibodies. J. Med. Virol. 21:321–327.

Yaegashi, N., Shiraishi, H., Tada, K., Yajima, A., and Sugamura, K. 1989. Enzyme-linked immunosorbent assay for IgG and IgM antibodies against human parvovirus B19: Use of monoclonal antibodies and viral antigen propagated in vitro. J. Virol. Meth. 26:171–182.

Yolken, R.H. and Leister, F.J. 1981. Enzyme immunoassays for measurement of cytomegalovirus immunoglobulin M antibody. J. Clin. Microbiol. 14:427–432.

16

Antiviral Drug Susceptibility Testing

Edgar L. Hill and M. Nixon Ellis

16.1 HISTORY

The increasing number of effective antiviral therapies for the treatment of several viral infections and the emergence of drug resistant virus strains underscores the need for the development of rapid methodologies to evaluate virus susceptibilities to these chemotherapeutic agents. Having knowledge of a particular clinical isolate's drug susceptibility profile will require consideration of alternate drugs or therapies and may help in the interpretation of other clinical information. Although the plaque reduction assay yields reproducible results and has been the standard technique to determine the susceptibility of viruses to antivirals, it is cumbersome to perform, costly in materials, labor intensive, and not readily amenable to the sensitivity testing of large numbers of isolates. This chapter will describe the materials and methods of a different cyto-pathogenic assay, the dye-uptake (DU) assay.

The DU assay was adapted from a method used to measure interferon activity

(Finter, 1969). The DU assay was developed to determine the in vitro sensitivities of large numbers of herpes simplex virus (HSV) clinical isolates to acyclovir (ACV) (McLaren et al, 1983). The preferential uptake of a vital dye (neutral red) by viable cells relative to virus-damaged cells forms the basis of the test. The extent of viral cytopathic effect (CPE) in different cultures is estimated by the relative amounts of dye retained by viable cells. The dye taken up by viable cells is eluted into a phosphate-alcohol buffer and measured colorimetrically. The drug concentration inhibiting 50% of the viral CPE is the inhibitory dose 50 (ID_{50}). This assay is readily adapted to test drugs against other cytopathic viruses. In addition, it has been automated to allow for screening of large numbers of isolates in a relatively short period of time.

16.2 MATERIALS

1. Maintenance and assay media. Eagle's minimal essential medium (EMEM) containing 5% heat inactivated (56 °C, 1 hour) fetal bovine serum (FBS), 0.075% sodium bicarbonate, 75 U/mL penicillin G, 75 μg/mL streptomycin, 2

Major portions of this chapter have been adapted from *The Manual of Clinical Microbiology, Fifth Edition*, with permission of the publisher, the American Society for Microbiology.

277

mM L-glutamine and 10 mM HEPES pH 6.5–7.0

2. Vero cells, continuous line of African green monkey kidney cells (ATCC CLL81)

3. Culture plates, 96-well flat-bottom (Costar No. 3596, Cambridge, MA)

4. Small, disposable, 13 × 100 mm sterile disposable test tubes with caps

5. Pipettes, 1.0, 5.0 and 10.0 mL sterile

6. Adjustable pipettor 20–200 μL (Rannin, Woburn, MA)

7. Adjustable 12-well multichannel pipettor (Costar)

8. Reservoir filling trough (Dynatech, Chantilly, VA), sterilized by autoclaving

9. Sterile pipet tips (Rannin)

10. Repeating pipettor (Eppendorf, Fremont, CA)

11. Sterile individually wrapped Combitips for repeating pipettor (Eppendorf)

12. Sterile sealing tape or plate sealers (Dynatech)

13. Sterile blotter papers (Dynatech)

14. Cornwall syringe, 1.0 and 2.0 mL, with 8-channel manifold, sterile

15. Neutral red dye (Sigma Chemical Co., St. Louis, MO)

16. Sodium Phosphate monobasic (Sigma)

17. Sodium Phosphate dibasic (Sigma)

18. 0.1 m phosphate buffer, pH 6.0; (Methods in Enzymology, Vol. 1)

19. Neutral red dye. 0.15 g of neutral red added to 100 mL phosphate buffer (0.1 M), filtered through a 0.45 μm filter, then through a 0.22 μm filter

20. Phosphate buffered saline (PBS) without calcium or magnesium (10× stock from Cell Culture Laboratories, Cleveland, OH)

21. Phosphate ethanol elution buffer. 1:1 mix of 0.1 M sodium phosphate monobasic and 95% ethanol

22. Ice bath

23. Water bath, 37 °C

24. Multichannel spectrophotometer for 96-well plates (Titertek Multiskan, Flow Laboratories, McLean, VA)

25. Autodiluter II (Dynatech)

26. Frozen virus specimens, including reference laboratory strains of herpes simplex virus types 1 and 2 (HSV-1, -2)

27. Hemacytometer

28. Filters (0.45 and 0.22 μM)

29. Ultrasonic cleaning bath (Sonicor Instrument Corp., Copiague, NY)

16.3 DYE UPTAKE ASSAY METHODS

16.3.1 Virus Infectivity Assays

1. Prepare suitable volumes (e.g., 1.8 mL) complete EMEM with 5% FBS in sterile tube with metal closures; keep on ice.

2. Rapidly thaw virus sample; briefly sonicate ($\approx$30) in ultrasonic cleaning bath to disrupt any virus aggregates; keep sample on ice.

3. Prepare tenfold dilution series of test virus in tubes containing EMEM: to tubes containing 1.8 mL of medium using an adjustable pipettor, add 0.2 mL of the original virus suspension into the first tube; make further serial dilutions up to 10^{-6} and hold on ice.

4. With the Dynatech Autodiluter II fitted with a fresh manifold, dispense 150 μL of medium containing 1.5×10^5 cell/mL in rows 1 through 9. All Dynatech dispensing steps can be done manually using the 12-well multichannel pipet and filling trough.

5. With the repeating pipet add 50 μL of the 10^{-6} dilution of virus to all wells of row 8 of the plate; add 50 μL of the 10^{-5} dilution of virus to all wells of row 7; add 50 μL of the appropriate dilution of virus to all wells of the other rows

Table 16.1 Virus Infectivity Assay Plate Arrangement

Contents	Row Numbers									
	1	2	3	4	5	6	7	8	9	10–12
Media and Cells 150 μL	+	+	+	+	+	+	+	+	+	Empty
Media 50 μL or Virus 50 μL	+	+	-1	-2	-3	-4	-5	-6	+	$-$

until the final virus dilution (10^{-1}) goes into all wells of row 3; add 50 μL of medium to all wells of rows 1, 2 and 9.

6. The virus infectivity plate should have the arrangement shown in Table 16.1.

7. Seal each plate with a sheet of sterile sealing film or plate sealer and replace the lid.

8. Incubate for 72 hours at 37 °C in a 5% CO_2 incubator.

9. After a 72-hour incubation, examine the plate for gross contamination and extreme pH changes of medium; check control wells in rows 1 and 2 for cell confluence; examine several wells of rows 3 and 4 for virus CPE; if CPE is absent from these rows do not continue.

10. If CPE is present in row 3, then, with the plate under a biosafety hood (to avoid producing an aerosol containing infectious virus), carefully remove the sealing film.

11. With a Cornwall syringe and manifold add 50 μL of a 0.15% solution of neutral red in 0.1 M phosphate buffer (pH 6.0) to each well.

12. Incubate the plate for 45 minutes at 37 °C in a 5% CO_2 incubator.

13. After incubation, briefly check several wells for the presence of neutral red crystals; if extensive crystallization has occurred, then high background readings may be obtained.

14. Aspirate off dye and medium.
    Fill wells to the top with PBS
    Aspirate off PBS

15. Using Dynatech Autodiluter II or Cornwall syringe and manifold add 150 μL of elution buffer (phosphate ethanol) to each well; gently rock the plate to ensure even elution of dye into buffer.

16. Determine optical density of the solution at 540 nm using a multichannel spectrophotometer designed for 96-well plates; the mean optical density (OD) of the cell control well is assigned a value of 100%, the mean of the control blank wells is assigned a value of 0% and the dilution of virus producing a 50% OD reading, i.e., 50% inhibition of cell growth is determined from a linear regression analysis of the data using a computer program. The titer of each virus pool is expressed as a 50% dye uptake [DU_{50}] value. This is the reciprocal of the dilution of virus producing a 50% reduction in neutral red dye uptake by the cells.

16.3.2 Virus Inhibition Assay

1. Prepare initial drug solution in EMEM; because 50 μL of drug is mixed in the well with 200 μL (1:5 dilution) of cell suspension and medium, the initial drug solution should be five times more concentrated than the highest concentration to be tested.

2. With the 12-well multichannel pipettor, add 50 μL of the initial drug solutions to all wells of rows 2, 3, and 4 of an empty microtiter plate.

3. Using a sterile Cornwall syringe (1 mL) and an 8-channel manifold with the Dynatech Autodiluter II, dispense 50 μL of medium and serially dilute (two-fold) the drug; flame the diluters to sterilize, fill the blot/rinse trays with H_2O.

   - Set dispensers for rows 1, 2, 4 and to 12
   - Set diluter for rows 4 to 11
   - Set blot cycle on
   - Place labeled microtiter plate in position
   - Press the RUN button

4. Prepare suspension of Vero cells (1.5×10^5 cells/mL) in EMEM (5% FBS); using the Dynatech Autodiluter II with sterile Cornwall syringe (2.0 mL) add 150 μL of the cell suspension to each well of the plate; if the Dynatech dual diluter is used, the medium addition, drug dilution, and cell seeding can be accomplished with one pass of the plate.

5. In vitro ID_{50} values generated by the DU assay are dependent on the virus challenge dose. The assay is considered valid when the virus challenge dose is between 10 and 100 DU_{50}/50 μL. The actual challenge dose is determined by running a back titration of the test virus in a separate microtiter plate.

To ensure that a valid challenge dose is used, run two drug plates (labeled "A" and "B") in the same assay. Using the DU_{50} titer determined in the virus infectivity titration, make two different virus challenge doses; make up the first in 6 mL of medium at a virus concentration of 50 DU_{50}/50 μL and designate as tube "A". From tube "A" take 0.6 mL and add it to 5.4 mL of medium. This 1:10 dilution is designated tube "B". Using the adjustable pipettor make a series of dilutions from 10^{-1} thru 10^{-3}, starting with 0.2 mL from tube "B" into 1.8 mL of medium. Use a fresh pipet tip for each dilution.

6. Starting with the back titration plate, inoculate the 10^{-3} dilution into all wells of row 7, the 10^{-2} dilution into all wells of row 6, and the 10^{-1} dilution into all wells of row 5. Next, inoculate tube "B" into all wells of row 4 of the back titration plate and all wells of rows 3 thru 12 of the "B" drug plate. Finally, inoculate tube "A" into all wells of row 3 of the back titration plate and all wells of rows 3 thru 12 of the "A" drug plate. Because the virus is inoculated beginning with the lowest virus concentration and ending with the highest virus concentration, the same Combitip can be used throughout. The number of replicates in each row may be eight or less (eight is recommended). In each inhibition assay, include standard laboratory strains (HSV-1 and -2) as internal controls. The back titration and

Table 16.2 Back Titration Plate Arrangement

Contents	Row Numbers								
	1	2	3	4	5	6	7	8	9–12
Cell 150 μL	+	+	+	+	+	+	+	+	Empty
Virus 50 μL	−	−	Tube A	Tube B	−1*	−2*	−3*	−	−

*These are 10^{-1}, 10^{-2}, and 10^{-3} dilution of Tube B.

Table 16.3 Virus Inhibition Assay Plate Arrangement

Contents and Order of Addition	Row Numbers											
	1	2	3	4	5	6	7	8	9	10	11	12
Highest Drug Conc. 50 μL	−	+	+	+	−	−	−	−	−	−	−	−
Medium μL	50	50	−	50	50	50	50	50	50	50	50	50
Drug Dilution	−	−	−	+	+	+	+	+	+	+	+	−
Cell 150 μL	+	+	+	+	+	+	+	+	+	+	+	+
Virus 50 μL	−	−	+	+	+	+	+	+	+	+	+	+

drug plates should have the arrangements depicted in Tables 16.2 and 16.3.

7. The inhibition plates should have the following format (see Table 16.3):

    row 1 = cell control
    row 2 = drug control
    row 3–11 = drug dilution serials
    row 12 = virus control

8. Seal the plate with sterile sealing tape or plate sealer; incubate the plate at 37 °C in a 5% CO_2 incubator.

9. After a 72-hour incubation, add neutral red dye solution to both drug plates and the back titration plate as described for the virus infectivity assay.

10. Use linear regression analysis of the data to determine the concentration of drug producing a 50% reduction of the viral CPE in relation to cell controls (0%) and virus controls (100%); this concentration of drug is the ID_{50}.

11. Determine the exact dose of the challenge virus by reading the back titration plate as described for virus infectivity assays; for valid assay the challenge dose should be between 10 and 100 $DU_{50}/50$ μL.

16.3.3 Applications of the Dye-Uptake Assay

Although the DU assay was developed for the determination of the sensitivity of HSV

clinical isolates to inhibition by acyclovir in Vero cells, it has been adapted for use with other drugs, viruses, and cells. Other drugs that have been screened against HSV in the DU assay include adenine arabinoside (Ara-A), iododeoxyuridine (IdU), phosphonoacetic acid (PAA), phosphonoformic acid (PFA), thymidine arabinoside (Ara-T), Ganciclovir (GCV), bromovinyldeoxyuridine (BVdU), and fluoroiodoaracytosine (FIAC). In addition, our laboratory has modified the assay to evaluate the combined action of drug combinations against HSV. The major requirements for the successful use of this system are a cytopathogenic virus and cells that form monolayers in 96-well, flat-bottom microtiter plates. The DU assay has been found to be reproducible, reliable, and quite suitable for use in the diagnostic virology laboratory (Brisebois et al, 1989).

16.3.4 Advantages of the Dye-Uptake Assay

The DU assay has several advantages over the standard plaque reduction (PR) assay, the most notable being the use of automated and semi-automated equipment. By linking the spectrophotometer to a computer one can analyze in vitro data from large numbers of clinical isolates with a reduced amount of time and effort. Our laboratory routinely assays 20 clinical isolates, plus HSV-1 and HSV-2 virus controls, in a day. In a typical week we can generate ID_{50} values for 60 different virus isolates. In microtiter plates (96-well), eight replicates of nine different drug concentrations are used, allowing for a wide range of drug concentrations and good statistical reproducibility. This format significantly reduces the amount of reagents used and significantly lowers the cost per test.

Perhaps the most significant advantage the DU assay has over the PR assay is its ability to detect small amounts of ACV-resistant virus in a heterogeneous virus population (Ellis et al, 1989). The ability to identify these viruses is important since ACV-resistant HSV are known to exist in natural virus populations never exposed to ACV (Parris and Harrington, 1982). Drug-resistant strains of HSV also have been recovered from normal and immunologically compromised patients receiving acyclovir (Sibrack et al, 1982; Burns et al, 1983; Ellis et al, 1987; Norris et al, 1988). In reconstruction experiments, the DU assay is able to detect small amounts (3% to 9%) of ACV-resistant virus in a mixed virus population (Ellis et al, 1989). The difference in sensitivity between the DU and PR assays may reflect the greater amount of input virus used in the DU assay—500 PFU/well compared to only 100–200 PFU/plate in the PR assay—or it may be the result of the different end points used in the assays. Another reason for the greater sensitivity of the DU assay over the PR assay may be the different overlay materials used. In the DU assay, the liquid overlay allows drug-resistant virus to spread over the monolayer and this amplified population of less sensitive virus is manifested by an increased ID_{50} value (McLaren et al, 1983). In the PR assay, extracellular spread of resistant virus is inhibited by either a solid overlay or a liquid overlay containing serum immunoglobin (Harmenberg et al, 1980; Hu and Hsiung, 1989).

The DU assay gives ID_{50} values that are three to five times greater than values obtained using the PR assay. This is a consistent result and the inclusion of reference viruses (both HSV-1 and -2) in each assay allows for comparisons of assays from week to week (McLaren et al, 1983). If the ID_{50} values of these reference strains are greater than two standard deviations from their mean values as determined in previous tests, then the entire assay is repeated (McLaren et al, 1983). Over the past eight years our laboratory has determined the ID_{50} values of over 2,500 HSV clinical isolates recovered from a variety of patients. We found that over 97% of these isolates had ACV ID_{50} values of ≤ 3.0 $\mu g/$mL. Therefore, with the DU assay we use this ID_{50} value as the cutoff between ACV-sensitive and ACV-resistant viruses (Barry et al, 1985).

16.3.5 Disadvantages and Technical Problems of the Dye-Uptake Assay

For a research laboratory with few isolates to test, the cost of the automated equipment needed for the DU assay is prohibitive. However, with the exception of the step involving spectrophotometric analysis, all other steps can be done manually (Brisebois et al, 1989). Technical problems with the equipment can be frustrating and can cause time-consuming delays. Occasionally, the neutral red dye will precipitate onto the monolayer, making an accurate determination of the amount of dye released from viable cells impossible. This problem can be avoided by maintaining the pH of the dye solution above 5.8 and by filtering the solution prior to use. Another pitfall is the overseeding of cells, which causes the monolayer to peel. This decreases the number of viable cells and can lead to poor results. This problem rarely occurs if the cells are seeded at a density of $2-3 \times 10^4$ per well.

Several factors affect the reliability of any assay that measures the inhibition of viral CPE. These include the virus and cells used, the challenge dose of virus used, and the gradient of drug concentrations that ensures a valid end point (Barry and Blum, 1983; Harmenberg, et al, 1980, 1985). In the DU assay most of these technical problems have been solved. However, the ID_{50} values generated by this assay are affected by the amount of virus used. It is necessary to determine the infectivity titer of each clinical isolate before the drug inhibition assay can be attempted. It is also useful to include a back titration plate for each test virus to determine the actual virus challenge dose used in the assay. In the DU assay a high challenge dose ($>100\ DU_{50}$) can cause an ACV-sensitive virus to have a high ID_{50} value. Alternatively, a low challenge dose ($<10\ DU_{50}$) of an ACV-resistant virus will give an artificially low ID_{50} value (McLaren et al, 1983). Challenge doses that fall within a range of 10 and 100 DU_{50}s have little effect on the ID_{50} values (McLaren et al, 1983). Hence, if the viral challenge dose is outside these limits, the test isolate is repeated. Because the infectivity titration of the isolate and the drug inhibition assay both require three days, the minimum time for determining the in vitro ID_{50} value for any particular isolate is approximately two weeks.

16.4 CONCLUSION

The factors affecting the choice of an appropriate antiviral susceptibility assay are different for each laboratory. In some instances the sheer number of isolates to be tested demands a so-called high "thru-put" assay. For other laboratories with few isolates to evaluate and a limited budget, a more traditional method may be satisfactory. Other factors that can influence this decision include the virus/cell system to be used, the incubation period, and the overall time needed to produce results.

The lack of standardization of these tests makes it very difficult to compare results from lab to lab. Hence, comparisons of in vitro data are only valid if the same assay, virus, and cell type are used in the test.

Correlation between in vitro sensitivity and in vivo response to therapy in humans has not been established; therefore, differences in absolute in vitro ID_{50} values between assay systems probably have little clinical relevance. Thus, the definition of a "sensitive" or "resistant" virus must be derived empirically for each assay system. A clearer understanding of the relationship between plasma drug level, in vitro sensitivity, and clinical response will emerge with continued use of antiviral drugs and subsequent sensitivity testing of large numbers of viruses by many laboratories.

REFERENCES

Barry, D.W. and Blum, M.R. 1983. Antiviral drugs: Acyclovir. In P. Turner and D.G. Shand (eds.), Recent Advances in Clinical Pharmacology. Edinburgh: Churchill Livingstone. p 57–80.

Barry, D.W., Nusinoff-Lehrman, S., Ellis, M.N., Biron, K.K., and Furman, P.A.

1985. Viral resistance, clinical experience. Scand. J. Infect. Dis. Suppl. 47:155–164.

Brisebois, J.J., Dumas, V.M., and Joncas, J.H. 1989. Comparison of two methods in the determination of the sensitivity of 84 herpes simplex virus (HSV) Type 1 and 2 clinical isolates to acyclovir and alpha-interferon. Antiviral Res. 11:67–76.

Burns, W.H., Saral, R., Santos, G.W., Laskin, O.L., Lietman, P.S., McLaren, C., and

Barry, D.W. 1982. Isolation and characterization of resistant herpes simplex virus after acyclovir therapy. Lancet 1:421–423.

Ellis, M.N., Keller, P.M., Fyfe, J.A., Martin, J.L., Rooney, J.F., Straus, S.E., Nusinoff-Lehrman, S., and Barry, D.W. 1987. Clinical isolate of herpes simplex virus type 2 that induces a thymidine kinase with altered substrate specificity. Antimicrob. Agents Chemother. 31:1117–1125.

Ellis, M.N., Waters, R., Hill, E.L., Lobe, D.C., Selleseth, D.W., and Barry, D.W. 1989. Orofacial infection of athymic mice with defined mixtures of acyclovir-susceptible and acyclovir-resistant herpes simplex virus type 1. Antimicrob. Agents Chemother. 33: 304–310.

Finter, N.B. 1969. Dye uptake methods for assessing viral cytopathogenicity and their application to interferon assays. J. Gen. Virol. 5:419–427.

Harmenberg, J., Wahren, B., and Öberg, B. 1980. Influence of cells and virus multiplicity on the inhibition of herpesviruses with acycloguanosine. Intervirology 14:239–244.

Harmenberg, J., Wahren, B., Sundqvist, V-A., and Levén, B. 1985. Multiplicity dependence and sensitivity of herpes simplex virus isolates to antiviral compounds. J. Antimicrob. Chemother. 15:567–573.

Hu, J.M. and Hsiung, G.D. 1989. Evaluation of new antiviral agents: I. *In vitro* perspectives. Antiviral Res. 11:217–232.

McLaren, C., Ellis, M.N., and Hunter, G.A. 1983. A colorimetric assay for the measurement of the sensitivity of herpes simplex viruses to antiviral agents. Antiviral Res. 3:223–234.

Norris, S.A., Kessler, H.A., and Fife, K.H. 1988. Severe, progressive herpetic whitlow caused by acyclovir-resistant virus in a patient with AIDS. J. Infect. Dis. 157:209–210.

Parris, D.S. and Harrington, J.E. 1982. Herpes simplex virus variants resistant to high concentrations of acyclovir exist in clinical isolates. Antimicrob. Agents Chemother. 22:71–77.

Sibrack, C.D., Gutman, L.T., Wilfert, C.M., McLaren, C., St. Clair, M.H., Keller, P.M., and Barry, D.W. 1982. Pathogenicity of acyclovir-resistant herpes simplex virus type 1 from an immunodeficient child. J. Infect. Dis. 146:673–682.

17

Nucleic Acid Hybridization

Attila T. Lörincz

17.1 INTRODUCTION

Viral detection is accomplished by many different and often unrelated methods. Choosing the most appropriate test for any particular situation is complicated by the fact that there is no universally superior procedure. The mainstay of viral diagnosis is virus isolation in cell culture. However, this approach is time-consuming and requires considerable expertise, making it vulnerable to challenge by alternative methods such as immunoassays and hybridization. For a number of viral agents which cannot be grown easily, such as the hepatitis, gastroenteritis, and papilloma viruses, direct detection of viral components can be used for diagnosis. The targets for direct immunoassays include protein or carbohydrate moieties. It is also possible to measure enzymatic activities associated with viral infections, such as the reverse transcriptases of retroviruses, or the DNA polymerase of hepatitis B virus. The genetic material of the virus forms the basis of hybridization tests. An exciting new strategy for detecting nucleic acids is the polymerase chain reaction (PCR). The sensitivity and potential specificity of PCR

is equal to, and often better than, methods such as immunoassays or cell culture. Nevertheless, technical barriers still prevent the routine application of PCR to viral diagnosis. Hybridization is a powerful method for detecting viruses even without the use of PCR amplification. Often there is sufficient viral DNA or RNA in clinical specimens to perform direct testing. In situ hybridization is well suited for detecting viruses in human tissues, including cytomegalovirus (CMV), herpes simplex viruses (HSV), human papillomaviruses (HPV) and others (Table 17.1). Dot blot and Southern blot hybridization require extracted nucleic acids, thus information on cell or tissue localization is lost. Nevertheless these methods are also very useful.

17.2 SPECIMEN HANDLING

DNA is a stable molecule which can survive for long periods and can be easily extracted from most clinical specimens (Sambrook et al, 1989). In contrast RNA is a labile molecule that requires careful handling; however, even RNA can be preserved intact more easily than some fastidious viruses and bacteria. The compo-

Table 17.1 Some Applications of Nucleic Acid Hybridization to Viral Detection*

Virus	Method	Reference
CMV	In situ	Danker et al, 1990
	In situ	Seyda et al, 1989
	In situ	Weiss et al, 1990
	Dot Blot	Kimpton et al, 1988
	Dot Blot	Agha et al, 1989
	Dot Blot	Spector et al, 1984
HSV	In situ	Schmidbauer et al, 1988
	In situ	Weiss et al, 1990
	Dot Blot	Redfield et al, 1983
VZV	Dot Blot	Seidlin et al, 1984
EBV	In situ	Seyda et al, 1989
	In situ	Eversole et al, 1988
	In situ	Sixbey et al, 1984
HBV	Southern Blot	Lie-Injo et al, 1983
	Dot Blot	Moestrup et al, 1985
	Dot Blot	Wu et al, 1986
	Dot and Southern Blot	Jenison et al, 1987
HBV	In situ	Burrell et al, 1984
	In situ, Southern Blot	Rijntjes et al, 1985
HIV	In situ	Seyda et al, 1989
	In situ	Wiley et al, 1986
HPV	Southern Blot	Lörincz et al, 1987
	Southern Blot	Ritter et al, 1989
	Dot Blot	Warford and Levy, 1989
	In situ	Richart and Nuovo, 1990
	In situ	Stoler and Broker, 1986
Adeno	In situ	Weiss et al, 1990
	Sandwich	Virtanen et al, 1983
HTLV	Southern Blot	Wong-Staal et al, 1983

*Abbreviations are as follows:
CMV = cytomegalovirus
HSV = herpes simplex virus
VZV = varicella-zoster virus
EBV = Epstein-Barr virus

HBV = hepatitis B virus
HIV = human immunodeficiency virus
HTLV = human T-cell leukemia virus
HPV = human papillomavirus
Adeno = adenovirus

sition of clinical specimens is heterogeneous and may consist of large amounts of mucus, proteins, microorganisms, etc. Thus the storage and extraction of each type of specimen must be optimized to yield the best quality nucleic acid. A universally acceptable state of storage and transport is frozen on dry ice. Most specimens yield suitable DNA and often even acceptable RNA after storage at −70 °C for several months. Transport and storage at 4 °C for a few days is often adequate for the recovery of DNA, but not RNA. Specialized sample transport media, commercially available for HPV testing, can stabilize DNA for several weeks at room temperature (Warford and Levy, 1989). Fixation of specimens in buffered formalin is suitable for subsequent analysis by in situ hybridization or PCR (Moench, 1987; Shibata et al, 1988). To perform dot blots or Southern blots, DNA must be extracted from the specimens. This is most commonly performed by treatment (large samples of solid tissues must be pulverized) with a digestion solution containing protease and detergent (Sambrook et al, 1989). The liberated DNA can be analyzed directly by dot blots or purified further with organic solvents followed by Southern blot analysis (Southern, 1975; Meinkoth and Wahl, 1984; Lörincz, 1989).

17.3 HYBRIDIZATION

17.3.1 Base-Pairing and Denaturation

DNA or RNA molecules can exist in either the single-stranded or the double-stranded state. Under the proper conditions, complementary single strands of DNA or RNA will pair to form a duplex. These double-stranded forms can be converted to their single-stranded forms by denaturation at highly alkaline pH or elevated temperatures. The hybridization of two complementary DNA strands is a highly specific process in which only the complementary strands can pair. This pairing process forms the basis of DNA-testing methodology. After hybridization, the duplexes that form must be distinguished from unreacted molecules. There are several ways of accomplishing this, the most popular being the immobilization of one of the reactants (target) to a solid support prior to hybridization, followed by the physical removal by washing the excess liquid phase reactant (probe) after hybridization is complete (Figure 17.1). Nylon filters and glass slides are the most commonly-used solid supports for DNA tests (Meinkoth and Wahl, 1984; Hames and Higgins, 1985; Lörincz, 1989).

Natural divergence has produced more than 60 types of HPV, distinguishable by differences in their nucleotide base sequence (Lörincz and Reid, 1989). Nevertheless, localized areas of homology exist between different HPV types. Similarly there are a large number of variants of the human immunodeficiency virus (HIV). In fact all viruses consist of families of more or less related members clustered about consensus sequences. Hybridization tests can be cross-reactive (related but divergent nucleotide sequences can be detected by stabilization of poorly-matched hybrids) or highly specific (very little reactivity with divergent viral types). The degree of cross-reactivity depends upon the relatedness of the nucleic acid strands and the "stringency" under which the reaction is conducted.

17.3.2 Controlling the Hybridization Test

Hybridization test specificity and sensitivity can be regulated by varying stringency via manipulation of the concentration of salt or other chemicals and by changing the temperature of the hybridization and wash solution (Table 17.2). The hybridization reaction is performed by incubating the filter in a defined solution containing the probes. Then unbound and non-specifically bound probes are removed by extensive washing of the filter (Figure 17.1). Reviews by Meinkoth and Wahl (1984); Hames and Higgins (1985); and Lörincz (1989) contain a more detailed treatment of this subject and of specific hybridization and wash solutions.

17.3.3 Signal Detection

Following washing, the filter is ready for signal detection. The method used depends on the types of label present on the probes employed in the test. Radioactive signals are usually detected by autoradiography (Figures 17.2 and 17.3). For filter tests, the most popular label is ^{32}P. For in situ hybridization tests, ^{35}S or ^{3}H are most often used.

Although frequently used for sensitive

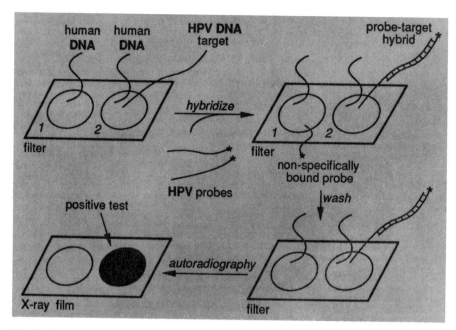

Figure 17.1 Basic principles of the dot blot. The specimen DNA is denatured with heat or alkali and immobilized onto the filter (by spotting and drying or with a filter manifold). Shown are two dots, one of which contains only human genomic DNA (dot number 1), and another of which contains human DNA plus HPV DNA (dot number 2). A solution of single-stranded HPV probes labeled with ^{32}P, biotin, or some other adduct is then incubated with the filter, and hybrids are allowed to form. Nonspecifically-bound and unbound probe are washed away, and the signal on the filter is developed. If the hybrids are radioactive, an autoradiograph is obtained. Alternatively, hybrids labeled with biotin can be detected with alkaline phosphatase. The final result shows a positive reaction in specimen 2 (Reproduced with permission, Lörincz A.T., 1989. Human papillomavirus testing. Diagnostics and Clinical Testing 27:28–37).

Table 17.2 Principal Controlling Factors of Hybridization Tests*

Temperature	Increase gives higher stringency; optimal hybridization is at Td −25 °C; denaturation of duplex occurs at Td.
Cations (Na$^+$)	Increased concentration gives lower stringency, favors hybridization rate (up to about 1.5 M), and enhances duplex stability.
Formamide	Increased concentration raises stringency (lowers Td) independent of temperature or salt.
Dextran Sulfate or Polyethylene Glycol	Accelerate hybridization rate.

*The Td is the denaturation temperature (temperature at which 50% of the nucleic acid duplexes are denatured). It can be estimated for any defined totally matched DNA-DNA duplex by: Td(in °C) = 81.5 + 16.6 (log M) + 0.41 (% GC) − 0.72 (% formamide), where M is the molarity of the sodium concentration and % GC is the percentage of deoxyguanosine plus deoxycytidine bases in DNA. DNA-RNA duplexes have a Td that is usually about 10 °C higher than for DNA-DNA; however this relationship is variable depending on the formamide concentration (with permission, Lörincz A.T., 1989. Human papillomavirus testing. Diagnostics and Clinical Testing 27:28–37).

Controls

High

Low

Negative

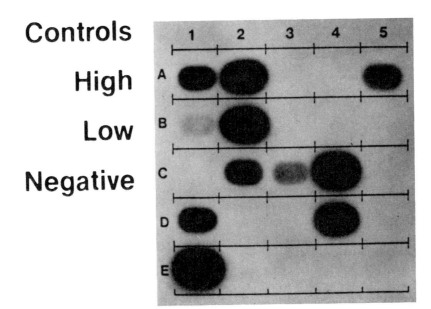

Figure 17.2 Example of a typical ViraPap® (Digene Diagnostics, Inc., Silver Spring, MD) dot blot result. Positions A1, B1 and C1 contain control samples. The other positions show various specimens of exfoliated cervical cells. Each of these specimens was obtained from a different woman and analyzed according to the recommended ViraPap® procedure. The ViraPap® kit uses essentially full-length ^{32}P-labeled RNA probes of HPV types 6, 11, 16, 18, 31, 33, and 35 to detect as little as 1 pg of HPV DNA (approximately 1×10^5 molecules of viral DNA) in cervical swab specimens. The high-positive control in position A1 contains 2.5×10^4 HeLa cells (at approximately 10 copies of HPV 18 per cell), the low-positive control (B1) contains 5×10^3 HeLa cells, and the negative control (C1) contains 1×10^5 HTB 31 cells (which lack HPV DNA).

detection assays, radioactive probes have the disadvantage of short useful-life, radiation hazards, and disposal problems. Nonradioactive detection systems have been developed as alternatives. Some use antibodies to haptenated probes, others use streptavidin and biotin, and yet others have enzymes directly conjugated to the DNA or RNA (Matthews and Kricka, 1988). When probes are labeled with biotin, the presence of the probe-target hybrids is detected using an enzyme such as alkaline phosphatase conjugated to streptavidin; the streptavidin binds tightly to the biotin and thus localizes the enzymatic activity to the position of the hybrids on the filter. Many different substrates are available for these enzymes. For alkaline phosphatase, a mixture of BCIP (5-bromo-4-chloro-3-indolyl phosphate) and

NBT (nitroblue tetrazolium) is commonly used. Recent reports indicate that newly developed chemiluminescent signalling systems may provide similar results to both ^{32}P and BCIP/NBT (Matthews et al, 1985).

Because of background problems, currently available nonradioactive detection systems perform poorly with dot blots of clinical specimens; however, they work well with Southern blots, where they are as sensitive and nearly as reproducible as ^{32}P-based tests. Several novel nonradioactive hybridization tests have been developed recently which do not employ filters. Typically hybridization is performed in solution followed by specific capture of the hybrids onto a solid support using antibodies or other avid intermolecular attractions. Alternatively, the probe may be immobilized

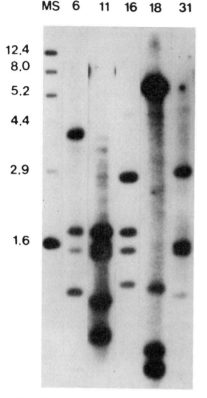

Figure 17.3 Southern blot showing *Pst*1 restriction endonuclease patterns of the five most prevalent HPV types detected in anogenital tissues. The lane labeled MS shows the molecular size markers (in Kb). The other five lanes show HPV DNAs from cervical biopsies detected by low-stringency hybridization (Td −37 °C) using a mixture of HPV types 6, 11 and 16 DNA probes; the numbers along the top indicate the HPV types detected in each specimen (Reproduced with permission, Lörincz A.T. et al., 1986. Characterization of human papillomaviruses in cervical neoplasias and their detection in routine clinical screening. Banbury Report 21:225–237).

on magnetic particles and used to remove the target from solution by hybridization. Such methods are less prone to background problems and do not require the use of radioactive probes. Our laboratory has applied a solution hybridization method using dioxetane-based chemiluminescence (Pollard-Knight et al, 1990) to the detection of HPV, hepatitis B virus (HBV), and chlamydia in crude clinical specimens. Prelimi-

nary results show that this approach is more sensitive and specific than dot blot hybridization employing ^{32}P-labeled probes. Another area where radiolabeled probes offer little or no advantages is in situ hybridization. It is now well recognized that nonradioactive in situ tests are as good as or better than most radioactive versions (Nuovo and Richart, 1989).

17.3.4 Probe Preparation

Construction of good probes is a prerequisite for any successful hybridization test. Probes can be either DNA or RNA, they can be short (oligonucleotides) or long, and they can be coupled to a wide array of adducts for use in subsequent detection. Kits to make DNA and RNA probes are available from several companies including Amersham (Arlington Heights, IL), Life Technologies (Gaithersburg, MD), and Promega Biotech (Madison, WI). One of the most commonly used procedures for preparing radioactively-labeled DNA probes is called random primer labeling (Feinberg and Vogelstein, 1983). In this method the input DNA is denatured and small random DNA oligonucleotide primers that hybridize to complementary regions on the template DNA act as the origins of synthesis for a DNA polymerizing enzyme, which makes a ^{32}P-labeled or biotinylated complementary copy of the template DNA strand. DNA to be used as a probe for clinical testing must always have the plasmid vector sequences removed by gel electrophoresis or by some other method, because specimens frequently have large numbers of bacteria harboring plasmids that may be cross-reactive with the vector sequences.

RNA probes are made in vitro using RNA polymerase encoded by certain bacteriophages (Little and Jackson, 1987). Plasmids have been constructed that contain specific phage RNA promoter sequences upstream of an inserted DNA sequence that is to be copied into the probe. In the presence of the phage RNA polymerase and appropriate substrates, the plasmids direct the synthesis of RNA probes complementary to these DNA regions. Carefully prepared RNA probes are practically free of sequences cross-reactive with bacterial plasmids. The use of RNase A to digest unhybridized regions of RNA make RNA probes equal to or superior to DNA probes.

17.4 SPECIFIC HYBRIDIZATION TESTS

Three basic DNA tests—the dot blot, the Southern blot, and the in situ hybridization test, are particularly important. In addition, PCR, which is not a detection test as such, but is a method for amplifying target DNA, has shown considerable promise to detect viruses. The PCR method is described in Chapter 19, thus it will not be dealt with in detail here.

17.4.1 Dot Blots

The dot blot is well suited for analysis of large numbers of specimens (Kimpton et al, 1988; Agha et al, 1989; Warford and Levy, 1989; Lörincz, 1989). Dot blots can be performed with crude clinical samples, or with nucleic acids purified from specimens (Figures 17.1 and 17.2). To avoid false positive results when DNA probes are employed, specimens must be extracted with phenol and chloroform prior to application to the filter. In contrast, ^{32}P-labeled RNA probes perform well with crude specimens and give results comparable to the Southern blot (Lörincz, 1989; Warford and Levy, 1989).

A limitation of the dot blot method is that it lacks the characteristic DNA banding patterns provided by Southern blots, which are responsible for this latter test's excellent specificity (Figure 17.3). Dot blots

also are unsuitable for low-stringency hybridization. But if adequate care is taken to optimize the sample preparation, hybridization, and washing conditions, and to avoid overloading with sample DNA, the dot blot can reliably provide specific and sensitive detection.

17.4.2 Southern Blots

In 1975, Southern's report of a procedure for faithfully reproducing on filters the images of DNA banding patterns in electrophoretic gels revolutionized the study of nucleic acids (Southern, 1975). Southern demonstrated that DNA fragments separated according to size in gels could be subsequently denatured and transferred to nitrocellulose filters. This DNA could then be hybridized with probes, thus allowing the detection of specific fragments of DNA (Figure 17.3). In trained hands, the Southern blot is an invaluable research tool; however, because of the skill required to perform the test, and the several days or more required to obtain results, this method is not easily adapted to clinical laboratories.

17.4.3 In Situ Hybridization

In situ hybridization is used to detect DNA or RNA targets within cells (Stoler and Broker, 1986; Moench, 1987; Eversole et al, 1988; Schmidbauer et al, 1988; and Weiss et al, 1990). Thin sections (3 to 5 μm) of formalin-fixed and paraffin-embedded biopsy specimens placed on specially-treated glass slides can be analyzed microscopically for target nucleic acids with the retention of good histological detail.

Prior to hybridization the sections are deparaffinized in xylene and rehydrated through graded concentrations of ethanol. The sections are usually treated with protease to make the cells more permeable. The cellular DNA is denatured by alkali or heat (RNA targets don't require denatur-

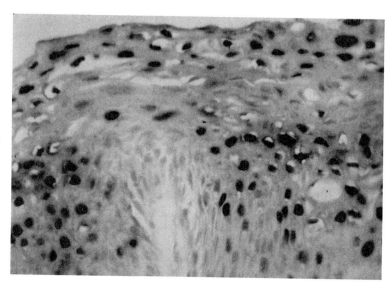

Figure 17.4 In situ hybridization of a cervical condyloma with biotin-labeled HPV 6 DNA (400× magnification). The tissue was sectioned on a microtome, and serial sections were taken for hybridization with various probes. Probe-target hybrids were revealed using alkaline phosphatase NBT-BCIP. Cells containing HPV 6 DNA are indicated by the darkly-staining nuclei.

ation) and allowed to hybridize with DNA or RNA probes. The slides are then washed to remove the unhybridized probes, and probe-target hybrids are detected as appropriate (Figure 17.4), depending on whether the probes were labeled with a radioisotope (detection with photographic emulsion) or with biotin or some other adduct (detection with BCIP/NBT). Robotic handling devices are now available for performing in situ hybridization tests (Brigati, 1988). Further developments along these lines should make in situ hybridization attractive for use in routine clinical testing laboratories.

17.5 PROBLEMS IN TEST PERFORMANCE AND INTERPRETATION: CHOOSING THE CORRECT TEST

The various detection tests each have specific useful characteristics. For example, the speed and simplicity of the dot blot make it useful for routine testing of large numbers of specimens. Southern blots can be used to verify the dot blot results and to search for new viruses. An investigator wishing to localize viral DNA or RNA to

specific cells or tissue regions can opt for in situ hybridization; however, this latter method cannot easily detect target DNA at low copy numbers. In situ hybridization can also be used to analyze paraffin-embedded tissues, samples that are poorly suited for dot blot or Southern blot analysis. Finally, if utmost sensitivity is needed, specimens can be subjected to PCR analysis.

All tests must be carefully monitored to detect erroneous results. False negatives can occur in several ways, e.g., the specimen is taken or stored improperly, or an error occurs in the technical performance of these tests, etc. Each set of analyses should be accompanied by positive and negative controls. Positive controls help to assess that signal generation is working properly. They consist of specimens containing known amounts of specific target DNAs at various concentrations, including the lower limit of detection desired (usually 1 pg of DNA or less). Negative controls are specimens known to be free of target DNA or that mimic as closely as possible true negative specimens. Several of these should be analyzed with each batch of tests; if any are positive, the entire set of test results is suspect.

17.6 APPLICATION OF HYBRIDIZATION TO VIRAL TESTING

In principle hybridization tests can be used to detect any organism with very high sensitivity and specificity (Zwadyk and Cooksey, 1987; Yolken, 1988). This is especially true of amplification tests such as PCR which can detect one or a few organisms in a specimen with a vast excess of other biomass. Furthermore, these tests work equally well on dead or viable organisms and will sometimes even work with partially degraded specimens. Thus, it would be futile to present a comprehensive list of all the applications of hybridization testing. Instead a few relevant examples will be chosen and discussed from among the viruses listed in Table 17.1.

17.6.1 CMV

CMV has a large DNA genome, some regions of which cross-react strongly with the human genome. It is essential that these regions be eliminated from the probes. Spector et al (1984) used dot blot hybridization (with the Eco R1 B and D fragments of the CMV AD169 strain as the probes) to detect CMV DNA in urine and buffy-coat specimens from neonates and bone marrow transplant recipients. DNA testing gave results faster than culture and in some cases positive DNA tests were obtained with immunosuppressed patients who had negative culture tests. Subsequently these patients were shown to have cytomegaloviral disease. Agha et al (1989) also used a dot blot method (using the Hind III O fragment of AD169) to study 148 urine specimens from 47 renal transplant and dialysis patients. The DNA test results were then compared with four other methods of CMV detection, including conventional tube cultures and immunofluorescence. The DNA test was more rapid and sensitive than

culture and was the most sensitive and specific of all the tests, correctly identifying CMV infection in 15 (32%) of the 47 patients. The next best test was immunofluorescence which detected CMV antigen in 12 (26%) of the 47 patients. A dot blot study (using the Hind III F, J, L and W fragments of AD169) by Kimpton et al (1988) correctly identified three culture positives of 35 urine specimens from neonates and bone marrow transplant recipients. In addition three of the 32 culture negatives were DNA-test positive.

In situ hybridization testing for CMV may be helpful in arriving at a definitive diagnosis when characteristic viral inclusions are not present in histological specimens of transplant patients suspected of active CMV infection. Weiss et al (1990) used in situ hybridization to detect CMV DNA in 26 (5%) of 477 lung or heart biopsies from 20 heart-lung recipients. In situ hybridization had a sensitivity of 85% and a specificity of 99% compared with histopathology. In contrast, culture was positive in only 68% of the patients diagnosed as CMV positive by both in situ hybridization and histopathology. In situ hybridization was positive for CMV DNA in four cases negative by histopathology. These cases probably represented true CMV infection because biopsies from other sites showed evidence of CMV infection. Thus from the above studies it appears that DNA testing for CMV has advantages of speed and sensitivity over culture and immunofluorescence and may be the test of choice for early diagnosis of CMV in immunosuppressed patients.

17.6.2 HPV

Hybridization testing is particularly well suited for detecting HPV. Ritter et al (1989) used the Southern blot to test cervicovaginal lavage specimens for HPV DNA from 191 women attending a colposcopy clinic. The performance of the DNA test was compared with the Pap smear using histopathological examination of colposcopically-directed biopsies as the "gold standard." The Pap smear detected 74% of the women with disease, and the DNA test detected 68%. The combined tests identified 89% of the women with cervical intraepithelial neoplasia (CIN). Lörincz et al (1987) used Southern blot hybridization to study the prevalence of various HPV types in 73 specimens of cervical cancer, in 45 low grade CIN, and in 33 high grade CIN. HPV types 6 or 11 were detected in only the low grade lesions; in contrast HPV types 16 and 18 were detected in 64% of the cancer specimens, in 48% of the high grade CIN and in 20% of the low grade CIN. Another 25% of the cancers were positive for other HPVs such as type 31.

Richart and Nuovo (1990) used a nonradioactive in situ test to analyze 70 specimens of low grade CIN. They found that 91% of these CIN were positive for one or more of 14 different HPV types, leading them to suggest that in situ hybridization may be a good method for correctly assigning histopathologically borderline specimens to the disease or the normal categories. The utility of hybridization for detecting and typing HPV is compelling. There is no culture system for HPV. Direct immunoassays are insensitive and cannot detect clinically important HPVs in a type-specific manner. Serology assays also do not appear to be particularly useful at present (Lörincz, 1989).

17.6.3 HBV

Most studies have shown a close correlation between the presence of hepatitis B virus e antigen (HBeAg) and the presence of HBV DNA in serum (Krogsgaard et al, 1985; Lok et al, 1985; Krogsgaard, 1988). Detection of HBV DNA in serum may be a marker of active viral replication in the liver. Krogsgaard et al (1985) reported that detection of HBV DNA in serum by dot blot upon admission to the hospital was positively correlated with clinical symptoms of short duration present in individuals prior to hospital admission. During follow-up of patients, HBV DNA was always cleared before HBsAg (surface antigen) and generally before HBeAg (Krogsgaard, 1988). However a few patients have been found to remain HBV DNA-positive after becoming HBeAg-negative indicating that under certain circumstances HBV DNA tests may be more sensitive than tests for HBeAg. Pontisso et al (1986) and Bortolotti et al (1986) reported that conversion from HBV DNA positive to negative (i.e., detectable to undetectable DNA) within one to two weeks after onset of symptoms showed a strong correlation with recovery. Conversely, persistence of circulating HBV DNA for more than eight weeks after onset of symptoms indicated the development of chronic HBV infection (Krogsgaard et al, 1985). Recent studies using α-interferon therapy demonstrated that a three to six month course of thrice weekly injections was effective in promoting clinical and biochemical remission of disease in 30% to 40% of patients with chronic hepatitis due to HBV (Hoofnagle, 1990). These studies also indicated that HBV DNA testing was a particularly efficacious monitoring modality. Rijntjes et al (1985) using in situ hybridization detected HBV DNA in formalin-fixed liver specimens, including three cases where the patients were not seropositive for HBV. Southern blots have also been used to demonstrate that HBV genomes are integrated into human DNA in hepatomas (Shafritz et al, 1981; Lie-Injo et al, 1983). In conclusion, HBV DNA analysis may be useful in certain situations such as monitoring antiviral therapy. Nevertheless, conventional serological methods remain the principal method for diagnosis of HBV-related conditions.

17.6.4 HSV

When using nucleic acid probes to detect HSV it is important to use only specific clones of HSV DNA from which sequences cross-reactive to human DNA have been eliminated. Redfield et al (1983) utilized the HSV-1 BamH1 A fragment to detect HSV-2 DNA in swabs of genital ulcers. Even though the HSV-1 probe was two to four-fold less sensitive for detection of HSV-2 DNA than for the fully complementary HSV-1 DNA, the sensitivity of the DNA test was 78% compared to isolation in cell culture, with a specificity of 100%. Schmid-bauer et al (1988) compared in situ hybridization with immunocytochemistry for diagnosis of HSV encephalitis. Both methods had comparable sensitivity (72%). However, in cases with a disease course longer than one month in situ hybridization was more sensitive.

17.7 PROS AND CONS OF NUCLEIC ACID TESTING

The full potential of nucleic acid testing has not yet been realized. However, the many apparent advantages of this detection modality are already appreciated by a substantial segment of the clinical laboratory community (Yolken, 1988). The hybridization test can be applied to the detection of virtually any organism, the major technical limitation being the identification of useful type-specific probes. Sensitivity is no longer an issue since the development of DNA and RNA amplification technologies. Culture is often too slow to help acute management of the patient. Also, cell culture can only detect the presence of viable organisms and in some situations is not even an option, e.g., HPV detection. Thus, there are obvious advantages of hybridization testing. The procedure is generally highly sensitive, specific, and relatively inexpensive. Furthermore, hybridization tests do not require viable (and potentially infectious) organisms. Often hybridization tests are superior to serology. Antibodies to disease agents cannot be detected until well into the course of the infection. Also, seropositivity persists well beyond the resolution of the acute infection. Measurement of specific immunoglobulin classes can provide information on the timing of acute infections; however, such tests can be error prone. In contrast, hybridization tests are extremely useful for quickly detecting acute phase infection. In situ hybridization is particularly useful in this regard in directly identifying the infectious agent in diseased tissues. Hybridization tests can be more sensitive and specific than immunodetection of viral antigens. The specimens can be subjected to relatively harsh treatment and still be acceptable for DNA analysis but not for immunoassays.

Unfortunately hybridization tests have been encumbered by many problems; in particular, sample preparation procedures, test formats, and the time required to run the test need improvement. Many of the tests employed in the past used radioactively-labeled probes which are clearly not practical for most routine laboratories. Non-radioactive probes have been available for years (Matthews and Kricka, 1988) but often did not offer high sensitivity or specificity.

Detection of RNA by slot blot procedures can be problematic as RNA is often rapidly degraded in clinical specimens. Rotbart et al, (1987), observed that the immediate addition of RNase inhibitors to specimens minimized the loss of RNA. Thus, improvements in technology should permit RNA detection as easily as DNA.

In conclusion, hybridization testing has a clear place in the routine detection of viruses and other disease agents. Researchers in academia and in the commercial sector are actively working on the remaining problems. Many companies are now selling, or will be making available shortly, DNA probe assays for detecting a variety of viral agents. These companies include Genprobe (San Diego, CA), Integrated Genetics (Framingham, MA), Digene Diagnostics (Silver Spring, MD), Abbott Laboratories (Abbott Park, IL), and Enzo Biochem (New York, NY), to name but a few (see also Zwadyk and Cooksey, 1987). The impact of hybridization testing on patient management is likely to grow considerably in the coming years.

REFERENCES

Agha, S.A., Mahmoud, L.A., Archard, L.C., Abd-Elaal, A.M., Selwyn, S., Mee, A.D., and Coleman, J.C. 1989. Early diagnosis of cytomegalovirus infection in renal transplant and dialysis patients by DNA-DNA hybridization assay. J. Med. Virol. 27:252–257.

Bortolotti, F., Bertaggia, A., Crivellaro, C., Armigliato, M., Alberti, A., Pontisso, P., Chemello, C., and Realdi, G. 1986. Chronic evolution of acute hepatitis type B: Prevalence and predictive markers. Infection 14:64–67.

Brigati, D.J., Budgeon, L.R., Unger, E.R., et al. 1988. Immunocytochemistry is automated: The development of a robotic workstation based upon the capillary action principle. J. Histotech. 11:165–183.

Burrell, G.J., Gowans, E.J., Rowland, R., Hall, P., Jilbert, A.R., and Marmion, B. 1984. Correlation between liver histology and markers of hepatitis B virus replication in infected patients: A study by in situ hybridization. Hepatology 4:20–24.

Danker, W.M., McCutchan, J.A., Richman, D.D., Hirata, K., and Spector, S.A. 1990. Localization of human cytomegalovirus in peripheral blood leukocytes by in situ hybridization. J. Infect. Dis. 161:31–36.

Eversole, L.R., Stone, C.E., and Beckman, A.M. 1988. Detection of EBV and HPV DNA sequences in oral "hairy" leukoplakia by in situ hybridization. J. Med. Virol. 26:271–277.

Feinberg, A.P. and Vogelstein, B. 1983. A technique for radiolabeling DNA restriction endonuclease fragments to high specific activity. Anal. Biochem. 132:6–13.

Hames, B.D. and Higgins, S.J. (eds.). 1985. Nucleic Acid Hybridization: A Practical Approach. Oxford, England: Oxford IRL Press Limited.

Hoofnagle, J.H. 1990. α Interferon therapy of chronic hepatitis B. Current status and recommendations. J. Hepatol. 11:S100–S107.

Jenison, S.A., Lemon, S.M., Baker, L.N., and Newbold, J.E. 1987. Quantitative analysis of hepatitis B virus DNA in saliva and semen of chronically infected homosexual men. J. Infect. Dis. 156:299–307.

Kimpton, C.P., Corbitt, G., and Morris, D.J. 1988. Detection of cytomegalovirus by dot blot DNA hybridization using probes labeled with ^{32}P by nick translation or random hexanucleotide priming. Mol. Cell. Probes. 2:181–188.

Krogsgaard, K. 1988. Hepatitis B virus DNA in serum. Applied molecular biology in the evaluation of hepatitis B infection. Liver 8:257–283.

Krogsgaard, K., Kryger, P., Aldershvile, J., Andersson, P., and Brechot, C. 1985. Hepatitis B virus DNA in serum from patients with acute hepatitis B. Hepatology 5:10–13.

Lie-Injo, L.E., Balasegaram, M., Lopez, C.G., and Herrera, A.R. 1983. Hepatitis B virus DNA in liver and white blood cells of patients with hepatoma. DNA 2:301–308.

Little, P.F.R. and Jackson, I.J. 1987. Application of plasmids containing promoters specific for phage-encoded RNA polymerases. In D.M. Glover (ed.), DNA Cloning III. A Practical Approach. Washington, DC: IRL Press. pp. 1–18.

Lok, A.S.F., Karayiannis, P., Jowett, T.P., Fowler, M.J.F., Farci, P., Monjardino, J., and Thomas, H.C. 1985. Studies of HBV replication during acute hepatitis followed by recovery and acute hepatitis progressing to chronic disease. J. Hepatol. 1:671–679.

Lörincz, A.T. 1989. Human papillomavirus testing. Diag. Clin. Test. 27:28–37.

Lörincz, A.T. and Reid, R. 1989. Association of human papillomavirus with gynecologic cancer. Cur. Op. Oncol. 1:123–132.

Lörincz, A.T., Temple, G.F., Kurman, R.J., Jenson, A.B., and Lancaster, W.D. 1987. Oncogenic association of specific human papillomavirus types with cervical neoplasia. J. Natl. Cancer Inst. 79:671–677.

Matthews, J.A. and Kricka, L.J. 1988. Analytical strategies for the use of DNA probes. Anal. Biochem. 169:1–25.

Matthews, J.A., Batki, A., Hynds, C., and Kricka, L.J. 1985. Enhanced chemiluminescent method for the detection of DNA dot-hybridization assays. Anal. Biochem. 151: 205–209.

Meinkoth, J. and Wahl, G. 1984. Hybridization of nucleic acids immobilized on solid supports. Anal. Biochem. 138:267–284.

Moench, T.R. 1987. In situ hybridization. Mol. Cell. Probes. 1:195–205.

Moestrup, T., Hansson, B.G., Widell, A., Blomberg, J., and Nordenfelt, E. 1985. Hepatitis B virus DNA in the serum of patients followed-up longitudinally with acute and chronic hepatitis B. J. Med. Virol. 17:337–344.

Nuovo, G.J. and Richart, R.M. 1989. A comparison of different in situ hybridization methodologies (biotin and ^{35}S based) for the detection of human papillomavirus DNA. Lab. Invest. 59:720–724.

Pollard-Knight, D., Simmonds, A.C., Schaap, A.P., Akhavan, H., and Brady, M.A.W. 1990. Non-radioactive DNA detection of Southern blots by enzymatically triggered chemiluminescence. Anal. Biochem. 185: 353–358.

Pontisso, P., Bortolotti, F., Schiavon, E., Chemello, L., Alberti, A., and Realdi, G. 1986. Serum Hepatitis B virus DNA in acute hepatitis B. Digestion 34:46–50.

Redfield, D.C., Richman, D.D., Albanil, S., Oxman, M.N., and Wahl, G.M. 1983. Detection of herpes simplex virus in clinical specimens by DNA hybridization. Diag. Microbiol. Infect. Dis. 1:117–128.

Richart, R.M. and Nuovo, G.J. 1990. Human papillomavirus DNA in situ hybridization may be used for the quality control of genital tract biopsies. Obstet. Gynecol. 75: 223–226.

Rijntjes, P.J.M., van Ditzhuijsen, T.J.M., van Loon, A.M., van Haelst, U.J.G.M., Bronkhorst, F.B., and Yap, S.J. 1985. Hepatitis B virus DNA detected in formalin-fixed liver specimens and its relation to serologic markers and histopathologic features of chronic liver disease. Am. J. Pathol. 120:411–418.

Ritter, D.B., Kadish, A.S., Vermund, S.H., Romney, S.L., Villari, D., and Burk, R.D. 1989. Detection of human papillomavirus deoxyribonucleic acid in exfoliated cervico-vaginal cells as a predictor of cervical neoplasia in a high risk population. Am. J. Obstet. Gynecol. 159:1517–1525.

Rotbart, H.A., Levin, M.J., Murphy, N.L., and Abzug, M.J. 1987. RNA target loss during solid phase hybridization of body fluids—quantitative study. Mol. Cell. Probes. 1:347–358.

Sambrook, J., Fritsch, E.F., and Maniatis, T. (eds.). 1989. Molecular Cloning: A Laboratory Manual. New York: Cold Spring Harbor Laboratory.

Schmidbauer, M., Budka, H., and Ambros, P. 1988. Comparison of in situ DNA hybridization (ISH) and immunocytochemistry for diagnosis of herpes simplex virus (HSV) encephalitis in tissue. Virchows Arch. A. Pathol. Anat. 414:39–43.

Seidlin, M., Takiff, H.E., Smith, H.A., Hay, J., and Straus, S.E. 1984. Detection of varicella-zoster virus by dot blot hybridization using a molecularly cloned viral DNA probe. J. Med. Virol. 13:53–61.

Seyda, M., Scheele, T., Neumann, R., and Krueger, G.R.F. 1989. Comparative evaluation of non-radioactive in situ hybridization techniques for pathologic diagnosis of viral infection. Path. Res. Pract. 184:18–26.

Shafritz, D.A., Schouval, D., and Sherman, H.J. 1981. Integration of hepatitis B virus DNA into the genome of liver cells in chronic liver disease and in hepatocellular carcinoma. N. Engl. J. Med. 305:1067–1073.

Shibata, D., Fu, Y.S., Gupta, J.W., Shah, K.V.,

Arnheim, N., and Martin, W.J. 1988. Detection of human papillomavirus in normal and dysplastic tissue by the polymerase chain reaction. Lab. Invest. 59:555–559.

Sixbey, J.W., Nedrud, J.G., Raab-Traub, N., Hanes, R.A., and Pagano, J.S. 1984. Epstein-Barr virus replication in oropharyngeal epithelial cells. N. Engl. J. Med. 310: 1225–1230.

Southern, E.M. 1975. Detection of specific sequences among DNA fragments separated by gel electrophoresis. J. Mol. Biol. 98:503–517.

Spector, S.A., Rua, J.A., Spector, D.H., and McMillan, R. 1984. Detection of human cytomegalovirus in clinical specimens by DNA-DNA hybridization. J. Infect. Dis. 150:121–126.

Stoler, M.H. and Broker, T.R. 1986. In situ hybridization detection of human papillomavirus DNAs and messenger RNAs in genital condylomas and a cervical carcinoma. Hum. Pathol. 17:1250–1258.

Virtanen, M., Palva, A., Laaksonen, M., Halonen, P., Soderlund, H., and Ranki, M. 1983. Novel test for rapid viral diagnosis: Detection of adenovirus in nasopharyngeal mucus aspirates by means of nucleic-acid sandwich hybridization. Lancet 1:381–383.

Warford, A.L. and Levy, R.A. 1989. Use of commercial DNA probes. Clin. Lab. Sci. 2:105–108.

Weiss, L.M., Movahed, L.A., Berry, G.J., and Billingham, M.E. 1990. In situ hybridization studies for viral nucleic acids in heart and lung allograft biopsies. Am. J. Clin. Pathol. 93:675–679.

Wiley, C.A., Schrier, R.D., Nelson, J.A., Lampert, P.W., and Oldston, M.B.A. 1986. Cellular localization of human immunodeficiency virus infection within the brains of acquired immune deficiency syndrome patients. Proc. Natl. Acad. Sci. USA. 83: 7089–7093.

Wong-Staal, F., Hahn, B., Manzari, V., Colombini, S., Franchini, G., Gelmann, E., and Gallo, R.C. 1983. A survey of a human leukaemias for sequences of human retrovirus. Nature 302:626–628.

Wu, J.C., Lee, S.D., Wang, J.Y., Ting, L.P., Tsai, Y.T., Lo, K.J., Chiang, B.N., and Tong, M.J. 1986. Analysis of the DNA of hepatitis B virus in the sera of Chinese patients infected with hepatitis B. J. Infect. Dis. 153:974–977.

Yolken, R.H. 1988. Nucleic acids or immunoglobulins: Which are the molecular probes of the future? Mol. Cell. Probes. 2:87–96.

Zwadyk, P. and Cooksey, R.C. 1987. Nucleic acid probes in clinical microbiology. CRC Crit. Rev. Clin. Lab. Sci. 25:71–103.

18

Application of Western Blotting for the Diagnosis of Viral Infections

Peter G. Medveczky

18.1 INTRODUCTION

The role of molecular biological techniques in the laboratory diagnosis of viral infections has increased in recent years. This methodology is more sensitive and specific than methods developed in the past and is especially important for the diagnosis of infections caused by agents that are difficult to propagate in tissue or cell culture. For example, techniques such as Western blotting (also referred to as immunoblotting) or the polymerase chain reaction are part of the growing panel of diagnostic procedures for human immunodeficiency virus (HIV), human T cell leukemia virus (HTLV) type I and II infections (Carlson, 1987; Centers for Disease Control, 1988a, 1988b, 1989; Gallo, 1986; Healey, 1989; Hirsch, 1990; Mullis, 1986; Saiki, 1988; The Consortium for Retrovirus Serology Standardization, 1988). This chapter discusses the principle, describes the methodology, and provides some practical clinical applications of Western blotting.

18.2 HISTORY AND PRINCIPLE OF WESTERN BLOTTING

The diagnosis of viral infection is often based on the detection of specific circulating antibodies to viral antigens in serum samples. Enzyme immunoassays (EIA) often are used for the diagnosis of viral diseases as well as for the screening of blood and blood products for viruses. Although EIA are very sensitive and highly specific, false positive reactions do occur. Given the medical and social significance of particular virus infections, e.g., HIV, HTLV, or hepatitis viruses, it is important that diagnostic tests for these virus infections should be as specific, accurate, and sensitive as possible. Although sensitivity and specificity of some EIA tests can be greater than 95%, e.g., the licensed EIA for HIV (Centers for Disease Control, 1988b; Petricciani, 1985), it is a standard laboratory practice to repeat a positive EIA test. According to guidelines adopted by the U.S. Public Health Service, if a second positive EIA result is obtained

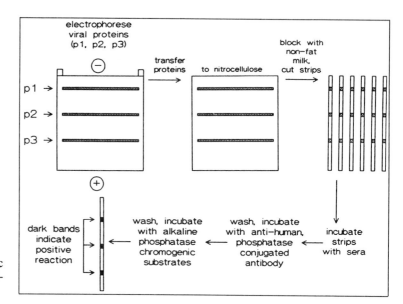

Figure 18.1 Schematic representation of Western blotting.

upon repeat testing the diagnosis must be confirmed by another assay, most often by Western blot (Centers for Disease Control, 1988a, 1988b, 1989; The Consortium for Retrovirus Serology Standardization, 1988).

Several "blotting" techniques have been developed. The initially described technique is referred to as Southern blotting, after its originator Edward Southern (Southern, 1975). It is a fundamental tool for the analysis of DNA fragments. An analogous method for analyzing RNAs is called Northern blotting (Alwine, 1977) as a molecular biologist's joke. The humor continued when the modification of the nucleic acid "blotting" methods for the study of proteins was developed; it is referred to jargonistically as "Western blotting" or more properly as "immunoblotting" (Burnette, 1981; Towbin, 1979).

These different blotting techniques share a common principle. Complex mixtures of macromolecules (DNA, RNA, or protein) are first separated by size in rectangular "slab" gels using electrophoresis. After separation the molecules are transferred ("blotted") onto the surface of a membrane, and the separated and immobi-

lized nucleic acid fragments or proteins are detected and/or identified on the membrane using specific molecular probes. For example, RNA blots can be probed with radioactive complementary DNA (cDNA) probes. The bound RNA hybridizes in situ with the labeled probe, and the reaction is then visualized by autoradiography (Alwine, 1977).

The basic approach to the use of Western blotting for the diagnosis of viral infections begins with purified virions which are disrupted by ionic detergent treatment, releasing viral proteins. As schematically shown in Figure 18.1, these virion proteins are then separated by electrophoresis in sodium dodecyl sulfate polyacrylamide gels (SDS-PAGE). These gels, approximately 1 mm thick, are formed in between two square glass plates, 14 × 14 cm being a typical size. Viral proteins are denatured by boiling in SDS/2-mercaptoethanol buffer, and a sample containing a few hundred μg protein is loaded across the top of the gel. After electrophoresis, the gel is placed on a nitrocellulose membrane of the same size and this gel/nitrocellulose unit is placed on top of several layers of wet filter paper. Filter papers are then laid on the gel and proteins are transferred ("blotted") to a membrane (nitrocellulose, nylon, etc.) by electrophoresis. Transfer is either done by "semi-dry" electroblotters when the gel/nitrocellulose/paper sandwich is placed between two rectangular metal or carbon electrodes (apparatus available from Hoefer Scientific Instruments Inc., San Francisco, CA; Fisher Inc., Springfield, NJ; or other distributors and manufacturers). In an older version of electroblotting, which is still the preferred method by some investigators, the sandwich is held together by a plastic device and is submerged in buffer in a large electrophoresis tank (Burnette, 1981; Towbin, 1979). In either case the transfer of proteins from the gel onto the membrane is mediated by electrical current so that the pattern of proteins obtained by SDS-PAGE is preserved during transfer. After transfer, the membrane is incubated with a buffer containing nonspecific proteins such as milk casein or serum albumin to block all unoccupied areas that could serve as binding sites. This blocking step prevents nonspecific adsorption of immunoglobulin proteins to the nitrocellulose filter during subsequent steps. The nitrocellulose sheet is then cut into several strips and each strip is ready to use for the detection of antiviral antibodies.

If a patient serum containing anti-HIV antibodies is incubated with a Western blot strip, the specific antibodies to the individual viral proteins will form stable complexes with the transferred protein species and the antibodies will remain bound to those antigens even after extensive washing. Typically, patients who have seroconverted have antibodies to several proteins of the viral agent (Centers for Disease Control, 1988a, 1988b, 1989; The Consortium for Retrovirus Serology Standardization, 1988).

After washing to remove unbound antibodies, the last step in the procedure is to visualize the patients' bound immunoglobulins. This can be achieved by using radio-labeled *Staphylococcus aureus* protein A, or a second antibody which is labeled with radioisotope or an enzyme. Clinical laboratories usually use safer non-radioactive methods such as alkaline phosphatase or horse radish peroxidase enzyme coupled anti-human antibodies. The anti-human immunoglobulin and the enzyme are coupled covalently in a manner that allows the activity of both molecules to remain intact.

Detection of patient antibody is accomplished by monitoring enzyme activity linked to the anti-human antibody (see Chapter 9). The enzyme activity can be demonstrated in situ by incubating the membrane with the appropriate chromogenic substrates. The product is insoluble in water after the reaction and develops a dark color at the site of the enzyme activity. During the last step of the Western blot procedure dark bands (blue for alkaline phosphatase, brown for peroxidase) corresponding to the physical location of the reactive viral proteins appear on the nitrocellulose filter.

In summary, the cascade of steps to perform a Western blot includes: electrophoretic separation of viral proteins, nitrocellulose immobilization of viral proteins, viral proteins capture patient's antibodies, patient's antibodies bind enzyme-labeled anti-human antibody, and detection of the presence of human antibody by enzyme reaction that produces visible dark bands at the site of these molecular complexes.

18.3 THE WESTERN BLOT PROCEDURE

The various aspects of purification of virions and viral proteins and details of SDS-PAGE will not be discussed here since clinical laboratories rarely have sufficient resources to perform these steps. We recommend the purchase of blotted viral proteins or Western blot kits available from several commercial sources (see below).

18.3.1 List of Materials Needed

18.3.1.a Instruments and Materials

Rocking platform

Vacuum aspirator with flasks

pH meter

Adjustable micropipettors (1–20 μL, 10–200 μL, and 100–1000 μL) and sterile tips

Pasteur and serological pipettes

Western blot incubation trays. The nitrocellulose strips are usually shipped in Western blot trays. If you wish to purchase separate trays two types are available: disposable or reusable. Reusable trays need thorough cleaning after exposure to reagents; disposable trays are more convenient although more costly in the long run. Before ordering trays, contact the manufacturer of the Western blot nitrocellulose strips you plan to purchase since dimensions of the strips and trays vary among different commercial sources.

Polaroid or other camera to photograph results for permanent record

18.3.1.b Reagents

Tris base (Trizma, Sigma product #T8524 or equivalent, Sigma Chemical Co., St. Louis, MO)

Hydrochloric acid, 5N

Sodium chloride (Sigma product #S9625 or equivalent)

Magnesium chloride, $6H_2O$ (Sigma product #M9272 or equivalent)

Ethylene diamine tetra acetate, di-sodium salt (EDTA, Sigma product #E5134 or equivalent)

Milli-Q or equivalent highly purified or double-distilled H_2O, autoclaved

Nonfat dried milk (from any local supermarket)

Nitrocellulose strips with blotted proteins (see below, Commercial Western blot kits)

Anti-human IgG, alkaline phosphatase conjugated (Promega, 2800 S. Fish Hatchery Rd., Madison, WI 53711-5305, 1-800-356-9526, product #P3821, or equivalent)

Tween 20 (Polyoxyethylenesorbitan monolaurate, Sigma product #P1379 or equivalent)

Nitro blue tetrazolium (NBT) powder, or NBT tablets (Sigma product #N6876, or #N5514 for tablets, or equivalent)

N,N-Dimethylformamide (Sigma, product #D8654)

5-bromo-4-chloro-3-indolyl phosphate (BCIP), p-toluidine salt, or BCIP tablets (Sigma product #B8503, or #B0274 for tablets, or equivalent)

18.3.1.c Solutions

TST buffer, (10mM Tris-HCl pH 8.0, 150mM NaCl, 0.02% Tween 20, 1 liter): Dissolve 1.21 g Tris base and 8.76 g NaCl in about 900 mL sterile Milli-Q H_2O, adjust pH to 8.0 with 5N HCl. Add 0.5 mL Tween 20 and fill with H_2O to 1 liter. Store at 4 °C.

TST buffer with 5% dry milk (10mM Tris-HCl pH 8.0, 150mM NaCl, 0.02% Tween 20, 50 mL): Dissolve 0.5 g nonfat dry milk in 50 mL TST buffer, store at −20 °C.

Alkaline phosphatase buffer (100 mM NaCl, 5 mM $MgCl_2$, 100 mM Tris-HCl pH 9.5, 100 mL): Dissolve 1.21 g Tris base, 0.10 g $MgCl_2$ $6H_2O$, and 0.58 g NaCl in about 90 mL sterile Milli-Q H_2O, adjust pH to 9.5 with 5N HCl. Fill with H_2O to 100 mL. Store at 4 °C.

NBT solution (2mL): Dissolve 100 mg NBT powder or tablets in 2 mL of 70% N,N-Dimethylformamide (mixture of 1.4

mL Dimethylformamide and 0.6 mL H_2O). Store in the dark at $-20\,°C$.

BCIP solution (1 mL): Dissolve 50 mg BCIP in 1 mL 100% N,N-Dimethylformamide. Store in the dark at $-20\,°C$.

Stop solution (5 mM EDTA, 50 mM Tris-HCl pH 7.5, 100 mL): Dissolve 1.21 g Tris base, 0.186 g EDTA, and 0.58 g NaCl in about 90 mL sterile Milli-Q H_2O, adjust pH to 7.5 with 5N HCl. Fill with H_2O to 100 mL. Store at $4\,°C$.

18.3.2 Step-by-Step Procedure

This description is for a procedure that uses an alkaline phosphatase-labeled second antibody for detection. Some minor modifications may be needed if another system is used (Sambrook et al, 1989).

1. *Check condition of equipment required, collect reagents, check expiration date of Western blot strips and sera (do not use beyond expiration date), and prepare solutions.*

2. *Incubation with primary antibody:*
   (a) Dilute serum 1:50 in TST buffer/5% dry milk. Prepare enough to allow use of 0.05 mL per cm^2 Western blot filter strip. As an example, if the area of the test strip is 5 cm^2 then a minimum of 0.25 mL diluted test serum should be prepared. Appropriate controls are essential in Western blotting; a positive and a negative control serum appropriately diluted must be included in all assays (manufacturers of Western blot kits offer such controls).
   (b) Place Western blot strips in multiwell tray using forceps and mark each well for future identification of each reaction.
   (c) Tilt tray at about a 30 degree angle and add 0.05 mL of TST buffer/5% dry milk per cm^2 filter strip to the bottom of each well. Slowly lower the tray to a horizontal position so that the strips

adsorb buffer gradually and evenly. Add diluted serum to the appropriate wells, close the lid, and incubate at room temperature for two hours on a rocking platform by gentle agitation.

3. *Wash filters five times with TST:*
   (a) Remove lid, tilt tray, and aspirate the liquid by Pasteur pipet connected to a vacuum flask. Do not allow filters to dry out!
   (b) Fill wells about half way with TST and wash with gentle agitation (rocking) for 7 minutes.
   (c) Remove TST and repeat this wash procedure four times for a total of five washes. Make sure that the buffer is completely removed after each wash.

4. *Incubation with secondary antibody:*
   (a) Dilute the anti-human IgG alkaline phosphatase conjugate according to the manufacturer's specification in TST buffer/5% dry milk.
   (b) Add the diluted serum to each well (0.1 mL of diluted serum per cm^2 filter strip).
   (c) Close the lid, and incubate at room temperature for one hour on a rocking platform by gentle agitation.

5. *Wash filters five times with TST as described in step 3.*

6. *Incubation with chromogenic substrates:* During the final period, prepare the chromogenic substrates; mix 66 μL NBT stock with 10 mL alkaline phosphatase buffer in a test tube then add 33 μL BCIP and mix well. Remove the washing buffer and add 0.1 mL chromogenic substrate solution per cm^2 filter strip. Incubate the tray with gentle agitation as before and monitor the development of dark blue bands.

7. *Stop reaction:* When strong bands are visible with the positive control serum and test samples but the negative control is still clear the reaction is stopped. Optimal incubation time with substrates

varies from a few minutes to half an hour and depends on the reagents and the titer of antibodies. To stop the reaction quickly aspirate the substrate solution and add 2 mL of the stop solution. Some experimentation is necessary to determine optimal conditions for each new set of reagents to be able to stop the reaction at the right moment. A timer should be used to monitor the incubation time with the chromogenic substrates in order to note optimal time for future reference. Over-incubation usually results in a high background and/or appearance of bands in the negative serum control. If this should occur the Western blot must be repeated with a shortened incubation period with the chromogenic substrate.

8. *Interpretation of the assay, special hints:* In general, a Western blot test is considered positive when the patient's serum is reactive with more than one viral antigen. For example, a licensed HIV test is interpreted positive when multiple bands are present, i.e., p24, p31, gp41, and gp160 (Centers for Disease Control, 1988a, 1988b, 1989; The Consortium for Retrovirus Serology Standardization, 1988). The question is how to identify these proteins? Most manufacturers offer sequentially numbered Western blot strips that have been originated from a single gel. The strips are numbered so that adjacent numbers refer to originally adjacent strips. The key to accurate results is to use a pair of adjacent nitrocellulose Western blot strips that originated from the same area of the SDS PAGE gel. One of these strips should be reacted with the positive control serum and the adjacent strip with the patient sample. When the test is complete strips should be aligned numerically according to the manufacturer's numbering and should be compared side by side and photographed. This method of analysis can help in the interpretation of dubious and nonspecific reactions.

18.4 ADVANTAGES AND DISADVANTAGES OF WESTERN BLOT ASSAY

It is obvious that the Western blot is more specific than EIA tests, since in the Western blot assay, antibodies to several antigens are detected simultaneously by using a group of electrophoretically-separated viral proteins. At present, Western blotting offers a very reliable confirmatory assay for HIV-1 infection (Centers for Disease Control, 1988b, 1989). A potential competitor for the Western blot assay has been reported for the diagnosis of HIV infection. This new technique, called recombinant-antigen immunoblot assay (RIBA-HIV216), utilizes a set of purified antigens produced by recombinant technology (Busch, 1991; Lillehoy, 1990; Oroszlan, 1985; Steimer, 1986; Truett, 1989). It remains to be determined whether this recombinant protein assay is more specific and/or more sensitive than the standard Western blot assay.

Western blot assays can produce nonspecific or so-called "indeterminate" reactions where often only one band is seen. These results are sometimes attributed, among other causes, to an underlying autoimmune disease which results in production of antibodies to cellular antigens (Healey, 1989). However, indeterminate status may precede a truly positive status and thus be indicative of an infection. Such results require a specimen to be collected from the patient at a later date for retesting. As demonstrated on Figure 18.2, an indeterminate reaction is observed at the early stage of HIV infection (day 14) with the antibody response directed only to a single HIV protein, p24. Gradually, the patient developed antibodies to several other viral

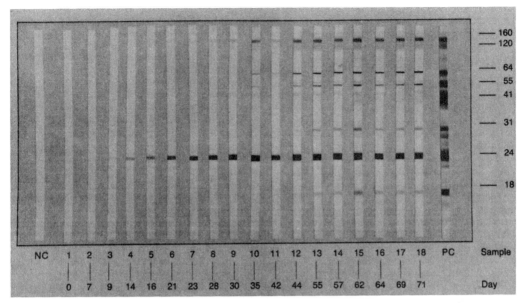

Figure 18.2 Demonstration of seroconversion after exposure to HIV by Western blotting. Western blot assay was performed using a Biotech Western Blot kit. Sera samples were taken from an individual at intervals as indicated. NC = negative control serum. PC = positive control serum. (Kindly provided by Boston Biomedica, Inc., Bridgewater, MA 02379.)

proteins and by day 35 is considered HIV positive.

Similarly, false negative reactions have been described (Kissler, 1987), but the occurrence of false negative reactions is relatively rare. Perhaps the main disadvantage of Western blot is that it requires experienced personnel, which limits its use to laboratories capable of performing more specialized services. Therefore, training personnel to perform this technique in experienced laboratories is highly recommended.

18.5 COMMERCIAL KITS AND NITROCELLULOSE STRIPS WITH BLOTTED PROTEINS

Nitrocellulose strips with blotted viral proteins or complete kits with all necessary reagents are available from commercial sources for HIV-1, HTLV-I, and Hepatitis C virus Western blot assays. Licensed HIV-1 kits are available from several distributors and manufacturers including:

Bio-Rad, Inc., 875 Alfred Nobel Dr., Suite D, Hercules, CA 94547, phone: 1-800-342-2072

Cambridge Biotech, 365 Plantation Street, Worcester, MA 01605, USA, phone: 508-797-5777

Organon-Teknika-Cappel, Inc., 100 Akzo Ave., Durham, NC 27704, phone: 1-800-523-7620

ORTHO Diagnostic Systems, Inc., Route 202, Raritan, NJ 08869, USA, phone: 1-800-322-6374

Some of these companies also sell hepatitis C virus, HTLV-I, and HTLV-II Western blot kits.

The information regarding commercial reagents was up-to-date in December 1991. For the list of currently available Western blot kits we recommend the yearly Guide to Biotechnology Products and Instruments published in Science magazine. Another useful commercial source of information is Linscott's Directory of Immunological and Biological Reagents. Linscott's Directory is a catalog of a large selection of biotechnology products and is available for $70 (as of January, 1992) at the following address:

4877 Grange Road, Santa Rosa, CA 95404, USA, phone: 707-544-9555.

REFERENCES

Alwine, J.C., Kemp, P.J., and Stark, G.R. 1977. Method for the detection of specific RNAs in agarose gels by transfer to diazobenzyl-methyl-paper and hybridization with DNA probes. Proc. Natl. Acad. Sci. USA 74: 5350–5354.

Burnette, W.H. 1981. Western blotting: Electrophoretic transfer of proteins from SDS-polyacrylamide gels to unmodified nitrocellulose and radiographic detection with antibody and radioiodinated Protein A. Anal. Biochem. 112:195–203.

Busch, M.P., El Amad, Z., McHugh, T.M., Chien, D., and Polito, A.J. 1991. Reliable confirmation and quantitation of human immunodeficiency virus type 1 antibody using a recombinant-antigen immunoblot assay. Transfusion 31:129–137.

Carlson, J.R., Yee, J., Hinrichs, S.H., Bryant, M.L., Gardner, M.B., and Pedersen, N.C. 1987. Comparison of indirect immunofluorescence and Western blot for the detection of anti-human immunodeficiency virus antibodies. J. Clin. Microbiol. 25:494–497.

Centers for Disease Control. 1988a. Licensure of screening tests for antibody to human T-lymphotropic virus type I. Morbid. Mortal. Weekly Rep. 37:736–747.

Centers for Disease Control. 1988b. Update: Serological testing for antibody to human immunodeficiency virus. Morbid. Mortal. Weekly Rep. 36:833–840.

Centers for Disease Control. 1989. Interpretation and use of the Western blot assay for serodiagnosis of human immunodeficiency virus type 1 infection. Morbid. Mortal. Weekly Rep. 38:(Suppl 7):1–7.

Gallo, D., Diggs, J.L., Shell, G.R., Dailey, P., Hoffman, M.N., and Riggs, F.L. 1986. Comparison of detection of antibody to the acquired immune deficiency syndrome virus by enzyme immunoassay, immunofluo-

rescence, and Western blot methods. J. Clin. Microbiol. 23:1049–1051.

Healey, D.S. and Howard, T.S. 1989. Activity to non-viral proteins on Western blot mistaken for reactivity to HIV glycoproteins. AIDS 3:545–546.

Hirsch, M.S. and Curren, J. 1990. Human immunodeficiency viruses. Biology and medical aspects in B.N. Fields and D.M. Knipe, (eds.). Virology, New York: Raven Press, pp. 1545–1570.

Kissler, H.A., Blauw, B., Spear, J., Paul, D.A., Falk, L.A., and Landay, A. 1987. Diagnosis of human immunodeficiency virus infection in seronegative homosexuals presenting with an acute viral syndrome. JAMA 258: 1196–1199.

Lillehoj, E.P., Alexander, S.S., Dubrule, C.J., Wikter, S., Adams, R., Taj, C., Manns, A., and Blattner, W.A. 1990. Development and evaluation of a human T-cell leukemia virus type I serological confirmatory assay incorporating a recombinant envelope polypeptide. J. Clin. Microbiol. 28:2653–2658.

Sambrook, J., Fritsch, E.F., and Maniatis, T. 1989. Molecular Cloning a Laboratory Manual. Cold Spring Harbor, NY: Cold Spring Harbor Laboratory Press, pp. 18.60–18.75.

Mullis, K.B., Faloona, F., Scharf, S.J., Saiki, R.K., Horn, G.T., and Erlich, H.A. 1986. Specific enzymatic amplification of DNA in vitro: The polymerase chain reaction. Cold Spring Harbor Symp. Quant. Biol. 51:263–273.

Oroszlan, S. and Copeland, T.D. 1985. Primary structure and processing of gag and env gene products of human T-cell leukemia viruses HTLV-I$_{CR}$ and HTLV-II$_{ATK}$. Curr. Top. Microbiol. Immunol. 115:221–233.

Petricciani, J.C. 1985. Licensed tests for antibody to human T-lymphotropic virus type III. Sensitivity and specificity. Ann. Intern. Med. 103:726–729.

Saiki, R.K., Gelfand, D.H., Stoffel, F., Scharf, S.J., Higuchi, R., Horn, G.T., Mullis,

K.B., and Erlich, H.A. 1988. Primer-directed enzymatic amplification of DNA with a ther-mostable DNA polymerase. Science 239:487–491.

Southern, E.N. 1975. Detection of specific sequences among DNA fragments separated by gel electrophoresis. J. Mol. Biol. 98:503–517.

Steimer, K.S., Higgins, K.W., Powers, M.A., Stephans, F.C., Gyenes, A., George-Nascimento, C., Luciw, P.A., Barr, P.J., Hallewell, R.A., and Sanchez-Pescador, R. 1986. Recombinant polypeptide from the endonuclease region of the acquired immune deficiency syndrome retrovirus polymerase (pol) gene detects serum antibodies in most infected individuals. J. Virol. 58:9–16.

Steimer, K.S., Puma, J.P., and Powers, M.A., 1986. Differential antibody responses of individuals infected with the AIDS-associated retroviruses surveyed using the viral core antigen p26gag expressed in bacteria. Virology 150:283–290.

The Consortium for Retrovirus Serology Standardization. 1988. Serological diagnosis of human immunodeficiency virus infection by Western blotting. JAMA 260:674–679.

Towbin, H., Staehelin, T., and Gordon, J. 1979. Electrophoretic transfer of proteins from polyacrylamide gels to nitrocellulose sheets: Procedure and some applications. Proc. Natl. Acad. Sci. USA 76:4350–4354.

Truett, M.A., Chien, D.Y., Calarco, T.L., DiNello, R.K., Polito, A.J. 1989. Recombinant immunoblot assay for the detection of antibodies to HIV. In P.A. Luciw, and K.S. Steimer (eds.). HIV Detection by Genetic Engineering Methods. New York: Marcel Dekker, pp. 121–141.

19

Polymerase Chain Reaction

Carleton T. Garrett, Kathleen Porter-Jordan, and
Suhail Nasim

19.1 GENERAL CONSIDERATIONS

19.1.1 Overview

One of the major advantages of DNA probe technology is that it makes use of nature's most specific marker of all living organisms, i.e., the order of the nucleotides comprising the organism's genome. The nucleotide sequence in the DNA of one species differs from that in another species. This is true when comparing different species of bacteria as well as when comparing the DNA in microorganisms and man. The polymerase chain reaction (PCR) is a technique which results in the in vitro amplification of DNA (Saiki et al, 1985; Mullis and Faloona, 1987). This action greatly enhances the detection of specific DNA sequences that may be present in clinical samples. PCR was developed by Dr. Kary Mullis of Cetus Corporation (Mullis and Faloona, 1987; Schaefer, 1991). Through close collaboration with his colleague, Dr. Henry Erlich, Dr. Mullis' idea was trans-

formed into a workable technical procedure (Saiki et al, 1985).

19.1.2 Problems Encountered Using DNA in Clinical Testing

Although DNA is a highly specific marker, detecting its presence in a clinical sample poses a number of problems (Table 19.1). First, the most sensitive detection methods currently utilize radioactive phosphorous (^{32}P). ^{32}P requires special handling and disposal procedures which complicate its use in the environment of a clinical laboratory. Second, even when using ^{32}P, 10,000 or more organisms may be necessary in a specimen in order to be detected (Cleary et al, 1984; Mullis and Faloona, 1987). Third, in the past it has been necessary to extensively purify DNA prior to analysis (Sambrook et al, 1989). This is time consuming and labor intensive. Fourth, detection of small mutations, such as point mutations, has required the isolation of the mutant

309

Table 19.1. Problems Associated with Using DNA in Clinical Test Protocols

1. The most sensitive detection methods use radioactive isotope (^{32}P)
2. Even with ^{32}P 10,000 or more sequences are required for detection
3. DNA extraction procedures are time consuming
4. Detection of small mutations may require cloning and sequencing

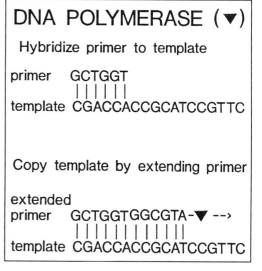

DNA POLYMERASE (▼)

Hybridize primer to template

primer   GCTGGT
           | | | | | |
template CGACCACCGCATCCGTTC

Copy template by extending primer

extended
primer   GCTGGTGGCGTA-▼ --›
           | | | | | | | | | | | |
template CGACCACCGCATCCGTTC

Figure 19.1 DNA Polymerase. The enzyme synthesizes a new strand of DNA using a pre-existing strand as a template. It commences adding complementary nucleotides beginning at the 3′ end of the primer.

gene by cloning (Scharf et al, 1986). The latter procedure requires special training and facilities.

19.2 POLYMERASE CHAIN REACTION—BASIC PRINCIPLES

19.2.1 DNA Polymerase

In order to understand the basic ideas which underlie PCR, it is necessary to review the properties and activities of the enzyme DNA polymerase. This enzyme synthesizes a new strand of DNA that is complementary to a preexisting strand of template DNA. The enzyme is active in the presence of the template, triphosphorylated deoxynucleotides, and a small fragment of the DNA termed a primer which attaches to the template forming a short double stranded region (Lehman, 1981; Saiki et al, 1985).

The primer, which is generally 15 to 20 nucleotides in length, attaches to the template through the process of hybridization or annealing, i.e., hydrogen bond formation between complementary bases in the primer and the template. When the primer has attached, the enzyme begins to extend the primer along the length of the template by sequentially joining complementary bases to each other (Figure 19.1).

19.2.2 Steps in the Polymerase Chain Reaction

PCR represents repeated cycles of DNA polymerase activity (Figure 19.2a). Each cycle contains three steps: denaturation, annealing, and extension.

19.2.2.a Denaturation

In order for PCR to work, the DNA in the sample must be denatured, i.e., the two strands must be separated. This is accomplished by using heat to break the hydrogen bonds between the complementary bases that hold the two strands of DNA together.

19.2.2.b Annealing

Next, the DNA strands and primers must hybridize or anneal to each other. The primers must be constructed so that one primer will anneal to one strand at the beginning of the target sequence while the other primer anneals to the other strand at the end of the target sequence.

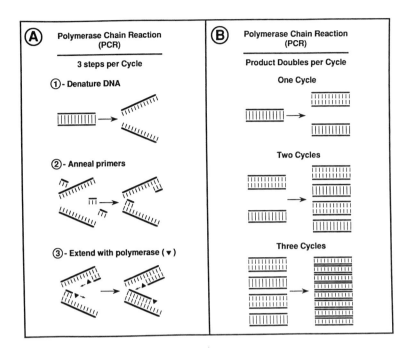

Figure 19.2 Polymerase Chain Reaction. Panel A–Denaturation, primer annealing, and extension comprise the steps of the PCR cycle. Panel B–The amount of product doubles with each PCR cycle.

19.2.2.c Extension

DNA polymerase and triphosphorylated deoxynucleotides then enter into the reaction. The polymerase extends the primers beginning with the first base after the primers. Through this extension of the primers, the DNA polymerase creates two new strands of DNA which are complementary to the target sequence.

19.2.3 Amplification by PCR

Thus, at the end of the first PCR cycle the original DNA target sequence has been converted into two copies. During the second PCR cycle, the two copies of the DNA target sequence are converted into four (Figure 19.2b). During the third cycle, the four are converted into eight. This geometric increase in the number of DNA target sequence copies continues for as many cycles as one runs the reaction, resulting after ten cycles in about a thousandfold increase, after 20 cycles in a millionfold increase, and after 30 cycles a billionfold increase (Oste, 1988).

19.3   ADVANTAGES DERIVING FROM IN VITRO AMPLIFICATION

19.3.1 Enhanced Sensitivity

By increasing the copy number of the target sequence, PCR overcomes many of the problems encountered when using DNA in clinical testing (Table 19.2). For example, PCR may eliminate the need to use radioactive isotopic detection methods. The advantage of radioactivity is that it is more sensitive than nonisotopic methods for detecting DNA target sequences. However, this higher sensitivity may not be necessary when the copy number of the target sequence is increased to very high levels (Corey and Chu-an, 1990). For example, PCR can eliminate the need to use radioactivity to detect the sickle cell point mutation (Chang and Kan, 1982; Chehab et al, 1987). At the sixth codon of the normal beta globin gene there is an MstII restriction endonuclease recognition site which is lost if the sickle cell mutation occurs (Chang and Kan, 1982). Digestion of DNA with the

Table 19.2 PCR Overcomes Many of the Problems Encountered Using DNA in Clinical Testing

1. PCR increases the number of DNA target sequences to very high levels (10^{+6}–10^{+9} copies) so that non-isotopic detection methods are feasible
2. DNA extraction can be greatly simplified
    only a short segment of DNA must be available to bind primers
    only a few copies of target sequence are required
    the extraction procedure can be reduced to a one step incubation
3. Detection of small mutations is relatively easy
    direct sequencing of PCR-generated DNA product removes need for cloning

sickle cell mutation by MstII produces a fragment of the beta globin gene which is longer than the one produced when the sickle cell mutation is not present (Chang and Kan, 1982). To detect the abnormally large fragment the digested DNA must be fractionated on an agarose gel, transferred onto a membrane, and hybridized with a radioactively-labeled probe to the beta globin gene (Chang and Kan, 1982). One million to ten million copies of the beta globin gene are present on the membrane from the digested DNA but this number is too small to be reliably detected using a nonradioactively-labeled probe. PCR increases the number of copies of the beta globin gene by one million to one billion-fold. This number is sufficient so that the gene can be directly visualized in an acrylamide or agarose gel using the nonradioactive procedure of staining with ethidium bromide (Chehab et al, 1987). Moreover, the amplified gene can be digested with MstII before being electrophoresed and the presence or absence of a sickle cell mutation determined quickly and without the use of radioactivity (Chehab et al, 1987).

19.3.2 Flexibility of Detection Methods

19.3.2.a Detection of an Amplified DNA Fragment of Known Molecular Weight

Since the sequence of the gene to be amplified by PCR is usually known, the size of the DNA fragment that will be generated by the PCR reaction is also known. To measure DNA fragment size, the PCR product is electrophoresed using an acrylamide or agarose (Nusieve) gel and examined using either ethidium bromide staining or, if radioactive nucleotides have been incorporated into the PCR product, by autoradiography. The level of certainty that one has, in fact, amplified the correct DNA sequence can be increased if a restriction endonuclease recognition site is present in the target sequence. In this case, digestion of the PCR product by the restriction enzyme should result in two DNA fragments of known molecular weight. This is easily verified by subjecting the digestion products to gel electrophoresis (Chehab et al, 1987; Jiang et al, 1989).

19.3.2.b Detection of Amplified Target DNA Through Oligonucleotide Hybridization

A second approach to detect amplified DNA utilizes oligonucleotide hybridization. Like primers, oligonucleotides are short (<50 bases long) sequences of single-stranded DNA. In this approach an oligonucleotide that is complementary to a portion of the sequence being amplified is attached to a radioactive or nonradioactive label and then hybridized to the PCR amplified target sequence. The formats usually adopted in this approach are:

Spot blot/dot blot—In this procedure a small portion of the amplified DNA is placed on a filter by either directly spotting it on the surface of the filter (spot blot) or drawing it through the filter under vacuum using a dot blot or slot blot apparatus (Abbott et al, 1988; Shibata et al, 1988; Vogelstein et al, 1988);

Liquid hybridization—An aliquot of the PCR product is hybridized in a solution with an oligonucleotide probe and then subjected to gel electrophoresis. Radioactive nucleotides are the usual method of labeling probes for this procedure, and the gel is then exposed to x-ray film for autoradiography. This approach has the advantage of providing information on both fragment size and homology (Cassol et al, 1989; Hsia et al, 1989; Porter-Jordan et al, 1990);

Southern blot detection—The PCR product is first electrophoresed in an agarose or acrylamide gel and transferred to a nylon membrane. The membrane is then hybridized with an appropriately labeled oligonucleotide probe (Cassol et al, 1989; Hsia et al, 1989).

In performing hybridization studies with oligonucleotide probes, one must be aware that the melting temperature of the probe, and hence the optimum hybridization temperature for the reaction, is highly sensitive to changes in the base composition, the length of the probe, and the composition of the medium in which the reaction takes place (Sambrook et al, 1989).

19.3.3 Simplifying DNA Purification Procedures

Use of PCR greatly reduces the problems associated with DNA extraction since just a few target sequences need be available to bind the primers to initiate the amplification reaction. Thus, DNA extraction can be reduced to a simple one-step process (Jeanpierre, 1987; Higuchi, 1989).

19.3.4 Increased Capability to Detect Genetic Changes

Detection of small mutations is also greatly simplified using PCR. One can amplify a portion of a gene by PCR and then directly sequence the amplified DNA without having to clone the gene (Nasim et al, 1990). If one desires to study the gene, it is a simple matter to insert the amplified DNA into a vector (Scharf et al, 1986; Thomas and Newbold, 1990). There is reason to believe that DNA sequencing may be of some clinical value in microbiology. For example, mutations in the *pol* gene of HIV-1 may predict resistance of the virus to AZT (Larder and Kemp, 1989).

19.4 LIMITATIONS OF PCR

19.4.1 Sequence Must Be Known

The sequence of the target gene must be known to enable preparation of primers used to amplify the target DNA (Oste, 1988). This limitation turns out to be of less importance than might otherwise be expected when applying PCR to clinical microbiology. This is due to the fact that sequence data are available for most clinically important microorganisms and these data are easily retrievable from government sponsored data bases such as GenBank (Intelligenetics, Inc., 700 E. El Camino Real, Mountain View, CA 94040). A related problem arises, however, from the fact that the genome of most microorganisms mutates over time. If the mutation occurs in the region of a primer it may prevent the primer from binding properly to the tem-

Table 19.3 PCR Limitations

1. Base sequence of the target gene is required (needed to make primers)
2. Contamination of laboratory and negative specimens by PCR product (amplicon)
3. Contamination from template fragments

Table 19.4 Amplicon Contamination

Spread by:
· mechanical pipettors
· gloves
· fingers
· clothing
· equipment
· reagents
· "virtually everything you might work with"

plate and thus interfere with the PCR. This phenomenon has been best studied in the case of HIV-1 where it has been determined that primers for several different genes within the virus must be used in order to ensure detection of the virus (Abbott et al, 1988).

19.4.2 Precautions to Prevent Contamination by Amplified Product

Amplified fragments of DNA, called amplicons, may be easily aerosolized into the work environment when sample tubes are opened following PCR or during any process or manipulation of the reaction products (Table 19.4). Therefore, special precautions must be taken since contamination of a negative sample by as few as one of these small DNA fragments may result in a false positive. Physical separation of pre-PCR and post-PCR work areas is a basic requirement. Care in opening tubes after PCR to minimize aerosolization must be taken by covering the tube with a chem-wipe when it is opened. Use of positive displacement pipettes is another standard practice in most laboratories performing PCR (Kwok and Higuchi, 1989; Oste, 1989).

19.4.3 Contamination from *In Vivo* Amplified DNA Fragments

Contamination can occur from a high background level of template DNA. This "high

level" is, however, actually so low that no one knew of its existence prior to PCR. It was detected in our laboratory during development of a PCR technique for the detection of human cytomegalovirus (hCMV) (Porter-Jordan et al, 1990; Porter-Jordan and Garrett, 1990). The problem can arise, however, in any situation in which there has been biological amplification. Autoclaving of biologically-amplified sequences leads to random shearing of the DNA which can still result in a positive signal through a process called "jumping PCR" even when the fragments are shorter than the DNA sequence one wishes to amplify (Paabo et al, 1989).

19.4.3.a Jumping PCR

PCR can be performed on overlapping randomly-sheared DNA fragments resulting in nearly complete recreation of the original sequence through jumping PCR (Paabo et al, 1989). Jumping PCR occurs when a primer combines with a fragment of DNA and during the first cycle extends to the end of that fragment [Figure 19.3(A)]. In the second cycle, that extended primer serves as a primer to an independent but overlapping randomly-sheared fragment of the sequence and is extended further. This process continues so long as there are overlapping fragments in sufficient abundance that during the time of the cycle a successful hybridization and subsequent extension can occur [Figure 19.3(A)]. Ultimately the primer extends completely across the target sequence [Figure 19.3(B)]. The regenerated complete sequence then acts as a copy for subsequent amplification.

The process of jumping PCR can be very helpful to archaeologists who have used this approach to study genes in degraded DNA obtained from Egyptian mummies and other long dead creatures (Paabo et al, 1989). However, jumping PCR can also be responsible for false positive results

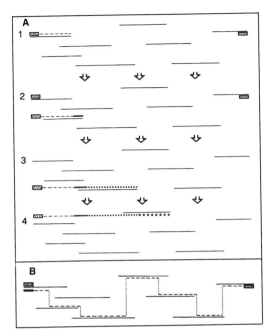

Figure 19.3 Jumping PCR. A. With each cycle the primer extends the length of an overlapping fragment of randomly-sheared DNA (See section 19.4.3.a for further details). B. After a sufficient number of cycles the primer extends to the region of the opposite primer. During the cycles in which it is extending (lag phase) no amplification occurs (Modified from Porter-Jordan and Garrett, 1990; printed with permission).

when PCR is performed in an environment in which the organism to be detected was previously cultured (Porter-Jordan and Garrett, 1990). When biological amplification of DNA, i.e., culturing a specific organism takes place in an environment for a sufficiently long time, fragments of DNA from the in vivo amplified organism can accumulate to a point that all reagents exposed to that environment will contain these DNA fragments. When PCR is performed with these reagents, all samples will be positive if a sufficient number of cycles are run (Porter-Jordan et al, 1990).

In jumping PCR a lag time occurs during the initial cycles of the PCR during which there is no amplification of the target sequence. In these early cycles the primer

is extending from one fragment to another [Figure 19.3(A)]. Amplification of the target sequence cannot occur until at least one intact copy of the target sequence is formed which extends between the two primers [Figure 19.3(B)]. However, after the entire sequence is regenerated, amplification can occur in the normal fashion (Porter-Jordan et al, 1990).

19.4.3.b Nested PCR

Nested PCR uses two sets of primers in which one set (the inner set) is located internal to the other set (the outer set). The outer set of primers amplifies a large sequence sometimes referred to as the outer nest. A portion of the sample from the outer nest is then transferred to a separate reaction vessel which contains the inner set of primers. These primers generate the final product called the inner nest (Mullis and Faloona, 1987).

Nested PCR can be useful in two sets of circumstances. First, the nested procedure boosts the amount of target sequence that is produced in comparison with the unnested PCR procedure thus increasing the sensitivity of the PCR assay (Mullis and Faloona, 1987; Porter-Jordan et al, 1990, Figure 19.4). Secondly, nested PCR can overcome false positive signals which arise from high background levels of randomly-sheared fragments of template DNA (Porter-Jordan et al, 1990, Figure 19.4). Nested PCR accomplishes this by interrupting the polymerase chain reaction prior to the creation of any completely regenerated sequences which might arise from overlapping background template DNA. The "jumping" process is interrupted during the transfer from the outer to inner nest while the amplification of intact sequences present in the original sample are enhanced by the transfer of amplified intact target sequences from the outer nest to the inner nest reaction.

A

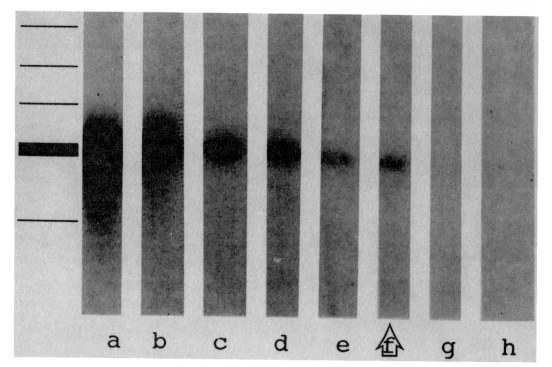

a   b   c   d   e   f   g   h

B

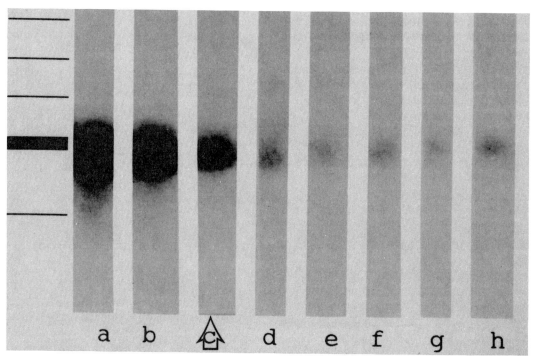

a   b   c   d   e   f   g   h

Figure 19.4 Detection of hCMV DNA by nested and non-nested PCR. Approximately 100,000 copies of Towne Strain hCMV DNA were diluted in ultra-pure water and then subjected either to a nested PCR procedure (20 cycles with the outer primer pair followed by 20 cycles with the inner primer pair–panel A) or to a non-nested procedure (35 cycles with the inner primer pair only–panel B). Samples were detected in a liquid hybridization reaction system. Results demonstrate that with a non-nested procedure a signal was detected in the blank control sample (lane h; panel B) which was not observed with the nested approach (lane h; panel A). Open arrows mark the highest dilution of Towne virus where the signal detected in the sample was clearly greater than that observed in a blank control sample (lane h in either panel A or B, respectively). The sensitivity using the nested approach was approximately 100 times greater than with the non-nested procedure. Lane a, 1:50 dilution of sample; lane b, 1:100; lane c, 1:500; lane d, 1:1000; lane e, 1:20,000; lane f, 1:50,000; lane g, 1:100,000; lane h, water. (Modified from Porter-Jordan et al, 1990; printed with permission)

19.5 SUMMARY

The polymerase chain reaction was first described at the end of 1985. Despite its relatively brief existence, PCR is currently at the forefront of much of the research effort in molecular biology. In addition, the technique is being applied in the clinical diagnostic laboratory, particularly in the field of microbiology and especially virology. The technique is highly sensitive and specific and is amenable to automation. A significant limitation to the use of this technique in a clinical setting is a potential occurrence of false positives due to the contamination of negative samples with PCR product (amplicon). A second potential source of contamination can occur through the presence of high background levels of template DNA. The latter is a serious concern wherever the sequence which is being detected by PCR has been amplified biologically through culture or cloning. Nested PCR is a method to enhance sensitivity and obviate problems arising from high background levels of template DNA. It is expected that within the next five years a number of PCR assays will become the "gold standard" to detect specific microorganisms.

REFERENCES

Abbott, M.A., Poiesz, B.J., Byrne, B.C., Kwok, S., Sninsky, J.J., and Ehrlich, G.C. 1988. Enzymatic gene amplification: Qualitative and quantitative methods for detecting proviral DNA amplified in vitro. J. Infect. Dis. 158:1158–1169.

Cassol, S.A., Poon, M.C., Pal, R., Naylor, M.J., Culver-James, J., Bowen, T.J., Russell, J.A., Krawetz, S.A., Pon, R.T., and Hoar, D.I. 1989. Primer-mediated enzymatic amplification of cytomegalovirus (CMV) DNA. Application to the early diagnosis of CMV infection in marrow transplant recipients. J. Clin. Invest. 83: 1109–1115.

Chang, J.C. and Kan, Y.W. 1982. A sensitive new test for sickle-cell anemia. N. Engl. J. Med. 307:30–32.

Chehab, F.F., Doherty, M., Cai, S., Kan, Y.W., Cooper, S., and Rubin, E.M. 1987. Detection of sickle cell anemia and thalassaemias. Nature 329:293–294.

Cleary, M.L., Chao, J., Warnke, R., and Sklar, J. 1984. Immunoglobulin gene rearrangement as a diagnostic criterion of B-cell lymphoma. Proc. Natl. Acad. Sci. USA 81:593–597.

Corey, L. and Chu-an, C. 1990. Nonisotopically labeled probes and primers. In M.A. Innis, D.H. Gelfand, J. Sninsky, and T.J. White (eds). PCR Protocols. San Diego, CA: Academic Press, pp. 99–112.

Higuchi, R. 1989. Rapid, efficient DNA extraction for PCR from cells or blood. Amplifications 2:1–3.

Hsia, K., Spector, D.H., Lawrie, J., Spector, S.A. 1989: Enzymatic amplification of human cytomegalovirus sequences by polymerase chain reaction. J. Clin. Microbiol. 27:1802–1809.

Jeanpierre, M. 1987. A rapid method for the purification of DNA from blood. Nucleic Acids Res. 15:9611.

Jiang, W., Kahn, S.M., Guillem, J.G., Lu, S-H., and Weinstein, I.B. 1989. Rapid detection of ras oncogenes in human tumors: Application to colon, esophageal, and gastric cancer. Oncogene 4:923–928.

Kwok, S. and Higuchi, R. 1989. Avoiding false positives with PCR. Nature 339:237–238.

Larder, B.A. and Kemp, S.D. 1989. Multiple mutations in HIV-1 reverse transcriptase confer high-level resistance to zidovudine (AZT). Science 246:1155–1158.

Lehman, I.R., 1981. DNA polymerase I of *Escherichia Coli*: In P.D. Boyer (ed.). The Enzymes, 3rd edition. New York: Academic Press, pp. 16–37.

Mullis, K.B. and Faloona, F.A. 1987. Specific synthesis of DNA in vitro via a polymerase-catalyzed chain reaction. Meth. Enzymol. 155:335–350.

Nasim, S., Mizuuchi, H., and Garrett, C.T. 1990. Use of the polymerase chain reaction (PCR) and direct DNA sequencing in the evaluation of genetic and neoplastic disease. Fresenius J. Anal. Chem. 337:120.

Oste, C. 1988. Polymerase chain reaction. Biotechniques 6:162–167.

Oste, C. 1989. Controlling false positives in PCR. Diag. Clin. Testing 27:50.

Paabo, S., Higuchi, R.G., and Wilson, A.C. 1989. Ancient DNA and the polymerase chain reaction. J. Biological Chem. 264: 9709–9712.

Porter-Jordan, K. and Garrett, C.T. 1990. Source of contamination in polymerase chain reaction assay. Lancet 1:1220.

Porter-Jordan, K., Rosenberg, E.L., Keiser, J., Gross, J.D., Ross, A.M., Nasim, S., and Garrett, C.T. 1990. Nested polymerase chain reaction assay for the detection of cytomegalovirus overcomes false positives due to contamination with fragmented DNA. J. Med. Virol. 30:85–91.

Saiki, R.K., Scharf, S., Faloona, F., Mullis, K.B., Horn, G.T., Erlich, H.A., and Arnheim, N. 1985. Enzymatic amplification of betaglobin genomic sequence and restriction site analysis for diagnosis of sickle cell anemia. Science 230:1350–1354.

Sambrook, J., Fritsch, E.F., and Maniatis, T. 1989. Molecular Cloning. A Laboratory Manual, 2nd edition. Cold Spring Harbor, NY: Cold Spring Harbor Laboratory Press, pp. 9.1–9.32, 11.45–11.48.

Schaefer, E. 1991. Biotechnology news: Cetus retains PCR patents. Nature 350:6.

Scharf, S.J., Horn, G.T., and Erlich, H.A. 1986. Direct cloning and sequence analysis of enzymatically amplified genomic sequences. Science 233:272–282.

Shibata, D., Martin, W.J., Appleman, M.D., Causey, D.M., Leedom, J.M., and Arnheim, N. 1988. Detection of cytomegalovirus DNA in peripheral blood of patients infected with human immunodeficiency virus. J. Infect. Dis. 158:1185–1192.

Thomas, R.F. and Newbold, J.E. 1990. Orienting inserts in recombinant DNA molecules by PCR. Amplifications 4:1,3–4.

Vogelstein, B., Fearson, E.R., Hamilton, S.R., Kern, S.E., Preisinger, A.C., Leppert, M., Nakamura, Y., White, R., Smits, A.M.M., and Bos, J.L. 1988. Genetic alterations during colorectal-tumor development. N. Engl. J. Med. 319:525–532.

Section 2

Viral Pathogens

20

Respiratory Viruses

Britt Åkerlind-Stopner and Maurice A. Mufson

20.1 INTRODUCTION

Among all infectious diseases, viral respiratory tract infections account for the predominate morbidity. Viruses belonging to different families representing several genera can infect the respiratory tract primarily and secondarily and produce a spectrum of diseases. Viruses which cause a primary infection of the respiratory tract, including influenza virus, parainfluenza virus (PIV) types 1, 2, 3, and 4, respiratory syncytial virus (RSV), coronavirus, adenovirus, and rhinoviruses, are discussed in this chapter (Table 20.1). Viruses which principally infect other organs, but also infect the respiratory tract include the coxsackie viruses, echoviruses, herpes simplex virus and varicella-zoster virus (Table 20.1).

Viral respiratory tract infections occur in epidemics or endemics worldwide. Influenza virus, PIV, and RSV occur epidemically. Adenoviruses, coronaviruses, and rhinoviruses occur endemically. Persons of all age groups are susceptible to viral respiratory tract infections, although children suffer about twice as many infections as adults (Table 20.2). These infections represent an important cause of a major number

of days lost from work and school, accounting for significant losses both financially and socially. Viruses which infect the respiratory tract are transmitted commonly by direct contact and less often by aerosols.

The severity of illness associated with viral respiratory tract infections ranges from mild upper respiratory tract illness, such as a common cold, to severe and life threatening lower respiratory tract illness, including pneumonia and bronchiolitis (Table 20.3). Virus infection alone or in combination with bacterial and mycoplasmal infections are associated with about 15% of community acquired pneumonias in adults (Figure 20.1) (Mufson et al, 1967). Virus as the sole pathogen, mainly influenza virus, accounts for about 10% of pneumonias in adults. Multiple factors contribute to the severity of illness, including the characteristics of the virus, inoculum size, and host factors, such as age, general health, and immune status. Low socioeconomic status and poor nutritional state also contribute to increasing susceptibility to infection and/or severity of disease.

Diagnosis of the etiology of a viral respiratory tract infection must be accomplished by conducting specific laboratory tests, as etiology cannot be determined

Table 20.1 Viruses Which Cause Respiratory Tract Infections

Family	Genus	Serotypes
Orthomyxoviridae	Influenzavirus	Influenza A, B, (C)
Paramyxoviridae	Paramyxovirus	Parainfluenza types 1, 2, 3, 4A, 4B
Pneumovirus	Respiratory syncytial virus	Subgroups A, B1, B2
Coronaviridae	Coronavirus	Types 229E and OC43; other strains
Adenoviridae	Adenovirus	Serotypes 1–40
Picornaviridae	Enterovirus	Coxsackievirus-A types 1–6, 8, 10, 21–22, 24
		Coxsackievirus-B types 4, 5
		Echovirus types 1–4, 6–9, 11, 16, 19–20, 22, 25
		Enterovirus type 68
	Rhinovirus	102 types classified; some unclassified
Herpesviridae	Herpes simplex virus	Serotypes 1, 2
	Varicella-zoster virus	Single serotype

with any confidence based on symptoms and signs alone (Table 20.4). One or both of two strategies can be employed: (1) direct identification of the virus either by isolation of the virus in cell cultures from nasal wash or oropharyngeal swab specimens, or detection of viral antigens in respiratory secretions and exfoliated cells using solid-phase enzyme immunoassay (ELISA), immunofluorescence (IF), or time-resolved fluoroimmunoassay (TR-FIA); and, (2) detection of a fourfold or greater antibody response during convalescence using any one of a number of serologic assays such as ELISA, complement fixation (CF), neutralization (NT), or hemagglutinin inhibition (HI) (as appropriate) or assays for high levels of IgM-specific antibody during the acute phase of illness (Table 20.4) (Hietala et al, 1988; Bruckova et al, 1989; Hierholzer et al, 1989). The newly-developed rapid diagnostic procedures based on

Table 20.2 Relative Importance of Viruses Among Differing Age Groups

Virus	Age Group*			
	Infants Children	Adolescent	Adult	Aged
Influenza virus	+++	++	++++	++++
Parainfluenza virus	++++	+	+	+
Respiratory syncytial virus	++++	+	+	++
Coronavirus	+++	+	+	−
Adenovirus	+++	++++	+	+
Coxsackie A virus	+	+	+	+
Coxsackie B virus	+	+	+	+
Echovirus	+	+	+	+
Enterovirus	+	+	+	+
Rhinovirus	++++	+++	++++	+++
Herpes simplex virus	++	+	+	+
Varicella-zoster virus	+	+	−	−

* + Minimal to + + + + major importance.
− No or negligible importance.

Table 20.3 Relative Importance of Viruses in Lower and Upper Respiratory Tract Disease

Virus	Site of Respiratory Tract Disease*	
	Upper	Lower
Influenza virus	+ + + +	+ + +
Parainfluenza virus	+ + +	+ + +
Respiratory syncytial virus	+ + +	+ + + +
Coronavirus	+	+
Adenovirus	+ + +	+ + +
Coxsackie A virus	+	−
Coxsackie B virus	+	+
Echovirus	+	+
Enterovirus	+	+
Rhinovirus	+ + + +	−
Herpes simplex virus	+	+
Varicella-zoster virus	−	+

* + Minimal to + + + + major importance.
  − No or negligible importance.

ELISA, TR-FIA, and IF not only provide speed, but also exhibit a high degree of sensitivity and specificity when compared with the "gold standard" of virus isolation

Figure 20.1 Venn diagram of the frequency of infection with viruses, mycoplasma, and bacteria in adults with pneumonia. Overall, about two thirds of adults with pneumonia had evidence of infection with viruses, mycoplasma, and bacteria, alone or in combination, and one third had no organism identified (as determined by isolation of bacteria from sputum and blood, virus isolation from sputum, and antibody responses to viruses and mycoplasma) (Mufson et al, 1967).

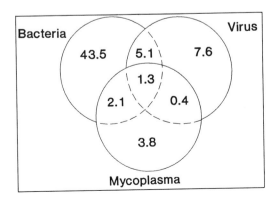

(Kellogg, 1991). A number of commercially available ELISA-based "kits" and monoclonal antibody-based IF can be employed for the rapid diagnosis of RSV or influenza viruses by the virology laboratory, in most instances obviating the need to attempt virus isolation in cell culture. These tests now represent the main approach to rapid diagnosis of viral respiratory tract infections. However, the virology laboratory should periodically run parallel assay with virus isolation as an institutional control for standardization of these two approaches to the diagnosis of virus infections.

Rapid identification of influenza virus and RSV infections can result in the use of specific anti-viral therapy early in the course of the illness. As few effective anti-viral drugs are currently available for the treatment of viral respiratory infections, rapid viral diagnosis can contribute to decisions on inappropriate antibiotic usage and provide information on emerging epidemic patterns.

20.2 INFLUENZA VIRUS

Of all viruses that infect the respiratory tract, influenza viruses cause the predominant number of serious acute respiratory tract illnesses, typically "the flu," and pneumonia. Influenza A virus was isolated in 1933, influenza B virus in 1940, and influenza C virus in 1951 (Smith et al, 1931; Francis, 1940; Taylor, 1951). Influenza A and B viruses belong to the family Orthomyxoviridae and the genus *Influenzavirus* (influenza C virus belongs to a separate genus). The virions are enveloped particles containing a single-stranded, negative sense segmented RNA genome (eights segments in influenza A and B viruses and seven segments in influenza C), which is surrounded by a helical capsid. The enveloped virion contains several structural proteins: the matrix (M) protein, nucleocapsid (NP) protein, and three large proteins that

Table 20.4 Recommended Procedures for Laboratory Diagnosis of Respiratory Viruses

Virus	Virus Isolation (Identification)	Antigen Detection	Antibody Response
Influenza Virus	cell cultures, eggs (TR-FIA, HI, HAdI, IF)	IF, ELISA	CF, HI, HAdI, ELISA-IgM
Parainfluenza Virus	cell cultures (HI, HAdI, IF)	IF, ELISA, TR-FIA	CF, HI, ELISA-IgM
Respiratory Syncytial Virus	cell cultures (TR-FIA, IF, ELISA)	IF, ELISA, TR-FIA	CF, ELISA-IgM
Coronavirus	cell cultures	PEI	CF, ELISA, IF, HI
Adenovirus	cell cultures (TR-FIA, HI, IF, NT)	IF, ELISA, TR-FIA	CF, ELISA, HI, NT
Rhinovirus	cell cultures (NT)	PROBE, ELISA	ELISA, NT

CF = complement fixation
ELISA = enzyme-linked immunosorbent assay
HAdI = hemadsorption-inhibition
HI = hemagglutinin inhibition
IF = immunofluorescence
NT = neutralization test
PEI = polyclonal enzyme immunoassay
PROBE = synthetic oligonucleotides as a probe for viral antigen
TR-FIA = time-resolved fluoroimmunoassay

function in RNA replication and transcription. The outer peplomers or spikes consist of two glycoproteins: the hemagglutinin (H), which is involved in the attachment of virus to cells and the initiation of infection, and the neuraminidase (N) exhibit, which facilitates release of the virus from cells (Wilson and Cox 1990). These two glycoproteins exhibit substantial antigenic variation, especially well documented among influenza A viruses (Table 20.5). The high rate of antigenic variation observed among the H and N glycoproteins of influenza A virus derives from its segmented genome; antigenic drift (due to mutations) reflected in minor variations in the antigenic characteristics of the H and N glycoproteins, and antigenic shift, due to major subgroup shifts of H or N or both, signal the impending development of a pandemic (Table 20.5). Such antigenic variation allows the virus to infect and produce disease among popula-

tions of persons who would otherwise possess immunity (Laver, 1985).

Epidemics caused by influenza viruses account for significant excess morbidity and mortality, epidemiologically speaking, especially among the elderly and persons of any age with pre-existing cardiac and pulmonary conditions. Influenza viruses are responsible for more severe morbidity and

Table 20.5 Predominant Influenza A Strains and Their H and N Subtypes for the Past Thirty Years

Common Name of First Strain	Subtype*	Years of Occurrence
"Spanish Flu"	H1N1	1918–1956
"Asian Flu"	H2N2	1956–1967
"Hong Kong Flu"	H3N2	1968–Present
"Russian Flu"	H1N1	1977–Present

* H = Hemagglutinin
  N = Neuraminidase

mortality than any of the other respiratory viruses (Figure 20.2). None of the other respiratory viruses approach the dominance of influenza viruses as a cause of life-threatening respiratory tract disease.

Influenza viruses comprise three types, A, B, and C. Only influenza A and B viruses are clinically important (influenza C virus infection is associated either with very minor or subclinical illness) (Hay et al, 1991). Influenza A and B viruses produce epidemics during the winter months in temperate climates, on a one- to three-year cycle for influenza A virus and on a four- to seven-year cycle for influenza B virus. Pandemics of influenza A virus occur at approximately ten-year intervals.

The incubation period of influenza virus infection is one to two days. The onset of disease begins abruptly, with sudden high fever, headache, myalgias, rhinitis, tracheobronchitis, pharyngitis, and a cough. The intensity of the signs and symptoms is characteristic of the "flu." In uncomplicated cases, the illness abates in about one week. Primary influenza virus pneumonia can complicate the course of illness, and secondary bacterial infection can develop, especially in elderly persons with pre-existing cardiovascular or pulmonary diseases (a high risk group). This is heralded by the onset of productive cough, with purulent sputum, new fever spikes, chest pain, and shortness of breath. *Streptococcus pneumoniae*, *Staphylococcus*, and *Hemophilus influenzae* are the most common causes of bacterial superinfections. In these cases, appropriate antibiotic treatment must be initiated promptly. Fatalities occur commonly in influenza-pneumonia, especially in persons in the high risk groups.

Diagnosis of influenza virus infection can be made either by isolation of the virus in cell cultures or detection of viral antigen in cells present in respiratory secretions. The specimen should be kept at 4 °C until processed. Influenza viruses do not pro-

duce cytopathic effects (CPE) in susceptible cultures of monkey kidney cells; growth of the virus must be detected by hemadsorption procedures, and isolates can be typed by hemadsorption-inhibition (HAdI) (Table 20.4). Most clinical isolates of influenza virus grow in primary rhesus or cynomologous monkey cell cultures, continuous rhesus kidney (LLC-MK2) cell cultures, or in Madin-Darby canine kidney (MDCK) in about two to five days (Minnich and Ray, 1987). Serotyping isolates using HI can be done the same day or by HAdI, a more sensitive technique, which requires sub-passage of the isolate and an additional two to three days of culture. Rapid diagnostic procedures have replaced virus isolation for confirming the diagnosis of influenza virus infection in the clinical virology laboratory. The common methods involve IF for detection of viral antigen in exfoliated cells of the respiratory tract, and ELISA or TR-FIA for antigen identification in respiratory secretions. Serologic diagnosis and typing of influenza virus can be complicated by the observation that the anamnestic response to infection is highest to the strain causing the primary infection, even when subsequently infected by other strains. This has been termed the "doctrine of original antigenic sin."

An immunologic response with detection of a fourfold or greater rise in serum antibody during convalescence also provides evidence of acute infection with influenza virus. The tests employed for serologic diagnosis are CF, HI, HAdI, and an ELISA-IgM. As these tests require acute-phase and convalescent-phase serum specimens tested together, they have limited use in establishing the diagnosis in individual cases within a time frame that permits therapeutic intervention, but they are useful to track epidemics in the community. IgM-specific and IgA-specific antibodies against influenza virus reach peak levels about 14 days after infection and IgG-spe-

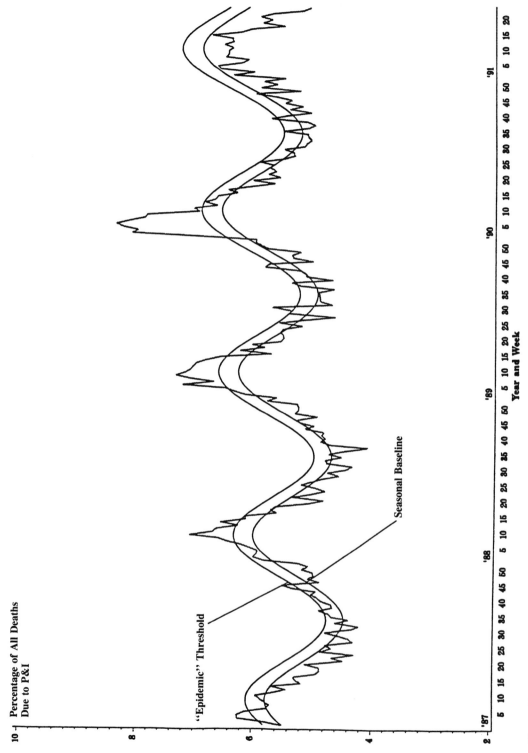

Figure 20.2 Pneumonia (P) and influenza (I) mortality for 121 U.S. cities for 1987 through early 1991. The influenza epidemics of the winters of 1988, 1989, and 1990 exceeded the "epidemic threshold" representing excess mortality from the disease. During 1990–1991, deaths from

Table 20.6A Influenza Vaccine Composition for 1991–1992

Virus	Subtype*	Quantity
A/Taiwan/1/86-like	H1N1	15 micrograms
A/Beijing/353/89	H3N2	15 micrograms
B/Panama/45/90-like	----	15 micrograms

* H = Hemagglutinin
 N = Neuraminidase

cific antibody reaches peak levels about four to seven weeks later.

Influenza virus vaccine is recommended for persons at high risk of serious disease, namely persons 65 years of age or older, persons of any age with underlying cardiovascular and pulmonary diseases, persons with chronic metabolic diseases, residents of nursing homes and chronic care facilities, and children and adolescents receiving chronic aspirin therapy, as well as persons who have the opportunity to transmit influenza virus infection to persons in these high risk groups (Table 20.6) (Heilman and La Montagne, 1990; CDC, 1991). Generally, influenza virus vaccines are at least 70% effective in preventing disease and have a duration of immunity of at least one year.

Amantadine (1-adamantanamine hydrochloride [Symmetrel]) is an effective prophylactic agent for influenza A virus infections only. Since it prevents attachment of the virus to cells, it should be started before the person becomes infected. The dose is one 100 mg tablet twice daily and should be taken for the time necessary

to protect the person from infection. Amantadine has also been used to treat the flu, and apparently shortens the duration of illness slightly (Heilman and La Montagne, 1990; CDC, 1991).

20.3 PARAINFLUENZA VIRUS

Among viruses that cause acute upper and lower respiratory tract diseases in infants and children, PIV rank after RSV as one of the two most important groups; a third important group in this regard is adenoviruses (Figure 20.3) (Belshe et al, 1983). First isolated from humans during 1956, there are five serotypes of PIV: 1, 2, 3, 4A and 4B. PIV belongs to the family and genus Paramyxoviruses and contain a negative single-stranded RNA genome encoding six mRNAs with each mRNA coding for one protein. PIV possess at least two envelope glycoproteins, the hemagglutinin-neuraminidase (HN) and the fusion (F) protein. HN mediates adsorption of the virion to cell receptors and the enzymatic cleavage

Table 20.6B Recommendations for Dosage of Influenza Vaccine for 1991–1992

Age Group	Type of Vaccine[a]	Dosage	Number of Doses	Route of Administration
6–35 months	Split virus only	0.25 mL	1 or 2[b]	Intramuscular
3–8 years	Split virus only	0.50 mL	1 or 2[b]	Intramuscular
9–12 years	Split virus only	0.50 mL	1	Intramuscular
12 years or more	Whole or split virus	0.50 mL	1	Intramuscular

[a] Administer split virus vaccines only to children. Such vaccines also called "subvirion" or "purified-surface-antigen."
[b] Two doses of vaccine recommended for children 9 years or older.
Data from CDC, 1991. Prevention and Control of Influenza. Morbid. Mortal Weekly Rep. 40:1–15.

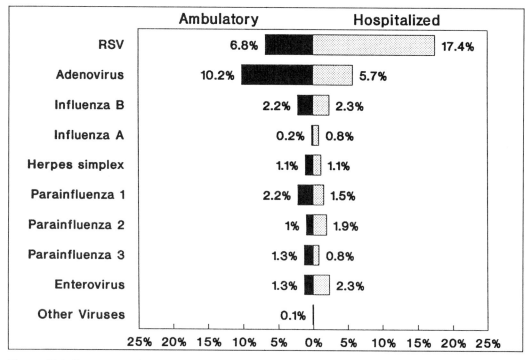

Figure 20.3 Frequency of virus infections among ambulatory and hospitalized infants and children with acute lower respiratory tract disease. Overall, about one fourth of ambulatory and one third of hospitalized infants and children had evidence of virus infection by virus isolation procedures (Belshe et al, 1983).

of sialic acid (neuraminic acid) residues of the virus that must be removed to prevent self-aggregation of virus particles during release from cells. The F glycoprotein facilitates penetration of the virus into cells. PIV infections are transmitted by direct contact and by aerosol droplets.

PIV types 1, 2, and 3 occur worldwide and among persons of all age groups. PIV4A and B are much less prevalent. PIV1 occurs in epidemics during the fall season usually in alternate years, while PIV2 occurs sporadically, and PIV3 occurs yearly in endemic areas. The incubation period is usually between two and six days. PIV3 and, to a lesser extent, PIV2 are the major etiological agents of croup (laryngotracheobronchitis) in infants and young children up to three years of age (Table 20.7). PIV3 causes pneumonia and bronchiolitis in infants and small children and the

incidence of these conditions to this virus is second to RSV. PIV4 occurs infrequently and is usually associated with mild symptoms of upper respiratory tract illness (rhinorrhea, sore throat, and cough). Adults infected with PIV usually experience mild symptoms of upper respiratory tract illnesses and only rarely does the infection progress to pneumonia.

PIV infections of the upper respiratory tract usually present as rhinitis, pharyngitis, and bronchitis, often associated with fever. Rhinorrhea, cough, and pharyngeal erythema are the usual findings. The infection can progress to croup with the typical "bark-like" cough. Although uncommon, cyanosis develops in the presence of marked airway obstruction. Persistent fever and productive cough signal lower respiratory tract involvement. Infants and children with bronchiolitis and pneumonia

Table 20.7 Relation of Parainfluenza Viruses to Respiratory Tract Syndromes in Infants and Children

Respiratory Tract Syndrome	Number in Group	Percent of Illnesses Associated with Indicated Serotype*		
		PIV1	PIV2	PIV3
Pneumonia	169	5.9	4.7	31.9
Bronchiolitis	120	6.7	4.2	23.4
Laryngotracheobronchitis	24	8.3	20.8	41.6

* All patients tested both by virus isolation and serologic response.
Data from Mufson et al, 1970.

manifest wheezing, tachypnea, retractions, and sometimes cyanosis.

Primary infection provides some measure of immunity, but it is neither complete nor long lasting. Immunity correlates well with the amount of secretory IgA present in the mucous membranes of the nose. Reinfections are very common. In one study among children with a primary serotype PIV3 infection, 17% became reinfected within nine months. However, second infections are rarely as severe as the primary infection.

A diagnosis of PIV infection can be made either by isolation of the virus in cell cultures or detection of viral antigens in specimens taken from the respiratory tract (Hierholzer et al, 1989). The virus is fragile and the specimen must be kept at 4 °C until processed. As PIV does not form lytic CPE in cell cultures, growth of the virus is revealed by hemadsorption procedures, and isolates can be typed using HAdI (Table 20.4). The most common methods for rapid diagnosis are IF for detection of viral antigen in exfoliated cells of the respiratory tract and antigen recognition in respiratory secretions using ELISA or TR-FIA (Julkunen, 1984; Hietala et al, 1988). Immunological response with detection of a fourfold or greater rise in serum antibody during convalescence also provides evidence of infection. The tests employed are CF, HI, or NT. Interpretation of serologic tests relative to the PIV type producing

infection may be complicated by the heterotypic antibody response seen in some infected individuals.

20.4 RESPIRATORY SYNCYTIAL VIRUS

Human respiratory syncytial virus, the most important etiological agent of respiratory tract disease in infants and children, was first isolated in 1956 from a symptomatic laboratory chimpanzee during an outbreak of illness resembling the common cold (Morris et al, 1956). Shortly afterward, the virus was recovered from a child with pneumonia and another child with croup and shown to be a human pathogen (Chanock and Frinberg, 1957; Chanock et al, 1957).

Until recently, RSV was classified in the genus Paramyxovirus. However, the virus lacks both a hemagglutinin and neuraminidase (Richman et al, 1971) and its nucleocapsid diameter is small (12 to 15 nm compared to 18 nm for the PIV) (Joncas et al, 1969). Therefore, RSV is now classified in a separate genus, Pneumovirus. RSV is an enveloped virus with a single-stranded, negative polarity, non-segmented RNA genome, which encodes for 10 distinct mRNAs (Collins and Wertz, 1983; Venkatesan et al, 1983; Collins et al, 1984, 1985; Sateke and Venkatesan 1984; Sateke et al, 1984) with each mRNA coding for a single

Table 20.8 Differentiation of Respiratory Syncytial Virus Subgroups on the Basis of Reactivity with Monoclonal Antibodies to the G Protein

Subgroup Specificity of Monoclonal Antibody	Epitope	Name of Monoclonal Antibody	Reaction with Indicated Subgroup of Respiratory Syncytial Virus*		
			A	B1	B2
A	G1	C793	+	+	+
	G2-6	B14, B18, B23, B109, B158, B17, B25, B119	+	−	−
B	G1	8188, 8305, 9177, 9273	−	+	+
	G2	8180, 9244, 9637, 9794	−	−	+

* + = Positive and − = Negative by ELISA, IF, and RIPA.
Data from Åkerlind et al, 1988.

protein. At least two virion proteins are structural enveloped proteins. The G protein, a heavily glycosylated glycoprotein (Wertz et al, 1985), is believed to mediate attachment to the cell (Belshe and Mufson, 1991). The fusion protein promotes membrane fusion. The application of monoclonal antibody technology to antigenic analysis of RSV permitted the recognition of two subgroups, A and B, based on their reactivity with a panel of monoclonal antibodies (Anderson et al, 1985; Mufson et al, 1985; Gimenez et al, 1986). The major difference between strains of subgroups A and B was found in the G glycoprotein. Heterogeneity also exists within the A and B subgroups (Åkerlind et al, 1988) (Table 20.8).

Epidemics of RSV occur yearly during the winter in temperate climates. The two subgroups can co-circulate in the same geographic area simultaneously (Åkerlind and Norrby, 1986) usually with a predominance of subgroup A (Mufson et al, 1991) (Figure 20.4). In the northern temperate zones the peak of the epidemic is usually January to April and in the tropics it is during the rainy seasons. The virus causes disease mainly among infants and children, with infants between 6 weeks and 6 months of age being the ones most often affected (Table 20.9). Among adults, RSV infections are less

common and result in mild upper respiratory illnesses. Elderly persons are a high risk group for developing serious RSV infections, especially among institutionalized adults. Although the rate of occurrence of serious RSV infections in adults is not as high as in infants and children, such RSV infections can be fatal in the elderly (Hart 1984; Falsey, 1990; Agius et al, 1990).

RSV infections are transmitted during close contact. The incubation period is about four to five days. No major clinical differences exist between illnesses caused by subgroup A and subgroup B strains (Table 20.9). RSV causes upper and lower respiratory tract illnesses. When the lower respiratory tract becomes involved (about 25% to 40% of infections), the patient can develop a cough, fever, sneezing, wheezing, hyperventilation, rhonchi, fine rales, and otitis media. The chest roentgenogram is usually normal, but often reveals hyperinflation and pulmonary infiltrates in more severe cases. Dyspnea, hyperexpansion, and tachypnea can develop, and apnea may occur. Often children with laryngotracheobronchitis, bronchiolitis, or pneumonia require hospitalization due to the need for intubation and ventilation. In an uncomplicated RSV infection, recovery occurs after 7 to 12 days. Generally, the mortality is less than 2% among children admitted to hospi-

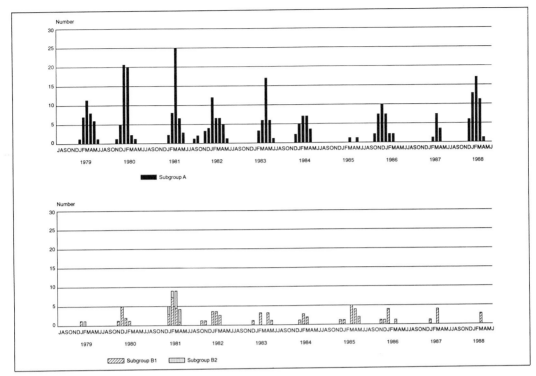

Figure 20.4 Temporal distribution of RSV subgroups A, B1, and B2 from July 1978 through June 1988 in Huntington, WV. RSV epidemics occur each winter in temperate climates starting early (November or December) and late (January or February) in alternating years (Data from Mufson et al, 1991).

tals due to the benefit of modern intensive care. Nevertheless, when a child has underlying cyanotic congenital heart disease the case fatality rate may be as high as 37%. Other pre-existing conditions that are associated with increased case fatality rates include bronchopulmonary disease, renal disease, and immunosuppression. RSV has been isolated from children who succumbed to sudden infant death syndrome (SIDS), but the role of RSV, if any, in these cases is not clear.

Table 20.9 Relation of Subgroup of Respiratory Syncytial Virus to Acute Respiratory Tract Illness in Infants and Children

Virus	Number (%) of Illnesses Associated with Indicated Subgroup*		
	A	B1	B2
Pneumonia	127 (39.8)	25 (34.2)	6 (46.1)
Bronchiolitis	97 (30.4)	11 (15.1)	2 (15.4)
Laryngotracheobronchitis	30 (9.4)	12 (6.4)	2 (15.4)
Upper respiratory tract illness or bronchitis	65 (20.4)	25 (34.3)	3 (23.1)

* Significantly fewer cases of bronchiolitis were associated with subgroup B strains (chi-square = 14.41, df = 6, P = 0.025). Data from Mufson et al, 1991.

Reinfections with RSV are common. Among school-age children the risk for RSV infection during each epidemic is about 20% to 40%. First infection with subgroup A strains is shown to provide some protection from a second infection with the homologous, but not the heterologous, subgroup of the virus (Mufson et al, 1987).

The diagnosis of RSV can be made either by isolation of the virus in cell cultures, by direct examination of respiratory secretions using either IF, ELISA, or TR-FIA, or by demonstration of a fourfold rise in antibody comparing acute and convalescent sera (Hughes et al, 1988; Masters et al, 1988; Waris et al, 1988; Kellogg, 1991). However, isolation of RSV in cell culture or direct detection of viral antigen is preferred to antibody detection for diagnosis of infection, since it yields information sooner. Since RSV replicates in the respiratory epithelia of the nasopharynx; nasopharyngeal swab or nasal wash specimens are the best source for virus. However, as RSV is very labile, specimens should be transported on wet ice to the laboratory and kept at 4 °C until processed. Freeze-thawing can cause substantial loss of infectivity. HEp-2 cells are the most commonly employed cell culture for isolation of RSV. The virus produces multinucleated giant cells, syncytia, in these cultures. The formation of CPE indicative of virus growth in tissue cultures usually takes about five to nine days in HEp-2 cells (or other susceptible cell cultures).

A number of commercial rapid diagnostic enzyme immunoassay (EIA) kits are available which identify RSV antigen in respiratory secretions. Generally, these EIA kits provide a high degree of sensitivity and specificity, although this varies somewhat with the individual EIA kit. The results are available in minutes to hours. TR-FIA procedures identify RSV antigen in nasopharyngeal secretions after only a one hour incubation at 37 °C.

Antibody response during convalescence detected using CF, NT, or ELISA methods also can be employed to diagnose an RSV infection. However, since these procedures require a serum specimen obtained 18 to 21 days after the onset of illness, this approach is more likely to be used in epidemiologic and research studies. Moreover, in children four to nine months of age, maternal antibodies are present and may complicate interpretation of the data.

Rapid diagnosis of RSV infection is important because of the availability of ribavirin (Virazole), a synthetic nucleoside, for the treatment of infants with serious RSV illness or infants with underlying diseases that place them at high risk for severe RSV illness. Ribavirin administered as a continuous aerosol at 20 mg/mL for 18 to 20 hours each day during the first several days of illness can hasten recovery.

20.5 ADENOVIRUS

Adenoviruses were isolated in 1953 (Rowe et al, 1953) from primary cell cultures of adenoids from children. Currently, 47 human adenovirus immunotypes have been identified that can be subdivided into six subgenera, designated subgroups A through F, based on differing classification schemes, including hemagglutination characteristics or percent of G+C content in their DNA (Table 20.10). Adenovirus particles contain a linear, double-stranded DNA-molecule, surrounded by an icosahedral capsid. They are non-enveloped. Adenovirus is not affected by the low pH in the stomach, the bile from the gallbladder or by the proteolytic enzymes secreted by the pancreas. They replicate well in the gastrointestinal tract.

Adenovirus infections occur worldwide. The transmission of adenovirus infection varies from sporadic to epidemic. Epidemics usually occur among children and among military recruits. The incuba-

Table 20.10 Classification of Adenoviruses According to Hemagglutination Characteristics and T Antigen Group

Subgenus	Serotypes[a]	Hemagglutination Pattern with Indicated Erythrocytes[b]		G+C DNA Percent
		Rhesus	Rat	
A	12, 18, 31	–	Poor or None	48
B	3, 7, 11, 14, 16, 21, 34, 35	Complete	–	51
C	1, 2, 5, 6	–	Partial[c]	58
D	8–10, 13, 15, 17, 19, 20, 22–30, 32, 33, 36–39, 42		Complete	58
E	4	–	Partial[c]	58
F	40, 41	–	Partial	52

[a] Candidate types 43–47 (subgenus D) not sufficiently characterized.

[b] – = Negative.

[c] Hemagglutination test requires adding a heterotypic antiserum to the diluent (type 4 antiserum for subgenus C and type 6 antiserum for subgenus E).

Hierholzer et al, 1988.

tion period ranges from five to eight days. The fecal-oral route, and to a lesser extent aerosols, account for the spread of infection in infants and children. Since adenoviruses are very stable, they can be transmitted in water and by fomites, for example ophthalmological instruments. The initial site of infection can be the conjunctivae, oropharynx, or the intestine, and subsequently, the infection spreads to the regional adenoid tissues. Although rare, adenovirus can be isolated from the lungs, liver, kidneys, and brain of fatal cases.

Adenoviruses are an important cause of acute upper and lower respiratory tract illnesses in infants and children (Figure 20.3) (Belshe, 1983). In industrialized countries, adenoviruses cause 2% to 5% of all acute respiratory tract diseases. Types 1, 2, 3, 5, and 7 account for seven eighths of adenovirus infections. Many military recruits who acquire adenovirus infection develop either upper respiratory disease, pharyngitis, or pneumonia.

Subgenus B contains types 3, 7, 11, 14, 16, 21, 34, and 35 (Table 20.10). Of the types in this subgenus, 3, 7, 14, and 21 cause epidemics of acute respiratory tract infection among children and military recruits (Table 20.11). Type 7 causes pneumonia among young children, with the potential for permanent lung damage. Sub-

Table 20.11 Common Adenovirus Serotypes Among Children, Adolescents, and Adults Associated with Specific Respiratory Diseases

Respiratory Tract Syndrome	Serotypes Associated with Illness in[a]		
	Children	Adolescents[b]	Adults
Pneumonia	1–3, 7	3, 4, 7	Rare
Acute Upper Respiratory Tract Disease	1–3, 5–7	3, 4, 7, 14, 21	3, 7
Pharyngoconjunctival Fever	3, 4, 7, 14	3, 4, 7, 14	3, 4, 7, 14
Epidemic Keratoconjunctivitis	–	–	8, 19, 37

[a] The most common types are shown; other types may be associated with these syndromes very infrequently or on a sporadic basis.

[b] Adolescents usually represented by military recruits.

genus C includes types 1, 2, 5, and 6, which are associated with respiratory tract diseases and fever among children younger than 5 years old. Subgenus D contains the largest number of serotypes, including several new candidate types (Hierholzer et al, 1988) (Table 20.10). Subgenus E, comprising only type 4, also occurs predominantly among military recruits, where it causes acute upper respiratory tract disease, pharyngitis, and pneumonia (Table 20.11) (Ylikoski and Karajalainen, 1989). Subgenus A and B have been associated with tumors in small laboratory animals, but not in humans. Other acute diseases caused by other adenoviruses include gastroenteritis, conjunctivitis, epidemic keratoconjunctivitis (especially type 8), acute hemorrhagic cystitis, pertussis-like syndrome, hepatitis, and latent infections in adenoid tissue.

Adenoviruses can be recovered from the conjunctivae, oropharynx, feces, or urine. The virus can be presumptively identified in cell culture by its characteristic CPE that develops two to five days after inoculation of susceptible tissue cultures. Immunologic verification can be made using CF or using IF and the strains serotyped using HI, NT, IF, or by DNA restriction analysis. Adenoviruses can be detected directly in respiratory secretions using TR-FIA, ELISA, and IF (Hietala et al, 1988; Bruckova et al, 1989). Type specific antibody responses can be detected by HI (for hemagglutinating strains), ELISA, and NT. A "family" antibody response to all adenoviruses can be detected by CF.

Live attenuated vaccines for types 4 and 7, consisting of these strains in an enteric coated capsule, are only recommended for and routinely administered to military recruits (Chaloner-Larsson et al, 1986). The vaccine causes a subclinical intestinal infection, resulting in the production of local antibody both in the respiratory and intestinal tracts and results in a significant reduction in illness among exposed recruits.

20.6 CORONAVIRUS

Coronaviruses, which comprise the single genus of the family Coronaviridae, were isolated and characterized in the early 1960s (Tyrrell and Bynoe, 1965; Hamre and Procknow, 1966). They are enveloped, pleomorphic particles 80 to 120 nm in diameter, with a single-stranded, positive-sense RNA genome. The prototype strains of coronavirus consist of three subgroups: (1) 229E and LP, (2) OC43 and OC38, and (3) B814 (Mufson, 1991). Purified virions of prototype strains 229E and OC43 possess four major proteins: two surface peplomeric (P) glycoproteins, 190Kd and 130Kd (the latter corresponds to hemagglutinin (H) in the hemagglutinating strains), nucleoprotein (N), and a membrane (M) protein (Jouvenne et al, 1990).

Coronavirus infections occur worldwide, usually appearing sporadically throughout the winter and spring. The virus is spread by large droplets via the respiratory route. Among children, coronavirus infection occurs commonly, one infection per child per year, which is about three times as often as in adults. These infections are associated with about 5% to 15% of upper respiratory tract illnesses. Coronaviruses cause mainly afebrile upper respiratory tract illnesses. They are infrequently associated with febrile illness and pneumonia, and rarely with otitis media (Arola et al, 1990a). The clinical features of upper respiratory tract illness consist of coryza, rhinorrhea, and nasal congestion, sore throat, and pharyngeal edema. Fever, headache, and cough are uncommon in mild upper respiratory tract illness. The illness lasts about one week.

The symptoms and signs of pneumonia associated with coronavirus infection in infants and children include fever, cough, restlessness, anorexia, rales, and pulmonary infiltrates consistent with a picture of viral pneumonia (Mufson, 1991). In adults, the clinical findings of coronavirus pneumonia are of a viral pneumonia which is clinically indistinguishable from other viral pneumonias.

Type specific antibody, usually only to a single prototype, is produced in response to infection and persists for long periods. Circulating and mucosal antibodies confer protection from infection and illness. In volunteers experimentally infected with prototype 229E, however, secretory IgA alone correlated with a shortened duration of virus shedding. Most adults possess circulating antibody to one or more prototype coronavirus strains. Nonetheless, reinfections are common. The presence of or change in antibody titer to one prototype was apparently not influenced by the level of antibody to the other prototype.

The laboratory diagnosis of coronavirus infection depends upon the detection of a fourfold or greater antibody rise during convalescence. The tests employed include ELISA, NT, CF, radioimmunoprecipitation, immunodiffusion, IF, and polyclonal enzyme immunoassay (PEI) (Arola et al, 1990a; Callow et al, 1990). Antibody to strain OC43 (and other hemagglutinating strains) can be assayed using HI. The assays of choice for routine diagnostic measurement of antibody response are ELISA, CF, NT, and HI (for hemagglutinating strains). Virus isolation, however, remains a research procedure. Coronavirus strain 229E grows in a narrow range of cell cultures. It can be isolated from respiratory secretions, for example, in the C-16 line of MRC-C cells (Callow et al, 1990).

20.7 RHINOVIRUS

Rhinoviruses are the "common cold virus." They cause more common cold (minor upper respiratory tract) illnesses than any other virus that infects the respiratory tract. Prevalent in all age groups, rhinoviruses account for one fourth to one half of upper respiratory tract infections. Morbidity from rhinovirus infections accounts for substantial absenteeism from schools and work, which is reflected in a considerable financial loss. The lost days annually amount to about 26 million days from schools and about 23 million from work. In the past few years, rhinoviruses have been associated with some cases of acute otitis media in children and with pneumonia in infants and children with a pre-existing underlying disease (Krilov et al, 1986; Arola et al, 1990b).

The search for a "common cold" virus led to the isolation of the first rhinovirus strains in the late 1950s (Pelon et al, 1957). Subsequently, during the next three decades, many studies revealed that rhinoviruses comprise a large group of antigenically distinct serotypes, of which 102 serotypes have been characterized and assigned numbers (Colonno et al, 1986; Hamparian et al, 1987). Rhinoviruses are small, naked viruses, containing a single molecule of RNA that has positive-sense polarity. Rhinoviruses, unlike the enteroviruses, lose infectivity upon exposure to mild acids, accounting for their failure to infect the gut. Rhinoviruses can be grouped into two receptor families, designated major and minor, based on utilization of cell attachment surface receptors. The major group, comprising 91 of 102 serotypes, utilizes intercellular adhesion molecule 1 (ICAM-1), a cell surface ligand (Staunton et al, 1989; Uncapher et al, 1991). The minor group of 10 serotypes utilizes a 120 Kd protein as yet not fully characterized. Serotype 87 alone requires sialic acid for attachment.

Rhinovirus infections occur year round, with peaks in the early spring and fall months in the temperate zones, and during the rainy period in the tropics. Rhinoviruses are spread person to person by direct contact with ill persons whose hands and fingers have become contaminated with infectious secretions and by aerosol.

Rhinoviruses replicate in the cells of mucous membranes of the nose. A direct correlation exists between the quantity of virus in secretions and the occurrence and severity of illness. Peak concentrations of virus are shed on the second and third day of the illness. As optimal replication of rhinoviruses takes place at 33 °C to 34 °C, the concentration of rhinovirus in the nose is 10 to 100 times higher than in pharyngeal secretions.

The usual incubation period of rhinovirus illness is one to four days, although it may be as long as seven days. Symptoms of the common cold illness include rhinorrhea, nasal obstruction, sneezing, sore throat, cough, and headache. Fever is uncommon. The illness usually lasts about one week. Secondary complications due to bacterial superinfection, e.g., sinusitis, rarely occur. Rhinoviruses cause otitis media in infants and children and these viruses, in the absence of other agents, can be recovered from middle ear fluid of about one in ten unselected otitis media illnesses (Arola et al, 1988). Rhinoviruses may be associated with acute lower respiratory tract diseases, in the absence of bacterial and fungal pathogens, in otherwise healthy infants under 2 months of age, as well as infants and children suffering from severe underlying diseases, such as malignancy, congenital heart disease, and respiratory tract abnormalities, (Krilov et al, 1986). Most of these children have a cough, fever, and wheeze; a chest x-ray reveals focal abnormalities.

Type-specific immunity develops following infection, characterized by IgG neutralizing antibodies in serum and secretions. However, long lasting immunity best correlates to the level and secretion of IgA antibodies from the nasal mucous membranes. Patients who recently have experienced a common cold caused by rhinovirus usually will not get infected by another rhinovirus within at least four weeks due to viral interference and local production of interferon.

Several novel assays have been developed recently for the rapid diagnosis of rhinovirus infection. These assays have been tested mainly using specimens from volunteers experimentally infected with rhinoviruses. However, their use in routine diagnosis of rhinovirus infection is moot, at least until safe and effective anti-viral therapy for rhinovirus infection becomes available. However, accurate diagnosis of rhinovirus should decrease the administration of unneeded antibiotics. At present, the diagnosis of rhinovirus illness is based on clinical presentation. However, when clinical findings dictate the need for a specific diagnosis of rhinovirus infection, new sensitive and specific assays for testing clinical samples including synthetic oligonucleotides as probes in a filter hybridization test, DNA probes with polymerase chain reaction amplification, or ELISA using a rabbit anti-rhinovirus hyperimmune serum as both the capture and test antibody (Bruce et al, 1989; 1990) could be utilized. Virus isolation in cell culture from nasopharyngeal or oropharyngeal swab specimens requires several days, but it is very sensitive especially when specimens are collected during the first few days of illness. Rhinoviruses grow well in fetal human fibroblast cells and primary human embryonic kidney cells. Organ cultures of nasal mucosa are especially sensitive to rhinoviruses, but this remains a research procedure.

Specific antibody to rhinovirus can be measured by macro and micro NT, plaque-reduction tests, and an ELISA using antibody-capture of rhinovirus antigen to detect IgA antibodies (Barclay and Al-Nakib, 1987). None of these tests are employed for routine diagnosis at this time.

REFERENCES

Agius, G., Dindinaud, G., Biggar, R.J., Peyre, R., Vaillant, V., Ranger, S., Poupet, J.Y., Cisse, M.F., and Castets, M. 1990. An epidemic of respiratory syncytial virus in elderly people: Clinical and serological findings. J. Med. Virol. 30:117–127.

Åkerlind, B. and Norrby, E. 1986. Occurrence of respiratory syncytial virus subtypes A and B strains in Sweden. J. Med. Virol. 19:241–247.

Åkerlind, B., Norrby, E., Örvell, C., and Mufson, M.A. 1988. Respiratory syncytial virus: Heterogeneity of subgroup B strains. J. Gen. Virol. 69:2145–2154.

Anderson, J., Hierholzer, J.C., Tsou, C., Hendry, R.M., Fernie, B.F., Stone, Y., and McIntosh, K. 1985. Antigenic characterization of respiratory syncytial virus strains with monoclonal antibodies. J. Infect. Dis. 151:626–633.

Arola, M., Ziegler, T., Ruuskanen, O., Mertsola, J., Nanto-Salonen, K., and Halonen, P. 1988. Rhinovirus in acute otitis media. J. Peds. 113:693–695.

Arola, M., Ruuskanen, O., Ziegler, T., Mertsola, J., Nanto-Salonen, K., Putto-Laurila, A., Viljanen, M.K., and Halonen, P. 1990a. Clinical role of respiratory virus infection in acute otitis media. Pediatrics 86:848–55.

Arola, M., Ziegler, T., Puhakka, H., Lehtonen, O.P., and Ruuskanen, O. 1990b. Rhinovirus in otitis media with effusion. Ann. Otol. Rhinol. Laryngol. 99:451–453.

Barclay, W.S. and Al-Nakib, W. 1987. An ELISA for the detection of rhinovirus specific antibody in serum and nasal secretion. J. Virol. Methods 15:53–64.

Belshe, R.B., Van Voris, L.P., and Mufson, M.A. 1983. Impact of viral respiratory diseases on infants and young children in a rural and urban area of southern West Virginia. Am. J. Epidemiol. 117:467–474.

Belshe, R.B. and Mufson, M.A. 1991. Respiratory syncytial virus. In R.B. Belshe (ed.), Textbook of Human Virology, 2nd edition. New York: Mosby-Year Book, Inc., pp. 388–407.

Bruce, C.B., Al-Nakib, W., Almond, J.W., and Tyrrell, D.A.J. 1989. Use of synthetic oligonucleotide probes to detect rhinovirus RNA. Arch. Virol. 105:179–187.

Bruce, C., Chadwick, P., and Al-Nakib, W. 1990. Detection of rhinovirus RNA in nasal epithelial cells by in situ hybridization. J. Virol. Methods 30:115–126.

Bruckova, M., Grandien, M., Pettersson, C.-A., and Kunzova, L. 1989. Use of nasal and pharyngeal swabs for rapid detection of respiratory syncytial virus and adenovirus antigens by enzyme-linked immunosorbent assay. J. Clin. Microbiol. 27:1867–1869.

Callow, K.A., Parry, H.F., Sergeant, M., and Tyrrell, D.A.J. 1990. The time course of the immune response to experimental coronavirus infection of man. Epidemiol. Infect. 105:435–446.

Chaloner-Larsson, G., Contreras, G., Furesz, J., Boucher, D.W., Krepps, D., Humphreys, G.R., and Mohanna, S.M. 1986. Immunization of Canadian Armed Forces personnel with live types 4 and 7 adenovirus vaccines. Can. J. Public Health 77:367–370.

Chanock, R.M. and Frinberg, L. 1957. Recovery from infants with respiratory illness of a virus related to chimpanzee coryza agent (CCA). II. Epidemiological aspects of infection in infants and young children. Am. J. Hyg. 66:291–300.

Chanock, R.M., Roizman, B., and Myers, R. 1957. Recovery from infants with respiratory illness of a virus related to chimpanzee coryza agent. I. Isolation, properties and characterization. Am. J. Hyg. 66:281–290.

Collins, P.L. and Wertz, G.W. 1983. cDNA cloning and transcriptional mapping of nine polyadenylated RNAs encoded by the genome of human respiratory syncytial virus. Proc. Natl. Acad. Sci. USA 80:3208–3212.

Collins, P.L., Huang, Y.T., and Wertz, G.W. 1984. Identification of a tenth mRNA of respiratory syncytial virus and assignment of polypeptides in the 10 viral genes. J. Virol. 49:572–578.

Collins, P.L., Anderson, K., Langer, S.J., and Wertz, G.W. 1985. Correct sequence of the major nucleocapsid protein mRNA of respiratory syncytial virus. Virology 146:69–77.

Colonno, R.J., Callahan, P.L., and Long, W.J. 1986. Isolation of a monoclonal antibody that blocks attachment of the major group of human rhinoviruses. J. Virol. 57:7–12.

Centers for Disease Control (CDC). 1991. Prevention and Control of Influenza: Recommendations of the Immunization Practices Advisory Committee (ACIP). Morbid. Mortal. Weekly Rep. 40:1–15.

Falsey, A.R. 1990. Serologic evidence of respiratory syncytial virus infection in nursing home patients. J. Infect. Dis. 162:568–569.

Francis, T. Jr. 1940. A new type of virus from epidemic influenza. Science 92:405–408.

Gimenez, H.B., Hardman, N., Keir, H.M., and Cash, P. 1986. Antigenic variation between human respiratory syncytial virus isolates. J. Gen. Virol. 67:863–870.

Hamparian, V.V., Colonno, R.J., Cooney, M.K., Dick, E.C., Gwaltney, J.M. Jr., Hughes, J.H., Jordan, W.S. Jr., Kapikian, A.Z., Mogabgab, W.J., Monto, A., Phillips, C.A., Rueckert, R.R., Schieble, J.H., Stott, E.J., and Tyrrell, D.A.J. 1987. A collaborative report: Rhinoviruses—extension of the numbering system from 89 to 100. Virology 159:191–192.

Hamre, D. and Procknow, J.J. 1966. A new virus isolated from the human respiratory tract. Proc. Soc. Exp. Biol. Med. 121:190–193.

Hart, R.J.C. 1984. An outbreak of respiratory syncytial virus infection in an old people's home. J. Infect. Dis. 8:259–261.

Hay, A.J., Belshe, R.B., Anderson, E.L., Gorse, G.J., and Westblom, T.U. 1991. Influenza Viruses. In R.B. Belshe, (ed.), Textbook of Human Virology, 2nd edition. New York: Mosby-Year Book, Inc., pp. 307–341.

Heilman, C. and La Montagne, J.R. 1990. Influenza: Status and prospects for its prevention, therapy, and control. Ped. Clin. N. Am. 37:669–688.

Hierholzer, J.C., Wigand, R., Anderson, L.J., Adrian, T., and Gold J.W. 1988. Adenoviruses from patients with AIDS: A plethora of serotypes and a description of five new

serotypes of subgenus D (types 43–47). J. Infect. Dis. 158:804–813.

Hierholzer, J.C., Bingham, P.G., Coombs, R.A., Johansson, K.H., Anderson, L.J., and Halonen, P.E. 1989. Comparison of monoclonal antibody time-resolved fluoroimmunoassay with monoclonal antibody capture-biotinylated detector enzyme immunoassay for respiratory syncytial virus and parainfluenza virus antigen detection. J. Clin. Microbiol. 27:1243–1249.

Hietala, J., Uhari, M., and Tuokko, H. 1988. Antigen detection in the diagnosis of viral infections. Scand. J. Infect. Dis. 20:595–599.

Hughes, J.H., Mann, D.R., and Hamparian, V.V. 1988. Detection of respiratory syncytial virus in clinical specimens by viral culture, direct and indirect immunofluorescence, and enzyme immunoassay. J. Clin. Microbiol. 26:588–591.

Joncas, J., Berthiaume, L., and Pavilanis, V. 1969. The structure of the respiratory syncytial virus. Virology 38:493–496.

Jouvenne, P., Richardson, C.D., Schrieber, S.S., Lai, M.M., and Talbot, P.J. 1990. Sequence analysis of the membrane protein of human coronavirus 229E. Virology 174:608–612.

Julkunen, I. 1984. Serological diagnosis of parainfluenza virus infections by enzyme immunoassay with special emphasis on purity of viral antigens. J. Med. Virol. 14:177–187.

Kellogg, J.A. 1991. Culture vs direct antigen assays for detection of microbial pathogens from lower respiratory tract specimens suspected of containing respiratory syncytial virus. Arch. Pathol. Lab. Med. 115:451–458.

Krilov, L., Pierik, L., Keller, E., Mahan, K., Watson, D., Hirsch, M., Hamparian, V., and McIntosh, K. 1986. The association of rhinoviruses with lower respiratory tract disease in hospitalized patients. J. Med. Virol. 19:345–352.

Laver, W.G. 1985. Immunochemistry of variants of influenza virus hemagglutinin and neuraminidase. Adv. Exp. Med. Biol. 185:149–174.

Masters, H.B., Bate, B.J., Wren, C., and Lauer,

B.A. 1988. Detection of respiratory syncytial virus antigen in nasopharyngeal secretions by Abbott diagnostics enzyme immunoassay. J. Clin. Microbiol. 26:1103–1105.

Minnich, L.L. and Ray, C.G. 1987. Early testing of cell cultures for detection of hemadsorbing viruses. J. Clin. Microbiol. 25:421–422.

Morris, J.A. Jr., Blount, R.E., and Savage, R.E. 1956. Recovery of cytopathogenic agent from chimpanzees with coryza. Proc. Soc. Exp. Biol. Med. 92:544–550.

Mufson, M.A. 1991. Coronavirus. In R.B. Belshe (ed.), Textbook of Human Virology, 2nd edition. New York: Mosby-Year Book, Inc., pp. 408–411.

Mufson, M.A., Chang, V., Gill, V., Wood, S.C., Romansky, M.J., and Chanock, R.M. 1967. The role of viruses, mycoplasmas, and bacteria in acute pneumonia in civilian adults. Am. J. Epidemiol. 86:526–544.

Mufson, M.A., Krause, H.E. Mocega, H.E., and Dawson, F.W. 1970. Viruses, *Mycoplasma pneumoniae* and bacteria associated with lower respiratory tract disease among infants. Am. J. Epidemiol. 91:192–202.

Mufson, M.A., Örvell, C., Rafnar, B., and Norrby, E. 1985. Two distinct subtypes of human respiratory syncytial virus. J. Gen. Virol. 66:2111–2124.

Mufson, M.A., Belshe, R.B., Örvell, C., and Norrby, E. 1987. Subgroup characteristics of respiratory syncytial virus strains recovered from children with two consecutive infections. J. Clin. Microbiol. 25:1535–1539.

Mufson, M.A., Åkerlind-Stopner, B., Örvell, C., Belshe, R.B., and Norrby, E. 1991. A single-season epidemic with respiratory syncytial virus subgroup B2 during 10 epidemic years, 1978 to 1988. J. Clin. Microbiol. 29:162–165.

Pelon, W., Mogabgab, W.J., Phillips, I.A., and Pierce, W.E. 1957. A cytopathogenic agent isolated from naval recruits with mild respiratory illness. Proc. Soc. Exp. Biol. Med. 94:262–267.

Richman, A.V., Pedreira, F.A., and Tauraso, N.M. 1971. Attempts to demonstrate hemagglutination and hemadsorption by respiratory syncytial virus. Appl. Microbiol. 21:1099–1100.

Rowe, W.P., Huebner, R.J., Gilmore, L.K., Parrot, R.H., and Ward, T.G. 1953. Isolation of a cytopathogenic agent from human adenoids undergoing spontaneous degeneration in tissue culture. Proc. Soc. Exp. Biol. Med. 84:570–573.

Sateke, M. and Venkatesan, S. 1984. Nucleotide sequence on the gene encoding respiratory syncytial virus matrix protein. J. Virol. 50:92–99.

Sateke, M., Elango, N., and Venkatesan, S. 1984. Sequence analysis of respiratory syncytial virus phosphoprotein gene. J. Virol. 52:991–994.

Smith, W., Andrewes, C.H., and Laidlaw, P.P. 1933. A virus obtained from influenza patients. Lancet 2:66–68.

Staunton, D.E., Merluzzi, V.J., Rothlein, R., Barton, R., Marlin, S.D., and Springer, T.A. 1989. A cell adhesion molecule, ICAM-1, is the major surface receptor for rhinoviruses. Cell 56:849–853.

Taylor, R.M. 1951. A further note on 1233 ("influenza C") virus. Arch. Gesamte. Virusforsch. 4:485–500.

Tyrrell, D.A.J. and Bynoe, M.L. 1965. Cultivation of a novel type of common cold virus in organ cultures. Br. Med. J. 1:1467–1470.

Uncapher, C.R., DeWitt, C.M., and Colonno, R.J. 1991. The major and minor group receptor families contain all but one human rhinovirus serotype. Virology 180:814–817.

Venkatesan, S., Elgano, N., and Chanock, R.M. 1983. Construction and characterization of cDNA clones for four respiratory syncytial virus genes. Proc. Natl. Acad. Sci. USA 80:1280–1284.

Waris, M., Halonen, P., Ziegler, T., Nikkari, S., and Obert, G. 1988. Time-resolved fluoroimmunoassay compared with virus isolation for rapid detection of respiratory syncytial virus in nasopharyngeal aspirates. J. Clin. Microbiol. 26:2581–2585.

Wertz, G.W., Collins, P.L., Huang, Y., Gruber, C., Levine, S., and Ball, L.A. 1985. Nucleotide sequence of the G protein gene of human respiratory syncytial virus reveals an unusual type of viral membrane protein. Proc. Natl. Acad. Sci. USA 82:4075–4079.

Wilson, I.A. and Cox, N.J. 1990. Structural basis of immune recognition of influenza virus hemagglutinin. Ann. Rev. Immunol. 8:737–771.

Ylikoski, J. and Karajalainen, J. 1989. Acute tonsillitis in young men: Etiological agents and their differentiation. Scand. J. Infect. Dis. 21:169–174.

21

Enteroviruses Including Hepatitis A Virus

Heinz Zeichhardt

21.1 INTRODUCTION

Enteroviruses comprise one genus in the *Picornaviridae* family. This family additionally contains the following genera: cardioviruses, rhinoviruses, and aphthoviruses (for classification and nomenclature of the viruses cf. Murphy and Kingsbury, 1990). The members of the enterovirus genus that infect humans are grouped together because of similar physicochemical properties and include the polioviruses, coxsackieviruses group A and B, echoviruses, and other enteroviruses (Table 21.1). Hepatitis A virus (HAV), also named enterovirus type 72, is still classified as an enterovirus although it has unique molecular and pathogenetical properties so that a reclassification is expected. All enteroviruses inhabit the human alimentary tract and most of them, except HAV, are able to infect the central nervous system (CNS). In addition, these viruses induce a variety of clinical syndromes. The reviews of Melnick et al (1979), Melnick (1982, 1990), Moore and Morens (1984), and Rueckert (1990) are recommended for extensive additional readings on enteroviruses.

Table 21.1 Serotypes of the Human Enteroviruses

Virus	Serotypes
Poliovirus	1–3
Coxsackievirus group A	1–24[a]
Coxsackievirus group B	1–6
Echovirus	1–33[b]
Other enteroviruses	68–71
Hepatitis A virus	[c]

[a] Coxsackievirus type A 23 is the same virus as echovirus type 9.
[b] Echovirus type 10 has been reclassified as reovirus type 1 and echovirus 28 as rhinovirus type 1A. Echovirus type 8 has been deleted because of identity with echovirus type 1.
[c] Formerly classified as enterovirus type 72. *See below.*

21.2 HISTORY OF VIRUS DISCOVERY

Crippling paralytic disease was recorded in ancient times (Melnick, 1982), however, a characterization of poliomyelitis was not

---

Note added in proof: Hepatitis A has recently been reclassified in the new genus "hepatovirus" (Francki, R.I.B., Fauquet, C.M., Knudson, D.L., and Brown, F. 1991. Classification and nomenclature of viruses. Fifth report of the International Committee on Taxonomy of Viruses. Archives of Virology, Supplementum 2, pp. 322–323).

reported until the turn of the 19th century. Poliomyelitis was established as a viral disease in 1909 when Landsteiner and Popper (1909) transmitted paralytic disease to monkeys by inoculating them with filtered stool from a patient with paralytic disease. During the next 40 years, animal inoculation was the method of choice for virus inoculation and study. Thus, in 1948, in Coxsackie, New York, a virus was isolated in suckling mice that were inoculated with a cell-free filtrate of stools obtained from two children suffering from paralysis (Dalldorf and Sickles, 1948). This virus, which could not be neutralized by antiserum against any of the three polioviruses, became the first member of the group A coxsackieviruses. The first of the group B coxsackieviruses was isolated in 1949 (Melnick et al, 1949). The major breakthrough for diagnosing and controlling poliomyelitis was the observation that poliovirus could be propagated in human embryonic tissues in culture (Enders et al, 1949). These tissue cultures allowed easy isolation of the viruses and were prerequisite for the development of vaccines, including both the inactivated vaccine of Salk and the attenuated (oral) vaccine of Sabin. After the introduction of tissue culture, the isolation of many other enteroviruses was possible. The echoviruses (enteric, cytopathic, human, orphan) were discovered in 1951 by Robbins et al (1951). These viruses often were isolated from stools of healthy children and, therefore, could not be related to a disease; hence, these "enteric viruses" were called "orphan viruses."

As early as the times of Hippocrates, a disease called infectious jaundice was described (for reviews, see Siegl, 1988; Hollinger and Ticehurst, 1990). However, the HAV was not isolated until the 1960s (Krugman et al., 1967). Initial studies suggested that only marmoset monkeys could be infected and it took until 1979 to cultivate hepatitis A virus in cell cultures (Provost and Hilleman, 1979).

21.3 STRUCTURAL, BIOPHYSICAL, BIOCHEMICAL, & BIOLOGICAL CHARACTERISTICS

21.3.1 Structure

Enteroviruses are small, spherical, naked viruses (Figure 21.1). The virus particles have icosahedral symmetry, a diameter of 27–30 nm, a buoyant density in CsCl of 1.34 g/ml, a molecular weight of 8.25×10^6 daltons, and a sedimentation coefficient of 156–160 S (for reviews, see Koch and Koch, 1985; Rueckert, 1990). Each virion contains one molecule of single-stranded RNA of about 7500 bases (molecular weight, 2.6×10^6 daltons), which serves as genetic information as well as viral messenger RNA ("positive" sense RNA). The genomes of several enteroviruses have been completely or at least partially sequenced (for review see Palmenberg, 1989). The polygenic RNA is monocystronic and translated into a large precursor protein. Approximately the first 740 nucleotides of the 5′ end of the RNA remain non-translated. The precursor is proteolytically cleaved into the four structural viral capsid proteins, VP1, VP2, VP3, VP4 (the molecular weights determined for poliovirus type are probably typical for all enteroviruses: VP1—33,500 daltons, VP2—30,000 daltons, VP3—26,400 daltons, VP4—7400 daltons), into VPg (molecular weight, 2400 daltons), and the functional proteins, proteases and RNA polymerase. VPg is covalently linked to the 5′ end of the RNA. The 3′ end of the RNA contains a poly-A-sequence. A virion contains 60 copies of each VP1, VP2, VP3, and VP4, and one copy of VPg. A virion may also contain a small number of copies of VPO, which is the uncleaved precursor of VP2 and VP4. The architecture of poliovirus and other picornaviruses has been investigated by X-ray cristallography (Hogle et al, 1985; Rossmann et al, 1985). VP1, VP2, and VP3 comprise the surface of the capsid as also shown by biochemical studies with poliovi-

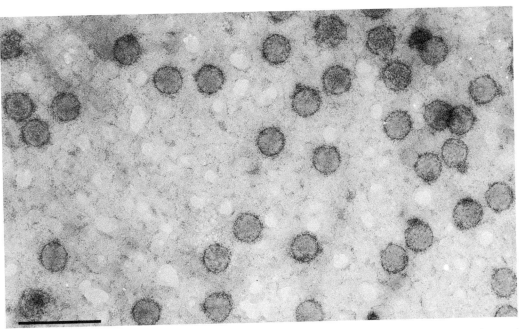

Figure 21.1 Electron micrograph of poliovirus type 1 particles, negatively stained with 0.5% uranyl acetate. Bar represents 100 nm.

rus (Wetz and Habermehl, 1979). A deep depression around the fivefold axis of the capsid, "canyon," is hypothesized to be the site of host cell attachment at the virus-specific receptor (see 21.3.4).

21.3.2 Antigenicity and Neutralization

Preparations of poliovirus contain two antigens that can be detected in complement fixation and precipitin tests: infective or "native" virus, called D (or N) antigen, and noninfective virus called C antigen or occasionally H (heated) antigen. It seems to be true for all enteroviruses that antigenic sites at the capsid surface determine the type-specific antigenicity that is best investigated in neutralization tests. For poliovirus three immunodominant antigenic sites associated with the binding of neutralizing antibodies have been localized to exposed regions of VP1, VP2, and VP3 on the capsid surface (for reviews, see Mosser et al, 1989; Rueckert, 1990). Recent data for HAV show that VP1 and VP3 are also involved in

the neutralization of this virus (Lemon and Ping, 1989). In contrast, VP2 of coxsackievirus type B3 was reported to comprise the major antigenic site(s) for neutralization (Beatrice et al, 1980).

Several mechanisms have been proposed for the neutralization (for reviews, see Dimmock, 1984; Mosser et al, 1989), each interfering with the entry of enteroviruses into their host cells (see 21.3.4). Neutralizing antibodies can (1) reduce the number of infectious units merely by aggregation, (2) create large immune complexes incapable of adsorbing to the cell surface, (3) induce a conformational change in the capsid, or (4) stabilize the capsid and thereby prevent virus uncoating. Stoichiometric analysis of the neutralization reaction revealed that binding of one polyclonal antibody or four monoclonal antibodies at a single virion is sufficient to neutralize infectivity (Icenogle et al, 1983; Wetz et al, 1986).

Several antigenic relationships between

enteroviruses have been observed. In neutralization tests poliovirus types 1 and 2 partially crossreact; coxsackievirus types A3 and A8, A11 and A15, A13 and A18, and echovirus types 1 and 8, 6 and 30, and 12 and 29 are antigenically related. The crossreactivity between several enteroviruses observed in the complement fixation test may be due to common antigenic sites of the virus proteins that are located in the interior of the capsid. These sites are accessible only when a soluble antigen is used. Such immunologic crossreactivity was recently confirmed by the immunoblot technique (Mertens et al, 1983). In addition, these antigenic relationships are reflected by limited homologies among the RNA and proteins of the different enteroviruses as studied by sequence alignments (Palmenberg, 1989 and personal communication). The following levels of homology between enteroviral capsid proteins were found: >80% between poliovirus types; >50% between polioviruses and coxsackieviruses; <20% between HAV and the other enteroviruses.

21.3.3 Reactivity to Chemical and Physical Agents and Virus Storage

All enteroviruses are resistant to low pH (pH 3) and several proteolytic enzymes, which is the prerequisite for virus passage through the stomach and duodenum. The viruses are resistant to several disinfectants, such as 70% alcohol, 5% lysol, or 1% quaternary ammonium compounds, to ether, deoxycholate, and various other detergents that destroy lipid-containing viruses. In general, enteroviruses are inactivated by the following chemicals (see Melnick, 1982; Moore and Morens, 1984; Hollinger and Ticehurst, 1990): formaldehyde (0.3%), HCl (0.1 N), free residual chlorine (0.3–0.5 ppm) and other halogens (free residual bromine or iodine, ≈0.5 ppm × 10 minutes contact time). Presence of organic matter with the virus may result in

protection against inactivation. For this reason 3% formaldehyde is recommended for disinfection. The following physical conditions are inactivating: drying, heat (50 °C for one hour in the absence of magnesium chloride), light (in the presence of vital dyes, such as neutral red, acridine orange, and proflavine). Hepatitis A virus has a higher stability than the other enteroviruses. Temperatures above 60 °C are necessary to destroy its infectivity within a short time (e.g., 85 °C for one minute).

Enteroviruses are stable for years when stored at −70 °C. Storage at −20 °C results in some loss in titer over months while enteroviruses in suspension stored at 4 °C usually will stay viable for weeks.

21.3.4 Replication in Cell Culture

The replication of all enteroviruses so far studied is a mechanism closely associated with cellular membranes. A replication cycle comprises the following steps (for reviews, see Crowell and Landau, 1983; Lentz, 1990; Rueckert, 1990):

Virus entry into the cell: Poliovirus, studied most thoroughly of all enteroviruses, enters its host cell by a phagocytosis mechanism (Dales, 1973). As recently shown, the entry is via receptor-mediated endocytosis (Zeichhardt et al, 1985). The virus adsorbs most likely via its "canyon" at the cell surface using specific receptors. The main receptor for poliovirus comprises a heterogeneous glycoprotein with a predominant species of 67-kDa which is a member of the immunoglobulin superfamily (Mendelsohn et al, 1989). In addition a protein of 100-kDa (Shepley et al, 1988) and glycoproteins of 23-/25-kDa and 50-kDa (Barnert et al, 1991) specifically bind poliovirus and may play a coregulating role in virus adsorption and entry. Virus-receptor-interaction can induce disintegration of the viral capsid; at least a subpopulation of the adsorbed virus, however, remains unaltered and is internalized into the cell via coated pits, coated vesicles, and endosomes (Zeichhardt et al, 1987; Willingmann et al, 1989). The site of uncoating (i.e., the release of the viral genome from the virus capsid) has been reported to occur at the cell surface membrane, in the cytoplasm and the endosome and/or lysosome. For poliovirus type 1, uncoating is most likely a pH-dependent process in endosomes and/or lysosomes (Habermehl et al, 1973; Zeichhardt et al, 1985).

Viral protein and RNA synthesis: Viral proteins are synthesized at the rough endoplasmic reticulum after release of VPg from the viral genome. After translation of a large precursor protein that is cleaved (most probably autocatalytically) into viral RNA polymerase, proteases and precursor of the capsid proteins, viral RNA synthesis takes place in a replicative complex at the smooth endoplasmic reticulum.

Morphogenesis of virus: The assembly of the virus takes place at the cytoplasmic membrane via defined capsid precursors; however, the association with rough or smooth membranes is still under investigation. There is likely to be a membrane-associated morphopoietic factor that facilitates the morphogenesis of the mature particle.

Virus release: The release of newly synthesized virus from the cell is not clearly understood. It has been shown for poliovirus that progeny virus is found in the supernatant prior to cytolysis. Further, only a small portion of the newly synthesized virus particles are mature infective virions. Ratios of infective virus to total virus particles of $1:10^1$–10^3 have been observed.

The replication cycle takes six to seven hours for poliovirus but is some hours longer for other enteroviruses. An exception is HAV for which replication times of more than four weeks were reported when using fecal specimens for primary inoculation. This was reduced to several days, or up to one week for virus inocula obtained after several passages (Siegl, 1988; Hollinger and Ticehurst, 1990).

Shortly after infection (one to three hours postinfection) most enteroviruses induce a pronounced inhibition of cellular RNA, protein and thereby cellular DNA synthesis (Diefenthal et al, 1973). Most enteroviruses are strongly cytolytic, that is, they induce a cytopathic effect (CPE), resulting in destruction of the cell by lysis. A typical example for such a CPE is shown in Figure 21.2 for poliovirus type 1 infecting a monolayer of HEp-2 cells (Zeichhardt et al, 1982). Most of the infected cells detach from the surface of the culture vessel. The remaining cells withdraw from adjacent cells, round up, and are attached to the substratum by long filopodia (Diefenthal and Habermehl, 1967). The microvilli at the cell surface merge and disappear. Ultrathin sections of poliovirus infected cells show drastic changes in the interior of the cell, such as vesicles arranged in clusters in the cytoplasm and a lobed nucleus with irregular distribution of condensed chromatin (Dales et al, 1965). Poliovirus induces characteristic mitotic changes and chromosomal aberrations. The early stage of replication induces enhancement of mitosis, later stages result in an arrest of mitosis in the metaphase (colchicine-like effect). Chromosomal damage is characterized by single chromatid breaks and pulverization (Habermehl et al, 1966; Bartsch et al, 1969).

21.4 EPIDEMIOLOGY

21.4.1 Mode of Transmission

Human enteroviruses have their reservoir only in humans (for growth and pathogenicity in animals, see Table 21.2). Enterovi-

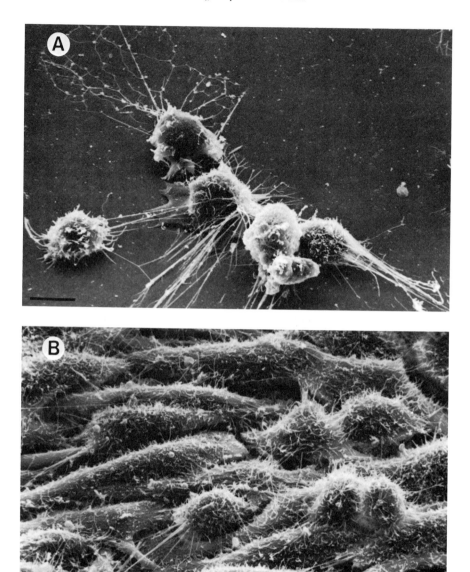

Figure 21.2 Scanning electron micrograph of HEp-2 cells infected with poliovirus type 1 (A) compared with mock-infected control cells (B). Bar represents 10 μm. (Zeichhardt, Schlehofer, Wetz, Hampl, Habermehl).

ruses can be isolated from the lower and/or upper alimentary tract and, therefore, can be spread both by the fecal–oral and respiratory routes (Melnick, 1982, 1990). In areas with poor sanitary conditions fecal–oral transmission is predominant. Transmission by respiratory routes can occur early in infection because the virus replicates in the upper respiratory tract. Sexual transmission of enteroviruses has not been reported, and blood transfusions and insect bites seem not to be responsible for virus

Table 21.2 Growth of Enteroviruses in Human and Monkey Kidney Cell Lines and Pathogenicity for Animals

Viruses	Human Cells	Monkey Kidney Cells	Pathogenicity	
			Mice	Monkeys
Polioviruses	+	+	[a]	+
Coxsackieviruses A	some[b]	some[c]	+[c]	some[d]
Coxsackieviruses B	+	+	+[c]	some
Echoviruses	some[b]	+[e]	some[f]	some
Enteroviruses 68–71	+	+/−	+/−	some
Hepatitis A virus	+[g]	+[g]		+

[a] Some strains of each type have been adapted to mice.

[b] Some strains grow preferentially in, or have been adapted to, human cell cultures. Coxsackievirus types A 11, 13, 15, 16, 18, 20, and 21 may be isolated directly in human cells.

[c] Coxsackievirus types A 7, A 9, and B strains grow readily in monkey kidney cells; some strains grow poorly in mice and fail to produce disease in these animals.

[d] Especially coxsackievirus type A 7.

[e] Echovirus type 21 is cytopathogenic for human epithelial cells but not for monkey kidney cells.

[f] Whereas the prototype and other strains of echovirus 9 are not pathogenic for mice, a number of other strains, especially after passage in monkey kidney cells, produce paralysis in mice (severe coxsackievirus type of myositis).

[g] Not used for routine isolation; propagation is difficult due to long replication times (several weeks) and low virus yield. Some adapted strains of hepatitis A virus have replication times of 3 days (see Hollinger and Ticehurst, 1990).

Modified according to Melnick et al, 1979.

transmission. Enteroviruses can be isolated from sewage, therefore, a fecal–water–oral route of transmission is possible. For HAV, transmission in clams and oysters from fecally contaminated waters is a common source of infection. Food borne acquisition of other enteroviruses has been noted. Nosocomial transmission of enteroviruses typically takes place in newborn nurseries and has been reported for several coxsackieviruses of group A and group B, echoviruses, and HAV.

21.4.2 Geographic, Seasonal, Socioeconomic, Sex, and Age Factors

Enteroviruses are found worldwide. Enterovirus infections characteristically take place during the summertime in areas of the north temperate zone. In tropical and semi-tropical areas enterovirus infections occur throughout the whole year. Persons of low socioeconomic status living in urban areas receive a greater exposure to enteroviruses and have a higher incidence of subclinical

infections than persons of higher socioeconomic status. This led to the paradox that poliomyelitis in the prevaccine era was a disease of "development" (Melnick, 1982), in other words, improvement in the hygienic and socioeconomic conditions was associated with lower subclinical exposure and an increase of incidence of severe illnesses. The improved hygiene might also have led to a decrease in mixed infections with other enteroviruses of patients infected with polioviruses. Mixed infections with more than one enterovirus at the same time can result in interference, leading to a suppression of replication of one of the viruses. Due to extensive vaccination, poliovirus infections are not currently a major problem in developed countries (Figure 21.3, see 21.8.), however, frequent endemics are observed in enclaves, usually religious groups who refuse vaccination.

Diseases due to enteroviruses occur more frequently in males than in females (male/female ratio of 1.5–2.5 : 1) (Moore and

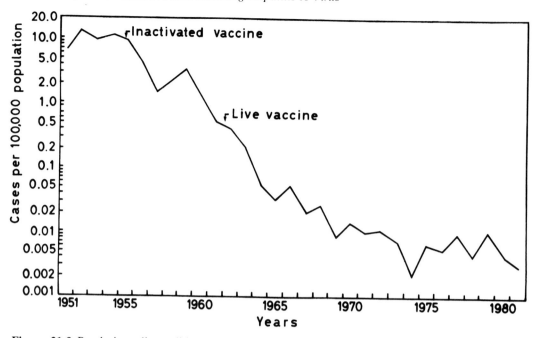

Figure 21.3 Paralytic poliomyelitis attack rates, United States, 1951–1981. (From CDC Poliomyelitis Surveillance Summary 1980–1981. Issued December 1982, p. 9. U.S. Dept. H.H.S., P.H.S.)

Morens, 1984). Generally, young children are the main transmitters of enteroviruses. Echovirus type 9 was found in 50% to 70% of children compared with 17% to 33% of adults. The age of initial infection decreases with bad hygienic conditions and low socio-economic status. It was reported for HAV that in Africa and Asia 90% of the children will be infected within the first eight years of life, however, in Scandinavia only 30% of the population will be infected by 30 to 40 years of age. Severity of disease is also related to age. Poliovirus infection in adults is more likely to lead to paralysis than in children, and infections with coxsackieviruses of group A and echoviruses are usually milder in children than in adults. In contrast coxsackieviruses of group B, can induce fulminant "viral sepsis," myocarditis, encephalitis, and death more likely in newborns than in older children and adults. A reason for the different susceptibility between children and adults for enterovirus infections might be found in a changing pattern of virus-specific receptors during development and aging.

21.4.3 Asymptomatic Infections

Infections with enteroviruses are very common, however, it should be reemphasized that the most common forms of infection are silent, mild, or subclinical. It has been reported that 90% to 95% of poliovirus infections are asymptomatic, whereas, only 0.1% to 1.0% of infections cause paralytic poliomyelitis. Asymptomatic infections are most common for polioviruses followed by echoviruses and coxsackieviruses (50%) (reviewed in Moore and Morens, 1984). The high incidence of inapparent infections with enteroviruses may be due to the virus passage of the gut. The cells of the epithelia of the gut normally have a high rate of turnover. Although 10^3 infective virus particles can be reproduced in one infected cell of the gut and consequently 10^6 to 10^9 viruses per gram can be detected in the

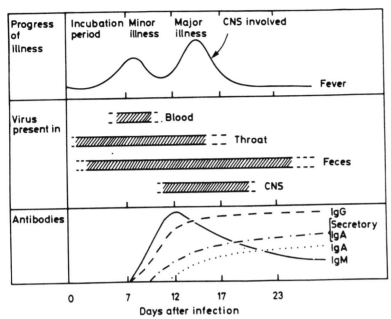

Figure 21.4 The course of infection with poliovirus (idealized). (According to Habermehl.)

stool of enterovirus infected persons, the virus-induced lysis of the cells might be without clinical consequence. Clinical symptoms occur only after massive infection of the gut epithelia.

21.5 PATHOGENESIS AND CLINICAL SYNDROMES

The mechanism of pathogenicity of enterovirus infections is the lytic infection of host cells resulting in severe cytopathic effects (see 21.3.4). Enteroviruses can lead to cyclic infections in their hosts with a viremia and subsequent transport of the virus to the target organs (spinal cord and brain, meninges, myocardium, skin, liver, etc.) (for reviews, see Melnick, 1982, 1990; Moore and Morens, 1984).

21.5.1 Polioviruses

The course of infection of polioviruses is the best understood of all enteroviruses and, therefore, presented as a typical example (Figure 21.4). The portal of entry of

polioviruses is the alimentary tract via the mouth. During the incubation period (6 to 20 days) poliovirus multiplies in the mucosal tissues of the pharynx, the lymphoid tissue (tonsils and Peyer's patches), and/or the gut. For this reason virus is spread via oral and fecal routes beginning shortly after infection. In most cases (90% to 95%) the virus infection will be apparent (i.e., the patient will be asymptomatic). However, virus can be spread to the draining lymph nodes, leading to a viremia characterized by a recovery of virus from the blood stream for a few days (days six to nine after infection). During this time the first nonspecific clinical symptoms (e.g., fever, malaise, sore throat, sometimes headache and vomiting), will be observed. In about 4% to 8% of poliovirus infections, the illness will not proceed and only take the form of a minor illness ("abortive poliomyelitis"). If poliovirus infects its target cells in the CNS, nonparalytic poliomyelitis (1% to 2%) and paralytic poliomyelitis (0.1% to 1.0%) will occur. In nonparalytic poliomyelitis patients have the same prodromal illness as those with minor illness followed after three to seven days by an illness similar to aseptic meningitis commonly accompanied by high fever, back pain, and muscle spasm. Paralytic poliomyelitis additionally comprises flaccid paralysis (involvement of the whole muscle) or paresis (involvement of only some muscle groups), which is due to spinal and/or bulbar damage. The bulbar poliomyelitis (ascendent infection) is less common than the spinal form and has a poor prognosis due to damage of cerebral nerve or vegetative centers. In the case of the spinal illness, recovery of motor function may occur to some degree after some months, however, remaining paralysis is permanent. The source of these severe clinical symptoms is the very specific extraintestinal target cell range of polioviruses; especially the anterior horn cells of the spinal cord, but also dorsal root ganglia, certain brainstem centers, cerebellum, spinal sensory columns, and occasionally the cerebral motor cortex. Histologic changes first observed are vascular engorgement, accompanied by perivascular infiltration with lymphocytes and also polymorphonuclear neutrophils, plasma cells, and microglia.

Certain factors increase the severity of disease, including very young and very old age, male sex, chronic undernutrition, corticosteroid treatment, physical exertion, hypoxia, cold, irradiation, tonsillectomy, pregnancy, adrenal-related endocrine changes, and possibly hypercholesterolemia (reviewed by Moore and Morens, 1984).

21.5.2 Coxsackie- and Echoviruses

Coxsackie- and echoviruses have a less specific extraintestinal target organ range than poliovirus and, therefore, can lead to a wider range of illness (Table 21.3). As described for polioviruses, coxsackie- and echoviruses multiply primarily in the pharynx and small intestine and are shed in the feces for up to one month and in respiratory secretions for several days. Generally speaking, coxsackieviruses and echoviruses, besides the alimentary tract, can infect the meninges, CNS, myocardium and pericardium, striated muscles, respiratory tract, and skin. Paralysis by coxsackie- and echoviruses are less common and less severe than paralysis induced by polioviruses. Coxsackieviruses usually are more pathogenic than echoviruses. Herpangina (vesicular pharyngitis) is only induced by coxsackieviruses of group A. Common colds and pneumonitis of infants are attributed to several serotypes of coxsackieviruses of group A and enterovirus 68. Epidemic myalgia (Bornholm disease) and pleurodynia are associated with most serotypes of coxsackieviruses of group B. Exanthemata with accompanying fever and pharyngitis are observed with several types of coxsackieviruses of group A and B and echoviruses. Aseptic meningitis very often accompanied by rashes are also induced by several coxsackie- and echoviruses. Meningoencephalitis (especially in children) is seen in infections with several coxsackieviruses of group B and with enterovirus 71 and encephalitis with several echoviruses. Myocarditis and pericarditis are more commonly caused by coxsackieviruses of group B in comparison to coxsackieviruses of group A and echoviruses. Acute hemorrhagic conjunctivitis is associated with enterovirus type 70. Coxsackieviruses of group B seem to be responsible for acute pancreatitis and coxsackievirus type B4 for diabetes. Several echoviruses and coxsackieviruses of group A have been recovered from feces mainly of children during epidemics of gastroenteritis, however, their strict role in epidemic diarrhea is still not completely solved. Some echo- and coxsackieviruses of group B are associated with hepatic disturbances or hepatitis, but HAV is the major source of picornavirus-induced hepatitis.

Table 21.3 Enterovirus Infections and Their Clinical Syndromes

Viruses	Types	Clinical Syndromes
Polioviruses types 1–3	1–3	Paralysis (slight to complete muscle weakness), aseptic meningitis, and undifferentiated febrile illness, particularly during the summer
Coxsackieviruses, group A, types 1–24	2, 3, 4, 5, 6, 8, 10	Vesicular pharyngitis (herpangina)
	10	Acute lymphatic or nodular pharyngitis
	2, 4, 7, 9, 10, 23	Aseptic meningitis
	7, 9	Paralysis (infrequently)
	2, 4, 9, 16, 23	Exanthem (macular rash)
	4, 5, 6, 9, 10, 16	Exanthem (vesicular rash; infrequently)
	5, 10, 16	A "hand-foot-and-mouth" disease
	9, 16	Pneumonitis of infants
	21, 24	"Common cold"
	4, 9	Hepatitis
	18, 20, 21, 22, 24	Infantile diarrhea
	24	Acute hemorrhagic conjunctivitis
Coxsackieviruses, group B, types 1–6	1, 2, 3, 4, 5	Pleurodynia
	1, 2, 3, 4, 5, 6	Aseptic meningitis
	2, 3, 4, 5	Paralysis (infrequently)
	1, 2, 3, 4, 5	Severe systemic infection in infants, meningoencephalitis, and myocarditis
	1, 2, 3, 4, 5	Epidemic myalgia (Bornholm disease)
	1, 2, 3, 4, 5	Pericarditis, myocarditis
	4, 5	Upper respiratory illness and pneumonia
	1, 3, 5	Exanthem (macular rash: infrequently)
	5	Hepatitis
	1, 2, 3, 4, 5, 6	Undifferentiated febrile illness
	1, 2, 4	Pancreatitis
	4	Diabetes
Echoviruses, types 1–34	all serotypes, except 12, 24, 26, 29, 32, 33, 34	Aseptic meningitis
	2, 4, 6, 9, 11, 30; possibly 1, 7, 13, 14, 16, 18, 31	Paralysis (infrequently)

(continued next page)

Table 21.3 (*continued*)

Viruses	Types	Clinical Syndromes
Echoviruses, types 1–34 (continued)	2, 6, 9, 19; possibly 3, 4, 7, 11, 14, 18, 22	Encephalitis, ataxia, or Guillain-Barré syndrome
	2, 4, 6, 9, 11, 16, 18; possibly 1, 3, 5, 7, 12, 14, 19, 20	Exanthem
	4, 9, 11, 20, 25; possibly 1, 2, 3, 6, 7, 8, 16, 19, 22	Respiratory disease
	different types have been recovered	Diarrhea (a consistent association has not been established)
	1, 6, 9	Epidemic myalgia (infrequently)
	1, 6, 9, 19	Pericarditis and myocarditis (infrequently)
	4, 9	Hepatic disturbances
Enteroviruses, types 68–71	68	Pneumonia and bronchiolitis
	70	Acute hemorrhagic conjunctivitis
	71	Aseptic meningitis
	71	Meningoencephalitis
	71	Hand-foot-and-mouth disease
Hepatitis A virus		Hepatitis

Modified according to Melnick et al, 1979.

21.5.3 Hepatitis A Virus

Hepatitis A virus seems to comprise a different mode of pathogenesis than the other enteroviruses (for reviews, see Siegl, 1988; Hollinger and Ticehurst, 1990). Hepatitis A virus infections have a long incubation period (four weeks) relative to other enteroviruses (Figure 21.5). So far, there is no evidence that HAV multiplies initially in the pharynx and/or the gut after uptake (fecal–oral route). This might be the reason that in contrast with other enteroviruses HAV cannot be recovered from the pharynx or gut shortly after infection. Virus can only be isolated from the feces of infected patients beginning two weeks after infection, which is about two weeks before the onset of clinical manifestations (jaundice). Virus recovery decreases shortly before the onset of illness and is mostly negative two to three weeks after onset. It is postulated that HAV is distributed in a viremic phase to its target organ (i.e., the liver). Virus multiplication takes place in hepatocytes and Kupffer cells. Hepatitis A virus gains access to the stool via the biliary system.

21.5.4 Chronic Infections

Chronic infections are not the main characteristic of enteroviruses; however, persistent infections with some serotypes of coxsackieviruses of group A and B have been observed (for review, see Melnick, 1990). Particularly coxsackie B viruses are thought to be associated with chronic cardiovascular disease, postviral fatigue syn-

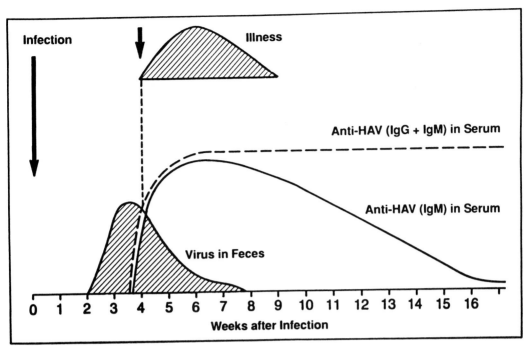

Figure 21.5 Hepatitis A virus can only be isolated from stools from two weeks after infection until the beginning of jaundice (four weeks after infection).

drome, and insulin-dependent diabetes mellitus.

21.5.5 Pregnancy

Pregnancy and infection with enteroviruses are an unsolved problem. In spite of several negative reports, maternal infections during the first trimester of pregnancy are suspected to result in the following anomalies in the infected fetus: coxsackievirus types B2 or B4, urogenital anomalies; coxsackievirus types B3 or B4, cardiovascular anomalies; and coxsackievirus type A9, digestive system malformations (cited in Moore and Morens, 1984; see also Melnick, 1990). Because the teratogenic risk of intrauterine infections with enteroviruses is lower by several orders of magnitude in comparison with infections with rubella virus, a recommendation of abortion is not generally accepted for women with proven enterovirus infection in the first trimester of pregnancy.

21.6 INCUBATION TIMES

All polioviruses, coxsackieviruses of group A and B, and echoviruses have incubation times ranging from 1 to 35 days, with an average of one to two weeks (reviewed by Moore and Morens, 1984). The shortest incubation period, 12 to 30 hours, has been reported for local infections of the eye by echovirus type 70. Hepatitis A virus has the longest incubation time. Usually, the incubation time is about four weeks with a range of 14 to 40 days.

21.7 IMMUNE RESPONSE

Humoral and secretory antibodies play the major role in the immunity to enterovirus infections (Figures 21.4 and 21.5). The role of cellular immunity in these infections is not yet well defined. For HAV, recent studies provide strong evidence that cytotoxic T-cells develop in the course of infection and are capable of lysing infected target cells under the restriction of the major histocompatibility complex (Kurane et al, 1985; Vallbracht et al, 1986). Humoral immunity is mediated by type-specific neutralizing IgG, IgM, and IgA, which prevent hematogenous spread of virus to the target organs. IgM appears first after infection (seven to ten days after infection with polioviruses, coxsackieviruses, and echoviruses, and four weeks after infection with HAV).

Virus-specific IgM persists for four weeks in 90% of infections. Virus-specific IgG and IgA appear a few days after IgM. IgG persists for years and, therefore, mediates the acquired humoral immunity. Production of antibodies in the CNS after poliovirus infection has been reported. Serum antibodies also may reach the CNS by crossing the blood–brain barrier due to breakdown of the integrity of the meninges. Secretory IgA is induced after two to four weeks after poliovirus infection and located mainly in nasopharyngeal and gut tissues. Secretory IgA prevents or limits the excretion of polioviruses in the alimentary tract.

21.8 VACCINATION

Of all enteroviruses only infections with polioviruses can presently be prevented by vaccination. Prerequisite for developing anti-poliomyelitis vaccines was the introduction of tissue culture for the production of virus in large quantity (Enders et al, 1949). Two vaccines have led to a dramatic decrease of poliomyelitis (from approximately 14 paralytic cases in 1952 to less than 0.01 cases in 1969 per 100,000 population) (Figure 21.3). The first vaccine was a formalin-inactivated vaccine developed by Salk and coworkers and first licensed in 1954 in the United States (Salk vaccine). This vaccine, administered intramuscularly, prevents poliovirus infections by inducing humoral neutralizing antibodies. At the same time,

live attenuated vaccines were developed by Sabin, Koprowski, Melnick, Cox, and others. This led to introduction of a live vaccine that is orally administered and consists of all three serotypes of poliovirus (Sabin vaccine). The viruses multiply in the gastrointestinal tract and give rise to a subclinical infection. The Sabin vaccine induces humoral IgG as does the Salk vaccine and additionally elicits secretory IgA in the gut. Therefore, Sabin vaccination prevents not only virus spread through the blood stream to the CNS, but also blocks primary virus multiplication in the gut. There is a possibility that one of the types of the trivalent vaccine will not multiply due to interference from one of the other types or from other enteroviruses during mixed infections. For this reason the vaccine is administered three times several weeks apart, in the north temperate zone, preferably not during the summer months when there is high frequency of other enterovirus infections (see 21.4.2). The live Sabin vaccine bears a small risk of developing clinical poliomyelitis including paralysis. Based on data from the United States, there is a risk of one case of poliomyelitis in vaccines per eight million doses of the live vaccine administered. A critical evaluation of both vaccines is given by Melnick (1982, 1990) and Moore and Morens (1984).

One reason for live vaccine associated poliomyelitis is changes in the genome of attenuated virus, often point mutations, that result in increased neurovirulence (for reviews, see Nomoto et al, 1989; Racaniello et al, 1989). Interestingly, many genome regions including the 5′ noncoding region, contribute to the phenotype of attenuation and neurovirulence. Therefore, several laboratories are presently investigating alternative, safe and efficacious vaccines against poliovirus (for review, see Melnick, 1990).

One strategy has been to prepare polypeptide vaccines representing the immunodominant antigenic sites of the three large capsid proteins that bind neutralizing antibodies (see 21.3.2). These peptides are either prepared by recombinant DNA technology or chemically synthesized. These vaccines should be extremely safe as they do not contain viral nucleic acid, however, several investigations have so far indicated that such peptides show rather low immunogenicity. Another approach is to develop live poliovirus by DNA recombinant technology that cannot mutate to increased neurovirulence. One example is the construction of a poliovirus chimera possessing immunodominant antigenic sites of poliovirus types 1 and 3 (Burke et al, 1988). Comparable investigations have been initiated for a hepatitis A vaccine (Siegl and Lemon, 1990). *Since going to press it has been reported that hepatitis A vaccines, most probably inactivated, can be expected in the near future (Melnick, 1990; Flehmig et al, 1990).*

Prophylaxis with immune serum globulin is usually recommended when the epidemiologic conditions are well known. Immune globulins against polioviruses or HAV from sera of convalescent patients are effective in preventing infections; however, they have to be administered within the first days after exposure.

So far enterovirus infection can be prevented only by prophylactic active or passive immunization and by interruption of the routes of virus transmission. Viral chemotherapeutic agents cannot as yet be utilized for treatment of humans. Antiviral compounds either affecting the virus-specific RNA polymerase such as guanidine and some benzimidazoles, or preventing viral uncoating by capsid stabilization have been proven to be potent in cell culture systems (Zeichhardt et al, 1987; Eggers, 1988; for reviews, see Eggers, 1982; Rossmann, 1989).

21.9 LABORATORY DIAGNOSIS

Methods used for laboratory diagnosis of enterovirus infections have been summarized elsewhere (Melnick et al, 1979; Melnick, 1985).

21.9.1 Virus Isolation and Identification

Specimens for virus isolation are usually stools and rectal swabs, throat swabs and washings, and cerebrospinal fluid (CSF). The successful isolation of virus from clinical specimens depends on the time postinfection, which relates to the pathogenesis of enterovirus infections. For example, virus is most readily isolated from the throat shortly after infection up to 15 days or more, from stools and rectal swabs up to four weeks after infection, and from the CSF during the time of manifestation of symptoms involving the CNS, usually two to three weeks after infection (Figure 21.4). Isolation from stools is most promising as virus concentrations in feces are higher than in other specimens (up to 10^6 to 10^9 virus particles per gram of feces). Viruses inducing vesicular rashes like coxsackievirus types A4, A5, A9, A10, A16, and enterovirus type 71 can be isolated also from the lesions. Virus isolation from blood is successful during viremia (days six to nine after infection), however, due to the short period of this phase virus isolation is generally not attempted from this source. In addition all specimens from target organs generally yield virus if a biopsy or autopsy specimen is taken during the clinical manifestation of disease.

Originally, pathologic lesions produced in mice were used for distinction between coxsackieviruses of groups A and B. More recently cell cultures are most commonly used for isolating most of the enteroviruses (Melnick et al, 1979; Melnick, 1985) (Table 21.2). Cells usually used for growth of polioviruses, coxsackievirus types A7, A9, and A16, coxsackieviruses of group B and echoviruses are human embryonic fibroblasts of the skin or lung, permanent human amnion cells, transformed human cell lines, such as HeLa and HEp-2 cells, and primary monkey cell lines, such as rhesus and African green monkey kidney cells. An exception is the growth of several types of coxsackieviruses of group A. Some of these viruses only replicate in a cell line derived from a human rhabdomyosarcoma cell line or only in newborn mice. Coxsackievirus types A1, A19, and A22 remain as types that require newborn mice for their cultivation (Melnick, 1982). Virus isolation in mice, however, is restricted to a few reference laboratories worldwide.

The appearance of the typical cytopathic effect in cell cultures (see 21.3.4) is proof of the presence of virus in a specimen. Neutralization tests are used for identification of the virus isolates. Most commonly, pools of internationally standardized hyperimmune equine antisera are used for typing the isolates. Typing is performed according to a pattern proposed by Lim and Benyesh-Melnick (1960). Neutralization tests are most favorable for this purpose as neutralizing antibodies are type-specific. In contrast, radio- or enzyme immunoassays, complement fixation tests, and hemagglutination inhibition tests for viruses with hemagglutinins (see 21.9.2) do not allow type-specific identification due to immunologic crossreactions (see 21.3.2.).

Direct virus visualization by electron microscopy is performed in specialized laboratories. Virus can be detected in preparations from stools due to high concentration of virus in feces. This is achieved by using the technique of negative staining. Immune electron microscopy allows a direct virus identification. Specimens containing virus are incubated with virus-specific antiserum and the resulting virus–antibody complexes are visualized. Immunofluorescence may be useful for directly typing enteroviruses in specimens obtained at biopsy or autopsy. However, this technique has not been used extensively.

Poliovirus strain characterization for differentiation of wild strains and vaccine-like viruses is performed in specialized laboratories usually by the following techniques: (1) determination of growth "markers" such as the reproductive capacity at elevated temperatures and plaquing capacities, (2) intratypic serodifferentiation of virus strains by polyclonal strain-specific antibodies (for review, see WHO, 1980). Additionally, monoclonal antibodies are used for this purpose (Minor et al, 1986; Horaud et al, 1987).

Oligonucleotide analysis ("fingerprinting") and nucleic acid sequencing are advanced methods for virus characterization both for differentiation of wild and vaccine viruses and studies of molecular epidemiology (Evans et al, 1985; Omata et al, 1986; Rico-Hesse et al, 1987). In situ nucleic acid hybridization has been introduced for detection of enteroviral RNA in myocardial tissue of patients with enterovirus associated heart disease (Kandolf, 1988; Rotbart, 1989). The polymerase chain reaction (PCR) is an additional technique that will detect enteroviral genomes in cells and tissues (Rotbart, 1990).

Hepatitis A virus can only be isolated from stools from two weeks after infection until the beginning of jaundice (four weeks after infection) (Figure 21.5). Identification of HAV can be performed using radio- or enzyme immunoassays; however, this is not a routine procedure as more than 50% of HAV-infected patients no longer excrete the virus when clinical symptoms appear (Figure 21.5). Due to the limited propagation of hepatitis A virus in cell culture and its protracted replication cycle (several days to four weeks) cell culture techniques are unfavorable for routine virus detection (for review, see Hollinger and Ticehurst, 1990).

21.9.2 Serologic Diagnosis

Serologic diagnosis in combination with virus identification is most favorable for confirming an infection with enteroviruses.

Documentation of a virus specific serologic response (fourfold or greater rise in antibody titer or a single high titer of virus-specific IgM) proves a recent enterovirus infection. Ideally, serum of a patient should be tested in the beginning of the illness and seven to ten days later (Figure 21.4). Neutralization tests using tissue culture are most commonly used for serologic diagnosis of infections with polioviruses, coxsackieviruses, and echoviruses, as neutralizing antibodies react with the virus in a type-specific manner (see 21.3.2 and 21.9.1). For those coxsackieviruses of group A not growing in cell culture, neutralization tests in mice can be performed.

The complement fixation test is another serologic procedure used to diagnose enteroviruses. It is of limited value because it allows only detection of group-specific antibodies. The same holds true for radio- and enzyme immunoassays. Enzyme immunoassays for coxsackieviruses have been reported (Katze and Crowell, 1980),

however, these tests have not been introduced into the routine laboratory so far. Theoretically, hemagglutination inhibition tests can be performed with some hemagglutinating strains of coxsackieviruses of group A and B, some echoviruses, and enterovirus type 68.

Detection of antibodies against HAV is performed in radio- or enzyme immunoassays. By applying antibodies against the different human immunoglobulins (IgG or IgM), acute hepatitis A infection can be differentiated from a previous infection (Figure 21.5).

ACKNOWLEDGMENT

Professor Dr. K.-O. Habermehl, Director, Institut für Klinische und Experimentelle Virologie, Freie Universität Berlin, is gratefully acknowledged for his comments and discussion, as well as the help of R. Joncker in preparing the manuscript.

REFERENCES

Barnert, H., Zeichhardt, H., and Habermehl, K.-O. 1992. Identification of 50- and 23-/25-kd HeLa cell membrane glycoproteins involved in poliovirus infection: Occurrence of poliovirus-specific binding sites on susceptible and nonsusceptible cells. Virology 186:533–542.

Bartsch, H.D., Habermehl, K.O., and Diefenthal, W. 1969. Correlation between poliomyelitisvirus-reproduction-cycle, chromosomal alterations and lysosomal enzymes. Arch. Ges. Virusforsch. 27:115–127.

Beatrice, S.T., Katze, M.G., Zajac, B.A., and Crowell, R.L. 1980. Induction of neutralizing antibodies by the coxsackievirus B3 virion polypeptide, VP2. Virology 104:426–438.

Burke, K.L., Dunn, G., Ferguson, M., Minor, P.D., and Almond, J.W. 1988. Antigen chimaeras of poliovirus as potential new vaccines. Nature (London) 332:81–82.

Crowell, R.L. and Landau, B.J. 1983. Receptors in the initiation of picornavirus infections. Comp. Virol. 18:1–42.

Dales, S. 1973. Early events in cell-animal virus interactions. Bact. Rev. 37:103–135.

Dales, S., Eggers, H.J., Tamm, I., and Palade, G.E. 1965. Electron microscopic study of the formation of poliovirus. Virology 26:379–389.

Dalldorf, G. and Sickles, G.M. 1948. An unidentified, filtrable agent isolated from the feces of children with paralysis. Science 108:61–63.

Diefenthal, W. and Habermehl, K.O. 1967. Die Bedeutung mikrokinematographischer Methoden in der Virologie. Res. Film 6:22–30.

Diefenthal, W., Habermehl, K.O., Lorenz, P.R., and Beneke, T. 1973. Virus-induced inhibition of host cell synthesis. Adv. Biosciences 11:127–148.

Dimmock, N.J. 1984. Mechanisms of neutralization of animal viruses. J. Gen. Virol. 65: 1015–1022.

Eggers, H.J. 1982. Benzimidazoles. Selective inhibitors of picornavirus replication in cell culture and in the organism. In P.E. Came, and L.A. Caliguiri (eds.), Handbook of Experimental Pharmacology 61. Berlin: Springer Verlag, pp. 377–417.

Eggers, H.J. 1988. Assay Systems: Testing of antiviral drugs in cell culture (in vitro). In E. De Clercq, R.T. Walker (eds.), Antiviral Drug Development, New York-London: Plenum Press, Series A: Life Sciences 143, pp. 139–148.

Enders, J.F., Weller, T.H., and Robbins, F.C. 1949. Cultivation of the Lansing strain of poliomyelitis virus in cultures of various human embryonic tissue. Science 109:85–87.

Evans, D.M.A., Dunn, G., Minor, P.D., Schild, G.C., Cann, A.J., Stanway, G., Almond, J.W., Currey, K., and Maizel, J.V. 1985. Increased neurovirulence associated with a single nucleotide change in a non-coding region of the Sabin type 3 poliovaccine genome. Nature (London) 314:548–550.

Flehmig, B., Heinricy, U., and Pfisterer, M. 1990. Prospects for a hepatitis A virus vaccine. In J.L. Melnick (ed.), Progress in Medical Virology, Vol. 37. Basel: Karger, pp. 56–71.

Habermehl, K.O., Diefenthal, R., and Diefenthal, W. 1966. Der Einfluß von Virusinfektionen auf den Ablauf der Zellteilung. Zbl. Bakt. Hyg., I. Orig. 199:273–314.

Habermehl, K.O., Diefenthal, W., and Buchholz, M. 1973. Distribution of parental viral constituents in the course of polioinfection. Adv. Biosciences 11:41–64.

Hogle, J.M., Chow, M., and Filman, D.J. 1985. Three-dimensional structure of poliovirus at 2.9 Å resolution. Science 229:1358–1365.

Hollinger, F.B. and Ticehurst, J. 1990. Hepatitis A Virus. In B.N. Fields, D.M. Knipe, R.M. Chanock, M.S. Hirsch, J.L. Melnick, T.P. Monath, B. Roizman (eds.), Virology, Second Edition, New York: Raven Press, pp. 631–667.

Horaud, F., Crainic, R., Van der Werf, S., Blondel, B., Wichowski, C., Akacem, O., Bruneau, P., Couillin, P., Siffert, O., and Girard, M. 1987. Identification and characterization of a continuous neutralization epitope (C3) present on type 1 poliovirus. Progr. Med. Virol. 34:129–155.

Icenogle, J., Shiwen, H., Duke, G., Gilbert, S., Rueckert, R., and Anderegg, J. 1983. Neutralization of poliovirus by a monoclonal antibody: Kinetics and stoichiometry. Virology 127:412–425.

Kandolf, R. 1988. The impact of recombinant DNA technology on the study of enterovirus heart disease. In M. Bendinelli, H. Friedman (eds.), Coxsackieviruses—A General Update. New York: Plenum Press, pp. 293–317.

Katze, M.G. and Crowell, R.L. 1980. Immunological studies of the group B coxsackieviruses by the sandwich enzyme-linked immunosorbent assay (ELISA) and immunoprecipitation. J. Gen. Virol. 50:357–367.

Koch, F. and Koch, G. 1985. The molecular biology of poliovirus. New York: Springer Verlag.

Krugman, S., Giles, J.P., and Hammond, J. 1967. Infectious hepatitis: Evidence for two distinctive clinical, epidemiological and immunological types of infection. JAMA 200: 365–373.

Kurane, I., Binn, L.N., Bankroft, W.H., and Ennis, F.A. 1985. Human lymphocyte responses to hepatitis A virus-infected cells: Interferon production and lysis of infected cells. J. Immunol. 135:2140–2144.

Landsteiner, K. and Popper, E. 1909. Übertragung der Poliomyelitis acuta auf Affen. Z. Immunitaetsforsch. Orig. 2:377–390.

Lemon, S.M. and Ping, L.H. 1989. Antigenic structure of hepatitis A virus. In B.L. Semler and E. Ehrenfeld (eds.), Molecular Aspects of Picornavirus Infection and Detection. Washington, D.C.: American Society for Microbiology, pp. 193–208.

Lentz, T.L. 1990. The recognition event between virus and host cell receptor: A target for antiviral agents. J. Gen. Virol. 71:751–766.

Lim, K.A. and Benyesh-Melnick, M. 1960. Typing of viruses by combinations of antiserum pools: Application to typing of enteroviruses (Coxsackie and echo). J. Immunol. 84:309–317.

Melnick, J.L. 1982. Enteroviruses. In A.S. Evans (ed.), Viral Infections of Humans, Epidemiology and Control. New York: Plenum Medical Books, pp. 187–251.

Melnick, J.L. 1985. Enteroviruses. In E.H. Lennette (ed.), Laboratory Diagnosis of Viral Infections, New York: Marcel Dekker, Inc., pp. 241–256.

Melnick, J.L. 1990. Enteroviruses: Polioviruses, coxsackieviruses, echoviruses, and newer enteroviruses. In B.N. Fields, D.M. Knipe, R.M. Chanock, M.S. Hirsch, J.L. Melnick, T.P. Monath, and B. Roizman (eds.), Virology, 2nd edition, New York: Raven Press, pp. 549–605.

Melnick, J.L. 1990. New picornavirus vaccines for hepatitis A, and lessons from the control of poliomyelitis by the prototype picornavirus vaccines. In J.L. Melnick (ed.), Progress in Medical Virology, Vol. 37. Basel: Karger, pp. 47–55.

Melnick, J.L., Shaw, E.W., and Curnen, E.C. 1949. A virus isolated from patients diagnosed as nonparalytic poliomyelitis or aseptic meningitis. Proc. Soc. Exp. Biol. Med. 71:344–349.

Melnick, J.L., Wenner, H.A., and Phillips, C.A. 1979. Enteroviruses. In E.H. Lennette and N.J. Schmidt (eds.), Diagnostic Procedures for Viral, Rickettsial and Chlamydial Infections. Washington, D.C.: American Public Health Association, pp. 471–534.

Mendelsohn, C.L., Wimmer, E., and Racaniello, V.R. 1989. Cellular receptor for poliovirus: Molecular cloning, nucleotide sequence, and expression of a new member of the immunoglobulin superfamily. Cell 56:855–865.

Mertens, T., Pika, U., and Eggers, H.J. 1983. Cross antigenicity among enteroviruses as revealed by immunoblot technique. Virology 129:431–442.

Minor, P.D., John, A., Ferguson, M., and Icenogle, J.P. 1986. Antigenic and molecular evolution of the vaccine strain of type 3 poliovirus during the period of excretion by a primary vaccinee. J. Gen. Virol. 67:693–706.

Moore, M. and Morens, D. 1984. Enteroviruses, including polioviruses. In R.B. Belshe (ed.), Textbook of Human Virology. Littleton, MA: PSG Publishing Company, pp. 407–483.

Mosser, A.G., Leippe, D.M., and Rueckert, R.R. 1989. Neutralization of picornaviruses: Support for the pentamer bridging hypothesis. In B.L. Semler and E. Ehrenfeld (eds.), Molecular Aspects of Picornavirus Infection and Detection. Washington, D.C.: American Society for Microbiology, pp. 155–167.

Murphy, F.A. and Kingsbury, D.W. 1990. Virus Taxonomy. In B.N. Fields, D.M. Knipe, R.M. Chanock, M.S. Hirsch, J.L. Melnick, T.P. Monath, and B. Roizman (eds.), Virology, Second Edition, New York: Raven Press, pp. 9–35.

Nomoto, A., Kawamura, N., Kohara, M., and Arita, M. 1989. Expression of the attenuation phenotype of poliovirus type 1. In B.L. Semler and E. Ehrenfeld (eds.), Molecular Aspects of Picornavirus Infection and Detection. Washington, D.C.: American Society for Microbiology, pp. 297–306.

Omata, T., Kohara, M., Kuge, S., Komatsu, T., Abe, S., Semler, B.L., Kameda, A., Itoh, H., Arita, M., Wimmer, E., and Nomoto, A. 1986. Genetic analysis of the attenuation phenotype of poliovirus type 1. J. Virol. 58:348–358.

Palmenberg, A.C. 1989. Sequence alignments of picornaviral capsid proteins. In B.L. Semler and E. Ehrenfeld (eds.), Molecular Aspects of Picornavirus Infection and Detection. Washington, D.C.: American Society for Microbiology, pp. 211–241.

Provost, P.J. and Hilleman, M.R. 1979. Propagation of human hepatitis A virus in cell culture in vitro. Proc. Soc. Exp. Biol. Med. 160:213–221.

Racaniello, V.R., La Monica, N., Moss, E.G., and O'Neill, R. 1989. Genetic analysis of neurovirulence, using a mouse model for poliomyelitis. In B.L. Semler and E. Ehrenfeld (eds.), Molecular Aspects of Picornavirus Infection and Detection. Washington,

D.C.: American Society for Microbiology, pp. 281–296.

Rico-Hesse, R., Pallansch, M.A., Nottay, B.K., and Kew, O.M. 1987. Geographic distribution of wild poliovirus type 1 genotypes. Virology 160:311–322.

Robbins, F.C., Enders, J.F., Weller, T.H., and Florentino, G.L. 1951. Studies on the cultivation of poliomyelitis viruses in tissue culture. V. The direct isolation and serologic identification of virus strains in tissue culture from patients with nonparalytic and paralytic poliomyelitis. Am. J. Hyg. 54: 286–293.

Rossmann, M.G., Arnold, E., Erickson, J.W., Frankenberger, E.A., Griffith, J.P., Hecht, H.J., Johnson, J.R., Kramer, G., Luo, M., Mosser, A.G., Rueckert, R.R., Sherry, B., and Vriend, G. 1985. Structure of a human common cold virus and functional relationships to other picornaviruses. Nature 317: 145–153.

Rossmann, M.G. 1989. Conformational adaptations by picornaviruses to antiviral agents and pH changes. In B.L. Semler and E. Ehrenfeld (eds.), Molecular Aspects of Picornavirus Infection and Detection. Washington, D.C.: American Society for Microbiology, pp. 139–154.

Rotbart, H.A. 1989. Human enterovirus infections: Molecular approaches to diagnosis and pathogenesis. In B.L. Semler and E. Ehrenfeld (eds.), Molecular Aspects of Picornavirus Infection and Detection. Washington, D.C.: American Society for Microbiology, pp. 243–264.

Rotbart, H.A. 1990. PCR amplification of enteroviruses. In M.A. Innis, D.H. Gelfand, J.J. Sninsky, and T.J. White (eds.), PCR protocols: A Guide to Methods and Applications. San Diego: Academic Press Inc., pp. 372–377.

Rueckert, R.R. 1990. Picornaviridae and their replication. In B.N. Fields, D.M. Knipe, R.M. Chanock, M.S. Hirsch, J.L. Melnick, T.P. Monath, and B. Roizman (eds.), Virology. New York: Raven Press, pp. 507–548.

Shepley, M.P., Sherry, B., and Weiner, H.L. 1988. Monoclonal antibody identification of a 100-kDa membrane protein in HeLa cells and human spinal cord involved in poliovirus attachment. Proc. Natl. Acad. Sci. USA 85:7743–7747.

Siegl, G. 1988. Virology of hepatitis A. In A.J. Zuckerman (ed.), Viral Hepatitis and Liver Disease. New York: Alan R. Liss, pp. 3–7.

Siegl, G. and Lemon, S.M. 1990. Recent advances in hepatitis A vaccine development. Virus Research, 17:75–92.

Vallbracht, A., Gabriel, P., Maier, K., Hartmann, F., Steinhardt, H.J., Müller, C., Wolf, A., Manncke, K.H., and Flehmig, B. 1986. Cell-mediated cytotoxicity in hepatitis A virus infection. Hepatology 6:1308–1314.

Wetz, K. and Habermehl, K.O. 1979. Topographical studies on poliovirus capsid proteins by chemical modification and crosslinking with bifunctional reagents. J. Gen. Virol. 44:525–534.

Wetz, K., Willingmann, P., Zeichhardt, H., and Habermehl, K.-O. 1986. Neutralization of poliovirus by polyclonal antibodies requires binding of a single IgG molecule per virion. Arch. Virol. 91:207–220.

WHO (World Health Organization). 1980. Poliovirus strain characterization: A WHO memorandum. Bull. WHO 58:727–730.

Willingmann, P., Barnert, H., Zeichhardt, H., and Habermehl, K.-O. 1989. Receptor-mediated endocytosis of poliovirus type 1: Recovery of structurally intact and infectious virus from HeLa cells until viral uncoating. Virology 168:417–420.

Zeichhardt, H., Schlehofer, J.R., Wetz, K., Hampl, H., and Habermehl, K.O. 1982. Mouse Elberfeld (ME) virus determines the cell surface alterations when mixedly infecting poliovirus-infected cells. J. Gen. Virol. 58:417–428.

Zeichhardt, H., Wetz, K., Willingmann, P., and Habermehl, K.O. 1985. Entry of poliovirus type 1 and Mouse Elberfeld (ME) virus into HEp-2 cells: Receptor-mediated endocytosis and endosomal or lysosomal uncoating. J. Gen. Virol. 66:483–492.

Zeichhardt, H., Otto, M.J., McKinlay, M.A., Willingmann, P., and Habermehl, K.-O. 1987. Inhibition of uncoating of poliovirus by disoxaril (WIN 51711). Virology 160: 281–285.

22

Rotavirus, Enteric Adenoviruses, Norwalk Virus, and Other Gastroenteritis Tract Viruses

Miguel O'Ryan, David O. Matson, and
Larry K. Pickering

22.1 INTRODUCTION

Viruses were first recognized as etiologic agents of acute diarrhea in the 1930s when a filterable agent that caused diarrhea in rabbits was discovered (Hodes, 1977). After the successful cultivation of enteroviruses in the late 1950s, a number of these viruses were identified in stool specimens from children with gastroenteritis (Ramos-Alvarez and Sabin, 1958; Bell et al, 1968). Subsequent studies failed to substantiate a role for enteroviruses as a cause of gastroenteritis because the rate of enterovirus detection was similar in children with and without diarrhea (Yow et al, 1970). Little progress in this field was achieved until 1970 when electron microscopy (EM) was applied to the study of stool samples. Using convalescent serum mixed with an acute stool from a patient with gastroenteritis from Norwalk, Ohio, virus-antibody complexes were directly visualized for the first time in a stool specimen from a patient with diarrhea (Kapikian et al, 1972). In the following two decades, the identification of a large number of viruses associated with gastroenteritis in children and adults was facilitated by

the use of EM and immunoelectron microscopy (IEM) techniques, the discovery of new cell lines for viral culture, and the development of antibody-based assays and genomic analysis. Most of these viruses would not replicate in standard cell lines. Table 22.1 lists the viruses that are currently considered to be pathogens of the intestinal tract. Both the classification scheme currently used and the number of viruses listed probably will change as more is learned about the molecular virology and epidemiology of enteric viruses.

22.2 ROTAVIRUSES

Rotaviruses are frequent pathogens in humans and other mammalian and avian hosts. Human rotaviruses (HRV) were first detected in 1973 (Bishop et al, 1973) by EM observation of biopsy specimens from the duodenum of children with acute diarrhea. Rotaviruses have been recognized as important pathogens in adults, particularly the elderly (Vollet et al, 1979; Wenman et al, 1979; Halvorsrud and Orstavik 1980; Marrie et al, 1982; Echeverria et al, 1983).

Table 22.1 Viruses Associated with Gastroenteritis

1. Rotavirus
2. Enteric adenoviruses (types 40 and 41)
3. Small round enteric viruses
   a. With characteristic structural features
      i. Small round structured viruses including Norwalk and Norwalk-like viruses
      ii. Astroviruses
      iii. Caliciviruses
   b. Featureless viruses
      i. Parvoviruses
      ii. Parvovirus-like viruses
4. Coronaviruses
5. Other viruses
   a. Pestivirus
   b. Breda virus

O'Ryan et al, 1990a.

22.2.1 Virus Structure

Rotaviruses are among the few human pathogens with a segmented genome (11 segments) of double-stranded RNA. The virus is 70 to 75 nm in diameter and is composed of an internal core and inner and outer capsids (Figures 22.1 and 22.2) (Estes and Cohen, 1989). The virus lacks an envelope. The genome is located within the inner core and each gene segment codes for separate viral proteins (VP). The VP can be part of the structure of the virus or it may be a nonstructural (NS) protein that functions in viral assembly or some other activity as described (Prasad et al, 1990) (Table 22.2 and Figure 22.1).

22.2.2 Antigenic Characteristics

Six distinct rotavirus groups (A through F) have been identified serologically based upon group specific antigens (Bridger, 1988; Desselberger, 1988). Three of these groups (A, B, and C) have been identified in humans. Group A rotaviruses are the most commonly-recognized cause of gastroenteritis in children. Group B rotaviruses

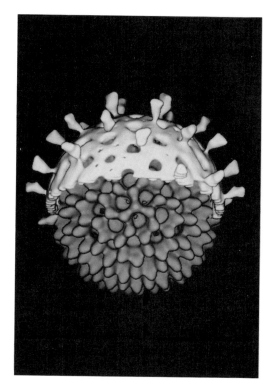

Figure 22.1 Three-dimensional representation of the inner and outer capsid of rotavirus. Image was obtained by computer processing of selected electron micrographs (Prasad et al, 1989). The spikes protruding from the outer capsid are formed by VP4. The matrix of the outer capsid is formed by VP7. The inner capsid is formed by VP6. (Courtesy of Dr. B.V.V. Prasad, Baylor College of Medicine, Houston, TX.)

have caused epidemics of cholera-like diarrhea in China and sporadic cases elsewhere (Hung et al, 1984; Brown et al, 1987; Nakata et al, 1987). Antibody surveys showed a low prevalence of Group B rotaviruses outside Southeast Asia. Group C rotaviruses have caused outbreaks of diarrhea in children in many countries, but have not been found in North America (Espejo et al, 1984; Bridger et al, 1986; Matsumoto et al, 1989; Penaranda et al, 1989a, 1989b). Only group A rotaviruses will be discussed further.

Group A rotaviruses can be divided into two subgroups (I and II) based on

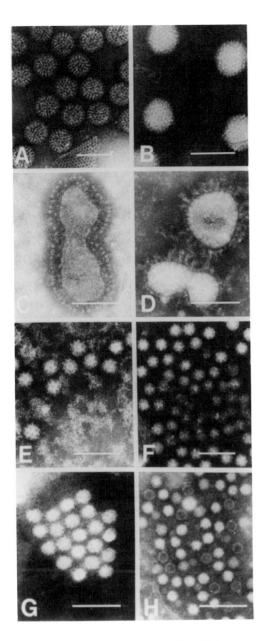

Figure 22.2 Electron micrographs of virus particles found in stool specimens from patients with gastroenteritis. These viruses were visualized following negative staining. Specific viruses and the original magnifications of the micrographs are as follows. A. Rotavirus (185,000x). B. Enteric adenovirus (234,000x). C. Coronavirus (249,000x). D. Breda virus (249,000). E. Calicivirus (250,000x). F. Astrovirus (196,000x). G. Norwalk-like virus (249,000x). H. Parvovirus (249,000x). The electron micrographs in panels

antigenic differences of epitopes on the major group antigen VP6 as defined by reaction with monoclonal antibodies (Estes and Cohen, 1989). The majority of the human group A rotaviruses are in subgroup II. The viruses in this subgroup also have in common a "long pattern" of RNA migration when the genome segments are separated by gel electrophoresis (Figure 22.3). Subgroup I viruses are less common than subgroup II viruses, and the RNA segments characteristically migrate with a "short pattern" (Figure 22.3). Viruses are further classified by serotypes based upon differences of neutralization epitopes located in the outer capsid. Fourteen serotypes have been described from both human and animal sources (Table 22.3) (Hoshino et al, 1985; Estes and Cohen, 1989; Taniguchi et al, 1990; Urasawa et al, 1990; Browning et al, 1991). Studies using monoclonal antibodies and genetic variants have demonstrated that VP7 is responsible for serotype-specific neutralization (Hoshino et al, 1985; Estes and Cohen, 1989). To date, seven serotypes have been isolated from humans (Table 22.3) (Estes and Cohen, 1989; Urasawa et al, 1990) including serotypes 1, 2, 3, 4, 8, 9, and 12.

The differentiation of types of VP4 has been less successful. Sera prepared in hyperimmunized animals do not discriminate different VP4 types in neutralizing assays. Monoclonal antibodies specific to different VP4 types are not well characterized. Gene sequencing data has suggested that at least 11 types of VP4 exist, five of which are known to infect humans (Estes and Cohen, 1989). Whether these VP4 (gene) types determine different antigenic types is as yet

C, D, and F–H were kindly provided by Dr. T. Flewett; the calicivirus photograph (panel E) was originally obtained from Dr. C.R. Madeley. The bar represents 100 nm. (Reprinted by permission of the publisher from Viral Infections of the Intestine by M.K. Estes and D.Y. Graham published in Gitnick (ed.), G. *Principles and Practice of Gastroenterology and Hepatology.* pp. 569. Copyright 1988 by Elsevier Science Publishing Co., Inc.)

Table 22.2 Rotavirus Gene Segments and Protein Products

Gene Segment	Protein Product	Comments
1	VP1	Inner core protein
2	VP2	Inner core protein; binds to RNA
3	VP3	Inner core protein
4	VP4 (cleaved to VP5 and VP8)	Outer capsid protein, hemagglutinin, neutralizing antigen, cleavage by protease enhances infectivity
5	NS 53	Function unknown
6	VP6	Major structural protein, inner capsid protein, group and subgroup epitopes
7	NS 34	Binds to RNA
8	NS 35	Possibly involved in RNA replication
9	VP7	Major outer capsid protein, glycoprotein, neutralization antigen
10	NS 28	Membrane protein of endoplasmic reticulum
11	NS 26	Possibly involved in RNA replication

Estes and Cohen, 1989.

unclear. A classification scheme based on antigenic variants of VP4 that describes four VP4 antigenic types (1A, 1B, 2, and 3), has been proposed (Gorziglia et al, 1990). Further studies will be required to determine the validity of this classification.

22.2.3 Epidemiology

Gastroenteritis caused by group A rotaviruses in humans most frequently occurs between 6 and 24 months of age (Kapikian et al, 1976; Rodriguez et al, 1977; Brandt et al, 1983). This virus group is responsible for 10% to 50% of the diarrheal episodes which require hospitalization in this age group. Incidence rates of rotavirus diarrhea from community-based studies range from 0.2 to 0.8 episodes per child per year (DeZoysa

and Feachem, 1985). In the day-care setting, the incidence of symptomatic rotavirus infection is about 0.5 episodes per child per year (Bartlett et al, 1988). Rotavirus outbreaks are common in closed environments such as day-care centers and hospitals where the virus can spread rapidly and involve a large number of children (Bartlett et al, 1988; Vial et al, 1988; O'Ryan et al, 1990a, 1990b).

A seasonal pattern has been shown in areas with a temperate climate (Kapikian et al, 1976; Brandt et al, 1983; Konno et al, 1983; Bartlett et al, 1988; LeBaron et al, 1990; Matson et al, 1990a). Rotavirus in these areas typically occurs during autumn and winter. In North America, the annual rotavirus season begins in late fall in Mex-

A   B

Figure 22.3 RNA electrophoresis of human rotaviruses in a polyacrylamide gel. Lane A contains a serotype 3 virus which has a "long" RNA profile so named by the migration of segments 10 and 11. Lane B contains a serotype 2 virus which displays a characteristic "short" RNA profile.

ico and moves across the continental United States from southwest to northeast, resulting in a peak of rotavirus activity in March and April in eastern Canada and the northeastern United States (LeBaron et al, 1990). Sporadic cases and rare outbreaks occur during the summer months. In tropical areas, this seasonal pattern is not seen. The reason for differences in the seasonal pattern between areas is unknown.

Information on the epidemiology of the various rotavirus serotypes has expanded rapidly in the past few years. Serotypes 1, 2, 3, and 4 circulate worldwide (Birch et al, 1988; Flores et al, 1988; Georges-Courbot et al, 1988; Gomez et al, 1990; Urasawa et al, 1989; Matson et al, 1990a; O'Ryan et al, 1990b). Several or all four of these serotypes appear to co-circulate in most urban regions each year. Several different patterns of temporal serotype variation have been observed. In Houston, a single serotype predominates as the cause of symptomatic and asymptomatic infection each rotavirus season (Matson et al, 1990a; O'Ryan et al, 1990b). The predominant serotype has changed every one to two years from 1979 through 1990. In a rural area of the north central United States, serotype 1 has predominated each year for eight consecutive rotavirus seasons (1981 through 1989, Matson et al, 1990a). Only serotype 1 has been detected in children hospitalized from many small towns in this region during this eight-year period. In Buenos Aires, Argentina, two serotypes predominated simultaneously as a cause of serious rotavirus infection during one season, after which time a switch to two other serotypes occurred (Gomez et al, 1990).

These findings indicate that the occurrence of rotavirus epidemics can be sustained by a single serotype or may involve multiple serotypes simultaneously. Studies in Houston demonstrated that the predominant serotype observed among children hospitalized with rotavirus gastroenteritis was also the serotype that caused symptomatic infection in the same seasons among children attending day-care centers (O'Ryan et al, 1990b). Changes of serotype

Table 22.3 Rotavirus VP7 Types

VP7 Type	Isolated From	Strains
1	Human, pig	WA, KU, K8, RV-4, M37, D, S12, Mont, C86, C60
2	Human, pig	DS-1, S2, RV-5, RV-6, HN-126, 1076, C134
3	Human, monkey, pig, dog, rabbit, mouse, horse, cat	Yo, Ito, P, M, AU-1, RV-3, W178, SA11, CV-1, Eq-H2, RRV
4	Human, pig	St Thomas 3, Hochi Gottfried
5	Pig, horse	OSU, TFR-41, EqH1, EE
6	Cow	NCDV, UK, WC3, 486, RF
7	Chicken, turkey	CH-2, Ty1
8	Human, cow	69M, B37
9	Human	W1-61, F45, AU32
10	Cow	B223
11	Pig	YM
12	Human	L26, L27
13	Horse	L338
14	Horse	F123

predominance are regional events likely to be limited by the extent of exposure and constraints on the length of the epidemic period. How children located in various regions, but with exposure to different serotypes, differ with respect to acquired protection against illness is unknown.

22.2.4 Pathophysiology

Information on the pathophysiology of rotavirus as well as other enteric viral infections has been obtained from animal studies and is supported by studies on biopsy specimens from seriously ill children and adult volunteers. Rotavirus primarily infects mature enterocytes located in the mid and upper villous epithelium (Davidson et al, 1975). The upper small intestine is most commonly involved, although lesions may extend to the distal ileum and colon (Phillips, 1988). The infected villous enterocytes are killed and sloughed as a result of which there seems to be an adjustment of the surrounding mucosal architecture which becomes stunted and flattened. This is followed by a blastic response in the crypt cells, which tends to involve most of the affected villi. Ischemia may also play an important role in the loss and stunting of villi (Stephen, 1988). During the recovery phase, the enteroblastic cells mature and reconstruct the villous structure. Because of the loss of mature enterocytes on the tips of the villi, the surface area of the intestine is reduced. Diarrhea that occurs may be a result of this decrease in surface area, disruption in epithelial integrity, disaccharide deficiency, and/or altered countercurrent mechanisms. Although both impaired absorption and increased fluid secretion have been documented in rotavirus diarrhea, it is not clear which of these mechanisms plays a greater role in the pathogenesis of this disease. Complete mucosal recovery has been found to occur three to eight weeks after infection (Davidson and Barnes, 1979).

22.2.5 Immunology

The mechanisms involved in the immune response against most of the viral enteric pathogens is not completely understood. Bishop and associates reported that newborns asymptomatically infected with rotavirus had less severe disease after a re-exposure to rotavirus later in life than children who did not have an early exposure (Bishop et al, 1983). Ogra and Karzon, (1980) showed that after a challenge with live attenuated poliovirus vaccine, both a systemic IgG and a mucosal IgA response occurred. These and similar observations as well as the known antigenic complexity of rotaviruses highlight a need for comprehensive studies of immune responses to rotavirus including systemic and mucosal cellular and humoral responses. Following a rotavirus infection a systemic rise in IgM, IgG, and IgA occurs as well as a rise in secretory IgA in intestinal fluid (Riepenhoff-Talty et al, 1981; Davidson et al, 1983; Grimwood et al, 1988; Coulson et al, 1990; Hjelt and Grauballe, 1990). The IgM response wanes after a few weeks while the serum IgG and IgA, as well as the IgA in intestinal fluid, may remain elevated for several months (Riepenhoff-Talty et al, 1981; Davidson et al, 1983; Hjelt et al, 1985; Grimwood et al, 1988; Losonsky et al, 1988). Data conflict as to whether pre-existing serum anti-rotavirus IgA antibody is sufficient to prevent illness by rotavirus infection (Riepenhoff-Talty et al, 1981; Hjelt et al, 1987).

A major goal over the past few years has been to determine which antigens are involved in protective immunity. Protective immunity to rotavirus is most probably directed against VP7 or VP4 (Ward et al, 1988; Matsui et al, 1989). Growth of rotavirus in tissue culture is inhibited by monoclonal antibodies to VP7 or VP4, but not by antibodies to other viral proteins (Estes and Cohen, 1989). Suckling mice orally inoculated with monoclonal antibodies against either of these two viral proteins are protected when subsequently challenged with rotavirus (Matsui et al, 1989). It is not clear which of these antigens plays a more important role in inducing protective immunity. Because some epitopes of VP7 are shared across serotypes, it is unclear whether the immune response to VP7 is predominantly homotypic or heterotypic. One report suggests that the humoral antibody response in humans may be homotypic in infants and small children and heterotypic in both older children and adults with a previous exposure (Green et al, 1990).

The importance of cell-mediated immunity in providing resistance to rotavirus is even less clear. Cytotoxic T-cell and helper T-cell activity has been detected in animal models after oral immunization with reoviruses (London et al, 1987). Rotavirus-specific cytotoxic T lymphocytes located in the intestinal mucosa of mice also can be induced by oral administration of rotavirus (Offit and Dudzik, 1989). Memory cells develop and are available for the rapid mobilization to the intestinal wall following a second exposure to the virus. Spleen cells used to reconstitute immunity in immuno-compromised, chronically infected mice are able to ablate the infection (Dharkul et al, 1990). Activation of mucosal lymphocytes is also likely in humans, but how this may correlate with protection is uncertain.

22.2.6 Transmission

Rotaviruses are most commonly transmitted from person-to-person by the fecal-oral route. Several studies have reported that spread from person-to-person is rapid, which in the absence of a common source such as a contaminated water supply, permits speculation that a respiratory route of spread may have occurred (Foster et al, 1980; Gurwith et al, 1981), although this has not been proven. Transmission within families, child care centers, and hospitals is common (Pickering et al, 1981; Rodriguez et al, 1982; Grimwood et al, 1983; O'Ryan et al, 1990a). Nosocomial spread may be facilitated by contact with contaminated hands of attendants who are not otherwise infected (Samadi et al, 1983) or by asymptomatic or ill children, staff, or parents infected with rotavirus (Flewett, 1983). Nosocomial outbreaks may be controlled by closing the affected ward or cohorting infected children and restricting staff movement between affected and unaffected wards (Flewett, 1983; O'Ryan et al, 1990a). Increased family size was reported to be a significant risk factor for acquiring rotavirus infection (Gurwith et al, 1983). Increased indoor crowding of infants and toddlers in diapers facilitates environmental contamination by rotavirus excreted in feces (Gurwith et al, 1983; Keswick et al, 1983). The child day-care environment has been shown to be contaminated with fecal coliform bacteria and rotavirus which may serve as vehicles for transmission of diseases (Van et al, 1991; Wilde et al, 1992). Waterborne outbreaks have been reported (Lycke et al, 1978; Sutmoller et al, 1982; Hopkins et al, 1984; Tao et al, 1984), but common source outbreaks of infection are uncommon. The importance of other factors in transmission of rotavirus such as temperature, relative humidity, domestic animals, and excretion by asymptomatic persons is unknown.

22.2.7 Clinical Manifestations

Rotavirus has a mean incubation period of two days with a range of one to three days in children, as well as in experimentally infected adults (Davidson et al, 1975; Kapikian et al, 1983). Excretion of rotavirus in stool can precede the onset of illness by several days and can continue for eight to ten days, even after symptoms of illness have abated (Pickering et al, 1988). The quantity of virus per gram of stool is highest shortly after the onset of illness (Nagayoshi et al, 1980; Vesikari et al, 1981b). The disease produced by rotavirus is usually self-limited, but symptomatic reinfections occur (Tallett et al, 1977; O'Ryan et al, 1990b). The mean duration of illness in immunocompetent hosts is five to seven days, but chronic infections can occur in immune-deficient children and disease may be more severe in malnourished individuals, as suggested by both clinical and experimental data (Saulsbury et al, 1980; Riepenhoff-Talty et al, 1989; Dagan et al, 1990; Uhnoo et al, 1990).

Rotavirus illness occurs in all age groups, but is most common in infants from 6 months to 2 years of age. Symptoms of illness produced by rotavirus vary according to age and also within an age group. In infants and young children rotavirus gastroenteritis usually has an abrupt onset characterized by explosive, watery diarrhea and is often associated with vomiting either before or after onset of diarrheal disease (Table 22.4). Other common clinical features among children with rotavirus gastroenteritis include isotonic dehydration and compensated metabolic acidosis. Dehydration occurs in 40% to 80% of patients and is usually less than 5% of body weight, but severe dehydration and death have been reported in children (Rodriguez et al, 1977; Ho et al, 1988) and adults (Marrie et al, 1982; Echeverria et al, 1983).

Patients with rotavirus diarrhea usually have watery stools which do not contain blood or fecal leukocytes. Mucus is occasionally seen (Pickering et al, 1977; Rodriguez et al, 1977). The finding of pale, fat-containing stools has been reported (Konno et al, 1977; Thomas et al, 1981), suggesting that rotavirus infection can impair the digestion of fat and alter bilirubin excretion. Concurrent respiratory tract

Table 22.4 Clinical Characteristics of 150 Children Hospitalized with Acute Diarrhea

Clinical Finding	Percent Having Each Clinical Finding	
	Rotavirus Infection Detected (72 Patients)	Rotavirus Infection Not Detected (78 Patients)
Vomiting	96	58
Fever (°C)		
37.9–39.0	46	29
39	31	33
None	23	38
Dehydration	83	40
Hypertonic	5	17
Isotonic	95	77
Hypotonic	0	6
Irritability	47	40
Lethargy	36	27
Pharyngeal erythema	49	32
Tonsillar exudate	3	3
Rhinitis	26	22
Red tympanic membrane with loss of landmarks	19	9
Rhonchi or wheezing	8	8
Palpable cervical lymph nodes	18	9

Rodriguez et al, 1977.

symptoms including pharyngitis and otitis media are frequently associated with rotavirus infection (Rodriguez et al, 1977; Goldwater et al, 1979; Lewis et al, 1979). Rotavirus has not been shown to infect the upper respiratory tract (Lewis et al, 1979; Kapikian et al, 1983), but was found in respiratory tract secretions obtained from four patients who were hospitalized with pneumonia (Santosham et al, 1983). An association of rotavirus with intussusception (Konno et al, 1978) has not been validated in prospective studies (Mulcahy et al, 1982; Nicolas et al, 1982). Rotavirus has been associated with, but not proven to be the cause of, several diseases or conditions

including Reye's syndrome and encephalitis (Salmi et al, 1978), aseptic meningitis (Wong et al, 1984), sudden infant death syndrome (Yolken and Murphy, 1982), inflammatory bowel disease (Blacklow and Cerkor, 1982), neonatal necrotizing enterocolitis (Rotbart et al, 1983), and Kawasaki's syndrome (Matsuno et al, 1983). Children with rotavirus diarrhea, when compared to children with diarrhea not caused by rotavirus, are more likely to have watery diarrhea, nausea, vomiting, abdominal pain, dehydration, and loss of appetite (Rodriguez et al, 1977), but so much overlap occurs between the groups that symptoms alone cannot be used to differentiate the cause of an episode of diarrhea.

There is increasing evidence that rotavirus may cause liver damage, especially in patients with immunodeficiency diseases. This has been shown experimentally in immunodeficient mice that developed hepatitis following infection with some rotavirus strains (Riepenhoff-Talty et al, 1990) and clinically in immunodeficient children who died after experiencing chronic rotavirus diarrhea and in whom rotavirus was detected in postmortem liver specimens (Gilger et al, 1992).

Infection of adults often occurs among those in close contact with young children (Kapikian et al, 1976; Kim et al, 1977; Pickering et al, 1981; Grimwood et al, 1983), and has been reported in adult travelers, military personnel, elderly persons in institutions, and hospitalized adults (Von-Bonsdorff et al, 1978; Vollet et al, 1979; Halvorsrud and Orstavik, 1980; Holzel et al, 1980; Marrie et al, 1982). Adults typically tend to be asymptomatic or have mild illnesses (Wenman et al, 1979). Asymptomatic rotavirus infections also are common in neonates (Christie et al, 1978; Bryden et al, 1982; Rodriguez et al, 1982; Bishop et al, 1983) and may be caused by less virulent strains (Gorziglia et al, 1987; Tam et al, 1990), a concept not yet proven. In children in day-care centers asymptomatic infections represent 40% to 50% of all rotavirus infections (Pickering et al, 1988; O'Ryan et al, 1990b).

22.2.8 Mixed Infection

Simultaneous infection of children by rotavirus and other viral, bacterial, and parasitic enteropathogens occurs, particularly in developing countries (Rodriguez et al, 1977; Pickering et al, 1978; Brandt et al, 1983; Van et al, 1992). Animal studies suggest that a mixed infection is more severe than infection caused by rotavirus alone (Runnels et al, 1986). This problem needs further study in humans.

22.2.9 Sequential Infection

The frequency of sequential rotavirus infections in individual children by different subgroups or serotypes is not clear. Recurrent rotavirus infections are common in children attending day-care centers (Bartlett et al, 1988; O'Ryan et al, 1990b) and in other settings where exposure is high. Second infections by the same and by different serotypes occur, but the clinical and immunologic significance of these reinfections remains to be determined (O'Ryan et al, 1990b).

22.2.10 Isolation Sites

Rotaviruses may be shed in stools in up to 10^{11} particles per gram of feces. Stools collected during the first several days of illness have the highest concentration of virus, whereas stools collected eight days or more after the onset of illness rarely contain virus (Vesikari et al, 1981b; Pickering et al, 1988). Stool specimens should be tested immediately, but may be held at 4 °C if delay is minimal. Otherwise, samples should be frozen at −70 °C until tested. For some methods of identification, such as ELISA, rectal swabs may be sufficient. For direct isolation of virus in cell culture, a rectal swab sample is usually inadequate

(Wyatt et al, 1983). Rotavirus recovery from other sites has been reported in a few cases (Santosham et al, 1983).

22.2.11 Stability

The stability of rotaviruses has been studied using simian SA-11 rotavirus as a model. The SA-11 strain is stable over a pH range of 3.5 to 10.0, and calcium is required for stability of infectious particles. Infectivity is lost by treatment with low concentrations (5 mM) of ethylenediamine tetraacetic acid (EDTA) or by heating at 50 °C for 15 minutes in 2 M $MgCl_2$. Rotavirus survival on environmental surfaces is enhanced by the presence of fecal material (Keswick et al, 1983). The presence of fecal material also protects viruses from the action of disinfectants. A solution of 70% ethanol or an ethanol-containing disinfectant is an effective solution for decontaminating surfaces (Brade et al, 1981), whereas other disinfectants, such as chlorhexidine gluconate and povidone-iodine are ineffective (Tan and Schnagl, 1981). Lysol, formalin, and an iodophor preparation also may be effective. Chlorine has been shown to rapidly inactivate SA-11 in water. However, human strains of rotavirus may be more resistant to the action of chlorine than animal strains (Butler and Harakeh, 1983). Rotaviruses have been found in treated drinking water and in sewage where the presence of solids such as fecal material may enhance their survival (Smith and Gerba, 1982; Deetz et al, 1984; Keswick et al, 1984). Likewise, rotavirus has been detected on surfaces in day-care centers (Keswick et al, 1983; Wilde et al, 1992) and undoubtedly could be found in the hospital environment.

22.2.12 Isolation Methods

Human rotaviruses are difficult to cultivate from clinical specimens. Even with optimal techniques and sample collection, many rotavirus samples fail to infect and grow in cell cultures. In 1980, the WA strain was successfully adapted to grow on serial passage in primary African green monkey kidney (AGMK) cell culture after 11 prior passages in newborn gnotobiotic piglets (Wyatt et al, 1980). Circumvention of the fastidious nature of human rotaviruses was accomplished by genetic reassortment during mixed infection with a temperature-sensitive mutant of a cultivatable bovine rotavirus. The human rotavirus gene 4, which is responsible for growth restriction in vitro, was replaced by the corresponding gene from the tissue culture adapted bovine rotavirus (Greenberg et al, 1981). Subsequently, direct isolation of numerous human rotaviruses has been successful without the cumbersome techniques initially utilized since investigators have become familiar with the growth requirements of these strains (Hasegawa et al, 1982; Kutsuzawa et al, 1982; Birch et al, 1983). Isolates may be recovered in roller tube cultures of MA104 cells by treating stool specimens with trypsin or other enzymes capable of appropriately cleaving VP4 before inoculation and by incorporating a small amount of trypsin or other enzymes, e.g., pancreatin in the maintenance medium.

A variety of cell lines have been tested for rotavirus isolation from clinical samples including MDBK, PK-15, BSC-1, LLC-MK2, BGM, CV-1, and MA104. In addition, primary cell cultures prepared from a variety of species are widely used. Primary AGMK cells and the MA104 cell line have proven to be the most useful. In particular, MA104 cells have been used with success for rotavirus isolation, passage, and plaque formation (Estes et al, 1983).

The following guideline for rotavirus culture is used in the Division of Molecular Virology, Baylor College of Medicine. Approximately 0.5 grams of stool are diluted in 500 μL of Tris-saline containing 2 mM $CaCl_2$. Two centrifugation steps at 31,000 xg yield a supernatant fluid free of most particles although it may be toxic to cell cultures. Fluorocarbon extraction is not required. The supernatant is treated with gentamicin and penicillin-streptomycin and incubated for one hour at 37 °C to eliminate bacteria. Virus is pre-activated with trypsin (10 μg/mL) or pancreatin (50 μg/mL) for 30 minutes at 37 °C. An aliquot of 100 μL virus suspension is added to a cell monolayer in each of two or three roller tubes and incubated for one hour at 37 °C. Thereafter, the virus suspension is decanted and 1.5 mL of medium (2× Eagle's minimum essential media) with 10 μg pancreatin/mL or 1 μg trypsin/mL is added. Tubes are incubated at 37 °C for four days and cells are observed daily for cytopathogenic effect (CPE). After four days, or earlier if CPE is noted, cells are frozen. Two or three passages may be needed to recover detectable virus. After the second or third passage, monitoring for rotavirus antigens by immunofluorescence, EM, or ELISA is recommended. If no virus is detected, further culturing of samples likely will be unsuccessful.

The development of plaque assays for HRV has provided an opportunity to study viral replication, genetic variation, and immune responses to specific serotypes (Urasawa et al, 1982; Wyatt et al, 1982; Estes et al, 1983). The plaque assays in the Molecular Virology Division at Baylor College of Medicine are performed in a cell monolayer under semisolid medium using 35 mm dishes or 6-well plates. The day before the assay, medium is decanted from cells and 3 mL/well of growth media (Medium 199) are added. If this is not possible, the cells are washed twice with a tris-buffer solution (TBS) before inoculation. Virus is pre-activated with 10-20 μg trypsin/mL and incubated at 37 °C for 30 minutes. At 4 °C virus is diluted tenfold (from 10^0 to 10^{-3}) in TBS or medium 199, containing penicillin-streptomycin, to a final volume of 1 or 2 mL for each dilution. The medium is decanted from the wells which are then inoculated with 0.2 mL of the appropriate dilution. Two wells per dilution are inoculated and incubated in a 5% CO_2 incubator at 37°C for 60 minutes, rotating the plates or dishes every 15 minutes; 3 mL of a 1:1 mixture of agarose and 2× medium 199 containing phenol red and also containing 3% glutamine, penicillin-streptomycin, 7.5% NaHCO3, pancreatin 4× (25 mg/mL) and DEAE-dextran (10 mg/mL, MW 500K) is added as overlay. Cells are stained two to three days later with agarose plus overlay solution without pancreatin or DEAE but with neutral red (1:1000) added; 2 mL/well of the solution is used for 35 mm dishes and 3 mL/well for 6-well plates. Plates are checked daily for plaque production. Plaques may be optimal for counting four to six days after inoculation.

Plaque reduction assays and immune peroxidase focus reduction assays have been developed and are currently in use to determine antibody responses to specific rotavirus serotypes. These assays are based on the serotype-specific neutralizing activity of serum obtained from individuals or animals infected with a particular rotavirus serotype (Gerna et al, 1984; Estes and Cohen, 1989).

22.2.13 Diagnostic Methods

The ability to study the epidemiology of rotaviruses and to identify characteristics of individual strains depends upon the availability of rapid, practical detection assays. Several methods of identification of rotavirus or rotavirus antigens in fecal extracts have been developed. The most widely used assays include EM, IEM, ELISA, latex agglutination, and gel electrophoresis.

22.2.13.a Electron Microscopy

Electron microscopy was the original diagnostic technique used for identification of rotavirus. It is now the reference method

for identification. Rotaviruses are readily recognized in stool specimens because of their characteristic 70 nm size and morphology (Figure 22.2). EM is also important in the detection of non-group A rotaviruses. These viruses have the characteristic morphology of rotavirus but do not react with group A specific immunologic reagents used in the currently available commercial assays. Because rotavirus groups and serotypes cannot be distinguished visually, the technique is limited to rotavirus identification. An advantage of EM is that other viruses (astrovirus, calicivirus, coronavirus, and adenovirus) may be visualized in stool specimens and dual virus infections can be detected.

22.2.13.b Immunoelectron Microscopy

Immunoelectron microscopy (IEM) has been employed for detection, identification and serotyping of rotaviruses and other viruses in feces (Brandt et al, 1981; Gerna et al, 1989). It is more sensitive than EM, and in some cases may be more useful because some of the enteric viruses which are approximately 30 nm in size are more readily identifiable by this technique. In this technique a serum containing high titer antibody to the virus being sought is heat inactivated at 56°C for 30 minutes and clarified by ultracentrifugation at 100,000 × g for 60 minutes. One tenth serum at an optimal dilution is added to 0.5 mL of fecal suspension and incubated at 37°C for one to two hours or overnight at 4°C. The volume is then increased to 5 mL and the virus antibody complexes are pelleted by ultracentrifugation for 90 minutes at 35,000 × g. The supernatant is discarded and the pellet is resuspended in water, transferred to coated grids, negatively stained, and examined by conventional techniques (see Chapter 5). Antibody virus complexes are easier to find than single virus particles. A disadvantage of IEM is that visualized par-

Table 22.5 Some of the Commercially Available Diagnostic Kits for Detection of Rotavirus

	Source
Monoclonal (MAb)-based EIA	
Rotaclone	Cambridge BioTech Corporation
Pathfinder	Sanofi Diagnostics Pasteur
Polyclonal (PAb)-based EIA	
Rotavirus EIA	International Diagnostic Laboratories
Wellcozyme	Wellcome Diagnostics
Rotazyme II	Abbott Laboratories
MAb and PAb EIA	
Testpack	Abbott Laboratories
Latex Agglutination Tests (PAbs)	
Rotastat	International Diagnostics Laboratories
Meritec Rotavirus	Meridian Diagnostics
Virogen Rotatest	Wampole Laboratories
Wellcome RV Latex	Wellcome Diagnostics

Marchlewicz et al, 1988; Brooks et al, 1989; Dennehy et al, 1990.

ticles are coated with antibody, which distorts virus structure. Coating of the grid with antibody before the direct application of the sample avoids this problem. This technique of solid phase IEM has been used for typing clinical HRV strains (Gerna et al, 1989).

22.2.13.c Immunoassays

Immunologic methods have become popular for detecting a wide variety of different organisms. In contrast to EM, IEM, and IF, this technique is relatively inexpensive, does not require special training, and can be applied to many samples simultaneously. Of these techniques, the best known and most widely used are enzyme-linked immunosorbent assay (ELISA) and latex agglutination (Morinet et al, 1984; Dennehy et al, 1988; Brooks et al, 1989; Kok and Burrell, 1989; Dennehy et al, 1990). Rotavirus ELISAs can detect 10^5 rotavirus particles per gram of stool, which is a tenfold greater sensitivity than EM.

Monoclonal antibodies have been shown to further increase the sensitivity and specificity of the ELISA (Cukor et al, 1984; Knisley et al, 1986; Pacini et al, 1988).

Many commercial kits are available for detection of rotavirus (Table 22.5). These kits have different degrees of specificity and sensitivity. Independent of the test used, it is important to evaluate each result in order to avoid false positive values. To assure specificity we recommend confirmation of a positive result by adding a blocking step with a polyclonal antibody (Pickering et al, 1988). This is especially important for samples obtained outside of the rotavirus season when the percentage of false positive results increases. Monoclonal antibody-based ELISAs that differentiate the most common rotavirus serotypes have been developed and are available as laboratory research tests and in one commercial kit (Coulson et al, 1987; Taniguchi et al, 1987; Birch et al, 1988; Matson et al, 1990a).

Latex agglutination tests have been shown to have a high specificity, although in general a lower sensitivity when compared to ELISA (Miotti et al, 1985; Doern et al, 1986). These tests may be used to rapidly screen children with diarrhea if samples are obtained early in the disease course when virus shedding is higher. The lower sensitivity may require that in certain patients negative specimens be confirmed by ELISA.

22.2.13.d Immunofluorescence

Immunofluorescent (IF) staining of cell cultures inoculated with fecal preparations permits identification of rotavirus (Birch et al, 1979; Yolken and Stopa, 1979). The sensitivity of the test can be increased by centrifuging inocula at 3000 × g for two hours to enhance virus-cell attachment (Banatvala et al, 1975) or by using LLC-MK2 cells grown in microtiter plates (Bryden et al, 1977). The IF assay appears to be as sensitive as EM but less sensitive than IEM. Rotavirus antigen also can be detected in infected cells or histologically-prepared tissue sections using an indirect immune peroxidase test or a peroxidase-anti-peroxidase test (Graham and Estes, 1979). These methods can also be applied to IEM (Altenburg et al, 1980; Petrie et al, 1982).

22.2.13.e Electropherotyping

Assessing the migration pattern of the 11 dsRNA genome segments of rotavirus following electrophoresis in polyacrylamide gels is called electropherotyping (Herring et al, 1982). The human rotaviruses show extensive diversity in their electrophoretic patterns (Rodger et al, 1981; Estes et al, 1983; Spencer et al, 1983; Dimitrov et al, 1984; Konno et al, 1984; Ruggeri et al, 1989; Gouvea et al, 1990a); electropherotyping is a valuable means of studying the epidemiology of infection. For example, because the number of different rotavirus electropherotypes within a geographic region at any given time is limited (Rodger et al, 1981; Dimitrov et al, 1984), electropherotyping would detect an outbreak of infection caused by more than one strain (Rodriguez et al, 1983; O'Ryan et al, 1990b). The prevalence of rotavirus electropherotypes changes over time within geographic regions (Espejo et al, 1980). Although many different electropherotypes may be detected within one serotype, common features have been identified. Almost all serotype 2 viruses have a "short" electropherotype, in which the 10 and 11 gene segments migrate a relatively short distance (Figure 22.3). Serotypes 1, 3, and 4, in contrast, display a "long" electropherotype (Kalica et al, 1981; Brown et al, 1988; Georges-Courbot et al, 1988).

Electropherotyping is useful for detecting non-group A rotaviruses, which do not react with the commercially available ELISAs. Although the RNA profiles of groups B through E rotaviruses differ markedly from the RNA profiles for group A rotaviruses, there is no classification scheme based on these electropherotype differences (Desselberger, 1988).

22.2.13.f Genetic Analysis

Radiolabeled RNA has been used in hybridization assays to detect rotavirus in fecal specimens (Flores et al, 1983). The sensitivity of this assay has been reported to be 10 to 100 times higher than that of ELISA and the specificity near 100%. Radiolabeled cDNA prepared from double-stranded RNA segments is highly sensitive and specific for detection of rotaviruses (Pedley and McCrae, 1984; Lin et al, 1985). Oligonucleotide probes for serotyping rotaviruses are being developed (Yamakawa et al, 1989; Sethabutr et al, 1990), and enzyme-conjugated synthetic probes have been developed (Bellinzoni et al, 1989). These latter probes have the advantage of not requiring the use of radioactive material (Olive and Sethi, 1989).

The polymerase chain reaction (PCR) has been applied to amplify rotavirus nucleic acid in stool samples and to serotype viruses based upon differences of the gene encoding VP7 (Gouvea et al, 1990b). This technique can be 1,000- to 10,000-fold more sensitive than ELISA or electrophoresis for virus detection (Wilde et al, 1990, 1992); techniques for removing inhibiting substances in some stool specimens are being standardized.

22.3 ENTERIC ADENOVIRUSES

22.3.1 Characteristics

The group classification of adenovirus (A-F) is based on differences in viral hemagglutinating or oncogenic properties, size of the structural proteins, and DNA homology (Wadell et al, 1984, 1986). Further differentiation into one of the 47 different serotypes is based upon neutralization assays. The enteric adenoviruses belong to group F. Adenoviruses are non-enveloped viruses that are 70 to 90 nm in diameter (Figure 22.2). The viral capsid has icosahedral symmetry and antenna-like projections emerging from each of the 12 vertices. The genome is a 33 to 45 kilobase (kb) linear double-stranded DNA (Van Loon et al, 1985).

The enteric adenovirus (EA) types 40 and 41 are second to rotavirus as a recognized cause of viral gastroenteritis in children (Uhnoo et al, 1984; Brandt et al, 1985; Madeley, 1986; Cruz et al, 1990; Kotloff et al, 1989). Adenoviruses from other groups (called non-enteric adenoviruses) are associated with diarrhea, although much less frequently and usually in association with symptoms of respiratory tract infection.

22.3.2 Epidemiology and Pathophysiology

Both EA types appear to be widespread and cause endemic diarrhea. Outbreaks have been reported in orphanages, hospitals, and day-care settings (Richmond et al, 1979; Chiba et al, 1983; Paerregard et al, 1990; Van et al, 1992). Seasonal shifts from one serotype to another have been shown in Holland, but not in Washington, D.C., where both serotypes were found to circulate simultaneously all year round (Wadell et al, 1988). Infection with EA appears to increase in the summer, although with a less marked seasonality than that exhibited by rotavirus.

EA are more likely to infect children than adults (Vesikari et al, 1981a). Antibody prevalence to EA has been shown to increase from 20% during the first 6 months of life to 50% or greater by the third to fourth year of life (Shinozaki et al, 1987). The mode of transmission is most probably fecal-oral (Isaacs et al, 1986). The incubation period is 3 to 10 days.

There is limited information on the pathophysiologic mechanisms involved in enteric adenovirus infections. Observations from a single human case suggest that the nuclei of villous epithelial cells are the site of virus concentration, and that microvilli structure is altered. How this affects intestinal function is not known (Phillips, 1988).

22.3.3 Clinical Manifestations

EA apparently cause a less severe but more prolonged illness than rotavirus. Diarrhea lasts from 6 to 9 days in most cases with a range of 4 to 23 days (Uhnoo et al, 1986; Hierholzer et al, 1988; Kotloff et al, 1989; Van et al, 1992). Vomiting and fever may precede or accompany the diarrhea. Persistent lactose intolerance has been reported in a few children infected with enteric adenovirus (Uhnoo et al, 1986). An association between enteric adenoviruses and coeliac disease has been postulated, but not proven (Kagnoff et al, 1984).

22.3.4 Diagnostic Methods

EM was the initial detection method for EA, but commercial assays that use monoclonal antibodies incorporated into an ELISA are now available. Hybridization assays have also been developed for adenoviruses, but they are less sensitive than antigen-detection techniques (Hammond et al, 1987). EA can be cultivated in special cell lines. Restriction enzyme analysis of DNA extracted from virus isolates is the definitive method for classifying individual isolates (Takiff et al, 1981; Wadell et al, 1984).

22.4 NORWALK VIRUSES

22.4.1 Characteristics

Norwalk virus is the best-studied member of a group of morphologically similar small round enteric viruses that bear little or no antigenic relation to each other. Many of these Norwalk-like viruses are known only from a single outbreak and have been named after the sites at which the outbreak occurred. They include Hawaii, Snow Mountain, Montgomery County, Taunton, Otofuke and Sapporo agents. None of these viruses have been passaged in cell culture as yet and each was initially detected by EM or IEM. The antigenic relationships among these viruses are not yet clear (Wyatt et al, 1974; Thornhill et al, 1977; Dolin et al, 1982; Okada et al, 1990). Norwalk, Hawaii, and Snow Mountain agents seem to be antigenically distinct, while the Montgomery County and Norwalk agents are probably antigenically related. The Marin County agent was initially thought to be Norwalk-like but is now considered to belong to one astrovirus type. The Ditchling, Cockle, and Paramatta agents, also initially thought to be related to Norwalk, differ in that they lack discernible surface features and a sharply delineated outer edge. These agents are also somewhat smaller, suggesting that they are more re-

lated to parvoviruses (Appleton, 1987). Further studies will be needed to classify these viruses as well as to determine their role in the pathogenesis of diarrheal disease.

Norwalk virus was discovered by Kapikian et al (1972) using IEM. Since then, IEM has been used to strengthen the association between viral particles and gastroenteritis in many outbreaks of diarrhea (Thornhill et al, 1977; Grohmann et al, 1980; Caul et al, 1982; Dolin et al, 1982; Kaplan et al, 1982; Morse et al, 1986; Sekine et al, 1989; Okada et al, 1990; Vial et al, 1990). The Norwalk virus particle is 27 nm in diameter, round, not enveloped, and has an irregular surface structure (Figure 22.2) (Dolin et al, 1972; Kapikian et al, 1972). When treated under laboratory conditions that have been applied to study other enteric viruses, the viruses in this group have been found to be fragile. The inability to passage the virus in cell culture has resulted in a lack of data on the physical properties of this agent and similar viruses. Serologic cross-reaction between calicivirus UK-4 and Norwalk virus was observed in two of three small outbreaks, suggesting that Norwalk virus may be a calicivirus (Cubitt et al, 1987a). The genome of Norwalk virus has been cloned and found to be a ssRNA of about 8 kb in length (Jiang et al, 1990).

22.4.2 Epidemiology

Serologic studies have demonstrated that there is a higher prevalence of antibodies to Norwalk virus among children in developing countries than among children in the United States and other developed countries (Blacklow et al, 1979; Greenberg et al, 1979a; Cukor et al, 1980), and that a high incidence of seroconversion to Norwalk virus occurs in the second and third years of life (Black et al, 1982). Serologic studies of families (Pickering et al, 1982) have demonstrated that Norwalk virus causes family outbreaks of diarrhea. Norwalk virus has a worldwide distribution, and in the United States 50% of adults have developed antibody by the time they are 40 years of age (Greenberg et al, 1979a).

Infection with Norwalk-like viruses is recognized to cause sporadic episodes and clusters of family illness, but more commonly is associated with large waterborne (Taylor et al, 1981; Baron et al, 1982; Kaplan et al, 1982a; Levine et al, 1990) or foodborne outbreaks (Kaplan et al, 1982b; Kuritsky et al, 1984; Riordan et al, 1984; Morse et al, 1986; Bean et al, 1990). Norwalk virus does not appear to be a zoonosis and, therefore, the sources of waterborne outbreaks probably are human excretors. Transmission of Norwalk virus occurs as a result of person-to-person spread and by both water (drinking water, ice, and swimming) (Levine et al, 1990) and contaminated food. Some of the foodborne outbreaks have been associated with shellfish, such as oysters (Grohmann et al, 1980; Gunn et al, 1982; Gill et al, 1983; Morse et al, 1986; Bean et al, 1990). It was assumed that the oysters were contaminated in the ocean by human sewage containing the virus. Foodborne outbreaks not involving shellfish have resulted from the direct contamination of salads and cake frosting by food handlers (Griffin et al, 1982; Pether and Caul, 1983; Kuritsky et al, 1984; Riordan et al, 1984).

22.4.3 Pathophysiology

Inoculation of Norwalk virus into chimpanzees resulted in excretion of virus in feces and in production of a Norwalk virus-specific immune response (Wyatt et al, 1978). Administration of the virus to other laboratory animals, including a number of primate species, has not resulted in illness (World Health Organization, 1980). When a human fecal filtrate containing the Norwalk virus was administered orally to human volunteers, approximately 50% developed illness (Blacklow et al, 1972; Johnson et al, 1990).

Table 22.6 Presence of Symptoms in 38 Outbreaks of Norwalk Virus Infection

Symptom	Number of Outbreaks Positive for Symptom	Percentage of Patients with Symptoms (Range)
Nausea	30	79 (51–100)
Vomiting	34	69 (25–100)
Diarrhea	34	66 (21–100)
Abdominal cramps	30	71 (17–90)
Headache	22	50 (17–80)
Fever	29	37 (13–71)
Chills	14	32 (5–74)
Myalgia	14	26 (11–73)
Sore throat	7	18 (7–32)

Kaplan et al, 1982a.

Pathologic examination of the intestines of adult volunteers infected with Norwalk virus showed abnormal histologic findings in the mucosa of the proximal small bowel, including mucosal inflammation, absorptive cell abnormalities, villous shortening, crypt hypertrophy, and increased epithelial cell mitosis. The gastric and colonic mucosa remain histologically normal (Agus et al, 1973; Meeroff et al, 1980). These abnormal findings persisted for at least four days after clinical symptoms ceased (Schreiber et al, 1973). However, the virus has not been detected within abnormal mucosal cells.

22.4.4 Clinical Manifestations

Much of what is known about the clinical aspects and immunologic properties of Norwalk virus has been the result of volunteer studies or of studies of large waterborne or foodborne outbreaks. In volunteers fed Norwalk virus, nausea, vomiting, diarrhea, and abdominal cramps lasted one to four days, if symptoms occurred (Table 22.6). Stool specimens generally do not contain blood or mucus. Vomiting is more frequent among children, whereas, diarrhea is more common among adults (Kaplan et al, 1982b). The vomiting may be caused by a decrease in gastric emptying, which has been demonstrated in adult volunteers (Meeroff et al, 1980). The illness may necessitate hospitalization and can be fatal in elderly, debilitated persons (Kaplan et al, 1982a). No prolonged illness or long-term side effects associated with Norwalk virus infection have been noted (Dolin et al, 1971; Blacklow et al, 1972; Johnson et al, 1990).

Transmission of Norwalk virus as measured by seroconversion occurs year-round, although in Bangladesh it is more frequent during the cool, dry season (Black et al, 1982). In the United States, outbreaks have been noted in elementary schools, colleges, camps, recreational areas, nursing homes, restaurants, cruise ships, and in cities with contaminated municipal water supplies (Greenberg et al, 1979b; Kaplan et al, 1982b; Reid et al, 1988; Ho et al, 1989). Common source outbreaks of gastroenteritis due to Norwalk virus may involve persons of all ages, occur during all seasons, and generally last no longer than one week. Longer outbreaks occur when new groups of susceptibles are introduced into the area (Kaplan et al, 1982a). The incubation period has a mean of 24 to 48 hours with a range of 4 to 77 hours. Maximal excretion occurs at the onset of illness and shortly thereafter (Thornhill et al, 1975). Excretion of low concentrations of virus may occur for weeks after the illness.

22.4.5 Isolation Sites

Norwalk virus has been identified in stool specimens and in vomitus (Greenberg et al, 1978, 1979c). Virus has not been detected from all volunteers fed Norwalk virus, including some who become ill. In situations where the confirmation of Norwalk virus infection is sought, stool specimens and paired sera should be collected and frozen at −70 °C until they can be processed.

22.4.6 Stability

Limited information is available on the stability of Norwalk virus. The virus is stable at −70 °C and resistant to acid treatment (pH 2.7 for three hours) and ether treatment, properties similar to other agents causing diarrhea (Dolin et al, 1972). The virus also may be resistant to heat treatment at 60 °C for 30 minutes (Cukor and Blacklow, 1984).

22.4.7 Isolation and Identification

Because none of the Norwalk virus agents can be passaged in cell culture or in a readily available laboratory animal model, the isolation of Norwalk virus from clinical samples is currently not possible. Identification of Norwalk-like agents is accomplished using EM, IEM, and RIA techniques. All serologic assays for viruses of the Norwalk group depend on reagents in critically short supply. Therefore, diagnosis of Norwalk virus infection is currently available only at reference laboratories.

Immunoelectron microscopy has been used successfully to visualize Norwalk virus and related agents in diarrheal stool specimens (Figure 22.2) (Grohmann et al, 1980; Kapikian et al, 1982; Okada et al, 1990). Due to the small size and amorphous surface of Norwalk virus particles, direct EM (in the absence of antibody-mediated clumping of the virus particles) is rarely adequate to detect or identify the agent in stool specimens. IEM using acute or convalescent sera can help establish the etiology of an illness by comparing the degree to and titer at which particles are coated with antibody.

The radioimmunoassay for Norwalk virus developed by Greenberg et al (1978) is more sensitive than IEM and is able to detect both soluble and particulate antigens. The assay utilizes acute and convalescent sera from volunteers. A modified assay detects antibodies to semi-purified

Norwalk virus from various outbreaks by employing a blocking step (Greenberg et al, 1978; Kapikian et al, 1982). Hybridization assays using gene probes and immunoassays using antisera to virus-derived recombinant proteins likely will be commercially available in the near future.

22.5 ASTROVIRUSES

22.5.1 Characteristics

Astroviruses are 20 to 30 nm in diameter and have a characteristic five to six pointed star appearance when visualized by EM (Figure 22.2). The virus genome is a positive-strand RNA of approximately 7,500 nucleotides that encodes four structural proteins (Kurtz and Lee, 1987). Five different antigenic types have been described using polyclonal rabbit antisera (Kurtz and Lee, 1984; 1987).

22.5.2 Epidemiology, Pathophysiology, and Clinical Manifestations

Gastroenteritis due to astrovirus occurs worldwide and has been associated with outbreaks of mild diarrhea in schools, pediatric wards, and nursing homes (Madeley et al, 1977; Madeley, 1979a; Konno et al, 1982). Illness occurs mainly in children and the elderly. A study from Bangladesh reported the detection of astrovirus in 8.6% of children less than 5 years of age with diarrhea compared to 2% of children without diarrhea. Co-infection with other pathogens occurred commonly (Herrman et al, 1991). This virus may account for 3% to 5% of hospital admissions for diarrhea (Ellis et al, 1984).

Astroviruses have been detected in the low villous epithelium of symptomatic humans, in surface epithelium of a flat biopsy, and in macrophages of the lamina propria (Phillips et al, 1998). In animals, astrovirus infected cells showed vacuolation followed by degeneration and cell death, which led to villous atrophy (Kurtz, 1988). As is true

for other enteric viruses, it remains to be determined which physiologic mechanisms might be impaired by this cytopathic effect.

Since more than 80% of adults have antibodies against the virus, they usually do not develop illness or develop mild symptoms characterized by fever, malaise and watery diarrhea that usually lasts three days. Vomiting is uncommon (Isaacs et al, 1986). Short-term monosaccharide intolerance and more prolonged cow's milk intolerance have been reported (Nazer et al, 1982), although these are probably uncommon events. Data suggest a peak incidence in winter (Grohmann, 1985), and an incubation period of one to four days (Konno et al, 1982).

22.5.3 Detection Methods

Astroviruses grow well in human embryo kidney cells in the presence of trypsin. EM, IEM, and immunofluorescence on cell culture can be used as detection methods. A monoclonal antibody-based ELISA had a sensitivity and specificity of more than 90% when compared to IEM (Herrman et al, 1990).

22.6 CALICIVIRUSES

22.6.1 Characteristics

Caliciviruses are 35 to 39 nm in diameter, with a distinct "Star of David" appearance by EM (Figure 22.2) (Madeley, 1979b). They have characteristic cup-shaped surface depressions arranged in icosahedral symmetry. The genome of human calicivirus has not been characterized, although animal caliciviruses contain a single-stranded RNA. Five antigenic types have been described (Cubitt et al, 1987b).

22.6.2 Epidemiology and Clinical Manifestations

Antibody prevalence studies indicate that virtually all children have been infected with this virus by five years of age (Cubitt and McSwiggan, 1987c; Nakata et al, 1988). Studies conducted in day-care centers have shown that calicivirus-associated diarrhea is widespread, although sporadic, and that there is a high proportion of asymptomatic infections (Matson et al, 1989; Matson et al, 1990b). Human calicivirus infections occur year-round although a seasonal predominance has been suggested in some studies (Cubitt, 1987b; Matson et al, 1989). Eighteen outbreaks have been reported since 1978, all in institutional settings (Matson, in press). Outbreaks involved the elderly in nursing homes in Japan and England (Cubitt et al, 1981; Humphrey et al, 1984; Gary et al, 1987) with attack rates ranging from 50% to 70%. Caliciviruses are transmitted by the fecal-oral route, and there is evidence that they may be a major cause of foodborne outbreaks of gastroenteritis (Cubitt, 1987b). The incubation period is approximately four days and the clinical symptoms are indistinguishable from those caused by rotavirus. Virus can be found in stools up to two weeks after the onset of symptoms, although in approximately half of the patients, viral excretion is complete after one week of illness (Cubitt, 1987b; Matson, 1989).

22.6.3 Diagnostic Methods

Caliciviruses can be detected in stools by EM, although the sensitivity of this method is low due to the small size of the virus. IEM and ELISA, which are more sensitive techniques than EM, are currently available only in research laboratories (Matson et al, 1989).

22.7 OTHER VIRUSES

A number of different viruses have been identified in stool samples from humans with diarrhea, but their role in causing gastroenteritis in humans remains unclear.

Coronaviruses are 180 to 200 nm in diameter and have characteristic petal-shaped surface projections that give the appearance of a "solar" corona when visualized by EM (Figure 22.2) (Caul et al, 1975; Garwes, 1982). These viruses are well-documented as a cause of gastroenteritis in animals (Garwes, 1982). In humans, coronaviruses cause the common cold. They have been found in stools of children with gastroenteritis, in newborns with necrotizing enterocolitis, and in the intestinal epithelium of a 13-year-old boy who died of diarrhea (Vaucher et al, 1982; Ettig and Altshuler, 1985; Gerna et al, 1985). They have also been identified in stool specimens from homosexuals and from people without diarrhea who live in crowded areas with a water supply that received inadequate treatment (Ashley and Caul, 1988). EM examination of samples suspected to contain human enteric coronaviruses must be interpreted with caution because pieces of sloughed intestinal epithelium may appear as coronavirus-like particle forms.

Parvoviruses and parvovirus-like agents are small (20 to 30 nm), featureless, round particles (Figure 22.2) (Caul and Appleton, 1982). Studies from Norway and England report this virus in stool specimens as the sole agent in a few children with gastroenteritis (Kjeldsberg, 1977; Oliver and Phillips, 1988).

Pestiviruses are 40 to 60 nm in diameter and belong to the Togaviridae. A monoclonal antibody-based ELISA detected this virus in stool specimens from 30 of 128 children with diarrhea of unknown etiology, in 1 of 28 children without diarrhea, and in 1 of 31 children with rotavirus diarrhea (Yolken et al, 1989). The diarrhea was mild and accompanied by respiratory tract symptoms. In 1984, a virus resembling the Breda virus of calf diarrhea was detected in stool specimens from 20 individuals with gastroenteritis (Figure 22.2) (Beards et al, 1984). A case of hemorrhagic enterocolitis associated with Breda virus infection has also been reported (Lacombe et al, 1988).

22.8 TREATMENT AND PREVENTION OF VIRAL GASTROENTERITIS

The cornerstone of the treatment of acute viral gastroenteritis is rehydration. Hospitalization of some patients may be required to assure adequate rehydration, correct acidosis, and to ensure adequate nutrition. There is no specific antiviral therapy for any of these viral enteropathogens.

Preventive measures are important to limit outbreaks of viral gastroenteritis, especially in crowded environments, such as day-care centers, hospitals, cruise ships, and camps. Adequate hygiene, frequent handwashing, and careful food preparation need to be enforced. Day-care center staff who prepare food should not change diapers. Soiled diapers should be disposed of in a way that eliminates further contact by care providers and children. Facilities should include designated diaper changing areas near sinks where paper towels and soap are available. Cohorting of children in an outbreak situation should be considered. Many of these procedures are dictated by local regulations, a list which should be sought from the regional health department.

Several vaccines against rotavirus have been developed and tested in clinical trials worldwide (Christy et al, 1988; Halsey et al, 1988; Gothefors et al, 1989; Kapikian et al, 1989; Lanata et al, 1989; Clark et al, 1990; Vesikari et al, 1990). The vaccines tested have included bovine serotype 6 viruses, rhesus serotype 3 viruses, and more recently multivalent vaccines. Reassortant viruses containing the gene for the human VP7 and the remaining ten genes from an animal rotavirus are also being studied. The protection conferred by the monovalent vaccines has varied from none to greater than 80% among different populations. Trials with the reassortant viruses are underway. However, the development of an effective vaccine will require a more complete understanding of the immunology of rotavirus infection.

22.9 SUMMARY

Enteric viruses are currently recognized as important etiologic agents of acute diarrhea. For some of these viruses (rotaviruses, enteric adenoviruses, Norwalk viruses, astroviruses, and caliciviruses) the association with disease has been clearly established, while for others (coronaviruses, some Norwalk-like viruses, parvoviruses, pestiviruses, and Breda viruses) the association is less clear and often based on single outbreaks. Future epidemiologic studies will be required to clarify this issue. Knowledge of the molecular biology of these viruses is increasing and will be helpful in expanding the diagnostic tools available to detect these enteric viruses. As more knowledge is gained at both the epidemiologic and basic molecular levels, new viruses causing enteric infections likely will be discovered and a clearer understanding of the relationship among the currently recognized viruses will be forthcoming.

REFERENCES

Agus, S.G., Dolin, R., Wyatt, R.G., Tousimis, A.J., and Northrop, R.S. 1973. Acute infectious nonbacterial gastroenteritis: Intestinal histopathology: Histologic and enzymatic alterations during illness produced by the Norwalk agent in man. Ann. Intern. Med. 7:18–25.

Altenburg, B.C., Graham, D.Y., and Estes,

M.K. 1980. Ultrastructural study of rotavirus replication in cultured cells. J. Gen. Virol. 46:75–85.

Appleton, H. 1987. Small round viruses: Classification and role in foodborne outbreaks. Ciba Foundation Symposium 128:108–125.

Ashley, C. and Caul, O.E. 1988. Human enteric coronaviruses, In M. Farthing (ed.), Viruses in the Gut. Proceedings of the Ninth BSG-SK and F International Workshop, Welwyn Garden City, Smith Kline & French, pp. 91–95.

Bantavala, J.E., Totterdell, B., Chrystie, I.L., and Woode, G.N. 1975. In vitro detection of human rotaviruses. Lancet 2:821.

Baron, R.C., Murphy, F.D., Greenberg, H.B., Davis, C.E., Biegman, D.J., Gary, W., Hughes, J.M., and Schonberger, L.B. 1982. Norwalk-related illness: An outbreak associated with swimming in a recreational lake and secondary person to person transmission. Am. J. Epidemiol. 115:162–163.

Bartlett, A.V. III, Reves, R.R., and Pickering, L.K. 1988. Rotavirus in infant-toddler daycare centers: Epidemiology relevant to disease control strategies. J. Pediatr. 113:435–441.

Bean, N.H., Griffin, P.M., Goulding, J.S., and Ivey, C.B. 1990. Foodborne disease outbreaks, 5-year summary 1983–1987. Centers for Disease Control M.M.W.R. 39:15–59.

Beards, G.M., Green, J.G., Hall, C., and Flewett, T.H. 1984. An enveloped virus in stools of children and adults with gastroenteritis that resembles the Breda virus of calves. Lancet 1:1050–1052.

Bell, E.J. and Grist, N.R. 1968. Viruses in diarrheal disease. Br. Med. J. 4:741–742.

Bellinzoni, R., Xi, J.A., Tanaka, T.N., Scodeller, E., and Estes, M.K. 1989. Rotavirus gene detection with biotinylated single-stranded RNA probes. Molec. Cell Probes 3:233–244.

Birch, C.J., Lehmann, N.I., Hawker, A.J., Marshall, J.A., and Gust, I.D. 1979. Comparison of electron microscopy, enzyme-linked immunosorbent assay, solid-phase radioimmunoassay, and indirect immunofluorescence for detection of human rotavirus antigen in faeces. J. Clin. Pathol. 32:700–705.

Birch, C.J., Rodger, S.M., Marshall, J.A., and Gust, I.D. 1983. Replication of human rotavirus in cell culture. J. Med. Virol. 11:241–250.

Birch, C.J., Heathm, R.L., and Gust, I.D. 1988. Use of serotype-specific monoclonal antibodies to study the epidemiology of rotavirus infection. J. Med. Virol. 24:45–53.

Bishop, R.F., Davidson, G.P., Holmes, I.H., and Ruck, B.J. 1973. Virus particles in epithelial cells of duodenal mucosa from children with acute nonbacterial gastroenteritis. Lancet 2:1281–1283.

Bishop, R.F., Barnes, G.L., Cipriani, E., and Lund, J.S. 1983. Clinical immunity after neonatal rotavirus infection. A prospective longitudinal study in young children. N. Engl. J. Med. 309:72–76.

Black, R.E., Greenberg, H.B., Kapikian, A.Z., Brown, K.H., and Becker, S. 1982. Acquisition of serum antibody to Norwalk virus and rotavirus and relation to diarrhea in a longitudinal study of young children in rural Bangladesh. J. Infect. Dis. 145:483–489.

Blacklow, N.R., Dolin, R., Fedson, D.S., DuPont, H., Northrop, R.S., Hornick, R.B., and Chanock, R.M. 1972. Acute infectious nonbacterial gastroenteritis: Etiology and pathogenesis. Ann. Intern. Med. 6:993–1008.

Blacklow, N.R., Cukor, G., Bedigian, M.K., Echeverria, P., Greenberg, H.B., Schreiber, D.S., and Trier, J.S. 1979. Immune response and prevalence of antibody to Norwalk enteritis virus as determined by radioimmunoassay. J. Clin. Microbiol. 10:903–909.

Blacklow, N.R. and Cukor, G. 1982. Viruses and gastrointestinal disease. In D.A.J. Tyrrell and A.Z. Kapikian (eds.), Virus Infections of the Gastrointestinal Tract. New York: Marcel Dekker, pp. 75–87.

Brade, L., Schmidt, W.A., and Gattert, I. 1981. [Relative effectiveness of disinfectants against rotaviruses (author's transl.)]. Zur relativen Wirksamkeit von Desinfektionsmitteln geganuber Rotaviren. Zentralbl. Bakteriol. Mikrobiol. Hyg. 174:151–159.

Brandt, C.D., Kim, H.W., Rodriguez, W.J., Thomas, L., Yolken, R.H., Arrobio, J.O., Kapikian, A.Z., Parrott, R.H., and Chanock, R.M. 1981. Comparison of direct electron microscopy, immune electron microscopy, and rotavirus enzyme-linked immunosorbent assay for detection of gastroenteritis viruses in children. J. Clin. Microbiol. 13:976–981.

Brandt, C.D., Kim, H.W., Rodriguez, W.J., Arrobio, J.O., Jeffries, B.C., Stallings, E.P., Lewis, C., Miles, A.J., Chanock, R.M., Kapikian, A.Z., and Parrott, R.H. 1983. Pediatric viral gastroenteritis during eight years of study. J. Clin. Microbiol. 18:71–78.

Brandt, C.D., Kim, H.W., Rodriguez, W.J., Arrobio, J.O., Jeffries, B.C., Stallings, E.P., Lewis, C., Miles, A.J., Gardner, M.K., and Parrott, R.H. 1985. Adenovirus and pediatric gastroenteritis. J. Infect. Dis. 151:437–443.

Bridger, J.C., Pedley, S., and McCrae, M.A. 1986. Group C rotaviruses in humans. Agriculture and Food Research Council Institute for Research on Animal Diseases, Compton, Nr. Newbury, Berkshire, United Kingdom. J. Clin. Microbiol. 23:760–763.

Bridger, J.C. 1988. Non-group A rotavirus, In M. Farthing (ed.), Viruses in the Gut. Proceedings of the Ninth BSG-SK and F International Workshop, Welwyn Garden City: Smith Kline & French, pp. 79–82.

Brooks, R.G., Brown, L., and Franklin, R.B. 1989. Comparison of a new rapid test (TestPack Rotavirus) with standard enzyme immunoassay and electron microscopy for the detection of rotavirus in symptomatic hospitalized children. J. Clin. Microbiol. 27:775–777.

Brown, M., Petric, M., and Middleton, P.J. 1984. Diagnosis of fastidious enteric adenovirus 40 and 41 in stool specimens. J. Clin. Microbiol. 20:334–338.

Brown, D.W., Beards, G.M., Chen, G.M., and Flewett, T.H. 1987. Prevalence of antibody to group B (atypical) rotavirus in humans and animals. J. Clin. Microbiol. 25:316–319.

Brown, D.W., Mathan, M.M., Mathew, M., Martin, R., Beards, G.M., and Mathan, V.I. 1988. Rotavirus epidemiology in Vellore, South India: Group, subgroup, serotype, and electrophoretype. J. Clin. Microbiol. 26:2410–2414.

Browning, G.F., Fitzgerald, F.A., Chalmers, R.M., and Snodgrass, D.R. 1991. A novel group A rotavirus G serotype: Serological and genomic characterization of equine isolate F123. J. Clin. Microbiol. 29:2043–2046.

Bryden, A.S., Davies, H.A., Thouless, M.E., and Flewitt, T.H. 1977. Diagnosis of rotavirus infection by cell culture. J. Med. Microbiol. 10:121–125.

Bryden, A.S., Thouless, M.E., Hall, C.J., Flewett, T.H., Wharton, B.A., Mathew, P.M., and Craig, I. 1982. Rotavirus infections in a special-care baby unit. J. Infect. Dis. 4:43–48.

Butler, M. and Harakeh, M.S. 1983. Inactivation of rotavirus in wastewater effluents by chemical disinfection. In M. Butler, A.R. Medlen, and R. Morris (eds.), Viruses and Disinfection of Water and Wastewater. Guildford, U.K.: University of Surrey Print Unit, pp. 282–289.

Caul, E.O., Paver, W.K., and Clark, S.K.R. 1975. Coronavirus particles in faeces from patients with gastroenteritis. Lancet 1:1972.

Caul, E.O. and Appleton, H. 1982. The electron microscopic and physical characteristics of small round human fecal viruses: An interim scheme for classification. J. Med. Virol. 9:257–265.

Chiba, S., Nakamura, I., Urasawa, S., Nakata, S., Taniguchi, K., Funinaga, K., and Nakao, T. 1983. Outbreak of infantile gastroenteritis due to type 40 adenovirus. Lancet 1:954–957.

Christy, C., Madore, H.P., Pichichero, M.E., Gala, C., Pincus, P., Vosefski, D., Hoshino, Y., Kapikian, A., and Dolin, R. 1988. Field trial of rhesus rotavirus vaccine in infants. Pediatr. Infect. Dis. J. 7:645–650.

Chrystie, I.L., Totterdell, B.M., and Banatvala, J.E. 1978. Asymptomatic endemic rotavirus infections in the newborn. Lancet 1:1176–1178.

Clark, H.G., Borran, F.E., and Plotkin, S.A. 1990. Immune protection of infants against rotavirus gastroenteritis by a serotype 1 reassortment of bovine rotavirus WC3. J. Infect. Dis. 161:1098–1104.

Coulson, B.S., Unicomb, L.E., Pitson, G.A., and Bishop, R.F. 1987. Simple and specific enzyme immunoassay using monoclonal antibodies for serotyping human rotaviruses. J. Clin. Microbiol. 25:509–515.

Coulson, B.S., Grimwood, K., Masendycz, P.J., Lund, J.S., Mermelstein, N., Bishop, R.F., and Barnes, G.L. 1990. Comparison of rotavirus immunoglobulin A coproconversion with other indices of rotavirus infection in a longitudinal study in children. J. Clin. Microbiol. 28:1367–1374.

Cruz, J.R., Caceres, P., Cano, F., Flores, J., Bartlett, A., and Toron, B. 1990. Adenovirus type 40 and 41 and rotaviruses associated with diarrhea in children from Guatemala. J. Clin. Microbiol. 28:1780–1784.

Cubitt, W.D., Moscovici, O.P., and Lebon, P.S. 1981. A new serotype of calicivirus associated with an outbreak of gastroenteritis in a residential home for the elderly. J. Clin. Pathol. 34:924–926.

Cubitt, W.D., Blacklow, N.R., Herrman, J.E., Nowak, N.A., Nakata, S., and Chiba, S. 1987. Antigenic relationships among human caliciviruses and Norwalk virus. J. Infect. Dis. 156:806–814.

Cubitt, W.D. 1987. The candidate caliciviruses. In G. Bock and J. Whelan (eds.), Novel Diarrhoea Viruses, Ciba Foundation Symposium. New York: Wiley, pp. 126–143.

Cubitt, W.D. and McSwiggan, D.A. 1987. Seroepidemilogical survey of the prevalence of antibodies to a strain of human calicivirus. J. Med. Virol. 21:361–368.

Cukor, G., Blacklow, N.R., Echeverria, P., Bedigian, M.K., Puruggan, H., and Basaca-Sevilla, V. 1980. Comparative study of the acquisition of antibody to Norwalk virus in pediatric populations. Infect. Immun. 29:822–823.

Cukor, G., Perron, D.M., Hudson, R., and Blacklow, N.R. 1984. Detection of rotavirus in human stools by using monoclonal antibody. J. Clin. Microbiol. 19:888–892.

Dagan, R., Bar-David, Y., Sarov, B., Katz, M., Kassis, I., Greenberg, D., Glass, R.I., Margolis, C.Z., and Sarov, I. 1990. Rotavirus diarrhea in Jewish and Bedouin children in the Negev region of Israel: Epidemiology, clinical aspects and possible role of malnu-

trition in severity of illness. Pediatr. Infect. Dis. J. 9:314–321.

Davidson, G.P., Bishop, R.F., Townley, R.R., and Holmes, I.H. 1975. Importance of a new virus in acute sporadic enteritis in children. Lancet 1:242–246.

Davidson, G.P. and Barnes, G.L. 1979. Structural and functional abnormalities of the small intestine in infants and young children with rotavirus enteritis. Acta. Paediatr. Scand. 68:181–186.

Davidson, G.P., Hogg, R.J., and Kirubakaran, C.P. 1983. Serum and intestinal immune response to rotavirus enteritis in children. Infect. Immun. 40:447–452.

Deetz, T.R., Smith, E.M., Goyal, S.M., Gera, C.P., Vollett, J.J., Tsai, L., DuPont, H.L., and Keswick, B.H. 1984. Occurrence of rota- and enteroviruses in drinking water in a developing nation. Water Res. 18:567–571.

Dennehy, P.H., Gauntlett, D.R., and Tente, W.E. 1988. Comparison of nine commercial immunoassays for the detection of rotavirus in fecal specimens. J. Clin. Microbiol. 26:1630–1634.

Dennehy, P.H., Gauntlett, D.R., and Spangenberger, S.E. 1990. Choice of reference assay for the detection of rotavirus in fecal specimens: Electron microscopy versus enzyme immunoassay. J. Clin. Microbiol. 28:1280–1283.

Desselberger, U. 1988. Molecular epidemiology of rotavirus. In M. Farthing (ed.), Viruses in the Gut. Proceedings of the Ninth BSG-SK and F International Workshops. Welwyn Garden City: Smith Kline & French, pp. 55–64.

DeZoysa, I. and Feachem, R.G. 1985. Interventions for the control of diarrhoeal diseases among young children: Rotavirus and cholera immunization. Bull. WHO 63:569–583.

Dharkul, T., Rott, L., and Greenberg, H.B. 1990. Recovery from chronic rotavirus infection in mice with severe combined immunodeficiency: Virus clearance mediated by adoptive transfer of immune CD8+T lymphocytes. J. Virol. 64:4375–4382.

Dimitrov, D.H., Graham, D.Y., Lopez, J., Muchinik, G., Velasco, G., Stenback, W.A., and Estes, M.K. 1984. RNA electropherotypes of human rotaviruses from North and South America. Bull. WHO 62:321–329.

Doern, G.V., Herrmann, J.E., Henderson, P., Stobbs-Walro, D., Perron, D.M., and Blacklow, N.R. 1986. Detection of rotavirus with a new polyclonal antibody enzyme immunoassay (Rotazyme II) and a commercial latex agglutination test (Rotalex): Comparison with a monoclonal antibody enzyme immunoassay. J. Clin. Microbiol. 23:226–229.

Dolin, R., Blacklow, N.R., DuPont, H., Formal, S., Buscho, R.F., Kasel, J.A., Chames, R.D., Hornick, R., and Chanock, R.M. 1971. Transmission of acute infectious nonbacterial gastroenteritis to volunteers by oral administration of stool filtrates. J. Infect. Dis. 123:307–312.

Dolin, R., Blacklow, N.R., DuPont, H.L., Buscho, R.F., Wyatt, R.G., Kasel, J.G., Hornick, R., and Chanock, R.M. 1972. Biological properties of Norwalk agent of acute infectious nonbacterial gastroenteritis. Proc. Soc. Exp. Biol. Med. 140:578–583.

Dolin, R., Reichman, R.C., Roessner, K.D., Tralka, T.S., Schooley, R.T., Gary, W., and Morens, D. 1982. Detection by immune electron microscopy of the Snow Mountain agent of acute viral gastroenteritis. J. Infect. Dis. 146:184–189.

Echeverria, P., Blacklow, N.R., Cukor, G.G., Vibulbandhitkit, S., Changchawalit, S., and Boonthai, P. 1983. Rotavirus as a cause of severe gastroenteritis in adults. J. Clin. Microbiol. 18:663–667.

Ellis, M.E., Watson, B., Mandal, B.K., Dunbar, E.M., Craske, T., Curry, A., Roberts, J., and Lomax, J. 1984. Micro-organisms in gastroenteritis. Arch. Dis. Child. 59:848–855.

Espejo, R.T., Munoz, O., Serafin, F., and Romero, P. 1980. Shift in the prevalent human rotavirus detected by ribonucleic acid segment differences. Infect. Immun. 27:351–354.

Espejo, R.T., Puerto, F., Soler, C., and Gonzalez, N. 1984. Characterization of a human pararotavirus. Infect. Immun. 44:112–116.

Estes, M.K., Palmer, E.L., and Obijeski, J.F. 1983. Rotaviruses: a review. Curr. Top. Microbiol. Immunol. 105:123–184.

Estes, M.K. and Cohen, J. 1989. Rotavirus gene structure and function. Microbiol. Rev. 53: 410–449.

Ettig, P.J. and Altschuler, G.P. 1985. Fatal gastroenteritis associated with coronaviruslike particles. J. Dis. Child. 139:245–248.

Flewett, T.H. 1983. Rotavirus in the home and hospital nursery [editorial]. Br. Med. J. (Clin. Res.) 287:568–569.

Flores, J., Boeggeman, E., Purcell, R.H., Sereno, M., Perez, I., White, L., Wyatt, R.G., Chanock, R.M., and Kapikian, A.Z. 1983. A dot hybridisation assay for detection of rotavirus. Lancet 1:555–558.

Flores, J., Taniguchi, K., Green, K., Perez-Schael, I., Garcia, D., Sears, J., Urasawa, S., and Kapikian, A.Z. 1988. Relative frequencies of rotavirus serotypes 1, 2, 3, and 4 in Venezuelan infants with gastroenteritis. J. Clin. Microbiol. 26:2092–2095.

Foster, S.O., Palmer, E.L., Gary, G.W. Jr., Martin, M.L., Herrmann, K.L., Beasley, P., and Sampson, J. 1980. Gastroenteritis due to rotavirus in an isolated Pacific island group: An epidemic of 3,439 cases. J. Infect. Dis. 141:32–39.

Garwes, D.J. 1982. Coronaviruses in animals. In D.A.J. Tyrrell and A.Z. Kapikian (eds.), Virus Infections of the Gastrointestinal Tract. New York: Marcel Dekker, pp. 315–359.

Gary, J.J., Wreghitt, T.G., Cubitt, W.D., and Elliott, P.R. 1987. An outbreak of gastroenteritis in a home for the elderly associated with astrovirus type 1 and human calicivirus. J. Med. Virol. 23:377–381.

Georges-Courbot, M.C., Beraud, A.M., Beards, G.M., Campbell, A.D., Gonzalez, J.P., Georges, A.J., and Flewett, T.H. 1988. Subgroups, serotypes, and electropherotypes of rotavirus isolated from children in Bangui, Central African Republic. J. Clin. Microbiol. 26:668–671.

Gerna, G., Battaglia, M., Milenesi, G., Passarani, N., Percivalle, E., and Cattaneo, E. 1984. Serotyping of cell culture-adapted subgroup 2 human rotavirus strains by neutralization. Infect. Immun. 43:722–729.

Gerna, G., Passarani, N., Battaglia, and Rondanelli, F.L. 1985. Human enteric coronaviruses: Antigenic relatedness to human coronavirus OC43 and possible etiologic role in viral gastroenteritis. J. Infect. Dis. 151:796–803.

Gerna, G., Passarani, N., Unicomb, L.E., Parea, M., Sarasini, A., Battaglia, M., and Bishop, R.F. 1989. Solid-phase immune electron microscopy and enzyme-linked immunosorbent assay for typing of human rotavirus strains by using polyclonal and monoclonal antibodies: A comparative study. J. Infect. Dis. 159:335–339.

Gilger, M.E., Matson, D.O., Conner, M.E., Rosenblatt, H.M., Hanson, L.G., Finegold, M.J., and Estes, M.K. 1992. Extraintestinal spread of rotavirus infections in children with immunodeficiency. J. Pediatr. 120: 912–917.

Gill, O.N., Cubitt, W.D., McSwiggan, D.A., Watney, B.M., and Bartlett, C.L.R. 1983. Epidemic of gastroenteritis caused by oysters contaminated with small round structured viruses. Br. Med. J. 287:1532–1534.

Goldwater, P.N., Chrystie, I.L., and Banatvala, J.E. 1979. Rotaviruses and the respiratory tract. Br. Med. J. 2:1551.

Gomez, J., Estes, M.K., Matson, D.O., Bellinzoni, R., Alvarez, A., and Grinstein, S. 1990. Serotyping of human rotaviruses in Argentina by ELISA with monoclonal antibodies. Arch. Virol. 112:249–259.

Gorziglia, M., Hoshino, Y., Buckler-White, A., Blumentals, I., Glass, R., Flores, J., Kapikian, A.Z., and Chanock, R.M. 1987. Conservation of amino acid sequence of VP8 and cleavage region of 84-kDa outer capsid protein among rotaviruses recovered from asymptomatic neonatal infection. Proc. Natl. Acad. Sci. 84:2062.

Gorziglia, M., Larraine, G., Kapikian, A.Z., Chanock, R.M., 1990. Antigenic relationships among human rotavirus as determined by outer capsid probe in VP4. Proc. Natl. Acad. Sci. USA 87:7155–7199.

Gothefors, L., Wadell, G., Juto, P., Taniguchi, K., Kapikian, A.Z., and Glass, R.I. 1989. Prolonged efficacy of rhesus rotavirus vaccine in Swedish children. J. Infect. Dis. 159:753–757.

Gouvea, V., Glass, R.I., Woods, P., Taniguchi, K., Clark, H.F., Forrester, B., and Fang, Z.Y. 1990a. Polymerase chain reaction amplification and typing of rotavirus nucleic acid from stool specimens. J. Clin. Microbiol. 28:276–282.

Gouvea, V., Ho, M.S., Glass, R., Woods, P., Forrester, B., Robinson, C., Ashley, R., Riepenhoff-Talty, M., Clark, F.H., Taniguchi, K., Meddix, E., McKellar, B., and Pickering, L. 1990b. Serotypes and electropherotypes of human rotavirus in the U.S.A. 1987–1989. J. Infect. Dis. 162:362–367.

Graham, D.Y. and Estes, M.K. 1979. Comparison of methods for immunocytochemical detection of rotavirus infections. Infect. Immun. 26:686–689.

Green, K.Y., Taniguchi, K., Mackow, E.R., and Kapikian, E.Z. 1990. Homotypic and heterotypic epitope-specific antibody responses in adult and infant rotavirus vaccines: Implications for vaccine development. J. Infect. Dis. 161:667–679.

Greenberg, H.B., Wyatt, R.G., Valdesuso, J., Kalica, A.R., London, W.T., Chanock, R.M., and Kapikian, A.Z. 1978. Solid-phase microtiter radioimmunoassay for detection of the Norwalk strain of acute nonbacterial, epidemic gastroenteritis virus and its antibodies. J. Med. Virol. 2:97–108.

Greenberg, H.B., Valdesuso, J., Kapikian, A.Z., Chanock, R.M., Wyatt, R.G., Szmuness, W., Larrick, J., Kaplan, J., Gilman, R.H., and Sack, D.A. 1979a. Prevalence of antibody to the Norwalk virus in various countries. Infect. Immun. 26:270–273.

Greenberg, H.B., Valdesuso, J., Yolken, R.H., Gangaroosa, E., Gary, W., Wyatt, R.G., Konno, T., Suzuki, H., Chanock, R.M., and Kapikian, A.Z. 1979b. Role of Norwalk virus in outbreaks of nonbacterial gastroenteritis. J. Infect. Dis. 139:564–568.

Greenberg, H.B., Wyatt, R.G., and Kapikian, A.Z. 1979c. Norwalk virus in vomitus. Lancet 1:55.

Greenberg, H.B., Kalica, A.R., Wyatt, R.G., Jones, R.W., Kapikian, A.Z., and Chanock, R.M. 1981. Rescue of noncultivatable human rotavirus by gene reassortment during mixed infection with ts mutants of a cultivatable bovine rotavirus. Proc. Natl. Acad. Sci. 78:420–424.

Griffin, M.R., Sorowiec, J.J., McCloskey, D.I., Greenberg, H., Capueno, B., Pierzynski, B., Quinn, M., Wojnalski, R., Parkin, W.E., and Gary, W.G. 1982. Foodborne Norwalk virus. Am. J. Epidemiol. 115:178–184.

Grimwood, K., Abbott, G.D., Fergusson, D.M., Jennings, L.C., and Allan, J.M. 1983. Spread of rotavirus within families: A community based study. Br. Med. J. (Clin. Res.) 287:575–577.

Grimwood, K., Lund, J.C., Coulson, B.S., Hudson, I.L., Bishop, R.F., and Barnes, G.L. 1988. Comparison of serum and mucosal antibody responses following severe acute rotavirus gastroenteritis in young children. J. Clin. Microbiol. 26:732–738.

Grohmann, G.S., Greenberg, H.B., Welch, B.M., and Murphy, A.M. 1980. Oyster-associated gastroenteritis in Australia. The detection of Norwalk virus and its antibody by immune electron microscopy and radioimmunoassay. J. Med. Virol. 6:11–19.

Grohmann, G. 1985. Viral diarrhoea in Australia. In S. Tzipori (ed.), Infectious Diarrhoea in the Young. Amsterdam, Excerpta Medica Elsevier, pp. 25–28.

Gunn, R.A., Janowski, H.T., Lieb, S., Prather, E.C., and Greenberg, H.B. 1982. Norwalk virus gastroenteritis following raw oyster consumption. Am. J. Epidemiol. 115:348–351.

Gurwith, M., Wenman, W., Hinde, D., Feltham, S., and Greenberg, H. 1981. A prospective study of rotavirus infection in infants and young children. J. Infect. Dis. 144:218–224.

Gurwith, M., Wenman, W., Gurwith, D., Brunton, J., Feltham, S., and Greenberg, H. 1983. Diarrhea among infants and young children in Canada: A longitudinal study in three northern communities. J. Infect. Dis. 147:685–692.

Halsey, N.A., Anderson, E.L., Sears, S.D., Steinhoff, M., Wilson, M., Belshe, R.B., Midthun, K., Kapikian, A.Z., Chanock, R.M., and Samorodin, R. 1988. Human-rhesus reassortant rotavirus vaccines: Safety and immunogenicity in adults, infants, and children. J. Infect. Dis. 158:1261–1267.

Halvorsrud, J. and Orstavik, I. 1980. An epidemic of rotavirus-associated gastroenteritis in a nursing home for the elderly. Scand. J. Infect. Dis. 12:161–164.

Hammond, G.W., Hannan, C., Yeh, T., Fischer, K., Mauthe, G., and Strauss, S.E. 1987. DNA hybridization for diagnosis of enteric adenovirus infection from directly spotted human fecal specimens. J. Clin. Microbiol. 25:1881–1885.

Hasegawa, A., Matsuno, S., Inouye, S., Kono, R., Tsurukubo, Y., Mukoyama, A., and Saito, Y. 1982. Isolation of human rotaviruses in primary cultures of monkey kidney cells. J. Clin. Microbiol. 16:387–390.

Herring, A.J., Inglis, N.F., Ojeh, C.K., Snodgrass, D.R., and Menzies, J.D. 1982. Rapid diagnosis of rotavirus infection by direct detection of viral nucleic acid in silver-stained polyacrylamide gels. J. Clin. Microbiol. 16:473–477.

Herrman, J.E., Nowak, N.A., Perron-Henry, D.M., Hudson, R.W., Cubitt, D.W., and Blacklow, N.R. 1990. Diagnosis of astrovirus gastroenteritis by antigen detection with monoclonal antibodies. J. Infect. Dis. 161:226–229.

Herrman, J.E., Taylor, D.N., Echeverria, A., Blacklow, N.R. 1991. Astroviruses as a cause of gastroenteritis in children. N. Engl. J. Med. 324:1757–1760.

Hierholzer, J.C., Wigand, R., Anderson, L.J., Adrian, T., and Gold, J.W. 1988. Adenoviruses from patients with AIDS: A plethora of serotypes and a description of five new serotypes of subgenus D (types 43-47). J. Infect. Dis. 157:804.

Hjelt, K., Grauballe, P.C., Schiotz, P.O., Andersen, L., and Krasilnikoff, P.A. 1985. Intestinal and serum immune response to a naturally acquired rotavirus gastroenteritis in children. J. Pediatr. Gastroenterol. Nutr. 4:60–66.

Hjelt, K., Grauballe, P.C., Paerregaard, A., Nielsen, O.H., and Krasilnikoff, P.A. 1987. Protective effect of pre-existing rotavirus-specific immunoglobulin A against naturally acquired rotavirus infection in children. J. Med. Virol. 21:39–47.

Hjelt, K. and Grauballe, C. 1990. Protective levels of intestinal rotavirus antibodies [letter]. J. Infect. Dis. 161:352–353.

Ho, M., Glass, R.I., Monroe, S.S., Madore, H.P., Stine, S., Pinsky, P.F., Cubitt, D.F., Ashley, C., and Caul, E.O. 1989. Viral gastroenteritis aboard a cruise ship. Lancet 2:961–964.

Hodes, H.L. 1977. Viral gastroenteritis. Am. J. Dis. Child. 131:729–731.

Holzel, H., Cubitt, D.W., McSwiggan, D.A., Sanderson, P.J., and Church, J. 1980. An outbreak of rotavirus infection among adults in a cardiology ward. J. Infect. Dis. 2:33–37.

Hopkins, R.S., Gaspard, G.B., Williams, F.P. Jr., Karlin, R.J., Cukor, G., and Blacklow, N.R. 1984. A community waterborne gastroenteritis outbreak: Evidence for rotavirus as the agent. Am. J. Public Health 74:263–265.

Hoshino, Y., Sereno, M.M., Midthun, K., Flores, J., Kapikian, A.Z., and Chanock, R.M. 1985. Independent segregation of two antigenic specificities (VP3 and VP7) involved in neutralization of rotavirus infectivity. Proc. Natl. Acad. Sci. 82:8701–8704.

Humphrey, T.J., Cruickshank, J.G., and Cubitt, W.D. 1984. An outbreak of calicivirus associated gastroenteritis in an elderly persons home: A possible zoonosis? J. Hyg. 93:293–299.

Hung, T., Wang, C.G., Fang, Z.Y., Chou, Z.Y., Chang, X.J., Lion, X.G., Chen, G.M., Yao, H.L., Chao, T.X., Ye, W., Den, S.S., and Chang, W. 1984. Waterborne outbreak of rotavirus diarrhoea in adults in China caused by a novel rotavirus. Lancet 1:1139–1142.

Isaacs, D., Day, D., and Crook, S. 1986. Childhood gastroenteritis: A population study. Br. Med. J. 293:545–546.

Jiang, X., Wang, K., Graham, D.Y., and Estes, M.K. 1990. Cloning and characterization of the Norwalk virus genome. Abstract, A.S.M. Annual Meeting; Anaheim, CA; May, 1990.

Johnson, P.C., Hoy, J., Mathewson, J.J., Eric-

sson, C.D., and DuPont, H.L. 1990. Occurrence of Norwalk virus infection among adults in Mexico. J. Infect. Dis. 162:389–393.

Kagnoff, M.F., Austin, R.K., Hubert, J.J., Bernardin, J.E., and Kasarda, D.O. 1984. Possible role for a human adenovirus in the pathogenesis of celiac disease. J. Exp. Med. 160:1544–1557.

Kalica, A.R., Greenberg, H.B., Espejo, R.T., Flores, J., Wyatt, R.G., Kapikian, A.Z., and Chanock, R.M. 1981. Distinctive ribonucleic acid patterns of human rotavirus subgroups 1 and 2. Infect. Immun. 33:958–961.

Kapikian, A.Z., Wyatt, R.G., Dolin, R., Thornhill, T.S., Kalica, A.R., and Chanock, R.M. 1972. Visualization by immune electron microscopy of a 27-nm particle associated with acute infectious nonbacterial gastroenteritis. J. Virol. 10:1075–1080.

Kapikian, A.Z., Kim, H.W., Wyatt, R.G., Rodriguez, W.J., Ross, S., Cline, W.L., and Parrott, R.H. 1974. Reovirus-like agents in stools: Association with infantile diarrhoea and development of serologic tests. Science 185:1049–1053.

Kapikian, A.Z., Kim, H.W., Wyatt, R.G., Cline, W.L., Arrobio, J.O., Brandt, C.D., Rodriguez, W.J., Sack, S.A., Chanock, R.M., and Parrott, R.H. 1976. Human reovirus-like agent as the major pathogen associated with winter gastroenteritis in hospitalized infants and young children. N. Engl. J. Med. 294:965–972.

Kapikian, A.Z., Greenberg, H.B., Wyatt, R.G., Kalica, A.R., and Chanock, R.M. 1982. The Norwalk group of viruses—Agents associated with epidemic viral gastroenteritis. In D.A. Tyrrell and A.Z. Kapikian (eds.), Virus Infections of the Gastrointestinal Tract. New York: Marcel Dekker, pp. 147–177.

Kapikian, A.Z., Wyatt, R.G., Levine, M.M., Yolken, R.H., Vankirk, D.H., Dolin, R., Greenberg, H.B., and Chanock, R.M. 1983. Oral administration of human rotavirus to volunteers: Induction of illness and correlates of resistance. J. Infect. Dis. 147:95–106.

Kapikian, A.Z., Flores, J., Midthun, K., Hoshino, Y., Green, K.Y., Gorziglia, M.,

Nishikawa, K., Chanock, R.M., Potash, L., and Perez-Schael, I. 1989. Strategies for the development of a rotavirus vaccine against infantile diarrhea with an update on clinical trials of rotavirus vaccines. Adv. Exp. Med. Biol. 257:67–89.

Kaplan, J.E., Gary, G.W., Baron, R.C., Singh, N., Schonberger, L.B., Feldman, R., and Greenberg, H.B. 1982a. Epidemiology of Norwalk gastroenteritis and the role of Norwalk virus in outbreaks of acute nonbacterial gastroenteritis. Ann. Intern. Med. 96: 756–761.

Kaplan, J.E., Goodman, R.A., Schonberger, L.B., Lippy, E.C., and Gary, W.G. 1982b. Gastroenteritis due to Norwalk virus: An outbreak associated with a municipal water supply. J. Infect. Dis. 146:190–197.

Keswick, B.H., Pickering, L.K., DuPont, H.L., and Woodward, W.E. 1983. Survival and detection of rotaviruses on environmental surfaces in day care centers. Appl. Environ. Microbiol. 46:813–816.

Keswick, B.H., Gerba, C.P., DuPont, H.L., and Rose, J.B. 1984. Detection of enteric viruses in treated drinking water. Appl. Environ. Microbiol. 47:1290–1294.

Kim, H.W., Brandt, C.D., Kapikian, A.Z., Wyatt, R.G., Arrobio, J.P., Rodriguez, W.J., Chanock, R.M., and Parrott, R.H. 1977. Human reovirus-like agent (HRVLA) infection: Occurrence in adult contacts of pediatric patients with gastroenteritis. JAMA 238:404–407.

Kjeldsberg, E. 1977. Small spherical viruses in faeces from gastroenteritis patients. Acta. Pathol. Microbiol. Immunol. Scand. 85: 351–354.

Knisley, C.V., Bednarz-Prashad, A.J., and Pickering, L.K. 1986. Detection of rotavirus in stool specimens with monoclonal and polyclonal antibody-based assay systems. J. Clin. Microbiol. 23:897–900.

Kok, T.W. and Burrell, C.J. 1989. Comparison of five enzyme immunoassays, electron microscopy, and latex agglutination for detection of rotavirus in fecal specimens. J. Clin. Microbiol. 27:364–366.

Konno, T., Suzuki, H., Kutsuzawa, T., Imai, A., Katsushima, N., Sakamoto, M., Kitaoka, S., Tsuboi, R., and Adachi, M. 1978. Human rotavirus infection in infants and young children with intussusception. J. Med. Virol. 2:265–269.

Konno, T., Suzuki, H., Ishida, N., Chiba, R., Mochizuki, K., and Tsunoda, A. 1982. Astrovirus-associated epidemic gastroenteritis in Japan. J. Med. Virol. 9:11–17.

Konno, T., Suzuki, H., Katsushima, N., Imai, A., Tazawa, F., Kutsuzawa, T., Kitaoka, S., Sakamoto, M., Yazaki, N., and Ishida, N. 1983. Influence of temperature and relative humidity on human rotavirus infection in Japan. J. Infect. Dis. 147:125–128.

Konno, T., Sato, T., Suzuki, H., Kitaoka, S., Katsushima, N., Sakamoto, M., Yazaki, N., and Ishida, N. 1984. Changing RNA patterns in rotaviruses of human origin: Demonstration of a single dominant pattern at the start of an epidemic and various patterns thereafter. J. Infect. Dis. 149:683–687.

Kotloff, K.L., Losonsky, G.A., Morris, J.G., Wasserman, S.S., Singh-Naz, H., and Levine, M.M. 1989. Enteric adenovirus infection and childhood diarrhea: An epidemiologic study in three clinical settings. Pediatrics 84:219–225.

Kuritsky, I.N., Osterholm, M.T., Greenberg, H.B., Korlatch, J.A., Godes, J.R., Hedberg, C.W., Forfang, J.C., and Kapikian, A.Z. 1984. Norwalk gastroenteritis: A community outbreak associated with bakery product consumption. Ann. Intern. Med. 100:519–521.

Kurtz, J.B. and Lee, T.W. 1984. Human astrovirus serotypes. Lancet 2:1405.

Kurtz, J.B. and Lee, T.W. 1987. Astroviruses: Human and animal. In G. Bock and J. Whelan (eds.), Novel Diarrhoea Viruses. Ciba Foundation Symposium 128. Chichester: Wiley, pp. 92–107.

Kurtz, J.B. 1988. Astroviruses, In M. Farthing (ed.), Viruses in the Gut. Proceedings of the Ninth BSG-SK and F International Workshop. Welwyn Garden City: Smith Kline & French, pp. 70–78.

Kutsuzawa, T., Konno, T., Suzuki, H., Kapikian, A.Z., Ebina, T., and Ishida, N. 1982. Isolation of human rotavirus subgroups 1 and 2 in cell culture. J. Clin. Microbiol. 16:727–730.

Lacombe, D., Lamouliatte, F., Billeud, C., and Sandler, B. 1988. Breda virus and hemorrhagic enteropathy. Reminder apropos of 1 case. Arch. Fr. Pediatr. 45:442.

Lanata, C.F., Black, R.E., del Aguila, R., Gil, A., Verastegui, H., Gerna, G., Flores, J., Kapikian, A.Z., and Andre, F.E. 1989. Protection of Peruvian children against rotavirus diarrhea of specific serotypes by one, two, or three doses of the RIT 4237 attenuated bovine rotavirus vaccine. J. Infect. Dis. 159:452–459.

LeBaron, C.W., Lew, J., Glass, R.I., Weber, J.W., and Ruiz-Palacios, G.M. 1990. The rotavirus study group. Annual rotavirus epidemic patterns in North America. JAMA 264:983–987.

Levine, W.C., Stephenson, W.T., and Craun, G.F. 1990. Waterborne disease outbreaks 1986–1988. Centers for Disease Control, M.M.W.R. 39:1–14.

Lewis, H.M., Parry, J.V., Davies, H.A., Parry, R.P., Mott, A., Dourmashkin, R.R., Sanderson, P.J., Tyrrell, D.A., and Valman, H.B. 1979. A year's experience of the rotavirus syndrome and its association with respiratory illness. Arch. Dis. Child. 54: 339–346.

Lin, M., Imai, M., Bellamy, A.R., Ikegami, N., Furuichi, Y., Summers, D., Nuss, D.L., and Deibel, R. 1985. Diagnosis of rotavirus infection with cloned cDNA copies of viral genome segments. J. Virol. 55:509–512.

London, S.D., Rubin, D.H., and Cebra, J.J. 1987. Gut mucosal immunization with reovirus serotype 1/L stimulates virus-specific cytotoxic T cell precursors as well as IgA memory cells in Peyer's patches. J. Exp. Med. 165:830–847.

Losonsky, G.A., Rennels, M.B., Lim, Y., Krall, G., Kapikian, A.Z., and Levine, M.M. 1988. Systemic and mucosal immune responses to rhesus rotavirus vaccine MMU 18006. Pediatr. Infect. Dis. J. 7:388–393.

Madeley, C.R., Cosgrove, B.P., Bell, E.J., and Fallan, R.J. 1977. Stool viruses in babies in Glasgow. 1. Hospital admissions with diarrhea. J. Hyg. Camb. 78:261–273.

Madeley, C.R. 1979a. Viruses in the stools. J. Clin. Pathol. 32:1–10.

Madeley, C.R. 1979b. Comparison of the features of astroviruses and caliciviruses seen in samples of feces by electron microscopy. J. Infect. Dis. 139:519–523.

Madeley, C.R. 1986. The emerging role of adenoviruses as inducers of gastroenteritis. Pediatr. Infect. Dis. 5:S63–S74.

Marchlewicz, B., Spiewak, M., and Lampinen, J. 1988. Evaluation of Abbott TESTPACK ROTAVIRUS with clinical specimens. J. Clin. Microbiol. 26:2456–2458.

Marrie, T.J., Lee, S.H., Faulkner, R.S., Ethier, J., and Young, C.H. 1982. Rotavirus infection in a geriatric population. Arch. Intern. Med. 142:313–316.

Matson, D.O., Estes, M.K., Glass, R.I., Bartlett, A.V., Penaranda, M., Calomeni, E., Tanaka, T., Nakata, S., and Chiba, S. 1989. Human calicivirus-associated diarrhea in children attending day care centers. J. Infect. Dis. 159:71–78.

Matson, D.O., Estes, M.K., Burns, J.W., Greenberg, H.B., Taniguchi, K., Urasawa, S. 1990a. Serotype variation of human group A rotaviruses in two regions of the United States. J. Infect. Dis. 162:605–614.

Matson, D.O., Estes, M.K., Tanaka, T., Bartlett, A.V., and Pickering, L.K. 1990b. Asymptomatic human calicivirus infection in a day care center. Pediatr. Infect. Dis. J. 9:180–186.

Matson, D.O. In press. Calicivirus infections. In R.D. Feigin and J.D. Cherry (eds.), Textbook of Pediatric Infectious Diseases, Philadelphia: W.D. Saunders Co.

Matsui, S.M., Offit, P.A., Vo, P.T., Mackow, E.R., Benfield, D.A., Shaw, R.D., Padilla-Noriega, L., and Greenberg, H.B. 1989. Passive protection against rotavirus-induced diarrhea by monoclonal antibodies to the heterotypic neutralization domain of VP7 and the VP8 fragment of VP4. J. Clin. Microbiol. 27:780–782.

Matsumoto, K., Hatano, M., Kobayashi, K., Hasegawa, A., Yamazaki, S., Nakata, S., Chiba, S., and Kimura, Y. 1989. An outbreak of gastroenteritis associated with acute rotaviral infection in schoolchildren. J. Infect. Dis. 160:611–615.

Matsuno, S., Utagawa, E., and Sugiura, A.

1983. Association of rotavirus infection with Kawasaki syndrome. J. Infect. Dis. 148:177.

Meeroff, J.C., Schreiber, D.S., Trier, J.S., and Blacklow, N.R. 1980. Abnormal gastric motor function in viral gastroenteritis. Ann. Intern. Med. 92:370–373.

Miotti, P.G., Eiden, J., and Yolken, R.H. 1985. Comparative efficiency of commercial immunoassays for the diagnosis of rotavirus gastroenteritis during the course of infection. J. Clin. Microbiol. 22:693–698.

Morinet, F., Ferchal, F., Colimon, R., and Perol, Y. 1984. Comparison of six methods for detecting human rotavirus in stools. Eur. J. Clin. Microbiol. 3:136–140.

Morse, D.L., Guzewich, J.J., Hanrahan, J.P., Stricof, R., Shayegani, M., Deibel, R., Grabau, J.C., Nowak, N.A., Herrmann, J.E., Cukor, G., and Blacklow, N.R. 1986. Widespread outbreak of clam and oyster-associated gastroenteritis. Role of Norwalk virus. N. Engl. J. Med. 314:678–681.

Mulcahy, D.L., Kamath, K.R., de Silva, L.M., Hodges, S., Carter, I.W., and Cloonan, M.J. 1982. A two-part study of the aetiological role of rotavirus in intussusception. J. Med. Virol. 9:51–55.

Nagayoshi, S., Yamaguchi, H., Ichikawa, T., Miyazu, M., Morishima, T., Ozaki, T., Isomura, S., Suzuki, S., and Hoshino, M. 1980. Changes of the rotavirus concentration in faeces during the course of acute gastroenteritis as determined by the immune adherence hemagglutination test. Eur. J. Pediatr. 134:99–102.

Nakata, S., Estes, M.K., Graham, D.Y., Wang, S.S., Gary, G.W., and Melnick, J.L. 1987. Detection of antibody to group B adult diarrhea rotaviruses in humans. J. Clin. Microbiol. 25:812–818.

Nakata, S., Estes, M.K., and Chiba, S. 1988. Detection of human calicivirus antigen and antibody by enzyme-linked immunosorbent assay (ELISA). J. Clin. Microbiol. 26:2001–2005.

Nazer, H., Rice, S., and Walker-Smith, J.A. 1982. Clinical associations of stool astrovirus in childhood. J. Pediatr. Gastroenterol. Nutr. 1:555–558.

Nicolas, J.C., Ingrand, D., Fortier, B., and Bricout, F. 1982. A one-year virological survey of acute intussusception in childhood. J. Med. Virol. 9:267–271.

Offit, P.A. and Dudzik, K.I. 1989. Rotavirus-specific cytotoxic T lymphocytes appear at the intestinal mucosal surface after rotavirus infection. J. Virol. 63:3507–3512.

Ogra, P.L. and Karzon, D.T. 1980. The role of immunoglobulins in the mechanisms of mucosal immunity to virus infection. Pediatr. Clin. N. Am. 17:385–400.

Okada, S., Sekine, S., Ando, T., Hayashi, Y., Murao, M., Yabuchi, K., Miki, T., and Ohashi, M. 1990. Antigenic characterization of small, round-structured viruses by immune electron microscopy. J. Clin. Microbiol. 28:1244–1248.

Olive, D.M. and Sethi, S.K. 1989. Detection of human rotavirus by using an alkaline phosphatase-conjugated synthetic DNA probe in comparison with enzyme-linked immunoassay and polyacrylamide gel analysis. J. Clin. Microbiol. 27:53–57.

Oliver, A.R. and Phillips, A.D. 1988. An electron microscopical investigation of faecal small round viruses. J. Med. Virol. 24:211–218.

O'Ryan, M.L. and Matson, D.O. 1990a. Viral gastroenteritis pathogens in the day care center setting. Semin. Pediatr. Infect. Dis. 1:252–262.

O'Ryan, M.L., Matson, D.O., Estes, M.K., and Pickering, L.K. 1990b. Molecular epidemiology of rotavirus in children attending day-care centers in Houston. J. Infect. Dis. 162:810–816.

Pacini, D.L., Brady, M.T., Budde, C.T., Connell, M.J., Hamparian, V.V., and Hughes, J.H. 1988. Polyacrylamide gel electrophoresis of RNA compared with polyclonal- and monoclonal-antibody-based enzyme immunoassays for rotavirus. J. Clin. Microbiol. 26:194–197.

Paerregard, A., Hjelt, A., Genner, J., Moslet, J., and Krasilnikoff, P.A. 1990. Role of enteric adenoviruses in acute gastroenteritis in children attending day care centers. Acta. Paediatr. Scand. 79:370–371.

Pedley, S. and McCrae, M.A. 1984. A rapid screening assay for detecting individual

RNA species in field isolates of rotaviruses. J. Virol. Methods 9:173–181.

Penaranda, M.E., Cubitt, W.D., Sinarachatanant, P., Taylor, D.N., Likanonsakul, S., Saif, L., and Glass, R.I. 1989a. Group C rotavirus infections in patients with diarrhea in Thailand, Nepal, and England. J. Infect. Dis. 160:392–397.

Penaranda, M.E., Ho, M.S., Fang, Z.Y., Dong, H., Bai, X.S., Duan, S.C., Ye, W.W., Estes, M.K., Echeverria, P., Hung, T., and Glass, R.I. 1989b. Seroepidemiology of adult diarrhea rotavirus in China, 1977 to 1987. J. Clin. Microbiol. 27:2180–2183.

Pether, J.V.S. and Caul, E.O. 1983. An outbreak of foodborne gastroenteritis in two hospitals associated with a Norwalk-like virus. J. Hyg. (London) 91:343–350.

Petrie, B.L., Graham, D.Y., Hanssen, H., and Estes, M.K. 1982. Localization of rotavirus antigens in infected cells by ultrastructural immunocytochemistry. J. Gen. Virol. 63: 457–467.

Phillips, A.D. 1988. Mechanisms of mucosal injury: Human studies. In M. Farthing (ed.), Viruses in the Gut. Proceedings of the Ninth BSG-SK and F International Workshops. Welwyn Garden City: Smith Kline & French, pp. 30–40.

Pickering, L.K., DuPont, H.L., Olarte, J., Conklin, R., and Ericsson, C. 1977. Fecal leukocytes in enteric infections. Am. J. Clin. Pathol. 68:562–565.

Pickering, L.K., Evans, D.G., DuPont, H.L., Vollet, J.J. III, and Evans, D.J. Jr. 1981. Diarrhea caused by Shigella, rotavirus, and Giardia in day-care centers: Prospective study. J. Pediatr. 99:51–56.

Pickering, L.K., DuPont, H.L., Blacklow, N.R., and Cukor, G. 1982. Diarrhea due to Norwalk virus in families. J. Infect. Dis. 146:116–117.

Pickering, L.K., Bartlett, A.V. III, Reves, R.R., and Morrow, A. 1988. Asymptomatic excretion of rotavirus before and after rotavirus diarrhea in children in day-care centers. J. Pediatr. 112:361–365.

Prasad, B.V.V., Burns, J.W., Mariette, E., Estes, M.K., and Chiu, W. 1990. Localization of VP4 neutralization sites in rotavirus by three-dimensional cryoelectron microscopy. Nature 343:476–479.

Ramos-Alvarez, M. and Sabin, A.B. 1958. Enteropathogenic viruses and bacteria. JAMA 167:147–156.

Reid, J.A., Caul, E.O., White, D.G., and Palmer, S.R. 1988. Role of infected food handler in hotel outbreak of Norwalk-like viral gastroenteritis: Implications for control. Lancet 2:321–323.

Richmond, S.J., Caul, E.O., Dunn, S.M., Ashley, C.R., Clarke, S.K., and Seymour, N.R. 1979. An outbreak of gastroenteritis in young children caused by adenoviruses. Lancet 1:1178–1181.

Riepenhoff-Talty, M., Bogger-Goren, S., Li, P., Carmody, P.J., Barrett, H.J., and Ogra, P.L. 1981. Development of serum and intestinal antibody response to rotavirus after naturally acquired rotavirus infection in man. J. Med. Virol. 8:215–222.

Riepenhoff-Talty, M., Uhnoo, I., Chegas, P., and Ogra, P.L. 1989. Effect of nutritional deprivation on mucosal viral infections. Immunol. Invest. 18:127–139.

Riepenhoff-Talty, M., Unhoo, I., Mueller, W.F., Chegas, W.P., Greenberg, H.B., and Ogra, P.L. 1990. Genetic determinants of rhesus rotavirus (RRV) associated with development of hepatitis: Role of viral protein 7 (VP-7) in hepatovirulence. Abstract 1065: p. 180A. Society for Pediatric Research, California, May 7–11.

Riordan, T., Craske, J., Roberts, J., Curry, A. 1984. Foodborne infection by a Norwalk-like virus (small round structured virus). J. Clin. Pathol. 37:817–820.

Rodger, S.M., Bishop, R.F., Birch, C., McLean, B., and Holmes, I.H. 1981. Molecular epidemiology of human rotaviruses in Melbourne, Australia, from 1973 to 1979, as determined by electrophoresis of genome ribonucleic acid. J. Clin. Microbiol. 13:272–278.

Rodriguez, W.J., Kim, H.W., Arrobio, J.O., Brandt, C.D., Chanock, R.M., Kapikian, A.Z., Wyatt, R.G., and Parrott, R.H. 1977. Clinical features of acute gastroenteritis associated with human rotavirus-like agent in infants and young children. J. Pediatr. 91: 188–193.

Rodriguez, W.J., Kim, H.W., Brandt, C.D., Fletcher, A.B., and Parrott, R.H. 1982. Rotavirus: A cause of nosocomial infection in the nursery. J. Pediatr. 101:274–277.

Rodriguez, W.J., Kim, H.W., Brandt, C.D., Gardner, M.K., and Parrott, R.H. 1983. Use of electrophoresis of RNA from human rotavirus to establish the identity of strains involved in outbreaks in a tertiary care nursery. J. Infect. Dis. 148:34–40.

Rotbart, H.A., Levin, M.J., Yolken, R.H., Manchester, D.K., and Jantzen, J. 1983. An outbreak of rotavirus-associated neonatal necrotizing enterocolitis. J. Pediatr. 103: 454–459.

Ruggeri, F.M., Marziano, M.L., Tinari, A., Salvatori, E., Donelli, G. 1989. Four-year study of rotavirus electropherotypes from cases of infantile diarrhea in Rome. J. Clin. Microbiol. 27:1522–1526.

Runnels, P.L., Moon, H.W., Matthews, P.J., Whipp, S.C., and Woode, G.N. 1986. Effects of microbial and host variables on the interaction of rotavirus and Escherichia coli infections in gnotobiotic calves. Am. J. Vet. Res. 47:1542–1550.

Salmi, T.T., Arstila, P., and Koivikko, A. 1978. Central nervous system involvement in patients with rotavirus gastroenteritis. Scand. J. Infect. Dis. 10:29–31.

Samadi, A.R., Huq, M.I., and Ahmed, Q.S. 1983. Detection of rotavirus in handwashings of attendants of children with diarrhoea. Br. Med. J. (Clin. Res.) 286:188.

Santosham, M., Yolken, R.H., Quiroz, E., Dillman, L., Oro, G., Reeves, W.C., and Sack, R.B. 1983. Detection of rotavirus in respiratory secretions of children with pneumonia. J. Pediatr. 103:583–585.

Saulsbury, F.T., Winkelstein, J.A., and Yolken, R.H. 1980. Chronic rotavirus infection in immunodeficiency. J. Pediatr. 97:61–65.

Schreiber, D.S., Blacklow, N.R., and Trier, J.S. 1973. The mucosal lesion of the proximal small intestine in acute infectious non-bacterial gastroenteritis. N. Engl. J. Med. 288: 1318–1323.

Sekine, S., Okada, S., Hayashi, Y., Ando, T., Terayama, T., Yabuuchi, K., Miki, T., and Ohashi, M. 1989. Prevalence of small round structured virus infections in acute gastro-enteritis outbreaks in Tokyo. Microbiol. Immunol. 33:207–217.

Sethabutr, O., Unicomb, L.E., Holmes, I.H., Taylor, D.N., Bishop, R.F., and Echeverria, P. 1990. Serotyping of human group A rotavirus with oligonucleotide probes. J. Infect. Dis. 162:368–372.

Shinozaki, T., Araki, K., Ushijima, H., and Fuji, R. 1987. Antibody response to enteric adenovirus types 40 and 41 in sera from people in various age groups. J. Clin. Microbiol. 25:1679–1682.

Smith, E.M. and Gerba, C.P. 1982. Development of a method for detection of human rotavirus in water and sewage. Appl. Environ. Microbiol. 43:1440–1450.

Spencer, E., Avendano, F., and Araya, M. 1983. Characteristics and analysis of electropherotypes of human rotavirus isolated in Chile. J. Infect. Dis. 148:41–48.

Stephen, J. 1988. Functional abnormalities in the intestine. In M. Farthing (ed.), Viruses in the Gut. Proceedings of the Ninth BSG-SK and F International Workshop. Welwyn Garden City: Smith Kline & French, pp. 41–44.

Sutmoller, F., Azeredo, R.S., Lacerda, M.D., Barth, O.M., Pereira, H.G., Hoffer, E., and Schatzmayr, H.G. 1982. An outbreak of gastroenteritis caused by both rotavirus and Shigella sonnei in a private school in Rio de Janeiro. J. Hyg. (London) 88:285–293.

Takiff, H.E., Strauss, S.E., and Garon, C.F. 1981. Propagation and in vitro studies of previously non-cultivable enteral adenoviruses in 293 cells. Lancet 2:832–834.

Tallett, S., MacKenzie, C., Middleton, P., Kerzner, B., and Hamilton, R. 1977. Clinical, laboratory, and epidemiologic features of a viral gastroenteritis in infants and children. Pediatrics 60:217–222.

Tam, J.S., Zeng, B.J., Lo, S.K., Yeung, C.Y., Lo, M., and Ng, M.H. 1990. Distinct population of rotaviruses circulating among neonates and older infants. J. Clin. Microbiol. 28:1033–1038.

Tan, J.A. and Schnagl, R.D. 1981. Inactivation of a rotavirus by disinfectants. Med. J. Aust. 1:19–23.

Taniguchi, K., Urasawa, T., Kobayashi, N., Gokrziglia, M., Urasawa, S. 1990. Nucleotide sequence of VP4 and VP7 genes of human rotaviruses with subgroup I specificity and long RNA pattern: Implication for new G serotype specificity. J. Virol. 64: 5640–5644.

Taniguchi, K., Urasawa, T., Morita, Y., Greenberg, H.B., and Urasawa, S. 1987. Direct serotyping of human rotavirus in stools by an enzyme-linked immunosorbent assay using serotype 1-, 2-, 3-, and 4-specific monoclonal antibodies to VP7. J. Infect. Dis. 155:1159–1166.

Tao, H., Guangmu, C., Changan, W., Aenli, Y., Zhaoying, F., Tungxin, C., Zinyi, C., Weiwe, Y., Xuejian, C., Shuasen, D., Xiaoguang, L., and Weicheng, C. 1984. Waterborne outbreak of rotavirus diarrhoea in adults in China caused by a novel rotavirus. Lancet 1:1139–1142.

Taylor, J.W., Gary, G.W., and Greenberg, H.B. 1981. Norwalk related gastroenteritis due to contaminated drinking water. Am. J. Epidemiol. 114:584–592.

Thomas, M.E., Luton, P., and Mortimer, J.Y. 1981. Virus diarrhoea associated with pale fatty faeces. J. Hyg. (London) 87:313–319.

Thornhill, T.S., Kalica, A.R., Wyatt, R.G., Kapikian, A.Z., and Chanock, R.M. 1975. Pattern of shedding of the Norwalk particle in stools during experimentally-induced gastroenteritis in volunteers as determined by immune electron microscopy. J. Infect. Dis. 132:28–34.

Thornhill, T.S., Wyatt, R.G., Kalica, A.R., Dolin, R., Chanock, R.M. and Kapikian, A.Z. 1977. Detection of immune electron microscopy of 26- to 27-nm virus-like particles associated with two family outbreaks of gastroenteritis. J. Infect. Dis. 135:20–27.

Uhnoo, I., Wadell, G., Svensson, L., and Johansson, M.E. 1984. Importance of enteric adenoviruses 40 and 41 in acute gastroenteritis in infants and children. J. Clin. Microbiol. 20:365–372.

Uhnoo, I., Olding-Stenkvist, E., and Kreuger, A. 1986. Clinical features of acute gastroenteritis associated with rotavirus, enteric adenoviruses, and bacteria. Arch. Dis. Child. 61:732–738.

Uhnoo, I.S., Freihorst, J., Riepenhoff-Talty, M., Fisher, J.E., and Ogra, P.L. 1990. Effect of rotavirus infection and malnutrition on uptake of a dietary antigen in the intestine. Pediatr. Res. 27:152–160.

Urasawa, S., Urasawa, T., and Taniguchi, K. 1982. Three human rotavirus serotypes demonstrated by plaque neutralization of isolated strains. Infect. Immun. 38:781–784.

Urasawa, S., Urasawa, T., Taniguchi, K., Wakasugi, F., Kobayashi, N., Chiba, S., Sakurada, N., Morita, M., Morita, O., Tokieda, M., Kawamoto, H., Minekawa, Y., and Ohseto, M. 1989. Survey of human rotavirus serotypes in different locales in Japan by enzyme-linked immunosorbent assay with monoclonal antibodies. J. Infect. Dis. 160:44–51.

Urasawa, S., Urasawa, T., Wakasugi, F., Kobayashi, N., Taniguchi, K., Lintag, I.C., Saniel, M.C., Goto, H. 1990. Presumptive seventh serotype of human rotavirus. Arch. Virol. 113:279–282.

Van, R., Morrow, A.L., Reves, R.R., and Pickering, L.K. 1991. Environmental contamination in child day-care centers. Am. J. Epidemiol. 133:460–470.

Van, R., Wun, C.C., O'Ryan, M.L., Matson, D.O., Jackson, L., Pickering, L.K. 1992. Outbreaks of human enteric adenovirus types 40 and 41 in Houston day care centers. J. Pediatr. 120:512–521.

Van Loon, A.E., Maas, R., Vaessen, R.T., Reemst, A.M., Sussenbach, J.S., and Rozijn, T.H. 1985. Cell transformation by the left terminal regions of the adenovirus 40 and 41 genomes. Virology 147:227–230.

Vaucher, Y.E., Ray, C.G., Minnich, L.L., Payne, C.M., Beck, D., and Lowe, P. 1982. Pleomorphic, enveloped, virus-like particles associated with gastrointestinal illness in neonates. J. Infect. Dis. 145:27–36.

Vesikari, T., Maki, M., Sarkkinen, H.J., Arstila, P.P., and Halonen, P.E. 1981a. Rotavirus, adenovirus, and non-viral enteropathogens in diarrhea. Arch. Dis. Child. 56:264–270.

Vesikari, T., Sarkkinen, H.K., and Maki, M. 1981b. Quantitative aspects of rotavirus excretion in childhood diarrhea. Acta. Pediatr. Scand. 70:717–721.

Vesikari, T., Rautanen, T., Varis, T., Beards, G.M., and Kapikian, A.Z. 1990. Rhesus rotavirus candidate vaccine. Clinical trial in children vaccinated between 2 and 5 months of age. Am. J. Dis. Child. 144:285–289.

Vial, P.A., Kotloff, K.L., and Losonsky, G.A. 1988. Molecular epidemiology of rotavirus infection in a room for convalescing newborns. J. Infect. Dis. 157:668–673.

Vial, P.A., Kotloff, K.L., Tall, B.D., Morris, J.G. Jr., and Levine, M.M. 1990. Detection by immune electron microscopy of 27-nm viral particles associated with community-acquired diarrhea in children. J. Infect. Dis. 161:571–573.

Vollet, J.J., Ericsson, C.D., Gibson, G., Pickering, L.K., DuPont, H.L., Kohl, S., and Conklin, R.H. 1979. Human rotavirus in an adult population with travelers' diarrhea and its relationship to the location of food consumption. J. Med. Virol. 4:81–87.

von Bonsdorff, C.H., Hovi, T., Makela, P., and Morttinen, A. 1978. Rotavirus infections in adults in association with acute gastroenteritis. J. Med. Virol. 2:21–28.

Wadell, G. 1984. Molecular epidemiology of human adenoviruses. Curr. Top. Microbiol. Immun. 110:191–220.

Wadell, G., Hammarskjold, M.L., Winberg, G., Varsanyi, T.M., and Sundell, G. 1986. Genetic variability of adenoviruses. Ann. N.Y. Acad. Sci. 354:16–42.

Wadell, G., Allard, A., Svennson, L., Uhnoo, I. 1988. Enteric adenoviruses, In M. Farthing (ed.), Viruses in the Gut. Proceedings of the Ninth BSG-SK and F International Workshop. Welwyn Garden City: Smith Kline & French, pp. 70–78.

Ward, R.L., Knowlton, D.R., Schiff, G.M., Hoshino, Y., and Greenberg, H.B. 1988. Relative concentrations of serum neutralizing antibody to VP3 and VP7 proteins in adults infected with a human rotavirus. J. Virol. 62:1543–1549.

Wenman, W.M., Hinde, D., Feltham, S., Gurwith, M. 1979. Rotavirus infection in adults. Results of a prospective family study. N. Engl. J. Med. 301:303–306.

Wilde, J., Eiden, J., and Yolken, R. 1990. Removal of inhibitory substances from human fecal specimens for detection of group A rotaviruses by reverse transcriptase and polymerase chain reaction. J. Clin. Microbiol. 28:1300–1307.

Wilde, J., Van, R., Pickering, L.K., Eiden, J., and Yolken, R. 1992. Rotaviruses in the day care environment—detection by reverse transcriptase polymerase chain reaction. J. Infect. Dis. In press.

Wong, C.J., Price, Z., and Bruckner, D.A. 1984. Aseptic meningitis in an infant with rotavirus gastroenteritis. Pediatr. Infect. Dis. 3:244–246.

World Health Organization Subgroup of the Scientific Working Group on Epidemiology and Etiology. 1980. Rotavirus and other viral diarrhoeas. Vol. 58. Geneva: World Health Organization, pp. 183–198.

Wyatt, R.G., Dolin, R., Blacklow, N.R., DuPont, H.L., Buscho, R.F., Thornhill, T.S., Kapikian, A.Z., and Chanock, R.M. 1974. Comparison of three agents of acute infectious nonbacterial gastroenteritis by cross-challenge in volunteers. J. Infect. Dis. 129:709–714.

Wyatt, R.G., Greenberg, H.B., Dalgard, D.W., Allen, W.P., Sly, D.L., Thornhill, T.S., Chanock, R.M., and Kapikian, A.Z. 1978. Experimental infection of chimpanzees with the Norwalk agent of epidemic viral gastroenteritis. J. Med. Virol. 2:89–96.

Wyatt, R.G., James, W.D., Bohl, E.H., Theil, K.W., Saif, L.J., Kalica, A.R., Greenberg, H.B., Kapikian, A.Z., and Chanock, R.M. 1980. Human rotavirus type 2: Cultivation in vitro. Science 207:189–191.

Wyatt, R.G., Greenberg, H.B., James, W.D., Pittman, A.L., Kalica, A.R., Flores, J., Chanock, R.M., and Kapikian, A.Z. 1982. Definition of human rotavirus serotypes by plaque reduction assay. Infect. Immun. 37:110–115.

Wyatt, R.G., James, H.D. Jr., Pittman, A.L., Hoshino, Y., Greenberg, H.B., Kalica, A.R., Flores, J., and Kapikian, A.Z. 1983. Direct isolation in cell culture of human rotaviruses and their characterization into four serotypes. J. Clin. Microbiol. 18:310–317.

Yamakawa, K., Oyamada, H., and Nakagomi,

O. 1989. Identification of rotaviruses by dot blot hybridization using an alkaline phosphatose conjugated synthetic oligonucleotide probe. Molec. Cell Probes 3:387–401.

Yolken, R.H., Wyatt, R.G., Zissis, G., Brandt, C.D., Rodriguez, W.J., Kim, H.W., Parrott, R.H., Urrutia, J.J., Mata, L., Greenberg, H.B., Kapikian, A.Z., and Chanock, R.M. 1978. Epidemiology of human rotavirus types 1 and 2 as studied by enzyme-linked immunosorbent assay. N. Engl. J. Med. 299:1156–1161.

Yolken, R.H. and Stopa, P.J. 1979. Enzyme-linked fluorescence assay: Ultrasensitive solid-phase assay for detection of human rotavirus. J. Clin. Microbiol. 10:317–321.

Yolken, R.H. and Murphy, M. 1982. Sudden infant death syndrome associated with rotavirus infection. J. Med. Virol. 10:291–296.

Yolken, R., Leister, F., Almeido-Hill, J., Dubovi, E., Reid, R., Santosham, M. 1989. Infantile gastroenteritis associated with excretion of pestivirus antigens. Lancet 1:517–519.

Yow, M.D., Melnick, J.L., Blattner, R.J., Stephenson, W.B., Robinson, N.M., and Burkhardt, M.A. 1970. The association of viruses and bacteria with infantile diarrhea. Am. J. Epidemiol. 92:33–39.

Viral Hepatitis

Mario R. Escobar

23.1 INTRODUCTION

23.1.1 Nature and Types of Hepatitis

Hepatitis represents an inflammatory disease that may result in necrosis of the liver. It is sometimes due to noninfectious causes including biliary obstruction, primary biliary cirrhosis, Wilson's disease, drug toxicity, and drug hypersensitivity reactions. More commonly, it may be attributed to an infectious process associated with a number of viruses. Occasionally, it may occur as a complication of leptospirosis, syphilis, tuberculosis, toxoplasmosis, or amebiasis (Jawetz et al, 1991).

Viral hepatitis refers more specifically to a primary infection of the liver by one of five etiologically-associated, but different, hepatotropic viruses. Hence, there are five types of hepatitis which include: type A (infectious hepatitis, epidemic jaundice, short incubation hepatitis); type B (serum hepatitis, homologous serum jaundice, long incubation hepatitis); type C (formerly non-A, non-B or NANB, post-transfusion hepatitis); type D (δ) hepatitis; and type E (enterically transmitted waterborne or epidemic hepatitis). With rare exceptions, the clinical manifestations of acute hepatitis seen in these five types are virtually identical. Other viruses that can cause sporadic hepatitis are yellow fever virus, human cytomegalovirus (CMV), Epstein-Barr virus (EBV), rubella virus, herpes simplex virus (HSV), varicella-zoster virus (VZV), and some enteroviruses. The histopathologic lesions produced by these viruses have many common features so that the etiologic agent cannot be distinguished on this basis.

Improved immunoassays for the detection of serologic markers of viral hepatitis, as well as the increase of our knowledge and understanding of the immunopathogenesis of viral infections, have advanced the field of viral hepatitis very rapidly during the last 20 years. The development of modern research techniques in molecular biology (e.g., molecular cloning) during the last few years has led to the discovery of new agents, such as the hepatitis C virus. This chapter presents information regarding the specific immunodiagnosis of viral hepatitis and a discussion of the clinical and pathologic sequelae in those individuals who do not fully recover from acute hepatitis.

23.1.2 Historical Review of the Agents of Viral Hepatitis

Certain clinical conditions characterized by jaundice have probably been grouped together as hepatitis since antiquity. However, it was not until the seventeenth and eighteenth centuries that epidemics of what would now be considered type A hepatitis were well documented. Many of these epidemics occurred during wars between nations when large concentrations of military personnel provided the epidemiologic setting for the spread of the disease. In 1947, the terms hepatitis A and hepatitis B were proposed to designate "infectious" and "serum" hepatitis, respectively. A large series of epidemiologic studies and experiments employing animals and human volunteers of all ages spanning from the mid-1940s to the early 1960s provided clear evidence that these two types of viral hepatitis were microbiologically and epidemiologically distinct (Krugman et al, 1962, 1967). Despite the numerous in vivo and in vitro studies, which resulted in a better understanding of the nature of viral hepatitis, none of the intensive efforts to identify the etiologic agents succeeded until later. The first indication of type B hepatitis came from Bremen, Germany in 1985; it involved shipyard workers who had been vaccinated against smallpox a few months earlier with glycerinated vesicular fluid of other vaccinated humans. Similar reports of parenterally transmitted viral hepatitis were reported in subsequent years. Although these cases occurred in different settings (e.g., diabetes, tuberculosis, and venereal disease clinics, as well as immunization and blood transfusion services), all had in common the use of contaminated materials and instruments (Zuckerman, 1976).

23.1.2.a Hepatitis A

The series of events that took place following the discovery of the Australia antigen by Blumberg (discussed in further detail below) were instrumental in elucidating the nature and identity of the etiologic agent of hepatitis A. Holmes et al (1969) conclusively demonstrated that marmoset monkeys were susceptible to infection with hepatitis A virus (HAV) and reported on the use of the marmoset model to perform the neutralization test as a measure of serum antibody (Holmes et al, 1973). Hepatitis A virus particles were observed by Feinstone et al (1973) in the feces of infected humans and by Provost et al (1975) in the livers of infected marmosets. Until the isolation of HAV in tissue culture by Provost et al (1979), the knowledge of this virus had evolved from immunologic identification of virus antigens. Many techniques have been developed for the detection of HAV in clinical specimens. Additionally, serologic tests have been devised to demonstrate the presence of antibodies to HAV. More recently, in vitro studies of HAV in a variety of cells in culture (Purcell et al, 1984), the availability of animal models (Dienstag et al, 1975; Mao et al, 1981), and the molecular cloning of the genome (Ticehurst et al, 1983) have improved greatly the prospects for the development of a live attenuated vaccine.

23.1.2.b Hepatitis B

In 1961, Blumberg et al (1964) performed a systematic analysis of the sera of multitransfused patients for isoprecipitins employing an agar gel double diffusion procedure. This study led to the discovery in 1962 (Blumberg, 1964) of a human lipoprotein polymorphism (AG system) demonstrating inherited antigenic differences among low density (β) lipoproteins. It was in 1963 when, serendipitously, a precipitating antibody was detected in the serum of a multitransfused patient with hemophilia, which reacted with only a single member of the 24 sera in the test panel, thus, revealing an antigen clearly different from those lipoprotein antigens previously identified. Because this unique antigen was detected in the serum obtained from an Australian aborigine, it was called the Australia antigen and designated by the symbol Au[1] (Blumberg, 1964). The epidemiologic significance, as well as the genetics and physical characteristics of the newly-discovered antigen, were rapidly investigated during the course of several years following this important discovery. However, confirmation of its association with acute viral hepatitis was not published until 1968 by investigators in Italy, Japan (Okochi and Murakami, 1968), and the United States (Blumberg et al, 1968) working independently. One of the best controlled clinical studies was carried out at Willowbrook (Krugman et al, 1979) where the new antigen was isolated from the sera of patients who suffered from "long-incubation-period serum hepatitis." Within a few years of this momentous discovery, the hepatitis B virus (HBV) was structurally characterized, its serologic determinants defined, and its etiologic role in acute and chronic hepatitis recognized. Although many questions concerning the host-virus interaction still remained unanswered, the accumulation of enough basic scientific information led to the successful development of an efficacious HBV vaccine within slightly over a decade following the identification of the virus (Szmuness et al, 1980). Moreover, the recently-discovered viruses that are structurally and immunologically related to human HBV and are also pathogenic for animals will provide important experimental models of hepatitis and hepatocellular carcinoma (Ganem, 1984).

23.1.2.c Hepatitis D

The detection of a distinct antigen by direct immunofluorescence in liver cell nuclei and in the serum of HBV carriers in southern Italy led Rizzetto et al (1977) to interpret their observation as the discovery of a new hepatitis agent which subsequently was called the (δ) agent, or hepatitis D virus (HDV). A fulminant form of hepatitis which occurred in the Sierra Nevada de Santa Marta in northern Colombia was described more than 50 years ago (Villanueva et al, 1988). This geographic area has a high prevalence of hepatitis D associated with HBV infection. This finding suggests that infections with this virus may have occurred much earlier than previously thought (Hoofnagle, 1985). This agent is unique in that it consists of an RNA genome smaller than the genome of conventional viruses but larger than that of viroids of plants and requires HBV infection for its replication. Since its discovery, many reports have been published dealing with all aspects of HDV infection. The availability in the last several years of commercial diagnostic kits and the more recent application of a probe for the direct measurement of δ RNA in clinical samples have opened new possibilities for a better understanding of the host-HDV and HBV-HDV interactions in both acute and chronic hepatitis.

23.1.2.d Hepatitis C and Hepatitis E: The New Types of Hepatitis Non-A, Non-B (NANB)

Non-A, non-B hepatitis causes over 90% of cases of posttransfusion hepatitis and 20% to 35% of sporadic viral hepatitis (Dienstag, 1983a, 1983b; Kuo et al, 1989). It was initially established, on the basis of clinical, epidemiologic, serologic, and electron microscopic observations, that the agents of hepatitis NANB were distinct from the known viruses responsible for acute and chronic hepatitis in humans. Infection with the agents of hepatitis A and B, as well as with other viruses also associated with hepatitis (e.g., CMV and EBV), has been excluded in a number of patients with acute hepatitis (Tabor et al, 1978). This exclusion was possible as a result of the development of sensitive serologic tests to diagnose infections by these other agents. It was also known that hepatitis could occur in individuals who had previously recovered from hepatitis B (Hoofnagle et al, 1977; Mosley et al, 1977). Studies in chimpanzees confirmed these findings and showed conversely that hepatitis B could be transmitted to animals that had recovered from hepatitis NANB (Tabor et al, 1978; Gerety et al, 1984). In contrast to an unconfirmed series of reports from the same laboratory (Hantz et al, 1980; Trepo et al, 1981), no HBV-like particles or serologic crossreactivity with HBV have been observed in hepatitis NANB (Hoofnagle et al, 1977). Analogous experiments were carried out to differentiate hepatitis A from NANB (Hoofnagle et al, 1977; Tabor et al, 1978; Khuroo, 1980).

After many years of confusion concerning the putative etiologic agent(s) of hepatitis NANB, the genome of the virus responsible for the majority of cases of this type of hepatitis, which is prevalent in the United States and western Europe, was finally isolated and characterized. Choo et al (1989) from the Chiron Corporation in California cloned an RNA from the serum of a chimpanzee that had been shown to be highly infectious in serial transmission studies in experimental animals. Since then, numerous studies have been published providing a rapid surge of knowledge about this agent now identified as the hepatitis C virus (HCV). Until more sensitive immunoassays become available, HCV antibodies have been detected so far in 0.7% to 1.4% of normal blood donors, in 50% to 90% of cases of posttransfusion hepatitis, in 50% to 60% of cases of sporadic hepatitis NANB, and in 60% to 90% of cases of suspected chronic hepatitis NANB. Furthermore, the majority of blood donors implicated in transmitting this type of hep-

atitis have tested positive for anti-HCV (Hoofnagle and Di Bisceglie, 1990).

Large waterborne outbreaks of hepatitis NANB were first reported from India and Pakistan about 12 years ago (Wong et al, 1980). Later, Balayan et al (1983) identified a similar non-enveloped 27 to 34 nm virus-like particle in the feces of patients during the incubation period of their disease. Inoculation of experimental animals with fecal material containing these agents was followed by the development of hepatitis and excretion of similar virus-like particles in their stools (Balayan et al, 1983). These findings were confirmed and extended by several groups of investigators (Gust et al, 1987; Kane et al, 1984; Kraczynski et al, 1988). Thus, this is the newest member of the hepatitis virus group and the disease it produces has been called "epidemic" or "enterically-transmitted" hepatitis NANB.

23.2 BRIEF DESCRIPTION OF THE CHARACTERISTICS OF THE VIRUSES

23.2.1 Hepatitis A Virus

It has been proposed recently that HAV should be classified as enterovirus 72. Hepatitis A virus is a 27 to 32 nm spherical naked particle with cubic symmetry. The viral capsid consists of 32 capsomeres arranged in an icosahedral conformation. The unit structure of the capsid antigen consists of four polypeptides: viral proteins 1 through 4 (VP1–VP4). The capsid surrounds a linear molecule of single-stranded RNA made up of about 8,100 nucleotides with a molecular weight of approximately 2.25×10^6 daltons (Ticehurst et al, 1983). It has been suggested that the RNA has positive polarity, that the 3′ end of the RNA is polyadenylated, and that the 5′ end has a small protein, the so-called "viral protein, genomic" (VPG), which may aid the virus in attaching to ribosomes, as is the case with the other enteroviruses (Ganem et al, 1984). Hepatitis A virus is stable to treatment with ether (20%), acid (pH 2.4), and heat (60 °C/hour). It is destroyed by autoclaving (120 °C/20 minutes), by boiling in water for five minutes, by dry heat (180 °C/hour), by ultraviolet irradiation (one minute at 1.1 watts), and by treatment with formalin (1:4000 for three days at 37 °C). Its infectivity can be preserved for at least one month after lyophilization and storage at 25 °C and 42% relative humidity or for years at −20 °C.

23.2.2 Hepatitis B Virus

Hepatitis B virus, also referred to as the "Dane particle," belongs to a new class called "Hepadna viruses." The virion is a complex 42-nm double-shelled particle. The outer surface, or envelope, contains hepatitis B surface antigen (HBsAg) and surrounds a 27-nm inner core that contains hepatitis B core antigen (HBcAg). Inside the core is the genome of HBV, a single molecule of circular DNA. One strand is complete, containing all of the genetic information for the production of HBsAg and HBcAg but is "nicked" (i.e., its 3' and 5' ends are not joined) (Hoofnagle, 1981). The complementary strand contains 50% to 90% of the complementary sequences, thus, leaving a single-stranded or gap region. The DNA molecule consists of 3,200 base pairs and has a molecular weight in excess of 2×10^6 daltons. It is the variable length of the gap region that results in genetically heterogeneous particles with a wide range of buoyant densities. It has been proposed that this unique structure of the viral core may be biologically relevant for integration of the DNA genome into the chromosome of the hepatocyte and thus would represent an important step in the eventual development of hepatocellular carcinoma (Hoofnagle, 1981). In addition, the HBV core also contains DNA-dependent DNA polymerase (which acts to complete the single-stranded region of the DNA), as well as hepatitis Be antigen (HBeAg). The latter can be detected in the serum of HBV infected individuals either as a single protein with a molecular weight of 19,000 daltons or complexed with serum immunoglobulin with a molecular weight of about 300,000 daltons. Its major clinical and epidemiologic significance relates to the finding that it appears to be a reliable serologic marker for the presence of high levels of HBV and thus a high degree of infectivity. Other markers with the same connotation include the Dane particle, DNA polymerase, or specific HBV-DNA. However, the procedures employed to demonstrate the presence of these markers are tedious and complicated in contrast to the use of a single enzyme immunoassay for detection of HBeAg.

HBsAg is the viral component found in the highest concentration in the serum of infected individuals (up to 10^{13} particles per milliliter). Its stability does not always coincide with that of the DNA particle. Nevertheless, both are stable at −20 °C for more than 20 years and stable after repeated freezing and thawing. Infectivity is not destroyed by incubation at 37 °C for 60 minutes and the virus remains viable after dessication and storage at 25 °C for one week. The Dane particle (but not HBsAg) is inactivated at higher temperatures (100 °C for one minute) or when incubated for longer periods (60 °C for ten hours) depending on the amount of virus present in the sample. HBsAg is stable at pH 2.4 for up to six hours with loss of viral infectivity. Sodium hypochlorite, 0.5% (e.g., 1:10 household bleach) destroys antigenicity within three minutes at low protein concentrations, but undiluted serum may require a higher bleach concentration (5%). HBsAg is not destroyed by UV irradiation of plasma or other blood products and viral infectivity may be retained after such treatment. Hepatitis B virus is unevenly distributed during Cohn ethanol fractionation of plasma. Most of the virus is contained in fraction I (fibrinogen, factor VIII) or III (prothrombin complex), whereas HBsAg is found in fractions II (gamma globulin) and IV (plasma protein).

23.2.3 Hepatitis D Virus

Hepatitis D virus is a 35 to 37 nm particle with a buoyant density of 1.24 to 1.25 g/mL. It contains HBsAg on the surface and δ antigen and a small RNA genome (with a molecular weight of 5.5×10^5) in the interior (Rizzetto et al, 1980). The particle consists of an organized structure that prevents hydrolysis of the internal RNA, thus protecting the infectivity of the virus (Hoyer et al, 1983). The δ antigen is distinct from the known antigenic determinants of HBV. It is localized to hepatocyte nuclei that have no HBcAg present. It has a buoyant density in cesium chloride of 1.28 g/mL and a molecular weight of 68,000 daltons (Jawetz et al, 1991). No homology with the HBV genome has been found using hybridization techniques. The δ agent is believed to be a defective virus that replicates only in HBV-infected cells. It appears to be resistant to treatment with EDTA, detergents, ether, nuclease, glycosidases, and acid; but partial to complete inactivation follows treatment with alkali, thiocyanate, guanidine hydrochloride, trichloracetic acid, and proteolytic enzymes.

23.2.4 Hepatitis C and Hepatitis E or Hepatitis NANB

Earlier clinical and epidemiologic studies as well as cross-challenge experiments in chimpanzees revealed at least two hepatitis NANB virus-like agents by immunoelectron microscopy (IEM) in human serum and liver tissue (Shimizu et al, 1979; Bradley et al, 1980). In the serum of patients with chronic hepatitis NANB these parti-

cles measured 37 nm, whereas, in the liver tissue of patients with acute NANB hepatitis, well-defined 27-nm particles were observed inside the hepatocyte nucleus. These agents were heat labile at 60 °C/10 hours (Gerety et al, 1984), and could be inactivated by chloroform and formalin (Bradley et al, 1983; Feinstone et al, 1983).

The major etiologic agent of transfusion-associated and community-acquired hepatitis C was recently cloned and found to consist of a positive-stranded RNA genome of approximately 10 kb in length with positive polarity and a single open reading frame (Choo et al, 1989). The nucleotide sequence has yet to be published, but characteristics of the genome and the size of the viral particle (30 to 60 nm by filtration studies) suggest that although HCV is distantly related to the flavivirus group, it appears to have some significant structural differences. According to studies in primates and consistent with previous reports for NANB, HCV has also been found to be sensitive to organic solvents such as chloroform and to have a low buoyant density (Hoofnagle and Di Bisceglie, 1990). Kaneko et al (1990) have recently reported on the genomic heterogeneity among HCV strains derived from cases in different countries and even among isolates from a single person. These investigators suggested that the apparent rapid evolution of this virus might create serious problems for viral pathogenesis, serological diagnosis, and vaccine design.

Molecular hybridization studies of HEV have suggested that this is a small RNA virus that probably belongs to the family of viruses called "caliciviruses" (Hoofnagle and Di Bisceglie, 1990). A molecular epidemiologic study of human fecal specimens collected from temporally and geographically distinct outbreaks revealed that the HEV clone (ET1.1), originally derived from a Burma isolate and passaged in the cynomolgus macaque, contains a single open reading frame encoding consensus amino acid residue recognized as an RNA directed-RNA polymerase in other plus-stranded RNA viruses. ET1.1 sequences were used to identify an overlapping set of molecular clones representing nearly the entire genome of HEV. Homologous sequences from a Mexican HEV isolate were also identified. However, the degree of sequence conservation along the genome varied (Reyes, 1990). It is to be expected that determination of the sequence of the full-length genome of HEV will undoubtedly help elucidate its genomic organization and contribute to our understanding of the pathogenesis of enterically-transmitted hepatitis NANB.

23.3 PATHOGENESIS OF VIRAL HEPATITIS

During its clinical course, viral hepatitis can lead to any one of four events: 1) typical, icteric hepatitis; 2) subclinical and anicteric hepatitis; 3) fulminant hepatitis; or 4) chronic hepatitis. The clinical manifestations of acute hepatitis for the five etiologic types of viral hepatitis are very similar. For each clinically-apparent case of typical acute (icteric) hepatitis, there are several subclinical cases that can be documented only by the detection of antibody to one of the hepatitis viruses when a history of hepatitis or jaundice is unrecognized. Fulminant hepatitis is a rare but dramatic outcome of viral hepatitis, which presents as an acute disease characterized by hepatic failure and symptoms of hepatic encephalopathy. Chronic hepatitis is a rare or questionable aftermath of HAV infection, but it does occur in about 10% of hepatitis B and 40% to 70% of hepatitis C cases. Chronic hepatitis D can generally develop in up to 40% of hepatitis B carriers. Hepatitis E, similar to hepatitis A, does not lead to chronic hepatitis or cirrhosis, and no carrier state of the infection has been demonstrated. The most striking clinical char-acteristic of hepatitis E is a high mortality rate (10%) in pregnant women which is ten times higher than that in nonpregnant women and men. The reason for the high fatality in pregnancy is unknown (Hoofnagle and Di Bisceglie, 1990).

23.3.1 Hepatitis A

The average incubation period for HAV infection is about one month. Hepatitis A virus infection rates are higher in close contact settings and antibody prevalence rates increase with age. Up to 50% of cases may be anicteric. Unlike HBV infection, a chronic carrier state does not occur and fatalities are very rare; infected patients recover fully and are immune to reinfection.

The pathogenesis of HAV infection is not fully understood. It appears that HAV enters the portal blood from the intestine and is transported to the liver. The specific cells involved in the primary site of infection have not been identified. In experimentally infected chimpanzees and marmosets, hepatitis A antigen has not been detected in intestinal cells during the acute illness (Mathiesen et al, 1977, 1978). However, this does not rule out the possibility of an earlier intestinal phase of replication. Viral antigen can be detected by immunofluorescence (IF) in the cytoplasm of chimpanzee and marmoset hepatocytes beginning approximately two weeks after inoculation and continuing for three to four weeks thereafter. This viral antigen is usually found about one week before liver enzyme elevations or histopathologic evidence of hepatitis. The ability of HAV to persist in humans is unresolved; however, viral antigen has been shown to persist in the hepatocytes of some marmosets for three to four months after inoculation with HAV (Mathiesen et al, 1978). Virus can no longer be detected in the feces of these animals after the appearance of serum antibodies. It cannot be ascertained, however, whether

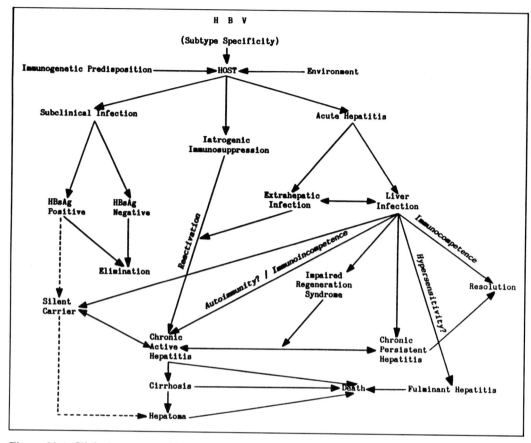

Figure 23.1 Clinical course and immunopathologic sequelae of hepatitis B virus infection.

or not infectious virus continues to be shed at a low level during the extended period when antigen is detected in the liver.

23.3.2 Hepatitis B

The incubation period of HBV is usually about three months, with a range between 45 and 180 days. Although HBV is transmitted predominantly by the percutaneous route, it is also recognized that it can be transmitted by other means, because HBsAg has been detected in virtually every type of body fluid. Spread of HBV by oral and/or genital contact has been demonstrated. Although HBV preferentially replicates in hepatocytes, it has been reported to replicate in nonhepatocytes and in lym-

phoid elements of hemopoietic tissue (Blum et al, 1984).

The natural history of the different forms of HBV infection, as well as the pathogenetic mechanisms responsible for the clinical course of hepatitis type B, are not clear. A recently revised diagram outlining the potentially immunopathologic sequelae of HBV infection is shown in Figure 23.1. HBV infection may result in a variety of syndromes, including subclinical infection with or without hepatitis B surface antigenemia (<50%), acute hepatitis with resolution of illness (30% to 40%), fulminant hepatitis or subacute hepatic necrosis with possible death within three months (1% to 3%), chronic active hepatitis frequently resulting in cirrhosis, chronic persistent hepatitis, or a silent carrier state with minimal (or absent) liver damage or even primary hepatocellular carcinoma (5% to 10% of adults after either symptomatic or subclinical acute hepatitis B).

No single immunologic alteration, as identified in laboratory testing, adequately explains the differences in outcomes in those infected with HBV. Differences in outcome may be related to immunomodulatory activities such as competition between cytolytic T cells and circulating antibody to HBcAg, serum and liver-derived immunoregulatory molecules, and immunoregulatory T cell influences (Dienstag, 1984).

23.3.2.a The Host

Certain host factors have been considered as potential pathogenic determinants of viral hepatitis. These include age, genetic background (Blumberg et al, 1970), sex, physiologic state, and the immune response of the host to viral antigens or autoantigens (Dudley et al, 1971).

Recent work suggests a link between a variety of human diseases and alleles within the major histocompatibility gene complex (HLA). Despite a weak association between HBV infection and some HLA phenotypes, the most convincing evidence for a genetic role in this disease is its very high prevalence in Down's syndrome and in certain Asian and African populations. However, environmental factors could just as easily explain these findings (Levy and Chisari, 1981). The clinical course that occurs in an individual following infection with HBV may be determined—at least in part—by three factors: 1) immunogenetic predisposition as related to the nature of the "match between the antigens of the virus and those of the host"; 2) immunocompetence as regards the humoral and cellular immune responses of the infected individual; and 3) iatrogenic immunosup-

pression occurring as a result of immunosuppressive drug therapy, which may lead to HBV reactivation.

23.3.2.b The Virus

The persistence of virus replication in silent carriers (Dudley et al, 1972), the exacerbation of biochemical abnormalities (e.g., liver enzymes, etc.) which follow the withdrawal of immunosuppressive therapy in patients with chronic active hepatitis B, and the occurrence of often severe acute hepatitis in HBsAg-positive cancer patients following withdrawal of cytotoxic chemotherapy (Dienstag, 1984), suggest that the immune response to HBV, and not the virus itself, is responsible for the liver damage observed in these patients. Nonetheless, viral persistence is often associated with chronic hepatitis (Melnick et al, 1977) suggesting that the virus may play the role of a trigger, if not a causative factor, in the pathogenesis of this disease. The possibility that virus strain, dose, and route of inoculation, or co-factors such as drugs, play a role in determining severity and chronicity has not been discounted and there are even suggestions that these factors are important (Barker and Murray, 1972; Schaefer et al, 1984).

23.3.3 Hepatitis D

The incubation period of hepatitis D varies from 2 to 12 weeks, with shorter incubation periods present in HBV carriers who are superinfected with the agent than in susceptible individuals who are simultaneously infected with both HBV and HDV. That is, the mechanism and outcome of the infection are different in the normal individual and in the individual who is already infected with HBV at the time of exposure to HDV. In normal individuals exposed to HDV (and simultaneously to HBV), the HBsAg necessary to rescue the pathogen is induced acutely by the HBV component of the infectious inoculum. In this context, a limiting factor for the expression of HDV is the short duration of the hepatitis B viremia (Smedile et al, 1981; Caredda et al, 1983). The clinical course of the HDV-associated disease depends on whether the carrier was asymptomatic or suffered from chronic active hepatitis B prior to infection with HDV. The silent carrier may present with an acute HBsAg-positive hepatitis with all the features of classic hepatitis B, except for the possible lack of anti-core IgM antibody (Farci et al, 1983). The disease is frequently severe, perhaps due to the explosive replication of HDV in the patient. This form accounts for a significant number of fulminant HBsAg-positive hepatitis cases in Europe and the United States (Smedile et al, 1982; Govindarajan et al, 1984a). Acute hepatitis D has a tendency to become chronic, which creates a diagnostic problem if the past HBsAg reactivity of the patient is unknown, as transition to chronicity with HDV could resemble progressive hepatitis B (Farci et al, 1983). Conversely, in patients with prior HBV-associated hepatitis, the primary infection with HDV may actually coincide with an apparent flare-up of the chronic hepatitis. Such exacerbations have been fatal in some cases (Raimondo et al, 1983). Liver damage in hepatitis D has been associated with a virus-mediated cytopathic mechanism, thus explaining the failure of conventional immunosuppressive therapy to alter the course of the disease (Rizzetto et al, 1983). The activity of chronic hepatitis D correlates with the presence but not with the amount of δ antigen in the liver. The immunologic features of HDV infection are still under investigation. However, it has been observed that children, who atypically develop mild cases of hepatitis B, often suffer severe and even fatal cases of hepatitis D (Popper et al, 1983). The same occurs in patients on renal dialysis (Smedile et al, 1982), suggesting that immunologic factors may be less important in the pathogenesis

of hepatitis D than in the pathogenesis of hepatitis B.

23.3.4 Hepatitis C and Hepatitis E or Hepatitis NANB

In general, hepatitis NANB has an incubation period ranging from 2 to 12 weeks postexposure. Two incubation periods are found within this range: a short period of two to four weeks which most likely corresponds to hepatitis E and a longer period of 8 to 12 weeks that would be more consistent with hepatitis C.

Three main features have been associated with the pathogenesis and clinical course of hepatitis C: 1) The characteristic pattern of enzyme elevations with marked spontaneous fluctuations in serum alanine aminotransferase levels. This enzyme pattern is rare in hepatitis B, except during periods of reactivation (Hoofnagle and Alter, 1984). 2) The histopathologic pattern characterized by fatty metamorphosis. It has been suggested that this feature is more typical of a cytopathic, rather than an immunologically-mediated, injury (Dienes et al, 1982). 3) The high incidence of chronicity with 50% to 70% of the patients developing chronic liver disease following the acute form of this type of hepatitis (Dienstag, 1983). Chronic hepatitis C is often mild and asymptomatic. Nonetheless, 10% to 25% of cases eventually progress to cirrhosis, and a proportion of these patients will die of hepatic failure or the complications of portal hypertension (Alter and Hoofnagle, 1984; Koretz et al, 1980; Realdi et al, 1982). The course of chronic hepatitis C can also be prolonged and insidious, and patients may not develop symptoms for many years after onset of the chronic infection. Cases of primary hepatocellular carcinoma are not limited to hepatitis B, but have also been attributed to infection with HCV (Okuda et al, 1984).

Large outbreaks of hepatitis E have been reported predominantly from underdeveloped areas in a number of countries including India, Pakistan, Nepal, Burma, Russia, Central Africa, Algeria, Peru, and Mexico. Imported, but not endemic, cases have been described from the United States (Gust and Purcell, 1987). Clinically, hepatitis E is a self-limited hepatitis that is often cholestatic (Gust and Purcell, 1987; Khuroo et al, 1981). Hepatitis E does not lead to chronic hepatitis or cirrhosis, and no carrier state of the infection has been identified.

23.4 PATHOLOGY OF VIRAL HEPATITIS

23.4.1 Hepatitis A

Ordinarily, acute HAV and HBV human infections cannot be distinguished by histopathologic criteria. Hepatitis A has no chronic stage, and the lesions found in chronic viral hepatitis do not apply to HAV infection. In the chimpanzee, certain differences between the hepatic lesions of HAV and HBV have been reported (Dienstag et al, 1975). In hepatitis A, the changes in the hepatocytes were primarily in the periportal areas and parenchymal histopathology was less severe than portal inflammation. In contrast, hepatitis B tended to involve the entire lobular parenchyma, but was predominantly centrilobular. The portal areas were less affected in HBV than in HAV infection. Although these experimental findings may mimic those observed in human hepatitis, the relatively smaller number of biopsies taken from hepatitis A patients is insufficient to support this assumption.

23.4.2 Hepatitis B

As described for hepatitis A, the histopathologic changes in the human liver during acute HBV infection have not been distinguished from those of acute HAV infection, except by special stains. Hepatitis B virus can be demonstrated in liver tissue by detection of HBsAg in infected hepatocytes (Phillips and Poucell, 1981). HBsAg has been detected by histochemical stains, such as aldehyde fuchsin (Shikata et al, 1974), aldehyde thionine (Shikata et al, 1974), modified orcein (Shikata et al, 1974), modified trichrome (Gubetta et al, 1977), Victorian blue (Tanaka et al, 1981), and resorcin fuchsin (Senba, 1982). Immunohistochemical procedures, such as IF and immunoperoxidase (Huang, 1975), as well as electron and IEM methods using immunoferritin and immunoperoxidase (Huang and Neurath, 1979) have been employed successfully for the histologic detection of HBV infection.

The histologic picture in hepatitis B varies according to the severity of the disease, ranging from loss of scattered individual hepatocytes and replacement by inflammatory cells, to massive necrosis of whole lobules of the liver (Peters, 1975). Characteristic lesions include acidophilic bodies (the dehydrated remnants of individual hepatocytes, sometimes with pycnotic nuclei), hepatocytolysis, "ballooned" hepatocytes that appear swollen and pale, excess lipofuscin pigment (resulting from breakdown of liver cells), and scattered or portal inflammation consisting of accumulations of Kupffer cells, lymphocytes, plasma cells, eosinophils, and fibroblasts. Chronic hepatitis may contain some of these elements, and in addition, varying degrees of fibrosis. Viral inclusions have not been reported in either acute or chronic viral hepatitis.

23.4.3 Hepatitis D

The morphologic changes seen in hepatitis NANB and in hepatitis D appear to be very similar with predominant cytotoxic rather than lymphocytotoxic effects (Popper, 1984).

Biopsy specimens of hepatitis B carriers who had hepatitis D demonstrated eosinophilic clumping of the cytoplasm of the hepatocyte, sometimes progressing to acidophilic bodies, and only rarely fine droplet steatosis. Lobular inflammatory cells were lymphocytes, mainly in the sinusoidal lumen, as well as macrophages with periodic acid-Schiff (PAS)-positive nonglycogenic granules in the perisinusoidal spaces, whereas, some parenchymal areas with significant hepatocellular alterations were free of inflammatory cells. By contrast, portal tracts were expanded and infiltrated by a large number of mononuclear cells (Rizzetto et al, 1983). These lesions in the acute stage were in distinct contrast to lesions of hepatitis B in chimpanzees. In hepatitis B, hepatocellular degeneration, including acidophilic bodies, was associated with many lymphocytes, often in close contact with normal and abnormal hepatocytes (Bianchi and Gudat, 1979). These cytotoxic lesions characteristic of hepatitis D have also been reported in studies of HDV infection in young South American Indians, chronic carriers of HBV, during an outbreak in Venezuela (Hadler et al, 1984). It was found that in the initial stage, within a few days after onset of symptoms, the liver exhibited numerous small fat droplets in hepatocytes that underwent focal necrosis, preceded by eosinophilic alterations and acidophilic bodies. Most of the inflammatory cells within the lobular parenchyma were PAS-positive macrophages, while the expanded portal tracts contained a large number of lymphocytes. Delta (δ) antigen was detected immunochemically in the nuclei of scattered hepatocytes. Subsequent stages of massive necrosis and postnecrotic collapse were not associated with detectable δ antigen. On the other hand, autopsy specimens obtained several months after onset of symptoms—where there was a transition to cirrhosis underway—contained large amounts of δ antigen. Finally, there is a similarity in the histopathologic features of hepatitis D and Labrea or "black fever," the etiology of which remains unknown (Andrade et al, 1983).

23.4.4 Hepatitis C and Hepatitis E or Hepatitis NANB

Non-A, non-B hepatitis can progress from the acute phase through the severe stages described for hepatitis B, but with a considerably higher tendency to become chronic. Until recently, more than 13 different particulate structures were associated with hepatitis NANB (Alter and Hoofnagle, 1984). These have shown great diversity in size and morphologic characteristics and none has been independently confirmed or shown to be aggregated by an antibody proven to be specific for the NANB agent. It is probable that most of these particles represent subcellular organelles, lipoproteins, or adventitious agents, but no definitive statement can be made until more reliable reagents become available for IEM (Alter and Hoofnagle, 1984). The histologic alterations found in human liver biopsy specimens, for the most part, are similar to those seen in cases of HBV infection, except that the lesions in NANB hepatitis are the result of virus-mediated cytotoxicity rather than immune-mediated cytotoxicity as is characteristic of hepatitis B by light microscopy, with the former usually being less severe (Popper, 1984). In this regard, there is a certain similarity between the lesion produced in hepatitis NANB and D (Popper, 1984). In addition there have been cytoplasmic tubular structures observed in thin-section electron micrographs of chimpanzee liver infected with human NANB agents. These cytoplasmic structures consist of double-unit membranes with an electron-dense center contiguous with the smooth endoplasmic reticulum. Also observed in chimpanzee liver were 20- to 27-nm intranuclear particles. The intranuclear and cytoplasmic ultrastructures were originally thought to be mutually exclusive and perhaps to represent the morphologic footprint of two NANB agents (Shimizu et al, 1979). Later studies, however, not only disproved their mutual exclusivity, but as previously reported (Schaff et al, 1984) the nuclear particles are neither associated with a second NANB agent nor are they specific for hepatitis NANB (Alter and Hoofnagle, 1984). Nevertheless, the cytoplasmic tubular structures constitute a valuable histologic marker of NANB infection even though only a single recent report noted their detection in liver biopsies of humans with hepatitis NANB (Watanabe et al, 1984).

The morphologic features of chronic hepatitis C were recently reported (Govindarajan et al, 1990) from a retrospective study of serologically-documented chronic HCV infection on the basis of liver biopsies which consistently revealed proliferation of atypical bile ducts including serpinginous growth patterns, hydropic changes of the epithelium, flattened epithelium, and rarely, hyperplastic changes. In addition, in all cases there was prominent periportal or periseptal sinusoidal fibrosis, and only occasionally perivenular fibrosis. In approximately 60% of the cases there were macrovesicular fatty changes with a random distribution. The authors suggested that morphologic features, such as fat, sinusoidal collagen, glycogen nuclei, and atypical ducts were unusual for other types of chronic hepatitis.

Although hepatitis E is at this time a well-established clinical entity and is considered the enterically-transmitted type of hepatitis NANB, there is a paucity of information regarding its pathology and pathogenesis. A recent report by Song et al (1990) indicates that the liver biopsies from a number of patients in this study revealed a predominant polymorphonuclear leukocyte response, marked canalicular and cytoplasmic cholestasis, and bile plugs seen in the middle of pseudoductules. However, repeat liver biopsies were normal in consonance with the limited clinical course of this type of hepatitis. Changes in hepatocyte ultrastructure were not diagnostic. No tubular structures or spongelike inclusion bodies were observed.

23.5 LABORATORY DIAGNOSIS OF VIRAL HEPATITIS

23.5.1 Histologic, Biochemical, and Hematologic Tests

Liver biopsy, when indicated, provides tissue for a histologic diagnosis of hepatitis. Tests for abnormal liver function, such as alanine aminotransferase (ALT; formerly SGPT) and bilirubin, supplement the clinical, pathologic, and epidemiologic data. Alanine aminotransferase levels in acute hepatitis range between 500 and 2000 units, and are almost never below 100 units; they are usually higher than those of serum aspartate transaminase (AST; formerly SGOT). A sharp elevation of ALT with a short duration (3 to 19 days) is more suggestive of HAV infection. Fluctuating levels of ALT are more characteristic of hepatitis C. Serum albumin levels are decreased and total serum globulin is increased. In many patients with hepatitis A, there is an abnormally high level of total IgM three to four days after ALT begins to rise. Hepatitis B patients have normal to slightly elevated IgM levels. Leukopenia is typical in the preicteric phase and may be followed by a relative lymphocytosis. Large atypical lymphocytes similar to those found in infectious mononucleosis may occasionally be present on smears prepared for differential white blood cell counts, but these generally do not exceed 10% of the total lymphocyte population.

23.5.2 Cell Culture for Virus Isolation

Of all the viruses associated with primary viral hepatitis only HAV has been isolated in cell culture (Provost et al, 1979). Its propagation was quickly confirmed by others and its host range in vitro was extended to other cells of primate origin, mainly primary, secondary, and continuous kidney cells (Provost et al, 1981; Binn et al, 1984) and diploid fibroblasts (Gauss-Müeller et al, 1981). Initially the virus grew poorly in cell culture with no cytopathic effects and was characterized by a long replicative cycle measured in weeks and low yield of virus and viral antigen. Staining of cultures with fluorescein-labeled anti-HAV revealed the presence of discrete fluorescent foci that slowly enlarged with time, suggesting that the virus was mostly cell-associated and not readily released into the tissue culture medium. It has been reported that the incubation period has been shortened to three weeks and the infectivity titers have been increased to 10^7 to 10^9 tissue culture 50% infectious doses per milliliter of lysed cells (Binn et al, 1984; Purcell et al, 1984). In addition, a rapidly replicating strain of HAV has been developed in Buffalo green monkey kidney cells, which can be passaged every three days. The rapidly replicating virus can be used to test for neutralizing anti-HAV and is as sensitive as commercially available radioimmunoassays (RIA) for anti-HAV antibody detection (Purcell et al, 1984), a finding consistent with that of Lemon and Binn (1983).

23.5.3 Serologic Markers and Immunodiagnosis in Viral Hepatitis

An algorithmic flow chart for the serodiagnosis of the various types of viral hepatitis is shown in Figure 23.2. Although it is self-explanatory, please note that dual infections may occasionally occur. Even though a final diagnosis may have been confirmed, serologic follow-up could include testing for additional serologic markers to identify HBV carriers and assess their degree of infectivity.

23.5.3.a Hepatitis A

The presence of HAV in stool and the humoral immune responses of the host from time of exposure up to 12 weeks post infection vis a vis the presence of clinical symptoms and liver enzyme elevation are illustrated in Figure 23.3. Initial evaluation of acute viral hepatitis should include tests for HAV and HBV. Assays used 15 years ago for anti-HAV included complement fixation, IEM, and immune adherence hemagglutination (Chernesky et al, 1984). Currently the most widely used immunoassay for IgM class anti-HAV employs anti-human IgM as a solid phase capture reagent, wherein reaction with a dilution of the patient's test serum results in the binding of a portion of all IgM molecules present in the specimen. In second and third reaction sequences, HAV and labeled anti-HAV probe are used to detect the presence of IgM anti-HAV, and either radioactive- or enzyme-conjugated anti-HAV probes may be utilized. These tests are sensitive, specific, and practical for diagnosing hepatitis A. The key for the specificity of the test is that the solid phase must not bind human IgG, because even minimum binding will lead to interpreting high levels of IgG anti-HAV as low levels of IgM (Decker et al, 1984). Anti-HAV IgM can also be detected with tests that measure total anti-HAV immunoglobulin after serum treatment with staphylococcal protein A (Bradley et al, 1979), but caution must be exercised because this treatment does not remove all subclasses of IgG. Positive results may be confirmed by treatment of the serum with 2-mercaptoethanol. Of course, tests for the detection of total anti-HAV immunoglobulins may also be used to assess past immunity in a single serum or for diagnostic purposes when paired sera show a fourfold or higher titer rise. Solid phase IgM-specific immunoassays are best because they are extremely accurate in measuring anti-HAV

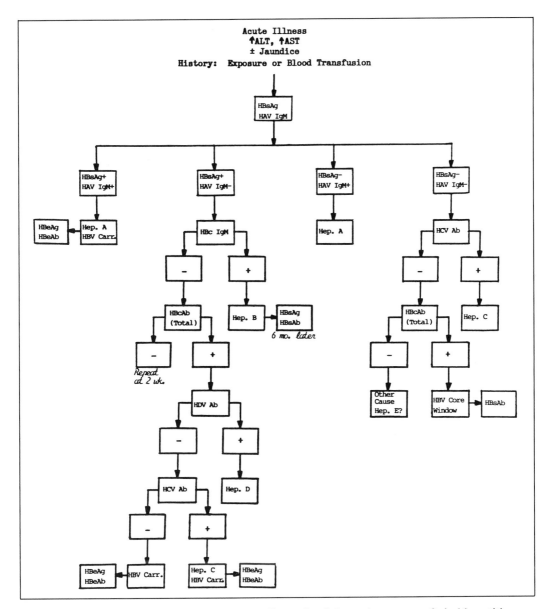

Figure 23.2 Algorithmic flow chart for the serodiagnosis of the various types of viral hepatitis.

IgM responses for up to four to six months (Chernesky et al, 1984). An alternate procedure for anti-HAV IgM detection using murine-derived monoclonal anti-HAV has been reported recently (Decker et al, 1984). It is more convenient than the conventional solid phase immunoassay described earlier and appears to have certain advantages for diagnostic effectiveness.

In most cases, it is impractical to test fecal specimens for HAV particles or antigens because most patients have few, if any, particles in their feces during the clinical phase of the illness, and what is present may be heavily coated by antibody, preventing serologic detection. Nonetheless, examination of feces from contacts of patients with hepatitis A, during an epidemic

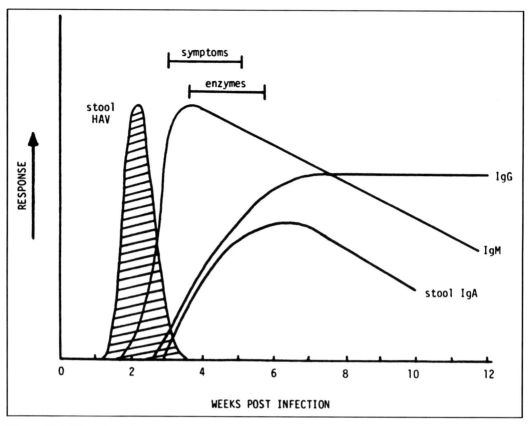

Figure 23.3 Virus detection and humoral immune responses in hepatitis A. (Reproduced from Chernesky et al, 1984)

outbreak, may be useful epidemiologically (Chernesky et al, 1984).

23.5.3.b Hepatitis B

The present unavailability of cell cultures for the isolation of HBV limits the laboratory diagnosis of hepatitis B to the detection of serologic markers by sensitive immunoassays. Third generation procedures including solid phase immunoassays using enzyme- or radioactive-labeled conjugates, latex agglutination, and hemagglutination tests are commercially available for the detection of at least six serologic markers of HBV infection. These markers include HBsAg, anti-HBs, anti-HBc IgG, anti-HBc IgM, HBeAg, and anti-HBe. Other serologic markers of HBV infection include

viral DNA polymerase (Kaplan et al, 1973), HBcAg (Rizzetto et al, 1981), and HBV DNA (Berninger et al, 1982). However, tests for the presence of these latter three markers are impractical for routine use and do not add significantly to the information provided by the immunoassays for HBeAg detection. This relatively complicated system of serologic markers of HBV infection may help establish the stage of the disease, degree of infectivity, prognosis, and immune status of the patient. The various phases of HBV infection as defined by the reactions obtained using these serologic markers are shown in Table 23.1. A typical serologic profile of HBV-associated markers after onset of acute hepatitis B is illustrated in Figure 23.4. A recent review deals

Table 23.1 Serodiagnostic Profiles of HBV Infection

Interpretation	Serological markers					
	HBsAg	HBeAg	IgM anti-HBc	Total anti-HBc	Anti-HBe	Anti-HBs
Acute infection						
Incubation period	+	+	−	−	−	−
Acute phase	+	+	+	+	−	−[b]
Early convalescent phase	+	−[a]	+	+	+	−
Convalescent phase	−	−	+	+	+	−
Late convalescent phase	−	−	−[c]	+	+	+
Long past infection	−	−	−	+[d]	+/−	+[d]
Chronic infection						
Chronic active hepatitis	+[e]	+/−[f]	+/−	+	+/−[f]	−[g]
Chronic persistent hepatitis	+	+/−	+/−	+	+/−	−
Chronic HBV carrier state	+	+/−	+/−	+	+/−	−
HBsAg immunization	−	−	−	−	−	+

[a] HBeAg "rarely" may persist for weeks to months after disappearance of HBsAg.

[b] Anti-HBs occasionally appears before the clearance of HBsAg in acute infection.

[c] IgM anti-HBc may persist for over 1 year after onset of acute infection when very sensitive assays are employed.

[d] Total anti-HBc and anti-HBs may be detected together or separately long after acute infection.

[e] HBsAg-negative chronic infection may occur where anti-HBc is the only detectable serological marker.

[f] Chronic infection is usually associated with liver disease more likely in the presence of HBeAg than of anti-HBe.

[g] Anti-HBs may be present in chronic infection more likely due to previous infection with a different HBsAg subtype than to antigen-antibody complexes.

Modified from Chernesky et al, 1984.

with the application and interpretation of tests for the detection of these serologic markers (Chernesky et al, 1984). This source of information is recommended for further details. Briefly, HBsAg appears prior to anti-HBc, and therefore, the presence of HBsAg with a negative test for anti-HBc indicates very early infection in nonvaccinated individuals. Testing for HBeAg and anti-HBe provides information about the relative infectivity of the patient. The presence of HBeAg correlates with abundant circulating Dane particles and/or elevated viral DNA polymerase activity. It has also been shown to have a strong correlation with an increased risk of transmitting infection upon accidental needlestick, and the transmission of HBV from mother to baby during the perinatal period, or following exposure in household settings. Conversely, a positive result for anti-HBe reflects a low number of Dane particles and

reduced risk of transmitting infection. Furthermore, serial HBeAg testing may be useful in assessing recovery from acute infection. Persistence of HBeAg by RIA at ten weeks or more after the onset of symptoms has the same prognostic value as the persistence of HBsAg at four to six months, and both may be useful in predicting development of the carrier state. Detection of anti-HBe indicates that recovery from infection is underway and complete resolution of infection is likely. Periodic testing (e.g., every four to six weeks) for HBsAg is always indicated when managing a patient with acute HBV hepatitis. When HBsAg has disappeared, anti-HBs testing is highly desirable to confirm immunologic recovery from infection and the presence of long-lasting immunity to reinfection. An enzyme immunoassay (EIA) for the detection of anti-HBc IgM has recently become available. This marker is very useful in the

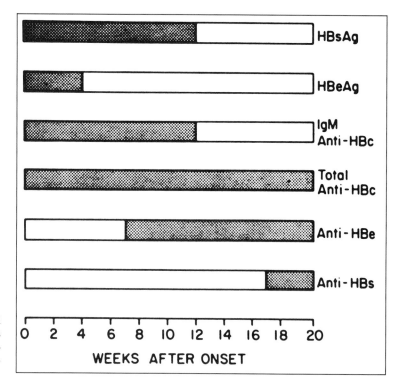

Figure 23.4 Serological profile of HBV markers in acute hepatitis B. (Reproduced from Chernesky et al, 1984)

differential diagnosis between: 1) acute hepatitis B and chronic HBsAg carrier state when HBsAg is undetectable; 2) acute hepatitis B with development of chronic HBsAg carrier state; and 3) non-B hepatitis in a healthy HBsAg carrier. In the latter the anti-HBc IgM titer is usually much lower (Frösner and Franco, 1984). High anti-HBc titers may be useful in the early diagnosis of fulminant hepatic necrosis (Gitlin, 1984). It has been suggested that the presence of anti-HBc among HBsAg-positive carriers places them in a higher risk group for the development of primary hepatocellular carcinoma (Beasley and Hwang, 1984).

23.5.3.c Hepatitis D

Initially, the detection of δ antigen in the nuclei (and occasionally in the cytoplasm) of hepatocytes was achieved by direct IF using either frozen unfixed or fixed embedded liver biopsy specimens (Rizzetto et al, 1977). Later, the δ antigen was visualized in

liver biopsy specimens by indirect immunoperoxidase staining (Govindarajan et al, 1984b). Two reports on the use of the EIA (Crivelli et al, 1981; Shattock and Morgan, 1984) for the detection of δ antigen and anti-δ led the way for the routine use of this assay in the clinical laboratory. The recent commercial availability of licensed RIA and EIA kits for the detection of antibodies to the δ agent will reduce the need for liver biopsy in many cases. Although detecting antigen in serum is possible, it is not a practical approach at this time.

Acute HBV and acute HDV coinfection is indicated when HBsAg, anti-HBc IgM, HBeAg, and anti-HD IgM are positive (Figure 23.5). A patient with this pattern will recover from HDV infection as soon as he is able to resolve the hepatitis B disease, because HDV depends on HBV for replication. Otherwise, the patient would also be very likely to become a chronic HDV carrier. Chronic HBV infection with superim-

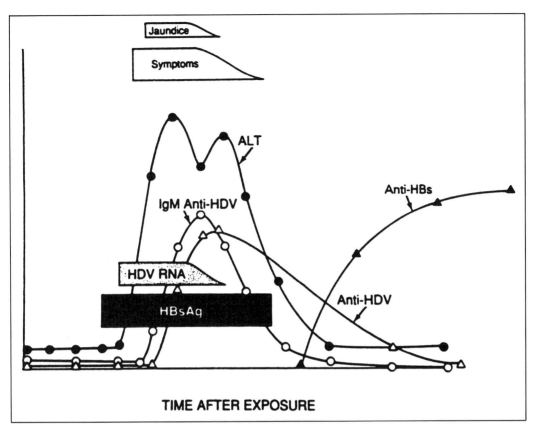

Figure 23.5 Clinical, biochemical, and serologic profile of a typical patient with acute hepatitis B and D coinfection. HDV hepatitis D virus; anti-HDV, antibody to HDV; HBsAg, hepatitis B surface antigen; anti-HBs antibody to HBsAg; ALT, alanine aminotransferase. (Reproduced with permission from Hoofnagle and Di Bisceglie, 1990)

posed HDV infection is indicated when HBsAg, anti-HBc IgG, and anti-HD IgM are positive (Figure 23.6). In both acute and chronic hepatitis D, HDV RNA should also be detectable concurrently with the presence of HBsAg in the blood. The clinical type of HDV infection in the presence of clinical symptoms of hepatitis can be further determined by titering the anti-HD antibodies. That is, an anti-HD titer of less than 10^2 is consistent with an acute infection and a titer of more than 10^3 with chronic infection. The same pattern above with the loss of anti-HD IgM in the absence of clinical symptoms is consistent with the recovery from HDV infection or onset of chronicity, again, depending on the an-

ti-HD titer. An anti-HD titer of less than 10^2 should indicate recovery; more than 10^3 should indicate the onset of chronic asymptomatic HDV infection. The disappearance of HBsAg and persistence of anti-HBc IgG and anti-HD IgG antibodies are compatible with the recovery from HBV and HDV infections. This can be further confirmed by demonstrating that anti-HBs antibodies are also present.

23.5.3.d Hepatitis C and Hepatitis E or Hepatitis NANB

Specific diagnosis of hepatitis NANB was hampered until recently because of the absence of defined serologic markers. Two

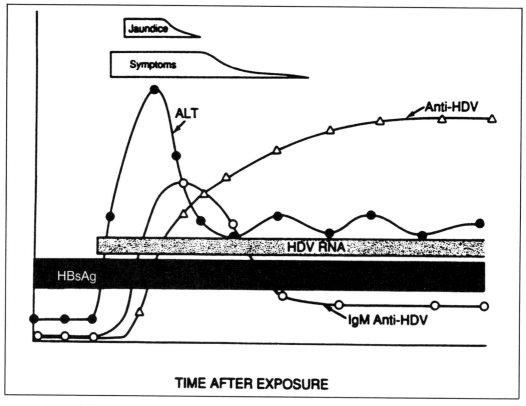

Figure 23.6 Clinical, biochemical and serologic profile of a typical patient with acute hepatitis D superinfection who develops chronic hepatitis D. HDV, hepatitis D virus; anti-HDV, antibody to HDV; HBsAg, hepatitis B surface antigen; anti-HBs, antibody to HBsAg; ALT, alanine aminotransferase. (Reproduced with permission from Hoofnagle and Di Bisceglie, 1990)

commercial EIA kits have already been licensed for diagnostic use. Their use for epidemiologic surveys and in clinical trials has resolved much of the confusion created in the literature in the past several years because results of those studies could not be confirmed or reproduced (Alter and Hoofnagle, 1984). These two kits are relatively sensitive for the detection of antibodies to the nonstructural proteins of the c-100-3 region (and also to the 5-1-1 region in one of these kits) of the HCV genome during chronic infection. However, both kits are insufficiently sensitive for the early diagnosis or during the acute phase of HCV infection because seroconversion may not be detectable until several months after exposure to the virus. Based on recent

evaluations of these tests (McFarlane et al, 1990; Weiner et al, 1990; Wong et al, 1990), it is recommended that "confirmatory" tests always be used; otherwise, positive results of first generation tests must be interpreted with caution. False positive results have been reported to occur in rheumatoid arthritis (Theilmann et al, 1990); hyperimmunoglobulinemia and autoimmune liver diseases (Ikeda et al, 1990) and, possibly, other immunologic disorders. Until we learn more about these tests, it is recommended that multiple freezing and thawing be avoided because storage conditions may also affect test results.

It is worthy of comment that the unlicensed "confirmatory" tests (e.g., recombinant neutralization and immunoblotting), which are commercially available at this time, are not truly confirmatory and require more sophistication. In addition, second generation assays should be more specific than the current tests and also more sensitive, particularly during early acute HCV infection (i.e., by the use of structural proteins or other epitopes of HCV). Precise and reliable immunoassays would: 1) contribute to the understanding of the pathogenesis and immune response in self-limited and chronic manifestations of HCV infection; 2) identify individuals who are immune to HCV in order to secure a dependable source of immune globulin; 3) monitor for viral replication and infectivity; and 4) facilitate the development of an effective HCV vaccine. The chimpanzee can be readily infected with HCV and is suggested as the ideal model for vaccine trials.

A reliable primate model of HEV infection has also been developed using cynomolgus monkeys to provide a readily available and dependable source of virus, viral antigens, and virus-specific antibodies for clinical and laboratory studies. The very recent development of a quantitative immunofluorescent assay for antigen and antibody has made it possible to determine HEV titers of acute and convalescent sera and to conduct detailed pathogenesis studies in experimentally infected primates. Studies are also well underway concerning the molecular cloning of the HEV genome and the production of recombinant HEV antigens that would be useful to the design of diagnostic immunoassays and the generation of prototype vaccines for the immunization of susceptible populations at high risk of contracting HEV infection (Bradley, 1989).

REFERENCES

Alter, J.H. and Hoofnagle, J.H. 1984. Non-A, non-B: Observations on the first decade. In G.N. Vyas, J.D. Dienstag, and J.H. Hoofnagle (eds.), Viral Hepatitis and Liver Disease. Orlando, FL: Grune & Stratton.

Andrade, Z.A., Santos, J.B., and Prata, A. 1983. Histopatologia de hepatite de labrea. Rev. Soc. Bras. Med. Trop. 16:31–40.

Balayan, M.S., Andzhaparidze, A.G., Savinskaya, S.S., Ketiladze, E.S., Braginsky, D.M., Savinov, A.P., and Poleschuk, V.F. 1983. Evidence for a virus in non-A, non-B hepatitis transmitted via the fecal oral route. Intervirology. 20:23–31.

Barker, L.F. and Murray, R. 1972. Relationship of virus dose to incubation time of clinical hepatitis and time of appearance of hepatitis associated antigen. Am. J. Med. Sci. 263: 27–33.

Beasley, R.P. and Hwang, L.Y. 1984. Epidemiology of hepatocellular carcinoma. In G.N. Vyas, J.L. Dienstag, and J.H. Hoofnagle

(eds.), Viral Hepatitis and Liver Disease. Orlando, FL: Grune & Stratton.

Berninger, M., Hammer, M., Hoyer, B., and Gerin, J.L. 1982. An assay for the detection of the DNA genome of hepatitis B virus in serum. J. Med. Virol. 9:57–68.

Bianchi, L. and Gudat, F. 1979. Immunopathology of hepatitis B. In H. Popper and F. Schaffner (eds.), Progress of Liver Diseases, Vol. VI. New York: Grune & Stratton.

Binn, L.N., Lemon, S.M., Marchwicki, R.H., Redfield, R.R., Gates, N.L., and Bancroft, W.H. 1984. Primary isolation and serial passage of hepatitis A virus strains in primate cell cultures. J. Clin. Microbiol. 20: 28–33.

Blum, H.E., Haase, A.T., and Vyas, G.N. 1984. Origin of replication of hepatitis B virus DNA in human liver. (Abst. 3B.32). In G.N. Vyas, J.L. Dienstag, and J.H. Hoofnagle (eds.), Viral Hepatitis and Liver Disease. Orlando, FL: Grune & Stratton, p. 633.

Blumberg, B.S. 1964. Polymorphisms of the serum proteins and the development of isoprecipitins in transfused patients. Bull. N.Y. Acad. Med. 40:377–386.

Blumberg, B.S., Sutnick, A.I., and London, W.T. 1968. Hepatitis and leukemia. Their relation to Australia antigen. Bull. N.Y. Acad. Med. 44:1566–1586.

Blumberg, B.S., Sutnick, A.I., and London, W.T. 1970. Australia antigen as a hepatitis virus: Variation in host response. Am. J. Med. 48:1–8.

Bradley, D.W., Fields, H.A., McCaustland, K.A., Maynard, J.E., Decker, R.H., Whittington, R., and Overby, L.R. 1979. Serodiagnosis of viral hepatitis A by a modified competitive binding radioimmunoassay for immunoglobulin M anti-hepatitis A virus. J. Clin. Microbiol. 9:120–127.

Bradley, D.W., Maynard, J.E., Cook, E.H., Ebert, J.W., Gravelle, C.R., Tisquaye, K.N., Kessler, H., Zuckerman, A.J., Miller, M.F., Ling, C., and Overby, L.R. 1980. Non-A, non-B hepatitis in experimentally infected chimpanzees: Cross-challenge and electronmicroscopic studies. J. Med. Virol. 6:185–201.

Bradley, D.W., Maynard, J.E., Popper, H., Cook, E.H., Ebert, J.W., McCaustland, K.A., Schable, C.A., and Fields, H.A. 1983. Posttransfusion non-A, non-B hepatitis: Physicochemical properties of two distinct agents. J. Infect. Dis. 148:254–265.

Bradley, D.W. 1989. Editor's Note. Update Testing in the Blood Bank, Ortho Diagnostic Systems, Inc., Raritan, NJ, Vol. 3, Issue No. 1.

Caredda, F., d'Arminio Monforte, A., Rossi, E., Farci, P., Smedile, A., Tappero, G., and Moroni, M. 1983. Prospective study of epidemic delta infection in drug addicts. In G. Verme, F. Bonino, and M. Rizzetto (eds.), Viral Hepatitis and Delta Infection. New York: Alan R. Liss.

Chernesky, M.A., Escobar, M.R., Swenson, P.D., and Specter, S.C. 1984. Laboratory diagnosis of hepatitis viruses. Washington, DC: American Society for Microbiology, Cumitech 18, pp. 1–12.

Choo, Q-L., Kuo, G., Weiner, A.M., Overby, L.R., Bradley, D.W., and Houghton, M. 1989. Isolation of a cDNA clone derived from a blood-borne non-A, non-B viral hepatitis genome. Science. 244:359–362.

Crivelli, O., Rizzetto, M., Lavarini, C., Smedile, A., and Gerin, J.L. 1981. Enzyme-linked immunosorbent assay for detection of antibody to the HBsAg-associated delta antigen. J. Clin. Microbiol. 14:173–177.

Decker, R.H., Dawson, G.J., and Mushahwar, I.K. 1984. Monoclonal antibodies to hepatitis A virus: Applications to the serologic detection of IgM anti-HAV. In G.N. Vyas, J.L. Dienstag, and J.H. Hoofnagle (eds.), Viral Hepatitis and Liver Disease. Orlando, FL: Grune & Stratton.

Dienes, H.P., Popper, H., Arnold, W., and Lobeck, H. 1982. Histologic observations in human hepatitis non-A, non-B. Hepatology. 2:562–571.

Dienstag, J.L., Feinstone, S.M., Purcell, R.H., Hoofnagle, J.H., Barker, L.F., London, W.T., Popper, H., Peterson, J.M., and Kapikian, A.Z. 1975. Experimental infection of chimpanzees with hepatitis A virus. J. Infect. Dis. 132:532–535.

Dienstag, J.L. 1983. Non-A, non-B hepatitis. Gastroenterology, 85:439–462, 743–768.

Dienstag, J.L. 1983a. Non-A, non-B hepatitis. I. Recognition, epidemiology, and clinical features. Gastroenterology. 85:439–462.

Dienstag, J.L. 1983b. Non-A, non-B hepatitis. II. Experimental transmission, putative virus agents and markers, and prevention. Gastroenterology. 85:743–768.

Dienstag, J.L. 1984. Immunologic mechanisms in chronic viral hepatitis. In G.N. Vyas, J.L. Dienstag, and J.H. Hoofnagle (eds.), Viral Hepatitis and Liver Disease, Orlando, FL: Grune & Stratton.

Dudley, F.J., Fox, R.A., and Sherlock, S. 1971. Relationship of hepatitis associated antigen (H.A.A.) to acute and chronic liver injury. Lancet 1:12.

Dudley, F.J., Fox, R.A., and Sherlock, S. 1972. Cellular immunity and hepatitis associated Australia antigen liver disease. Lancet 1:723–726.

Farci, P., Smedile, A., Lavarini, C., Piantino, P., Crivelli, O., Caporaso, N., Toti, M., Bonino, F., and Rizzetto, M. 1983. Delta hepatitis in inapparent carriers of hepatitis

B surface antigen. Gastroenterology 85: 669–673.

Feinstone, S.M., Kapikian, A.Z., and Purcell, R.H. 1973. Hepatitis A: Detection by immune electron microscopy of a virus-like antigen associated with acute illness. Science 182:1026–1028.

Feinstone, S.M., Mihalik, K.B., Kazmimura, T., Alter, H.J., London, W.T., and Purcell, R.H. 1983. Inactivation of hepatitis B virus and non-A, non-B hepatitis by chloroform. Infect. Immun. 41:816–821.

Frösner, G.G. and Franco, E. 1984. New developments and open questions in the serology of hepatitis A and B. In G.N. Vyas, J.L. Dienstag, and J.H. Hoofnagle (eds.), Viral Hepatitis and Liver Disease. Orlando, FL: Grune & Stratton.

Ganem, D. 1984. Animal models in hepatitis research. In G.N. Vyas, J.L. Dienstag, and J.H. Hoofnagle (eds.), Viral Hepatitis and Liver Disease. Orlando, FL: Grune & Stratton.

Gauss-Müeller, V., Frösner, G.G., and Deinhardt, F. 1981. Propagation of hepatitis A virus in human embryo fibroblasts. J. Med. Virol. 7:233–239.

Gerety, R.J., Tabor, E., Schaff, A., Seto, B., and Coleman, W.G. Jr. 1984. Non-A, non-B hepatitis agents. In G.N. Vyas, J.L. Dienstag, and J.H. Hoofnagle (eds.), Viral Hepatitis and Liver Disease. Orlando, FL: Grune & Stratton.

Gitlin, N. 1984. The treatment of viral hepatitis: Uncomplicated acute viral hepatitis and fulminant viral hepatic necrosis. In G.N. Vyas, J.L. Dienstag, and J.H. Hoofnagle (eds.), Viral Hepatitis and Liver Disease. Orlando, FL: Grune & Stratton.

Govindarajan, S., Chin, K.P., Redeker, A.G., and Peters, R.L. 1984a. Fulminant B. Viral hepatitis: Role of delta agent. Gastroenterology 86:1717–1720.

Govindarajan, S., Lim, B., and Peters, R.L. 1984b. Immunohistochemical localization of the delta antigen associated with hepatitis B virus liver biopsy sections embedded in araldite. Histopathology 8:63–67.

Govindarajan, S., Kanel, G.C., McHutchinson, J.G., Redeker, A.G., and Kuo, G. 1990.

Morphologic features of chronic HCV infection—liver biopsy study. (Abst. #387) In The 1990 International Symposium on Viral Hepatitis and Liver Disease, p. 146.

Gubetta, L.L., Rizzetto, M., Crivelli, O., Verme, G., and Arico, S. 1977. A trichrome stain for the intrahepatic localization of the hepatitis B surface antigen (HBsAg). Histopathology 1:227–288.

Gust, I.D. and Purcell, R.H. 1987. Report of a workshop: Waterborne non-A, non-B hepatitis. J. Infect. Dis. 156:630–635.

Hadler, S.C., De Monzon, M., Ponzetto, A., Anzola, E., Rivero, D., Mondolfi, A., Bacho, A., Francis, D.P., Gerber, M.A., and Thung, M.A. 1984. An epidemic of severe hepatitis due to delta virus infection in Yupca Indians of Venezuela. Ann. Intern. Med. 100:339–344.

Hantz, O., Vitvitski, L., and Trepo, C. 1980. Non-A, non-B hepatitis: Identification of hepatitis B-like virus particles in serum and liver. J. Med. Virol. 5:73–86.

Holmes, A.W., Wolfe, L., Rosenblate, H., and Deinhardt, F. 1969. Hepatitis in marmosets: Induction of disease with coded specimens from a human volunteer study. Science 165:816–817.

Holmes, A.W., Deinhardt, F., Wolfe, L., Frösner, G.G., Peterson, D., Casto, B., and Conrad, M. 1973. Specific neutralization of human hepatitis type A in marmoset monkeys. Nature (London) 243:419–420.

Hoofnagle, J.H., Gerety, R.J., Smallwood, L.A., and Barker, L.F. 1977. Subtyping hepatitis B. Antigen and antibody by radioimmunoassay. Gastroenterology 72:290–296.

Hoofnagle, J.H. 1981. Serological markers of hepatitis B virus infection. Ann. Rev. Med. 32:1–11.

Hoofnagle, J.H. and Alter, H.J. 1984. Chronic viral hepatitis. In G.N. Vyas, J.L. Dienstag, and J.H. Hoofnagle (eds.), Viral Hepatitis and Liver Disease. Orlando, FL: Grune & Stratton.

Hoofnagle, J.H. 1985. Hepatitis. In G. Mandell, R. Douglas, and J. Bennett (eds.), Principles and Practice of Infectious Disease. New York: John Wiley and Sons.

Hoofnagle, J.H., and Di Bisceglie, A.M. 1990. Antiviral Therapy of Viral Hepatitis. In G.J. Galasso, R.J. Whitney, and T.C. Merigan (eds.), Antiviral Agents and Viral Diseases of Man. New York: Raven Press.

Hoyer, B., Bonino, F., and Ponzetto, A. 1983. Properties of delta-associated ribonucleic acid. In G. Verme, F. Bonino, and M. Rizzetto (eds.), Viral Hepatitis and Delta Infection, New York: Alan R. Liss.

Huang, S.N. 1975. Immunohistochemical demonstration of hepatitis B core and surface antigens in paraffin sections. Lab. Invest. 33:88–95.

Huang, S.N. and Neurath, A.R. 1979. Immunohistologic demonstration of hepatitis B viral antigens in liver with reference to its significance in liver injury. Lab. Invest. 40:1–17.

Ikeda, Y., Toda, G., Hashimoto, N., and Kurokawa, K. 1990. Antibody to superoxide dismutase, autoimmune hepatitis and antibody tests for hepatitis C virus. Lancet 335:1346.

Jawetz, E., Melnick, J.L., Adelberg, E.A., Brooks, G.F., Butel, J.S., and Ornston, L.N. 1991. Medical Microbiology, Chap. 37 (Hepatitis Viruses). Norwalk, CT: Appleton & Lange, pp. 451–468.

Kane, M.A., Bradley, D.W., Shresta, S.M., Maynard, J.E., Cook, E.H., Mishra, R.P., and Joshi, D.D. 1984. Epidemic non-A, non-B hepatitis in Nepal. Recovery of a possible etiologic agent and transmission studies in marmosets. JAMA 252:3140–3145.

Kaneko, S., Uhoura, M., Murakami, S., Hattori, N., and Kobayashi, K. 1990. Sequence analysis of hepatitis C virus genomes isolated from 5 patients with chronic non-A, non-B hepatitis. (Abst. #357) In The 1990 International Symposium on Viral Hepatitis and Liver Disease, p. 138.

Kaplan, P.M., Greenman, R.L., Gerin, J.L., Purcell, R.H., and Robinson, W.S. 1973. DNA polymerase associated with human hepatitis B antigen. J. Virol. 12:995–1005.

Khuroo, M.S. 1980. Study of an epidemic of non-A, non-B hepatitis: Possibility of another human hepatitis virus distinct from posttransfusion non-A, non-B type. Am. J. Med. 68:818–824.

Khuroo, M.S., Teli, M.R., Skidmore, S., Lofo, M.A., and Khuroo, M.I. 1981. Incidence and severity of viral hepatitis in pregnancy. Am. J. Med. 70:252–255.

Koretz, R.L., Stone, O., and Gitnick, G.L. 1980. The long-term course of non-A, non-B post-transfusion hepatitis. Gastroenterology 79:893–898.

Kraczynski, K., Bradley, D.W., and Kane, M.A. 1988. Virus associated antigen and epidemic non-A, non-B hepatitis and specific antibodies in outbreaks and in sporadic cases of NANB hepatitis. Hepatology. 8:1223.

Krugman, S., Ward, R., and Giles, J.P. 1962. The natural history of infectious hepatitis. Am. J. Med. 32:717–728.

Krugman, S., Giles, J.P., and Hammond, J. 1967. Infectious hepatitis: Evidence for two distinct clinical, epidemiological and immunological types of infection. JAMA 200:365–373.

Krugman, S., Overby, L.R., Mushahwar, I.K., Ling, C., Frösner, G., and Deinhardt, F. 1979. Viral hepatitis type B. Studies on natural history and prevention re-examined. N. Engl. J. Med. 300:101–107.

Kuo, G., Choo, Q-L., Alter, H.J., Gitnick, G.L., Redeker, A.G., Purcell, R.H., Miyamura, T., Dienstag, J.L., Alter, M.J., Stevens, C.E., Tegtmeier, G.E., Bonino, F., Colombo, M., Lee, W.-S., Kuo, C., Berger, K., Shuster, J.R., Overby, L.R., Bradley, D.W., and Houghton, M. 1989. An assay for circulating antibodies to a major etiologic virus of human non-A, non-B hepatitis. Science. 244:362–364.

Lemon, S.M. and Binn, L.N. 1983. Serum neutralizing antibody response to hepatitis A virus. J. Infect. Dis. 148:1033–1039.

Levy, G.A. and Chisari, F.V. 1981. The immunopathogenesis of chronic HBV induced liver disease. In H.C. Thomas (ed.), Immunopathology of Hepatitis B Virus Infection. New York: Springer International.

MacFarlane, I.G., Smith, H.M., Johnson, P.J., Bray, G.P., Vergani, D., and Williams, R. 1990. Hepatitis C virus antibodies in chronic-active hepatitis: Pathogenic factor of false-positive results. Lancet 335:754–757.

Mao, J.S., Go, Y.Y., Huang, H.Y., Yu, P.H.,

Huang, B.Z., Ding, Z.S., Chen, N.L., Yu, J.H., and Xie, R.H. 1981. Susceptibility of monkeys to human hepatitis A virus. J. Infect. Dis. 144:55–60.

Mathiesen, L.R., Feinstone, S.M., Purcell, R.H., and Wagner, J.A. 1977. Detection of hepatitis A antigen by immunofluorescence. Infect. Immun. 18:524–530.

Mathiesen, L.R., Feinstone, S.M., Wong, D.C., Skinhoej, P., and Purcell, R.H. 1978. Enzyme-linked immunosorbent assay for detection of hepatitis A antigen in stool and antibody to hepatitis A antigen in sera: Comparison with solid phase radioimmunoassay, immune electron microscopy and immune adherence hemagglutination assay. J. Clin. Microbiol. 7:184–193.

Melnick, J.L., Dreesman, G.R., and Hollinger, F.B. 1977. Viral hepatitis. Sci. Am. 44:237.

Mosley, J.W., Redeker, A.G., Feinstone, S.M., and Purcell, R.H. 1977. Multiple hepatitis viruses in multiple attacks of acute viral hepatitis. N. Engl. J. Med. 296:75–78.

Okochi, K., and Murakami, S. 1968. Observations on Australia angtigen in Japanese. Vox Sang. 15:376–385.

Okuda, H., Obata, H., Motoike, Y., and Hisamitsu, T. 1984. Clincopathological features of hepatocellular carcinoma: Comparison of seropositive and seronegative patients. Hepatogastroenterology 31:64–68.

Peters, R.L. 1975. Viral hepatitis: A pathologic spectrum. Am. J. Med. Sci. 270:17–31.

Phillips, J.J. and Poucell, S. 1981. Modern aspects of the morphology of viral hepatitis. Hum. Pathol. 12:1060–1084.

Popper, H., Thung, S.N., Gerber, M.A., Hadler, S.C., De Monzon, M., Ponzetto, A., Anzola, E., Rivera, D., Mondolfi, A., Bracho, A., Francis, D.P., Gerin, J.L., Maynard, J.E., and Purcell, R.H. 1983. Histologic studies of severe delta agent infection in Venezuelan Indians. Hepatology 3:906–912.

Popper, H. 1984. Summary of workshop on pathology. In G.N. Vyas, J.L. Dienstag, and J.H. Hoofnagle (eds.), Viral Hepatitis and Liver Disease. Orlando, FL: Grune & Stratton.

Provost, P.J., Ittensohn, Q.L., Villarejos, V.M., and Hilleman, M.R. 1975. A specific complement-fixation test for human hepatitis A employing CR326 virus antigen. Diagnosis and epidemiology. Proc. Soc. Exp. Biol. Med. 148:962–969.

Provost, P.J. and Hilleman, M.R. 1979. Propagation of human hepatitis A virus in cell cultures *in vitro*. Proc. Soc. Exp. Biol. Med. 160:213–221.

Provost, P.J., Giesa, P.A., McAleer, W.J., and Hilleman, M.R. 1981. Isolation of hepatitis A virus *in vitro*. Proc. Soc. Exp. Biol. Med. 167:201–206.

Purcell, R.N., Feinstone, S.M., Daemer, R.J., Ticehurst, J.R., and Baroudy, B.M. 1984. Approaches to the development of hepatitis A vaccines: The old and new. In R.M. Chanock and R.A. Lerner (eds.), Cold Spring Harbor Symposium Modern Approaches to Vaccines: Molecular and Chemical Basis of Virus Virulence and Immunogenicity, pp. 59–63.

Raimondo, G., Longo, G., and Squadrito, G. 1983. Exacerbation of chronic liver disease due to hepatitis B surface antigen after delta infection. Br. Med. J. 286:845.

Realdi, G., Alberti, A., Rugge, M., Rigoli, A.M., Tremolada, F., Schivazappa, L., and Ruol, A. 1982. Long-term follow-up of acute and chronic non-A, non-B post-transfusion hepatitis: Evidence of progression to liver cirrhosis. Gut 23:270–275.

Reyes, G.R. 1990. Molecular cloning of the hepatitis E virus (HEV). 1990. (Abst. #566A) In The 1990 International Symposium on Viral Hepatitis and Liver Disease, p. 200.

Rizzetto, M., Canese, M.G., Airco, S., Crivelli, O., Bonino, F., Trepo, C., and Verme, G. 1977. Immunofluorescence detection of a new antigen-antibody system (δ/Anti-δ) associated to the hepatitis B virus in the liver and in the serum of HBsAg carriers. Gut 18:997–1003.

Rizzetto, M., Hoyer, B., and Canese, M.G. 1980. Delta antigen. The association of delta antigen with hepatitis B surface antigen and ribonucleic acid in the serum of delta-infected chimpanzees. Proc. Soc. Natl. Acad. Sci. USA 77:6124–6128.

Rizzetto, M., Shih, J.W.K., Verme, G., and Gerin, J.L. 1981. A radio-immunoassay for HBcAg in the sera of HBsAg carriers: Serum HBcAg, serum DNA polymerase activity, and liver HBcAg immunofluorescence as markers of chronic liver disease. Gastroenterology 80:1420–1427.

Rizzetto, M., Verme, G., Recchia, S., Bonino, F., Farci, P., Arico, S., Calzia, R., Picciotto, A., Columbo, M., and Popper, H. 1983. Chronic HBsAg hepatitis with intrahepatic expression of delta antigen. An active and progressive disease unresponsive to immunosuppressive treatment. Ann. Intern. Med. 98:437–441.

Schaefer, R.L., Wolfe, R.L., Steinfeld, C.M., Kawas, E.E., Martin, D.A., Hinds, M.W., Johnson, R.H., Hayes, J.R., Jacobs, J., Pendleton, J. Jr., Redeker, A.G., Roberto, R.R., and Chin, J. 1984. Fulminant hepatitis B among parenteral drug abusers—Kentucky, California. Morbid. Mortal. Wkly. Rep. 33:70–77.

Schaff, Z., Tabor, E., Jackson, D.R., and Gerety, R.J. 1984. Ultrastructural alterations in serial liver biopsy specimens from chimpanzees experimentally infected with a human non-A, non-B hepatitis agent. Virchows Arch. B. Cell. Path. 45:301–312.

Senba, M. 1982. Staining methods for hepatitis B surface antigen (HBsAg) and its mechanism. Am. J. Clin. Pathol. 77:312–315.

Shattock, A.G. and Morgan, B.M. 1983. Sensitive enzyme-immunoassay for the detection of delta antigen and anti-delta using serum as a delta antigen source. J. Med. Virol. 12:73–82.

Shikata, G., Uzawa, T., Yoshiwara, N., Akatsuka, T., and Yamazaki, S. 1974. Staining methods of Australia antigen in paraffin section. Jpn. J. Exp. Med. 44:25–36.

Shimizu, Y.K., Feinstone, S.M., Purcell, R.H., Alter, H.J., and London, W.T. 1979. Non-A, non-B hepatitis: Ultrastructural evidence for two agents in experimentally infected chimpanzees. Science 205:197–200.

Smedile, A., Dentico, P., Zanetti, A., Sagnelli, E., Nordenflet, E., Actis, G., and Rizzetto, M. 1981. Infection with HBV-associated delta (δ) agent in HBsAg carriers. Gastroenterology 81:992–997.

Smedile, A., Farci, P., Verme, G., Czredda, F., Cargnel, A., Caporaso, N., Dentico, P., Trepo, C., Opolon, P., Gimson, A., Vergani, D., Williams, R., and Rizzetto, M. 1982. Influence of delta infection on severity of hepatitis B. Lancet 2:945–947.

Song, D.Y., Zhuang, H., Wang, X.C., Liu, K.M., Hao, F.M., Chen, D.G., Jia, Q., Li, Z., Luo, K.X., and Ai, Z.Z. 1990. Enterically transmitted non-A, non-B hepatitis (ET-NANB) in Hetian City: A report of 562 cases. (Abst. #574A) In The 1990 International Symposium on Viral Hepatitis and Liver Disease, p. 202.

Szmuness, W., Stevens, C.E., Harley, E.J., Zang, E.A., Oleszko, W.R., William, D.C., Sadovsky, R., Morrison, J.M., and Kellner, A. 1980. Hepatitis B vaccine: Demonstration of efficacy in a control clinical trial in a high-risk population in the United States. N. Engl. J. Med. 303:833–841.

Tabor, E., Drucker, J.A., Hoofnagle, J.H., April, M., Gerety, R.J., Seeff, L.B., Jackson, D.R., Barker, L.F., and Pineda-Tamondong, G. 1978. Transmission of non-A, non-B hepatitis from man to chimpanzee. Lancet 1:463–466.

Tanaka, K., Mori, W., and Suwa, K. 1981. Victoria blue-nuclear fast red stain for HBs antigen detection in paraffin section. Acta Pathol. Jpn. 31:93–98.

Theilmann, L., Blazek, M., Goeser, T., Gmelin, K., Lommerell, B., and Feihn, W. 1990. False positive anti-HCV tests in rheumatoid arthritis. Lancet 335:1346.

Ticehurst, J.R., Racaniello, V.R., Baroudy, B.M., Baltimore, D., Purcell, R.H., and Feinstone, S.M. 1983. Molecular cloning and characterization of hepatitis A virus cDNA. Proc. Natl. Acad. Sci. USA 80:5885–5889.

Trepo, C., Vitvitski, L., and Hantz, O. 1981. Non-A, non-B hepatitis virus: Identification of a core antigen-antibody system that cross reacts with hepatitis B core antigen and antibody. J. Med. Virol. 8:31–47.

Villanueva, A., Fraser, P., Garcia, R., Ospino, P., Ariza, P., and O'Brien, T. 1988. Fulminant hepatitis in the Sierra Nevada: A superinfection of delta virus on hepatitis B virus. In A.J. Zuckerman (ed.), Viral Hepatitis and Liver Disease, New York: Alan R. Liss, Inc.

Watanabe, S., Reddy, K.R., Jeffers, L.J., Dickinson, G.M., O'Connell, M., and Schiff, E.R. 1984. Electron microscopic evidence for non-A, non-B hepatitis markers and virus-like particles in immunocompromised humans (Abst. 3A.29). In G.N. Vyas, J.L. Dienstag, and J.H. Hoofnagle (eds.), Viral Hepatitis and Liver Disease. Orlando, FL: Grune & Stratton, p. 620.

Weiner, A.J., Truett, M.A., Rosenblatt, J., Han, J., Quan, S., Polito, A.J., Kuo, G., Choo, Q.L., Houghton, M., Agius, G.,

Page, E., and Nelles, M.J. 1990. HCV testing in low-risk population. Lancet 336:695.

Wong, D.C., Purcell, R.H., Sreenivasan, M.A., Prasad, S.R., and Pavri, K.M. 1980. Epidemic and endemic hepatitis in India. Evidence for non-A/non-B hepatitis virus etiology. Lancet 2:876–878.

Wong, D.C., Diwan, A.R., Rose, L., Guerin, J.L., Johnson, R.G., Polito, A., and Purcell, R.H. 1990. Nonspecificity of anti-HCV test for sero-epidemiological analysis. Lancet 336:750–751.

Zuckerman, A.J. 1976. Twenty-five centuries of viral hepatitis. Rush-Presbyt. St. Luke's Med. Bull. 15:57–82.

Rabies

George R. Anderson

24.1 HISTORY

Rabies has challenged and plagued humans for more than 25 centuries. Whenever the struggle appeared to be favoring mankind, this highly resourceful virus has found a mode of adaptation necessary for survival.

Diseases similar or identical to rabies were described by the Greeks and Romans between the years 300 and 500 B.C. The first isolation of the virus is credited to Zinke, who in 1804 used virus-infected saliva to transmit the disease to a normal dog (Zinke, 1804). Negri (1903) described the presence of intracytoplasmic inclusion bodies in the neurons of rabid animals. Negri's findings provided the basis upon which the diagnosis of rabies rested for almost three quarters of a century.

In the latter part of the nineteenth century, Louis Pasteur and a few of his scientific contemporaries, working with crude techniques and limited means of communication, generated much of the information that has formed the basis of our knowledge about rabies.

Dr. George Newman (1904) stated, "Although rabies was mentioned by Aristotle and has been studied by a large number of workers, the additions of Pasteur have been greater than all other additions to our knowledge of the disease put together." This same author referred to the comments of Dr. Rose Bradford, who stated that "Pasteur has established, a) that the virus was not only in the saliva but also in the central and peripheral nervous system—yet absent from the blood, b) that the disease was most readily inoculated in the nervous system, c) that by suitable means the virus could be attenuated, and d) that by means of an attenuated virus, preventive and even curative methods might be adopted."

Webster and Clow (1936) and Kanazawa (1936) reported the first isolation of rabies virus in cell culture. That same year, Galloway and Elford (1936) determined the size of rabies virus to be in the range of 100 to 150 nm.

Three research contributions between 1945 and 1980 deserve mention as part of the historical perspective of rabies. The first was the adaptation of the Flury strain of rabies virus to the developing chick embryo (Koprowski and Cox, 1948). Through the process of adaptation (40 to 50 passages), the virus lost its virulence for dogs but not its ability to infect dogs or induce protection against challenge with "street" (wild type) virus. This work set the stage for the development of attenuated rabies vaccines suitable for use in dogs. The attenuated rabies vaccines have proven to be highly efficacious, with the result that the domestic dog is no longer a significant reservoir for the dissemination of rabies to humans in developed countries.

The second major contribution of this era was the development and specific application of fluorescent antibody procedures for use in the diagnosis of rabies by Goldwasser and Kissling (1958). This procedure not only facilitated diagnosis but opened new vistas for studying the location and the distribution of the virus in tissues and cell cultures.

The third major contribution was made by Wiktor and colleagues, (1964, 1966), who generated critical information regarding the adaptation and cultivation of rabies virus in cell culture. This work provided the essential groundwork for the development of tissue culture-derived vaccines and for further elucidation of the structure, replication, and composition of the virus.

In the period from 1985 to 1990, the focus of rabies research has been directed at the molecular composition of the virus (Wunner et al, 1988), at the refinement and simplification of diagnostic procedures, and at the development of a variety of "high output," low cost vaccines that can be used efficiently in third world countries and in wildlife populations in all countries where rabies is endemic.

24.2 CHARACTERISTICS OF THE VIRUS

The rabies virus is a bullet-shaped RNA virus that matures at the plasma membranes of the host cell (Murphy, 1975). The virus consists of a central helical nucleocapsid and an outer membrane or envelope. Small filamentous projections or spikes arise from the surface of the envelope. The rabies virus contains five structural proteins: a glycoprotein (G), a membrane protein (M), a large protein (L), a phosphoprotein (N), and a nucleoprotein (NP). The nucleocapsid is composed of nucleoprotein (NP) and a single strand of RNA. The envelope contains two non-glycosylated forms of the M protein (m_1 and m_2), the glycoprotein (G) and lipid. Tordo et al (1988) have completed work on the structure of the rabies viral genome through a series of cloning and sequencing experiments. They reported that the L protein encodes for amino acid sequences that correspond to the viral RNA-dependant RNA polymerase. The L protein appears to be required for RNA synthesis.

Rabies virus can be disrupted by a combination of physical and chemical treatments (Sokol, 1975). Sodium deoxycholate solubilizes the lipid component of the membrane coat and breaks the virus particle into nucleocapsid and envelope components. The glycoprotein surface projections of the envelope can be released by treatment of the virion with a nonionic detergent such as Nonidet P-40 (NP-40). Further separation of the individual components can be accomplished by density gradient centrifugation and column chromatography. The protective antigen, a hemagglutinin, and at least one and possibly two soluble antigens are associated with the virion (Cox et al, 1980). The protective antigen and hemagglutinin are derived primarily from envelope glycoprotein, the soluble antigens are derived from the nucleocapsid.

The glycoprotein (G) component is responsible for the induction of neutralizing antibody (Wiktor et al, 1973). Dietzschold et al (1983) described the isolation of a soluble rabies glycoprotein antigen (G_s). The G_s and G antigens were found to have identical antibody binding characteristics but the G_s antigen, unlike the G antigen, is a weak immunogen. Certain fixed virus strains contain two classes of glycoprotein:

Table 24.1 Lyssa Subgroup Viruses

Virus	Principal Species Involved	Disease Occurrence
Rabies	Bats, foxes, dogs, cats, skunks, wolves, raccoons, cattle	Worldwide
Lagos bat	Bats	Nigeria, Central Africa
Mokola	Man, shrews	Nigeria
Duvenhage	Man	South Africa
Kotonkan	Mosquito	Nigeria
Obodhiang	Mosquito	Sudan

one class with three glycosylated peptides susceptible to trypsin digestion; and other class with only one glycosylated tryptic peptide.

The hemagglutinin is apparently composed of aggregated or reassociated aggregates of glycoprotein. The reaggregated glycoprotein is highly immunogenic and induces protection against a rabies challenge.

Recent work has demonstrated that the nucleoprotein component of the virus may play a key role in activating the T lymphocyte subsets that are necessary for cytolysis, B cell activation, and the release of lymphokines (Dietzschold et al, 1987).

The rabies virus is classified as a member of the Lyssa subgroup of Rhabdoviruses. Presently five other viruses are included in this subgroup (Table 24.1). The degree of relatedness of the six viruses in the Lyssa subgroup was determined by cross neutralization, cross protection, and complement fixation tests. Rabies virus is most closely related to the Lagos bat and Mokola viruses. The latter two viruses would appear to serve as bridge viruses between rabies and the Kotonkan and Obodhiang viruses.

The work of Shope on the viruses of the Lyssa serogroup (Tignor and Shope, 1972; Shope, 1975) and that of Wiktor et al (1980) with nucleocapsid antigens from a variety of rabies virus strains suggests that some degree of heterogeneity exists among rabies virus strains. Rabies vaccines for animal and human use have been made from a wide variety of strains with evidence of cross protection against heterologous challenge strains.

Rabies virus is inactivated by heat, sunlight, tissue enzymes, ultraviolet irradiation, and a number of chemical agents including formalin, quaternary ammonium compounds, strong acids, 70% ethanol, and lipid solvents. Virus infectivity may be retained for one or two weeks at room temperature (20 °C) and is preserved for extended periods of time by low temperature storage (−70 °C or below), by freeze drying, or by storage of infected tissue in 50% glycerol.

Beta propiolactone (BPL) and the aziridines or their derivatives inactivate rabies virus with minimum alteration of the envelope protein (Larghi and Nebel, 1980). BPL has been used extensively for the inactivation of rabies virus in the preparation of vaccines and virus antigens for use as diagnostic reagents. Gamble et al (1980) demonstrated the inactivation of rabies virus in brain tissue in situ by subjecting mice to 1.26 megarads of radiation. This procedure reduces the risk associated with the handling of brains or brain tissues from animals harboring rabies virus.

24.3 PATHOGENESIS

The fact that the rabies virus is capable of infecting such a wide range of animal species allows it to propagate and survive in nature. Rabies is one of the few infectious diseases in which the exact time and site of exposure (infection) in man is usually known. It would seem that having this specific information would facilitate our understanding of the pathogenesis of this disease. Unfortunately, this is not the case. A detailed review of the literature suggests that there is still confusion regarding the mechanism involved in the spread of the virus from point of entry—usually muscle, connective tissue, or abraded skin—to the central nervous system (CNS). The early studies in animals indicated that the virus gained access to the CNS via peripheral nerve pathways and not through hematogenous spread. These early investigations have been supported by the work of Dean et al (1963), who demonstrated that the progress of rabies virus from a leg or foot wound could be blocked by amputation or the use of local anesthetics.

Murphy (1975) demonstrated virus budding from the plasma membranes of myocytes, supporting the concept that the virus replicates locally prior to its migration to the central nervous system. The virus was then observed to progress to neuromuscular and neurotendinal spindles. Baer (1975) reported that removal of the perineurium, epineurium, and perineurial epithelium in the legs of rats prior to virus inoculation did not prevent the development of clinical rabies, though in some cases the appearance of symptoms was delayed. When the nerve fasciculus was removed from the leg infected with rabies virus, the development of rabies was prevented. The progression of the virus was not inhibited following demyelination or the disintegration of axons. Baer postulated that the virus moves passively through the tissue spaces between Schwann's cells or the interstitial nerve spaces within the nerve bundle.

Ceccaldi et al (1989) demonstrated that axonal transport of rabies virus is delayed following the injection of colchicine, a mitotic inhibitor, in a peripheral nerve near the site of inoculation of the virus.

The studies of Watson et al (1981) further clarified events at the site of inoculation. Sequential histochemical and immunofluorescent staining procedures of the same tissue sections identified the axon terminal of the neuromuscular junction (NMJ) and the presence of virus at the NMJ. A portion of the nerve at the NMJ is not protected by perineurial sheath. Watson postulated that the virus moves from the point of entry in the body into the extracellular space and then gains access to the unsheathed terminal of the motor axon. The virus then traverses the motor axon to the spinal cord. Viral antigen was detected consistently at the NMJ within one hour post inoculation but by 72 and 95 hours viral antigen was detected only rarely at this site.

Rabies is found in the United States and Canada most commonly in bats, skunks, foxes, and raccoons and less commonly in dogs, cats, cattle, horses, sheep, and humans. In South and Central America, rabies is found commonly in vampire bats, cattle, dogs, and cats. In Western Europe, the fox has become the dominant reservoir of rabies. It is rare to find rabies in wild rodents or wild rabbits; in fact, it is sufficiently rare that many laboratories no longer examine rodents or rabbits for evidence of rabies except in extenuating circumstances, e.g., where there is an unprovoked attack.

Using a pool of monoclonal antibodies Smith et al (1986) categorized 231 rabies virus isolates from terrestrial animals. These isolates separated the virus into four distinct antigenic groups which reflected enzootics based on species and geographical location, e.g., red foxes from the northeastern United States and eastern Canada, raccoons from the southeastern and mid-Atlantic United States, skunks from the south central and north central United States. Four different reaction patterns were also found from bat isolates collected throughout the United States. This work should be of value in tracking future rabies outbreaks in terrestrial animals. All of the standard laboratory animal species are susceptible to rabies, with the hamster showing the highest degree of susceptibility. There is variability in resistance or susceptibility between species, between strains within a species, and between individual animals within the same strain. Lodmell (1983) demonstrated that resistance to challenge with street virus within inbred strains of mice was dominant and not controlled solely by the major histocompatibility (H-2) locus. The degree of susceptibility or resistance of individual mice may profoundly influence the course of the clinical disease. Resistant mice may show few if any signs of clinical rabies even though the challenge virus is found in their spinal cord and brain five to seven days after peripheral inoculation. In the resistant animals, clinical symptoms are rare as the virus does not appear to induce malfunction of the motor nerves. The incubation period in susceptible mice is relatively short. These animals become paralyzed and prostrate, and death ensues rapidly.

The incubation period from exposure to development of overt clinical rabies is extremely variable. In humans the incubation period may be as short as 13 days or as long as two years. However, the vast majority of human rabies cases develop within 20 to 60 days following exposure. The incubation period of rabies in animals in their natural habitat is less well defined but appears to approximate that described for humans. The incubation period in experimental animals is influenced by a number of factors, including the route of injection, the species and strain of animal, the quantity of virus injected, the strain of virus, and the vaccination history of the challenged animal. All warm-blooded animals and some cold-blooded animals are susceptible to rabies. There are, however, only a relatively few animals that serve as a reservoir for rabies and are of true concern to the clinical virologist.

Rabies manifests itself in many forms. Domestic and wild animals infected with street virus most often become agitated and aggressive and will attempt to bite inanimate objects or other animals. The symptoms may evolve to that more typical of the paralytic form of the disease. Laboratory animals infected with street virus may show some agitation, e.g., jumping and aggressiveness, but more often develop the paralytic or "dumb" form of the disease. In this form the animal develops tremors and becomes unsteady, developing a slowly ascending paralysis. Prior to death the animal is usually prostrate and nonresponsive. Fixed virus tends to induce the paralytic form of the disease, though exceptions have been noted.

Since the time of Pasteur, two basic types of rabies virus have been recognized: the wild or naturally occurring virus referred to as street virus and laboratory or passaged virus that is referred to as fixed virus. Fixed virus is derived from street virus. Fixed virus has been passaged in animals a sufficient number of time to result in a virus whose pathogenic "behavior" is modified from that of street virus.

Laboratory animals infected with fixed virus strains generally show symptoms within five to seven days following intracerebral inoculation and usually die within ten days. After peripheral challenge with fixed virus, virus can be demonstrated in the cord in about 72 hours and in the brain in about 96 hours with symptoms and death occurring in four to seven days. When street virus is injected intramuscularly infection progresses more slowly in laboratory animals than it does with fixed virus, with virus appearing in the brain six to nine days following experimental infection. Clinical symptoms appear about 10 to 15 days after inoculation with street virus. French authors, early in this century, coined the term "septinevrite" to characterize the general dissemination of rabies virus throughout the nervous system of an infected animal (VanRooyen and Rhodes, 1948). Septinevrite confirms the affinity of the virus for all of the nerve cells in the body and reflects the centrifugal spread of the virus throughout the body, including the salivary glands and respiratory tract mucosa. Whatever the mechanism for the transfer of virus from nerve cell to saliva, the fact remains that the involvement of the salivary gland is the essence of rabies pathogenesis, for without salivary gland involvement the disease would have to develop an alternative mode of spread between individual animals or it would gradually disappear. Among certain wildlife species rabies may be transmitted on occasion by inhalation or ingestion, but these routes of transmission play a minor role in the overall picture.

Vaughan et al (1963, 1965) investigated the excretion of wild rabies virus in dogs and cats. Twenty-three of 26 infected cats showing clinical symptoms excreted virus in the saliva. The three cats that did not excrete the virus developed the paralytic form of the disease. The earliest that rabies virus was detected in the saliva of cats was one day before the onset of clinical symptoms. With one exception, salivary excretion of virus continued until death. In contrast, rabies virus was found in the saliva of only 24 of 54 infected dogs. In dogs, virus excretion began as early as three days before onset of illness. Once detected, salivary excretion of the virus in the dog continued until death.

The studies of Vaughan et al are important to the clinical virologist because they confirm three assumptions that have been paramount in determining the risk of infection following exposure:

1. If a domestic animal does not develop clinical rabies within five to seven days after biting a human or another animal, the chances of virus excretion at the time of bite are negligible.
2. Not all animals that develop clinical rabies excrete virus in their saliva.
3. The bite of a cat is potentially more dangerous than that of a dog considering that 90% of rabid cats were virus excretors as opposed to only 45% of rabid dogs.

24.4 THE ISOLATION AND CULTIVATION OF RABIES VIRUS

Rabies virus may be isolated from the brain, spinal cord, or salivary glands of infected animals. The brain of a suspected animal is usually the tissue of choice for the isolation and/or demonstration of rabies virus or virus antigens. The hippocampus (Ammon's horn), the cerebellum, and the thalamus are the sites of the brain usually selected for the isolation of the virus. The salivary glands are often overlooked as a source of virus even though the old adage "that an animal bites with its mouth not its brain" still applies. Unfortunately, the salivary glands are more difficult to locate and identify particularly in very small animals.

The adult mouse is used most commonly for the primary isolation of rabies virus because it is susceptible, available, and easy to handle. Hamsters and neonatal mice are highly susceptible to rabies virus and, when available, may be used in addition to or in lieu of adult mice. These animals (hamsters and neonatal mice) are valuable for detecting attenuated rabies viruses or confirming that inactivated rabies vaccines are innocuous.

Fluorescent antibody methodology has supplanted animal inoculation in the public health diagnostic laboratory for the diagnosis of rabies. The only situation for which animal inoculation is still useful is for those tissues where cell structure is sufficiently disturbed (damaged) that it is difficult to recognize and properly characterize fluorescent foci.

Rabies virus will propagate in a variety of cell cultures, including primary hamster kidney cells, human diploid cells, rhesus monkey diploid cells, chick embryo fibroblasts, and mouse embryo brain cells. Smith et al (1978) reported the isolation of street virus strains in a chick embryo-related (CER) cell line with an efficiency rate that parallels mouse inoculation. Wild or street virus strains of rabies virus need to be adapted to tissue culture before consistent high titer virus production occurs. Adaptation can be accomplished by rapid serial passage, i.e., by the harvest, transfer, and subculture of infected cells, or by the mixing and co-cultivation of infected cells with noninfected cells.

The BHK 21 continuous cell line, derived from baby hamster kidney, has become a standard cell line for virology laboratories working with rabies virus (Wiktor and Clark, 1975). Rabies virus adapts easily to this cell line and once adaptation has occurred, this cell line will consistently support the production of high levels of virus. The BHK 21 cells have been used for antigen production, virus growth studies, neutralization assays, and the production of virus for physical and chemical studies.

Virus-infected cells are harvested by treatment of monolayers with trypsin or the use of rubber scraping devices. Release of infectious virus and/or the viral protective antigen can be obtained by cycles of freezing and rapid thawing. Ultrasound has also been used for this purpose but appears to offer little or no advantage over the freeze-thaw method.

Viruses adapted for growth in primary hamster kidney cells or in rhesus diploid cells reach maximum titers in four to five days post inoculation. However, protective antigen is released and then accumulates in the culture fluid beyond the time of maximum virus growth. The optimal time for antigen harvest is eight to ten days post inoculation.

Fernandes et al (1964) were able to establish chronic rabies infection in rabbit endothelial cell cultures. These cells, observed over an extended period, displayed no demonstrable alteration in cell metabolism or cell replication during the period of persistent virus infection.

24.5 VIRUS IDENTIFICATION

In 1903, Negri described the appearance of eosinophilic staining inclusion bodies in the cytoplasm of nerve cells from animals infected with rabies virus. These inclusion bodies are referred to as Negri bodies. Negri bodies occur most frequently in the hippocampus (Ammon's horn), the cerebellum, and the base of the brain. They may also be found in other parts of the brain, the spinal cord, and in ganglia of the CNS. For most of this century the diagnosis of rabies was based on the identification of Negri bodies in impression smears made from sections of brain tissue. Negri body identification is still the most economical and rapid means of diagnosis of rabies in animals; however, inclusion body identification is the least sensitive method to demonstrate virus and is generally unsuitable for use in animals infected with the fixed or adapted strains of rabies virus.

The histopathology of rabies infection is neither peculiar to, nor characteristic of, the disease except for the presence of Negri bodies. Very little histopathology is noted in the brain during the course of the disease. At autopsy, leukocytic infiltration, perivascular cuffing, and mononuclear cuffing of nerve cells are the most common neuropathological features of rabies. The white blood cell infiltration and cuffing of nerve cells is not necessarily a phenomenon limited to rabies infection and may reflect generalized degenerative changes rather than virus-induced pathology. By contrast specific virus-induced pathologic changes are observed in other infections of the CNS, e.g., the equine encephalitides.

Laboratories in some parts of the world still use the conventional Negri body technique for the diagnosis of rabies. This technique should be replaced by more accurate and sensitive immunoassay procedures that do not require extensive training of personnel or a sizeable equipment expenditure.

The fluorescent antibody (FA) test is the universal test of choice for the identification of rabies antigen in infected tissues (Gardner and McQuillin, 1980). This test has the advantage of being easy to perform, with the end result obtainable in two to four hours. The FA test is based on the principle that high titered rabies antibody previously conjugated with a fluorescent dye will bind to rabies antigen present in animal tissues or infected cell cultures.

There are some points that should be taken into consideration with the use of the FA test for the diagnosis of rabies:

1. The individual performing the test should have normal color vision.
2. It is essential to use positive and negative antigen controls to make certain that the FA conjugate is working properly.
3. The FA-conjugated antibody or the FA-stained impression smear should receive minimal exposure to natural light.
4. Several areas of the brain and/or salivary glands should be tested for the presence of rabies antigen.
5. If rabies antigen is not detected in a brain impression smear by FA after careful examination, it can be assumed that the saliva of the suspect rabid animal was noninfectious.
6. The test will not distinguish between wild and fixed strains of rabies virus.
7. Euthanasia can be performed immediately on any suspect animal for FA ex-

amination (with the Negri body test it is desirable to allow the animal to develop symptoms or die before examining the brain).

8. Decomposed brain is unsuitable for FA examination.
9. Stained smears should not be left under ultraviolet light for an extended period before examination, because this causes a decrease in fluorescence.

Impression smears for the FA diagnosis of rabies may be made from a number of sites in the brain, from salivary gland tissue, and even from the surface of the cornea. The corneal smear is a noninvasive procedure that has potential application in establishing a diagnosis in humans.

It is not uncommon for a virology laboratory to receive severely damaged or decomposed brains for examination for rabies. Decomposed brains should never be examined by conventional histopathologic or immunologic procedures. The chance for error is too great. Rarely, a decomposed brain may be inoculated into mice in lieu of in vitro examination. Even this approach is fraught with uncertainly, for the virus may have been destroyed by the action of proteolytic enzymes. Fresh or nondecomposed damaged brain (e.g., damaged by gunshot) may be examined by FA with the provision that a negative finding with brain material from a highly suspect animal must be interpreted with caution and may require the use of the mouse inoculation test.

A rabies tissue culture infection test (RTCIT) described by Rudd and Trimarchi (1989) would appear to offer a reasonable alternative to the mouse inoculation test (MIT). In the RTCIT test supernatant fluid from a 10% brain tissue suspension is added to neuroblastoma cells grown on plastic plates. The cells are then incubated at 34 °C for four days in a humidified chamber containing 5% CO_2 in air. The presence of rabies virus antigen is detected by FA. Brain tissues from 2,112 FA-negative animals were found negative by RTCIT and MIT, whole brain tissues from 278 FA-positive animals were found positive by RTCIT and MIT. Seventy-two specimens tested by RTCIT were categorized as unsatisfactory mainly due to brain decomposition.

A rabies enzyme immunoassay (REIA) has been proposed as an alternative to FA for the detection of rabies antigen in animal brain tissue. In 1989, 2,290 brain tissue specimens from a variety of animals were tested by both REIA and FA (Bourhy et al, 1989). When compared to FA the sensitivity of the REIA test was 95% with a specificity of 99.8%. In the REIA procedure, an antinucleocapsid IgG is used to capture rabies antigen, this is subsequently reacted with enzyme-labeled anti-IgG and then substrate to indicate the presence of virus. This technique would appear to have great potential for use in surveys or as a replacement for the Negri body test in laboratories lacking FA equipment.

An ELISA immunocapture assay, using a monoclonal antiglycoprotein antibody was reported as an effective in vitro technique to evaluate potency during vaccine processing. The presence of soluble glycoprotein interferes with the accuracy of this procedure in the presence of a polyclonal antiglycoprotein rabies antibody (Perrin et al, 1990).

24.5.1 Detection of Anti-Rabies Antibody

Unlike so many other viral diseases, the detection and quantitation of antibody has not played an important role in the diagnosis, treatment, or surveillance of rabies infection in animals or man. This is changing rapidly. In the last few years, several relatively simple, sensitive, low-cost testing procedures have made rabies antibody testing practical. In addition, the development of a number of promising innovative new rabies vaccines for use in humans has

resulted in a need for procedures to qualitate and quantitate rabies antibodies.

Detectable rabies antibody appears 14 to 18 days after the onset of clinical symptoms and may rise to very high levels prior to death. The presence of circulating serum antibody may be the best and only criterion to confirm the diagnosis of rabies in nonvaccinated humans prior to death.

Antibody analysis following rabies vaccination is the simplest and most efficient way of determining vaccine efficacy and the immune status of an individual. Every pre-exposure vaccination program should include a provision for the periodic determination of antibody against rabies in the serum of vaccinated individuals.

Currently, the most widely used test for detecting rabies antibody is the rapid fluorescent focus inhibition test (RFFIT), developed by Smith et al (1973). The RFFIT is a virus neutralization test performed in cell culture. There is a high degree of correlation between the RFFIT and the antibody-mediated rabies virus neutralization test performed in mice. The RFFIT test is sensitive, specific, and test results can be reported within 48 hours, whereas the results of the neutralization test in mice require 10 to 14 days for completion and reporting.

Three other in vitro procedures have promise in disease surveillance and in vaccine efficacy studies.

Budzko et al (1983) reported the adaptation of immunoadherence hemagglutination (IAHA) for use in the detection and measurement of anti-rabies antibody. This procedure is simple to perform, low cost, and offers a quick turnaround time (under 4 hours). The IAHA test has also been used to detect rabies antibody in household and stray dogs (Bota et al, 1987). Eighty-nine percent of household dogs and 38% of strays were IAHA antibody positive. IAHA detects complement binding antibody. The data obtained by IAHA correlate well with RFFIT, although the IAHA test is not as sensitive as RFFIT for detecting low levels of anti-rabies antibody.

Antanasiu et al (1977), Nicholson and Prestage (1982), and Perrin and Sureau (1987) quantitated anti-rabies antibody in human sera by application of enzyme-linked immunosorbent assay (ELISA). The antibody values obtained by the ELISA test appeared to correlate well with those obtained by an in vitro neutralization test. The results of the ELISA test are available within 48 hours. The fact that the ELISA test has become a universal tool in the clinical laboratory makes it highly attractive to the clinical virologist and the epidemiologist for use in the detection and quantitation of rabies antibody and in seroprevalence studies.

A dot immunobinding assay (DIA) that employs whole rabies virus antigen dotted on nitrocellulose sheets offers another simple, low-cost assay for use in serodiagnosis (Herberling et al, 1987). FA and DIA antibody test results were identical on 55 pre- and post-vaccination sera from dogs and humans.

24.6 CONSIDERATIONS FOR THE CLINICAL VIROLOGIST

Few, if any, diseases produce a greater degree of hysteria than rabies. There are many misconceptions and myths surrounding this disease. Therefore, it is very important for the clinical virologist to be fully informed about rabies so that facts and knowledge supercede misconceptions and myths.

Rabies is not transmitted through intact skin and rarely through intact mucosa. Humans do not acquire rabies by touching animal skin, animal excreta, or a variety of fomites, as long as there is no break in the skin.

Most animals that bite are not rabid. The bite of a rabid animal does not always result in overt clinical rabies nor does anti-rabies treatment after exposure guarantee protection against clinical rabies.

The bite of a rabid carnivore is more serious than that of an herbivore, as the saliva of the carnivore contains hyaluronidase and other spreading factors. Rabies has rarely been transmitted to man from a herbivore (cow, horse, sheep) in the United States (VanRooyen and Rhodes, 1948). However, in Middle Eastern countries, donkey and camel bites are an important mode of rabies transmission to man.

The location of a bite is of paramount consideration in determining risk and whether to institute treatment. Facial or finger bites pose a greater risk than bites incurred on the trunk or other locations on the extremities. Multiple bites increase risk over a single bite. The degree of mortal risk has been "classified" by Webster according to the location and the severity of the bite (VanRooyen and Rhodes, 1948).

Clinical virologists, laboratorians, veterinarians, animal control officers, and others with a high risk of contact with rabid animals, the tissues of rabid animals, or infected tissue cultures should be vaccinated against rabies. A sufficient number of injections of rabies vaccine should be given to prime the immune system to recognize and respond to rabies antigen. Priming is confirmed by the development of serum neutralizing antibody of the IgG class (Turner, 1978). Usually, three injections of a potent rabies vaccine of tissue culture origin spaced at 0, 7, and 21 or 28 days are sufficient to stimulate the immune system to react to the rabies vaccine. A single booster at six months following the primary vaccine series will reinforce the primary series so that an additional booster will not be needed for at least one year. Once the primary series is completed, the requirement for additional boosters will depend, in part, on the degree of risk of exposure to rabies virus. For example, persons who work with rabies virus in a laboratory should have their antibody titer determined every six months and, if necessary, be given a booster to maintain adequate antibody levels. One half (½) of an International Unit (IU) of rabies antibody is considered as minimal for protection against rabies (Centers for Disease Control, 1991). Others, such as animal control officers, should have their antibody titer determined every two years and receive a booster to maintain adequate antibody levels. If the means to determine antibody titers is not available, then these workers should be given a booster at the appropriate intervals.

In humans, domestic animals, and wild animals, rabies may mimic other diseases. Even though human rabies is a truly rare disease, and although rabies is found less frequently in domestic animals in Western countries, one should never dismiss this disease from clinical consideration (Fishbein et al, 1988). Cardinal clinical signs of rabies, such as dysphagia, oculomotor paralysis, excessive salivation, and hyperesthesia or hypalgesia, may be overlooked. Many veterinarians have sought for foreign bodies in the mouths of rabid animals.

Warrell et al (1988) have reported the use of skin biopsy in the diagnosis of clinical rabies. Punch biopsies of skin including hair follicles were taken from the neck or leg and examined by IF. The biopsy specimens confirmed rabies in nine of ten patients with rabies positive (FA) brain smears. Corneal smears were negative in all nine cases found positive by biopsy and brain smear.

Currently there is only one type of rabies vaccine that is being distributed nationwide for use in humans. This is an inactivated vaccine derived from rabies virus-infected human diploid (HDRV) cells. HDRV induces an excellent antibody response following a preexposure primary series of two or three injections (Anderson et al 1980). Treatment failures and some inadequate antibody responses have been reported from Africa following the intradermal inoculation of this vaccine (Centers for Disease Control, 1983). The reason for these treatment and antibody response failures has not been fully ascertained, but there is a suggestion that the problem may be related to the use of the vaccine in individuals immunosuppressed by other debilitating diseases or by drug treatment.

The human diploid vaccine can trigger severe hypersensitivity type reactions in a small percentage of individuals following a booster injection (Centers for Disease Control, 1984). Modified human albumin, the stabilizing component in the vaccine preparation, is suspected to be the principal sensitizing allergen in the vaccine.

The intradermal route of administration of the human diploid vaccine has been proposed by a number of investigators as a cost reduction alternative to the conventional intramuscular route (Morrison et al, 1987; Fishbein et al, 1987; Bernard et al, 1987). Our experience with a rabies vaccine of duck embryo origin indicated that this procedure should be restricted to those skilled in intradermal inoculation. However, when performed properly this procedure is time-consuming and may be less acceptable to some vaccine recipients than intramuscular administration.

An inactivated absorbed rabies vaccine (RDRV) derived from rhesus diploid cells is now licensed for use in the United States (Centers for Disease Control, 1988). This vaccine has been shown to induce high levels of serum antibody when administered intramuscularly in a primary series of two or three doses (Berlin et al, 1982, 1983). Following adequate primary stimulation, a booster injection of this vaccine stimulates antibody recall in practically all recipients. Ninety-eight percent of recipients who received a three-dose series of RDRV were found to have >0.1 IU of antibody 18 to 24 months later (Berlin, 1990). So far, adverse reactions following both primary and booster injection of the rhesus diploid cell vaccine have been minimal. Levenbook et al (1988) were unable to induce experimental allergic encephalomyelitis in guinea pigs and rats injected with RDRV combined with myelin basic protein, suggesting that RDRV does not induce autoimmunity.

Rabies immune globulin (RIG) of human origin is available for use in post-exposure situations. RIG should be used in conjunction with vaccine and should be administered as quickly as possible following exposure (Rubin et al, 1973; Aoki et al, 1989).

Possibly the most effective and probably the most overlooked method of treatment for persons bitten by an animal or cut by a contaminated instrument or bone fragment is local treatment at the site of the wound. It is important to thoroughly cleanse a bite wound using a soap solution, a detergent solution, or 50% to 70% ethyl alcohol. If rabies antiglobulin or antiserum is available, it should be infused in and around the wound site. Local treatment must be administered within one hour of the time of injury in order to remove or neutralize virus prior to its dissemination from the wound site.

24.7 FUTURE APPROACHES TO RABIES PREVENTION

Vaccinia and at least one other pox virus have been shown to be suitable vectors for the expression of the rabies glycoprotein (Esposito et al, 1988; Tolson et al, 1988). Recently Prevec et al (1989) described the insertion of the rabies virus glycoprotein gene into the DNA of adenovirus type 5. The recombinant adenovirus containing the rabies virus glycoprotein was shown to be capable of inducing neutralizing antibodies in dogs and mice when the vaccine was administered orally, intranasally, or via a parenteral route. Recombinant vaccines that utilize a pox or adenoviruses should have application in the large scale vaccination of wildlife (Baer, 1988) but will proba-

bly face considerable resistance as vaccine candidates for human use.

Ajjan and Pilet (1989) reported the results of a double blind study in French veterinary students comparing a human diploid cell vaccine and a vaccine derived from Vero cells (a continuous cell line derived from African green monkey kidney). All 72 recipients of the Vero cell derived vaccine seroconverted. Rabies antibody was still demonstrable 21 months later in 43 of these vaccinees. It was concluded from these studies that the Vero cell vaccine compared favorably with the rabies antibody stimulating potential of HDCV. Vero cell derived vaccine could serve as a source for use in vaccine production in developing countries, possibly on a regional basis (Halstead, 1988).

Interferon production has been demonstrated in tissue culture cells infected with high multiplicities of rabies virus. Interferon has been shown to inhibit rabies virus infection and its early induction may be critical in protecting individuals who are infected prior to receipt of any rabies vaccine. Perrin et al (1988) examined the use of interleukin 2 (IL 2) in experimental rabies. IL 2 appears to have some role in protection but this role has not been clarified.

Joffret et al (1991) vaccinated mice against rabies and found a positive correlation between the induction of IL 2 and protection. The production of IL 2 appears to reflect an increase in activity of CD4+CD8− cells and cells bearing IL 2 receptors. Vaccines that induce high levels of IL 2 may trigger the early cell-mediated response that appears to be essential for the success of post-exposure vaccination.

Intravenous rabies immune globulin (IRIG) may soon replace the intramuscular product in post-exposure prophylaxis. The level of rabies neutralizing antibody following injection of IRIG was shown to be two and one half times that of an intramuscular product of similar potency. With the intravenous product the antibody is placed directly into the blood stream where it is more readily available to interact at the sites where virus has been deposited (Aoki et al, 1989).

In western countries costs for rabies treatment are high whether the treatment is administered on a pre-exposure or post-exposure basis. There is a need to improve vaccine production technology to enhance the efficiency of generating protective antigens. A recombinant rabies vaccine using a bacterial vector may offer the best chance to reach this goal. Even here it will be essential that the vaccine contain the principal antigens necessary to stimulate the development of protection against challenge by street virus. Conserved nucleotide sequences derived from both the rabies virus glycoprotein and the nucleoprotein appear to be important for inducing protection (Ertl et al, 1989).

There is a tendency to recommend a full treatment schedule (local cleansing of the wound, RIG, and at least five doses of the vaccine) in the United States following an animal bite even when the actual need for treatment is questionable. State epidemiologists should be asked to provide more definitive information on the distribution of the disease and in assessing risk. Careful analysis of the incidence of rabies in the area and the biting incident may preclude the need for treatment.

In the third world keeping costs and procedures to a minimum is of paramount importance (Bogel et al, 1986). Thus, in third world countries or in other countries where health "dollars" are in short supply, there may be a tendency to use therapy as a means to combat this disease, i.e., using antiserum of equine origin and using a vaccine derived from suckling mouse brain, because these products may cost one tenth that of RIG and tissue culture-derived virus vaccines.

Twenty-one different rabies vaccines are being marketed for animal use in the United States in 1990 (Centers for Disease Control, 1990). Some of these vaccines provide three year protection, others only one year protection. With some vaccines the dosage is varied with the size of the animal, others are not. Boosters are recommended either every three years or annually depending on the vaccine. It seems inconceivable that a dog or cat would need a rabies inoculation every year or even every three years for the rest of its life. At some point the immune mechanism of these animals becomes sufficiently primed to rabies virus antigens to respond very efficiently to the incursion of wild rabies virus without continuing vaccine boosters. Tetanus toxoid booster inoculations in humans are spaced ten years apart after age 7. We have observed that, after a priming of a series of tetanus toxoid, a lengthy rest of the immune system before boostering will result in higher levels of antitoxin than that obtained by frequently repeated boosters.

When a human is bitten by a domestic animal, little heed is often paid to the fact that the animal has been vaccinated against rabies. Physicians, public health officials, and even some veterinarians seem to have little confidence that current animal rabies vaccines protect against rabies. This reaction is not surprising when it is so difficult to decode all the variables affecting the success or failure of animal vaccination. There is a need to either report or generate seroprevalence and/or protection data to determine whether the current rabies vaccines work well in animals. The introduction of low-cost serologic procedures (REIA, IAHA, and DIA) should make it easier to obtain these data.

24.7.1 Unique Features of the Disease

The "Early Death Syndrome" and the failure of high levels of anti-rabies serum antibody to guarantee protection in previously vaccinated individuals following postexposure treatment are two unusual features of this disease. A number of investigators have reported the occurrence of early deaths in experimental animals who received their first rabies vaccine or a vaccine series immediately following virus challenge. Challenged vaccinated animals who do not survive, often die three to four days earlier than the non-vaccinated challenge controls. The early death phenomenon has been reported in mice, guinea pigs, rhesus monkeys, and humans (Blancou et al, 1980; Porterfield, 1981; Prabhakar and Nathanson, 1981). Various hypotheses have been presented to explain this unusual feature of rabies, i.e., that vaccine virus may initially inhibit natural defense mechanisms; that rabies vaccine antigen may combine with small quantities of preformed anti-rabies antibody thereby making the antibody unavailable to act against wild virus; or that following antigen stimulation, blocking antibodies may prevail over neutralizing antibody. VanRooyen and Rhodes in their 1948 edition, ''Virus Diseases of Man,'' referred to a statement from the work of Proca and Bobes: ''that antirabies treatment may actually bring about death from rabies after a shorter incubation period than in untreated persons.'' A hypothesis was offered that fixed vaccine virus inhibited defense mechanisms and accelerated the passage of street virus. So the ''Early Death Syndrome'' in rabies is really nothing new. It has just taken us 40 years to catch up with what our predecessors already knew.

The second unique feature of rabies worthy of mention is the failure of antibody to protect a certain percentage of animals who have been exposed to rabies virus and then vaccinated following exposure. In our laboratories we have observed clinical disease and death in monkeys and guinea pigs whose serum contained high levels of rabies antibody prior to the appearance of clinical symptoms. These animals were vaccinated following virus exposure in experiments that were designed to simulate what happens most commonly with humans, i.e., the attempt to generate immunity de novo in the face of an exposure.

Wiktor was unable to protect a group of mice when vaccine was administered 24 hours after infection even though these animals had a mean antibody titer of 1000 seven days after infection (Wiktor, 1978). None of these animals had demonstrable antirabies cell-mediated immunity at seven days postinfection. In animals vaccinated at the time of exposure or prior to exposure, cell-mediated immunity was demonstrable at seven days after infection and the majority of these animals survived. These studies indicate that rabies-neutralizing antibody alone is not sufficient to guarantee protection in postexposure treatment situations. It would be desirable for any vaccine intended for postexposure use to be able to stimulate cell-mediated immunity, humoral immunity, and interferon to ensure the development of adequate protection.

Rabies itself is a unique disease. It presents to us, in a single entity, a researcher's delight. The virus can establish a symbiotic relationship with certain hosts. Cell-mediated immunity may be as important in rabies as it is in cancer. The majority of warm-blooded animals are susceptible to this disease. Some host cells are modified sufficiently by the virus to no longer be recognized as ''self.''

One hundred years after Louis Pasteur, the excitement that he faced is still there. We have conquered smallpox, we have almost conquered measles, yet rabies continues to remain the elusive abominable snowman.

REFERENCES

Ajjan, N. and Pilet, C. 1989. Comparative study of the safety protective value, in pre-exposure use, of rabies vaccine cultivated on human diploid cell (HDCV) and of the new vaccine grown on Vero cells. Vaccine 7:19–21.

Anderson, L., Winkler, G., Haflin, B., Keelyside, R., D'Angels, L., and Deitch, M. 1980. Clinical experience with a human diploid cell rabies vaccine. JAMA 244:781–784.

Aoki, F., Rubin, M., Freisen, A., Bowman, J. and Saunders, J. 1989. Intravenous human rabies immunoglobulin for post-exposure prophylaxis: Serum rabies neutralizing antibody concentrations and side effects. J. Biol. Stndz. 17:91–104.

Antanasiu, P., Savy, V., and Perrin, P. 1977. Epreuve immunoenzymatique pour la détection rapide des anticorps antirabique. Ann. Microbiol. 128A:489–498.

Baer, G. 1975. Pathogenesis to the central nervous system. In G. Baer (ed.), The Natural History of Rabies. New York: Academic Press, 181–198.

Baer, G. 1988. Oral Rabies Vaccination. Rev. Inf. Dis. 10:5644–5648.

Berlin, B., Mitchell, J., Burgoyne, G., Oleson, D., Brown, W., Goswick, C., and McCullough, N. 1982. Rhesus diploid rabies vaccine (adsorbed). A new rabies vaccine I. Results of initial clinical studies of pre-exposure vaccination. JAMA 247:1726–1728.

Berlin, B., Mitchell, J., Burgoyne, G., Oleson, D., Brown, W., Goswick, C., and McCullough, N. 1983. Rhesus diploid vaccine (adsorbed). A new rabies vaccine II. Results of clinical studies simulating prophylactic therapy for rabies exposure. JAMA 249:2663–2665.

Berlin, B. 1990. Rabies vaccine adsorbed: neutralizing antibody titers after three-dose pre-exposure vaccination. AJPH 80:476–478.

Bernard, K., Mallonee, J., Wright, J., Reid, F., Makintubee, S., Parker, R., Dwyer, D., and Winkler, W. 1987. Pre-exposure immuniza-tion with intradermal human diploid cell rabies vaccine, JAMA 257:1057–1063.

Blancou, J., Andral, B., and Andral, L. 1980. A model in mice for the study of the early death phenomenon after vaccination and challenge with rabies virus. J. Gen. Virol. 50:433–435.

Bogel, K. and Motschwiller, E. 1986. Incidence of rabies and post-exposure treatment in developing countries. Bull. W.H.O. 64:883–887.

Bota, C., Anderson, R., Goyal, S., Charamella, L., Howard, D., and Briggs, D. 1987. Comparative prevalence of rabies antibodies among household and unclaimed/stray dogs as determined by immune adherence haemagglutination assay. Int. J. Epidemiol. 16:472–476.

Bourhy, H., Rollin, P., Vincent, J., and Sureau, P. 1989. Comparative field evaluation of the fluorescent-antibody test, virus isolation from tissue culture, and enzyme immunodiagnosis for rapid laboratory diagnosis of rabies. J. Clin. Microbiol. 27:519–523.

Budzko, D., Charamella, L., Jelinek, D. and Anderson, G. 1983. Rapid detection of rabies antibodies in human serum. J. Clin. Microbiol. 17:481–484.

Ceccaldi, P., Gillet, J., and Tsiang, H. 1989. Inhibition of the transport of rabies virus in the central nervous system. J. Neuropath. Exp. Neurology 48:620–630.

Centers for Disease Control. 1983. Human rabies—Kenya. Morbidity and Mortality Weekly Report. 32:494–495.

Centers for Disease Control. 1984. Systemic allergic reactions following immunization with human diploid cell rabies vaccine. M.M.W.R. 33:185–187.

Centers for Disease Control. 1988. Rabies vaccine adsorbed: A new rabies vaccine for use in humans. M.M.W.R. 37:221–223.

Centers for Disease Control. 1990. Compendium of animal rabies control. M.M.W.R. 39:1–8.

Centers for Disease Control. 1991. Rabies Prevention. United States, 1991. M.M.W.R. 40:1–16.

Cox, J., Dietzschold, B., Weiland, F., and Schneider, L. 1980. Preparation and characterization of rabies virus hemagglutinin. Infect. Immun. 30:572–577.

Dean, D., Baer, G., and Thompson, W. 1963. Studies on the local treatment of rabies infected wounds. Bull. W.H.O. 28:477–486.

Dietzshold, B., Wiktor, T., Wunner, W., and Varrichio, A. 1983. Chemical and immunological analysis of the rabies soluble glycoprotein. Virology 124:330–337.

Dietzschold, B., Wang, H., Rupprecht, C., Celis, E., Tollis, M., Ertl, H., Herber-Katz, E., and Koprowski, H. 1987. Induction of protective immunity against rabies by immunization with rabies virus ribonucleoprotein. Proc. Natl. Acad. Sci. USA 84: 9165–9169.

Ertl, H., Dietzschold, B., Gore, M., Otvos, L., Larson, J., Wunner, W., and Koprowski, H. 1989. Induction of rabies virus-specific T-helper cells by synthetic peptides that carry dominant T-helper cell epitopes of the viral ribonucleoprotein. J. Virol. 63:2885–2892.

Esposito, J., Knight, J., Shaddock, J., Novembre, R., and Baer, G. 1988. Successful oral rabies vaccination of raccoons with racoon poxvirus recombinants expressing rabies virus glycoprotein. Virology 165:313–316.

Fernandes, M., Wiktor, T., and Kiprowski, H. 1964. Endosymbiotic relationship between animal viruses and host cells. A study of rabies virus in tissue culture. J. Exp. Med. 1220:1099–1116.

Fishbein, D., Pacer, R., Holmes, D., Ley, A., Yager, P., and Tong, T. 1987. Rabies pre-exposure prophylaxis with human diploid cell rabies vaccine: A dose response study. J. Inf. Dis. 156:50–55.

Fishbein, D. and Baer, G. 1988. Animal rabies: Implications for diagnosis and human treatment. Ann. Int. Med. 109:935–937.

Galloway, I. and Elford W. 1936. Size of virus of rabies ("fixed" strain) by ultrafiltration analysis. J. Hyg. (Lond.) 36:532–535.

Gamble, W., Chappell, W., and George, E. 1980. Inactivation of rabies diagnostic reagents by gamma radiation. J. Clin. Microbiol. 12:676–678.

Gardner, P.S. and McQuillin, J. 1980. Rapid

Virus Diagnosis. London: Butterworth, pp. 174–184.

Goldwasser, R. and Kissling, R. 1958. Fluorescent antibody staining of street and fixed virus rabies antigens. Proc. Soc. Exp. Biol. Med. 98:219–223.

Halstead, S. 1988. Tissue culture-based rabies vaccines: Vaccine production technology transfer. Rev. Inf. Dis. 10:9764–9765.

Heberling, R., Kalter, S., Smith, J., and Hildebrand, D. 1987. Serodiagnosis of rabies by dot immunobinding assay. J. Clin. Microbiol. 25:1262–1264.

Joffret, M.L., Zanetti, C., Morgeaux, S., Leclevc, D., Sureau, P., and Perrin, P. 1991. Appraisal of rabies vaccine potency by determination of *in vitro*, specific interleukin 2 production. Biologicals 19:113–123.

Kanazawa, K. 1936. Sur La Culture in Vitro du virus del la rage. Jpn. J. Exp. Med. 14:519–522.

Koprowski, H. and Cox, H. 1948. Studies on chick-embryo adapted rabies virus I—cultural characteristics and pathogenicity. J. Immunol. 60:533–554.

Larghi, O. and Nebel, A. 1980. Rabies virus inactivation by binary ethylenimine: New method for inactivated vaccine production. J. Clin. Microbiol. 11:120–122.

Levenbook, I., Elisberg, B., and Driscoll, B. 1988. Rhesus diploid rabies vaccine (adsorbed): Neurological safety in guinea pigs and Lewis rats. Vaccine 4:3–5.

Lodmell, D. 1983. Genetic control of resistance to street rabies virus in mice. J. Exp. Med. 157:451–460.

Morrison, A., Hunt, E., Atuk, N., Schwartzman, J., and Wenzel, R. 1987. Rabies pre-exposure prophylaxis using intradermal human diploid cell vaccine: Immunologic efficacy and cost-effectiveness in a university medical center and a review of selected literature. Am. J. Med. Sci. 293:293–297.

Murphy, F. 1975. Morphology and morphogenesis. In G. Baer (ed.), The Natural History of Rabies. New York: Academic Press, 33–61.

Negri, A. 1903. Beitrag zum Studium der Aetiologie der Tollwuth. Ztschr. fur Hyg. und Intekhonski. 43:507–528.

Newman, G. 1904. Pasteur's treatment for rabies. In Bacteriology and Public Health. London: John Murray, 420.

Nicholson, K. and Prestage, H. 1982. Enzyme-linked immunosorbent assay: A rapid reproducible test for the measurement of rabies antibody. J. Med. Virol. 9:43–49.

Perrin, P. and Sureau, P. 1987. A collaborative study of an experimental kit for rapid rabies enzyme immunodiagnosis (RREID) Bull. W.H.O. 65:489–493.

Perrin, P., Joffret, M.L., Leclerc, C., Oth, D., Sureau, P., and Thibodeau, L. 1988a. Interleukin 2 increases protection against experimental rabies. Immunobiol. 177:199–209.

Perrin, P., Morgeaux, S., and Sureau, P. 1990. In vitro rabies vaccine potency by ELISA: Advantages of the immunocapture method with a neutralizing anti-glycoprotein monoclonal antibody. Biologicals 18:321–330.

Porterfield, J. 1981. Antibody-mediated enhancement of rabies virus. Nature 290:542.

Prabhakar, B. and Nathanson, N. 1981. Acute rabies death mediated by antibody. Nature 290:590–591.

Prevec, L., Campbell, J., Christie, B., Belbeck, L., and Graham, F. 1989. A recombinant human adenovirus vaccine against rabies. J. Inf. Dis. 161:27–30.

Rubin, R., Sikes, K., and Gregg, M. 1973. Human rabies immune globulin clinical trials and effects of serum antigamma globulins, JAMA 224:871–874.

Rudd, R. and Trimarchi, C. 1989. Development and evaluation of an in vitro virus isolation procedure as a replacement for the mouse inoculation test in rabies diagnosis. J. Clin. Microbiol. 27:2522–2528.

Shope, R. 1975. Rabies virus antigenic relationships. In G. Baer (ed.), The Natural History of Rabies, New York: Academic Press, 141–152.

Smith, A., Tignor, G., Emmons, R., and Woodie, J. 1978. Isolation of field rabies virus strains in CER and murine neuroblastoma cell cultures. Intervirology 9:359–361.

Smith, J., Yager, R., and Baer, G. 1973. A rapid tissue culture test for determining rabies neutralizing antibody. In M. Kaplan and H. Koprowski (eds.), Laboratory Techniques in Rabies, 3rd ed. Geneva, Switzerland: WHO Publications, 354–357.

Smith, J., Reid-Sanden, F., Roumillat, L., Trimarchi, C., Clark, K., Baer, G., and Winkler, W. 1986. Demonstration of antigenic variation among rabies virus isolates by using monoclonal antibodies to nucleocapsid proteins. J. Clin. Microbiol. 24:573–580.

Sokol, F. 1975. Chemical composition and structure of rabies virus. In G. Baer (ed.), The Natural History of Rabies. New York: Academic Press, 79–113.

Tignor, G. and Shope, R. 1972. Vaccination and challenge of mice with viruses of the rabies serogroup. J. Inf. Dis. 125:322–327.

Tolson, N., Charlton, K., Casey, G., Knowles, M. Rupprecht, C., Lawson, K., and Campbell, J. 1988. Immunization of foxes against rabies with recombinant virus expressing the rabies virus glycoprotein. Arch. Virol. 102:297–301.

Tordo, N., Poch, O., Ermine, A., Keith, G., and Rougeon, F. 1988. Completion of the rabies virus genome sequence determination: Highly conserved domains among the L(polymerase) proteins of unsegmented negative-strand RNA viruses. Virology 165: 565–575.

Turner, G. 1978. Immunoglobulin (IgG) and (IgM) antibody responses to rabies vaccine. J. Gen. Virol. 40:595–604.

VanRooyen, C. and Rhodes, A. 1948. The Rabies Virus. London: Thomas Nelson, 811–833.

Vaughan, J., Gerhardt, P., and Newell, K. 1965. Excretion of street rabies virus in the saliva of dogs. JAMA 193:363–368.

Vaughan, J., Gerhardt, P., and Paterson, J. 1963. Excretion of street rabies virus in saliva of cats. JAMA 184:705–708.

Warrell, M., Looareesuwan, S., Manatsathit, S., White, N., Phuapradit, P., Vejjajiva, A., Hoke, C., Burke, D., and Warrell, D. 1988. Rapid diagnosis of rabies and post-vaccinal enphalitides. Clin. Exp. Immunol. 71:229–234.

Watson, H., Tignor, G., and Smith, A. 1981. Entry of rabies virus into the peripheral nerves of mice. J. Gen. Virol. 56:371–382.

Webster, L. and Clow, A. 1936. Propagation of

rabies virus in tissue culture and the successful use of culture as an antirabic vaccine. Science 84:487–488.

Wiktor, T. 1966. Dynamics of rabies virus infection in tissue culture. In Symposia Series in Immunobiological Standardization, 1. New York: Karger, 65–80.

Wiktor, T. 1978. Cell mediated immunity and post-exposure protection from rabies by inactivated vaccines of tissue culture origin. Dev. Biol. Stand. 40:255–264.

Wiktor, T. and Clark, H. 1975. Growth of rabies virus in cell culture. In G. Baer (ed.), The Natural History of Rabies. New York: Academic Press, 155–179.

Wiktor, T., Fernandes, M., and Koprowski, H. 1964. Cultivation of rabies virus in human diploid cell strain WI-28. J. Immunol. 93: 353–366.

Wiktor, T., Gyorgy, H., Schlumberger, H., Sokol, F., and Koprowski, H. 1973. Antigenic properties of rabies virus components, J. Immunol. 110:269–276.

Wiktor, T., Flamand, A., and Koprowski, H. 1980. Use of monoclonal antibodies in diagnosis of rabies virus infection and differentiation of rabies and rabies related viruses. J. Immunol. Methods 1:43–66.

Wunner, W., Larson, J., Dietzshold, B., and Smith, D. 1988. Molecular biology. Biology of rabies viruses. Rev. Inf. Dis. 10:5771–5784.

Zinke, G. 1804. Neue Ansichten der Hundswuth, ihrer Ursachen und Folgen. nebst einer sichern Behandlungsart der von Follen Thieren gebissenen. Menschen Gabler. Jena 16,212.

Arboviruses

Robert E. Shope

25.1 HISTORY

The early white settlers in both Africa and the Americas were familiar with diseases we now know were caused by arboviruses. Yellow fever virus was responsible for a clearly identified epidemic in the Yucatan in 1648 and induced a much feared illness along the major rivers and in the seaports of the New World throughout the days of sailing ships. *Aedes aegypti* mosquitoes were the urban vectors of yellow fever. This mosquito also transmitted dengue virus, the cause of dengue fever and dengue hemorrhagic fever and shock syndrome. Dengue fever was rampant in the southern United States until the 1920s, when populations of the vector mosquito were controlled. Both dengue and yellow fever continue to occur in tropical America and Africa, although yellow fever can be prevented by vaccination. Today, dengue hemorrhagic fever and shock syndrome (Halstead, 1980) is a major, lethal, epidemic disease of children in Southeast Asia, and appeared for the first time in the New World in Cuba in 1981.

In modern times the primary clinical manifestation of life-threatening arboviral disease in North America has been encephalitis. Three mosquito-borne viruses that cause human encephalitis were discovered during the 1930s. Western equine encephalitis (WEE) virus was isolated in 1930 from horses (Meyer et al, 1931) and in 1938 was associated with encephalitis in humans in California; since then it has been isolated regularly in the plains and plateaus of the western United States and Canada. Eastern equine encephalitis (EEE) was isolated in 1933 (TenBroeck and Merrill, 1933) and was associated with encephalitis in horses; it was subsequently recovered in 1938 from people along the Atlantic coast and more recently along the Gulf coast and in upper New York State and Michigan. In 1933, St. Louis encephalitis (SLE) virus caused an epidemic of encephalitis in St. Louis and surrounding communities with 1095 reported cases (Cumming, 1935). Endemic (rural) SLE is prevalent each year in much of the western United States, and urban epidemics occur every seven to ten years in widely scattered loci in cities such as Chicago, Philadelphia, and Houston. The last major epidemic, in 1975, was responsible for 1815 reported cases of SLE.

California encephalitis virus was isolated in 1943 from *Aedes* mosquitoes in

California and was later associated serologically with three pediatric encephalitis cases in California (Hammon and Reeves, 1952). Not until 1964, however, was the full significance of the California group viruses realized. In that year a virus closely related to California encephalitis virus was isolated from the stored brain of a child who had died in 1960 in LaCrosse, Wisconsin (Thompson et al, 1965). Starting in the early 1960s, the LaCrosse (LAC) subtype has been associated in the United States with about 30 to 140 cases per year of California group encephalitis. Two other closely related California group viruses, snowshoe hare (SSH) and Jamestown Canyon (JC), have been etiologically associated with a small number of encephalitis cases in the United States and Canada since 1980 (Fauvel et al, 1980; Grimstad et al, 1982). Another mosquito borne virus, Venezuelan equine encephalitis (VEE), induced encephalitis in Florida, where three endemic cases were recorded (Ehrenkranz et al, 1970). Similarly, according to Centers for Disease Control (CDC) reports, the tick transmitted Powassan (POW) virus has caused only ten reported cases of human encephalitis in the United States and Canada since it was first isolated from the brain of a child in Powassan, Ontario in 1958.

Colorado tick fever (CTF) is another arboviral syndrome, prevalent in the western mountain region of the United States. The disease was known as mountain fever as early as 1855, but the etiologic virus isolated from the blood of febrile patients was not described until 1944 (Florio et al, 1944). Colorado tick fever is a tick borne virus that causes diphasic fever, muscle aches, malaise and, occasionally, hemorrhagic or central nervous system (CNS) complications in children. It is most common in campers, hikers, and other persons coming in contact with *Dermacentor andersoni* ticks.

Detailed reviews are recommended for SLE (Monath, 1980), for California group encephalitis (Calisher and Thompson, 1983), for alphaviruses and flaviviruses (Schlesinger, 1980), for bunyaviruses (Kolakofsky, 1991), and for arboviruses (Monath, 1988).

25.2 BIOCHEMICAL, SEROLOGIC, AND EPIDEMIOLOGIC CHARACTERISTICS OF ARBOVIRUSES

Over 400 different serotypes of arboviruses are known, of which at least 90 have been incriminated in human disease. Arboviruses encountered in patients in North America belong to one of four genera: *Alphavirus* (family Togaviridae), *Flavivirus* (family Flaviviridae), *Bunyavirus* (family Bunyaviridae), and *Orbivirus* (family Reoviridae). Table 25.1 lists those human disease arboviruses indigenous to or apt to be encountered in travellers to North America.

Eastern, western, and Venezuelan encephalitis viruses are alphaviruses. These are single-stranded, spherical RNA viruses, 40 to 70 nm in diameter. The RNA genome has positive (sense) polarity. The viruses have a lipid envelope into which are inserted virus specified glycoprotein spikes that endow the particle with the property of hemagglutination. The virus multiplies in the cytoplasm of vertebrate and invertebrate cells and matures by budding through the plasma membrane. There are several complexes of closely related viruses within the *Alphavirus* genus. Eastern equine, WEE, and VEE viruses are serologically related but each belongs to a different complex and thus is easily distinguished. The hemagglutination inhibition (HI) test and the enzyme-linked immunosorbent assay (ELISA) demonstrate the crossreactions; the neutralization test is relatively specific.

The clinical virologist must understand the epidemiology in order to entertain a rational index of suspicion when considering a diagnosis. Eastern equine encephalitis

Table 25.1 The Distribution and Natural Cycles of Arboviral Diseases Found in North America

Disease	Disease Distribution	Virus	Vector	Vertebrate	Annual Number of U.S. Cases Reported to CDC
Eastern encephalitis	Atlantic and Gulf coasts	Eastern encephalitis	Mosquito	Birds, horses, pheasants	0–14
	Upper New York State, Michigan				
Western encephalitis	Western United States and Canada	Western encephalitis	Mosquito	Birds	0–133
Venezuelan encephalitis	Florida	Venezuelan encephalitis	Mosquito	Forest & swamp rodents	0–1
St. Louis encephalitis	North America	St. Louis encephalitis	Mosquito	Birds	15–1815
Powassan encephalitis	U.S. and Canada	Powassan	Tick	Woodchuck, skunk, other small mammals	0–1
LaCrosse encephalitis[a]	North Central and Northeast U.S.	LaCrosse	Mosquito	Chipmunk, squirrel	30–160
Colorado tick fever	Rocky Mountain states	Colorado tick fever	Tick	Ground squirrel	At least 200[b]
Dengue fever	Caribbean, in travelers to U.S.	Dengue, types 1–4	Mosquito	Humans	24 confirmed in 1990[b]

[a] California encephalitis, snowshoe hare encephalitis, and Jamestown Canyon encephalitis are caused by viruses closely related to LaCrosse virus; the diseases are known collectively as California group encephalitis.

[b] Not reportable diseases.

virus is maintained in a transmission cycle between birds and *Culiseta melanura* mosquitoes. These are swamp breeding mosquitoes found along the Gulf and Atlantic coasts, in upper New York State, and in Michigan. The distribution of the vector presumably accounts for the distribution of the disease. It is not known how the virus survives the winter, but virus appears each year in the vicinity of swamps, first infect-ing wild birds. This leads to an amplification of the transmission cycle. *Culiseta melanura* feeds almost exclusively on birds. It is thus necessary to implicate other mosquitoes such as *Aedes* and *Coquillettidia*. These mosquitoes transmit virus from birds to horses and pheasants, which develop encephalitis. About two weeks later, encephalitis appears in people. There are commercial vaccines for horses for EEE,

WEE, and VEE viruses; if these vaccines have been used, the disease in horses will not be available as an indicator to alert the physician of impending illness in people. Cases of EEE occur early in the summer along the Gulf coast, but not until late June and lasting into October in more northern climates.

Western equine encephalitis virus is also maintained in a cycle of birds and mosquitoes, the vector being *Culex tarsalis*. This mosquito occurs in the western part of the United States. It breeds in ground pools, especially in irrigated areas and river flood-plains. Sporadic cases of WEE are detected between June and October each year and range as far north as Canada. Epidemics are usually associated with heavy irrigation, excess runoff from snowpack, or flood conditions in river valleys.

Endemic VEE virus in the United States is maintained in a cycle of tropical *Culex* (*Melanoconion*) spp. mosquitoes in the south Florida swamps. The vertebrate hosts are swamp rodents such as the cotton rat. Recognized encephalitis is rare. When it occurs it is usually in older persons who have entered the swamp to fish or for other recreation. In 1971, a major epizootic of VEE in horses moved from Mexico into Texas where it caused encephalitis in horses and people. The virus differed antigenically from the endemic VEE of Florida. All transmission was eliminated after a campaign to vaccinate horses and to kill mosquitoes by aerial insecticide spraying. Venezuelan equine encephalitis infection has not been detected in Texas since.

St. Louis encephalitis and Powassan viruses are flaviviruses. These are spherical, positive sense single-stranded RNA viruses that are 40 to 50 nm in diameter, somewhat smaller than alphaviruses. They have a lipid envelope into which are inserted virus specified glycoprotein spikes, which are responsible for the property of hemagglutination. The virus multiplies in the cytoplasm of vertebrate and invertebrate cells but, unlike alphaviruses, particles do not bud from the plasma membrane; they mature in association with endoplasmic reticulum. There are several complexes of closely related flaviviruses. St. Louis encephalitis and POW viruses are serologically related but each belongs to a different complex and, thus, is easily distinguished. The HI test and the ELISA demonstrate the crossreactions; the neutralization test is quite specific.

The epidemiology of SLE virus in rural North America is similar to that of WEE. Transmission is maintained in a bird-*Culex tarsalis* cycle with sporadic cases occurring each summer, somewhat later than cases of WEE because SLE virus requires warmer temperatures to replicate efficiently in the mosquito. Unlike WEE, SLE virus has alternate vectors involved in urban outbreaks. St. Louis encephalitis epidemics follow unusually warm and dry periods in urban centers. The vectors are *Culex pipiens* complex mosquitoes, which breed in water with high organic content such as city storm sewers and poorly draining sewage. This mosquito takes blood-meals from birds and humans. Cases of SLE are prevalent from mid-summer until frost. In Florida, SLE virus is transmitted by *Culex nigripalpus*.

LaCrosse, California encephalitis (CE), snowshoe hare, and Jamestown Canyon viruses are bunyaviruses. These are spherical, negative polarity single-stranded RNA viruses. The RNA genome consists of three segments. The viruses replicate in the cytoplasm and mature by budding into smooth surfaced vesicles in association with the Golgi apparatus. They have a lipid envelope into which are inserted spikes made up of two glycoproteins. The spikes confer the ability to agglutinate red blood cells. The particles are 90 to 100 nm in diameter. California encephalitis is the type virus of the California serogroup. LaCrosse and snowshoe hare viruses are varieties of

California encephalitis virus and crossreact extensively in serologic tests; Jamestown Canyon is a subtype of California encephalitis virus and is easily distinguished serologically (Bishop and Shope, 1979).

Encephalitis caused by LaCrosse virus may occur somewhat earlier in the summer than other arbovirus encephalitides because this virus is transmitted transovarially in *Aedes triseriatus*, its mosquito vector. Because the virus passes via the egg, the mosquito can infect vertebrate hosts as soon as the adult emerges and takes its first bloodmeal in the early summer. The insect continues to transmit until frost. This mosquito lays its eggs in water-containing tree holes in the deciduous hardwood forests of Wisconsin, Minnesota, Iowa, Michigan, Ohio, Indiana, Illinois, New York, and to a lesser extent, neighboring states. In recent years it has also adapted to using tires and other peridomestic water containers. Children presenting with encephalitis give a history of living near the woods, of having tire swings or discarded tires in the yard, or of camping or hiking in forests. The mosquito feeds on tree squirrels and chipmunks that become viremic and may serve as amplifying hosts of the virus. The disease is focal and endemic because of the ecology of the vector.

Colorado tick fever virus is an orbivirus. It is a spherical particle of 65 to 80 nm diameter. The genome consists of 12 segments of double-stranded RNA. Replication is in the cytoplasm. The particles lack an envelope. Colorado tick fever virus has no close serologic relatives in North America.

Colorado tick fever infects *Dermacentor andersoni* ticks inhabiting the western mountain states, including Oregon, Washington, California, Idaho, Nevada, Montana, Wyoming, Utah, Colorado, New Mexico, and South Dakota, as well as western Canada. The immature tick feeds on ground squirrels and other small rodents with a CTF viremia and they serve as amplifying hosts. The tick remains infected transtadially and the adult tick transmits to people. Disease is most common during April and May at lower altitudes and June and July at higher elevations, because the adult tick is most active during these months.

25.3 PATHOGENESIS

Arboviruses gain entry through the skin by the bite of an infected arthropod. Knowledge of the initial events of infection is superficial. The mosquito saliva enters the dermis and at times enters the small capillaries directly when the mosquito's proboscis threads the vessel. It is presumed that the virus replicates initially in the dermal tissues, including the capillary endothelium, although it is also possible that virus is transported directly in the blood to primary target organs. Replication also occurs in the regional lymph nodes and from there the blood is seeded, inducing a secondary viremia, which in turn carries virus to infect muscle and connective tissue cells. This viremia is often of very high titer and is accompanied by fever, leukopenia, and malaise. It is during this viremic phase that an arthropod may feed and become infected. The period between infection and viremia (intrinsic incubation period) is usually short, from one to three days. Viremia may last two to five days. Colorado tick fever viremia is of much longer duration because immature red blood cells are infected and virus remains in the blood cells for two to six weeks.

The vast majority of human arboviral infections are either asymptomatic or self-limited febrile illnesses. Antibody is produced and it complexes with and neutralizes circulating virus. The process is accompanied by complete recovery and leads to the presence of life-long antibody. Occasionally, however, an infected person develops encephalitis. The mechanism of

Table 25.2 Clinical Features of Some Arboviral Encephalitides of North America

Disease	Age	Onset	Seizures	Case Fatality	Sequelae
Eastern encephalitis	Severe in children and older adults	Abrupt	75%	About 50%	Severe in about 30%
Western encephalitis	Severe in children	Preceded by fever	Common in infants	5%–10% under 1 year	Common and severe in children
St. Louis encephalitis	Severe & higher attack rate >50 years	Often abrupt	About 50%	3%–20%	Uncommon
Powassan encephalitis	Severe in children	Gradual		About 10%	Frequent and severe
LaCrosse encephalitis	Children	Abrupt	>50%	0.5%	Seizures later in life in 6%–13%
Jamestown Canyon encephalitis	Children & adults	Preceded by fever	About 10%	Low	Uncommon

entry of virus into the central nervous system is not completely understood. Nor is it understood why one person develops encephalitis and another apparently similar individual does not. Virus may reach the brain by seeding of cerebral capillaries during viremia, then by direct invasion of the brain parenchyma through the capillary walls. Alternatively, certain neural cells such as the olfactory neurons are exposed directly to circulating blood; viremia may seed these nerve endings and the virus may pass directly to the olfactory lobe of the brain. Regardless of the mechanism, it is important to note that the process of seeding the brain and productive infection of brain cells takes time. By the time the patient presents with encephalitis, serum antibody is usually detectable as is antibody in the cerebrospinal fluid (CSF). At this stage of infection, viremia has ceased and diagnosis is made by serologic assay.

The clinical laboratory findings and histopathology of arboviral encephalitis are often not helpful in arriving at an etiologic diagnosis. A definitive diagnosis can be made only in the virus diagnostic laboratory. The histopathology is characterized by perivascular cuffing, neuronal chromatolysis, cell shrinkage, and neuronophagia. Eastern equine encephalitis brain lesions are unusually necrotizing and are associated with high lymphocyte counts and modestly elevated protein levels in the CSF. Central nervous system infections with other arboviruses have CSF cell counts <500 and normally slightly elevated CSF protein.

25.4 CLINICAL DESCRIPTION

The arboviral encephalitides are not readily distinguished on clinical grounds, but the age of the patient has predictive value (Table 25.2). Western equine, Powassan, and LaCrosse encephalitis have the highest attack rates and are also more severe in the pediatric age-group. St. Louis encephalitis is a more common and severe disease in people over 50 years. Fever, malaise, and often severe headache usually precede neurologic signs. LaCrosse encephalitis patients often present with seizures, which are sometimes uncontrollable with antiseizure medication (Chun, 1983). Abnormal reflexes, paresis, and paralysis are common, but the prognosis with conservative treatment is good. Eastern encephalitis patients develop severe damage to the CNS, manifest by rapid progression to drowsiness and coma, convulsions, spasticity, periorbital or facial edema, high fever, and death as early as three to five days after onset (Feemster, 1957). Western encephalitis is characterized by seizures (90% of infants), lethargy or coma, restlessness, stiff neck, and fever (Kokernot et al, 1953). Remission may be sudden. St. Louis encephalitis usually has a benign course with fever, headache, and uneventful recovery. Individuals over 55 years of age, however, have severe disease with convulsions, paralysis, abnormal reflexes, confusion, and stiff neck. Again, remission may be sudden after an illness of three to ten days. Urinary tract symptoms including frequency, urgency, and incontinence are found in about 25% of patients, sometimes preceding the onset of encephalitic symptoms. The syndrome of inappropriate secretion of antidiuretic hormone has also been described as a common complication of SLE (White et al, 1969).

The differential diagnosis of arboviral encephalitis includes cerebrovascular accidents, neoplasia, and a variety of parasitic, bacterial, mycotic, and spirochetal diseases. Among the viral causes of encephalitis are enteric viruses, mumps, measles, lymphocytic choriomeningitis, rabies, and herpes simplex viruses. Herpes simplex is important to differentiate, because its onset with temporal lobe seizures may closely mimic arboviral encephalitis (especially EEE and LaCrosse types). The diagnosis of herpes encephalitis requires brain biopsy and, for best results, treatment with vidarabine or similar drugs should be started early. In Summer–Fall cases (especially in children) however tests to detect antibody to arbovirus should be conducted prior to biopsy.

Colorado tick fever is a self-limited febrile disease of persons exposed to ticks in the Rocky Mountain area of the United States. In addition to fever, patients have malaise and muscle aches and pains (Goodpasture et al, 1978). The fever may be diphasic and is occasionally accompanied by a maculopapular or petechial rash. Leukopenia is consistently present on days two to six of fever (Anderson et al, 1985). Patients recover by the ninth or tenth day with weakness but without long-term effects of the disease. Rare complications are encephalitis, myocarditis, or a hemorrhagic syndrome. The disease is usually mild in children; however, adults can be quite ill.

25.5 LABORATORY PROCEDURES

25.5.1 Collection of Specimens

Details of laboratory procedures for arboviruses are described by Shope and Sather (1979). Specimens should be collected aseptically and refrigerated at 4 °C for 24 hours or less until tested, or frozen at −60 °C or colder if they are to be stored for longer periods. Virus is rarely found in the serum or plasma of encephalitis cases, but virus is readily isolated during the first three or four days of systemic illnesses such as dengue fever; in CTF the virus is in the erythrocytes for as long as six weeks. Acute phase blood and CSF should also be taken for antibody detection. If a diagnosis is not made acutely, then a second serum should be collected one week or later. At necropsy, virus is present in the brain in cases of encephalitis, and may be recovered from the spleen, liver, and kidney in systemic illnesses.

25.5.2 Isolation of Virus

Intracerebral inoculation of mice one to three days of age is the isolation method of choice for those arboviruses causing human encephalitis and for CTF. Alternate isolation systems include Vero, BHK-21, and primary chicken or duck embryo cells. The C6/36 clone of *Aedes albopictus* mosquito cells is highly susceptible to the encephalitis viruses and to dengue and yellow fever viruses, but these cells do not always develop cytopathic effect and the virus usually must be detected by immunofluorescence or subculture into another host.

Colorado tick fever is readily isolated from patients' red blood cells for two weeks and sometimes for as long as six weeks after onset of fever. For cases of arbovirus encephalitis, the viremic period has passed before the patient is seen by the physician. Virus may still be isolated from brain tissue at biopsy or autopsy. At autopsy, tissue from multiple sites in the brain should be sampled, because arboviruses may replicate better in one part of the brain than another. It is rarely possible to isolate virus from the CSF.

Isolation of virus takes time. Mice inoculated with EEE, WEE, or VEE viruses die in less than 48 hours; mice inoculated with LaCrosse and other California group viruses die within 48 to 72 hours; and mice inoculated with SLE and Powassan viruses die in four to seven days. The time for appearance of plaques or cytopathic effect in cell culture is usually shorter than the incubation period in mice.

Arboviruses are stable at −70 °C, or lyophilized and stored at −20 °C. Lyophilized viruses may be transported for short periods at room temperature without loss of titer.

Virus is identified by complement fixation, HI, immunofluorescence, neutralization, or ELISA.[1] An antigen is made by suspending infected mouse brain in veronal buffer (for complement fixation) or in borate saline buffer pH 9.0 (for HI). Alternatively, the brain is weighed and suspended 5% w/v in 8.5% sucrose and extracted twice with cold acetone. The resulting precipitate is dried, rehydrated, and used as antigen in the complement fixation, HI, or ELISA methods. For confirmation of the identification, mice are immunized with infected mouse brain of the unknown virus and the resulting antibody is used in reciprocal tests with the antigen of the unknown, and the antigen of and antibodies to reference viruses known to be in the geographic area of exposure. Type-specific monoclonal antibodies reactive in the immunofluorescence test are available from reference centers for the four types of dengue virus.

25.5.3 Detection of Antigen and Antibody

Antigen detection offers a more rapid diagnosis than virus isolation and is the method of choice for CTF virus. Colorado tick fever antigen is readily demonstrated by immunofluorescence in red blood cells taken within the first two weeks of infection, and even as late as six weeks (Emmons et al, 1969). The immunofluorescence test is also used for the rapid diagnosis of arbovirus encephalitis when brain biopsy or autopsy material is available. Either frozen sections or impression smears may be used. Antigen detection by ELISA may also be employed, although this technique is not in general use.

IgM antibody detection is the method of choice for diagnosis of arbovirus encephalitis. In most cases of encephalitis, onset of CNS signs and symptoms coincides with or follows development of antibody. The viremic phase of the infection has terminated and it is usually no longer possible to isolate virus except in fatal cases or by brain biopsy. With the IgM antibody capture ELISA, a reliable presumptive diagnosis can be made within the first five days after onset of CNS signs in most patients. In a study of 29 LaCrosse encephalitis patients, 83% were diagnosed on the day of

---

[1]Reference hyperimmune mouse ascitic fluids are available from the CDC, Vector-Borne Viral Diseases Laboratory, Box 2087, Fort Collins, CO 80522, or from the Yale Arbovirus Research Unit, Box 3333, New Haven, CT 06510.

hospital admission using the IgM capture ELISA with acute phase sera (Jamnback et al, 1982).

The IgM antibody capture ELISA utilizes anti-μ chain antibody to coat the solid phase. This antibody captures the patient's serum IgM or CSF IgM. Anti-μ chain antibody captures total IgM (i.e., arboviral infection-specific as well as other IgM). Other IgM is found in serum, but usually not in substantial quantities in CSF. The color intensity of the IgM antibody capture ELISA depends on the proportion of arboviral-specific IgM, not total IgM. Therefore ELISA using CSF is more sensitive than that of serum. The ELISA indicator system utilizes the arboviral antigen and a conjugated mouse or rabbit specific antiarboviral serum. An analogous method using either ELISA or radioimmune assay has been applied to flaviviral encephalitis (Heinz et al, 1981; Burke et al, 1982). The antibody capture IgM ELISA has not been thoroughly tested with all arboviruses, but it has proved reliable in experience to date.

Many public health and hospital laboratories still rely on demonstration of a fourfold or greater rise in antibody titer using paired acute and convalescent phase sera to diagnose arboviral encephalitis. This classic procedure is not completely satisfactory. Not only is the diagnosis delayed, but a study of LaCrosse encephalitis cases showed that the complement fixation test alone diagnosed only 50%, the HI test alone 79%, and the neutralization test alone 85% (Calisher and Bailey, 1981). The neutralization test, although the most reliable, is expensive and not widely used.

The IgM antibody capture ELISA is rapid. When the differential diagnosis includes herpes encephalitis, the ELISA for arboviral encephalitis should be performed before the brain is biopsied, to preclude an unnecessary invasive procedure.

25.6 TRENDS

Summertime acute encephalitis in the United States continues to be an enigma. More than 50% of the reported encephalitis cases are still undiagnosed. Their Summer–Fall seasonality is reason to believe that they may be mosquito borne. The clinical virologist is urged to sample these cases. Already, in the 1980s, snowshoe hare and Jamestown Canyon viruses have emerged as causes of encephalitis. Additional viruses will almost certainly be linked in the future. Table 25.1 shows the annual numbers of reported cases of arboviral encephalitis in the past decade. It is probable that there is marked underreporting. The clinical virologist can help obtain a truer picture.

Imported dengue cases are also of public health concern. In 1990, the CDC recorded 24 serologically confirmed cases; three of these were in states that had *Aedes aegypti* (CDC, 1991). So far, dengue virus transmission has not become established in the southern United States, but to alert authorities to such an event, it is critical that clinical virologists continue surveillance, including laboratory testing of specimens for virus isolation and for serologic confirmation where dengue is suspected.

REFERENCES

Anderson, R.D., Entringer, M.S., and Robinson, W.A. 1985. Virus-induced leukopenia: Colorado tick fever as a human model. J. Infect. Dis. 151:449–453.

Bishop, D.H.L. and Shope, R.E. 1979. Bunyaviridae. In H. Fraenkel-Conrat and R.R. Wagner (eds.), Comprehensive Virology. Vol. 14. New York: Plenum, 1–156.

Burke, D.S., Nisalak, A., and Ussery, M.A. 1982. Antibody capture immunoassay detection of Japanese encephalitis virus immunoglobulin M and G antibodies in cerebrospinal fluid. J. Clin. Microbiol. 16: 1034–1042.

Calisher, C.H. and Thompson, W.H. (eds.). 1983. California Serogroup Viruses. New York: Alan R. Liss.

Calisher, C.H. and Bailey, R.E. 1981. Serodiagnosis of LaCrosse virus infections in humans. J. Clin. Microbiol. 13:344–350.

Centers for Disease Control. 1991. Imported dengue–United States. M.M.W.R. 40:519–520.

Chun, R.W.M. 1983. Clinical aspects of LaCrosse encephalitis: Neurological and psychological sequelae. In C.H. Calisher and W.H. Thompson (eds.), California Serogroup Viruses. New York: Alan R. Liss.

Cumming, H.S. (ed.). 1935. Report on the St. Louis Outbreak of Encephalitis. Public Health Bulletin No. 214. Washington, D.C.: U.S. Treasury Department, Public Health Service.

Ehrenkranz, N.J., Sinclair, M.C., Buff, E., and Lyman, D.O. 1970. The natural occurrence of Venezuelan equine encephalitis in the United States. N. Engl. J. Med. 282:298–302.

Emmons, R.W., Dondero, D.V., Devlin, V., and Lennette, E.H. 1969. Serologic diagnosis of Colorado tick fever. A comparison of complement-fixation, immunofluorescence, and plaque-reduction methods. Am. J. Trop. Med. Hyg. 18:796–802.

Fauvel, L.M., Artsob, H., Calisher, C.H., Davignon, L., Chagnon, A., Skvorc-Ranko, R., and Belloncik, S. 1980. California group virus encephalitis in three children from Quebec: Clinical and serological findings. Can. Med. Assoc. J. 122:60–64.

Feemster, R.F. 1957. Equine encephalitis in Massachusetts. N. Engl. J. Med. 257:701–704.

Florio, L., Stewart, M.O., and Mugrage, E.R. 1944. The experimental transmission of Colorado tick fever. J. Exp. Med. 80:165–188.

Goodpasture, H.C., Poland, J.D., Francy, D.B., Bowen, G.S., and Horn, K.A. 1978. Colorado tick fever: Clinical, epidemiologic, and laboratory aspects of 228 cases in Colorado

in 1973–1974. Ann. Intern. Med. 88:303–310.

Grimstad, P.R., Shabino, C.L., Calisher, C.H., and Waldman, R.J. 1982. A case of encephalitis in a human associated with a serologic rise to Jamestown Canyon virus. Am. J. Trop. Med. Hyg. 31:1238–1244.

Halstead, S.B. 1980. Immunological parameters of togavirus disease syndromes. In R.W. Schlesinger (ed.), The Togaviruses. New York: Academic Press, 107–173.

Hammon, W.F.McD. and Reeves, W.C. 1952. California encephalitis virus. A newly described agent. Calif. Med. 77:303–309.

Heinz, F.X., Roggendorf, M., Hormann, H., Kunz, C., and Deinhardt, F. 1981. Comparison of two different enzyme immunoassays for detection of immunoglobulin M antibodies against tick-borne encephalitis virus in serum and cerebrospinal fluid. J. Clin. Microbiol. 14:141–146.

Jamnback, T.L., Beaty, B.J., Hildreth, S.W., Brown, K.L., and Gundersen, C. 1982. Capture immunoglobulin M system for rapid diagnosis of LaCrosse (California encephalitis) virus infections. J. Clin. Microbiol. 16:577–580.

Kokernot, R.H., Shinefield, H.R., and Longshore, W.A. Jr. 1953. The 1952 outbreak of encephalitis in California: Differential diagnosis. Calif. Med. 79:73–77.

Kolakofsky, D. (ed.). 1991. Bunyaviridae. New York: Springer-Verlag, 256.

Meyer, K.F., Haring, C.M., and Howitt, B. 1931. The etiology of epizootic encephalomyelitis in horses in the San Joaquin Valley. Science 74:227–228.

Monath, T.P. (ed.). 1980. St. Louis Encephalitis. Washington, D.C.: American Public Health Association.

Monath, T.P. (ed.). 1988. The Arboviruses: Epidemiology and Ecology. Boca Raton FL: CRC Press.

Schlesinger, R.W. (ed.). 1980. The Togaviruses. New York: Academic Press.

Shope, R.E. and Sather, G. 1979. Arboviruses. In E.H. Lennette and N.J. Schmidt (eds.), Diagnostic Procedures for Viral, Rickettsial and Chlamydial Infections. Washington, D.C.: American Public Health Association, 767–814.

TenBroeck, C. and Merrill, M. 1933. A serological difference between eastern and western equine encephalomyelitis virus. Proc. Exp. Biol. Med. 31:217–220.

Thompson, W.H., Kalfayan, B., and Anslow, R.O. 1965. Isolation of California encephalitis group virus from a fatal human illness. Am. J. Epidemiol. 81:245–253.

White, M.G., Carter, N.W., Rector, F.C., and Seldin, D.W. 1969. Pathophysiology of epidemic St. Louis encephalitis. I. Inappropriate secretion of antidiuretic hormone. Ann. Intern. Med. 71:691–702.

Papovaviruses

Keerti V. Shah

26.1 INTRODUCTION

Papovaviruses are small, naked, icosahedral viruses that have a double-stranded DNA genome and that multiply in the nucleus. The family *Papovaviridae* consists of two subfamilies, papillomavirus (wart virus) and polyomavirus. Viruses of both subfamilies are widely distributed in nature (Table 26.1). Humans are hosts to at least 60 different human papillomaviruses (HPV types 1 to 60) and to polyomaviruses BK virus and JC virus. Viruses of each subfamily have a common evolutionary origin. All members of a subfamily are immunologically related and share some nucleotide sequences, but there is no evidence of a relationship between the papillomaviruses and polyomaviruses.

Papillomaviruses are larger in size than polyomaviruses (virion diameter of 55 nm compared with 45 nm) and have larger genome (8×10^3 base pairs compared with 5×10^3 base pairs). All of the genetic information in papillomaviruses is located on one strand. In contrast, the genetic information in polyomaviruses is about equally divided between the two strands. Viruses of the two subfamilies also differ biologically. Papillomaviruses infect surface epithelia and produce warts at the site of multiplication on the skin or the mucous membrane. On the other hand, polyomaviruses, after initial multiplication at the site of entry in the respiratory or the gastrointestinal tract, reach internal organs (kidney, lung, brain) following viremia. Although viruses of both subfamilies can transform cells and produce tumors experimentally, only papillomaviruses are associated with naturally occurring tumors.

26.2 HUMAN PAPILLOMAVIRUSES

The infectious nature of human warts has been suspected for many centuries (Rowson and Mahy, 1967). Its viral etiology was established in 1907 by experimental transmission of warts from person to person by inoculation of a cell-free extract of wart tissue. The virus was visualized in the 1950s soon after electron microscopy came into general use and, on the basis of morphologic similarities and nuclear site of multiplication, was grouped with polyomavirus of mice and vacuolation agent (SV40)

This work was supported by National Institutes of Health grants CA-13478 and PO1 AI-16959.

Table 26.1 Natural Hosts of Papovaviruses

Host	Papillomavirus	Polyomavirus
Man	Human papillomaviruses types 1–60	BK virus and JC virus
Monkeys	Rhesus papillomavirus	Simian virus 40 of macaques; Simian agent 12 of baboons; lymphotropic papovavirus of African green monkeys
Cattle	Bovine papillomaviruses types 1–6	Bovine polyomavirus
Rabbit	Cottontail rabbit papillomavirus	Rabbit kidney vacuolating virus
Rodents	*Mastomys natalensis* papillomavirus	Polyoma and K viruses of mice; hamster papova virus
Birds	Chaffinch papillomavirus	Budgerigar fledgling disease virus
Other	Horse, dog, sheep, European elk, deer	—

of rhesus monkey to form the papova (*pap-illoma*, *polyoma*, and *vacuolating agent*) group (Melnick, 1962). Warts, because of their characteristic histopathologic features, have been recognized at many different sites in humans (skin, genital tract, respiratory tract, oral cavity) and in many mammalian species. However, papillomaviruses still cannot be grown in culture. The existence of a large number of distinct human papillomaviruses became evident only after the development of recombinant DNA technology, which permitted the

cloning of viral genomes from different sites and the comparison of these genomes (Orth et al, 1977b; zur Hausen, 1980). Different genotypes of HPV were associated with lesions of specific morphology and at specific anatomic sites (Howley, 1990) (Table 26.2).

26.2.1 Characteristics of the Virus

The virion is nonenveloped and has a diameter of 55 nm, icosahedral symmetry, and 72 capsomers. The viral genome is a dou-

Table 26.2 Human Papillomaviruses[a]

Main Site of Recovery	HPV Type(s)
Skin of patients	1–4, 7[b], 10, 26–29, 38, 41, 46, 48, 49
Skin of patients with epidermodysplasia verruciformis	5, 8, 9, 12, 14, 15, 17, 19–25, 36, 37, 47, 50
Genital tract	6, 11, 16, 18, 31, 33–35, 39, 40, 42–45, 51–56, 58
Oral cavity	13, 32
Larynx	30, 57

[a] HPV-1 through HPV-58. Many of the higher-numbered types have not yet been fully characterized.

[b] HPV-7 is recovered largely from butcher's warts.

ble-stranded, circular DNA molecule with 8 × 10^3 base pairs and a molecular weight of 5.2 × 10^6 daltons. Complete nucleotide sequences are known for many HPV types. All of the open reading frames in papillomavirus DNA are located on only one of the two strands, indicating that only one strand carries the genetic information. Detailed physical maps have been constructed for most of the HPV genomes. Few functions have been localized directly on these maps because simple biological assays for viral multiplication and transformation are not yet available. However, by alignment of HPV with bovine papillomavirus (BPV) (for which transformation assays are available), the genome can be divided into early and late regions. The early region contains eight open reading frames (E1-E8) and the late region two open reading frames (L1-L2).

The genomes of all papillomaviruses share some nucleotide sequences. DNA hybridization performed under conditions of low stringency (which allow for the detection of weakly homologous regions), shows extensive crosshybridization between human papillomaviruses and between human and animal papillomaviruses (Heilman et al, 1980). Surprisingly, HPV-1 and HPV-2 do not show a greater degree of relatedness with one another than they do with viruses of bovine or rabbit origin. In DNA hybridization tests performed under conditions of high stringency, when only strongly homologous regions are detected, HPV do not crosshybridize with animal papillomaviruses. Among the HPV, viruses can be subgrouped together on the basis of significant crosshybridization under conditions of high stringency (Pfister, 1990). An HPV genome that displays less than 50% duplex formation with all other known HPV genotypes in liquid hybridization tests performed under stringent conditions is designated a new genotype. Within each genotype, subtypes are identifiable on the basis of variations in restriction enzyme digest patterns.

The viral capsid proteins consist of a major polypeptide of approximately 57 k and a number of minor polypeptides with molecular weights between 43 and 53 k. Purified virions contain four histones of host origin. The virion surface displays type-specific antigenic determinants; immune sera prepared against the whole virus particle show virtually no crossreactivity between different viral types. Conversely, some of the genus-specific determinants, shared by all viral types, are probably located internally; immune sera prepared against disrupted virions are broadly crossreactive (Jenson et al, 1980).

26.2.2 Pathogenesis and Disease Potential

Human papillomaviruses infect only epithelia of skin and mucous membranes. The virus probably infects cells of the basal layer of the epithelium which undergo proliferation and form the wart. Histologically, a wart is localized epithelial hyperplasia with a defined boundary and an intact basement membrane. All layers of the normal epithelium are represented in the wart. The prickle cell layer is irregularly thickened, the granular layer contains foci of koilocytotic cells, and the cornified layer displays hyperkeratosis. The viral capsid antigen and viral particles are found only in the nuclei of cells of the differentiated, nondividing, superficial layers of the wart. In the infected cell, the multiple copies of the viral genome are present in an unintegrated state.

Warts vary widely in their appearance, morphology, site of occurrence, and pathogenic potential. The clinical spectrum of warts ranges from transient, barely noticed, self-limiting, benign tumors of the skin to infections that lead to malignancies of the

Table 26.3 Viral Genotypes Most Frequently Recovered from Skin Warts

Type of Wart	Virus Genotype
Deep plantar wart	HPV-1
Common wart	HPV-2, HPV-4
Flat wart	HPV-3, HPV-10

skin or of the genital tract. Many factors determine the clinical significance of papillomavirus infection, as described below.

26.2.2.a Location of Lesion

This is best exemplified by laryngeal papilloma. Although the tumors are benign, they may cause life-threatening respiratory obstruction because of their location on the vocal cord.

26.2.2.b Genotype of Virus

There is considerable correlation between genotype of the infecting virus and the morphology and site of the lesion (Tables 26.3 and 26.4). For example, almost all flat warts of the skin yield HPV-3 or HPV-10. Most deep plantar warts are caused by HPV-1 and common warts by HPV-2. Virus types HPV-6 and HPV-11 are recovered from most of the genital warts (condylomas). Oncogenic potential is also correlated with viral genotype. In the genital tract,

Table 26.4 Viral Genotypes Most Frequently Recovered from Patients with Epidermodysplasia Verruciformis

Type of Lesion	Virus Genotype
Flat skin warts	HPV-3, HPV-10
Red-brown (macular) plaques[a]	HPV-5, 8, 9, 12, 14, 15, 17, 19, 20, 21, 22, 23, 24, 25, 36, 47, 50

[a] Lesions with HPV-5, 8, or 14 have a potential for malignant conversion to squamous cell carcinoma.

HPV-16 and HPV-18 are strongly associated with malignancies and HPV-6 and HPV-11 with benign warts. In the rare dermatologic disorder epidermodysplasia verruciformis (EV), lesions caused by HPV-5, HPV-8, and HPV-13 have a greater tendency to convert to malignancy than lesions caused by several other types (Orth et al, 1977a).

26.2.2.c Host Factors

Warts tend to increase in size and numbers in conditions associated with immunologic impairment, especially T-lymphocyte deficiency. This immunologic impairment may be subtle, as in pregnancy, or gross, as in organ transplant recipients, patients receiving anticancer therapy, and AIDS patients.

Papillomavirus infection is acquired in a variety of ways: through skin abrasions (skin warts), by sexual intercourse (genital warts), during passage through an infected birth canal (juvenile-onset laryngeal papilloma), and probably in other ways (e.g., papillomas of the oral cavity by autoinoculation or by oral sex).

26.2.3 Clinical Types of Warts

26.2.3.a Skin Warts

There are many morphologic types of warts and each type may have preferred locations on the skin (Bunney, 1982). Common warts are found on hands, and there are generally multiple lesions. The warts are characteristically dome-shaped, with numerous conical projections (papillomatosis) that give their surface a velvety appearance. Deep plantar warts (on the bottom surface of the foot) generally occur singly, and have a highly thickened corneal layer (hyperkeratosis). Flat warts (with little or no papillomatosis) almost always occur as multiple warts and are found most often on arms and face and around the knees. The thread-like filiform warts occur most often on the face and neck.

Skin warts are transmitted by direct contact with an infected individual or indirectly by contact with contaminated objects. The incubation period is difficult to estimate but may be as short as one week or as long as several months. As a rule, warts in an otherwise healthy individual are few in number and small in size, but a large number of warts may develop in immunodeficient individuals or in apparently normal persons. Most warts regress within two years, probably as a result of cell-mediated immune responses. Treatment or excision of one wart often results in regression of the remaining warts. This may result from a "triggering" of an immune response due to immune-competent cells, which come in contact with antigens that are released as a result of treatment.

Warts are most prevalent in children and young adults. At any one time, as many as 10% of school children may have warts at some site. It is not known if the reduced prevalence in the older population represents acquired immunity, reduced exposure, or both. The incidence of warts in the general population is believed to be increasing. Recreational activity in which bare skin may be exposed to virus contaminated objects (for example, swimming in communally used pools) increases the risk of acquiring warts, especially plantar warts (Bunney, 1982). Types of HPV most frequently recovered from skin warts are HPV-1, -2, -3, -4, and -10 (Table 26.3).

28.2.3.b Epidermodysplasia Verruciformis

Epidermodysplasia verruciformis is a rare, life-long disease in which a patient is unable to resolve the wart virus infection (Jablonska et al, 1972). Most patients exhibit defects of cell-mediated immunity. The disease probably has a genetic basis (Lutzner, 1978). Patients frequently give a history of parental consanguinity and, despite the rarity of the disease, multiple cases occur in some families. It is postulated that EV patients have an inherited immunologic defect as a result of homozygosity for a rare recessive autosomal gene. The nature of the presumed genetic defect is not known.

The onset of the disease occurs in infancy or childhood. The patient develops multiple, disseminated, polymorphic wart-like lesions that tend to become confluent. The warts are of two clinical types: flat warts and red or reddish-brown macular plaques resembling pityriasis versicolor. The warts contain abundant amounts of viral particles, viral antigen, and viral DNA. The flat warts of EV patients yield HPV-3 and -10, the same genotypes that are recovered from flat warts of normal individuals. However, a bewildering variety of viral genotypes are recovered from the macular plaques of EV patients (Orth et al, 1980) (Table 26.4). It is unclear how EV patients acquire these infections because these genotypes are seldom encountered in normal populations.

In about 33% of the cases, multiple foci of malignant transformation arise in the reddish-brown plaques, especially in lesions occurring in areas exposed to sunlight. Histologically, the tumors may be in situ (bowenoid) or invasive squamous cell carcinoma. The tumors grow slowly and are generally nonmetastasizing. The malignant cells contain multiple copies of viral DNA (HPV-5, -8, or -14) but no viral particles or capsid antigen. Human papillomavirus DNA is also recovered from metastatic tumor cells (Ostrow et al, 1982).

The carcinomas occurring in EV patients illustrate how several factors working in concert result in papillomavirus-induced malignancy. Viruses of specific genotypes infecting an immunologically impaired host produce malignant transformation in lesions that are exposed to sunlight.

26.2.3.c Genital Warts (Anogenital Warts, Condyloma, Genital Papilloma)

Papillomavirus infection of the genital tract occurs predominantly in young adults and in sexually promiscuous populations. It is increasing in incidence over the past 20 years and is now among the most common sexually transmitted diseases. In the United States, an estimated 946,000 individuals consulted private physicians for genital warts in 1981, compared with an estimated 169,000 in 1966 (Centers for Disease Control, 1983). The number of comparable consultations for genital herpes in 1981 was 295,000, or about 31% of that for condyloma. In the United Kingdom, the annual incidence of genital warts per 100,000 population rose from about 30 in 1971 to 50 in 1978. In sexually transmitted disease clinics, genital warts account for about 4% of patient visits, compared with 24% of visits for gonorrhea; however, in a population-based study in Rochester, Minnesota, the incidence rate for genital warts was about one half that for gonorrhea (Chuang et al, 1984). In the United States and Canada, 1.3% to 1.6% of routinely collected Papanicolaou smears show cytopathologic evidence of HPV infection (Reid et al, 1980; Meisels et al, 1982).

The incubation period for condylomas is estimated to be between three weeks and eight months, with an average of 2.8 months (Oriel, 1971). About 66% of the sexual partners of condyloma patients develop the disease. Condylomas may be papillary (condyloma acuminatum) or flat (condyloma planum). The most frequent sites for papillary (or exophytic) condylomas are the penis, around the anus, and on the perineum in the male and the vaginal introitus, vulva, the perineum, and around the anus in the female. On the cervix, flat condylomas are far more frequent than papillary condylomas (Meisels et al, 1982). The flat lesion on the cervix was not recognized to be due to papillomavirus infection until the late 1970s. It is now known to be the most common manifestation of genital HPV infection in the female. The lesion is generally seen only by a careful colposcopic examination and is confirmed by cytology and histopathology.

In a large number of infected individuals, condylomas occur at more than one site in the genital tract. Condylomas may increase in number and size during pregnancy and regress after delivery. Immunosuppressed populations—for example, patients with acquired immunodeficiency syndrome (AIDS)—have a high prevalence of condylomas. The closely related HPV-6 and -11 are responsible for a large majority of the condylomas (Gissmann et al, 1983). Viral genotypes that are primarily associated with skin warts are rarely found in genital warts. Many genital warts regress with time but some may persist for long periods. They may cause local irritation and itching, become infected, and cause severe physical and psychological difficulties for the patient if they enlarge in size or increase in numbers. The presence of condylomas during pregnancy is a risk factor for the transmission of HPV from mother to newborn during birth and for the consequent development of respiratory papilloma in the offspring.

26.2.3.d Respiratory Papilloma

This is a chronic, rare, and a recurrent disease in which benign viral papillomas in the respiratory tract may become life-threatening because of their location. The vocal cords in the larynx are the site most often affected, although the disease may occur at other locations (e.g., trachea) without laryngeal involvement. The most common presenting symptom is hoarseness of voice or change of voice. The papillomas may produce respiratory distress and obstruction, especially in children. The disease tends to recur following surgical removal of the papilloma and patients may require frequent operations, sometimes as often as every two to four weeks. Surgery may lead to dissemination of disease to other sites, for example, to the lungs. Malignant conversion of papilloma is rare and is usually associated with a history of previous radiation therapy.

The highest risk of onset of respiratory papilloma is under the age of 5 years. About 33% to 50% of the cases occur by that age, and about 33% of the cases have onset of illness in adult life (Mounts and Shah, 1984). The viral types recovered from both juvenile- and adult-onset disease are HPV-6 and -11, the viruses that are responsible for genital warts (Mounts et al, 1982; Gissmann et al, 1983). The transmission of virus in juvenile-onset cases probably occurs during the process of birth, in the course of fetal passage through an infected birth canal (Hajek, 1956). Mothers of patients with laryngeal papilloma frequently give a history of genital warts during pregnancy. The risk of acquiring laryngeal papilloma for children born to mothers with active genital papillomavirus infection is estimated to be between 1:100 and 1:1000. Cesarean delivery prior to rupture of membrane very likely reduces the risk of virus transmission. It is not known if transmission of virus in adult-onset disease occurs intrapartum, with the virus remaining latent for many years, or if the infection is acquired in other ways, for example, by oral contact with infected genitalia.

26.2.3.e Warts at Other Sites

Several morphologic types of warts occur in the oral cavity. They have been described as common warts, flat warts, condylomas, or respiratory papillomas on the basis of their clinical and histologic features. HPV-6 and HPV-11 are the viruses recovered most frequently from these lesions. A clinically well-defined entity, focal epithelial hyperplasia has been de-

scribed only in the oral mucosa. The condition occurs with a high frequency in American Indians in North and South America but it has also been seen in other races and in many parts of the world. Clinically, there are discrete, multiple, elevated nodules on the oral mucosa (lips, buccal mucosa, tongue), which may persist for many years and have the histologic appearance of warts. These lesions are associated with HPV types 13 and 32 (Pfister, 1990).

Papillomavirus particles have been found in a small proportion of esophageal papillomas but the infecting genotype has not been identified.

26.2.3.f Cervical Dysplasia (Cervical Intraepithelial Neoplasia (CIN) and Cancer of the Cervix

Cervical dysplasias are a progressive spectrum of abnormalities of the cervical epithelium which precede the development of cervical cancer. Numerous studies have been performed to determine if papillomavirus infection of the cervix is benign and self-limiting, or if it has the potential to progress to severe dysplasia and invasive cancer. These studies have been prompted by a number of considerations: 1) Cancer of the cervix has the epidemiologic characteristics of a sexually transmitted disease; 2) A majority of lesions previously diagnosed as mild cervical dysplasia are now recognized as flat cervical condylomas; 3) The cervix is one of the most common sites of papillomavirus infection and many condylomas are located in the transformation zone where cervical cancer originates; and 4) Condylomas and severe dysplasias are often found side by side in histological sections.

Condylomas, dysplasias, and genital cancers have been examined for evidence of productive papillomavirus infection as well as for the presence of papillomavirus-related DNA sequences. The frequency of

Table 26.5 Association of HPV with Cervical Cancers

Cancer	HPV Type(s)
Squamous cell carcinoma	
Strong association	16, 18
Moderate association	31, 33, 35, 45, 51, 52, 56
Weak association	6, 11, 42, 43, 44
Adenocarcinoma	16, 18

productive virus infection decreases as the lesion becomes progressively more severe. Viral capsid antigen has been demonstrated in about 50% to 70% of condylomas, in 10% to 45% of dysplasias and in virtually no lesions of carcinoma-in-situ or invasive cervical cancer (Guillet et al, 1983; Kurman et al, 1983).

Examination of pre-cancerous lesions and invasive cancer for HPV genomic sequences has provided compelling evidence for an etiological association between HPV and cervical cancer. HPV DNAs are found in a large majority of invasive cancers as well as in precursor or pre-cancerous lesions. There is a preferential association of some HPV types with invasive cancer (Table 26.5). While all HPV types found in the genital tract are associated with some form of mild dysplasia and/or with subclinical infections, HPV-16 and HPV-18 predominate in invasive cancers. The viral genome is found in the cancer cells themselves and is present in both the primary and metastatic tumors. The viral genome is extrachromosomal in dysplasias but is integrated in all HPV-18-associated, and many HPV-16-associated, invasive cancers (Cullen et al, 1991). The linearization of the viral genome occurs most frequently by a break in the E1/E2 region, thus leading to the deregulation of the viral genes E6 and E7. The E6/E7 genes, which code for the transforming functions of HPV, are always expressed in HPV-associated cancers. Experimental data indicate that the E6 and E7 proteins of the oncogenic HPV may exert oncogenic effect by complexing with and inactivating tumor suppressor proteins p53 and RB105, respectively (Scheffner et al, 1991).

HPV-16 is the virus most predominantly associated with cervical cancers in worldwide studies, and accounts for about 50% of the cancers. HPV-18 is associated with 10% to 20% of cervical cancers and is not distributed uniformly in different geographic areas. There is suggestive evidence that HPV-18-associated cancers progress more rapidly than HPV-16-associated cancers.

26.2.4 Diagnosis

Clinically, a papillary wart is seldom misdiagnosed as something else, but other dermatologic conditions (e.g., molluscum contagiosum, plantar corns, skin tags) may be mistaken for warts. Histologic examination of the tissue generally establishes the diagnosis of a wart, but does not assist in identification of the genotype of the infecting virus. No serologic tests are available for virus identification. Human papillomaviruses cannot be grown in culture and there is no other source that can be used to regularly obtain viral antigens of known genotypes.

26.2.4.a Tests for Viral Antigen

A broadly crossreactive genus-specific antiserum is available, which is capable of recognizing capsid antigen of all human and animal papillomavirus by immunoperoxidase or immunofluorescence tests (Jenson et al, 1980). Tests can be performed on sections of routinely collected, formalin-fixed, paraffin-processed tissues, as well as on exfoliated cells (Gupta et al, 1983). The viral antigen is present in the nuclei of cells of the superficial layers of the epithelium. For detection of virus, an immunologic test for viral capsid antigen is considerably more sensitive than demonstration of viral particles by electron microscopy (Ferenczy et al, 1981). However, the antigen is not detectable in at least 25% of histologically-confirmed warts. In antigen-positive tissues the number of cells displaying antigen is variable, ranging from only one or two cells to a large number of cells in the section. Only a proportion of cytologically-affected cells exhibit antigen. Warts at different sites differ markedly with respect to their yield of viral particles and patterns of antigen distribution (Braun et al, 1983). Viral particles and antigen are abundant in some plantar and common warts, but are scarce in genital tract and laryngeal papillomas. In the genital tract, the antigen prevalence decreases as the lesion progresses toward malignancy. Genotype-specific antisera for individual viral genotypes are not available. The identification of the viral genotype in a tissue requires DNA hybridization.

26.2.4.b Identification of Viral Genotypes

The DNA hybridization methods employed for viral genotype identification, and their advantages and disadvantages, are listed in Table 26.6. Hybridization by the Southern transfer method utilizing DNA extracted from fresh or fresh-frozen tissue (Gissman et al, 1983) provides the most trusted method of genotype diagnosis and also permits identification of viral subtypes. Hybridization tests, however, can also be performed in situ on paraffin sections of formalin-fixed tissues (Beckmann et al, 1985; Gupta et al, 1985). These tests, although they are less sensitive than the Southern transfer method, permit retrospective diagnosis of routinely collected material and correlation of viral genotype with pathologic characteristics of the tissues. Hybridization of cells placed on a nitrocellulose filter and denatured in situ provides a simple but erratic method suitable for testing of large numbers of speci-

mens collected by a noninvasive technique (Wagner et al, 1984). Almost all of the HPV DNA have been cloned in plasmid vectors. The viral DNA probes for the hybridization tests are prepared using either radio-labeled or biotin-labeled nucleotides. As a result, ^{32}P-labeled probes have been employed in filter hybridizations and ^{35}S-labeled, ^{3}H-labeled and biotin-labeled probes for in situ hybridizations of paraffin sections and cells fixed on slides.

All papillomavirus DNA share some conserved nucleotide sequences. In order to make a specific diagnosis, therefore, hybridizations are performed under conditions of high stringency (i.e., at high effective temperature; for example, at Tm $-20\,°C$). Under these conditions, regions of weak homology are not detected and crosshybridization will occur only between closely related viruses. Conversely, hybridization under conditions of low stringency (e.g., at Tm $-35\,°C$) permits the detection of weakly homologous regions, and is useful for screening of tissue DNA for papillomavirus sequences.

The most sensitive and highly specific method for HPV diagnosis employs polymerase chain reaction (PCR) technology. In the most widely used version of this method (Manos et al, 1989), consensus primers, which are capable of amplifying a large number of HPV types, are used to amplify a portion of the L1 gene of HPV. The PCR products are hybridized with a generic probe to identify any HPV type and with type-specific probes to identify specific types. The technique can be used for both fixed tissues and fresh tissues and is capable of detecting 10^1 to 10^2 copies of the HPV genome, as compared to the other techniques which require 10^5 to 10^6 copies of the genome.

26.2.5 Treatment

Most skin warts and genital warts regress spontaneously. The patient seeks treatment

Table 26.6 DNA Hybridization Techniques for Identification of Papillomavirus Genotypes

Hybridization Method	Procedure	Advantages	Disadvantages
Southern transfer	Extract DNA from fresh or fresh-frozen tissue or cells, fractionate on gel, denature and transfer to filter, hybridize	Sensitive; permits identification of subtype and detailed studies of genome in tissue	Requires unfixed, fresh or carefully stored tissue; lengthy protocol; not suitable for large studies
DNA dot blot hybridization	Extract DNA from fresh or fresh-frozen tissue or cells, denature, spot denatured DNA on filter, hybridize	Sensitive; omits gel fractionation; simpler than Southern transfer	Requires unfixed, fresh, or carefully stored tissue; does not permit subtype identification or detailed study of genome in tissue
Cells denatured in situ on filter	Place cells on filter, denature, hybridize	Suitable for large studies	Requires unfixed, fresh, or carefully stored cells; gives erratic results
In situ hybridization of paraffin-processed tissues (or cell smears)	Deparaffinize sections, denature cells, hybridize	Permits retrospective study of routinely collected material; allows identification of cell types harboring genome	Labor intensive; does not permit subtype identification or detailed study of genome in tissue
Hybridization following polymerase chain reaction (PCR)	Amplify a portion of genome by PCR, test product by hybridization	Highly sensitive and specific	Potential for false-positive diagnosis if there is even minor contamination

for cosmetic reasons, pain, discomfort, and disability depending on the location and size of warts. The most difficult problems for therapy are posed by children with recurrent laryngeal papilloma, patients with EV, pregnant women with genital warts, and warts in immunocompromised individuals. There is no "one-time" treatment for all warts (Bunney, 1982). The therapies in use include application of caustic agents such as podophyllin and salicylic acid, cryotherapy, surgical removal, antimetabolites such as 5-fluorouracil applied in a cream or a solution, immunotherapy with "autogenous vaccines," and treatment with interferon. Both laryngeal papillomas and genital warts are reported to respond to interferon therapy (Haglund et al, 1981; Schonfeld et al, 1984), but recurrence after cessation of therapy is not uncommon.

26.3 HUMAN POLYOMAVIRUSES

The first conclusive evidence of human infection with polyomaviruses was obtained in the mid-1960s when polyomavirus particles were consistently demonstrated by electron microscopy in the enlarged nuclei of oligodendrocytes in the affected areas of brains of patients with progressive multifocal leukoencephalopathy (PML) (Zu Rhein, 1969). In 1971, JC virus (JCV), the causative agent of PML, was isolated from a PML brain in primary human fetal glial cell cultures. In the same year, another polyomavirus, BK virus (BKV), was isolated in Vero cell cultures from the urine of a renal transplant recipient (Padgett and Walker, 1976; Gardner, 1977). Subsequent studies have shown that in the immunocompetent host, both JCV and BKV persist in the kidney following clinically inapparent primary infection in childhood, and that they are reactivated in a variety of conditions that impair cell-mediated immune responses. Almost all of the pathologic effects of BKV and JCV infections occur in immunodeficient individuals.

Between 1955 and 1961, millions of people were inadvertently exposed to simian virus 40 (SV40), an oncogenic polyomavirus of Asian macaques, which had contaminated inactivated (Salk) poliovirus vaccines and experimental live poliovirus vaccines prepared from virus pools grown in primary rhesus kidney cultures. There is no persuasive evidence of any ill effect attributable to SV40 in these vaccines (Shah and Nathanson, 1976).

26.3.1 Characteristics of the Virus

The virion is nonenveloped and has a diameter of 44 nm, icosahedral symmetry, and 72 capsomers. The viral genome is a double-stranded, circular DNA molecule with 5×10^3 base pairs and a molecular weight of 3.2×10^6 daltons. Each of the two DNA strands carries about 50% of the genetic information. Complete nucleotide sequences, as well as detailed physical and physiologic maps, are known for both JCV and BKV genomes. There is extensive nucleotide sequence homology between BKV and JCV throughout their genomes with the highest conservation in the late region,

which codes for the capsid proteins (Howley, 1980).

Both JCV and BKV hemagglutinate human erythrocytes. The capsid consists of three virus-specified proteins (VP1, VP2, and VP3) and three cellular histones (VP4, VP5, VP6). The major capsid protein, VP1, accounts for more than 70% of the virion mass and has a molecular weight of 39 to 44 k. Cells infected with or transformed by the viruses express T antigens, which are coded by the "early" regions of the genomes and are not part of the viral capsid. T antigens are required for the initiation of viral DNA synthesis. BK virus codes for a large T antigen (86 to 97 k) and a small t antigen (17 k).

Despite the extensive nucleotide sequence homology between the two genomes, JCV and BKV can be readily distinguished from one another by immunologic and DNA hybridization tests. Antibodies to the two viruses in human sera display minimal or no crossreactivity in neutralization, hemagglutination-inhibition (HI), or enzyme-linked immunosorbent assays (ELISA). Both JCV and BKV share with other polyomaviruses genus-specific immunologic determinant(s) that are physically located on VP1 but are internal to the virion surface (Shah et al, 1977). Antibodies against the genus-specific determinant(s), prepared by immunization with disrupted capsids, react with all human and animal polyomaviruses.

Both JCV and BKV transform cells in tissue culture and are oncogenic in laboratory animals. JC virus transforms hamster brain cells and human amnion cells, whereas BKV transforms cells of hamster, rat, rabbit, monkey, and mouse origin. Both viruses are oncogenic for newborn hamsters. JC virus also produces cerebral neoplasm in owl and squirrel monkeys, and provides the only model of a primate central nervous system (CNS) tumor caused by a virus (London et al, 1983).

26.3.2 Pathogenesis and Disease Potential

26.3.2.a In Immunocompetent Hosts

Primary infections with JCV and BKV occur in childhood. Infection with JCV is acquired at a later age than BKV infection. In the United States, 50% of children develop antibodies to JCV by the age of 10 to 14 years and to BKV by the age of 3 to 4 years. Infection, in healthy children, is most often subclinical. Serologic studies suggest that primary BKV infection may be associated with mild upper respiratory disease but BKV has not been isolated from respiratory secretions (Goudsmit et al, 1982). An occasional case of cystitis in an otherwise normal child may occur as a result of primary BKV infection (Padgett and Walker, 1983).

The viruses persist in the kidney following primary infection. Viral genomes can be demonstrated in cadaver kidney tissues (McCance, 1983). It is likely that after multiplication at the site of entry, the viruses reach the kidney by a process of viremia, and infection of the kidney is associated with transient viruria. However, JCV and BKV have been recovered very rarely from blood, urine, or from any other site in healthy, immunocompetent children.

26.3.2.b In Immunocompromised Hosts

Most of the infections in immunocompromised hosts are the result of reactivation of viruses latent in the kidney and are evidenced by viruria. Conditions in which viruses are reactivated include pregnancy, diabetes, organ transplantation, antitumor therapy, and immunodeficiency diseases. Unchecked virus multiplication after primary infection of immunodeficient individuals may lead to pathologic consequences.

26.3.2.c Progressive Multifocal Leukoencephalopathy

This is a rare, fatal, subacute demyelinating disease of the central nervous system (CNS) that results from JCV infection of oligodendrocytes in the brain (Johnson, 1982; Walker and Padgett, 1983). It occurs as a complication of a wide variety of conditions associated with T-cell deficiencies. These conditions include lymphoproliferative disorders, such as Hodgkin's disease, chronic lymphocytic leukemia, and lymphosarcoma; chronic diseases, such as sarcoidosis and tuberculosis; primary immunodeficiency diseases; prolonged immunosuppressive therapy as, for example, in renal transplant recipients and patients with rheumatoid arthritis, systemic lupus erythematosus, and myositis; and AIDS. Most cases of PML occur in middle age or later life, but the disease is being increasingly identified in younger patients, e.g., in children with primary immunodeficiency diseases, in renal transplant recipients, and in AIDS patients. Cases of progressive multifocal leukoencephalopathy (PML) in the older patients are most likely the result of reactivation of latent JCV. In the younger patient, it is possible that unchecked primary JCV infection may lead to PML.

Progressive multifocal leukoencephalopathy has unique pathologic features. The affected area of the brain contains foci of demyelination, which have at their edges enlarged oligodendrocytes. The nuclei of the oligodendrocytes are two to three times their normal size, basophilic, and they may contain basophilic or eosinophilic inclusion bodies. The centers of the demyelinating foci contain macrophages and ''reactive'' astrocytes. Most lesions also have bizarre, giant astrocytes with hyperchromatic pleomorphic nuclei. Inflammation is minimal or absent. Neurons are unaffected. The characteristics of these lesions and the occurrence of PML in immunodeficient individuals led Richardson (1961) to propose that the key event in the pathogenesis of PML was infection of oligodendrocytes with a common virus, which atypically infected these cells when immune defenses were impaired, and that demyelination was a result of destruction of these cells, which are normally responsible for the formation and maintenance of myelin sheaths. This hypothesis proved correct. The nuclei of affected oligodendrocytes contain abundant numbers of JCV particles. Progressive multifocal leukoencephalopathy is caused by an atypical course of JCV in an immunocompromised host.

Clinically, PML has an insidious onset, and may occur at any time in the course of the underlying illness. The signs and symptoms point to a multifocal involvement of the brain. Impaired speech and vision and mental deterioration are common early features of the disease. The patient remains afebrile and headache is uncommon. The cerebrospinal fluid (CSF) remains normal. As a rule, the disease is progressive, resulting in death within three to six months after onset. Paralysis of limbs, cortical blindness, and sensory abnormalities occur in later stages. A few patients may survive for years with stabilization of the condition and even apparent remission. A longer survival time is thought to be associated with a more marked inflammatory response in the brain.

The diagnosis of PML can be conclusively established by pathologic examination of a biopsy or at postmortem. Macroscopically, the brain shows foci of demyelination that may vary widely in size and may become confluent and necrotic in the advanced stages of disease. The lesions are most frequent in the subcortical white matter. The cerebrum is almost always affected. Microscopically, the presence of enlarged oligodendrocyte nuclei around the foci of demyelination is diagnostic. These altered nuclei contain abundant amounts of JCV particles, antigen, and DNA. JC virus particles or antigen are not found in normal

brains or in nondiseased areas of PML brains. Small amounts of viral DNA may be recovered from extraneural sites such as kidney, liver, lymph node, and spleen (Grinnell et al, 1983).

In recent years, noninvasive techniques, such as computed tomographic scan or nuclear magnetic resonance imaging of the brain, have provided effective means for the diagnosis of PML.

26.3.2.d Renal and Bone Marrow Transplant Recipients

About 33% to 50% of these organ transplant recipients excrete one or both of these viruses in urine in the posttransplant period (Gardner, 1977; Arthur et al, 1983). The duration of viruria varies from a few days to several months. BK virus infection is more common than JCV infection. Most infections are due to reactivation of latent viruses. The frequency of infection is higher in recipients with a history of diabetes (Hogan et al, 1980). The risk of BKV infection is increased if a kidney from a seropositive donor is transplanted into a seronegative recipient (Andrews et al, 1983). The infections rarely lead to severe pathologic consequences and it is unclear if they result in loss of renal function or rejection of allografts. Some cases of ureteral obstruction, an uncommon and late complication in renal transplantation, have been ascribed to JCV and BKV infections. In addition, BK virus reactivation appears to be responsible for hemorrhagic cystitis in bone marrow transplant recipients (Arthur et al, 1986).

26.3.2.e Primary Immunodeficiency Diseases

BK virus has been isolated from the urine of patients with primary immunodeficiency diseases. A fatal end result of BKV infection has been reported. A 6-year-old boy with hyperimmunoglobulin M deficiency developed massive BKV viruria, tubulo-interstitial nephritis with viral inclusions in the lesions, and irreversible renal failure (Rosen et al, 1983).

26.3.2.f Pregnancy

Both viruses are reactivated in some women during normal pregnancy. In a prospective study, cytopathology in cells obtained from urine sediment suggested JCV and BKV infections in 3.2% of pregnant women. This was most frequently observed in the last trimester of pregnancy (Coleman et al, 1980). In another study, 16% of the women showed an antibody rise to one or the other virus during pregnancy. All the infections were reactivations of latent viruses in antibody-positive individuals (Andrews et al, 1983). It has been reported that fetal sera may have BKV-specific IgM, indicating transplacental transmission of the virus; these observations have not been confirmed (Andrews et al, 1983).

26.3.2.g Role in Human Malignancies

JC virus and BKV are oncogenic for laboratory animals and they transform cultured cells. These viruses as well as SV40, therefore, have been investigated for their roles in human malignancies (Howley, 1983). Tumors of the urinary tract and the CNS have received special attention. In one instance, multifocal gliomas corresponded topographically to lesions of PML demyelination, suggesting that the tumors arose in these lesions. There are sporadic reports of finding viral genome or viral antigen in individual human tumors, especially in meningiomas, but these observations have not been confirmed. A reproducible and consistent association of JCV or BKV with any human malignancy has not been demonstrated.

26.3.3 Diagnosis

Evidence of multifocal brain disease in an immunocompromised individual suggests the possibility of PML. The unique histopathologic features of PML are seen by light microscopy. Except for two cases in 1971 in which SV40 was isolated from PML brains, all other cases have yielded JCV. The virus can be specifically identified by an immunoperoxidase test of frozen sections or paraffin sections of the affected tissue using a monospecific anti-JCV serum. Alternatively, the viral genome in the lesion can be identified by hybridization of the total DNA extracted from the affected tissue (Grinnell et al, 1983), by in situ hybridization of paraffin sections with a JCV probe or by detection of JCV by PCR technology (Arthur et al, 1989). Serologic studies are not helpful in the diagnosis of PML. JC virus antibodies are present at the onset of the disease and they do not show any marked increase as the disease progresses. Viral antibodies are not detected in the CSF.

In conditions other than PML, cytomorphology of the urinary tract epithelial cells suggestive of virus excretion in the urine is often the first indication of virus infection. Virus-infected cells are enlarged and their nuclei contain single, large, basophilic inclusions that may occupy the whole nucleus (Coleman, 1975; Kahan et al, 1980; Traystman et al, 1980). There are no cytoplasmic inclusions. Differential diagnosis includes cells infected with cytomegalovirus (CMV) and sometimes urothelial cancer cells. Cytomegalovirus-infected cells may have both nuclear and cytoplasmic inclusions; the nuclear inclusions are small, surrounded by a clear peripheral zone (halo), and are either basophilic or eosinophilic. The malignant cell nucleus has rough-textured chromatin in contrast to the structureless inclusion in the virus-infected nucleus.

The cytologic abnormalities are not

always present or clear-cut during viruria, and these abnormalities do not distinguish between JCV and BKV infection. JC virus grows best in primary human fetal glial cells and BKV grows best in primary human embryonic kidney or human diploid fibroblast cells. Both viruses also grow in primary urothelial cell cultures collected from infant urine (Beckmann and Shah, 1983). Isolation of virus in tissue culture is inefficient and it may take several weeks before a specific diagnosis is possible. Human fetal glial cells or urinary tract epithelial cells are difficult to obtain. An enzyme-linked immunosorbent assay for detection of viral antigens in urine and DNA dot blot hybridization assays for identification of viral genomes in cells from the urinary tract offer the prospect of rapid diagnosis of JCV and BKV (Arthur et al, 1983). However, the most effective method of diagnosing BKV and JCV employs PCR technology (Arthur et al, 1989).

26.3.4 Treatment

Attempts to treat PML have not been successful although some remissions have been reported with the use of nucleic acid base analogs, adenine arabinoside, and cytosine arabinoside (Walker and Padgett, 1983). It would be useful, when possible, to reduce or discontinue immunosuppressive therapy. No attempts have been made to treat urinary infections.

REFERENCES

Andrews, C., Daniel, R., and Shah, K. 1983. Serologic studies of papovavirus infections in pregnant women and renal transplant recipients. In J.L. Sever and D.L. Madden (eds.), Polyomaviruses and Human Neurological Diseases. New York: Alan R. Liss.

Arthur, R.R., Dagostin, S., and Shah, K.V. 1989. Detection of BKV and JCV in urine and brain tissue by the polymerase chain reaction. J. Clin. Microbiol. 27:1174–1179.

Arthur, R.R., Shah, K.V., Baust, S.J., Santos, G.W., and Saral, R. 1986. Association of BK viruria with hemorrhagic cystitis in recipients of bone marrow transplants. N Engl. J. Med. 315:230–234.

Arthur, R., Shah, K., Yolken, R., and Charache, P. 1983. Detection of human papovaviruses BKV and JCV in urines by ELISA. In J.L. Sever and D.L. Madden (eds.), Polyomaviruses and Human Neurological Diseases. New York: Alan R. Liss.

Beckmann, A. and Shah, K. 1983. Propagation and primary isolation of JCV and BKV in urinary epithelial cell cultures. In J.L. Sever and D.L. Madden (eds.), Polyomaviruses and Human Neurological Diseases, New York: Alan R. Liss.

Beckmann, A.M., Myerson, D., Daling, J.R., Kiviat, N.B., Fenoglio, C.M., and McDougall, J.K. 1985. Detection of human papillomavirus DNA in carcinomas by in situ hybridization with biotinylated probes. J. Med. Virol. 16:265–273.

Braun, L., Farmer, E., and Shah, K. 1983. Immunoperoxidase localization of papillomavirus antigen in cutaneous warts and Bowenoid papulosis. J. Med. Virol. 12:187–193.

Bunney, M. 1982. Viral Warts: Their Biology and Treatment. Oxford: Oxford University Press.

Centers for Disease Control 1983. Condyloma acuminatum—United States, 1966–1981. M.M.W.R. 32:306–308.

Chuang, T.-Y., Perry, H.O., Kurland, L.T., and Ilstrup, D.M. 1984. Condyloma acuminatum in Rochester, Minn., 1950–1978. I. Epidemiology and clinical features. Arch. Dermatol. 120:469–483.

Coleman, D.V. 1975. The cytodiagnosis of human polyomavirus infection. Acta Cytol. 19:93–96.

Coleman, D., Wolfendale, M., Daniel, R.,

Dhanjal, N., Gardner, S., Gibson, P., and Field, A. 1980. A prospective study of human polyomavirus infection in pregnancy. J. Infect. Dis. 142:1–8.

Cullen, A.P., Reid, R., Campion, M., and Lorincz, A.T. 1991. Analysis of the physical state of different human papillomavirus DNAs in intraepithelial and invasive cervical neoplasia. J. Virol. 65:606–612.

Durst, M., Gissmann, L., Ikenberg, H., and zur Hausen, H. 1983. A papilloma-virus DNA from a cervical carcinoma and its prevalence in cancer biopsy samples from different geographic regions. Proc. Natl. Acad. Sci. USA 80:3812–3815.

Ferenczy, A., Braun, L., and Shah, K.V. 1981. Human papillomavirus (HPV) in condylomatous lesions of cervix. A comparative ultrastructural and immunohistochemical study. Am. J. Surg. Pathol. 5:661–670.

Gardner, S. 1977. The new human papovaviruses: Their nature and significance. In A.P. Waterson, (ed.), Recent Advances in Clinical Virology. New York: Livingstone.

Gissmann, L., Wolnik, L., Ikenberg, H., Koldovsky, U., Schnurch, H., and zur Hausen, H. 1983. Human papillomavirus types 6 and 11 DNA sequences in genital and laryngeal papillomas and in some cervical cancers. Proc. Natl. Acad. Sci. USA 80:560–563.

Goudsmit, J., Wertheim-van Dillen, P., van Strein, A., and van der Noordaa, J. 1982. The role of BK virus in acute respiratory tract disease and the presence of BKV DNA in tonsils. J. Med. Virol. 10:91–99.

Grinnell, B., Padgett, B. and Walker, D. 1983. Distribution of nonintegrated DNA from JC papovavirus in organs of patients with progressive multifocal leukoencephalopathy. J. Infect. Dis. 147:669–675.

Guillet, G., Braun, L., Shah, K., and Ferenczy, A. 1983. Papillomavirus in cervical condylomas with and without associated cervical intraepithelial neoplasia. J. Invest. Dermatol. 81:513–516.

Gupta, J.W., Gupta, P.K., Shah, K.V., and Kelly, D.P. 1983. Distribution of human papillomavirus antigen in cervicovaginal smears and cervical tissues. Int. J. Gynecol. Pathol. 2:160–170.

Gupta, J., Gendelman, H.E., Naghashfar, Z.,

Gupta, P., Rosenshein, N., Sawada, E., Woodruff, J.D., and Shah, K. 1985. Specific identification of human papillomavirus type in cervical smears and paraffin sections by in situ hybridization with radioactive probes: A preliminary communication. Int. J. Gynecol. Pathol. 4:211–218.

Haglund, S., Lundquist, P.G., Cantell, K. Strander, H. 1981. Interferon therapy in juvenile laryngeal papillomatosis. Arch. Otolaryngol. 107:327–332.

Hajek, E. 1956. Contribution to the etiology of laryngeal papilloma in children. J. Laryngol. 70:166–168.

Heilman, C.A., Law, M.F., Israel, M.A., and Howley, P.M. 1980. Cloning of human papillomavirus genomic DNAs and analysis of homologous polynucleotide sequences. J. Virol. 36:395–407.

Hogan, T., Borden, E., McBain, J., Padgett, B., and Walker, D. 1980. Human polyomavirus infections with JC virus and BK virus in renal transplant patients. Ann. Intern. Med. 92:373–378.

Howley, P. 1980. Molecular biology of SV40 and the human polyomaviruses BK and JC. In G Klein (ed.), Viral Oncology. New York: Raven Press.

Howley, P. 1983. Papovaviruses: Search for evidence of possible association with human cancer. In L.A. Phillips (ed.), Viruses Associated with Human Cancer. New York: Marcel Dekker.

Howley, P.M. 1990. Papillomavirinae and their replication. In B.N. Fields, and D.M. Knipe (eds.), Virology, 2nd ed. New York: Raven Press, 1625–1650.

Jablonska, S., Dabrowski, J., and Jakubowicz, K. 1972. Epidermodysplasia verruciformis as a model in studies on the role of papovaviruses in oncogenesis. Cancer Res. 32:583–589.

Jenson, A., Rosenthal, J., Olson, C., Pass, F., Lancaster, W., and Shah, K. 1980. Immunological relatedness of papilloma viruses from different species. J. Natl. Cancer Inst. 64:495–500.

Johnson, R. 1982. Progressive multifocal leukoencephalopathy. In Viral Infections of the Nervous System. New York: Raven Press.

Kahan, A., Coleman, D., and Koss, L. 1980. Activation of human polyomavirus infection—detection by cytologic technics. Am. J. Clin. Pathol. 74:326–332.

Kremsdorf, D., Favre, M., Jablonska, S., Obalek, S., Rueda, L.A., Lutzner, M.A., Blanchet-Bardon, C., Van Voost Vader, P.C., and Orth, G. 1984. Molecular cloning and characterization of the genomes of nine newly recognized human papillomavirus types associated with epidermodysplasia verruciformis. J. Virol. 52:1013–1018.

Kurman, R., Jenson, A., and Lancaster, W. 1983. Papillomavirus infection of the cervix. II. Relationship to intraepithelial neoplasia based on the presence of specific viral structural proteins. Am. J. Surg. Pathol. 7:39–52.

London, W.T., Houff, S.A., McKeever, P.E., Wallen, W.C., Sever, J.L., Padgett, B.L., and Walker, D.L. 1983. Viral-induced astrocytomas in squirrel monkeys. In J.L. Sever and D.L. Madden (eds.), Polyomavirus and Human Neurological Diseases. New York: Alan R. Liss.

Lutzner, M.A. 1978. Epidermodysplasia verruciformis: Autosomal recessive disease characterized by viral warts and skin cancer: A model for viral oncogenesis. Bull. Cancer 65:169–182.

Manos, M.M., Ting, Y., Wright, D.K., Lewis, A.J., Broker, T.R., and Wolinski, S.M. 1989. The use of the polymerase chain reaction amplification for the detection of genital human papillomaviruses. Cancer Cells 7:209–214.

McCance, D. 1983. Persistence of animal and human papovaviruses in renal and nervous tissues. In J.L. Sever and D.L. Madden (eds.), Polyomaviruses and Human Neurological Disease. New York: Alan R. Liss.

Meisels, A., Morin, C., and Casas-Cordero, M. 1982. Human papillomavirus infection of the uterine cervix. Int. J. Gynecol. Pathol. 1:75–94.

Melnick, J. 1962. Papova virus group. Science 135:1128–1130.

Mounts, P. and Shah, K. 1984. Respiratory papillomatosis: Etiological relation to genital tract papillomaviruses. Prog. Med. Virol. 29:90–114.

Mounts, P., Shah, K.V., and Kashima, H. 1982. Viral etiology of juvenile- and adult-onset squamous papilloma of the larynx. Proc. Natl. Acad. Sci. USA 79:5425–5429.

Oriel, J. 1971. Natural history of genital warts. Br. J. Vener. Dis. 47:1–13.

Orth, G., Breitburd, F., Favre, M., and Croissant, O. 1977a. Papilloma viruses: Possible role in human cancer. In H.H. Hiatt, J.D. Watson, and J.A. Winsten (eds.), Origins of Human Cancer. Cold Spring Harbor Conferences on Cell Proliferation, Vol. 4. New York: Cold Spring Harbor Laboratory.

Orth, G., Favre, M., and Croissant, O. 1977b. Characterization of a new type of human papillomavirus that causes skin warts. J. Virol. 24:108–120.

Orth, G., Favre, M., Breitburd, F., Croissant, O., Jablonska, S., Obalek, S., Jarzabek-Chorzelska, M., and Rzesa, G. 1980. Epidermodysplasia verruciformis: A model for the role of papilloma viruses in human cancer. In M. Essex, G. Todaro, and H. zur Hausen (eds.), Viruses in Naturally Occurring Cancers. Cold Spring Harbor Conferences on Cell Proliferation, Vol. 7. New York: Cold Spring Harbor Laboratory.

Ostrow, R., Bender, M., Nhmura, M., Seki, T., Kawashima, M., Pass, F., and Faras, A. 1982. Human papillomavirus DNA in cutaneous primary and metastasized squamous cell carcinomas from patients with epidermodysplasia verruciformis. Proc. Natl. Acad. Sci. USA 79:1634–1638.

Padgett, B., and Walker, D. 1976. New human papovaviruses. Prog. Med. Virol. 22:1–35.

Padgett, B., and Walker, D. 1983. BK virus and nonhemorrhagic cystitis in a child. Lancet 1:770.

Pfister, H. 1990. General introduction to papillomaviruses. In H. Pfister (ed.), Papillomaviruses and Human Cancer. Boca Raton: CRC Press, 1–9.

Reid, R., Laverty, C., Coppleson, M., Isarangkul, W., and Hills, E. 1980. Noncondylomatous cervical wart virus infection. Obstet. Gynecol. 55:476–483.

Richardson, E. 1961. Progressive multifocal leukoencephalopathy. N. Engl. J. Med. 265:815–823.

Rosen, S., Harmon, W., Krensky, A., Edelson, P., Padgett, B., Grinnell, B., Rubino, M., and Walker, D. 1983. Tubulo-interstitial nephritis associated with polyomavirus (BK type) infection. N. Engl. J. Med. 308:1192–1196.

Rowson, K.E.K. and Mahy, B.W.J. (1967). Human papova (wart) virus. Bacteriol. Rev. 31:110–131.

Schonfeld, A., Schattner, A., Crespi, M., Hahn, T., Levavi, H., Yarden, O., Shoham, J., Doerner, T., and Revel, M. 1984. Intramuscular human interferon-B injections in treatment of condylomata acuminata. Lancet 1:1038–1042.

Scheffner, M., Munge, K., Byrne, J., and Howley, P. 1991. The state of the p53 and vetinoblastoma genes in human cervical carcinoma cell lines. Proc. Natl. Acad. Sci. USA 88:5523–5527.

Shah, K. and Nathanson, N. 1976. Human exposure to SV40: Review and comment. Am. J. Epidemiol. 103:1–12.

Shah, K., Ozer, H., Ghazey, H., and Kelly, T. Jr. 1977. Common structural antigen of papovaviruses of the simian virus O-polyoma subgroup. J. Virol. 21:179–186.

Traystman, M.D., Gupta, P.K., Shah, K.V., Reissig, M., Cowles, L.T., Hillis, W.D., and Frost, J.K. 1980. Identification of viruses in the urine of renal transplant recipients by cytomorphology. Acta Cytol. 24:501–510.

Wagner, D., Ikenberg, H., Boehm, N., and Gissmann, L. 1984. Identification of human papillomavirus in cervical smears by deoxyribonucleic acid in situ hybridization. Obstet. Gynecol. 65:767–772.

Walker, D. and Padgett, B. 1983. Progressive multifocal leukoencephalopathy. In H. Fraenkel-Conrat and R.R. Wagner (eds.), Comprehensive Virology, Vol. 18. New York: Plenum Press.

Zu Rhein, G. 1969. Association of papova-virions with a human demyelinating disease (progressive multifocal leukoencephalopathy). Prog. Med. Virol. 11:185–247.

zur Hausen, H. 1980. Papilloma viruses. In J. Tooze (ed.), DNA Tumor Viruses. New York: Cold Spring Harbor Laboratory.

Herpes Simplex Viruses

Laure Aurelian

27.1 INTRODUCTION

The herpesviruses are a unique group of viruses with a host range that includes invertebrates and vertebrates of all species examined thus far. The word herpes has been in the medical vocabulary for at least 25 centuries. In Greek it means "to creep" and it was originally used to define "an animal that goes on all fours" (herpeton). In the Hippocratic corpus it was used to describe an assortment of cutaneous lesions including descriptions compatible with herpes simplex or herpes zoster lesions. In the early 19th century six clinical entities were delineated, including facial and genital herpes. However, herpes was not considered to be a communicable disease at that time, probably because of the idiosyncratic appearance of symptoms in conjunction with disparate well-defined febrile illnesses. It took 100 years to demonstrate that material derived from human herpetic lesions passed through a filter with pores small enough to retain bacteria, produces a serially transmissible keratoconjunctivitis in rabbits (Luger and Lauda, 1921) and can cause a variety of diseases in humans. Now almost everyone knows that

herpes virus causes sexually transmitted recurrent disease in man; the word herpes usually refers to herpes simplex virus type 1 (HSV-1) and herpes simplex virus type 2 (HSV-2). However, there are other members included in the human herpesvirus group and these agents will be discussed in the succeeding chapter.

27.2 CHARACTERISTICS OF HERPES SIMPLEX VIRUSES

Herpesviruses exhibit a unique morphology that allows rapid identification by electron microscopy. However, differentiation among the herpesviruses by visual examination is very difficult because the morphology is typical of all group members and the morphological differences are slight. The herpesviruses are enveloped, ether-sensitive, and consist of four morphologic elements. The protein shell (capsid) is approximately 100 nm in diameter. It has an icosahedral symmetry and is composed of 162 capsomers. It encloses the core, an electron opaque structure measuring approximately 77.5 nm in diameter that contains the virus DNA. The outer covering of the virus is a lipid-containing membrane (envelope) that is acquired when the capsids bud through the inner nuclear membrane of an infected cell. Cell proteins are not detectable in the virion envelope. The electron-dense amorphous material located between the capsid and the envelope, called the tegument, is composed of virus proteins whose properties and functions are largely unknown.

Herpes simplex virus (HSV) virions contain approximately 30 to 35 proteins. The precise number has been difficult to determine because of resolution problems, including contamination with cellular proteins that are difficult to separate from the virions and the presence in the virions of both precursor and product forms of the same protein. Type A capsids, found primarily in the nucleus, have an electron-translucent core and are therefore considered to lack an internal core. The empty capsids consist of five proteins, namely VP5, VP19c, VP23, VP24 and a 12kDa protein (Cohen et al, 1980). Recent studies suggest that VP5 is a component of both pentameric and hexameric capsomers (Schrug et al, 1989). It appears to be linked to VP19c by a disulfide bridge (Zweig et al, 1979). However, VP19c has strong binding affinity for DNA, and it has been suggested that it is an internal capsid protein involved in DNA packaging (Braun et al, 1984). Seven HSV glycoproteins have been identified [glycoprotein B (gB), gC, gD, gE, gG, gH, and gI] in infected cells and all except gI are present in the virion envelope (Marsden, 1987). A putative eighth glycoprotein, the product of gene US5 (McGeoch et al, 1985) remains to be identified. The HSV-2 glycoprotein gG is the only known HSV glycoprotein that is cleaved to yield a secreted product. The nature and location of the cleavage event are unknown (Weldon et al, 1990).

HSV-1 and HSV-2 DNAs are linear double-stranded molecules of approximately 100×10^6 daltons and have a base composition of 67 and 69 guanine plus cy-

tosine moles %, respectively (Kieff et al, 1971). The nucleic acids of HSV-1 and HSV-2 share approximately 47% to 50% base sequence homology (Kieff et al, 1972), and the buoyant density of the DNA in cesium chloride gradients is 1.726 g/cm^3 for HSV-1 and 1.728 g/cm^3 for HSV-2 (Kieff et al, 1971). The density of the HSV virion has been calculated by measuring the position of the virus band at equilibrium in density gradients. Densities ranging from 1.253 to 1.285 g/cm^3 have been reported, but differences in virus strains, choice of gradient, and cells in which the virus is grown may account for observed discrepancies (Roizman and Roane, 1961).

HSV DNA has a very intriguing structure. It consists of two covalently-linked components that represent the long (L) region (82%) and the short (S) region (18%) of the HSV genome. Each of these two components consists of unique (U) sequences that are bracketed by inverted repeated sequences (Roizman, 1979; Hayward and Sugden, 1986). The HSV DNA can form four isomeric arrangements (Sheldrick and Berthelot, 1975) due to the ability of the L and S components to invert relative to each other. Therefore, the DNA isolated from virions consists of four populations that are approximately equimolar in concentration and differing in the orientations of the U_L and U_S components. The U_L and U_S segments structurally resemble prokaryotic DNA sequences capable of excision and insertion into the same or different DNAs. The generation of inverted forms of viral DNA and the capacity of the virus to cause nonproductive infection hinge on the ability of the U_L and U_S components to insert and excise into viral and host DNA, respectively. In this latter capacity viral DNA bears some homology to eukaryotic cell DNA (Hayward and Sugden, 1986).

Antigenic and biologic markers capable of differentiating between HSV-1 and HSV-2 are summarized in Table 27.1. Ad-

Table 27.1 Properties of HSV Serotypes

Property	HSV-1	HSV-2
Site of infection	Primarily nongenital	Primarily genital[a]
Transmission	Primarily nonsexual	Primarily sexual
Biochemical properties		
% G + C content of viral DNA	67	69
Homology of viral DNA	Approximately 50%	
Viral proteins by electrophoretic mobility	Some differences	
Antigenic properties	Mostly cross-reactive with intratypic variation	
Biologic properties		
Pock size CAM[b]	Small	Large
Plaque size CE cells[c]	None or small	Yes or large
CPE in other cells	Tight adhesion of rounded cells	Loose aggregation, propensity for syncytia formation
Growth cycle		
Titers in RK[b]	6×10^7	8×10^5
Particle to PFU ratio in RK cells	36	2000
% enveloped virus in RK cells[c]	38	6.8
Microtubules[d]	No	Yes
Sensitivity to		
Temperature (37 °C)		
log loss/hr	0.07	0.27
IUdR or BUdR	Sensitive	Relatively insensitive
Interferon	Relatively resistant	Sensitive
Neurotropism	More	Less
Virulence ($\log_{10}$ PFU/LD$_{50}$)[e]	4.0	0.5

[a] Frequency of genital HSV-1 infection differs in various population groups.
[b] CAM: Chorioallantoic membrane of fertilized eggs; CE: chick embryo cells; RK: rabbit kidney cells.
[c] EM analysis reveals a high proportion of non-enveloped virus particles in HSV-1 infected cells.
[d] Formation of unique microtubule structures in HSV-2 infected cells described by EM studies.
[e] Done in mice. Large variation among HSV-2 isolates.

ditionally, analyses of restriction endonuclease digests of HSV-DNA have revealed that epidemiologically unrelated isolates of the same serotype are not identical and analysis of restriction enzyme digests has become the best criterion for identifying HSV isolates. Differences include occasional deletions and the presence or absence of restriction endonuclease cleavage sites.

27.3 REPLICATIVE CYCLE

To understand the multipotentiality of HSV and its ability to cause a wide spectrum of disease in humans, a basic knowledge of the HSV replication cycle is required. The events in the HSV-1 growth cycle require 8 to 16 hours from infection to the end of the growth phase. The replicative cycle of HSV-2 is somewhat longer; however, this varies with the host cell type used. The initial stages appear to be similar to those of other viruses and include: 1) attachment to and penetration of the cell membrane, 2) release of DNA or of a nucleoprotein complex, and 3) migration to the nucleus where HSV DNA synthesis and transcription of messenger RNA (mRNA) occur (Roizman, 1980).

The two HSV serotypes have a broad host range suggesting that the receptors for HSV are ubiquitous cell surface components. Initial interaction was shown to involve virion binding to the widely distributed cell surface heparan sulfate proteoglycans (WuDunn and Spear, 1989). It is generally believed, however, that this only facilitates interaction with another cellular receptor with which virus must interact before it can penetrate by fusion of the virus envelope with the cell membrane. Furthermore, virion interaction with the cell surface may induce the expression of specific cellular genes important for cell penetration (Preston, 1990). In a recent study, HSV-1 penetration was blocked by inhibitors of basic fibroblast growth factor (FGF) receptor, suggesting that HSV-1 may be bound and internalized by the high affinity FGF receptor (Kaner et al, 1990). There is evidence to suggest that HSV-1 and HSV-2 may bind different cell receptors.

The virion component involved in attachment to the cell surface is still unknown. Glycoproteins are on the virion surface and are therefore likely candidates. gH, gE, gG, gI, and the putative US5 gene product are not essential for infectivity in tissue culture (Weber et al, 1987) and therefore are presumably not involved in attachment. Of the remaining three glycoproteins (gB, gC, and gD) that are essential for virus growth in tissue culture gB (Cai et al, 1988) and gD (Ligas and Johnson, 1988) are dispensible for attachment but may be required for cell penetration. Indeed, mutants that do not produce gB or gD attach to cells but are unable to penetrate into the cytoplasm (Weber et al, 1987). Possibly, when one glycoprotein is missing another one may fulfill its function. Non-glycosylated protein(s) may also be involved in attachment as attachment can occur in the absence of glycosylation (Kuhn et al, 1988).

After attachment, fusion between the viral envelope and the cell membrane is required in order to liberate the nucleocapsid into the cytoplasm of the cell. This is followed by controlled disassembly of the nucleocapsid and release of the viral DNA. Expression of viral genes occurs once the DNA reaches the cell nucleus. The mRNA migrates to the cytoplasm and is translated into various HSV-specific proteins which are then transported back to the nucleus. Expression is highly regulated. Three classes of genes have been defined based on the timing of their expression (Honess and Roizman, 1974; Roizman, 1980; Hayward and Sugden, 1986). The α- or immediate early (IE) genes are expressed before any other viral products. Five IE proteins have been identified and of these, at least four have regulatory functions. The expression of the IE genes is enhanced by a virion protein (VP16 or Vmw65) that is a component of the tegument. The β- or early (E) proteins include enzymes involved in viral DNA replication. Their synthesis requires prior synthesis of an IE protein and peaks at 5 to 7 hours postinfection. The γ- or late proteins are primarily structural components of the virus. Their synthesis requires the presence of functional IE and E proteins and they are made at increasing rates until 15 to 18 hours postinfection. Normal levels of late gene expression depend on viral DNA replication. DNA replication and its encapsidation by structural proteins occur in the nucleus. The capsid then buds through the inner nuclear membrane, which adds the envelope to the virus particle. The particles travel through the endoplasmic reticulum and the Golgi apparatus to the cell surface and exit the cell into extracellular spaces and fluids. Budding, as an alternative mechanism of egress, may take place at the cell membrane.

Infection does not always result in efficient virus replication. Cells that support virus replication (permissive) usually support the production of as many as 1 to 2 × 10^5 virions per cell. It is accompanied by the inhibition of cellular macromolecular synthesis manifested by a series of cytopathic events. These include damage to host chromosomes, margination of the chromatin, development of an early basophilic Feulgen-positive inclusion body that contains viral DNA, and a late eosinophilic intranuclear inclusion body (type A) devoid of virus material and representing a "scar" of virus infection. Most cells eventually become rounded and are destroyed. Viral glycoproteins are found on the surfaces of infected cells as well as in the envelopes of the virions. Cell surface changes include the acquisition of a receptor for the Fc domain of immunoglobulin [a function assigned to HSV glycoproteins gE and gI acting as a complex (Johnson et al, 1988)], a receptor for C3b, a fragment of the third component of complement [a function assigned to gC (Friedman et al, 1986)], and serotype-specific and cross-reactive antigenic determinants that are recognized in cytolytic immune reactions that can abort the release of infectious virus from cells.

27.4 PATHOGENESIS OF HERPES SIMPLEX VIRUS INFECTION

HSV infects neonates, children, and adults, and is able to produce a wide spectrum of diseases (Table 27.2). The primary (or initial) infection is generally in the form of a localized lesion. The severity varies among individuals, ranging from asymptomatic to a severe systemic illness that sometimes is fatal. Some of the factors that can influence severity are the age and gender of the patient, host genetic factors, immune competence, associated illnesses, and the virulence of the infecting strain. The effects of gender (more severe in women than men) may be related to hormonal influences but this has not been studied in any detail. Recurrent skin disease, the hallmark of HSV pathogenesis, is usually localized to the site of the primary lesion or close to it. With the exception of severely immuno-

Table 27.2 Spectrum of Diseases Caused by Herpes Simplex Viruses

Stomatitis
Herpes labialis
Genital lesions
Primary herpetic dermatitis
Eczema herpeticum
Traumatic herpes
Acute herpetic rhinitis
Keratoconjunctivitis
Keratitis
Chorioretinitis
Neonatal herpes
Meningitis
Encephalitis
Pharyngitis
Herpetic hepatitis
Pneumonitis
Urethritis
Arthritis
Disseminated rash
Autonomic system dysfunction
Peptic ulcer
Erythema multiforme

compromised patients, recurrent herpes is relatively benign producing fewer and smaller lesions that heal more quickly. Systemic symptoms are rare, and the duration of viral shedding is shorter. A prodrome often signals a recurrence and is characterized by a tingling sensation that may precede lesions by a few hours to two days at the site at which the lesions will appear. This sensation may be accompanied by radiating radicular pain.

The initial HSV-1 disease is generally gingivostomatitis, a serious infection of the gums, tongue, mouth, lip, facial area and pharynx, seen primarily in children from 1 to 3 years of age, and often accompanied by high fever, malaise, myalgias, swollen gums, irritability, inability to eat and cervical lymphadenopathy. Reactivated HSV is associated with the development of mucosal ulcerations or with lesions at the mucocutaneous junction of the lip presenting as small vesicles that last four to seven days and are often referred to as herpes labialis, facialis, febrilis, cold sores, or fever blisters. The virus may also be excreted in the saliva from asymptomatic individuals. Serologic studies suggest that most HSV-1 infections are asymptomatic but they may be followed by subsequent clinically symptomatic recurrent disease.

HSV-1 and HSV-2 were isolated from the posterior pharynx in 11% of cases of pharyngitis. Concomitant lesions of the tongue, buccal mucosa and gingiva are seen in one third of cases. Acute herpetic rhinitis is a primary infection of the nose recognized by the appearance of tiny vesicles in the nostrils usually associated with fever and enlarged cervical lymph nodes. Other HSV-1-induced diseases of the skin include primary herpes dermatitis (a generalized vesicular eruption), eczema herpeticum (usually a manifestation of a primary infection in which the skin is the portal of entry), and traumatic herpes (resulting from traumatic breaks due to burns or abrasions in the normal skin of a susceptible child).

Herpetic whitlow is an occupational hazard (dentists, hospital personnel, wrestlers) resulting from infection of broken skin (often on fingers) in contact with virus on another individual. HSV-1 infections of the eye can lead to keratoconjunctivitis which, in its most serious form, can cause blindness. Chorioretinitis is a manifestation of disseminated HSV infection that may occur in neonates or in patients with AIDS.

The genitourinary tract is the primary site of HSV-2 infections. In the female, the infection is manifested by vesicles on the mucous membranes of the labia and the vagina. Cervical involvement is common. Severe forms result in ulcers that cover the entire area surrounding the vulva. Symptoms of primary infection include itching, pain, and lymphadenopathy. Systemic symptoms are more common in women. They often accompany the appearance of primary lesions and include fever, headache, photophobia, malaise, and generalized myalgias. As a rule they are not seen in recurrent disease. HSV infections of the genitourinary tract in women are easily mistaken for more common urinary tract infections as the major symptoms—dysuria, urinary retention, urgency and frequency, pain, and discharge—are similar. Therefore, when a patient presents with these symptoms, particularly in the absence of typical HSV lesions in the vulvar, vaginal, or cervical regions, cultures should be taken for both HSV and common gramnegative organisms.

In males, common sites of infection with HSV-2 are the shaft of the penis, the prepuce and the glans penis. Urethritis is probably the main local extension accompanied by a watery discharge often resulting in dysuria. Symptomatic urethritis is rare in recurrent disease, but virus often can be cultured from the urethra. Anal and rectal infections occur in both sexes. They are particularly seen in homosexual men and in women, as a result of anal intercourse.

The central nervous system (CNS) is also a target of HSV infection and when this happens the disease is usually severe. HSV encephalitis is the most commonly reported viral CNS infection accounting for 10% to 20% of all cases. The estimated incidence is about 2.3 cases per million population per year. The age distribution is biphasic with one peak at 5 to 30 years of age and a second peak above 50 years of age. HSV-1 accounts for 95% of cases. In children and young adults encephalitis usually results from a generalized primary infection. Virus enters the CNS presumably through neurotropic spread by way of the olfactory bulb. In these cases virus can be isolated from other organs as well as the CNS. In adults encephalitis is often preceded by recurrent lesions. Thus, there is a level of uncertainty as to the contribution of reinfection with exogenous virus vs reactivation of endogenous virus. CNS infections are rare in adults. However, meningitis due to HSV-2 infection of the CNS has been described. Furthermore, immunocompromised adults can develop a severe generalized disease similar to neonatal herpes; this is occasionally responsible for herpetic hepatitis. Differentiation of HSV encephalitis from other viral encephalitides and focal CNS diseases is difficult. In most cases antibody levels in cerebrospinal fluid (CSF) are increased. However, this is rarely seen before ten days of illness and therefore it is not helpful for the early diagnosis essential for initiation of therapy. HSV was isolated from CSF in 0.5% to 3% of patients with aseptic meningitis. It is generally seen in association with genital HSV-2 infections. Recurrent bouts of aseptic meningitis related to HSV reactivation have been reported, but it has been difficult to repeatedly isolate HSV from the CSF of these patients.

Neonates appear to have the highest frequency of visceral and CNS involvement of all HSV infected patients. Skin lesions are the most commonly recognized features of the disease and at least 70% of untreated neonatal HSV cases will lead to disseminated disease. Almost all infections due to HSV-2 are acquired by passage through an infected birth canal. However, congenitally-acquired infections have been described, mostly in association with primary HSV-2 infection of the mother during pregnancy.

Among recipients of bone marrow transplants HSV pneumonitis appears to account for 6% to 8% of cases of interstitial pneumonia. Mortality due to HSV pneumonia in immunosuppressed patients is above 80%. Therefore early diagnosis is essential for initiation of antiviral chemotherapy. HSV has been isolated from 40% of patients with acute respiratory distress syndrome. Other uncommonly reported complications of HSV include monoarticular arthritis, adrenal necrosis, idiopathic thrombocytopenia, and glomerulonephritis. In the immunocompromised patient, generalized HSV with involvement of adrenal glands, pancreas, small and large intestine, and bone marrow have been reported. More recently HSV was associated with erythema multiforme (Brice et al, 1989) and peptic ulcer (Lohr et al, 1990) based on the detection of HSV antigen and DNA at the sites of the lesions using immunoperoxidase (IP) staining and polymerase chain reaction (PCR) assays, respectively.

27.5 EPIDEMIOLOGY OF HERPES SIMPLEX VIRUS INFECTIONS

HSV infections are found worldwide (Guinan et al, 1985). Early seroepidemiologic studies, using assays that cannot differentiate between antibodies to the two serotypes, demonstrated that in almost all the populations studied, more than 90% had antibodies to HSV by the fourth decade of life even though only 10% to 15% of all primary infections produce clinical illness. Since much of the humoral immune response to HSV is to type-common antigenic determinants it is difficult to detect

HSV-2 antibodies in patients with prior HSV-1 infections and HSV-1 antibodies in those with prior HSV-2 infections. Several serologic assays, such as neutralization, indirect immunofluorescence (IF), passive hemagglutination, radioimmunoassay (RIA), and enzyme-linked immunosorbent assay (ELISA) have been developed that can measure, with relative selectivity, serum antibodies to the two HSV serotypes. Recent studies using such assays suggest that first exposure to HSV-1 may occur in many western industrialized middle class populations at an older age than the first 18 months after birth as previously thought. Infection with HSV-2 occurs at puberty and correlates with sexual activity.

Based on the number of patient visits to private practitioners and sexually transmitted disease (STD) clinics it has been estimated that in the United States there are at least 2 to 3×10^5 new cases of genital HSV infection per year. Considering 2×10^5 new infections and a conservative 2:1 overall estimate of recurrent to new infections the total that is generated, including new and recurrent infections is 6×10^5 per year. This amounts to a cumulative build up of 400,000 cases per year which after ten years will result in a prevalence of approximately 4 million cases per year.[1] However, the true annual incidence of genital herpes is probably much higher. Overall occurrence estimates have ranged from 2 to 20 million episodes each year.

HSV infections occur throughout the year. The incubation period ranges from 1 to 26 days (median, 6 to 8 days). Contact with active ulcerative lesions results in transmission. Additionally, salivary excretion of HSV-1 has been reported in 2% to 9% of adults and 5% to 8% of asymptomatic children, and HSV-2 has been isolated from the genital tract of 0.3% to 5.4% of males and 1.6% to 8% of asymptomatic females

---

[1] Source: National Disease and Therapeutic Index (NDTI).

attending sexually transmitted disease clinics. Asymptomatic virus shedding, originally considered an improbable source of infection, is becoming increasingly recognized as a significant reservoir for virus transmission although its true incidence is difficult to assess due to the short and intermittent nature of the episodes. Virus titers in skin lesions are 100 to 1000 times higher than those found in salivary or genital secretions from asymptomatic subjects. Therefore, the efficiency of transmission is likely to be significantly higher during symptomatic periods of viral excretion than during asymptomatic periods. As already stated, HSV-2 can be transmitted from mother to fetus primarily during passage through an infected birth canal, and this can result in a severe and often fatal generalized herpetic infection of the neonate. This infection can be prevented by delivery by Cesarean section. A higher frequency of orogenital sexual practices may be contributing to the recently increased incidence of HSV-1 infection in the genital tract.

27.6 HERPES SIMPLEX VIRUS-INDUCED LATENCY

Latency is defined as virus persistence in the infected host in a repressed state that is compatible with cell survival. It is often followed by subsequent episodes of virus reactivation and clinical symptoms. Latency poses a number of fascinating questions: 1) what tissues harbor the latent infection, 2) how is latency established and maintained, 3) what mechanisms underlie virus reactivation, and 4) what mechanisms regulate the development of recurrent disease symptoms?

Infectious virus can only rarely be recovered from sensory or autonomic nervous system ganglia. However, maintenance and growth of the neural cells in tissue culture results in the production of infectious virions and subsequent permissive infection of susceptible cells—a process called cocultivation. Using this process the trigeminal and sacral dorsal root (S2–S4) ganglia were respectively identified as the most acknowledged sites for latent HSV-1 and HSV-2 infections. Studies of the temporal appearance of infectious virus in different tissues suggest that virus replicates at the peripheral inoculation site in cells of the epidermis and dermis. Whether or not clinically apparent lesions develop, sufficient viral replication may occur to permit infection of either sensory or autonomic nerve endings. Virus, or more likely nucleocapsids, are then thought to be transported intraxonally to the nerve-cell bodies in ganglia. However, it is unclear whether latency always results from peripheral mucosal infection or whether viral replication occurs in ganglia prior to establishment of latency. In humans, the time from the inoculation of virus in peripheral tissue to the spread of virus to the ganglia is unknown. In mice and guinea pigs, HSV can be recovered from ganglia by one to two days after peripheral inoculation. However, after resolution of the primary disease, infectious virus can no longer be recovered in the ganglia.

Viral DNA found in the ipsilateral trigeminal ganglion and brain stem of mice inoculated with HSV-1 on the cornea is in a different physical state from that observed during acute infection or that present in virions, in that it lacks detectable termini. This can be interpreted to indicate that the DNA is integrated into multiple sites in the host cell genome or it is present in an extrachromosomal form. However, as viral and cellular DNA present in murine brain stems can be separated by centrifugation on buoyant density gradients, it seems that the viral DNA persists in an extrachromosomal state, the most likely physical form of which would be a "circular episome." Viral genomes with similar properties are also present in latently infected human trigeminal ganglia in which it was estimated that

there are 0.01 to 0.4 HSV genome equivalents per latently infected cell. Because neurons constitute only 10% of the cells in neural tissues the actual number is probably larger; it was calculated to be 20 copies per latently infected neuron (For review see Stevens, 1989; Aurelian, 1989a).

The mechanisms by which latency is maintained and by which various stimuli cause reactivation of HSV infection are unknown and it is unclear why certain individuals experience recurrences and others do not. Information is beginning to emerge regarding the viral gene products that govern these aspects. For example, thymidine kinase-negative HSV mutants have a diminished ability to establish latency in experimentally infected mice, or as suggested by recent studies, a diminished ability to be reactivated from latently infected cells. Viral ribonucleotide reductase is another function that appears to be required for virus growth in non-replicating cells, such as neurons. Its expression may be required at least for virus reactivation. Expression of the IE protein IE110 also may be involved in the establishment and the reactivation of latent virus (Leib et al, 1989).

One region of the HSV-1 genome, termed latency associated transcript (LAT), is transcribed in latently infected sensory neurons. The transcripts map to the long terminal repeat region of the viral genome and are largely confined to the nucleus. The major transcript (50% to 90% of the total, depending on the virus strain or the experimental animal studied) is approximately 2.2 kilobases (kb) in length. Another transcript (10% to 50% of the total) is a spliced derivative of the first and is about 1.5kb long. Both are transcribed from the DNA strand opposite that encoding the transcript for IE110. They initiate 3' to the IE110 gene and extend approximately one third of the distance into the transcribed region of this gene. Despite intensive searches, virus-encoded proteins have not been reproducibly demonstrated in neurons latently infected with HSV. Theoretically, LAT could function in the establishment, maintenance, or reactivation of the latent state. In the rabbit eye model, LAT was shown to facilitate virus reactivation (Hill et al, 1990). However, using deletion mutants that do not synthesize LAT it was shown that LAT expression is not absolutely required for any of the phases in the latency cycle. Whether some HSV strains are more apt to cause more severe mucocutaneous disease, or disease that is more frequently reactivated, is unknown (Stevens, 1989).

People who do reactivate HSV, do so after nonspecific stimuli that include fever due to bacterial or viral infections, exposure to ultraviolet irradiation (i.e., sunlight), stress, and possibly hormonal irregularities. The stimuli seem to have very little in common. Virus reactivation results in the transport of viral genomes in some unknown form, presumably by reverse axonal transport, to the body surface where replication can occur in the dermal and epidermal surfaces. The development of clinically apparent recurrent lesions is probably influenced by the amount of virus, the nature of the virus, virus-cell interactions, and the rapidity of the host's immune response in clearing the virus. We proposed that transient suppression of virus-induced cell-mediated immunity allows unimpeded growth of the reactivated ganglionic virus in the skin, thereby reaching the levels required for lesion development (For review see Aurelian, 1989b). Supporting this interpretation HSV specific T-cell responses are inhibited or at least significantly decreased during prodrome and recrudescence. Decrease is mediated by suppressor T lymphocytes (Ts) and soluble suppressor factor(s). Lymphokine activities that are decreased at these times include interferon and interleukin-2, both of which contribute to the containment of HSV infections by enhancing the cytotoxic activity of natural killer cells.

Prostaglandin E_2 (PGE$_2$) may provide a stimulus required for the transient immunosuppression associated with recurrent HSV disease. Increased levels of PGE$_2$ in the skin were observed in experimental animals for at least four days after skin treatment with a stimulus that causes recurrent disease. Treatment of the skin with a stimulus that reactivates ganglionic virus but fails to produce recurrent disease did not increase the levels of PGE$_2$ in the skin and HSV specific responses of T cells obtained at recrudescence were restored by treatment with the prostaglandin inhibitor indomethacin. Possibly in those individuals who ultimately experience recurrent disease, primary infection is characterized by priming for Ts development upon re-exposure to reactivated ganglionic virus. This priming may involve virus interaction with skin antigen presenting (Langerhans) cells (for review see Aurelian, 1989b).

27.7 DIAGNOSIS OF HERPES SIMPLEX VIRUS INFECTIONS

Factors to be considered in choosing an appropriate diagnostic test for HSV infection include sensitivity and specificity as well as speed and availability. For instance, rapid diagnosis is necessary when a woman with a history of genital herpes is entering labor, since virus may be present in the vaginal canal and/or the cervix, even in the absence of symptoms. Speed is also essential when certain complications, such as systemic dissemination of the disease, are suspected and therapy must be initiated early in the course of the disease. There are two possible approaches to the detection of HSV infection: 1) demonstration of the presence of infectious virus or viral antigen and 2) detection of HSV antibodies (Table 27.3).

Table 27.3 Methods for the Diagnosis of HSV Infections

Type of infection	Method	Advantages	Disadvantages	Cost	General Availability
Skin, Mucocutaneous	Isolation	High sensitivity/specificity	Slow; complex	High	Variable
	Cytology (Tzanck)	Rapid, simple	Poor-fair sensitivity/specificity	Low	Yes
	Antigen detection *IF, IP*	Good sensitivity, rapid, simple	Poor-fair specificity	Low	Yes
	RIA	Good sensitivity	Inconvenient	High	Variable
	ELISA	Good sensitivity, specificity, rapid, simple	Variable results	Low	Yes
Keratitis	Isolation	High sensitivity/specificity	Slow, complex	High	Variable
	Antigen detection *IF, IP*	Good sensitivity, rapid	Poor-fair specificity	Low	Yes
Encephalitis (Biopsy or CSF)	Isolation	High sensitivity/specificity	Slow, complex	High	Variable
	Antigen detection *IF, IP*	Good specificity	Not completely reliable	Low	Yes
Previous infection	Serology *neutralization*	High sensitivity/specificity	Slow, complex (variable results with different procedures)	High	Variable
	ELISA	Rapid, good specificity/sensitivity	Variable results	Low	Variable
	Spot blot	Rapid, good specificity	Fair sensitivity; medium speed	Low	No

Abbreviations: *IF* = immunofluorescent antibody staining; *IP* = immunoperoxidase antibody staining; *ELISA* = enzyme-linked immunosorbent assay; *RIA* = radioimmunoprecipitation.

27.7.1 Virus Isolation

Demonstration of the presence of infectious virus by virus culture is the most definitive method for the diagnosis of HSV infection. HSV can be isolated from oral and genital lesions, ocular samples, brain biopsies and less commonly from CSF. Skin vesicles are punctured and the vesicular fluid is taken on a swab; also the base of the lesion is swabbed and both may be used to inoculate cell cultures. For uterine cervical samples the exocervix should be swabbed. Virus can also be cultured from throat swabs and washings, bronchial/alveolar washings, and various tissues. For best results specimens should be inoculated within one hour after collection. Otherwise, they should be put in transport medium and refrigerated or frozen (at $-70\,°C$) in order to preserve the specimen. The transport medium routinely used in our laboratory is Eagle's minimal essential medium (MEM) with Earle's salts and L-glutamine supplemented with 0.8% bovine serum albumin Fraction V, 10mM HEPES buffer and antibiotics (100 μg/mL penicillin/streptomycin, 5 μg/mL amphotericin).

When inoculated into appropriate cell cultures, HSV causes a characteristic CPE consisting of rounded refractile cells. It is sometimes accompanied by the development of multinucleated giant cells (depending on the virus strain and the cell type; also see Table 27.1 for serotype properties). CPE can be seen within 18 hours, but it generally takes two to three days (depending on the sensitivity of the cells and the amount of virus inoculated) and cultures are routinely kept for 14 days. If hematoxylin-eosin stain is applied, the cells will exhibit eosinophilic intranuclear inclusions. The major drawback of the culture procedure is the length of time required to observe the CPE. Since this is directly affected by the amount of virus in the inoculum, best results are obtained when the lesion is still in the vesicular or early ulcerative stage (days one to three post onset of recurrent lesions) at which time mean virus titers are $\log_{10}$ plaque forming units [PFU] $\pm$ SEM = 2.5 $\pm$ 0.5. By day four post lesion onset virus titers are often much lower ($\log_{10}$PFU $\pm$ SEM = 1.3 $\pm$ 1) and therefore the probability of successful virus isolation is greatly reduced.

A wide variety of cell lines are available for virus culture. Primary cultures are generally more sensitive than established lines. The most sensitive are rabbit kidney cells and human embryo cells. Those most commonly used at present are human lung fibroblasts (MRC-5), human lung carcinoma cells (A549), human epidermoid carcinoma (HEp-2), human embryonic kidney and African green monkey kidney (Vero).

Results may be confounded by the presence in the specimens of other viruses that cause a similar CPE. For instance, adenovirus is a common isolate from ocular specimens. To address this question at least two cell types that differ in their ability to support the growth of these viruses should be routinely used for virus culture. In our laboratory specimens are routinely cultured on MRC-5, A549, HEp-2, and Rhesus monkey cells. HSV grows in MRC-5 cells; adenovirus does not. Additionally, while adenovirus grows in A549 cells, its CPE is first seen on days four to five as compared to days one to three for HSV. Ultimately, however, virus identification can only be done using specific antisera in an appropriate immunologic test such as immunofluorescence. This approach can also be used for serotyping of the isolates, a task made much easier by the recent development of serotype-specific monoclonal antibodies.

The restriction endonuclease mapping technique, which is not yet commercially available, shows great promise both for typing of HSV and for identification of strain variations within each type. Restriction endonucleases, which are commercially available, cleave DNA at specific sites. The number and size of the cleavage products or DNA fragments generated differ between the two HSV serotypes. These fragments are separated by slab gel electrophoresis and visualized by chemical staining or by autoradiography if the viral DNA was labeled with ^{32}P while grown in culture. Restriction endonuclease analysis of DNA can be used to differentiate between HSV-1 and HSV-2 and to differentiate between various isolates within each serotype. This provides valuable epidemiologic information regarding transmission such as in the identification of common source outbreaks of HSV.

HSV readily infects experimental animals such as hamsters, mice, rats, guinea pigs, rabbits, and embryonated chicken eggs. Newborn mice are particularly sensitive to HSV. When inoculated with the virus intracerebrally or intraperitoneally, the animals will develop encephalitis within a short time. Upon examination of sections stained with hematoxylin and eosin, mouse brain tissue will exhibit Cowdry type A inclusion bodies. The HSV serotypes can be recognized successfully using the chorioallantoic membranes of embryonated eggs on which HSV-1 produces well-defined pinpoint pocks and HSV-2 produces easily recognizable, large, clear pocks within three to four days post inoculation. Differential sensitivity to iododeoxyuridine or bromodeoxyuridine has also been used to define the HSV serotype; HSV-1 is sensitive, HSV-2 relatively insensitive (Table 27.1).

27.7.2 Antigen Detection

Direct antigen detection is an alternative approach to virus culture. The basic principle is that antigens in the lesion will interact with added anti-HSV antibodies, producing a complex that may be detected by means of a variety of methods such as IF, IP, ELISA, or RIA. Sensitivity and specificity can be increased by antigen amplification through short term (16 to 24 hour) growth in tissue culture prior to IF detection, a procedure known as the shell vials assay. Antigen detection assays are rapid and relatively inexpensive, but there still is some uncertainty as to their sensitivity and specificity. Recently available monoclonal antibodies directed against specific HSV antigens have improved the sensitivity of these assays and provided the necessary tools to differentiate between the two HSV serotypes.

For direct antigen detection, lesion scrapings are placed on a glass slide and fixed with acetone. If monoclonal antibodies are used, fixation may have to be in ethanol or methanol since some epitopes are acetone-sensitive. After washing with phosphate-buffered saline (PBS), the cells are stained with anti-HSV serum that has been conjugated to a detector molecule such as fluorescein isothiocyanate (30 minutes at 37 °C in a humid atmosphere). This is the direct IF staining assay. If HSV is present in the lesion, the fluorescein-labeled antibodies will bind, and examination under a fluorescent microscope will show brilliant patchy green areas within cells. The success of direct IF depends on having a sufficient number of infected cells in the sample. The sensitivity of the method is about 65% to 70% of that of virus culture and it depends on the titers of virus in the lesions. In direct comparison, we found that the sensitivity of the IF assay is 70% and 53.8% of that of virus culture when specimens are collected from primary and recurrent lesions respectively. This correlates with mean titers of total recoverable virus ($\log_{10}$PFU $\pm$ SEM) of 4.0 $\pm$ 1.0 and 2.5 $\pm$ 0.5 respectively in primary and recurrent lesions. Specificity may be a problem particularly in the presence of vesicle fluid, mucus, or other body secretions that may nonspecifically bind protein.

The indirect IF assay uses unconjugated anti-HSV serum followed by 30 minutes at 37 °C with the appropriate fluorescein-conjugated anti-immunoglobulin.

This technique can also utilize the high affinity of biotin for avidin, an egg white protein. Anti-HSV antibody conjugated to biotin is added to the culture, followed by the addition of avidin-fluorescein conjugate. The indirect (sandwich) technique is more sensitive than the direct IF assay but it may be somewhat less specific. Samples prepared as in the IF test can also be stained by the IP procedure. Anti-HSV serum is incubated with the sample, followed by the addition of anti-immunoglobulin conjugated to the enzyme horseradish peroxidase. The sample is then treated with diamino-benzidine, a substrate for the enzyme. If HSV is in the sample, it will bind anti-HSV antibodies, which in turn will bind the enzyme-conjugated "anti-antibodies." Thus, an enzyme reaction occurs that is detected by observation of reddish-brown areas using a light microscope. The IP and IF assays are equally sensitive and specific in direct comparison. The advantage of the IP assay is that it requires an ordinary light microscope instead of a fluorescent microscope and the slides can be kept indefinitely.

Routine specimen collection (Figure 27.1) should include vesicle fluid swabs and scrapings for both culture and antigen detection assays. Fluid is obtained from the vesicles by puncture with a sterile needle. The lesions are then swabbed and placed in virus transfer medium for inoculation of cell cultures. Next, the base of the vesicle is scraped and these scrapings are placed into a drop of PBS on slides. After drying, the slides with the tissue scrapings are fixed and stained. This is the in situ assay in which the tissue is directly assayed for viral antigen. If IF is used the indirect assay is preferable. Commercially prepared and standardized anti-HSV serum can be used. The presence of HSV antigen in the scrapings can be determined with known negative cell cultures and HSV-infected cell cultures as controls. The medium containing the lesion swab and vesicular fluid is

used to inoculate appropriate cell cultures: 1) on coverslips in shell vials for antigen amplification prior to staining, and/or 2) in tubes for virus culture. Low speed centrifugation ($700 \times g$; 40 minutes; room temperature) of inoculated shell vials increases virus adsorption onto the cell monolayers thereby enhancing the sensitivity of the assay (Gleaves et al, 1985). The duplicate cultures serve to confirm the results. Some direct comparison studies reported identical detection rates by virus culture and shell vials assays (Gleaves et al, 1985; Salmon et al, 1986; Espy and Smith, 1988). Others (Seal et al, 1991) found that the sensitivity of the shell vial assay was only 66.2%, a finding consistent with our observations. However, the advantage of the shell vial assay is its relative rapidity (16 to 24 hours as compared to 4 to 14 days for culture).

When a diagnosis is needed for HSV infections of the cornea, the foregoing approach can be used after modification. Specifically, the procedure is as follows. The cornea is anesthetized; specimens are obtained by scraping the corneal epithelium with a sterile scalpel blade and then transferring the specimens to sterile PBS on glass slides. The blade is washed in transport medium and a swab of the affected area is placed in the virus transport medium. Standard procedures for the indirect IF test are followed to obtain a preliminary diagnosis while awaiting confirmation from virus culture. Antigen detection (by IF) has been found to be the most effective method currently in use to diagnose eye disease due to HSV infection.

ELISA is a rapid method for the diagnosis of HSV infections that is versatile and adaptable to automation. Its sensitivity is comparable to that of the other immunoassays. Its major drawback is the wide variation in results obtained with different commercially available kits. These include systems in which anti-viral antibody is coated onto polystyrene microtiter wells (to

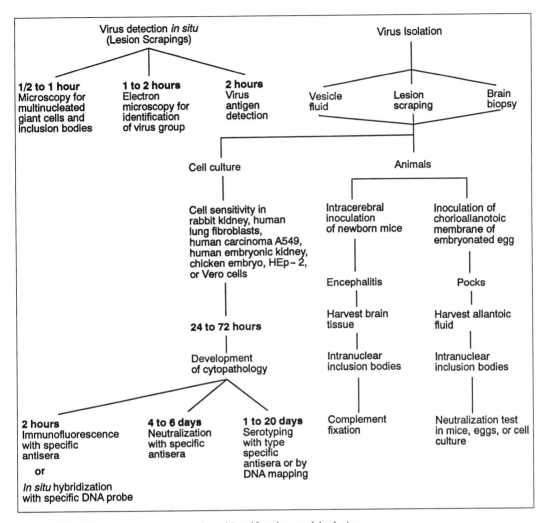

Figure 27.1 Flow chart for herpesvirus identification and isolation.

bind the HSV antigen in the test specimen) and other systems in which the test specimen is directly attached to the microtiter well. In both systems the antigen containing wells are treated with anti-HSV serum (produced in a different species than the coated antibody, if this system is used). Finally, antiglobulin (against the second anti-HSV antibody) that is conjugated to an enzyme, such as alkaline phosphatase, is added. Addition of substrate results in the formation of a soluble colored reaction product that can be read quantitatively in a spectrophotometer.

Since the available ELISA procedures differ in sensitivity and specificity, the assays should be standardized in direct comparisons prior to being adopted. Variables that affect the ELISA for antigen detection include the use of different buffers and binding of the antigen to various plastics. In direct comparison studies (Aurelian, 1982), optimal results were obtained using polyvinyl plates exposed to the virus in PBS (or distilled H_2O) for two hours and washed with PBS-tween. Irreproducible results and/or higher background levels were observed with polystyrene plates, when carbonate buffer, 1% glutaraldehyde, or Ca^{++}/Mg^{++} free PBS were used and when reactions were incubated overnight. Additionally, restandardization must be performed for each new lot of microtiter plates as they vary in their ability to bind the antigen. Finally, reactivity was affected by the quality of the anti-HSV serum. Antiserum prepared in guinea pigs with virus grown in syngeneic cells gave rise to a reproducible, highly specific reaction. Direct comparison to virus isolation showed that when performed under optimal conditions, the direct ELISA had a false-negative level of 6.2% and a false-positive level of 8.3%. Tissue culture amplification increased the sensitivity of the assay (Aurelian, 1982). A similar commercially available ELISA (trademark Herpcheck) had an essentially similar sensitivity/specificity level by comparison to virus culture (6.1% false positives; 12.5% false negatives).

The methodology employed in immunoassays can also use a radioisotope (viz. ^{125}I) as the detector molecule (RIA). The binding of the isotope is quantitated in a gamma counter after completion of the reaction. This method has several disadvantages: radioactive reagents are relatively unstable, there are working hazards, precautions are required, and expensive equipment is needed. However, radioimmunoassay is highly sensitive and does not have the disadvantage of "background" enzymes in sera that can lower the specificity of ELISA.

Spot blotting is the most recent addition to the commercially-available antigen detection assays. The patient sample is collected as described for all other antigen assays. Following appropriate cell lysis the sample is placed on a membrane that captures the HSV antigen, if present. Following thorough washing to remove excess material the membrane is treated with peroxidase conjugated HSV antibody and a dye solution to generate a colorimetric re-

action that is read by comparison to membranes containing positive and negative controls. The assay is rapid and simple. However, its specificity and sensitivity depend on the specific monoclonal antibodies that are sold with the kit and the experimental conditions that have been predetermined by the vendor. The sensitivity of these assays is generally around 70% and it can be increased by culture amplification. Culture confirmation is advisable.

27.7.3 DNA Technology

In situ hybridization with radioactively-labeled or biotinylated DNA probes is a new detection method for samples inoculated onto shell vials. It is claimed to be as sensitive and specific as the antigen detection or virus culture procedures (Espy and Smith, 1988). However, results will differ with different probes and therefore culture confirmation is advisable. A novel approach that has achieved recent prominence and is based on recent advances in molecular biology is the polymerase chain reaction (PCR) used to amplify genomic DNA for cloning and sequencing. It is a powerful and relatively simple analytical tool of high specificity that has been rapidly replacing standard cloning protocols for the identification of homologous genomic DNA sequences. It has been used to identify the presence of HSV DNA sequences in tissue from erythema multiforme (Brice et al, 1989) and is a potentially powerful method for HSV diagnosis in infected tissues where a relatively small proportion of cells may be infected or when virus levels are low, such as in the CSF.

27.7.4 Cytology

Cytologic examination of tissue scrapings is somewhat useful for screening, but the insensitivity of this method, as well as its inability to distinguish HSV types, argues against its use at a time when other, equally rapid, tests are available. In direct comparison, virus isolation detected HSV in 84% of the cases, compared with 57% by indirect IF or IP and only 40% by cytologic examination.

27.8 DIAGNOSIS OF ENCEPHALITIS DUE TO HERPES SIMPLEX VIRUS

Mortality rates due to HSV encephalitis can be as high as 70% in untreated cases. Survivors are often scarred for life with serious physical and neurologic abnormalities. As a result, there is an increasing need for fast, accurate diagnosis of the agent responsible for the encephalitis. Because chemotherapy is now available to treat encephalitis due to HSV, the need for an excellent diagnostic method is great. Virus culture takes too long and may delay prompt, appropriate treatment. Therefore, it cannot be the only diagnostic tool employed.

Burr-hole exploration of the temporal lobe can be used to obtain brain biopsy specimens in order to differentiate between brain tumor, brain abscess, and viral encephalitis. Using IF tests on specimens removed at brain biopsy, a diagnosis can be made in three hours compared with 24 to 48 hours required for virus isolation. The indirect IF test on impression preparations from brain biopsies provides the most rapid and sensitive method of diagnosis. To do this, the surface of the biopsy is pressed several times onto the surface of one or more microscopic slides. When dry, the specimens are fixed in cold acetone for one minute. Longer fixation or use of different fixatives will yield suboptimal results. The specimens are then stained using the direct or indirect IF test, although some reports indicate that the indirect IF test allows more flexibility.

HSV antigens can also be found in cells from the CSF of suspected encephalitis patients. Cells in CSF can be separated using a Shandon-Elliott "cytocentrifuge" and then spread onto microscope slides. The cells are fixed in cold acetone and then stained using a conventional, indirect IF test. By this procedure, more than 92% of the cases have been correctly diagnosed. However, this technique is not routinely used because interpretation is difficult and the cell counts in the CSF early in the disease are often too low.

A final method for rapid diagnosis of HSV encephalitis is the indirect IP method. With this method, brain biopsy specimens are teased and separated into a cell suspension, placed on slides, and fixed in acetone. The specimens are treated with anti-HSV antiserum for 45 minutes at 35 °C and reacted with the anti-species immunoglobulin conjugated to peroxidase for 45 additional minutes at 35 °C. After thorough washing the specimens are stained immediately with Kaplow's reagent, washed, and counterstained with safranin for 30 seconds. The specimens are dehydrated through alcohol to xylene and mounted in Permount. The cells that are positive for HSV antigens will show blue granules indicating sites of peroxidase activity. This procedure has been described as reliable, rapid, and specific. Its advantages over the IF are easier interpretation of results, use of cell suspensions rather than frozen sections, permanent preparations, and the use of a light microscope.

27.9 SEROLOGIC DIAGNOSIS

Serodiagnosis of HSV infections in a patient presenting with visible lesions is based on the demonstration of a significant rise (fourfold or greater) in serum antibody levels over the course of the illness. This is done by taking paired serum samples. The first, or acute sample, is taken as early in the course as possible. The second, or convalescent, sample is taken two to three weeks later, which is the time required for development of antibodies. Many factors may confound the results of serodiagnosis. Because of the considerable cross-reactivity between the two serotypes, it is difficult to detect HSV-2 antibodies in patients who already have high titers of antibodies to HSV-1 and vice versa. Furthermore, demonstrable increases in antibody levels are not seen in patients with recurrent disease. Thus, serodiagnosis is not commonly used in diagnostic virus laboratories. When it is done, its value is limited to: 1) documentation of primary HSV infection in which a clear-cut rise in titer is definitive, or 2) as evidence for prior infection that may constitute an indication for prophylactic therapy, for instance in the immunocompromised patients. Type-specific serodiagnostic techniques, which are not readily available at this time, would have more practical applications. They would be particularly useful in evaluating past HSV-2 infections from a transmission or prevalence standpoint. Distinguishing between types may also be of some value in predicting recurrences, since it has been established that genital HSV-2 infection may be more likely to recur than HSV-1 infection.

A number of serologic assays are available. The complement fixation test was the most commonly used method for routine measurements. It is a relatively complex assay of limited sensitivity and specificity, and it cannot differentiate between the two HSV serotypes. Accordingly, it recently has been replaced by ELISA with antigen-coated plates for antibody detection. This assay is much easier to perform and is commercially available. Its limitations are similar to those discussed earlier for the antigen detection ELISA, and it generally does not differentiate between antibodies to the two HSV types. More recently, methods were developed using purified glycoproteins and polypeptides from HSV-1 and HSV-2 prepared on immunoaffinity columns containing bound monoclonal antibodies. One such method employs the ELISA technique. Another is the Western blot assay in which HSV proteins are subjected to polyacrylamide gel electrophoresis. Proteins on the gel are transferred by electrophoresis to a sheet of nitrocellulose paper that is incubated with test serum and probed with peroxidase-labeled protein A. Detection is carried out by color development.

Measurement of neutralizing antibodies in serum is less frequently used as it has the disadvantages of greater complexity, a longer time to run, and expense. There are a number of neutralization procedures described in the literature, and all have good sensitivity and specificity. The most commonly used test is the microneutralization. Test serum and virus are incubated and then inoculated onto cell cultures. If neutralizing antibody is present in the serum, virus is inactivated, and CPE is not observed. The antibody titer is defined as the maximum dilution of serum that causes 50% reduction in the number of cells showing CPE. The microneutralization test can distinguish between HSV-1 and HSV-2. However, there is a certain amount of cross-reactivity between the antigens used in this assay, and antibody to one HSV type will be associated with a neutralizing titer to the heterologous antigen.

The kinetics of neutralization assay is more complex, but it has the advantage of providing a much better differentiation between the two antibodies. A specific modification used in our laboratory [termed multiplicity analysis kinetics of neutralization assay (MAKNA) (Aurelian et al, 1970; Sheridan and Aurelian, 1983)] is even capable of defining increases in serum antibody titers to one HSV serotype in patients infected with both serotypes. This assay is based on the relative distribution of the antibodies to HSV type-specific and type-common determinants in the patient serum, between a known mixture of HSV-1 and HSV-2. In direct comparisons there was a close correlation (above 90%) between antibody typing with this assay and the type of the HSV isolated from these patients. Additionally, there was a very good correlation between the results of the MAKNA assay and clinical findings. In a series of 133 patients with genital HSV infections, 31 were considered to have primary infection on clinical criteria. Of these, 28 were positive for both HSV serotypes in the acute phase. The titer of antibody to HSV-2 increased in the convalescent serum from 20 of the 28 patients while the titer of HSV-1 antibody remained unchanged in all 28. This is consistent with a diagnosis of primary HSV-2 infection in 20 patients. In the eight remaining patients the titer to HSV-1, but not HSV-2, increased in the convalescent serum, consistent with a diagnosis of primary HSV-1 infection. In 100 patients that were diagnosed as having recurrent HSV based on clinical criteria, titers of antibody to both HSV serotypes were similar in both the acute and convalescent sera consistent with recurrent disease. Discordant findings were obtained for only three patients who were diagnosed clinically as having a primary infection but had similar HSV-1 and HSV-2 antibody titers in the acute and convalescent sera (Sheridan and Aurelian, 1983).

Other recently developed assays for detection of type-specific antibody include IF, ELISA, and indirect hemagglutination. The HSV-2-specific glycoprotein gG was recently used to detect specific antibody to HSV-2 (Lee et al, 1985). However, the sensitivity of this assay is relatively low; it misses those patients who develop antibody to type-specific determinants on HSV-2 proteins other than gG. Elevated levels of HSV-specific serum IgA determined with the indirect IF assay (geometric mean titer = 380.03) were recently described as the means for rapid diagnosis of the asymptomatic virus-shedding patient (Guiglielmino et al, 1989).

27.10 IMMUNITY AND VACCINE DEVELOPMENT

The significant rise in the incidence of HSV-2 infections has evoked public desire for vaccine development. However, progress has been impeded by our relatively poor understanding of the role of virus-specific immunity in protection against infection. Recent efforts using subunit vaccines containing different glycoproteins have met with limited success. Other studies have used vaccinia virus recombinants containing different HSV glycoproteins. Most were shown to induce virus-specific neutralizing antibody and some protected mice from fatal HSV-1 challenge. However, they failed to protect from cutaneous disease due to high dose HSV challenge (Martin and Rouse, 1987).

In an effort to develop a vaccine that protects against cutaneous HSV disease, recent studies in our laboratory sought to determine the role of temporal regulation of HSV gene expression on the induction of protective immunity (Wachsman et al, 1989a, 1989b). We used two vaccinia recombinants that express the HSV-1 glycoprotein gD (gD-1) under the control of the late vaccinia virus 11K promoter (VP254) or the early vaccinia virus promoter (Pi) (VP176). We found that VP254 and VP176 induce equal titers of HSV-specific neutralizing antibodies and protect mice from cutaneous HSV-2 challenge at 14 days postimmunization. However, temporal regulation of gD-1 expression has a significant effect on long-term protective immunity against neural and cutaneous HSV disease.

In the mouse and guinea pig models of HSV infection, protection mediated by VP176 correlated with its ability to elicit a stronger T-cell response as determined by HSV-specific DTH ($P < 0.01$) and lymphoproliferation ($P < 0.005$) (Table 27.4). The failure of VP254 to induce HSV specific T-cell immunity was observed despite

Table 27.4 Effect of VP176 or VP254 Immunization on Primary and Recurrent Cutaneous HSV-2 Disease

Disease, parameter	Immunization		
	VP176	VP254	VTK$^-$79
Primary disease			
No. positive/no. inoculated	2/17[a]	7/8	9/9
Duration (days ± SE)*	14 ± 1.0	10.9 ± 1.8[b]	15.1 ± 1.2
Maximal lesion score[c]	4	2.7 ± 0.3	3.4 ± 0.24
Maximal footpad Δ mm[d]	ND	0.3 ± 0.1	0.5 ± 0.1
HSV-2 titer at inoculation site (PFU/mL)[e]	3×10^1	ND	1×10^5
Recurrent disease			
No. positive/no. inoculated	2/17[a]	4/8	6/9
Duration (days ± SE)*	9.5 ± 0.5	10.8 ± 2.3	9.2 ± 1.2
Maximal lesion score[c]	3.0 ± 1.0	2.5 ± 0.5	2.8 ± 0.3
Maximal footpad Δ mm[d]	0.2	0.2 ± 0.1	0.3 ± 0.1
Episodes per animal	1	2–2.5	2–2.5

* Parameters listed below refer only to positive animals.

[a] Same animals developed primary and recurrent disease. P < 0.0005 by χ^2 compared with VTK$^-$79 controls.

[b] P < 0.025 by one-tailed student's t test compared with VTK$^-$79 controls.

[c] Scoring criteria for disease severity are: 0 = negative; 1 = slight erythema or healing vesicles; 2 = moderate erythema with swelling; 3 = severe erythema with swelling and small vesicles; 4 = severe erythema with swelling and large vesicles.

[d] Difference in millimeters between the size of the infected and uninfected footpads.

[e] Mean titers for groups of five animals.

Hartley strain guinea pigs were given a primary and secondary immunization with VP176, VP254, or VTK$^-$79 (2×10^6 PFU). They were challenged with HSV-2 (5×10^5 PFU) 7 days later and followed for the development of primary and recurrent cutaneous disease (Wachsman et al, 1989b).

very high immunizing doses (2×10^6 PFU), and in the presence of very strong vaccinia-specific T-cell responses, and it correlated with the failure of VP254 to protect from HSV-2 disease (7 of 8 immunized animals developed HSV-2 lesions). However, while vaccinia immunized mice died within 20 days postchallenge, 58% of those immunized with VP254 survived (Wachsman et al, 1989b), consistent with partial protection.

These findings indicate that recombinants expressing a foreign gene under the control of an early (but not late) vaccinia virus promoter elicit protective T-cell responses. Since: 1) Ia$^+$ epidermal cells (EC) act as antigen-presenting cells (APC) in HSV-2-induced T-cell proliferation, and 2) impairment of this function is associated with immune suppression and increased disease severity (Aurelian, 1989b), we considered the involvement of EC in the induction of HSV-specific immunity following immunization with vaccinia virus recombinants. We found that EC infected with VP176 (10 or 25 PFU/cell) but not mock-infected EC, EC infected with 10 or 25 PFU/cell of VP254, or EC infected with vaccinia virus had antigen presenting capacity for HSV-immune lymphocytes. EC infected with either recombinant could present antigen to vaccinia-immune lymphocytes. The failure of VP254-infected EC to present antigen to HSV-immune lymphocytes correlates with the absence of HSV-specific lymphoproliferative responses in VP254-immunized mice and with impaired processing of the gD-1 precursor protein in

VP254-infected EC. VP176-infected EC, on the other hand, expressed the product form of gD-1 (57kDa), had accessory cell function for HSV-immune lymphocytes, and induced HSV-specific protective immunity.

One possible interpretation of the failure to detect HSV-specific T-cell responses (and protection) in VP254-immunized animals is that the mature gD-1 is not expressed in the APC, thereby impairing antigen presentation. An alternative interpretation is that restricting major histocompatibility-complex antigen must be synthesized simultaneously with the foreign antigen in order to achieve effective T-cell recognition, and this does not occur in VP254-infected APC in which cellular protein synthesis is inhibited before the synthesis of the gD precursor. These findings underscore the importance of HSV-specific T-cell responses for cutaneous and long-term protection and emphasize the role of regulatory aspects of viral gene expression and antigen presentation.

27.11 UNUSUAL FEATURES OF HERPES SIMPLEX VIRUSES

The herpes simplex viruses have several properties that distinguish them from other animal viruses. Perhaps the most unique feature of this complex group of viruses is their low species-specificity for replication. These viruses will replicate in almost any type of cell and in many different animals. Along with nonspecificity for replication and probably because of it, HSV is responsible for a wide spectrum of diseases, from gingivostomatitis to keratoconjunctivitis to encephalitis. As new and highly sophisticated molecular technology approaches (such as PCR) are made available to the diagnostic laboratory, and by virtue of the relatively high proportion of immunosuppressed patients provided by the AIDS epidemic, the association of HSV with additional clinical conditions is also becoming increasingly evident.

Another property characteristic of HSV is the ability to become latent. Before the recent recognition of HIV, the herpesviruses were the only viruses known to cause latent infection. However, the mechanism of HSV latency is probably distinct from that of HIV. HSV has an extremely high preference for nerve cells and can remain in a quiescent state for many years without causing clinical disease. Biological as well as external stimuli can trigger the reactivation of HSV and various types of disease can result. The transient immunosuppression associated with recurrent disease is another unusual property of the herpesviruses that is of particular significance in HIV-infected individuals in whom these viruses may contribute to AIDS pathogenesis (Aurelian, 1990). Both latency and immunosuppression render this group of viruses a particularly difficult challenge for vaccination.

The final unique property of HSV is the ability to cause neoplastic transformation in vitro. Herpes simplex viruses enter a state of latency with no apparent damage to the host. However, the HSV gene that causes neoplastic transformation (oncogene) could induce the expression of repressed cellular genes that cause unregulated cell growth, i.e., neoplastic transformation. Recent studies indicate that the HSV-2 oncogene behaves as a growth factor receptor protein kinase, the expression of which is required for transformation (Chung et al, 1989, 1990). Significantly, the HSV-2 oncogene has good (65%) homology to a cellular gene that is overexpressed in human cervical cancer (Smith et al, 1991). Indeed, HSV-2 has been associated with cervical cancer in numerous studies, including some very recent ones (Kjaer et al, 1988), and the E2 gene of human papillomavirus type 16 (that is also associated with cervical cancer) was shown to activate the HSV-2 oncogene (Wymer and Aurelian, 1990). Immunohistochemistry with antibody to an antigenic determinant (LA-1) within the HSV-2 oncogene (homologous to the cervical cancer gene) has potential diagnostic/prognostic value for the management of the patient with early (preinvasive) cervical lesions (Aurelian et al, 1989; Aurelian, 1990).

27.12 TREATMENT OF HSV INFECTIONS

The ideal HSV therapy would both reduce the severity of the primary disease and prevent the establishment of latency, thereby preventing recurrent disease. In reality, there is no known agent that can prevent latency, presumably because it is already established by the time symptomatic disease appears. Therefore, the major goals for the treatment of primary HSV disease are to reduce: 1) the time to complete healing of the lesions and resolution of clinical symptoms, 2) the likelihood of complications and, should they occur, their severity, and 3) the time of virus shedding and therefore the likelihood of transmission. Some of the agents that have been used to treat HSV are listed in Table 27.5. While many are without any or little scientific basis, they often remain popular because clinical manifestations are to some degree influenced by a placebo effect. Topical inactivators are meant to destroy the virus at the point of contact. They are not effective and some may even prolong healing time, act as skin irritants, or be downright dangerous. For example, photodynamic inactivation with proflavine or neutral red increases HSV oncogenicity. The rationale for the use of immunotherapeutic agents is that they nonspecifically stimulate cell-mediated immunity but their effectiveness is at best minimal. The therapeutic potential of HSV vaccines is predicated on the definition of protective

Table 27.5 Agents That Have Been Used To Treat HSV

Topical Inactivators
Ether/chloroform
Nonoxynol-9
Photodynamic inactivation
Laser treatment

Immunotherapy
Unrelated vaccines: (smallpox, BCG)
HSV vaccine (Lupidon-G)
Isoprinosine
Levamisole
Interferon
Interferon inducers

Others
Co-trimoxazole
Griseolfulvin
Corticosteroid creams
Lithium
Chlorpromazine
DMSO
Glutaraldehyde
Lysine
Zinc
Salicylic acid
Vitamins B_{12}, C, E

Specific Antivirals
Idoxuridine
Vidarabine
Acyclovir

immunity which is still poorly understood. As for the miscellaneous treatments, none have convincingly shown any effect on the course of primary HSV or the frequency or severity of recurrences. Some may be toxic at the levels needed to inhibit HSV. Others, such as the corticosteroid creams, may prolong healing time because of their immunosuppressive effect.

The most logical approach to antiviral therapy is the use of specific agents that selectively destroy viruses or virus-infected cells and are nontoxic to uninfected cells, i.e., agents that have a high therapeutic index. Such agents would have no (or more realistically, few) short-term and long-term adverse effects and would accumulate preferentially in the infected cells. Additionally, ideal antiviral agents would not allow for the selection of resistant virus strains, and could be administered in such a manner that they easily reach their target. Synthetic nucleoside analogs are structurally related to endogenous pyrimidine or purine bases. Accordingly, they specifically inhibit the activity of viral enzymes and function as false substrates for DNA synthesis and are good candidates for antiviral agents.

Two synthetic nucleoside analogs (idoxuridine and vidarabine) were already available for the treatment of HSV infections before 1982. Both inhibit HSV replication in vitro. Idoxuridine is licensed for the treatment of HSV keratitis. However, it is immunosuppressive, and is potentially mutagenic. Vidarabine is available in parenteral and ophthalmic preparations for the treatment of HSV encephalitis, neonatal herpes, and herpetic keratitis. In 1982, a new synthetic purine nucleoside analog, acyclovir, was licensed for the treatment of primary genital HSV infections. It is currently available in topical, intravenous, and oral formulations and is the only agent used in the treatment of HSV infections. At therapeutic doses, acyclovir interferes with viral DNA synthesis while leaving cell DNA synthesis largely unaffected. HSV thymidine kinase (TK), which has physicochemical and biologic properties that are different from those of the host cell enzyme, mistakenly recognizes acyclovir as an endogenous nucleoside and phosphorylates it. By contrast, the cellular TK phosphorylates acyclovir at a rate 10^6-fold slower. Acyclovir monophosphate is phosphorylated twice more by host cell enzymes. Acyclovir triphosphate, the active form of the drug, is recognized by viral

DNA polymerase much more readily than by the host cell DNA polymerase. It blocks DNA synthesis by acting as an inhibitor of polymerase activity as well as its substrate, becoming incorporated into the growing DNA chain and causing termination of chain growth. As DNA synthesis is required for acyclovir to function, the drug cannot destroy the virus during the latent period.

Acyclovir has been shown to shorten the duration of HSV symptoms and reduce virus shedding in immunocompromised patients and in patients with primary genital HSV (Whitley et al, 1984). Intravenous and oral acyclovir also prevent reactivation of HSV in seropositive patients undergoing immunosuppressive chemotherapy for acute leukemia or transplantation (Saral et al, 1981; Ambinder et al, 1984). Oral acyclovir has been shown to speed the healing and resolution of symptoms of primary and recurrent episodes of genital HSV infection. However, the subsequent rates of recurrence are not affected. Routine use for recurrent episodes of disease, especially for mild episodes, is therefore not recommended. Long-term daily suppressive therapy may be useful in reducing the frequency of reactivation in patients with very frequent recurrent episodes (Mertz et al, 1984; Reichman et al, 1984). Daily administration of two to five 200-mg capsules of oral acyclovir for four to six months appears to be safe. However, additional data are required on the drug's long-term toxicity. Furthermore, resistant HSV strains emerge with relatively high frequency. Long-term suppressive therapy with oral acyclovir does not eliminate ganglionic latency, and reactivation of disease occurs after therapy is discontinued. Data are not available on the use of oral acyclovir in the treatment of primary or recurrent gingivostomatitis.

Ongoing research and development of new antiviral chemotherapy is focused on antisense technology. This approach is particularly attractive as it has the potential to specifically inhibit viral genes that are critical for replication or other viral functions such as the establishment of latency, or oncogenesis without affecting cellular genes. Such an antisense oligonucleotide that inhibits HSV growth in vitro has recently been described (Kulka et al, 1989).

Its ultimate clinical significance remains to be determined.

ACKNOWLEDGMENT

The studies in our laboratory were supported by grant AI-22192 from the National Institute of Allergy and Infectious Diseases.

REFERENCES

Ambinder, R.F., Lietman, P.S., Burns, W.H., and Saral, R. 1984. Prophylaxis: A strategy to minimise antiviral resistance. Lancet 2:1154–1155.

Aurelian, L., Royston, I., and Davis, H.J. 1970. Antibody to genital herpes simplex virus: Association with cervical atypia and carcinoma in situ. J. Natl. Cancer Inst. 45:455–464.

Aurelian, L. 1982. Herpes simplex virus diagnosis: Antigen detection by ELISA and flow microfluorometry. Diagn. Gyn. Obstet. 4:375–388.

Aurelian, L. 1989a. The pathogenesis of herpes simplex virus infections: Latency. In J.W. Gorrod, O. Albano, S. Papa, (eds.), Molecular Aspects of Human Disease, Vol. 1. Ellis Horwood Publishers, 253–268.

Aurelian, L. 1989b. Herpes simplex. In S. Specter, M. Bendinelli, and H. Friedman, (eds.), Virus-Induced Immunosuppression. N.Y.: Plenum Publishing Corp., 73–100.

Aurelian, L. (ed.). 1990. Herpesviruses, the Immune System, and AIDS. Developments in Medical Virology, Kluwer Academic Publishers.

Aurelian, L., Terzano, P., Smith, C.C., Chung, T., Shamsuddin, A., Costa, S., and Orlandi, C. 1989. Amino-terminal epitope of herpes simplex virus type 2 ICP10 protein as a molecular diagnostic marker for cervical intraepithelial neoplasia. Cancer Cells 7:187–191.

Aurelian, L., Costa, S., Terzano, P., and Orlandi, C. 1990. Sexually transmitted viruses and cervical cancer: The LA-1 oncogene hypothesis. Rech. Gynecol. 2:64–70.

Braun, D.K., Batterson, W., and Roizman, B. 1984. Identification and genetic mapping of a herpes simplex virus capsid protein which binds DNA. J. Virol. 50:645–648.

Brice, S.L., Krzemien, D., Weston, W.L., and Huff, J.C. 1989. Detection of herpes simplex virus DNA in cutaneous lesions of erythema multiforme. J. Inv. Dermatol. 93:183–187.

Cai, W., Person, S., Warner, S.C., Zhou, J., and DeLuca, N.A. 1987. Linker-insertion nonsense and restriction-site deletion mutations of the gB glycoprotein gene of herpes simplex virus type 1. J. Virol. 61:714–721.

Chung, T.D., Wymer, J.P., Smith, C.C., Kulka, M., and Aurelian, L. 1989. Protein kinase activity associated with the large subunit of herpes simplex virus type 2 ribonucleotide reductase (ICP10). J. Virol. 63:3389–3398.

Chung, T.D., Wymer, J.P., Kulka, M., Smith, C.C., and Aurelian, L. 1990. Myristylation and polylysine-mediated activation of the protein kinase domain of the large subunit of herpes simplex virus type 2 ribonucleotide reductase (ICP10). Virology 179:168–178.

Cohen, G.H., Ponce de Leon, M., Deggelmann, H., Lawrence, W.C., Vernon, S.K., and Eisenberg, R.J. 1980. Structural analysis of the capsid polypeptides of herpes simplex virus types 1 and 2. J. Virol. 34:521–531.

Espy, M.J. and Smith, T.F. 1988. Detection of herpes simplex virus in conventional tube cell cultures and in shell vials with a DNA probe kit and monoclonal antibodies. J. Clin. Microbiol. 26:22–24.

Friedman, H.M., Glorioso, J.C., Cohen, G.H., Hastings, J.C., Harris, S.L., and Eisenberg, R.J. 1986. Binding of complement component C3b to glycoprotein C of herpes simplex virus type 1: Mapping of gC-binding sites and demonstration of conserved C3b binding in low-passage clinical isolates. J. Virol. 60:470–475.

Gleaves, C.A., Wilson, D.J., Wold, A.D., and Smith, T.F. 1985. Detection and serotyping of herpes simplex virus in MRC-5 cells by use of centrifugation and monoclonal antibodies 16 h postinoculation. J. Clin. Microbiol. 21:29–32.

Guglielmino, S.P.P., Pinizzotto, M.R., Furneri, P.M., Corbino, N., Castro, A., and Cianci, S. 1989. Serum IgA antibodies in HSV asymptomatic genital infections. J. Med. Virol. 27:210–214.

Guinan, M.E., Wolinsky, S.M., and Reichman, R.C. 1985. Epidemiology of genital herpes simplex virus infection. Epidemiol. Rev. 7:127–146.

Hayward, G.S. and Sugden, B. 1986. Herpesviruses: I. Genome structure and regulation. II. Latent and oncogenic infections by human herpesviruses. Cancer Cells 4:59–93.

Hill, J.M., Sedarati, F., Javier, R.T., Wagner, E.K., and Stevens, J.G. 1990. Herpes simplex virus latent phase transcription facilitates in vivo reactivation. Virology 174:117–125.

Honess, R.W. and Roizman, B. 1974. Regulation of herpesvirus synthesis. I. Cascade regulation of the synthesis of three groups of viral proteins. J. Virol. 14:8–19.

Hyman, R. 1980. Comparison of herpesvirus genome. In F. Rapp, (ed.), Oncogenic Herpesviruses. Boca Raton, FL: CRC Press, I:1–18.

Johnson, D.C., Frame, M.C., Ligas, M.W., Cross, A.M., and Stow, N.D. 1988. Herpes simplex virus immunoglobulin G Fc receptor activity depends on a complex of two viral glycoproteins, gE and gI. J. Virol. 62:1347–1354.

Kaner, R.J., Baird, A., Mansukhani, A., Basilico, C., Summers, B.D., Florkiewicz, R.Z., and Hajjar, D.P. 1990. Fibroblast growth factor receptor is a portal of cellular entry for herpes simplex virus type 1. Science 248:1410–1413.

Kieff, E.D., Bachenheimer, S.L., and Roizman, B. 1971. Size, composition and structure of the deoxyribonucleic acid of herpes simplex virus subtypes 1 and 2. J. Virol. 8:125–132.

Kieff, E.D., Hoyer, B., Bachenheimer, S.L., and Roizman, B. 1972. Genetic relatedness of type 1 and type 2 herpes simplex viruses. J. Virol. 9:738–745.

Kjaer, S.K., de Villiers, E.M., Haugaard, B.J., Christensen, R.B., Teisen, C., Moller, K.A., Poll, P., Jensen, H., Vestergaard, B.F., Lynge, E., and Jensen, O.M. 1988. Human papillomavirus, herpes simplex virus and cervical cancer incidence in Greenland and Denmark. A population-based cross-sectional study. Int. J. Cancer 41:518–524.

Kuhn, J.E., Eing, B.R., Brossmer, R., Munk, K., and Braun, R.W. 1988. Removal of N-linked carbohydrates decreases the infectivity of herpes simplex virus type 1. J. Gen. Virol. 69:2847–2858.

Kulka, M., Smith, C.C., Aurelian, L., Fishelevich, R., Meade, K., Miller, P., and Ts'o, P.O.P. 1989. Site specificity of the inhibitory effects of oligo(nucleoside methylphosphonate)s complementary to the acceptor splice junction of herpes simplex virus type 1 immediate early mRNA 4. Proc. Natl. Acad. Sci. USA 86:6868–6872.

Lee, F.K., Coleman, R.M., Pereira, L., Bailey, P.D., Tatsuno, M., and Nahmias, A.J. 1985. Detection of herpes simplex virus type 2-specific antibody with glycoprotein G. J. Clin. Microbiol. 22:641.

Leib, D.A., Coen, D.M., Bogard, C.L., Hicks, K.A., Yager, D.R., Knipe, D.M., Tyler, K.L., and Schaffer, P.A. 1989. Immediate-early regulatory gene mutants define different stages in the establishment and reactivation of herpes simplex virus latency. J. Virol. 63:759–768.

Ligas, M.W. and Johnson, D.C. 1988. A herpes simplex virus mutant in which glycoprotein D sequences are replaced by β-galactosidase sequences binds to but is unable to penetrate into cells. J. Virol. 62:1486–1494.

Lohr, J.M., Nelson, J.A., and Oldstone, M.B.A. 1990. Is herpes simplex virus associated with peptic ulcer disease? J. Virol. 64:2168–2174.

Luger, A. and Lauda, E. 1921. Transmissibility of herpetic keratitis in man to the cornea of rabbit. Wien. Klin. Wochenschr. 34:132.

Marsden, H.S. 1987. Herpes simplex virus glycoproteins and pathogenesis. In W.C. Russell and J.W. Almond, (eds.), Molecular Basis of Virus Disease. Cambridge: Cambridge University Press, 259–288.

Martin, S. and Rouse, B.T. 1987. The mechanism of antiviral immunity induced by a vaccinia virus recombinant expressing herpes simplex virus type 1 glycoprotein D: Clearance of local infection. J. Immunol. 138:3431–3437.

McGeoch, D.J., Dolan, A., Donald, S., and Rixon, F.J. 1985. Sequence determination and genetic content of the short unique region in the genome of herpes simplex virus type 1. J. Mol. Biol. 181:1–13.

Mertz, G.J., Critchlow, C., Benedetti, J., Reichman, R.C., Dolin, R., Connor, J.D., Redfield, D.C., Savoia, M.C., Richman, D.D., and Tyrrell, D.L. 1984. Double-blind placebo-controlled trial of oral acyclovir in first-episode genital herpes simplex virus infection. JAMA 252:1147–1151.

Preston, V.G. 1990. Herpes simplex virus activates expression of a cellular gene by specific binding to the cell surface. Virology 176:474–482.

Reichman, R.C., Badger, G.J., Mertz, G.J., Corey, L., Richman, D.D., Connor, J.D., Redfield, D.C., Savoia, M.C., and Bryson, Y. 1983. Treatment of recurrent herpes simplex infections with oral acyclovir: A controlled trial. JAMA 251:916–921.

Roizman, B. and Roane, P.R. Jr. 1961. A physical difference between two strains of herpes simplex virus apparent on sedimentation in cesium chloride. Virology 15:75–79.

Roizman, B. 1979. The organization of the herpes simplex virus genomes. Annu. Rev. Genet. 13:25–27.

Roizman, B. 1980. Structural and functional organization of herpes simplex virus genomes. In F. Rapp, (ed.), Oncogenic Herpesviruses. Boca Raton, FL: CRC Press, I:19–51.

Salmon, V.C., Turner, R.B., Speranza, M.J., and Overall, J.C. Jr. 1986. Rapid detection of herpes simplex virus in clinical specimens by centrifugation and immunoperoxidase staining. J. Clin. Microbiol. 23:683–686.

Saral, R., Burns, W.H., Laskin, O.L., Santos, G.W., and Lietman, P.S. 1981. Acyclovir prophylaxis of herpes-simplex-virus infections: A randomized, double-blind, controlled trial in bone-marrow-transplant recipients. N. Engl. J. Med. 305:63–67.

Schrag, J.D., Prasad, B.V.V., Rixon, F.J., and Chiu, W. 1989. Three-dimensional structure of the HSV-1 nucleocapsid. Cell 56:651–660.

Seal, L.A., Toyama, P.S., Fleet, K.M., Lerud, K.S., Heth, S.R., Moorman, A.J., Woods, J.C., and Hill, R.B. 1991. Comparison of standard culture methods, a shell vial assay, and a DNA probe for the detection of herpes simplex virus. J. Clin. Microbiol. 29:650–652.

Sheldrick, P. and Bethelot, N. 1975. Inverted repetitions in the chromosome of herpes simplex virus. Cold Spring Harbor Symp. Quant. Biol. 39:667–678.

Sheridan, J.F. and Aurelian, L. 1983. Immunity to herpes simplex virus type 2. V. Risk of recurrent disease following primary infection: Modulation of T-cell subsets and lymphokine (LIF) production. Diagn. Immunol. 1:245–256.

Smith, C.C., Wymer, J.P., Luo, J., and Aurelian, L. 1991. Genomic sequences homologous to the protein kinase region of the bifunctional herpes simplex virus type 2 protein ICP10. Virus Genes, 5:215–225.

Stevens, J.G. 1989. Human Herpesviruses: A consideration of the latent state. Microb. Rev. 53:318–332.

Wachsman, M., Luo, J.H., Aurelian, L., Perkus, M.E., and Paoletti, E. 1989. Antigen-presenting capacity of epidermal cells infected with vaccinia virus recombinants containing the herpes simplex virus glycoprotein D, and protective immunity. J. Gen. Virol. 70:2513–2520.

Wachsman, M., Aurelian, L., Smith, C.C., Perkus, M.E., and Paoletti, E. 1989. Regulation of expression of herpes simplex virus

(HSV) glycoprotein D in vaccinia recombinants affects their ability to protect from cutaneous HSV-2 disease. J. Infect. Dis. 159:625–634.

Weber, P.C., Levine, M., and Glorioso, J.C. 1987. Rapid identification of nonessential genes of herpes simplex virus type 1 by Tn5 mutagenesis. Science 236:576–579.

Weldon, S.K., Su, H.K., Fetherston, J.D., and Courtney, R.J. 1990. In vitro synthesis and processing of herpes simplex virus type 2 gG-2, using cell-free transcription and translation systems. J. Virol. 64:1357–1359.

Whitley, R.J., Levine, M., Barton, N., Herchey, B.J., Davis, G., Keeney, R.E., Whelchel, J., Diethelm, A.G., Kartus, P., and Soong, S.Y. 1984. Infections caused by herpes simplex virus in the immunocompromised host: Natural history and topical acyclovir therapy. J. Infect. Dis. 150:323–329.

WuDunn, D. and Spear, P.G. 1989. Initial interaction of herpes simplex virus with cells is binding to heparan sulfate. J. Virol. 63:52–58.

Wymer, J.P. and Aurelian, L. 1990. Papillomavirus trans-activator protein E2 activates expression from the promoter for the ribonucleotide reductase large subunit from herpes simplex virus type 2. J. Gen. Virol. 71:1817–1821.

Zweig, M., Heilman, C.J., and Hampar, B. 1979. Identification of disulfide-linked protein complexes in the nucleocapsids of herpes simplex virus type 2. Virology 94:442–450.

28

Cytomegalovirus, Varicella-Zoster Virus, Epstein-Barr Virus, and Human Herpesvirus Type 6

David Paar and Stephen E. Straus

28.1 INTRODUCTION

This chapter reviews the biological, clinical, and diagnostic features of cytomegalovirus (CMV), varicella-zoster virus (VZV), Epstein-Barr virus (EBV) and human herpesvirus-6 (HHV-6) infections. HHV-6 was first described in 1986 and, although data are rapidly accumulating, less is known about this virus than the other three. Although these viruses are structurally related, they are each quite unique. They possess little or no antigenic or sequence homology, and each of their genomes is organized somewhat differently as well. Nonetheless, all four agents are extremely common causes of disease in humans.

Except for HHV-6, these viruses are known to transform cells to varying degrees. They establish latent infections, have the potential for clinical or subclinical reactivation, and are of special concern when they affect neonate and immunodeficient hosts. In recent years, all of these viruses have been recovered with increasing frequency from patients infected with human immunodeficiency virus. Although the morbidity that can be attributed to HHV-6 is unclear at the present time, the other three viruses are significant pathogens. Indeed, CMV is the most common viral pathogen responsible for morbidity and mortality in acquired immunodeficiency syndrome (AIDS) patients. Thus, with the availability of effective means of treating or preventing serious infections with CMV, VZV, and EBV, prompt and accurate diagnosis is more important than ever. The diagnoses of infections with these viruses, which rest largely on clinical grounds, can be supported by specific serologic determinations and confirmed by an ever-expanding array of tests for the detection and identification of viruses (Schmidt and Emmons, 1989). Different laboratories may adapt diagnostic tests for their own use. Clinicians will need to become familiar with the methods employed in their clinical laboratories.

28.2 CYTOMEGALOVIRUS

28.2.1 Historical Perspectives

Cytomegalovirus was first recognized by the pathoanatomic changes it induces in renal and salivary gland tissues during the course of congenital infection. In 1881, Ribbert reported enlarged "protozoan-like" cells in the kidney of a stillborn infant (Ribbert, 1904). Similar changes, as well as intranuclear and cytoplasmic inclusions, were reported in the salivary glands of 12% of infants who died from a variety of causes (Farber and Wolbach, 1932). Three groups of investigators reported in quick succession the propagation in cell culture of the agent, variously termed salivary gland virus, cytomegalic inclusion disease virus and, finally, cytomegalovirus (Rowe et al, 1956; Smith, 1956; Weller et al, 1957). CMV is associated with a wide range of clinical syndromes affecting individuals of all ages with normal or impaired immune systems (Falloon, 1990). CMV is the most common cause of life-threatening opportunistic viral infection in organ transplant patients and in patients with AIDS (Apperley et al, 1988; Jacobson and Mills, 1988).

28.2.2 Biology and Pathogenesis of Infection

CMV strains appear to fall into two or three closely related groups (Weller et al, 1960). A uniform serotyping convention has not been agreed upon and, because the strains share sufficient antigenicity, it is practical to dismiss these differences. CMV is a typical human herpesvirus, having an icosahedral nucleocapsid and a glycoprotein-bearing lipid envelope. It has the largest genome complement (150 megadaltons) of any of the seven known human herpesviruses. Although epithelial cells, leukocytes, and fibroblasts can be infected with CMV in vivo, only fibroblasts readily support CMV replication in vitro.

Compared with herpes simplex virus (HSV) and VZV, replication of CMV proceeds very slowly in vitro. The replicative processes of CMV, like those of all herpesviruses, involve an orderly cascade of biochemical events that include immediate early (α), early (β), and late (γ) gene transcription, DNA synthesis, and assembly and release of progeny virions. There is little cell-free CMV, and infection is spread predominantly to contiguous cells.

There are no good animal hosts for human CMV, so much of our thinking about the pathogenesis of CMV disease derives from careful analyses of naturally infected patients as well as from studies of model infections with murine and guinea pig CMVs in their natural hosts. These models simulate human CMV infections in that they are transmissible in utero and by transfusion or transplantation; virus persists in leukocytes and salivary tissues; and disease is more severe in immune suppressed animals (Brodsky and Rowe, 1958; Mannini and Medearis, 1961; Bia et al, 1983; Klotman et al, 1990).

CMV causes disease in at least three, and possibly four, ways: by the direct destruction of the cells it infects; by provoking immunopathologic responses; by its ability to persist in cells and reactivate; and possibly by transforming cells. Examples of diseases associated with direct CMV-induced cytopathic changes are CMV retinitis, gastroenteritis, and pneumonitis. In each of these, one finds infected cells with intranuclear inclusions. In fact, the hallmark of such CMV infections that distinguishes them from infections due to other herpesviruses is the development of perinuclear cytoplasmic inclusions in addition to the typical herpetic intranuclear inclusions (Hanshaw, 1968) (Figure 28.1). CMV also has a propensity for massive enlargement of the affected cell, hence the name for the classical congenital CMV infection, cytomegalic inclusion disease.

Some complications of CMV infections may reflect immunopathologic processes. Examples include certain late occurring hematologic, neurologic, and renal problems such as hemolytic anemia or thrombocytopenia, transverse myelitis, and nephritis. Shanley and colleagues, in a series of studies in the murine model, have suggested a role for immunopathologic pro-

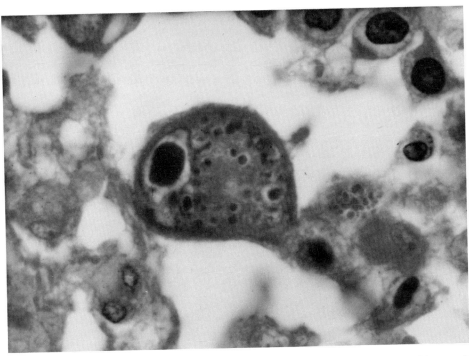

Figure 28.1 Biopsy specimen of CMV-infected lung tissue showing enlarged cells with cytoplasmic and intranuclear inclusions (hematoxylin and eosin stain). (Courtesy of Dr. William Travis.)

cesses in the development of CMV pneumonia (Shanley et al, 1982).

CMV, like all other herpesviruses, induces a lifelong infection. In situ nucleic acid hybridization studies have demonstrated CMV nucleic acids in the lymphocytes and monocytes (Schrier et al, 1985) and the arterial walls (Hendrix et al, 1989) of patients without evidence of active CMV disease. Thus lymphocytes, monocytes, and possibly arterial endothelial or smooth muscle cells are sites of CMV latency. Additionally, virus can be shed from the pharynx or in the urine for years after primary infection (Diosi et al, 1969; Hanshaw, 1978). Because of this persistent infection and because CMV can transform cells in vitro (Stevenson and Macnab, 1989), attempts have been made to etiologically link CMV infection with human tumors, particularly classical (European and African) and AIDS-associated Kaposi's

sarcoma. However, molecular analyses have failed to consistently demonstrate CMV DNA sequences in human tumor biopsies of any type. Thus, there is no conclusive evidence for the role of CMV in human neoplastic diseases (Mach et al, 1989). Similarly, although CMV DNA sequences have been detected in the arterial walls of patients with atherosclerosis, there remains inadequate evidence to link this virus with atherogenesis.

Immune responses to CMV are complex and involve both humoral and cellular mechanisms. As with other herpesviruses, the cellular mechanisms are most critical. Agammaglobulinemic and neutropenic patients without T-cell deficits are not at an increased risk of suffering serious CMV infections, which is in direct contrast to patients having cellular T lymphocyte immune deficiencies, especially organ transplant recipients and AIDS patients who may succumb to progressive visceral CMV disease (Suwansirikul et al, 1977; Ho, 1990).

CMV-specific IgG and IgM antibodies, as well as responses by natural killer and antibody-dependent killer cells, appear following infection with CMV; however, the ability to recover from serious CMV infections in bone marrow transplant recipients correlates best with the development of cytotoxic T lymphocyte responses to CMV. Fatal CMV infections are often seen in patients who fail to develop cytotoxic responses despite the presence of rising antibody titers (Quinnan et al, 1981, 1982; Rook, 1988). The determinants of protective immunity to CMV are glycoprotein (g) complexes found on the virion envelope. The gA/gB complex induces neutralizing antibodies and cytotoxic responses in animals and man (Gonczol et al, 1990), while the p86 glycoprotein and a 47 to 52 kD glycoprotein induce neutralizing antibodies in animals (Rasmussen et al, 1985; Furlini et al, 1987).

28.2.3 Epidemiology

CMV is transmissible by several different means, all requiring intimate contact with virus-bearing material. Virus is shed in urine, saliva, semen, breast milk, and cervical secretions, in addition to being carried in circulating white blood cells (Diosi et al, 1969; Lang and Hanshaw, 1969; Land and Kummer, 1972; Montgomery et al, 1972; Hanshaw, 1978; Stagno et al, 1980). CMV can be spread transplacentally, during birth by direct contact with infected cervical secretions, by accidental or inadvertent needlestick exposure and transfusions, and by transplantation of infected organs or tissues. In individuals from developing nations (Krech and Jung, 1971) and in promiscuous homosexual men (Drew et al, 1982), the seroprevalence of antibodies to CMV approaches 100%. In individuals in industrialized nations, about 50% to 70% are

seropositive by middle age (Steen and Elek, 1965; Krech and Jung, 1971). An increased incidence of CMV infection among children in day-care centers, as well as an increased rate of infection in their caretakers and parents, has been well documented (Pass et al, 1982, 1986; Adler, 1989). Congenital CMV infection occurs in approximately 1% of live births, but only a small percentage of infected infants are symptomatic. The earlier in fetal life the infection occurs, the greater is the risk of clinically-apparent congenital infection (Hanshaw et al, 1976). The risk of acquiring CMV from transfused blood is approximately 3% per unit transfused (Armstrong et al, 1976). Nearly all recipients of renal, liver, heart, heart-lung, and bone-marrow transplants develop CMV infection (Ho, 1990). Many of these infections represent reactivation of the recipient's own virus and are mild or subclinical, but primary infection by virus in the donor tissue or transfused blood carries a high risk of morbidity and mortality (Pass et al, 1979). The incubation period for CMV infection is approximately four to eight weeks.

28.2.4 Clinical Features

CMV infection is truly protean in its clinical expression (Weller, 1971). Most infections are subclinical. Congenital infection is classically associated with hepatosplenomegaly, thrombocytopenia, hemolytic anemia, chorioretinitis, and encephalitis. It is the leading cause of mental retardation and congenitally acquired deafness in the United States (Porath et al, 1990). In children and adults, symptomatic CMV infection is characterized by fever, heterophile agglutinin-negative mononucleosis, and subclinical hepatitis (Klemola et al, 1970). Pharyngitis and lymphadenopathy are uncommon.

CMV infection is most severe in neonates and in those with impairments in cellular immunity, particularly AIDS patients and transplant recipients on immuno-suppressive therapy. In these groups, CMV results in pneumonitis, retinitis, hepatitis, pancreatitis, meningoencephalitis, and gastrointestinal infections including esophagitis, gastritis, and colitis (Apperley and Goldman, 1988; Jacobson and Mills, 1988).

28.2.5 Prevention and Therapy

The prevention and management of CMV infections is a rapidly evolving discipline. The use of CMV seronegative blood products and organs has proven effective in preventing CMV infection in neonates and transplant recipients (Forbes, 1989). A live, attenuated CMV vaccine induces humoral and cellular immunity in healthy individuals and provides protection against experimentally induced CMV infection in this group (Plotkin et al, 1989). In CMV seronegative renal transplant recipients vaccinated prior to transplantation, CMV infection produced less serious disease than in nonvaccinated transplant recipients. Moreover, vaccination is associated with prolonged allograft survival at one and five years post-transplant (Brayman et al, 1988). Despite these successes, candidate CMV vaccines have not been tested sufficiently and are not close to approval for routine use. The prophylactic administration of intravenous gamma globulin or CMV hyperimmune globulin to renal, bone marrow, and heart transplant recipients may reduce the morbidity, though not the incidence, of CMV infection in these groups (Snydman et al, 1987; Winston et al, 1987; Metselaar et al, 1990). Prophylactic administration of intravenous acyclovir to bone marrow transplant recipients reduces the risk of CMV infection (Meyers et al, 1988).

Established CMV infections are difficult to treat. Interferon, vidarabine, and acyclovir are clinically ineffective alone or in combination. Ganciclovir, a derivative of acyclovir, and foscarnet, an inorganic pyrophosphate analogue, have shown some-what more promise in the treatment and suppression of CMV infections in selected settings. For example, CMV retinitis and gastrointestinal disease in AIDS patients respond to ganciclovir, but chronic maintenance therapy is necessary to prevent relapse (Jacobson and Mills, 1988). Renal and heart transplant recipients and AIDS patients with CMV pneumonia or other CMV syndromes may respond to ganciclovir therapy (Erice et al, 1987; Keay et al, 1987; Jacobson and Mills, 1988), but the data are not as compelling as for retinitis in AIDS patients. The combination of ganciclovir and intravenous immunoglobulin improves survival in bone marrow transplant recipients with pneumonia (Reed et al, 1988; Schmidt et al, 1988). Foscarnet is effective for CMV retinitis in AIDS patients, although chronic maintenance therapy is necessary to prevent relapse (Polis et al, 1990). Unfortunately, both ganciclovir and foscarnet possess risks of significant toxicity, and they must be administered intravenously, rendering them less than ideal agents. With the growing number of AIDS patients who require treatment for months or years to suppress recurrent CMV disease, there is a pressing need for better alternatives.

28.2.6 Diagnosis

Isolation of virus in tissue culture is the gold standard in establishing the diagnosis of CMV infection. Virus can be recovered from body fluids (blood, urine, saliva, tears, milk, semen, stool, and vaginal or cervical secretions) and tissues obtained by biopsy or at autopsy. All samples should be transported and held at 4 °C and processed within a few hours of collection. Avoid freezing, because CMV is susceptible to freezing. If a specimen must be frozen, add an equal volume of 70% sorbitol, and keep the specimen at or below −70 °C.

Clean voided urine is treated with antibiotics, centrifuged, and the sediment inoculated into culture tubes. Most other fluid specimens are collected sterilely, transported to the laboratory on ice, and inoculated promptly onto cell cultures.

Blood is collected into a tube with an anticoagulant, preferably citrate, and transported on ice. Leukocytes are isolated either by centrifugation and withdrawn for buffy coat culture or separated with a Ficoll-Hypaque gradient, resuspended in medium, and inoculated into tubes containing cell cultures.

Tissues from biopsy or autopsy are placed into medium containing antibiotics for transport to the laboratory. In the laboratory, the tissues are minced with sterile instruments and cells are dispersed by incubation with an EDTA/trypsin solution for several hours. After centrifugation, the cells are resuspended in medium and inoculated onto cell cultures.

CMV can be grown in cell culture employing commercially available human fibroblasts. The traditional technique is often referred to as tube culture, since tubes containing monolayers of cells are used. The tubes are maintained on rotating drums or racks and fed periodically with fresh tissue culture media. Growth of virus is slow and, if detected, may take as long as three to six weeks. Eventually, the virus induces a cytopathologic effect (CPE) which is reminiscent of the gladiolus flower (Figure 28.2). The CPE is initially focal, but gradually spreads to cover most of the monolayer. Although CMV-induced CPE has a fairly typical appearance, virus growth can be confirmed by staining the infected monolayer with fluorescent or immunoperoxidase-labeled antibodies.

For the acutely ill patient who may have CMV disease, a far more rapid means of establishing the diagnosis than that afforded by tube culture diagnosis is imperative. The shell vial technique of virus isolation is rapid, sensitive, specific, and simple enough that it is currently used by most large hospital or commercial laboratories. After routine sample preparation as

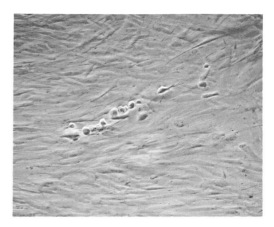

Figure 28.2 Early cytopathic changes induced by CMV in WI-38 cells.

described above, the clinical specimen is inoculated into a small vial containing media overlying a cover slip on which a monolayer of fibroblasts has been grown. The vial is centrifuged at low speed for 20 minutes to deposit infected cells and cell debris onto the monolayer. The vial is incubated for 18 to 24 hours. The coverslip is removed, rinsed, and stained with fluorescent-labeled antibodies directed against early antigens of CMV. The appearance of cells fluorescing with the appropriate pattern and intensity confirms the presence of CMV.

The application of CMV-specific monoclonal antibodies (Mabs) to the diagnosis of CMV has been useful in several ways. Using fluorescent techniques, Mabs have been used to stain pulmonary biopsy specimens that are infected with CMV. The staining of urinary sediment and bronchoalveolar lavage (BAL) specimens with CMV Mabs is less sensitive. Finally, Mabs are useful in confirming the presence of CMV in tube and shell vial cultures as described above.

Other rapid diagnostic tests such as nucleic acid hybridization, amplification of viral DNA by the polymerase chain reaction (PCR), and electron microscopy have not been adapted as yet for routine use in the diagnostic laboratory. PCR promises increased sensitivity and may be amenable to automation.

CMV infection can also be diagnosed serologically either by demonstration of seroconversion or by detecting the presence of IgM antibodies. Complement fixation, indirect immunofluorescence, latex agglutination, enzyme immunoassay, indirect hemagglutination, and radioimmunoassay techniques are available for these determinations. Rising titers of IgG antibodies develop after one to two weeks of illness. IgM antibodies usually develop after the first week of illness and may persist for nine months. In immunocompromised patients, the ability to mount IgG or IgM antibody responses may be impaired, and serologic diagnosis in this setting is not as reliable (Apperley and Goldman, 1988; Drew, 1988; Schmidt and Emmons, 1989).

28.3 VARICELLA-ZOSTER VIRUS

28.3.1 Historical Perspectives

Varicella (or chickenpox) is a familiar exanthematous disease that afflicts most humans during childhood or adolescence (Weller, 1983). A unique property of the etiologic agent for varicella is its ability to cause a distinct clinical syndrome, zoster (or shingles), upon reactivation. In 1892, after observing cases of varicella in children who were in contact with patients with zoster, von Bokay first suggested that varicella and zoster are related infections (von Bokay, 1909). Classic experiments that confirmed this theory involved direct inoculation of zoster vesicle fluids into children, with the resulting development of varicella (Kundratitz, 1925). Based on these and other observations, Garland (1943) and, later, Hope-Simpson (1965) postulated that zoster represents reactivation of varicella virus that had persisted in a latent form in the sensory ganglia.

VZV was first cultivated in monolayer cultures of human cells by Weller (1953).

Immunofluorescence and biochemical studies subsequently have proven that the agents recovered from varicella and zoster infections are essentially identical and represent herpesviruses (Weller and Witton, 1958; Weller et al, 1958; Hyman, 1981). The hypothesis that zoster represents reactivation of virus that had remained latent, since the primary varicella infection was finally proven by direct molecular analysis of VZV DNA recovered from the sequential varicella and zoster infections of the same person (Straus et al, 1984).

28.3.2 Biology and Pathogenesis of Infection

VZV is a typical herpesvirus, consisting of double-stranded linear DNA of approximately 80 megadaltons which resides within an icosahedral nucleocapsid and that is surrounded by a lipid envelope bearing viral-specific glycoproteins (Hyman, 1981). VZV has a narrow host range, growing best in human fibroblasts or epithelial cell lines, moderately well in simian cells, and poorly (if at all) in other mammalian cell lines. The replicative cycle of VZV is similar to that described previously for CMV.

Following primary infection, VZV spreads to local lymphoid tissues and then through the bloodstream to its many target organs: the skin, liver, lungs, brain, etc. Through mechanisms that have not been defined as yet, latent VZV may reactivate. Based upon the finding of VZV in satellite cells of sensory ganglia, it is presumed that the virus must first spread to adjacent nerves and then descend to the periphery. The ensuing reactivation infection, zoster, is generally confined to a single cutaneous dermatome (Brunell, 1979) but, in the setting of cellular immune impairment, it may spread to the bloodstream to cause syndromes similar to those seen in more severe or more generalized varicella.

As with CMV, some clinical manifestations of VZV infections could be immunopathologically mediated. These include varicella-associated erythema multiforme as well as some hematologic and neurologic complications of varicella or zoster.

VZV is thought to be a transforming agent (Gelb et al, 1980), although no human malignancies have been epidemiologically associated with VZV infection. While it is uncertain whether transforming infection occurs in vivo, the virus does establish a lifelong latent infection. It is believed that during varicella infection, sensory nerve roots become infected either by contiguous spread from cutaneous epithelial cells or by direct viremic spread to the nerve ganglia. Initial studies of human ganglia suggested that VZV persists in neurons, but more recent work indicates that the virus remains latent in satellite cells of sensory ganglia (Gilden et al, 1983; Hyman et al, 1983; Croen et al, 1988). The lifelong carriage of VZV in satellite cells is in direct contrast to the latency of HSV in neurons and suggests a model of disease reactivation that could explain the major clinical and biological differences between recurrent HSV and VZV infections (Straus, 1989).

Humoral immunity can be passively transferred with zoster immune globulin, conferring partial protection against primary varicella infection. Otherwise, cellular immunity is of paramount importance in maintaining infection in a latent state, and in determining the duration and severity of a VZV infection.

28.3.3 Epidemiology

Varicella is a highly contagious illness, with about 80% to 90% of susceptible household contacts acquiring the infection following direct contact with lesions or inhalation of infectious virus in aerosols (Weller, 1979). The incubation period is between 11 and 20 days in nearly all cases (Preblud et al, 1984). Except in isolated communities, about 90% of individuals contract varicella before adulthood. Patients are infectious two days before onset of rash and until all lesions have crusted. Viremia has been demonstrated from one to five days before the onset of the rash (Asano et al, 1985).

Zoster develops in about 10% to 20% of normal individuals, with an increasing risk with advancing age. Patients with cellular immune impairment, i.e., those with lymphoproliferative malignancies, AIDS, and transplant recipients, are at a greatly increased risk of developing zoster. Recurrent zoster is infrequent in the immunocompetent host but is well documented in AIDS patients (Jacobson et al, 1990).

28.3.4 Clinical Features

Varicella is an easily recognizable infection in its classic form (Brunell, 1979). Most infections are clinically apparent, with a brief prodrome of fever, chills, and myalgias followed by the development of vesicular lesions surrounded by erythema. The lesions appear first on the head or trunk and are distributed centrifugally. The average case involves some 100 to 500 lesions appearing over three to five days, with the gradual drying, crusting, and healing of lesions over 10 to 14 days. Visceral complications such as meningoencephalitis, pneumonitis, and hepatitis are uncommon except in immunodeficient patients, especially leukemic children, for whom the mortality rate from progressive varicella infection is approximately 7%. Pregnant women with varicella have a higher rate of complications, particularly pneumonia, than other adults. The congenital varicella syndrome characterized by limb hypoplasia, cutaneous scarring, cortical atrophy, chorioretinitis, and other anomalies occurs most commonly subsequent to a first trimester maternal infection (Paryani and Arvin, 1986).

Zoster is also easily recognizable, but mild, well localized infections can be mistaken for HSV infections or vice versa. The emergence of lesions is occasionally heralded by neuralgia within the affected dermatome. Grouped vesicular lesions on an erythematous base evolve and coalesce over several days. A few scattered lesions outside of the dermatome are commonly observed, but frank cutaneous or visceral dissemination is a morbid event that is essentially seen only in transplant recipients, patients with solid or hematologic malignancies (Schuchter et al, 1989), and AIDS patients (Cohen and Grossman, 1989). Post-herpetic neuralgia is the major complication of zoster in the normal host and may continue for weeks to months. The incidence of post-herpetic neuralgia increases with increasing age (Ragozzino et al, 1982).

In AIDS patients, VZV infections may become severe, chronic, and recurrent. In one small series, all children infected with human immunodeficiency virus (HIV) who developed primary varicella also developed pneumonitis. New skin lesions developed for up to six weeks and often recurred despite suppressive therapy with oral acyclovir. Relapsing varicella-like lesions were more frequent in those with lower CD4 counts (Jura et al, 1989).

In adults, early HIV infection is associated with an increased incidence of zoster (Friedman-Kien et al, 1986). In adults with AIDS, zoster recurs in the same or different dermatomes. Progressive skin lesions are characterized by epidermal necrosis with extensive ulcerations and black eschars. Verrucous hyperkeratotic lesions may evolve even in the face of acyclovir treatment. Some virus strains recovered from such treated lesions have proven drug resistant (Gilden et al, 1988; Cohen and Grossman, 1989; Jacobson et al, 1990).

28.3.5 Prevention and Therapy

A live, attenuated varicella vaccine has been shown to be immunogenic in healthy children, healthy adults, and children with leukemia; it decreases the attack rate of varicella in children with leukemia, and protective immunity persists for at least six years (Gershon et al, 1989). The vaccine is nearing approval for universal childhood vaccination. Administration of varicella-zoster immune globulin (VZIG) can prevent or lessen clinical symptoms of varicella in susceptible, exposed individuals if given within 96 hours of exposure (Centers for Disease Control, Immunization Practices Advisory Committee, 1984).

Antiviral therapy of varicella and zoster is most critical in the immunocompromised patient, with the aim of therapy being the prevention of virus dissemination. VZV isolates are generally susceptible in vitro to vidarabine, acyclovir, ganciclovir, and foscarnet. In controlled studies, leukocyte interferon, vidarabine, and acyclovir decreased the duration and severity of infection and the rate of dissemination in immunocompromised patients (Straus et al, 1988), but interferon is not used clinically, and vidarabine is not as effective as acyclovir. In the normal host, acyclovir modestly shortens the duration of varicella and zoster but there is little compelling reason to treat symptomatic patients except in the setting of early zoster in the V_1 nerve root distribution, which otherwise possesses a risk of ocular and cerebrovascular involvement (Cobo et al, 1986). In AIDS patients, acyclovir speeds healing of cutaneous lesions and is presumably effective against disseminated disease. Strains of VZV isolated from AIDS patients that are resistant to acyclovir retain sensitivity to vidarabine and foscarnet in vitro (Jacobson et al, 1990).

Post zoster-associated neuralgia is difficult to prevent or treat. Many studies have addressed the prevention of this condition by treatment with antiviral agents and corticosteroids (Elliott, 1964; Eaglstein et al, 1970; Esmann et al, 1987). A comprehensive review of the literature (Schmader and Studenski, 1989) showed that adenosine monophosphate and idoxuridine in di-

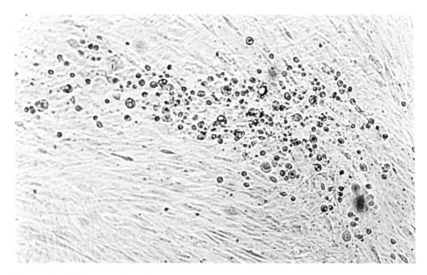

Figure 28.3 Cytopathic changes characteristic of VZV replication in human embryonic fibroblasts (Flow 5000 cell line). (Courtesy of J. Felser.)

methyl sulfoxide (DMSO) had marginal clinical benefit, but these drugs have limited clinical applicability. The treatment of established post-herpetic neuralgia is equally difficult to treat. Analgesics, anticonvulsants, tricyclic antidepressants, neuroleptics, surgical methods, and transcutaneous electrical nerve stimulation (TENS) have been used with varying success (Robertson and George, 1990). A recent report suggests that topical capsaicin may be effective therapy in the management of post-herpetic neuralgia (Bernstein, 1989). Unfortunately, the treatment itself is uncomfortable, severely limiting its acceptance.

28.3.6 Diagnosis

The diagnoses of varicella and zoster are predominantly made on clinical grounds, but can be, and at times must be, confirmed by one of several laboratory techniques. VZV can be recovered from vesicles fairly easily. It has also been cultured from blood, CSF, and tissues obtained by biopsy or at autopsy. Human diploid fibroblasts are the cell lines of choice for isolating VZV, but the virus has been grown in a variety of human cell types. The cell cultures are observed microscopically for CPE, which usually occurs from day 3 to day 14. VZV CPE is characterized by the emergence of small and slowly enlarging foci of rounded and swollen refractile cells (Figure 28.3). A presumptive diagnosis can be made based on the typical CPE and confirmed by staining with fluorescent antibodies.

For culture, vesicular lesions should be aspirated gently through a small gauge needle into a small volume syringe containing a diluent (viral transport medium, cell culture medium, or skim milk). The aspirate is inoculated into cell culture at the bedside or immediately transported to the laboratory for inoculation. If immediate inoculation is not performed, the aspirate can be transferred into a vial containing viral transport medium and kept at 4 °C for 24 hours or frozen indefinitely at −70 °C. Even with the most efficient and expeditious techniques, virus is still not recovered from about 20% of clear vesicles. The virus is quite fragile and only accumulates to low titers. Blood cultures are not routinely performed. For recovery of virus from biopsy or autopsy specimens, the tissues are homogenized, suspended in tissue culture medium, and inoculated into cell culture.

VZV-specific antigens can be detected in clinical specimens by fixing tissue sections or imprints of lesion scrapings and staining with fluorescein or immunoperoxidase-labeled polyclonal or monoclonal antibodies. Other techniques for demonstrating VZV antigens, not used in the routine diagnostic laboratory, include gel diffusion, countercurrent immunoelectrophoresis (CIE), and enzyme immunoassay (EIA). Electron microscopy can demonstrate viral particles in clinical specimens, but also is not used in routine diagnosis. Detection of viral nucleic acids can be achieved by various nucleic acid hybridization techniques, but these remain research tools. Because healthy persons should not shed VZV, PCR holds promise as a diagnostic tool.

Serology is used primarily to assess immunity to VZV rather than to establish a diagnosis. Complement fixation (CF) techniques are not sensitive, since CF antibodies decline rapidly after convalescence. Other more useful techniques include fluorescent antibody-to-membrane antigen (FAMA), direct and indirect immunofluorescence, anticomplement-immunofluorescence (ACIF), and enzyme immunofluorescent assay (EIA). The FAMA test is considered the most sensitive serologic tool and the best indicator of prior infection with VZV. Any of these can be performed in routine diagnostic laboratories or reference laboratories (Schmidt and Emmons, 1989).

28.4 EPSTEIN-BARR VIRUS

28.4.1 Historical Perspectives

The discovery of EBV and its association with infectious mononucleosis (IM) was serendipitous. Epidemiologic investigations suggested that mononucleosis had an infectious etiology (Hoagland, 1955), but serious attempts in the 1940s and 1950s to recover an agent or to experimentally transmit the infection were unsuccessful (Evans, 1947). In 1958, however, Burkitt described an unusual lymphoma in African children with a predilection for head and neck involvement (Burkitt, 1958). A variety of climatologic and geographic features of the regions in which the tumor was most prevalent suggested an infectious etiology. Efforts to identify a viral agent led to the discovery, reported by Epstein, Achong, and Barr (1964), of a herpesvirus-like particle in Burkitt tumor biopsy specimens. The Henles (1966) determined that Burkitt's patients have antibodies that react in an indirect immunofluorescence assay with antigens within the lymphomatous tissues. As an appropriate control, they chose nonreacting serum from workers in their laboratory. In 1967, one of their technicians developed classical acute IM, after which her serum reacted positively with Burkitt's tissue antigens (Henle et al, 1968). This occurrence represented the first clue that EBV may be associated with IM. Through a series of studies in college students, it was determined that heterophile-positive infectious mononucleosis occurred in EBV-seronegative patients and was followed by seroconversion (Niederman et al, 1970). Subsequently, the virus has been recovered from throat washings of acutely infected individuals by propagation of the virus in lymphocytes obtained from fetal cord blood (Chang and Golden, 1971). EBV is now known to be a ubiquitous agent that not only is associated with acute IM, but also with a variety of other aggressive lymphoproliferative disorders that arise as complications in patients with congenital or acquired immunodeficiency states (Schooley and Dolin, 1990).

28.4.2 Biology and Pathogenesis

EBV is a unique herpesvirus (Epstein and Achong, 1979). Although it exhibits typical herpesvirus-like structural and biochemical properties, its exceedingly narrow host range is remarkable. Two strains of EBV are presently recognized. Isolates similar to HR1 virus primarily induce lytic, productive infections, while isolates similar to the B95-8 strain are predominantly transforming. Most clinical isolates are of the latter type (Hinuma et al, 1967; Miller and Lipman, 1973).

In vitro, EBV can be propagated in B lymphocytes (Pattengale et al, 1973) and, to some extent, in epithelial cells as well (Sixbey et al, 1983). EBV induces a blastogenic transformation and immortalization of B cells. The viral genome remains predominantly in the form of superhelical circular episomes, but integration of DNA into the host chromosome also occurs (Adams, 1979). Cells transformed by EBV express only ten of the virus' 80 or more genes (Littler et al, 1986; Dillner and Kallin, 1988). The receptor on the B lymphocyte for EBV is the receptor for the C3d fragment of complement (Fingeroth et al, 1984). This molecule also serves as an EBV receptor on the surface of pharyngeal epithelial and uterine cervical cells (Young et al, 1986; Sixbey et al, 1987). Replicating EBV has been demonstrated in desquamated epithelial cells in pharyngeal washings (Sixbey et al, 1984), suggesting that epithelial cells in the oropharynx are primarily infected with EBV and that B lymphocytes become secondarily infected during their passage through pharyngeal lymphoid tissue while virus is being shed.

The EBV-transformed B cell clones persist for life. The potential for unlimited growth of these lines can be realized in vitro, indicating that in vivo immune mechanisms must be continuously vigilant to prevent unbridled B-cell lymphoproliferation. Quiescent EBV is known to reactivate intermittently with shedding of virus in pharyngeal secretions and renewed expression of antibodies to selected early viral antigens. Current information suggests that most such reactivation events are clinically silent except in severely immunodeficient patients.

Most data indicate that IM is largely a disease provoked by immunopathologic responses to primary EBV infection. It is uncertain whether any EBV-associated disorder reflects direct cytopathic changes, except perhaps in oral hairy leukoplakia. Lymphoproliferative and epithelial malignancies associated with EBV are relatively common in certain parts of the world, revealing this virus' strong transforming and growth promoting potential.

28.4.3 Epidemiology

In developing nations, EBV infection is nearly universal during early childhood (Evans, 1974). In industrialized nations, at least 50% of all EBV infections are delayed until late adolescence and adulthood, and it is in this setting that classical IM is recognized (Sawyer et al, 1971). The infection is most likely to be spread by exchange of infected saliva, as EBV can be recovered frequently from the oropharynx of seropositive patients (Chang et al, 1973). The virus is not easily communicable, however, so that outbreaks within enclosed communities or families are not problematic. Blood transfusions account for a small percentage of EBV infections (Gerber et al, 1969). That the virus can replicate in uterine cervical cells and can be cultured from the uterine cervix suggests that the potential for sexual transmission exists (Young and Sixbey, 1988).

28.4.4 Clinical Manifestations

In children, EBV infection is generally asymptomatic. With college-age patients, 33% to 75% of infections are clinically recognizable and typically take the form of IM. This is a self-limited illness lasting two to four weeks, characterized by fever, sore throat, lymphadenopathy, and hepatosplenomegaly (Schooley and Dolin, 1990). There is an increase in the number and proportion of circulating lymphocytes, with the emergence of a substantial number of atypical, antigen-reactive T lymphocytes (Pattengale et al, 1974). Infection of B lymphocytes by EBV induces polyclonal activation, resulting in the transient appearance of heterophile antibodies and autoimmune manifestations.

The wide variety of less common complications of IM includes splenic rupture, hemolytic anemia or thrombocytopenia, hepatitis, and neurologic manifestations, including Guillain-Barré syndrome, Bell's palsy, and encephalitis (Schooley and Dolin, 1990). IM is rarely fatal, but such an outcome is predictable in young boys with an X-linked immunodeficiency syndrome (Purtilo et al, 1975).

EBV has been associated with African Burkitt's lymphoma, nasopharyngeal carcinoma (Zur Hausen et al, 1970) and, less convincingly, with other intra-oral, parotid, and thymic tumors (Young and Sixbey, 1988). The investigation of organ allograft recipients provides convincing evidence that EBV is involved in the development of lymphoproliferative disorders. The incidence of lymphoproliferative disorders in transplant recipients is many times that observed in the general population. These lymphoproliferations have the histological appearance of non-Hodgkins lymphomas. Most of these neoplasms are composed of polyclonal B cells and, thus, are not true malignancies, although transformation to true monoclonal malignancies occurs (List et al, 1987). Epstein-Barr nuclear antigen (EBNA) and EBV DNA can be demonstrated in neoplastic tissue. Although the histology of transplant-associated non-Hodgkin's lymphoma is similar to that of AIDS-associated non-Hodgkin's lymphoma, EBV DNA cannot consistently be detected in biopsy specimens of the AIDS-associated neoplasia, casting some doubt

on the etiologic role of EBV in this setting (Kaplan et al, 1989).

EBV is associated with a variety of other disorders in AIDS patients. Viral DNA and proteins are found in most biopsies of hairy leukoplakia, an intra-oral lesion that responds to acyclovir therapy (Greenspan, 1985). In children with AIDS, EBV DNA has been demonstrated in CNS lymphoma tissues and in lung biopsies in which there is histologic evidence of chronic lymphocytic interstitial pneumonitis (Andiman et al, 1985).

28.4.5 Therapy

IM is treated symptomatically with rest, hydration, and antipyretics. High dose, rapidly tapered corticosteroid therapy is used to treat airway obstruction and hemolysis associated with IM. Although EBV is sensitive to acyclovir and ganciclovir in vitro, clinical benefit is not generally apparent when EBV-associated syndromes are treated with these drugs. The one exception is hairy leukoplakia, which responds to acyclovir therapy, but the lesion often recurs when therapy is stopped (Resnick et al, 1988).

28.4.6 Diagnosis

EBV can be detected in saliva, throat wash material, and peripheral blood mononuclear cells by transformation assays using human umbilical cord lymphocytes. Blood is collected in the delivery room from the cut end of an umbilical cord, and the lymphocytes are separated on a Ficoll-Hypaque gradient. These lymphocytes are suspended in tissue culture medium and must be used within five days of collection. The cord lymphocytes and clinical specimens are then mixed together in glass tubes with cell culture media and are maintained for six to eight weeks with periodic replenishment of the medium. The tubes containing the cells are examined microscopically.

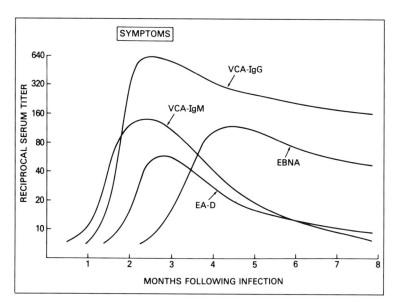

Figure 28.4 Typical kinetics of appearance and persistence of antibodies to EBV-specific antigens after a primary infection. Individuals with recent infection have IgM and IgG antibodies to VCA. Antibodies to early antigens (EA-D) develop in 70% to 80% of patients and persist for several months. Several weeks after acute infection, antibodies to EBNA appear and persist for life.

Transformation is indicated by a rapid increase in the number of cells, many of which become enlarged and grow in clusters. Infection is confirmed by anti-complementary immunofluorescence staining for EBNA. This assay is too complicated to be performed in routine diagnostic laboratories and remains a research tool.

Serology is the most common way to confirm a clinical diagnosis of IM. Nonspecific antibodies (heterophile antibodies) and antibodies directed against specific viral antigens can be detected. Heterophile antibodies represent a non specific response of B cells to EBV infection and are directed against a multitude of non-viral antigens. The ability of certain of these antibodies to agglutinate or lyse sheep, beef, goat, and horse RBCs has been adapted in the preparation of commercial "spot" tests that use RBC antigens in slide agglutination tests. The majority of tests in patients with IM due to EBV will be positive in the first week of clinical illness. If the initial test is negative, the test should be repeated in subsequent weeks of illness, since there may be a three to four week delay before a rise in heterophile antibody titers can be demonstrated.

Antibody directed against specific viral antigens have been detected, traditionally, by immunofluorescence or anti-complement immunofluorescence. More recently, serologic kits based upon enzyme immunoassays entered routine clinical use for detection of EBV antibodies. The pattern of EBV antibody response to primary infection is indicated in Figure 28.4. The diagnosis of current or recent infection can be made by detection in a single acute-phase serum of IgM antibody to viral capsid antigen (VCA). High titers of IgG antibody to VCA, the presence of antibody to early antigen (EA), and the absence of antibodies to EBNA in a single serum are suggestive of current or recent infection, but are also seen after primary infection in patients with cellular immune deficiency states.

Epstein-Barr nuclear antigen is expressed in all cells that contain EBV DNA. EBNA can be detected in circulating lymphocytes of patients with acute IM and in tissues from patients with EBV-associated lymphoproliferative disorders. Anti-complement immunofluorescent antibody techniques are used to detect this antigen complex.

Research tools that are used to detect

EBV nucleic acids in tissues include nucleic acid hybridization with various techniques and DNA amplification by PCR (Schmidt and Emmons, 1989; Jiwa et al, 1990).

28.5 HUMAN HERPESVIRUS-6 (HHV-6)

28.5.1 Historical Perspectives

Salahuddin et al (1986) reported the isolation in cell culture of a unique herpesvirus from peripheral blood mononuclear cells of six patients with lymphoproliferative disorders, including two HIV-infected patients. Because the virus initially appeared to grow only in B lymphocytes, it was called human B-lymphotrophic virus (HBLV). Subsequently, the virus was shown to infect T lymphocytes, megakaryocytes, and glioblastoma cells (Ablashi et al, 1987, 1988). Detailed molecular (Josephs et al, 1986) and ultrastructural analysis (Salahuddin et al, 1986; Biberfeld et al, 1987) confirmed that HBLV was, in fact, a unique herpesvirus and it was renamed human herpesvirus-6, or HHV-6.

HHV-6 was an orphan virus not associated with a disease until 1988, when Yamanishi et al (1988) reported convincing evidence that it was the etiologic agent of roseola infantum (exanthem subitum). Other investigators have corroborated this work (Asano et al, 1989; Yoshiyama et al, 1990). Seroepidemiologic data show that infection with HHV-6 is quite common, with up to 95% of adolescent and adult populations having antibody to this virus (Levy et al, 1990).

28.5.2 Biology and Pathogenesis

HHV-6 is a double-stranded DNA virus with an icosahedral nucleocapsid. Electron microscopic analyses indicate that, like other herpesviruses, the nucleocapsid is surrounded by an amorphous tegument and a protein-spiked envelope. The diameter of the enveloped virion is 160 to 200 nm (Biberfeld et al, 1987). The HHV-6 genome is 170 kilobases (Josephs et al, 1988a) and is more closely related to the genome of CMV than to those of other herpesviruses (Lawrence et al, 1990). It also shares some DNA sequence homology with Marek's disease virus, an avian herpesvirus (Kishi et al, 1988).

The exact mode of transmission of HHV-6 has not been defined; however, since the virus has been isolated frequently from the saliva of healthy individuals (Levy et al, 1990) and since viremia is known to occur (Salahuddin et al, 1986; Yamanishi et al, 1988), respiratory droplet and close intimate contact with exchange of body fluids seem likely modes of transmission. In this regard, transmission of HHV-6 via hepatic transplantation has been documented in one report (Ward et al, 1989), and needlestick transmission was suggested in another report (Kirchesch et al, 1988).

Whatever the mode of transmission, viremia occurs and, presumably, there is dissemination to various organs. Evidence of viral infection of tissues has been demonstrated by antigen detection in the cervical lymph nodes of patients with histiocytic necrotizing lymphadenitis and lymphoma (Eizuru et al, 1989); by the identification of viral nucleic acids by Southern hybridization, in situ hybridization, and PCR employing tissue obtained from lymphoid malignancies and various lymphoproliferations (Buchbinder et al, 1988); and by virus isolation in cell culture and antigen detection in kidney allografts of transplant patients (Asano et al, 1989; Okuno et al, 1990).

The site and nature of HHV-6 latency are unknown. Reactivation does occur and, as noted above, is manifested by asymptomatic salivary shedding in healthy people and by viremia in the immunocompromised. Potential, though unproven, sites of latency are blood mononuclear cells and organs of the lymphoreticular system.

28.5.3 Epidemiology

Initial seroepidemiologic reports suggested that the prevalence of antibodies to HHV-6 was relatively low except in immunocompromised individuals. With more sensitive techniques, however, it has become apparent that HHV-6, like most of the other herpesviruses, is an ubiquitous agent. The highest infection rate occurs in children under the age of 3 years; immunity is then maintained in up to 95% of the adult population until age 40, when antibodies begin to decline with increasing age. This suggests that repeated exposures to this virus or virus reactivation may be less frequent in adults than is the case with most other herpesviruses. Although there is not a specific population, i.e., the immunocompromised, with an increased seropositivity to HHV-6, mean antibody titers may be higher in those individuals who have a greater degree of immunosuppression (Levy, 1990).

28.5.4 Clinical Features

Roseola infantum is a common childhood disease that occurs between 6 months and 3 years of age. It is characterized by the abrupt onset of high fever in a child who looks otherwise well except for occasional mild coryza, pharyngitis, and cervical lymphadenopathy. After three to five days, as the fever falls, a maculopapular rash evolves on the trunk and neck which lasts for hours to days and may recur repeatedly for weeks, especially with exposure to heat (Gershon, 1989). In 1988, Yamanishi et al reported the isolation of HHV-6 from the blood of four infants obtained during the febrile phase of roseola infantum. In addition, seroconversion to HHV-6 was demonstrated in these four infants. Other investigators have substantiated these findings (Knowles and Gardner, 1988; Yoshiyama et al, 1990).

An infectious mononucleosis-like syndrome that occurs in children and young adults has also been attributed to acute infection with HHV-6. It is characterized by atypical lymphocytosis based upon examination of peripheral blood smear and one or more of the following: fever, mild pharyngitis, cervical lymphadenopathy, splenomegaly, and mild elevations in hepatic transaminases. One patient was asymptomatic, but the atypical lymphocytosis was noted on a pre-operative blood specimen. Serology for acute infection with EBV and CMV was negative. Acute HHV-6 infection was suggested by the presence of IgM and rising titers of IgG (Niederman et al, 1988; Steeper et al, 1990).

Because HHV-6 is a lymphotropic virus that can frequently be isolated from the blood of patients with AIDS (Salahuddin et al, 1986; Biberfield et al, 1987), work has been done to determine whether it may contribute to the pathogenesis of AIDS in HIV-1-infected individuals. Lusso et al (1989) co-infected peripheral blood mononuclear cells with both HIV-1 and HHV-6 in vitro. Co-infection with HHV-6 accelerated the appearance of cytopathic changes induced by HIV-1, increased the expression of HIV-1-associated antigens on CD4 positive lymphocytes, and accelerated the death of CD4 lymphocytes. In addition, co-infection with HHV-6 was shown to activate HIV-1 transcription from the retroviral long terminal repeat sequence, the genetic element responsible for the initiation of viral replication. These findings indicate that HHV-6 may facilitate the depletion of CD4 lymphocytes and, thus, enhance the progression to AIDS in individuals who are infected with HIV-1; however, one clinical study reported that the prevalence of antibodies to HHV-6 was actually lower in AIDS patients as compared to patients with lymphadenopathy-associated syndrome and seronegative homosexual men (Spira et al, 1990).

HHV-6 viremia and increased titers of antibodies to HHV-6 have been associated with other disease states in which the role of HHV-6 is even less well defined than in those described above. For example, HHV-6 viremia occurs in patients with lymphoproliferative disorders, and viral DNA has been detected by various molecular techniques in clinical specimens from patients with EBV-positive Burkitt's lymphoma, B-cell lymphomas, nodular histiocytic lymphoma, and small cleaved cell lymphoma (Josephs et al, 1988b; Krueger et al, 1988, 1989). Perhaps of greater significance is the reported demonstration of HHV-6 DNA by in situ and Southern hybridizations in tissues exhibiting pre-malignant atypical polyclonal lymphoproliferations (Krueger et al, 1988). This raises the possibility that persistent infection with HHV-6 may induce lymphoid hyperplasia and, thus, enhance the susceptibility of lymphoid tissues to malignant transformation.

Chronic fatigue syndrome (CFS) is another disorder to which links to HHV-6 have been drawn. It is characterized by persistent or relapsing fatigue that reduces daily activity by 50% and lasts longer than six months (Holmes et al, 1988). Since some of these patients have relatively elevated antibody titers against HHV-6 (Krueger et al, 1989), and since there are unsubstantiated claims that HHV-6 has been isolated from the blood of some of these patients more readily than from controls (Ablashi, personal communication), it has been suggested that chronic infection with HHV-6 is responsible for CFS. The bulk of data regarding CFS argues, alternatively, that nonspecific immunological abnormalities that have been observed in some CFS patients may permit secondary viral reactivation and elevation of antibody levels, neither of which would contribute to the illness.

28.5.5 Therapy

Roseola infantum, the only disease proven to be caused by HHV-6, does not require therapy other than supportive measures. It is unclear whether specific inhibition of HHV-6 could benefit patients with AIDS or lymphoproliferative disorders in which roles for HHV-6 are proposed. HHV-6 is resistant in vitro to acyclovir and ganciclovir, but appears to be sensitive to levels of foscarnet and ganciclovir that are achievable in the serum of patients treated with these drugs (Burns and Sandford, 1990).

28.5.6 Diagnosis

Infection with HHV-6 is confirmed by culture and serology. At the present time, virus isolation in cell culture remains a research tool, while serology is performed in reference laboratories and, increasingly, in commercial laboratories. IgG and IgM antibodies to HHV-6 are detected by indirect immunofluorescent techniques.

For cell culture, fresh peripheral blood mononuclear cells from patients with suspected HHV-6 infection are grown in primary culture in RPMI 1640 supplemented with 20% fetal bovine serum and hydrocortisone. Gradually a small number of large, refractile cells which are mono- or binucleated transiently appear. These primary cultures are then passaged into freshly isolated, phytohemagglutinin-stimulated human leukocytes from umbilical cord blood, adult peripheral blood, bone marrow, or spleen. Within two to four days, large refractile cells similar to those in primary cultures appear (Figure 28.5). These cells become the predominant cells in culture and survive for 8 to 12 days. The presence of HHV-6 in these cultures must then be confirmed by in situ hybridization

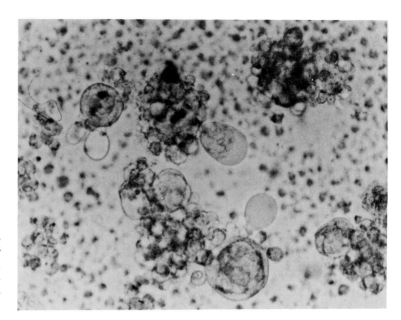

Figure 28.5 Human umbilical-cord mononuclear cells infected with HHV-6. Large rounded cells are infected with HHV-6 (courtesy of D. Ablashi).

or indirect immunofluorescent staining (Salahuddin, 1986). The primary isolates can also be passaged into T-cell lines, B-cell lines, megakaryocyte lines, and glioblastoma cell lines (Ablashi et al, 1987).

ACKNOWLEDGMENT

We thank H. Smith, J. Felser, W. Travis, and D. Ablashi for photographs of infected cells and histopathology specimens.

REFERENCES

Ablashi, D.V., Josephs, S.F., Buchbinder, A., Hellman, K., Nakamura, S., Llana, T., Lusso, P., Kaplan, M., Dahlberg, J., Memon, S., Imam, F., Ablashi, K.L., Markham, P.D., Kramarsky, B., Krueger, G.R.F., Wong-Staal, F., Salahuddin, S.Z., and Gallo, R.C. 1988. Human B-lymphotropic virus (human herpesvirus-6). J. Virol. 21:29–48.

Ablashi, D.V., Salahuddin, S.Z., Josephs, S.F., Imam, F., Lusso, P., and Gallo, R.C. 1987. HBLV (or HHV-6) in human cell lines. Nature 329:287.

Adams, A. 1979. The state of the virus genome in transformed cells and its relationship to host cell DNA. In M.A. Epstein and B.G. Achong (eds.), The Epstein-Barr virus. Berlin: Springer-Verlag, 155–183.

Adler, S.P. 1989. Cytomegalovirus and child day care: Evidence for an increased infection rate among day-care workers. N. Engl. J. Med. 321:1290–1296.

Andiman, W.A., Martin, K., Rubinstein, A., Pahwa, S., Eastman, R., Katz, B.Z., Pitt, J., and Miller, G. 1985. Opportunistic lymphoproliferations associated with Epstein-Barr viral DNA in infants and children with AIDS. Lancet 1:1390–1393.

Apperley, J.F. and Goldman, J.M. 1988. Cytomegalovirus: Biology, clinical features and methods for diagnosis. Bone Marrow Transplant. 3:253–264.

Armstrong, J.A., Tarr, G.C., Youngblood, L.A., Dowling, J.N., Saslow, A.R., Lucas, J.P., and Ho, M. 1976. Cytomegalovirus infection in children undergoing open-heart surgery. Yale J. Biol. Med. 49:83–91.

Asano, Y., Itakura, N., Kiroishi, Y., Hirose, S.,

Nagai, T., Ozaki, T., Yazaki, T., Yamanishi, K., and Takahashi, M. 1985. Viremia is present in incubation period in nonimmunocompromised children with varicella. J. Pediatr. 106:69–71.

Asano, Y., Yoshikawa, T., Suga, S., Yazaki, T., Hata, T., Nagai, T., Kajita, Y., Ozaki, T., and Yoshida, S. 1989. Viremia and neutralizing antibody response in infants with exanthem subitum. J. Pediatr. 114:535–539.

Asano, Y., Yoshikawa, T., Suga, S., Yazaki, T., Hirabayashi, S., Ono, Y., Tsuzuki, K., and Oshima, S. 1989. Human herpesvirus-6 harbouring in kidney. Lancet 2:1391.

Bernstein, J.E. 1989. Capsaicin in postherpetic neuralgia. Med. Times 117:113–116.

Bia, F.J., Griffith, B.P., Fong, C.K.Y., and Hsiung, G.D. 1983. Cytomegaloviral infections in the guinea pig: Experimental models for human disease. Rev. Infect. Dis. 5:177–195.

Biberfeld, P., Kramarsky, B., Salahuddin, S.Z., and Gallo, R.C. 1987. Ultrastructural characterization of a new human B lymphotropic DNA virus (human herpesvirus-6) isolated from patients with lymphoproliferative disease. J. Natl. Cancer Inst. 79:933–941.

Brayman, K.L., Dafoe, D.C., Smythe, W.R., Barker, C.F., Perloff, L.J., Naji, A., Fox, I.J., Grossman, R.A., Jorkasky, D.K., Starr, S.E., Friedman, H.M., and Plotkin, S.A. 1988. Prophylaxis of serious cytomegalovirus infection in renal transplant candidates using live human cytomegalovirus vaccine. Interim results of a randomized controlled trial. Arch. Surg. 123:1502–1508.

Brodsky, I. and Rowe, W.P. 1958. Chronic subclinical infection with mouse salivary gland virus. Proc. Soc. Exp. Biol. Med. 99:654–655.

Brunell, P.A. 1979. Varicella-zoster virus. In G.L. Mandell, R.G. Douglas Jr., and J.E. Bennett (eds.), Principles and Practices of Infectious Diseases. New York: John Wiley and Sons, 1295–1306.

Buchbinder, A., Josephs, S.F., Ablashi, D., Salahuddin, S.Z., Klotman, M.E., Manak, M., Krueger, G.R.F., Wong-Staal, F., and Gallo, R.C. 1988. Polymerase chain reaction amplification and in situ hybridization for the detection of human B-lymphotropic virus. J. Virol. 21:191–197.

Burkitt, D. 1958. A sarcoma involving the jaws in African children. Br. J. Surg. 46:218–223.

Burns, W.H. and Sandford, G.R. 1990. Susceptibility of human herpesvirus-6 to antivirals in vitro. J. Infect. Dis. 162:634–637.

Centers for Disease Control. 1984. Varicella-zoster immune globulin for the prevention of chickenpox: Recommendations of the Immunization Practices Advisory Committee. Ann. Intern. Med. 100:859–865.

Chang, R.S. and Golden, H.D. 1971. Transformation of human leukocytes from throat washings from infectious mononucleosis patients. Nature 234:359–360.

Chang, R.S., Lewis, J.P., and Abildgaard, C.F. 1973. Prevalence of oropharyngeal excretors of leukocyte-transforming agents among a human population. N. Engl. J. Med. 289:1325–1329.

Cobo, L.M., Foulks, G.N., Liesegang, T., Lass, J., Sutphin, J.E., Wilhermus, K., Jones, D.B., Chapman, S., Segreti, A.C., and King, D.H. 1986. Oral acyclovir in the treatment of acute herpes zoster ophthalmicus. Ophthalmology 93:763–770.

Cohen, P.R. and Grossman, M.E. 1989. Clinical features of human immunodeficiency virus-associated disseminated herpes zoster virus infection—a review of the literature. Clin. Exp. Dermatol. 14:273–276.

Croen, K.D., Ostrove, J.M., Dragovic, L.J., and Straus, S.E. 1988. Patterns of gene expression and sites of latency in human nerve ganglia are different for varicella-zoster and herpes simplex viruses. Proc. Natl. Acad. Sci. USA 85:9773–9777.

Dillner, J. and Kallin, B. 1988. The Epstein-Barr proteins. Adv. Can. Res. 50:95–158.

Diosi, P., Moldovan, E., and Tomescu, N. 1969. Latent cytomegalovirus infection in blood donors. Br. Med. J. 4:660–662.

Drew, W.L. 1988. Diagnosis of cytomegalovirus infection. Rev. Infect. Dis. 10:S468–S474.

Drew, W.L., Miner, R.D., Ziegler, J.L., Gullett, J.H., Abrams, J.I., Conant, M.A., Huang, E., Groundwater, J.R., Volberding, P., and Mintz, L. 1982. Cytomegalovirus and Kaposi's sarcoma in young homosexual men. Lancet 2:125–128.

Eaglstein, W.H., Katz, R., and Brown, J.A. 1970. The effects of early corticosteroid therapy on the skin eruption and pain in herpes zoster. JAMA 211:1681–1683.

Eizuru, Y., Minematsu, T., Minamishima, Y., Kikuchi, M., Yamanishi, K., Takahashi, M., and Kurata, T. 1989. Human herpesvirus-6 in lymph nodes. Lancet 1:40.

Elliott, F.A. 1964. Treatment of herpes zoster with high doses of prednisone. Lancet 2:610–621.

Epstein, M.A. and Achong, B.G. 1979. The Epstein-Barr Virus. Berlin: Springer-Verlag.

Epstein, M.A., Achong, B.G., and Barr, Y.M. 1964. Virus particles in cultured lymphoblasts from Burkitt's lymphoma. Lancet 1:702–703.

Erice, A., Jordan, M.C., Chace, B.A., Fletcher, C., Chinnock, B.J., and Balfour, H. 1987. Ganciclovir treatment of cytomegalovirus disease in transplant recipients and other immunocompromised hosts. JAMA 257:3082–3087.

Esmann, V., Kroon, S., Peterslund, N.A., Ronnie-Rasmussen, J.O., Geil, J.P., Fogh, H., Petersen, C.S., and Danielsen, L. 1987. Prednisolone does not prevent post-herpetic neuralgia. Lancet 2:126–129.

Evans, A.S. 1947. Experimental attempts to transmit infectious mononucleosis to man. Yale J. Biol. Med. 20:19–26.

Evans, A.S. 1974. New discoveries in infectious mononucleosis. Mod. Med. 1:18–24.

Farber, S. and Wolbach, S.B. 1932. Intranuclear and cytoplasmic inclusion ('protozoan-like bodies') in the salivary glands and other organs of infants. Am. J. Pathol. 8:123–236.

Falloon, J. and Masur, H. 1990. Cytomegalovirus and the immunosuppressed patient. In G.J. Galasso, R.J. Whitley, and T.C. Merigan (eds.), Antiviral Agents and Viral Diseases of Man, 3rd ed. New York: Raven Press, 669–689.

Fingeroth, J.D., Weis, J.J., Tedder, T.F., Strominger, J.L., Biro, P.A., and Fearon, D.T. 1984. Epstein-Barr virus receptor of human B lymphocytes is the C3d receptor CRZ. Proc. Natl. Acad. Sci. USA 81:4510–4514.

Forbes, B.A. 1989. Acquisition of cytomegalovirus infection: An update. Clin. Microbiol. Rev. 2:204–216.

Friedman-Kien, A.E., Lafleur, F.L., Gendler, E., Hennessey, N.P., Montagna, R., Halbert, S., Rubinstein, P., Krasinski, K., Zang, E., and Poisey, B. 1986. Herpes zoster: A possible early clinical sign for development of acquired immunodeficiency syndrome in high-risk individuals. J. Am. Acad. Dermatol. 14:1023–1028.

Furlini, G., Gonczol, E., Szokan, G., Ianacone, J., and Plotkin, S.A. 1987. Monoclonal antibodies directed to two groups of viral proteins neutralizing human cytomegalovirus in vitro. Hybridoma 6:321.

Garland, J. 1943. Varicella following exposure to herpes zoster. N. Engl. J. Med. 228:336–337.

Gelb, L.D., Huang, J.J., and Wellinghoff, W.J. 1980. Varicella-zoster virus transformation of hamster embryo cells. J. Gen. Virol. 51:171–177.

Gerber, P., Walsh, J.H., Rosenblum, E.N., and Purcell, R.H. 1969. Association of EB virus infection with the post-perfusion syndrome. Lancet 1:593–596.

Gershon, A.A. 1989. In G.L. Mandell, R.G. Douglas Jr., and J.E. Bennett (eds.), Principles and Practices of Infectious Diseases, 3rd ed. New York: Churchill Livingston, 2170–2171.

Gershon, A.A., Steinberg, S.P., and the Varicella Vaccine Collaborative Study Group of the National Institute of Allergy and Infectious Diseases. 1989. Persistence of immunity to varicella in children with leukemia immunized with live attenuated varicella vaccine. N. Engl. J. Med. 320:892–897.

Gilden, D.H., Murray, R.S., Wellish, M., Kleinschmidt-DeMasters, B.K., and Vafai, A. 1988. Chronic progressive varicella-zoster virus encephalitis in an AIDS patient. Neurology 38:1150–1153.

Gilden, D.H., Vafai, A., Shtram, Y., Becker, Y., Devlin, M., and Wellish, M. 1983. Varicella-zoster virus DNA in human sensory ganglia. Nature 306:478–480.

Gonczol, E., Ianacone, J., Ho, W., Starr, S., Meignier, B., and Plotkin, S. 1990. Isolated gA/gB glycoprotein complex of human cytomegalovirus envelope induces humoral and cellular immune-responses in human volunteers. Vaccine 9:130–136.

Greenspan, J.S., Greenspan, D., Lennette, E.T., Abrams, D.I., Conant, M.A., Petersen, V., and Freese, U.K. 1985. Replication of Epstein-Barr virus within the epithelial cells of oral "hairy" leukoplakia, and AIDS-associated lesion. N. Engl. J. Med. 313:1564–1571.

Hanshaw, J.B. 1968. Cytomegaloviruses. In Virology, Monograph No. 3. New York: Springer-Verlag, 1–23.

Hanshaw, J.B. 1978. Congenital cytomegalovirus. In J.S. Remington and J.O. Klein (eds.), Viral Disease of the Fetus and Newborn. Philadelphia: W.B. Saunders, 97–152.

Hanshaw, J.W., Scheiner, A.P., Moxley, A.W., Gaev, L., Abel, V., and Scheiner, B. 1976. School failure and deafness after "silent" congenital cytomegalovirus infections. N. Engl. J. Med. 295:468–470.

Henle, G. and Henle, W. 1966. Immunofluorescence in cells derived from Burkitt's lymphoma. J. Bacteriol. 91:1248–1256.

Henle, G., Henle, W., and Diehl, V. 1968. Relation of Burkitt's tumor-associated herpes-type virus to infectious mononucleosis. Proc. Natl. Acad. Sci. USA 59:94–101.

Hendrix, M.G.R., Dormans, P.H.J., Kitslaar, P., Bosman, F., and Bruggeman, C.A. 1989. The presence of cytomegalovirus nucleic acids in arterial walls of atherosclerotic and nonatherosclerotic patients. Am. J. Pathol. 134:1151–1157.

Hinuma, Y., Kohn, M., Yamaguchi, J., Wudarski, D.J., Blakely, J.R., and Grace, J.T. 1967. Immunofluorescence and herpes type virus particles in the P₃HR-1 Burkitt lymphoma clone. J. Virol. 1:1045–1051.

Ho, M. 1990. Cytomegalovirus. In R. Mandell, G. Douglas, and J.E. Bennett (eds.), Principles and Practice of Infectious Diseases, 3rd ed. New York: Churchill Livingstone, 1159–1172.

Hoagland, R.J. 1955. The transmission of infectious mononucleosis. Am. J. Med. Sci. 229:262–272.

Holmes, G.P., Kaplan, J.E., Gantz, N.M., Komaroff, A.L., Schonberger, L.B., Straus, S.E., Jones, J.F., Dubois, R.E., Cunningham-Rundles, C., Pahwa, S., Tosato, G., Zegans, L.S., Purtilo, D.T., Brown, N., Schooley, R.T., and Brus, I. 1988. Chronic fatigue syndrome: A working case definition. Ann. Int. Med. 108:387–389.

Hope-Simpson, R.E. 1965. The nature of herpes zoster: A long-term study and a new hypothesis. Proc. R. Soc. Med. 58:9–20.

Hyman, R.W. 1981. Structure and function of the varicella-zoster virus genome. In A.J. Nahmias, W.R. Dowdle, and R.F. Schinazi (eds.), The Human Herpesviruses: An Interdisciplinary Perspective. New York: Elsevier Press, 63–71.

Hyman, R.W., Ecker, J.R., and Tenser, R.B. 1983. Varicella-zoster virus RNA in human trigeminal ganglia. Lancet 2:814–816.

Jacobson, M.A., Berger, T.G., Fikrig, S., Becherer, P., Moohr, J.W., Stanat, S.C., and Biron, K.K. 1990. Acyclovir-resistant varicella-zoster virus infection after chronic oral acyclovir therapy in patients with the acquired immunodeficiency syndrome (AIDS). Ann. Int. Med. 112:187–191.

Jacobson, M.A. and Mills, J. 1988. Serious cytomegalovirus disease in the acquired immunodeficiency syndrome (AIDS): Clinical findings, diagnosis, and treatment. Ann. Int. Med. 108:585–594.

Jiwa, M., Steenbergen, R.D.M., Zwaan, F.E., Kluin, P.M., Raap, A.K., and van der Ploeg, M. 1990. Three sensitive methods for the detection of cytomegalovirus in lung tissue of patients with interstitial pneumonitis. Am. J. Clin. Pathol. 93:491–494.

Josephs, S.F., Ablashi, D.V., Salahuddin, S.Z., Kramarsky, B., Fronza, B.R. Jr., Pellett, P., Buchbinder, A., Memon, S., Wong-Staal, F., and Gallo, R.C. 1988a. Molecular studies of HHV-6. J. Virol. Methods 21:179–190.

Josephs, S.F., Buchbinder, A., Streicher, H.W., Ablashi, D.V., Salahuddin, S.Z., Guo, H.-G., Wong-Staal, F., Cossman, J., Raffeld, M., Sundeen, J., Levine, P., Biggar, R., Krueger, G.R.F., Fox, R.I., and Gallo, R.C. 1988b. Detection of human B-lymphotropic virus (human herpesvirus

6) sequences in B cell lymphoma tissues of three patients. Leukemia 2:132–135.

Josephs, S.F., Salahuddin, S.Z., Ablashi, D.V., Schachter, F., Wong-Staal, F., and Gallo, R.C. 1986. Genomic analysis of the human B-lymphotropic virus (HBLV). Science 234:601–603.

Jura, E., Chadwick, E.G., Josephs, S.H., Steinberg, S.P., Yogev, R., Gershon, A.A., Krasinski, K.M., and Borkowsky, W. 1989. Varicella-zoster virus infections in children infected with human immunodeficiency virus. Pediatr. Infect. Dis. J. 9:586–590.

Kaplan, L.D., Abrams, D.I., Feigal, E., McGrath, M., Kahn, J., Neville, P., Ziegler, J., and Volberding, P.A. 1989. AIDS-associated non-Hodgkin's lymphoma in San Francisco. JAMA 261:719–724.

Keay, S., Petersen, E., Icenogle, T., Zeluff, B.J., et al. 1988. Ganciclovir treatment of serious cytomegalovirus infection in heart and heart-lung transplant recipients. Rev. Infect. Dis. 10:S563–S572.

Kirchesch, H., Mertens, T., Burkhardt, U., Kruppenbacher, J.P., Höffken, A., and Eggers, H.J. 1988. Seroconversion against human herpesvirus-6 (and other herpesviruses) and clinical illness. Lancet 2:273–274.

Kishi, M., Harada, H., Takahashi, M., Tanaka, A., Hayashi, M., Nonoyama, M., Josephs, S.F., Buchbinder, A., Schachter, F., Ablashi, D.V., Wong-Staal, F., Salahuddin, S.Z., and Gallo, R.C. 1988. A repeat sequence, GGGTTA, is shared by DNA of human herpesvirus-6 and Marek's disease virus. J. Virol. 62:4824–4827.

Klemola, E., von Essen, R., Henle, G., and Henle, W. 1970. Infectious mononucleosis-like disease with negative heterophil agglutination test. Clinical features in relation to Epstein-Barr virus and cytomegalovirus antibodies. J. Infect. Dis. 121:608–614.

Klotman, M.E., Henry, S.C., Greene, R.C., Brazy, P.C., Klotman, P.E., and Hamilton, J.D. 1990. Detection of mouse cytomegalovirus nucleic acid in latently infected mice by in vitro enzymatic amplification. J. Infect. Dis. 161:220–225.

Knowles, W.A. and Gardner, S.D. 1988. High prevalence of antibody to human herpesvirus-6 and seroconversion associated with rash in two infants. Lancet 2:912–913.

Krech, U. and Jung, M. 1971. Age distribution of complement-fixing antibodies in Tanzania, 1970. In U. Krech and M. Jung (eds.), Cytomegalovirus Infections of Man. Basel: Karger, 27–28.

Krueger, G.R.F., Ablashi, D.V., Salahuddin, S.Z., and Josephs, S.F. 1988. Diagnosis and differential diagnosis of progressive lymphoproliferation and malignant lymphoma in persistent active herpesvirus infection. J. Virol. Methods 21:255–264.

Krueger, G.R.F., Manak, M., Bourgeois, N., Ablashi, D.V., Salahuddin, S.Z., Josephs, S.S., Buchbinder, A., Gallo, R.C., Berthold, F., and Tesch, H. 1989. Persistent active herpes virus infection associated with atypical polyclonal lymphoproliferation (APL) and malignant lymphoma. Antican. Res. 9:1457–1476.

Kundratitiz, K. 1925. Experimentelle übertragung von herpes zoster auf den menschen und die bezienhungen von herpes zoster zu varicellen. Monatsschr. Kinderheilkd. 29: 516–522.

Lang, D. and Kummer, J.F. 1972. Demonstration of cytomegalovirus in semen. N. Engl. J. Med. 287:756–758.

Lang, D. and Hanshaw, J.B. 1969. Cytomegalovirus infection and the post-perfusion syndrome: Recognition of primary infection in four patients. N. Engl. J. Med. 280:1148–1149.

Lawrence, G.L., Chee, M., Craxton, M.A., Gompels, U.A., Honess, R.W., and Barrell, B.G. 1990. Human herpesvirus-6 is closely related to human cytomegalovirus. J. Virol. 64:287–299.

Levy, J.A., Ferro, F., Greenspan, D., and Lennette, E.T. 1990. Frequent isolation of HHV-6 from saliva and high seroprevalence of the virus in the population. Lancet 1:1047–1050.

List, A.F., Greco, A., and Vogler, L.B. 1987. Lymphoproliferative diseases in immunocompromised hosts: The role of Epstein-Barr virus. J. Clin. Oncol. 5:1673–1689.

Littler, E., Zeuthen, J., McBride, A.A., Sorensen, E.T., Powell, K.L., Walsh-Arrand,

J.E., and Arrand, J.R. 1986. Identification of an Epstein-Barr virus-coded thymidine kinase. EMBO J. 6:1959–1966.

Lusso, P., Ensoli, B., Markham, P.D., Ablashi, D.V., Salahuddin, S.Z., Tschachler, E., Wong-Staal, F., and Gallo, R.C. 1989. Productive dual infection of human CD4$^+$ T lymphocytes by HIV-6 and HHV-6. Nature 337:370–373.

Mach, M., Stamminger, T., and Jahn, G. 1989. Human cytomegalovirus: Recent aspects from molecular biology. J. Gen. Virol. 70: 3117–3146.

Mannini, A. and Medearis, D.N. 1961. Mouse salivary gland virus infections. Am. J. Hyg. 73:329–343.

Metselaar, H.J., Balk, A.H.M.M., Mochtar, B., Rothbarth, P.H., and Weimar, W. 1990. Cytomegalovirus seronegative heart transplant recipients: Prophylactic use of anti-CMV immunoglobulin. Chest 97:396–399.

Meyers, J.D., Reed, E.C., Shepp, D.H., Thornquist, M., Dandliker, P.S., Vicary, C.A., Flournoy, N., Kirk, L.E., Kersey, J.H., Thomas, E.D., and Balfour, H.H. 1988. Acyclovir for prevention of cytomegalovirus infection and disease after allogenic marrow transplantation. N. Engl. J. Med. 318:70–75.

Miller, G. and Lipman, M. 1973. Release of infectious Epstein-Barr virus by transformed marmoset leukocytes. Proc. Natl. Acad. Sci. USA 70:190–194.

Montgomery, R., Youngblood, L., and Medearis, D.N. Jr. 1972. Recovery of cytomegalovirus from the cervix in pregnancy. Pediatrics 49:524–531.

Niederman, J.C., Evans, A.S., Subrahmanyan, M.S., and McCollum, R.W. 1970. Prevalence, incidence and persistence of EB virus antibody in young adults. N. Engl. J. Med. 282:361–365.

Niederman, J.D., Liu, C.-R., Kaplan, M.H., and Brown, N.A. 1988. Clinical and serological features of human herpesvirus-6 infection in three adults. Lancet 2:817–818.

Okuno, T., Higashi, K., Shiraki, K., Yamanishi, K., Takahashi, M., Kokado, Y., Ishibashi, M., Takahara, S., Sonoda, T., Tanaka, K., Baba, K., Yabuuchi, H., and Kurata, T.

1990. Human herpesvirus-6 infection in renal transplantation. Transplantation 49: 519–522.

Paryani, S.G. and Arvin, A.M. 1986. Intrauterine infection with varicella-zoster virus after maternal varicella. N. Engl. J. Med. 314:1542–1546.

Pass, R.F., August, A.M., Dworsky, M., and Reynolds, D.W. 1982. Cytomegalovirus infections in a day-care center. N. Engl. J. Med. 307:477–479.

Pass, R.F., Hutto, C., Ricks, R., and Cloud, G.A. 1986. Increased rate of cytomegalovirus infection among parents of children attending day-care centers. N. Engl. J. Med. 314:1414–1418.

Pass, R.F., Whitley, R.J., Diethelm, A.G., Whelchel, J.D., Reynolds, D.W., and Alford, C.A. Jr. 1979. Outcome of renal transplantation in patients with primary cytomegalovirus infection. Transplant. Proc. 11:1288–1290.

Pattengale, P.K., Smith, R.W., and Gerber, P. 1973. Selective transformation of B lymphocytes by EB virus. Lancet 2:93–94.

Pattengale, P.K., Smith, R.W., and Perlin, E. 1974. Atypical lymphocytes in acute infectious mononucleosis. Identification by multiple T and B markers. N. Engl. J. Med. 291:1145–1148.

Plotkin, S.A., Starr, S.E., Friedman, H.M., Gonczol, E., and Weibel, R.E. 1989. Protective effects of Towne cytomegalovirus vaccine against low-passage cytomegalovirus administered as a challenge. J. Infect. Dis. 159:860–865.

Polis, M.A., Palestine, A.G., deSmet, M.D., Baird, B., Davey, R.T., Rubin, B., Falloon, J., Kovacs, J.A., Zunick, K., Zurlo, J.J., Mellow, B., Whitcher, S., Nussenblatt, R.B., Masur, H., and Lane, H.C. 1990. Toxicity and anti-retroviral activity of foscarnet in a randomized controlled clinical trial for the treatment of cytomegalovirus retinitis. Clin. Res. 38:280A.

Porath, A., McNutt, R.A., Smiley, L.M., and Weigle, K.A. 1990. Effectiveness and cost benefit of a proposed live cytomegalovirus vaccine in the prevention of congenital disease. Rev. Infect. Dis. 12:31–40.

Preblud, S.R., Orenstein, W.A., and Bart, K.J.

1984. Varicella: Clinical manifestations, epidemiology, and health impact in children. Pediatr. Infect. Dis. 3:505–509.

Purtilo, D.T., Cassel, C.K., Yang, J.P.S., Stephenson, S.R., Harper, R., Landing, B.H., and Vawter, G.F. 1975. X-linked recessive progressive combined variable immunodeficiency (Duncan's disease). Lancet 1:935–940.

Quinnan, G.V., Kirmani, N., Esber, E., Saral, R., Manischewitz, J.F., Rogers, J.L., Rook, A.H., Santos, G.W., and Burns, W.H. 1981. HLA-restricted cytotoxic T lymphocyte and nonthymic cytotoxic lymphocyte responses to cytomegalovirus infection of bone marrow transplant recipients. J. Immunol. 126:2036–2041.

Quinnan, G.V., Kirmani, N., Rook, A.H., Manischewitz, J.F., Jackson, L., Moreschi, G., Santos, G.W., Saral, R., and Burns, W.H. 1982. Cytotoxic T cells in cytomegalovirus: HLA-restricted T-lymphocyte and non-T-lymphocyte cytotoxic responses correlate with recovery from cytomegalovirus infection in bone-marrow-transplant recipients. N. Engl. J. Med. 307:7–13.

Ragozzino, M.W., Melton, L.J., Kurland, L.T., Chu, C.P., and Perry, H.O. 1982. Population-based study of herpes zoster and its sequelae. Medicine 61:310–316.

Rasmussen, L., Mullenax, J., Nelson, M., and Merigan, T.C. 1985. Human cytomegalovirus polypeptides stimulate neutralizing antibody in vivo. Virology 145:186–190.

Reed, E.C., Bowden, R.A., Dandliker, P.S., Lilleby, K.E., and Meyers, J.D. 1988. Treatment of cytomegalovirus pneumonia with ganciclovir and intravenous cytomegalovirus immunoglobulin in patients with bone marrow transplants. Ann. Int. Med. 109:783–788.

Resnick, L., Herbst, J.S., Ablashi, D.V., Atherton, S., Frank, B., Rosen, L., and Horwitz, S.N. 1988. Regression of oral hairy leukoplakia after orally-administered acyclovir therapy. JAMA 259:384–388.

Ribbert, H. 1904. Ubeer protozoenartige zellen in der niere eines syphilitischen neugebordnen und in der parotis von kindern. Zbl. Allg. Path. Pathol. Anat. 15:945–948.

Robertson, D.R.C. and George, C.F. 1990.

Treatment of post-herpetic neuralgia in the elderly. Br. Med. Bull. 46:113–123.

Rook, A.H. 1988. Interactions of cytomegalovirus with the human immune system. Rev. Infect. Dis. 10:S460–S467.

Rowe, W.P., Hartley, J.W., Waterman, S., Turner, H.C., and Huebner, R.J. 1956. Cytopathogenic agent resembling human salivary gland virus recovered from tissue culture of human adenoids. Proc. Soc. Exp. Biol. Med. 92:418–424.

Salahuddin, S.Z., Ablashi, D.V., Markham, P.D., Josephs, S.F., Struzenegger, S., Kaplan, M., Halligan, G., Biberfeld, P., Wong-Staal, F., Kramarsky, B., and Gallo, R.C. 1986. Isolation of a new virus, HBLV, in patients with lymphoproliferative disorders. Science 234:596–601.

Sawyer, R.N., Evans, A.S., Niederman, J.C., and McCollum, R.W. 1971. Prospective studies of a group of Yale University freshmen. I. Occurrence of infectious mononucleosis. J. Infect. Dis. 123:263–270.

Schmader, K.E. and Studenski, S. 1989. Are current therapies useful for the prevention of postherpetic neuralgia? A critical analysis of the literature. J. Gen. Intern. Med. 4:83–89.

Schmidt, G.M., Kovacs, A., Zaia, J.A., Horak, D.A., Blume, K.G., Nadamanee, A.P., O'Donnell, M.R., Snyder, D.S., and Forman, S.J. 1988. Ganciclovir/immunoglobulin combination therapy for the treatment of human cytomegalovirus-associated interstitial pneumonia in bone marrow allograft recipients. Transplantation 46:905–907.

Schmidt, N.J. and Emmons, R.W. (eds.). 1989. Diagnostic Procedures for Viral, Rickettsial and Chlamydical Infections, 6th ed. Washington, DC: American Public Health Assn., 347–367; 390–400; 425–446.

Schooley, R.T. and Dolin, R. 1990. In G.L. Mandell, R.G. Douglas, and J.E. Bennett (eds.), Principles and Practice of Infectious Diseases, 3rd. ed. New York: Churchill Livingston, 1172–1185.

Schrier, R.D., Nelson, J.A., and Oldstone, M.B.A. 1985. Detection of human cytomegalovirus in peripheral blood lymphocytes in a natural infection. Science 230:1048–1050.

Schuchter, L.M., Wingard, J.R., Piantadosi, S.,

Burns, W.H., Santos, G.W., and Saral, R. 1989. Herpes zoster infection after autologous bone marrow transplantation. Blood 74:1424–1427.

Shanley, J.D., Pesanti, E.L., and Nugent, K.M. 1982. The pathogenesis of pneumonitis due to murine cytomegalovirus. J. Infect. Dis. 146:388–396.

Sixbey, J.W., Davis, D.S., Young, L., Hutt-Fletcher, L., Tedder, T.F., and Rickinson, A.B. 1987. Human epithelial cell expression of an Epstein-Barr virus receptor. J. Gen. Virol. 68:805–811.

Sixbey, J.W., Lemon, S.M., and Pagano, J.S. 1986. A second site for Epstein-Barr virus shedding: The uterine cervix. Lancet 2:1122–1124.

Sixbey, J.W., Nedrud, J.G., Raab-Traub, N., Hanes, R.A., and Pagano, J.S. 1984. Epstein-Barr virus replication in oropharyngeal epithelial cells. N. Engl. J. Med. 310:1225–1230.

Sixbey, J.W., Vesterinen, E.H., Nedrud, J.G., Raab-Traub, N., Walton, L.A., and Pagano, J.S. 1983. Replication of Epstein-Barr virus in human epithelial cells infected in vitro. Nature 306:480–483.

Smith, M.G. 1956. Propagation in tissue cultures of a cytopathogenic virus from human salivary gland virus (SGV) disease. Proc. Soc. Exp. Biol. Med. 92:424–430.

Snydman, D.R., Werner, B.G., Heinze-Lacey, B., Berardi, V.P., Tilney, N.L., Kirkman, R.L., Milford, E.L., Cho, S.I., Bush, H.L., Levey, A.S., Strom, T.B., Carpenter, C.B., Levey, R.H., Harmon, W.E., Zimmerman, C.E., Shapiro, M.E., Steinman, T., LoGerfo, F., Idelson, B., Schroter, G.P.J., Levin, M.J., McIver, J., Leszczynski, J., and Grady, G.F. 1987. Use of cytomegalovirus immune globulin to prevent cytomegalovirus disease in renal-transplant recipients. N. Engl. J. Med. 317:1049–1054.

Spira, T.J., Bozeman, L.H., Sanderlin, K.C., Warfield, D.T., Feorino, P.M., Holman, R.C., Kaplan, J.E., Fishbein, D.B., and Lopez, C. 1990. Lack of correlation between human herpesvirus-6 infection and the course of human immunodeficiency virus infection. J. Infect. Dis. 161:567–570.

Stagno, S., Reynolds, D.W., Pass, R.F., and

Alford, C.A. 1980. Breast milk and the risk of cytomegalovirus infections. N. Engl. J. Med. 302:1073–1076.

Steen, H. and Elek, S.D. 1965. The incidence of infection with cytomegalovirus in a normal population: A serologic study in greater London. J. Hyg. 63:79–87.

Steeper, T.A., Horwitz, C.A., Ablashi, D.V., Salahuddin, S.Z., Saxinger, C., Saltzman, R., and Schwartz, B. 1990. The spectrum of clinical and laboratory findings resulting from human herpesvirus-6 (HHV-6) in patients with mononucleosis-like illnesses not resulting from Epstein-Barr virus or cytomegalovirus. Am. J. Clin. Pathol. 6:776–783.

Stevenson, K. and MacNab, J.C.M. 1989. Cervical carcinoma and human cytomegalovirus. Biomed. Pharmacother. 43:173–176.

Straus, S.E. 1989. Clinical and biological differences between recurrent herpes simplex virus and varicella-zoster virus infections. JAMA 262:3455–3458.

Straus, S.E., Ostrove, J.M., Inchauspé, G., Felser, J.M., Freifeld, A., Croen, K.D., and Sawyer, M.H. 1988. Varicella-zoster virus infections: Biology, natural history, treatment, and prevention. Ann. Int. Med. 108:221–237.

Straus, S.E., Reinhold, W., Smith, H.A., Ruyechan, W.T., Henderson, D.K., Blaese, R.M., and Hay, J. 1984. Endonuclease analysis of viral DNA from varicella and subsequent zoster virus infections in the same patient. N. Engl. J. Med. 311:1362–1364.

Suwansirikul, S., Rao, N., Dowling, J.N., and Ho, M. 1977. Clinical manifestations of primary and secondary CMV infection after renal transplantation. Arch. Intern. Med. 137:1026–1029.

von Bokay, J. 1909. Über den ätiologischen zusammenhaug der varisellen mit gervissen fällen von herpes zoster. Wien. Klin. Wochenschr. 22:1323–1326.

Ward, K.N., Gray, J.J., and Efstathiou, S. 1989. Brief report: Primary human herpesvirus-6 infection in a patient following liver transplantation from a seropositive donor. J. Med. Virol. 28:69–72.

Weller, T.H. 1953. Serial propagation in vitro of

agents producing inclusion bodies derived from varicella and herpes zoster. Proc. Soc. Sys. Br. Med. 83:340–346.

Weller, T.H. 1971. The cytomegalovirus: Ubiquitous agents with protean clinical manifestations. N. Engl. J. Med. 285:203–214, 267–274.

Weller, T.H. 1979. Varicella-herpes zoster virus. In A.S. Evans (ed.), Viral Infection of Humans. Epidemiology and Control. New York: Plenum, 457–480.

Weller, T.H. 1983. Varicella and herpes zoster. Changing concepts of the natural history, control, and importance of a not-so-benign virus. N. Engl. J. Med. 309:1362–1368, 1434–1440.

Weller, T.H., Hanshaw, J.B., and Scott, D.E. 1960. Serologic differentiation of viruses responsible for cytomegalic inclusion disease. Virology 12:130–132.

Weller, T.H., Macauley, J.D., Craig, J.M., and Wirth, P. 1957. Isolation of intranuclear inclusion producing agents from infants with illnesses resembling cytomegalic inclusion disease. Proc. Soc. Exp. Biol. Med. 94:4–12.

Weller, T.H. and Witton, H.M. 1958. The etiologic agents of varicella and herpes zoster: Serologic studies with the viruses as propagated in vitro. J. Exp. Med. 108:869–890.

Weller, T.H., Witton, H.M., and Bell, E.J. 1958. The etiologic agents of varicella and herpes zoster: Isolation, propagation, and cultural characteristics in vitro. J. Exp. Med. 108:843–868.

Winston, D.J., Ho, W.G., Lin, C., Bartoni, K., Budinger, M.D., Gale, R.P., and Champlin, R.E. 1987. Intravenous immune globulin for prevention of cytomegalovirus infection and interstitial pneumonia after bone marrow transplantation. Ann. Int. Med. 106:12–18.

Yamanishi, K., Okuno, T., Shiraki, K., Takahashi, M., Kondo, T., Asano, Y., and Kurata, T. 1988. Identification of human herpesvirus-6 as a causal agent for exanthem subitum. Lancet 1:1065–1067.

Yoshiyama, H., Suzuki, E., Yoshida, T., Kajii, T., and Yamamoto, N. 1990. Role of human herpesvirus-6 infection in infants with exanthema subitum. Pediatr. Infect. Dis. J. 9:71–74.

Young, L.S. and Sixbey, J.W. 1988. Epstein-Barr virus and epithelial cells: A possible role for the virus in the development of cervical carcinoma. Cancer Sur. 7:507–518.

Young, L.S., Sixbey, J.W., Clark, D., and Rickinson, A.B. 1986. Epstein-Barr virus receptors on human pharyngeal epithelia. Lancet 1:240–242.

Yeager, A.S., Grumet, F.C., Hafleigh, E.B., Arvin, A.M., Bradley, J.S., and Prober, C.G. 1981. Prevention of transfusion-acquired cytomegalovirus infections in newborn infants. Fetal Neonat. Med. 98:281–287.

Zur Hausen, H., Schulte-Holthausen, H., Klein, G., Henle, W., Henle, G., Clifford, P., and Santesson, L. 1980. EBV DNA in biopsies of Burkitt's tumors and anaplastic carcinoma of the nasopharynx. Nature 228:1056–1058.

29

Poxviruses

James H. Nakano*

29.1 DESCRIPTION OF DISEASES

29.1.1 Smallpox (Variola)

Smallpox is believed to have originated sometime after 10,000 B.C. in some agricultural settlement in Asia or Africa (Hopkins, 1983). Supporting evidence that the disease had existed in antiquity was the discovery of the mummy of Ramses V of Egypt, who died at the age of 40 in 1157 B.C. with lesions resembling those of smallpox on the surface of his face, neck, shoulders, and arms (Hopkins, 1983).

Smallpox was endemic in 33 countries in 1967, but because of the intensified effort for global smallpox eradication by the World Health Organization (WHO), the world saw the last case of endemic smallpox, which occurred in Merka, Somalia, in October 1977. Two cases of smallpox, with one death, occurred in Birmingham, England, in August 1978, but these cases were laboratory-associated, rather than the result of endemic infection.

Perhaps the first tangible evidence for the existence of the etiologic agent for

smallpox was the description of structures which are now known as elementary bodies in infections of variola and vaccinia virus reported by Buist in 1887 (cited by Smadel, 1948). This was described also by Paschen in 1906 (cited by Smadel, 1948). The first variola virus isolated was alastrim, a strain also known as variola minor virus, on chorioallantoic membranes of embryonated chicken (Torres and de Castro Teixeira, 1935).

Smallpox is transmitted to humans by person-to-person contact or by fomites. The incubation period for human infection ranges from 7 to 17 days, with an average of 12 days. The virus enters via the upper respiratory tract, and the prodromal period is characterized by fever (38.8 °C to 39.4 °C), chills, headache, backache, vomiting, pain in the limbs, and prostration. The virus invades the lymph glands and is carried through the bloodstream to the internal organs, where the virus reproduces and is shed into the bloodstream. The skin eruption appears on the third or fourth day, as the fever subsides. The eruption develops through the stages of macule, papule, vesicle, and pustule within five or six days. The distribution of the eruption characteristically involves the face and the limbs. If a

* James Hiroto Nakano died February 9, 1990. He is revered as a leader in the smallpox eradication effort of the Centers for Disease Control.

person is to recover, the lesions usually begin to dry up at about the tenth day, and scabs will be shed, almost completely, after the third week.

Two types of smallpox, identifiable only clinically during smallpox outbreaks, are variola major and variola minor. Variola major, which prevailed on the Asian subcontinent, was severe, with case fatality rates of 15% to 40% (WHO, 1972). Variola minor apparently did not exist until the 19th century (Hopkins, 1983). It is also known as amaas, Kaffir pox, or alastrim (Mardsen, 1948); it first occurred in Southern Africa and the West Indies and spread into Brazil, North American, and Europe (Hopkins, 1983). It was variola minor that existed in Ethiopia and Somalia, where the last case of endemic smallpox was seen in 1977.

29.1.2 Human Monkeypox

Monkeypox was first discovered as a disease entity in captive monkeys in 1958 (von Magnus et al, 1959), but it was not until 1970 in Zaire that it also was discovered to be a disease entity in humans. This occurred two years after the last case of smallpox was recorded in the area. It has been called human monkeypox to differentiate it from monkeypox in monkeys. From 1970 (when it was first discovered to infect humans) until April 1986, 345 cases have been verified by laboratory testing. Cases of human monkeypox have occurred in Liberia, Ivory Coast, Sierra Leone, Nigeria, Benin (from Nigeria), Cameroon, and Zaire, with most of the cases occurring in Zaire. Although all the sources of monkeypox virus infection in humans are still unknown, one source may be wild monkeys, because many human patients had direct contact with monkeys before the onset of the disease. Also, a number of species of African monkeys captured in the wild were found to have specific monkeypox antibody determined by a radioimmunoassay adsorption test (Hutchinson et al, 1977). How-

ever, because many other individuals with human monkeypox had no contacts with monkeys, other animals such as rodents are suspected as being sources of the disease. In 1985, we isolated monkeypox viruses from a sick squirrel with lesions, identified as Funisciurus anerythrus, captured near a village where a human case of monkeypox was found (Khodakevich et al, 1986).

The disease is clinically similar to smallpox; in fact, it is so similar that the final diagnosis would require laboratory testing if smallpox were still prevalent. The disease begins with two to four days of prodromal illness with fever and prostration before the eruption. The course of development of lesions is similar to that of smallpox. However, many patients with monkeypox, unlike those with smallpox, show prominent submandibular, cervical, and inguinal lymphadenopathy. Patients usually recover in two to six weeks. The transmission rate of monkeypox is about 3.3% and the fatality rate is about 15% (compared with 15% to 40% for variola major, and less than 1% for variola minor) and the overall transmission rate is about 3.3% to known susceptible contacts (Breman et al, 1980). (Compared with 25% to 40% for variola). Undoubtedly, human monkeypox currently is the most important orthopoxvirus disease, because its clinical manifestations are so similar to those of smallpox. Because its transmissibility to susceptible contacts has been much lower than that of smallpox, however, it is not a serious public health problem at the present time. Due to the appearance of two tertiary cases of human monkeypox in 1983 for the first time and the sudden increase in the number of cases since then, the disease is receiving more concern than previously anticipated by WHO.

29.1.3 Vaccinia

Vaccinia is a "man-made" disease as long as smallpox vaccination is practiced, because vaccination is the only way in which vaccinia virus is introduced into the human population.

The origin of the virus is unknown, but its existence became known sometime after Jenner's report in 1798 (Jenner, 1798) on smallpox vaccination. Some believe that it evolved from serial passages of cowpox virus on the skin of calves; some claim that it evolved from serial passages of variola virus on the skin of calves; and still others claim that it is a genetic hybrid of cowpox and variola virus. Recently, however, it was suggested that it originated from horsepox virus (Baxby, 1981), which is now apparently extinct. Vaccinia infection in humans can cause:

1. Erythema multiforme, a macular and erythematous rash, which appears 7 to 14 days after smallpox vaccination and is caused by an allergic reaction to vaccine components
2. Generalized vaccinia, a benign disease with multiple lesions appearing on the body of vaccinees whose antibody production was delayed, but adequate
3. Congenital vaccinia, a severe disease of the fetus following primary vaccination of pregnant women
4. Progressive vaccinia (vaccinia gangrenosa or vaccinia necrosum), a very serious infection following a vaccination in individuals apparently with a defective immune system
5. Postvaccinial encephalitis, a serious disease manifested by meningeal signs, ataxia, muscular weakness, paralysis, lethargy, coma, and convulsion
6. Eczema vaccinatum, a serious local or disseminated infection in individuals with eczema or a history of eczema.

Because of the problems accompanying smallpox vaccination, most countries have discontinued vaccination after the declaration of worldwide smallpox eradication in 1980 by WHO. In March 1983, WHO reported that 155 of its 160 member states and associated members had officially discontinued vaccination. In Egypt, revaccination has been discontinued, but primary vaccination continues. In France, primary vaccination has been discontinued, but revaccination continues. In the United States, vaccination of military personnel continues and does contribute to producing cases of vaccinia, especially in the family contacts of vaccinees. In 1983, the Advisory Committee for Immunization Practice, which meets annually at the Centers for Disease Control (CDC), recommended vaccination only for those individuals who come in contact with orthopoxviruses (including viruses of smallpox, monkeypox, vaccinia, and cowpox) because of their occupation, and that the distribution of smallpox vaccine to vaccinate the general civilian population in the United States be discontinued.

In laboratories in which orthopoxviruses are handled, three levels of requirements for smallpox vaccination are recommended according to the viruses involved and with the assumption that the viruses are handled in a class 2-type biosafety cabinet:

1. In a laboratory with smallpox virus, every person entering the special biocontained laboratory should be vaccinated with smallpox vaccine annually. Only two laboratories currently are in this category; one is at CDC and the other at the Research Institute of Virus Preparations, Moscow, USSR.
2. In a laboratory with monkeypox, vaccinia, and cowpox viruses, the laboratory need not be bio-contained, but entry to the area should be restricted. Every person entering the laboratory should be vaccinated every three years. These persons include laboratory workers, service persons, and visitors. Although monkeypox virus is a class 2 security-level agent (class 3 when inoc-

ulated into susceptible animals), it does cause systemic infection similar to smallpox in individuals who are not protected by smallpox vaccination, and it has caused a number of secondary and some tertiary transmissions.

3. In a laboratory with vaccinia and cowpox viruses, but without monkeypox virus, only those who are working with the viruses need to be vaccinated every three years with smallpox vaccine. Service persons and visitors need not be vaccinated, but an unrestricted entry of these persons into the laboratory is not recommended during the time when the viruses, especially in high concentration, are being handled.

Laboratories in the third category are most numerous in the United States with only a few laboratories in the second category.[1]

29.1.4 Cowpox

Cowpox was known to Edward Jenner, who used this virus in his early smallpox vaccine. It is endemic in Great Britain and in Western Europe, but there is no evidence that it prevailed in the United States. Reports of cowpox in humans before the 1930s in the United States were in patients who were infected by cows suffering with either vaccinia virus infection (previously transmitted from humans) or pseudocowpox (parapox, milker's nodule) virus infection. Of course, the confusion was understandable because the distinction between vaccinia and cowpox viruses was not made until 1939 (Downie et al, 1939). Furthermore, indirect evidence that indicates

cowpox virus never existed in the United States is that in recent times in Europe, poxviruses similar to cowpox virus have been isolated from zoo animals such as the giant anteater (Marennikova et al, 1976), the family Felidae, including lions, black panthers, cheetahs, pumas, jaguars, ocelots (Marennikova et al, 1975; Baxby et al, 1979), okapis (Zwart et al, 1971), and also domestic cats (Thomsett et al, 1978). In the United States, however, not only have we not seen cowpox virus infection in carnivores in zoos and domestic cats, we have not encountered cowpox virus in the poxvirus laboratory which was established at CDC in 1966. Therefore, it is my belief that cowpox virus was never transported from Europe to the United States and it never existed in the United States.

Cowpox virus infection in humans is transmitted by direct contact with infections on the skin of the udder and teats of cows. The lesions in humans are found on the fingers, with reddening and swelling which develop into papules that become vesicular in four to five days and heal in two to four weeks.

Although the disease was always believed to be transmitted to humans by direct contact with infected cows, it is now believed that rats or other rodents (Baxby, 1977) are an important vector for disease transmission. This mode of transmission can explain how zoo animals and domestic cats can be infected and, thus, how domestic cats can transmit the disease to humans.

29.1.5 Buffalopox

Poxvirus disease in buffalo, in the past, has been caused by vaccinia, variola, and cowpox viruses, but a unique poxvirus which is now known as buffalopox virus (Mathew, 1976; Singh and Singh, 1967; Baxby and Hill, 1971) caused outbreaks in buffaloes in India in the 1960s and 1970s. During an outbreak of the disease in buffaloes, it is transmitted to humans by direct contact

---

[1] Because smallpox vaccine has not been distributed for use with the general civilian population since May 1983, laboratories in categories 2 and 3 can request a supply of vaccinia from the Centers for Disease Control, Division of Host Factors, Clinical Medicine Branch, Atlanta, Georgia 30333, telephone number 404-329-3356.

with an infected buffalo. The lesions are localized on the fingers, hands, and sometimes on the face of humans. No generalized infection has been seen so far and no person-to-person transmission has been reported. The patient recovers from the infection, with the scabs falling off in about two weeks.

29.1.6 Whitepox

Six isolates of whitepox virus are known today and were isolated from nonhuman primates and rodents (cited by Nakano, 1979). The first two were isolated from kidney cell cultures of two cynomolgus monkeys in a laboratory in the Netherlands in 1964; the third was isolated from a chimpanzee captured in Zaire in 1971; the fourth from a sala monkey in Zaire in 1973; the fifth from a rodent (Mastomys) in Zaire in 1974; and the sixth from another rodent (Heliosciurus) in Zaire in 1975. The latter four were isolated from wild animals captured during investigations of human monkeypox cases. Whitepox virus cannot be differentiated from variola virus by biological methods or by DNA analysis (Esposito et al, 1978; Dumbell and Archard, 1980). Although we do not know how this virus may affect humans because it was never isolated from humans, we do know from our investigation at CDC that it can cause a smallpox-like or a monkeypox-like disease in African green monkeys (*Cercopethecus aethiops*) (Nakano, 1977). When whitepox virus first became known, investigators theorized that wild animals might be a reservoir for variola virus. However, it has been fairly well established (Dumbell and Kapsenberg, 1983) that at least the first two strains isolated from kidney cell cultures of the two cynomolgus monkeys in 1964 are products of laboratory contamination of a variola virus strain from Vellore, India, which was present in the laboratory at that time. Furthermore, the whitepox viruses from the chimpanzee and the sala monkey

were indistinguishable by the biological marker test (Dumbell and Archard, 1980) from Harvey, the international reference strain for variola major virus. Whitepox virus DNA from the sala monkey and DNA from the Harvey strain were indistinguishable in every respect; the virus DNA from the chimpanzee was indistinguishable except for a minor difference (Esposito et al, 1978) when endonuclease-digested DNA was examined by electrophoresis.

29.1.7 Tanapox

Downie et al (1971) first reported epidemics of tanapox in humans (1957 and 1962) in the Wapakoma tribe which lives in Kenya along the Tana River. Tanapox was also found in the northeast area of Zaire (Lisala area) in 1975 by a WHO surveillance team that was looking for smallpox and human monkeypox. The diagnosis was verified by the CDC Poxvirus Laboratory. In Zaire, the disease is called the "river smallpox" (WHO investigators M. Szczeniowski and J. Jezek, personal communication).

Tanapox in humans as described by Downie et al (1971) can begin with a febrile period of three to four days, and can include backache, severe headache, and prostration, or, as found in many cases in Zaire, the disease may cause very few clinical signs. Lesions seen on uncovered areas of the skin are few, usually one or two. Very few patients examined are found to have more than ten lesions. The lesions appear on the skin of the upper arms, face, neck, or trunk and start as papules similar to those of smallpox and then become vesicles. Fluid, however, is difficult to extract from these vesicles. The lesions become umbilicated without pustulation and usually heal in two to four weeks. The healing time, however, may extend to seven weeks. The disease is believed to be transmitted by mosquitoes; the source of the infection is unknown, but monkeys are suspected.

29.1.8 Milker's Nodule

Dr. Jenner called milker's nodule "spurious coxpox," and he knew that it did not provide immunity against cowpox or smallpox; thus he suggested that it is unrelated to cowpox and smallpox. This disease, as first reported in the United States in 1940 (Becker, 1940), primarily affects cattle and is known as pseudocowpox. In cows, milker's nodule produces lesions on the skin of the udder and teats, and in calves it produces lesions on the lips and nose. The infection may spread to the head, trunk, and limbs. Milker's nodule is transmitted to humans by direct contact with an infected animal, and the lesion(s) is usually located on the abraded skin area of the fingers and hands. The lesion(s) starts as an erythematous papule five to seven days after exposure and becomes a firm, elastic, bluish-red, and semi-globular nodule, which measures from 1 to 2 cm and has a central depression. The lesion(s) flattens as it heals and disappears in four to six weeks. The infection sometimes causes the regional lymph nodes to swell.

29.1.9 Bovine Papular Stomatitis

Bovine papular stomatitis (BPS) is a mild infection in calves and is manifested by lesions occurring on the muzzle, margin of the lips, and buccal mucosa. It affects cows and bulls of all ages, but calves seem to be affected most often. Although the calves are infected in the buccal area, cows are not always infected on the teats and the udders as found in pseudocowpox (paravaccinia). Also unlike the pseudocowpox, BPS is found more often in beef animals than in dairy animals. The infection is transmitted to humans, manifested by lesions on the hands and arms, and heals in three to four weeks. Although there is a great deal of similarity between BPS virus and milker's nodule virus, the question of whether they are the same virus is still unresolved.

29.1.10 Orf

Orf, also known as contagious ecthyma, contagious pustular stomatitis, contagious pustular dermatitis, or sore mouth, is an infection of sheep and goats which is transmitted to humans by direct contact with an infected animal. The infection in sheep was first reported in 1887 and is prevalent worldwide (Tripathy et al, 1981). Infection sites are usually on the fingers, hands, and arms, but are sometimes on the face and neck. Only very occasionally is infection generalized. The lesions develop after an incubation period of three to six days through maculopapular stages. A red center then develops in the lesions surrounded by a white ring and a red halo. A nodular stage with red and weeping surface follows, often with a central umbilication. The lesions become granulomatous or papillomatous in three to four weeks, and some become ulcerated and superinfected with bacteria. Healing occurs in four to seven or more weeks.

29.1.11 Molluscum Contagiosum

Molluscum contagiosum (MC) was recognized in 1817 and is found worldwide. Two types of this disease are known to occur (Brown et al, 1981). One, found in childhood, manifests itself with lesions on the face, trunk, and limbs and is transmitted by direct contact from skin to skin or by fomites. This type is common in the tropics. The second, found in young adulthood, manifests itself with lesions located mostly in the lower abdominal area, pubis, inner thighs, and genitalia and is transmitted by sexual contact. The incubation time ranges from one week to six months. The lesions begin as pimples and become umbilicated papules which are pale pink to white, measuring 2 to 8 mm. The surface area over the lesions often appears tightly stretched with a slight central depression. A semi-solid caseous material can be expressed and used

Table 29.1 Taxonomic and Morphologic Classification of Poxviruses of Human Infections

Orthopoxvirus	Unclassified Virus	Pararpoxvirus
Brick-Shaped Morphology	Brick-Shaped Morphology[a]	Ovoid Morphology
1. Variola	1. Tanapox	1. Milker's nodule (pseudocowpox, paravaccinia)
2. Monkeypox (human and monkey)	2. Molluscum contagiosum	2. Bovine papular stomatitis
3. Vaccinia		3. Orf (contagious ecthyma, contagious pustular stomatitis, contagious pustular dermatitis, sore mouth)
4. Cowpox		
5. Buffalopox		
6. Whitepox		

[a] Prominent tubules for viruses of tanapox and molluscum contagiosum; envelope for tanapox virus.

for examination. The disease is self-limiting, but may last from several months to several years.

29.2 DESCRIPTION OF VIRUSES

29.2.1 Physical Characteristics and Classification

Poxviruses that cause human infections are relatively large, brick-shaped, or ovoid virions which possess an external coat containing lipid and tubular or globular protein structures, enclosing two bilateral bodies and a core which contains the genome. The genome is a single molecule of double-stranded DNA with a molecular weight of 130 to 240 $\times$ 10^6. The G + C content of orthopoxviruses is 35% to 40% and of parapoxviruses is about 63%. As shown in Table 29.1, the orthopoxvirus group includes viruses of variola, monkeypox, vaccinia, cowpox, buffalopox, and whitepox; the parapoxvirus group includes viruses of

milker's nodule, BPS, and orf; and the unclassified group includes viruses of tanapox and MC.

There are two morphologic groups: one group, as illustrated by Figure 29.1 a and b, is brick-shaped and includes the viruses of the orthopoxvirus group and the unclassified group (Table 29.1), and the second group, as illustrated by Figure 29.2 a and b, is ovoid or elongated and includes the viruses of the parapoxvirus group (Table 29.1). The surface tubules of these viruses are characteristic in that they are uniquely arranged in a parallel and criss-crossing pattern (Figure 29.2). The sizes of the viruses in the orthopoxvirus and the unclassified groups range from 140 to 230 $\times$ 210 to 380 nm, and the size of the viruses in the parapoxvirus group range from 120 to 160 $\times$ 250 to 310 nm.

Although cursory comparison of viruses in the *orthopoxvirus group* and the *unclassified group* shows no discernible

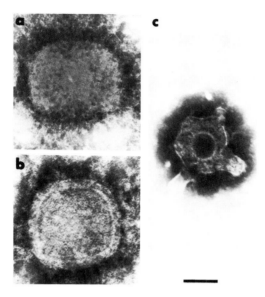

Figure 29.1 A. "M" form of variola virus from Somalia. B. "C" form of variola virus from Somalia. C. a typical varicella virus with envelope. The line represent 100 nm. See the text for the differentiation of "M" and "C" forms.

morphologic difference by electron microscopy, I found that tanapoxvirus and MC virus in field specimens (lesion materials) can, in certain situations, be differentiated from the orthopoxviruses. This is discussed under "electron microscopy" in this chapter.

Note, as shown in Figure 29.1, that an

"M" form of poxvirus is a viral particle into which a negative stain had not penetrated; therefore, when examined by electron microscopy, only the surface resembling that of a "mulberry" is seen. A "C" form is a poxvirus particle into which a negative stain had penetrated; therefore, the particle appears to be "capsulated." The "M" form is found more often in vesicular specimens (wet), and the "C" form in scabs (dry).

29.2.2 Description of Antigens

Viruses of the genus *Orthopoxvirus* are so closely related antigenically that there is no easy, routine serologic method that can differentiate these viruses. However, because of this close antigenic relationship, an antiserum against one virus can be used to identify any virus in this group as an orthopoxvirus. For example, an antivariola virus rabbit serum cannot identify an isolate as variola virus, but can identify the isolate as an orthopoxvirus.

Viruses of milker's nodule, BPS, and orf in the genus *Parapoxvirus* are not antigenically related to any of the orthopoxviruses and neither of the two groups are related to tanapoxvirus or MC virus. By the complement-fixation test, however, one may be able to see one way cross-reactivity

Figure 29.2 Electron microscope photograph of a parapoxvirus. A. "M" form; B. "C" form.

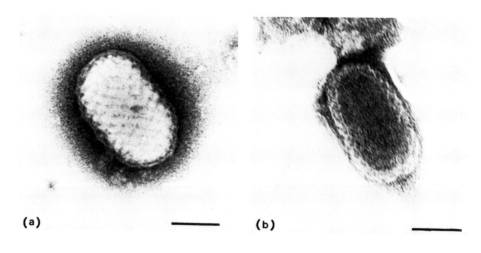

(a)   (b)

between orthopoxviruses (monkeypox and vaccinia) and tanapoxvirus.

Antigens of orthopoxviruses, parapoxvirus, tanapoxvirus, and MC virus are composed of two types: the structural proteins of virions which are the viral nucleoproteins and the viral surface tubule proteins, and the soluble antigens which are released from infected cells during the course of infection. An additional antigen, a hemagglutinin, which is produced only by orthopoxviruses, is derived from the plasma membrane of the infected cells and is separate from the soluble antigen and the structural proteins. A small percentage of extracellular vaccinia virus particles are found enveloped. The envelope surrounding these particles also contains a hemagglutinin (Payne and Norrby, 1976), which is composed of glycoprotein (Payne, 1979).

29.3 COLLECTION AND HANDLING OF SPECIMENS

For each of the poxvirus infections described in this chapter it is in the skin lesions that are the specimen of choice for virus isolation. These poxviruses are relatively stable and therefore remain viable after several weeks of storage even without refrigeration. We have been able to isolate viruses from specimens of smallpox and monkeypox even when the specimens were received at CDC six weeks after they were collected in Africa. A sufficient amount of specimen should be submitted to permit effective testing. Because specimens are examined by electron microscopy at CDC they should not be diluted with any "holding" fluid prior to shipment.

Clinical specimens for virus isolation during *orthopoxvirus* infections, should be from skin lesions collected at macular-papular, vesicular-pustular, and crusting stages. During the macular-papular stage, the lesions are scraped onto a slide with a scalpel blade reaching deeply into the lower

epithelial layer. During the vesicular-pustular stage, the fluid can be collected by capillary tubes, swabs, or on slides, noting especially to collect the materials from the base of the vesicles where virus is found in higher concentration. During the crusting stage, no fewer than three crusts (preferably ten) are collected and placed in a screw-capped vial and processed. For vaccinia and cowpox, only fragments of crusts may be available for collection.

In *parapoxvirus* infections, vesicular fluid, if available, and crusts should be collected. A biopsy from the periphery of the lesions is occasionally useful when crusts are not available.

For *tanapoxvirus*, specimens similar to those for parapoxvirus are collected.

For *molluscum contagiosum*, expelled materials from the lesions are collected on swabs or slides.

29.4 METHODS FOR ISOLATION

29.4.1 Chicken Embryo Chorioallantoic Membrane

The procedures for the preparation of the chorioallantoic membrane (CAM) for inoculating viruses and harvesting the membranes are described in Chapter 3 of this book. Fertile chicken eggs must be incubated at 38°C to 39°C for 11 to 13 days. The eggs incubated less than 11 days can be insusceptible to viruses, and those incubated more than 13 days have embryos that are too large to enable use of the CAM.

Although viruses of parapox, tanapox, and MC do not grow on the CAM, orthopoxviruses, especially viruses of smallpox, monkeypox, vaccinia, cowpox, and buffalopox, grow well. The use of the CAM is the best method to isolate these viruses, because not only does it isolate orthopoxviruses, but the viruses can also be identified according to the characteristic morphology of the pocks that each virus produces on the CAM. Because these orthopoxviruses cannot routinely be differentiated serologically, this method of identification is very useful.

29.4.1.a Smallpox Virus

Smallpox virus pocks are about 1 mm in diameter 72 hours postinoculation. They are grayish-white, opaque, convex, raised above the CAM surface, round, regular, with a smooth outer edge, are not hemorrhagic, and all are nearly the same size. The pocks appear like a fried sunny side-up egg with an opaque area with a halo.

29.4.1.b Monkeypox Virus

Monkeypox virus pocks are about the same size as those of smallpox virus at 72 hours but they are flat and ridged along the periphery and not raised above the CAM surface. Many pocks have a crater in the center that appears as a punched-out hole, and many are hemorrhagic when incubated below 35 °C. The hemorrhagic appearance in monkeypox virus pocks is caused by the deposition of red blood cells in the surface cell layer of the CAM, not in the pock itself.

29.4.1.c Vaccinia Virus

Vaccinia virus pocks at 72 hours measure 3 to 4 mm in diameter, are flattened with central necrosis and ulceration, and are sometimes hemorrhagic. Certain strains of vaccinia virus produce hemorrhagic pocks. In the pocks with a hemorrhagic appearance, the red blood cells are deposited in the surface cell layer of the CAM, like those found with monkeypox virus pocks.

29.4.1.d Cowpox Virus

Cowpox virus pocks at 72 hours measure 2 to 4 mm in diameter, are flattened and rather round, and have a bright red central

area which is caused by the red blood cells deposited in the pock proper, unlike those deposited on the surface cell layer of the CAM for monkeypox virus pocks.

29.4.1.e Buffalopox Virus

Buffalopox virus pocks at 48 hours measure about 1.6 mm in diameter and characteristically produce two types of pocks. One type is white and raised with little or no hemorrhage and the other is flat and gray.

29.4.1.f Whitepox Virus

Whitepox virus pocks at 72 hours are exactly like those of smallpox virus.

29.4.2 Cell Cultures

The use of cell culture for isolating orthopoxviruses is warranted as a back-up method for CAM inoculation, because the CAM can manifest periodic unpredictable "insusceptibility" for these viruses.

29.4.2.a Orthopoxviruses

Orthopoxviruses can be isolated in all human and nonhuman primate cells, e.g., human embryonic diploid cells, LLC-MK$_2$ (stable rhesus monkey kidney cells), and Vero (stable African green monkey kidney cells). Most of these viruses can grow in cells of other animals such as rabbit, mouse, and hamster. Smallpox virus does require adaptation in rabbit kidney cells before the virus can attain satisfactory growth.

29.4.2.b Smallpox Virus and Whitepox Virus

From clinical specimens, these produce a cytopathic effect (CPE) within one to three days in LLC-MK$_2$ or Vero cells with rounding of the cells and the formation of hyperplastic foci seen as small plaques 1 to 3 mm in diameter.

29.4.2.c Monkeypox, Vaccinia, and Cowpox Viruses

These viruses produce CPE in nonprimate cells, and in LLC-MK$_2$ or Vero cells in one to three days by fostering cell fusion and thereby forming foci. This is followed by the formation of plaques in two to three days. The plaques produced by these viruses are much larger, measuring 2 to 6 mm, than those made by smallpox virus. The plaques usually show cytoplasmic bridging. Although these three viruses can be differentiated from smallpox virus by the large plaques they produce, they cannot be differentiated from one another based on their plaque characteristics.

A continuous line of pig embryonic kidney (PEK) cells can be used to differentiate monkeypox virus from smallpox, whitepox, vaccinia, and cowpox viruses because monkeypox virus cannot grow in PEK, but the other four viruses can (Marennikova et al, 1972).

29.4.2.d Buffalopox Virus

Little work has been done to characterize the CPE that buffalopox virus produces in cell cultures, but it grows well in primary monkey kidney cells, primary rabbit kidney cells, hamster kidney cells, and human amnion cells (Mathew, 1976; Baxby and Hill, 1971).

29.4.2.e Orf Virus

Orf virus, a parapoxvirus, grows well in ovine cells such as embryonic ovine kidney and ovine testis. We found that some strains of orf virus can be isolated from clinical specimens in primary rhesus monkey kidney (PRMK) cells. Once isolated,

the virus can be passaged in other primate cell lines such as LLC-MK$_2$. Orf virus from human infections can also be isolated in bovine cells, but that from ovine infections cannot be isolated in bovine cells. The virus produces CPE in three to six days.

29.4.2.f Milker's Nodule

Milker's nodule virus can be isolated in bovine cells such as bovine embryonic lung and calf testis. Most can also be isolated in ovine cells. Once isolated, the virus can be grown in primate cells such as human diploid cells and LLC-MK$_2$.

29.4.2.g Tanapoxvirus

Tanapoxvirus was first isolated from humans in human thyroid cells (Downie et al, 1971). Once isolated, the virus grows in WI-38, primary vervet monkey kidney cells, Vero cells, HEP-2, and primary patas monkey kidney cells (Downie et al, 1971). We have been growing the virus in human embryonic lung fibroblast cells, primary African monkey kidney cells, and LLC-MK$_2$ at the CDC. Of 145 positive cases of tanapox determined using electron microscopy, only 20 isolations have been made by using LLC-MK$_2$ cells incubated at 33 °C instead of at 35 °C to 37 °C (Nakano, 1982). The virus produces CPE in six to ten days.

29.4.2.h Molluscum Contagiosum Virus

The MC virus has been virtually impossible to grow in any cell culture system.

29.5 METHODS FOR IDENTIFICATION

29.5.1 Electron Microscopy

Electron microscopy (EM) is the best method for the identification of the viruses in smallpox and human monkeypox. Its reliability at CDC is rated at 98.6% compared with 89% for that of viral isolation by CAM (Nakano, 1982). However, all of the specimens for smallpox and human monkeypox received at CDC were sent from Africa or other countries overseas, and the time required for the specimens to arrive in Atlanta was several weeks to two months. Therefore, if each specimen were sent in sufficient amount and received fresh, the reliability of EM and CAM could have been very close to 100%.

Electron microscopy is advantageous in that results in most cases can be obtained within one hour after receipt of the specimen.

29.5.1.a Orthopoxviruses

Orthopoxviruses cannot be differentiated by EM because they are similar in size and morphology. Use of CAM can, however, prevent mistaking chickenpox for smallpox and human monkeypox. As shown in Figure 29.1, smallpox (Figure 29.1a) and human monkeypox (Figure 29.1b) are similar but the herpesvirus of varicella shown in Figure 29.1c is distinctly different from orthopoxviruses.

In our experience, the possibility of finding poxvirus by EM for smallpox and monkeypox was 98.6%; however, for finding herpesvirus in patients with chickenpox it was about 60% to 70%. Based on this experience, if a patient is suspected of smallpox or monkeypox and no poxvirus or herpesvirus is found by EM, the patient is considered not to have smallpox or monkeypox, but may have chickenpox or another disease.

When a poxvirus (brick-shaped), as stated in Table 29.1, is seen by EM for suspected cases of vaccinia, cowpox, or buffalopox, a tentative diagnosis can be made because of the characteristic clinical picture and the clinical history.

29.5.1.b Parapoxviruses

Parapoxviruses are alike in size and morphology and, therefore, they cannot be differentiated from each other (Figure 29.2 a and b). These viruses have tubules on their surface which are arranged in parallel and form a criss-cross pattern. These viruses are easily differentiated from those of orthopoxviruses, tanapoxviruses, and MC viruses by EM.

29.5.1.c Tanapoxvirus

Because tanapox lesions are quite characteristic, observing brick-shaped poxvirus by EM can confirm the disease. Although these virions are morphologically similar to those of orthopoxviruses, they can often be differentiated from those of orthopoxviruses. About 90% of the specimens from tanapox patients reveal viral particles that are enveloped when examined by EM (Table 29.1). Furthermore, the surface tubules on tanapoxvirus appear to be more pronounced than those found on viruses of smallpox and monkeypox and other orthopoxviruses when the virions are found in field specimens. Although viruses of smallpox, monkeypox, and vaccinia can show prominent tubules when they are grown in cell cultures or when they are "cleaned" during a virus concentration procedure, generally, prominent tubules are not seen on virus particles in field specimens. Vaccinia virus is occasionally seen with an envelope when it is grown in tissue culture, but we have never seen it enveloped in specimens taken from patients.

29.5.1.d Molluscum Contagiosum

Molluscum contagiosum lesions are characteristic and, therefore, may not require laboratory assistance for diagnosis. Although MC virions cannot be differentiated from virions of orthopoxviruses on the basis of shape and size, most MC virions do have the prominent surface tubules similar to those on the surface of tanapox virions. Therefore, most MC virions can be differentiated from those of orthopoxviruses. Note again that the comparison is made on the basis of virions seen only in field specimens.

Four other methods which can identify the poxviruses mentioned in this chapter are described only briefly, because they cannot be used with the speed and reliability shown by the EM.

29.5.2 Agar Gel Precipitation

Agar gel precipitation (AGP) test has been used extensively for the preliminary identification of smallpox virus antigen. The detailed procedure for AGP has been described (Nakano, 1979). The test identifies orthopoxviruses as a group and, therefore, does not differentiate smallpox, human monkeypox, vaccinia, cowpox, and buffalopox viruses from one another. At CDC, although the efficiency of detecting smallpox by this method has been about 72% (Nakano, 1982), I believe that this percentage could have been much higher if fresh specimens were tested. This is a good substitute if the EM method is not available, because AGP is easily installed for routine use, can yield a positive result in about two hours, and can yield a verified negative in 24 hours.

The AGP can be used for the identification of all poxviruses. However, its reliability for the identification of virus groups other than orthopoxvirus group is unknown.

29.5.3 Immunofluorescence

Immunofluorescence (IF) testing by the direct or indirect method has been used to diagnose smallpox, but we found that the method can give false-positive results when a field specimen has been stored unfrozen for more than seven days. The test identifies orthopoxviruses as a group and cannot differentiate one orthopoxvirus from another. The test has been successfully used also for the antigenic detection of parapoxviruses, tanapoxvirus, and MC virus. However, the reliability of identification of these viruses has not been tested.

29.5.4 Complement Fixation

The complement-fixation (CF) test for the identification of viral antigen is sensitive and was useful in earlier days for the laboratory identification of smallpox virus antigen. Like the other serologic tests, it identifies orthopoxviruses as a group. This test has been used also to detect the presence of parapoxviruses, tanapoxvirus, and MC virus, but unless the lesion materials are "cleaned" by treatment with ether, an anticomplementary reaction often occurs, which makes the results useless.

29.5.5 Stained Smears

Stained smears of lesion materials were often used in the past to look for inclusion bodies present in specimens collected from smallpox patients but not present in specimens from chickenpox patients. The use of EM made this method obsolete, but it can still be used when EM is not available.

Table 29.2 lists seven methods which can identify poxvirus antigens. Although enzyme-linked immunosorbent assay (ELISA) and radioimmunoassay (RIA) are not listed in the table, ELISA has been used for identifying virus antigens of variola, vaccinia, monkeypox, and tanapoxviruses. Except for tanapoxvirus, these tests can only identify each virus as belonging to a group.

Table 29.2 Tests to Identify Poxvirus Antigens

Viruses	EM	CAM	TC	AGP	FA	CF	Stained Smears
Smallpox	+	+	+	+	+	+	+
Monkeypox	+	+	+	+	+	+	+
Vaccinia	+	+	+	+	+	+	+
Cowpox	+	+	+	+	+	+	+
Camelpox	+	+	+	+	+	+	+
Whitepox	+	+	+	+	+	+	+
Milker's Nodule	+		+	$+^a$	+	+	
BPS	+		+	$+^a$	+	+	
Orf	+		+	$+^a$	+	+	
Tanapox	+			$+^a$	+	+	
MC	+			$+^a$	$+^a$	$+^a$	+

Abbreviations: EM = electron microscopy; CAM = choriallantoic membrane of embryonated chicken, 11 to 13 days old; TC = tissue culture (cell culture); AGP = agar gel precipitation test; FA = fluorescent antibody test; CF = complement-fixation test; BPS = bovine papular stomatitis; MC = molluscum contagiosum.

^{a} The test can be used, but its reliability is uncertain.

29.6 SEROLOGIC METHODS FOR ANTIBODY ASSAY

For quick and accurate diagnosis of poxvirus infections, virologic specimens are preferred over serum specimens; however, such a specimen at times cannot be collected because the patient was seen too late in the course of the infection. Thus, a blood specimen for antibody assays must be collected and becomes very important for the diagnosis of an infection. Blood specimens are also important in an epidemiologic survey to determine the extent of a past infection in a population. A list of serologic methods for antibody assays of various poxvirus infections is given in Table 29.3.

29.6.1 Hemagglutination Inhibition Test

As shown in Table 29.3, the hemagglutination inhibition antibody assay can be performed only for smallpox, monkeypox, vaccinia, cowpox, and buffalopox, because only viruses in the orthopoxvirus group produce hemagglutinin.

A standard microtiter method as de-

scribed in Chapter 14 can be used. Virtually any virus in the orthopoxvirus group can be used as the hemagglutinin (viral antigen) for the HI test; however, because of its easy availability and its fairly high hemagglutinin content, vaccinia virus traditionally has been used. (See Nakano, 1979 for details of preparation.)

A suspension of chicken erythrocytes is used as a standard reagent, but since erythrocytes from only about 50% of the chickens will agglutinate with vaccinia hemagglutinins, erythrocytes must be pretested. (See Nakano, 1979 for details of preparing chicken erythrocyte suspension.)

Sera stored for a long time with inadequate refrigeration and sera from blood samples collected after death may contain nonspecific HI factor. These sera must be treated with periodate (Nakano, 1979). Occasionally sera will contain a nonspecific hemagglutinin which must be removed by absorbing with 50% chicken erythrocytes suspension before the HI test. (See reference by Nakano, 1979, for the procedure.)

Hemagglutination inhibition antibody is detectable within four to seven days after

Table 29.3 Serologic Methods in Assaying Poxvirus Antibodies

Poxvirus Infection	HI	NT	IFA	ELISA	RIA	RIAA	CF	AGP
Smallpox	+	+	+	+	+	+	+	+
Monkeypox (human and monkey)	+	+	+	+	+	+	+	+
Vaccinia	+	+	+	+	+	+	+	+
Cowpox	+	+	+	+	+		+	+
Buffalopox	+	+	+	+	+		+	+
Tanapox		+	+	+			+	+
Milker's nodule		+	+	+			+	+
Bovine papular stomatitis		+	+				+	+
Orf		+	+	+			+	+
Molluscum contagiosum		+[a]	+[a]				+[a]	+[a]

Abbreviations: HI = hemagglutination inhibition; NT = neutralization test; IFA = indirect fluorescent antibody; ELISA = enzyme-linked immunosorbent assay; RIA = radioimmunoassay; RIAA = radioimmunoassay-adsorption; CF = complement fixation; AGP = agar gel precipitate.

[a] No routine test can be performed.

infection in patients with smallpox or human monkeypox. For a patient with a dependable clinical history of either of these diseases, an HI titer of greater than 40 in blood specimens collected four weeks after onset of the infection may be diagnostically significant, provided that the serum was treated to eliminate nonspecific hemagglutinin and HI factor. The HI test is also suitable to measure antibody response in other orthopoxvirus infections, in addition to smallpox and monkeypox.

Although the HI test is useful in measuring a response to smallpox vaccination in individuals who are vaccinated for the first time or who are being revaccinated after several years (more than five years), it is not useful for a large number of individuals who are vaccinated every three years and for a greater number of individuals who are vaccinated every year, because the viral replication at the site of inoculation is far less in the latter two groups of individuals. Although HI antibody titer was traditionally believed to be short-lived, we have found HI titers of 10 to 20 after three or more years in many patients who had human monkeypox. The HI test identifies

orthopoxvirus group antibody and not specific virus antibody.

29.6.2 Neutralization Test

As shown in Table 29.3, an assay for neutralizing antibody assay can be performed for the ten infections listed. For the orthopox group (smallpox, monkeypox, vaccinia, cowpox, and buffalopox), the assay can be done by a method using tissue culture or CAM, but for infections caused by other viruses that do not grow on CAM (tanapox, milker's nodule, BPS, orf, and MC), only the tissue culture method can be used.

Procedures for neutralization test (NT) using tissue culture are found in Chapter 12 and in Nakano (1979) but for procedures using CAM see reference by Boulter (1957).

The live-virus antigen used for orthopoxvirus NT is monkeypox virus, and the neutralizing antibody assay is based on 50% plaque reduction, in which the plaques formed under a liquid medium are counted at 44 to 48 hours after the virus-serum mixture is inoculated on cell cultures. Because of the close antigenic relationship of the viruses within the orthopoxvirus group, as demonstrated by the HI test, NT cannot identify the causative poxvirus from the antibody assay.

In a patient with smallpox, human monkeypox, or vaccinia, and probably with other orthopoxvirus infections, neutralizing antibody is detected in the latter part of the first week or during the second week after infection and the antibody persists for a number of years. Orthopoxvirus NT antibodies can vary from 500 to 2000 and, therefore, a positive control serum with an average titer of 1000 should be included in each test run.

In many cases in which laboratory diagnosis of smallpox or human monkeypox is dependent upon serologic results, only one serum specimen is collected at some time after the onset of rash. If this specimen has a titer of less than 500, a definite diagnosis cannot be made, but if the titer is greater than 1000, the patient probably had the infection.

The neutralizing antibody assay for parapox is the 50% plaque reduction technique using tissue cultures such as human embryonic lung fibroblast cells or LLC-MK$_2$ ovine or bovine cells. Specific live virus is used to assay its corresponding NT antibody, e.g., orf virus must be used to assay antibody for orf virus, etc. Although the NT for parapoxviruses has been used successfully in the laboratory, its reliability is unknown because its usage has been limited. We do know that a convalescent-phase serum should be collected at four to five weeks after onset of the infection, however, and any positive result in conjunction with clinical disease confirms the diagnosis. Because the three viruses are antigenically related, the specific etiologic agent cannot be identified unless the animal contact of the patient is identified (e.g., orf virus for sheep, milker's nodule virus for milking cows and their calves, and BPS virus for beef cattle and calves.)

The technique for tanapox is again 50%

plaque reduction with tissue culture choices of primary African green monkey kidney cells, human embryonic fibroblast cells, or LLC-MK$_2$, and with live tanapox virus. Any positive result again confirms the diagnosis of tanapox. Because very few blood specimens collected from cases of human tanapox within two weeks after onset of the infection were positive, we recommend that the blood specimens be collected at four to five weeks after onset of the infection.

Routine NT for MC is not available because the virus does not grow well in any cell system or any known laboratory animal.

29.6.3 Indirect Fluorescent Antibody Test

As shown in Table 29.3, indirect fluorescent antibody (IFA) test can be used for the ten infections listed. Procedure for IFA is described in Chapter 7, but the method used at the CDC differs from the usual in that the infected cells on the test slides are not fixed with acetone or methanol but merely dried thoroughly, leaving the slides with the infected cells overnight in a class II safety cabinet with the air-ventilation switch on.

To detect antibodies to orthopoxviruses, IFA has not been used extensively for smallpox, but has been used for monkeypox. Cells infected with monkeypox virus are the antigen used to detect antibody in the specimen. An IFA titer of greater than 32 in a serum specimen from a case of suspected monkeypox virus is probably diagnostic of the infection if the patient had no previous smallpox vaccination. The monkeypox virus antibody measured by IFA test does not become detectable in many cases until the second week after onset of the infection. The IFA test was not suitable for serologic survey of monkeypox; the IFA titers begin to decrease after six months and may not be detectable at one or one and a half years.

The IFA test for antibody to parapoxvirus in our experience is more sensitive than the CF test. Although the viruses show some cross-reaction, we have used orf virus for orf antibody assay and milker's nodule virus for milker's nodule and BSP antibody assays. Any IFA positive result is diagnostic of these diseases since these diseases are not commonly found in humans.

The convalescent-phase serum should be collected about five weeks after onset of an infection since those collected at two to three weeks are often negative. Again, as described previously, the etiologic agent can be determined only after the patient's animal contact is known.

The IFA test for antibodies to tanapoxvirus is often negative for serum collected 12 to 15 days after the infection and, therefore, the serum should be collected at five to six weeks after onset of the infection.

No routine IFA test is presently available for MC since cells infected with MC virus cannot be produced.

29.6.4 Enzyme-Linked Immunosorbent Assay

An indirect ELISA test was adapted at CDC from the indirect ELISA test described by Voller et al (1976) to detect antibody to orthopoxviruses using monkeypox virus as the antigen. An ELISA test using selective adsorption has been developed by S.S. Marennikova at the Research Institute for Viral Preparations, Moscow, USSR, that can identify specific monkeypox antibody (personal communication).

An ELISA, test used at CDC, has aided in the diagnosis and serologic survey of human monkeypox cases in Africa. An ELISA titer greater than 160 in a serum from a patient with suspected monkeypox who had no previous smallpox vaccination and whose blood had been collected two to

six weeks after onset of the illness is diagnostic for human monkeypox.

The ELISA test has also been adapted at CDC for antibody assays of parapoxviruses (orf and milker's nodule) and tanapoxvirus. For parapoxviruses, the reliability of the test remains to be determined. But for tanapox, the ELISA test, for many patients, is capable of detecting antibody at an earlier time after onset of the disease than the IFA or the NT.

The ELISA test is not available for the assay of MC antibody since the virus virtually cannot be grown in a laboratory.

29.6.5 Radioimmunoassay

The RIA for orthopoxvirus antibody assay was reported by Ziegler et al (1975) and was used at CDC in diagnosing smallpox, vaccinia, and human monkeypox. Among the serologic tests for poxvirus antibody, RIA is probably the most sensitive in detecting antibodies to orthopoxviruses. However, for the detection of antibodies to smallpox and monkeypox in the very early stage of these illnesses, HI appears to be better than the RIA. Because of its high sensitivity, RIA titers less than 100 are questionable. Radioimmunoassay titers of 3000 to 20,000 or more can be found in a blood specimen collected within four to six weeks after onset of human monkeypox.

29.6.6 Radioimmunoassay-Adsorption Test

The radioimmunoassay-adsorption test reported by Hutchinson et al (1977) has been used at CDC mainly for the identification of specific antibodies to monkeypox virus, but it has also been used for the identification of specific antibody of vaccinia and variola viruses. The RIA test is first used to screen sera with titers greater than 500. After a high-titered serum is adsorbed by normal CAM and vaccinia virus antigen, the residual antibody in the serum is reacted against

CAM, vaccinia virus, monkeypox virus, and variola virus. The reaction pattern produced determines whether the serum contains specific antibody to monkeypox, vaccinia, or variola viruses. The test often cannot identify a specific antibody when a blood specimen was collected too early (three to four weeks) after onset of a disease. Also, the test sometimes cannot identify specific antibodies to monkeypox virus if the patient with monkeypox was vaccinated against smallpox at some time before

contracting the illness. The test has been useful in identifying cases of monkeypox in a serologic survey and in a situation in which no virologic specimens could be collected.

29.6.7 Complement-Fixation Test and Agar Gel Precipitation for Antibody Assay

The use of these two tests has been discontinued at CDC.

REFERENCES

Baxby, D. 1977. Is cowpox misnamed? A review of ten human cases, 1977. Br. Med. J. 1:1379–1380.

Baxby, D. 1981. Jenner's Smallpox Vaccine. The Riddle of the Origin of Vaccinia Virus. London: Heinemann Educational Books.

Baxby, D., Ashton, D.G., Jones, D., Thomsett, L.R., and Denham, E.M.H. 1979. Cowpox virus infection in unusual hosts. Vet. Rec. 109:175.

Baxby, D. and Hill, B.J. 1971. Characteristics of a new poxvirus isolated from Indian buffaloes. Archiv. ges. Virusforsch. 35:70–79.

Becker, F.T. 1940. Milker's nodules. JAMA 115:2140–2144.

Boulter, E.A. 1957. The titration of vaccinial neutralizing antibody on chorioallantoic membranes. J. Hyg. 55:50–52.

Breman, J.G., Kalisa-Ruti, Steniowski, M.V., Zanotto, E., Gromyko, A.I., and Arita, I. 1980. Human monkeypox 1970–1979. Bull. WHO 58:165–182.

Brown, S.T., Nalley, J.F., and Kraus, S.J. 1981. Molluscum contagiosum. Sex. Transm. Dis. 8:227–233.

Downie, A.W. 1939. A study of the lesions produced experimentally by cowpox virus. J. Pathol. Bacteriol. 48:361–379.

Downie, A.W., Taylor-Robinson, C.H., Count, A.E., Nelson, G.S., Mason-Bahr, P.E.C., and Matthews, T.C.H. 1971. Tanapox: A

new disease caused by a poxvirus. Br. Med. J. 1:363–368.

Dumbell, K.R. and Archard, L.C. 1980. Comparison of whitepox (b) mutants of monkeypox virus with parental monkeypox and with variola-like viruses isolated from animals. Nature 286:29–32.

Dumbell, K.R. and Kapsenberg, J.G. 1983. Laboratory investigation of two "whitepox" viruses and comparison with two variola strains from southern India. Bull. WHO 60:3281–3287.

Esposito, J.J., Obijeski, J.F., and Nakano, J.H. 1978. Orthopoxvirus DNA: Strain differentiation by electrophoresis of restriction endonuclease fragmented virion. DNA. Virology 89:53–66.

Hopkins, D.R. 1983. Princes and Peasants: Smallpox in History. Chicago: University of Chicago Press.

Hutchinson, H.D., Ziegler, D.W., Wells, D.E., and Nakano, J.H. 1977. Differentiation of variola, monkeypox and vaccinia antisera by radioimmunoassay. Bull. WHO 55:613–623.

Jenner, E. 1798. An inquiry into the causes and effect of the variolae vaccinae, a disease known by the name of the cow pox. Sampson Low No. 7, Soho, London.

Khodakevich, L., Jezek, Z., and Kinzanka, K. 1986. Isolation of monkeypox virus from wild squirrel infected in nature. Letters to the Editor, Lancet 1:98–99. (No. 8472, 11 January 1986).

Marennikova, S.S., Seluhina, E.M., Maltseva,

N.N., Cimiskjan, K.L., and Macevic, G.R. 1972. Isolation and properties of the causal agent of a new variola-like disease (monkeypox) in man. Bull. WHO 46:599–661.

Marennikova, S.S., Maltseva, N.N., Korneeva, V.I., and Garanina, V.M. 1975. Pox infection in carnivora of the family Felidae. Acta Virol. 19:260.

Marennikova, S.S., Maltseva, N.N., and Korneeva, V.I. 1976. Pox in giant anteater due to agent similar to cowpox virus. Br. Vet. J. 132:182–186.

Mardsen, J.P. 1948. Variola minor, a personal analysis of 13,686 cases. Bull. Hyg. 30:735–746.

Mathew, T. 1976. Comparative studies on the propagation of poxvirus in chick embryo with special reference of buffalopox virus. Kerala J. Vet. Sci. 1:48–56.

Nakano, J.H. 1977. Comparative diagnosis of poxvirus diseases. In E. Kurstak and C. Kurstak (eds.), Comparative Diagnosis of Viral Diseases. Vol. I, Part A. Human and Related Viruses. New York: Academic Press, 289–339.

Nakano, J.H. 1979. Poxviruses. In E.H. Lennette and N.J. Schmidt (eds.), Diagnostic Procedures for Viral, Rickettsial and Chlamydial Infections, 5th Ed. Washington, D.C.: American Public Health Association, 257–308.

Nakano, J.H. 1982. Human poxvirus diseases and laboratory diagnosis. In L.M. de la Maza and E.M. Peterson (eds.), Medical Virology. New York: Elsevier Science Publishing, 125–147.

Payne, L.G. and Norrby, E. 1976. Presence of haemagglutinin in the envelope of extracellular vaccinia virus particles. J. Gen. Virol. 32:63–72.

Payne, L.G. 1979. Identification of the vaccinia hemagglutinin polypeptide from a cell system yielding large amounts of extracellular enveloped virus. J. Virol. 31:147–155.

Singh, I.P. and Singh, S.B. 1967. Isolation and characterization of the aetiologic agent of buffalopox. J. Res. Ludhiana 4:440–448.

Smadel, J.E. 1948. Smallpox and vaccinia. In T.M. Rivers (ed.), Viral and Rickettsial Infections of Man. Philadelphia: J.B. Lippincott, 314–336.

Thomsett, L.R., Baxby, D., Denham, E.M.H. 1978. Cowpox in domestic cats. Vet. Rec. 108:567.

Torres, C.M. and de Castro Teixeira, J. 1935. Culture du virus l'alastrim sus les membranes de l'embryon de poulet. Compt. Rend. Seances Soc. Biol. Filiales 118:1023–1024.

Tripathy, D.N., Hanson, L.E., and Crandell, R.A. 1981. Poxviruses of veterinary importance: Diagnosis of infections. In E. Kurstak and C. Kurstak (eds.), Comparative Diagnosis of Viral Diseases. Vol. III. Vertebrate Animals and Related Viruses, Part A. DNA Viruses. New York: Academic Press, 268–348.

Voller, A., Bidwell, D., and Bartlett, A. 1976. Microplate enzyme immunoassays for the immunodiagnosis of virus infection. In N.E. Rose and H. Friedman (eds.), Manual of Clinical Immunology. Washington, D.C.: American Society of Microbiology, 506–512.

Von Magnus, P., Andersen, E.K., Petersen, K.B., and Birch-Anderson, A. 1959. A pox-like disease in cynomolgus monkeys. Acta Pathol. Microbiol. Scand. 46:156–176.

World Health Organization. 1972. Expert Committee on Smallpox Eradication. WHO Tech. Rep. Ser. No. 493:24–27.

Ziegler, D.W., Hutchinson, H.D., Koplan, J.P., and Nakano, J.H. 1975. Detection by radioimmunoassay of antibodies in human smallpox patients and vaccinees. J. Clin. Microbiol. 1:311–317.

Zwart, P., Gispen, R., and Peters, J.C. 1971. Cowpox in okapis: Okapia Johnsloni at Rotterdam Zoo. Br. Vet. J. 127:20–24.

30

Parvoviruses

Stanley J. Naides

30.1 INTRODUCTION

The family *Parvoviridae* is comprised of small, single-stranded DNA viruses. Parvo derives from the Latin "parvus," meaning small. The family is composed of three genera. The genus *Parvovirus* is comprised of those members of the family which infect mammalian hosts and are autonomous in their ability to replicate in host cells. The genus *Dependovirus* consists of those members of the family that require the presence of a helper virus such as adenovirus or herpesvirus in order to replicate. The dependoviruses were formerly known as adeno-associated viruses. The third genus, *Densovirus*, is composed of those autonomously replicating insect viruses previously known as densonucleosis viruses because of characteristic nuclear changes in the host cell occurring during infection.

Prior to 1975, parvoviruses had not been documented to infect humans. In 1975, Yvonne Cossart and her colleagues reported the discovery of parvovirus-like particles in human serum that was being screened for hepatitis B surface antigen (HB$_s$Ag). They were comparing sensitivity

and specificity of HB$_s$Ag detection by electrophoresis versus reverse passive hemagglutination (RPHA) using 3,219 sera received in a routine clinical laboratory (Vandervelde et al, 1974). Three sera were found to be positive for HB$_s$Ag by electrophoresis, but not by what were considered more sensitive RHPA or radioimmunoassay tests (R1A). Two of these three sera were positive for HB$_s$Ag by electron microscopy. One of these three sera was included in a control panel of sera for HB$_s$Ag tests. This serum was labelled No. 19 in panel B.

Cossart and her colleagues then realized that the sera that were positive in electrophoresis assays for HB$_s$Ag, but negative in more sensitive RPHA and RIA tests, contained a new viral antigen. The antibody source for the detection of HB$_s$Ag by the electrophoresis tests was human serum. This serum contained antibodies against HB$_s$Ag as well as antibodies against the new antigen. The antibody source used for detection of HB$_s$Ag in the RPHA and RIA tests were hyperimmune antisera that was raised in animals using a purified preparation of HB$_s$Ag. Thus, these animal antisera contained no antibodies against the

547

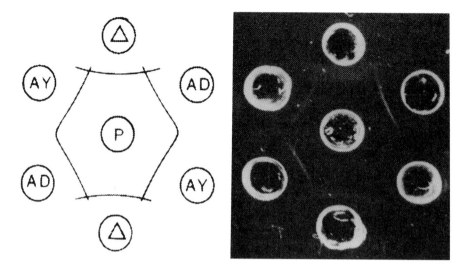

Figure 30.1 Ouchterlony gel diffusion test showing newly discovered B19 antigen (△) non-identity with HB antigen subtypes ad and ay, using anti-B19 antibody positive antiserum P (with permission from Cossart et al, 1975).

new antigen. The non-identity of the new antigen to HB$_s$Ag was demonstrated in Ouchterlony gel diffusion tests (Figure 30.1) (Cossart et al, 1975). Electron microscopy of the serum containing the new antigen revealed spherical particles and empty shells typical of the *Parvoviridae*. On cesium chloride gradients, the antigen banded at a bouyant density of 1.36 to 1.40, also typical of *Parvoviridae*. The antigen was initially identified in the serum from nine healthy blood donors, one patient with acute hepatitis, and one recent renal transplant recipient. Convalescent sera from four of these individuals demonstrated loss of virus with concomitant seroconversion. While Cossart was unable to associate the presence of virus with a specific illness, initial studies found 30% of adults had antibodies against this virus. Cossart noted the similarity of their new antigen, subsequently labelled B19, to particles in feces described in patients with acute gastroenteritis (Paver et al, 1973; Pattison, 1988).

30.2 CHARACTERISTICS OF PARVOVIRUS B19

Parvovirus B19 is a non-enveloped, icosahedral virus which on electron microscopy may appear spherical. The particle measures approximately 23 nm in diameter, although a range of particles, from 20.5 nm to 25 nm, has been described. The parvovirus B19 genome is a 5.6 kilobase, single-stranded DNA molecule characterized by imperfect palindromes at both the 3′ and 5′ ends which can form terminal hairpin loops (Summers et al, 1983; Cotmore and Tattersall, 1984; Astell and Blundell, 1989). While most autonomous parvoviruses possess unique sequences at each terminus, sequencing studies revealed that B19 differs in that its termini are inverted terminal repeats (Deiss et al, 1990). In this respect, B19 resembles adeno-associated virus (AAV) of the sister genus *Dependovirus* (Berns and Hauswirth, 1984). As a result of this structure, B19 is able to package either a plus or a minus DNA strand in a given virion in approximately equal numbers during replication. A single copy of the genome is encapsidated.

B19 replication follows a modified rolling hairpin model characteristic of the autonomous parvoviruses (Tattersall and Ward, 1976). The imperfect palindrome at the 3' end of the molecule forms a region of double-stranded DNA that primes 3' elongation during replication. As the 3' complementary strand (C_1) elongates, it unfolds the 5' end of the parental strand (V_1). This represents the monomeric replicative form. The 3' end of C_1 which is now complementary to the parental 5' V_1 hairpin, is able to fold back on itself and continue elongation. As C_1 elongates, a complementary strand (V_2) with the same sequence as the parental virus (V_1) is made. As 3' end elongation continues, a second complementary strand (C_2) is synthesized. The resulting form is the dimeric replicative form. A postulated "nickase" reduces the replicative forms to progeny virus. Nickase reduction results in two distinct configurations of the distal 375 nucleotide palindromes which are inverted complements of each other. These alternate configurations of the terminal hairpins have been found in all parvoviruses analyzed so far, and are referred to as "flip" and "flop" (Astell, 1990; Carter et al, 1990). While parvovirus B19 is autonomous in that it does not require a helper virus, productive replication is still restricted to rapidly dividing cells where cellular factors required for viral replication are found in the nuclei. Parvovirus B19, like other parvoviruses, replicates and assembles in the cell nucleus.

Nearly full-length DNA from clones of two viral isolates have been sequenced (Shade et al, 1986; Blundell et al, 1987). The virus employs a somewhat simple coding strategy. Interposed between the palindromes, a single strong promoter at map unit 6 initiates transcription of the left-handed non-structural protein gene region,

as well as the right-hand structural gene region (Blundell et al, 1987). The non-structural protein NS1 is approximately 74,000 daltons in size and is encoded between nucleotides 435 to 2,448. NS1 is thought to provide the nickase activity necessary to reduce replicative forms to progeny virus. NS1 may also play a role in assembly of viral DNA into mature viral capsids. Cotmore and Tattersall (1989) have identified NS1 on the external surface of mature B19 viral particles covalently linked to virion DNA.

Structural proteins VP1 and VP2 are encoded by nucleotides 2444–4786 and 3125–4786, and are 84,000 and 58,000 daltons, respectively (Cotmore et al, 1986; Ozawa and Young, 1987). Both structural proteins are encoded in the same open reading frame, but VP2 results from an alternatively spliced transcript. Messenger RNA for VP2 is initiated at an ATG codon at nucleotides 3125 to 3127 (Ozawa et al, 1987a). Comparing the two sequenced isolates, 37 nucleotide changes were located within the large open reading frame, including the VP1 and VP2 genes, of which eight nucleotides caused non-conservative changes in seven amino acids. One of these two isolates sequenced was the original B19-WI isolate described by Cossart and her colleagues. The other isolate, B19-AU, was derived from a teenager with homogeneous sickle cell anemia in aplastic crisis (Blundell et al, 1987). Digestion of B19 isolates with panels of restriction endonucleases has revealed strain variation and sequence drift, but these were not associated with differences in clinical syndromes (Mori et al, 1987).

30.3 PATHOGENESIS OF VIRUS INFECTION

Much of the early knowledge of the natural history of B19 infection is based upon studies of experimental infection in normal volunteers conducted by Mary Anderson and her colleagues at the Common Cold Research Unit, Harvard Hospital, in Salisbury, England (Anderson et al, 1985a). Plasma containing parvovirus B19 was obtained from a normal blood donor. The presence of other infectious agents was ruled out by inoculation of samples of this plasma into an array of bacterial and viral screening tests. The plasma was diluted in Hank's buffer saline containing 0.2% bovine serum albumin, and 0.5 mL of diluted plasma was inoculated into each nostril of a volunteer. Volunteers with pre-existent anti-B19 antibodies did not develop viremia. Virus was first detected six days after inoculation in previously seronegative individuals. Peak titers of virus were reached on days eight and nine post-inoculation. Viremia was present up to seven days. Virus was detected in nasal washes and gargle specimens between days seven and eleven post inoculation, during the same period as the viremia. Virus was not detected in urine or fecal specimens from any of the volunteers, except for one woman whose urine was contaminated with menstrual blood. High titer IgM antibody to B19 developed during the second week after inoculation. Anti-B19 IgG antibody began to develop at the end of the second week or early in the third week after inoculation. In one individual with a trace amount of pre-existing anti-B19 IgG, increased levels of IgG antibody were detected at the time of a mild IgM antibody response nine to ten days after inoculation. None of the volunteers with a significant anti-B19 IgG antibody level prior to inoculation developed a subsequent IgM antibody response.

The clinical illness associated with experimental B19 infection was biphasic. In some individuals the viremia was asymptomatic, but in others it was associated with a flu-like illness consisting of malaise, myalgia, and/or transient fever. Several individuals experienced headache, pruritis, and chills. The second phase of the illness began towards the end of the second week post-inoculation and was characterized by rash, arthralgia, and arthritis.

Hematologic alterations associated with B19 infection were noted by Anderson and her colleagues. There was an absolute areticulocytosis from the period of peak viremia until several days after the onset of the antibody response. The hemoglobin fell during the week after onset of areticulocytosis, but the decrease in hemoglobin was not clinically significant. Neutropenia was detected as early as day eight post-inoculation, and persisted through the second week following infection. Lymphocyte and platelet counts were transiently depressed during the second week postinoculation. No significant trends were observed in the monocyte, basophil, or eosinophil counts.

The incubation period between inoculation and onset of symptoms in natural infections can be inferred to range from 6 to 18 days based upon the volunteer studies and the epidemiologic studies of B19 outbreaks; however, studies of some outbreaks have suggested that the incubation period may be as long as 28 days. Individuals are infectious during the period of viremia. Since many patients will not present until the onset of either rash or joint symptoms, most patients will no longer be infectious at the time of presentation. Onset of rash, polyarthralgia, or polyarthritis are temporally associated with development of an anti-B19 antibody response which, as we have seen, results in clearance of viremia and cessation of viral shedding (Joseph, 1986). Anti-B19 IgM antibody may be present up to two to three months, but declines thereafter. Anti-B19 IgG antibody response is long-lived. Anti-B19 IgG antibody seroprevalence in the adult population ranges between 40% to 60%. Acquisition of antibody to B19, suggesting infection, is accelerated after 5 years of age when many children first enter school (Anderson et al, 1986).

30.4 CLINICAL MANIFESTATIONS OF PARVOVIRUS B19 INFECTION

After Cossart's report, parvovirus B19 was a virus in search of a disease. While screening serum in a clinical laboratory for evidence of parvovirus B19 infection, Pattison and his colleagues observed six patients with sickle cell disease in aplastic crisis who had evidence of a recent B19 infection (Pattison et al, 1981). Aplastic crisis in sickle cell disease was first described in 1950 as an acute fall in hemoglobin associated with cessation of new erythrocyte formation. Following areticulocytosis lasting approximately seven to ten days, a brisk reticulocyte rebound occurred with eventual return of hemoglobin to baseline levels. While the etiology of aplastic crisis in sickle cell anemia and other hemolytic anemias was not known, an infectious agent was considered to be likely because the crisis usually occurred only once in any given individual and occasionally occurred in outbreaks within chronic hemolytic anemia clinic patient populations. Shortly after Pattison's observation, Serjeant and his colleagues confirmed that B19 was the cause of an epidemic of aplastic crisis in Jamaica between 1979 and 1981 (Serjeant et al, 1981). Subsequent studies demonstrated that B19 infection may cause aplastic crisis in chronic hemolytic anemia regardless of the anemia's underlying etiology. In short order, B19-induced aplastic crisis also was found to occur in individuals with hereditary spherocytosis, alpha and beta thalassemias, pyruvate kinase deficiency, glucose-6-phosphate dehydrogenase deficiency, pyrimidine 5'-nucleotidase deficiency, hereditary stomatocytosis, autoimmune hemolytic anemia, and HEMPAS (hereditary erythrocytic multinuclearity associated with a positive acidified (HAMS) test (Pattison et al, 1981; Duncan et al, 1983; Rao et al, 1983; Davidson et al, 1984; Evans et al, 1984;

Green et al, 1984; Kelleher et al, 1984; Bertrand et al, 1985; Summerfield and Wyatt, 1985; Tsukada et al, 1985; Lefrere et al, 1986a–d, 1986f; Saarinen et al, 1986; Takahashi et al, 1986; West et al, 1986; Hanada et al, 1988; Rappaport et al, 1989; Rechavi et al, 1989; Goldman et al, 1990; Mabin and Chowdhury, 1990). All these patients shared the need for brisk reticulocytosis to maintain their hemoglobin level in the face of shortened erythrocyte survival and during the period of areticulocytosis, individuals usually required transfusion support. Examination of the bone marrow revealed growth arrest at the giant pronomoblast stage of erythrocyte developments. Marginal intranuclear inclusions may be seen and represent accumulated B19 virus (Burton, 1986; Caul et al, 1988; Knisely et al, 1988).

Since the turn of the century when pediatricians enumerated the rashes of childhood, the fifth rash—erythema infectiosum, or fifth disease—was a well-described clinical entity. In May 1983, an outbreak of a rash illness with features of erythema infectiosum occurred in a primary school in North London and was investigated by Mary Anderson and her colleagues (Anderson et al, 1984). The majority of the children had the typical rash of classic erythema infectiosum which presents with bright red "slapped cheeks." The rash occurs on the cheeks in the majority of children and is also seen on the torso and extremities. The exanthem is frequently lacy or reticular in pattern but may be macular, maculopapular, and occasionally vesicular or hemorrhagic. It is pruritic in approximately half the cases. In the majority of children, the rash may recur after initial resolution. Recurrence is usually precipitated by sun exposure, hot bath, or physical activity. Recrudescence may occur for weeks or months following the acute infection, but during episodes of rash recrudescence, children are not infectious. While infection in children may be asymptomatic, when symptoms do occur they tend to be mild. These include sore throat, headache, fever, cough, anorexia, vomiting, diarrhea, and arthralgia (Brandrup and Larsen, 1976; Cramp and Armstron, 1976; Lauer et al, 1976; Shneerson et al, 1980; Anderson et al, 1983; Andrews et al, 1984; Clark, 1984; Okabe et al, 1984; Mynott, 1985; Nunoue et al, 1985; Plummer et al, 1985; Shiraishi et al, 1985; Chorba et al, 1986; van Elsacker-Niele and Anderson, 1987; Mansfield, 1988; Naides et al, 1988b). At the time of presentation, children with a rash usually have anti-B19 IgM antibodies (Shiraishi et al, 1985).

A number of uncommon dermatologic manifestations of B19 infection have been reported. A vesiculopustular eruption has been seen in adult patients with fifth disease. It is unusual because it combines the features of morbilliform and vesiculopustular lesions. Sub-epidermal edema, ballooning necrosis of the dermis, and a lymphohistiocytic infiltrate may be noted. The infiltrate may contain unusual binucleate giant cells. The vesicle itself may contain a neutrophilic infiltrate. Extravasation of erythrocytes into the dermis eventually gives the vesiculopustules a hemorrhagic appearance (Naides et al, 1988b). Purpura may be seen in the absence of thrombocytopenia (Lefrere et al, 1985; Mortimer et al, 1985b; Shiraishi et al, 1989). Some patients may have purpura as a result of thrombocytopenia (Kilbourne et al, 1987; Lefrere et al, 1989). The clinical appearance may suggest Henoch-Schönlein purpura (Lefrere et al, 1985; 1986e).

B19 infection may be associated with paresthesias in the fingers. Rarely, progressive arm weakness may occur as may numbness of the toes. In such instances, nerve conduction studies may show a mild slowing of nerve conduction velocities and decreased amplitudes of motor and sensory potentials (Faden et al, 1990).

Erythema infectiosum may also be seen in adults not previously infected. In adults, the rash tends to be more subtle and the bright red "slapped cheeks" absent. When adults are symptomatic, they tend to have a more severe flu-like illness in which polyarthralgia and joint swelling are more prominent. For example, Ager and his colleagues observed adult involvement in an outbreak of erythema infectiosum in Port Angeles, Washington in 1961–62. Subjects were identified by the presence of a typical rash. Only 5.1% of children under ten years of age had joint pain, and 2.8% joint swelling. In children infected during their adolescent years, joint pain and swelling occurred in 11.5% and 5.3%, respectively. However, in the 20 year or older age group, joint pain occurred in 77.2% and joint swelling 59.6% (Ager et al, 1966).

White and his colleagues were the first to demonstrate that B19 could cause a chronic rheumatoid-like arthropathy (White et al, 1985). Sera were collected from individuals presenting at an "early synovitis" clinic at the Royal National Hospital for Rheumatic Diseases, Bath, England, beginning in mid-1979. Primary care physicians were invited to refer patients to the clinic as soon as possible after the onset of acute joint symptoms. Traditionally, referral would wait until it became clear that joint symptoms had become chronic, usually a period of three months or more. Sera from 153 patients with early synovitis were retrospectively tested when tests for parvovirus B19 became available. Nineteen patients had evidence of a recent B19 infection. Although 49 of the cohort were men, all 19 patients with evidence of a recent B19 infection were women. Eighteen of the infected patients presented with acute, moderately severe symmetric polyarthritis that usually started in the hands or knees, and within 24 to 48 hours involved wrists, ankles, feet, elbows, and shoulders. The cervical spine was involved in two cases and the lumbosacral spine in one case. In three cases, the upper limbs alone were affected. All patients complained of joint pain, stiffness, and variable swelling. The acute polyarthropathy usually improved within two weeks, but completely resolved in only two cases. In 17 cases, symptoms persisted for more than two months, and in three cases, for more than four years. Thirteen of the nineteen B19 infected patients reported an influenza-like illness with malaise, fever, gastrointestinal symptoms, and/or rash. Two thirds of the patients had episodic flares, but remained symptomatic between flares. One third of the patients had episodic flares, but were symptom free between flares. There was no associated long-term joint damage (White et al, 1985). The distribution of joint involvement and its symmetry may suggest a diagnosis of a rheumatoid arthritis (Reid et al, 1985; White et al, 1985; Woolf et al, 1989; Naides et al, 1990). Many patients experience morning stiffness. About half of the patients with chronic B19 arthropathy meet the criteria of the American Rheumatism Association for diagnosis of rheumatoid arthritis (Arnett et al, 1988; Silman, 1988; Naides et al, 1990). Patients may have a transient expression of autoantibodies during acute infection, including rheumatoid factor, anti-DNA antibodies, and anti-lymphocyte antibodies (Luzzi et al, 1985; Semble et al, 1987; Naides and Field, 1988; Sasaki et al, 1989; Soloninka et al, 1989). While initial reports suggested that chronic B19 arthropathy was associated with the major histocompatibility antigen HLA DR4, as is seen in classic erosive rheumatoid arthritis, subsequent studies have demonstrated no increased association with DR4 (Klouda et al, 1986; Woolf et al, 1987).

During a regional outbreak of erythema infectiosum in Scotland, six pregnant women were found to have serologically documented B19 infection. Four delivered normal term infants. Two aborted grossly hydropic fetuses with anemia during the second trimester. There was evidence of fetal leukoerythroblastic reaction with eosinophilic changes in hematopoietic cell nuclei, hepatitis, and excess iron pigment in the liver. Hybridization with radio-labelled B19 DNA probes demonstrated viral DNA in fetal liver, heart, thymus, kidney, adrenal, and placental tissues (Anand et al, 1987). Additional observations have confirmed the association of maternal B19 infection with fetal hydrops (Brown et al, 1984; Knott et al, 1984; Mortimer et al, 1985a; Bond et al, 1986; Gray et al, 1986; Anand et al, 1987; Carrington et al, 1987; Clewley et al, 1987; Matsunaga et al, 1987; Woernle et al, 1987; Anderson and Hurwitz, 1988; van Elsacker-Niele et al, 1989; Knisely, 1990; Salimans, 1990).

The fetus is similar to an individual with chronic hemolytic anemia in that red cell survival is only 45 to 70 days and fetal red cell mass increases 34-fold during the second trimester (Gray et al, 1987). In effect, the B19 infected fetus develops B19-induced aplastic crisis resulting in a high output cardiac failure. The fetus becomes hydropic with soft tissue edema, ascites, pleural effusions, and in some cases, polyhydramnios. Ultrasound examination reveals a characteristic picture of fluid retention. During the first trimester and early second trimester when normal values for maternal serum alpha fetoprotein are available, a rise in this marker may herald ultrasound evidence of fetal hydrops (Carrington et al, 1987). Until a report by Woernle and his colleagues in 1987, fetal B19 infection was thought to be uniformly fatal. However, Woernle reported four pregnant women with anti-B19 IgM positive serology of whom one delivered a still-born hydropic fetus whose tissues were positive for B19 DNA by nucleic acid hybridization. While the other three IgM positive women gave birth to normal offspring, one of the neonates had anti-B19 IgM antibody positive cord serum. A second apparently healthy neonate was anti-B19 IgM negative, but IgG positive; the anti-B19 IgG antibody persisted in the infant's serum at nine months of age, confirming that it was of fetal origin consistent with an in utero infection (Woernle et al, 1987). While microopthalmia with abnormal lens development has been reported in an abortus, congenital anomalies have not been a common feature of B19 infection (Weiland et al, 1987; Kinney et al, 1988). However, recent reports suggest that occasionally abortuses may show evidence of developmental anomalies and that survivors of B19 infection in utero may demonstrate evidence of a congenital syndrome characterized by anemia, thrombocytopenia, and cardiac and hepatic dysfunction (Naides et al, 1988a). Fetal viral cardiomyopathy as a mechanism of fetal hydrops has been reported (Naides and Weiner, 1989). Hepatic dysfunction in neonates as well as adult B19 infection has been described (Naides, 1987; Naides et al, 1988a; Metzman et al, 1989). Whether developmental anomalies represent a direct viral effect or the indirect effect of severe illness during gestation remains to be determined. Fetuses have now been successfully treated in utero during the fetal aplastic crisis by in utero transfusion with excellent salvage and outcome (Peters and Nicolaides, 1990; Soothill, 1990; Sahakian et al, 1991).

Beginning in 1987, a series of reports noted that patients with congenital or acquired immune deficiency were unable to clear B19 viremia. These included Nezelof's syndrome, prior chemotherapy for lymphoproliferative disorders, immunosuppressive therapy for transplantation, or AIDS (Kurtzman et al, 1987, 1988, 1989b; Graeve et al, 1989; Young et al, 1989; Frickhofen and Young, 1989, 1990; Chrystie et al, 1990; de Mayolo and Temple, 1990; Frickhofen et al, 1990; Rao et al, 1990; Naides et al, 1991b). In the immune competent host, the IgM antibody response may last two months or more. Anti-B19 IgM antibody and acute phase IgG antibody (less than one week postinoculation) recognize determinants on VP2. In convalescent serum, anti-B19 IgG antibody recognizes

determinants on the VP1 structural protein (Kurtzman et al, 1989). B19 VP1 and VP2 are products of alternate transcription of the same open reading frame, and VP1 contains an additional 227 N-terminal amino acids not present in VP2 (Shade et al, 1986). VP1 therefore contains unique determinants not present in the truncated form represented by VP2. These determinants may be in the unique non-overlapping N-terminal region or, alternatively, represent conformational differences in the sequences shared between the two proteins. Western blot analysis of serum from individuals with congenital immune deficiency, prior chemotherapy, or AIDS, demonstrated the absence of convalescent anti-B19 IgG antibodies directed against VP1. These sera were unable to neutralize B19 virus in experimental bone marrow in vitro culture systems. While this work suggests that neutralizing determinants are unique to VP1 (since antibodies from immune deficient patients recognized VP2), recent work with synthetic peptides suggests that neutralizing determinants may be found on VP2 (Kurtzman et al, 1989a; Sato et al, 1991a, 1991b). Neutralizing activity to B19 is found in commercially available pooled immunoglobulin since seroprevalence of anti-B19 IgG antibodies in the adult population is approximately 50% (Anderson et al, 1986; Frickhofen et al, 1990; Naides et al, 1991b). In the absence of neutralizing antibodies to B19, virus persists in the bone marrow and causes chronic or intermittent suppression of one or more hematopoietic lineages. For example, B19 is a major cause of red cell aplasia in individuals with human immunodeficiency virus (HIV-1) infection. Young and his colleagues have demonstrated the utility of intravenous immunoglobulin in the treatment of B19 associated bone marrow suppression and B19 persistence (Frickhofen et al, 1990). A preliminary report suggests a role for intramuscular immunoglobulin therapy in the treatment of concurrent B19 infection in

AIDS patients (Naides et al, 1991b). However, immunoglobulin therapy may not be universally successful in clearing B19 persistence (Bowman et al, 1990).

Measures for infection control in the out-patient setting are limited to avoiding exposure for high risk groups. Unfortunately, most infections result from exposure to index cases during the period of viremia which is either asymptomatic or characterized by non-specific influenza-like symptoms. In the hospitalized patient, infection control measures are important in avoiding exposure of patients at risk and hospital personnel. Outbreaks of B19 infection amongst hospital staff have been documented (Bell et al, 1989). The following infection control measures are employed by the University of Iowa Hospitals and Clinics. Patients with fifth disease are isolated with secretion precautions for 24 hours after onset of rash, arthralgia, or arthritis. Secretion precautions require wearing gowns and gloves when handling body fluids or secretions, e.g., saliva, nasal aspirates, urine, stool, or blood. Patients who are likely to be viremic are isolated with secretion precautions. However, patients with hemolytic anemia, pregnancy, or immune compromise are not permitted to share a room with a potentially viremic patient. Isolation is continued for the duration of the illness and in most cases for the entire period of hospitalization. Mothers of infected newborns are placed in isolation, under secretion precautions for the duration of the hospital stay. Mothers may visit neonates in the nursery but must follow infection control procedures that include careful hand-washing and avoidance of contact with other infants. Household contacts of viremic patients are isolated with secretion precautions from day 7 until day 18 after contact. Throughout the period of hospitalization, employees, visitors, and patients who are pregnant are separated from persons who have fifth disease, those who are likely to be viremic, or who are

household contacts with either of these. While these measures have been universally accepted for infection control, some centers have also recommended respiratory precautions for viremic patients (Anderson et al, 1989). Respiratory precautions require separation of the patient in a single room and wearing of a mask by staff during all patient contact.

Special care should be taken when handling B19 in the laboratory. Laboratory acquisition of infection has been suggested with the most likely source being aerosolization during centrifugation, resuspension of viral pellets, and during washing stages of immunoassays (Cohen et al, 1988). Exposure of B19 virus to ultraviolet light does decrease infectivity, but it also reduces antigenicity of virus preparations used in serologic assays. It would appear prudent to survey laboratory personnel for their serological status and to caution individuals who are pregnant, immunocompromised, or who have hemolytic anemia from directly working with B19 virus or in situations that could lead to their direct exposure to the virus.

30.5 PRESENCE OF B19 IN VARIOUS TISSUES

In the immunocompromised patient, virus may be found continuously or intermittently in serum during the period of viremia. It is presumed that the virus may also be found in the bone marrow of the individuals during this period, representing a persistent infection. In fetal infections, virus may be found in serum, amniotic fluid, and fetal ascites, as well as an array of body tissues including liver, spleen, kidney, thymus, bone marrow, heart, and placenta. The time after infection of the prospective mother when B19 can be recovered from fetal tissues is not entirely clear, and may depend on the severity of infection, gestational age of the fetus at the time of infection, and whether transfusion therapy for the fetus is attempted. In apparently immune competent individuals with chronic B19 arthropathy, B19 DNA has been found in bone marrow aspirates and synovium (Foto et al, 1990; Naides et al, 1991a). Whether mature virus may be isolated from these tissues remains to be determined.

30.6 PREFERRED SITES FOR VIRUS ISOLATION

Virus is most easily isolated from serum obtained from individuals with aplastic crisis or persistent infection particularly in immunocompromised individuals. Virus may be found in bone marrow aspirates during the period of areticulocytosis in individuals with aplastic crisis, or in those with persistent virus infection. In immunodeficient individuals with persistent infection and in chronic B19 arthropathy, B19 may be detected in bone marrow even in the absence of detectable viremia (Rao et al, 1990; Foto et al, 1990). In fetal infection, virus has been isolated from cord serum, ascites, or amniotic fluid (Naides and Weiner, 1989). Small, round, parvovirus-like particles have been found in stool from individuals with gastroenteritis, but B19 has not been isolated from such specimens nor has it been reported in stool from individuals with known B19 infection. Rather, the small, round, parvovirus-like particles found in stool may be a related human parvovirus with sequence homology similar to B19 (Oliver and Phillips, 1988; Turton et al, 1990). Detection of B19 DNA in urine has been reported in one case of neonatal B19 infection (Naides et al, 1988a).

30.7 VIRUS STABILITY

B19 is not affected by ether, chloroform, 0.25% sodium deoxycholate, RNase, mi-

crococcal DNase, potassium iodide, or heating at 45 °C for 30 minutes. It is readily inactivated when heated at 56 °C for 5 minutes, by treatment with 1 mg/mL proteinase K, or by treatment with 0.05 N NaOH. Prior treatment of B19 with either 0.05 N HCl or glycine-HCl, pH 2.8 is about 75% effective in inhibiting B19 growth in bone marrow culture (Young et al, 1984). B19 may be stored frozen at −85 °C for indefinite periods without loss of virion infectivity. Special handling is not required although repetitive freeze-thawing may result in some loss of virus titer.

B19 may survive the dry heat processing of factor VIII and IX concentrates at 80 °C for 72 hours (Lyon et al, 1989). Solvent-detergent inactivation of factor VIII concentrate with tri-(n-butyl) phosphate detergents will not inactivate B19, and terminal dry heating of these preparations at 100 °C for 10 to 30 minutes or more has been recommended (Rubinstein and Rubinstein, 1990).

30.8 PROPAGATION OF B19 IN VITRO

Routine culture of parvovirus B19 in vitro is not available. All continuous cell lines tested failed to support B19 growth, including erythroleukemic cell lines such as K562 and HEL. Only very recently has a permissive cell line been established. The first reported growth of parvovirus B19 in primary bone marrow suspension cultures that were supplemented with erythropoietin was by Ozawa et al (1986). Optimally, the bone marrow should be derived from individuals with hemolytic anemia, e.g., sickle cell anemia, in whom the erythroid precursor pool is increased. The input to output ratio of virus was only approximately 1:50 for culture supernatants and 1:200 for total cultures (Ozawa et al, 1987b). Recently, propagation of B19 in primary fetal liver cell culture and human cord blood culture have been described (Brown et al, 1991; Zhou et al, 1991). Takahashi and his colleagues propagated B19 in bone marrow from a patient with chronic myeloid leukemia in erythroblastic crisis but were unable to establish continuous cell lines (Takahashi et al, 1989). Very recently, Komatsu and his colleagues reported propagation of B19 in a megakaryoblastic leukemia continuous cell line conditioned in erythropoietin (Komatsu et al, 1991; Shimomura et al, 1992). The difficulty in culturing B19 has led investigators to seek a readily renewable antigen source through recombinant DNA technology.

30.9 LABORATORY DIAGNOSIS OF B19 INFECTION

A number of approaches have been used to diagnose parvovirus B19 infection. In practice, several methodologies are used to confirm a diagnosis.

30.9.1 Electron Microscopy

Electron microscopy provides morphologic identification of the virus. Standard touch preparations that are negatively stained are made by allowing a liquid sample to dry on an electron microscopy grid that had been previously coated with a thin layer of plastic. After the sample has dried, it is stained with phosphotungstic acid or uranyl acetate, both providing electron dense material that accumulates around the particle to give a base relief appearance, or negative image.

In the absence of endogenous antibody to B19, incubation of the sample with specific antiserum or monoclonal antibody to B19 may cause aggregation of viral particles which can be visualized by electron microscopic examination (see Chapter 5). Cohen and his colleagues mixed the test specimen with an equal volume of anti-B19 positive serum. The mixture was then incubated for one hour at room temperature before being diluted 1:25 in phosphate buffered saline, centrifuged at 40,000 × g for one hour, and the pellet negatively stained with phosphotungstic acid (Cohen, 1988).

A combined pseudoreplica-immunochemical staining technique offers the advantage of preserving the morphology while providing a specific serologic diagnosis (Naides and Weiner, 1989). This approach may be useful even in the presence of endogenous B19 antibodies that cause virus to aggregate, since antigenic sites for binding exogenous anti-B19 antibodies may still be available. Virus in various body fluids may be examined by this technique. A 25 μL sample is allowed to absorb into an agarose block leaving the viral particles on the surface. The agarose is layered or coated with plastic which, after hardening, is floated off the agarose, inverted, and applied to a support grid. The pseudoreplica may then be negatively stained with phosphotungstic acid or uranyl acetate. Immune electron microscopy (IEM) may be performed on pseudoreplicated samples prior to negative staining. The samples for IEM are applied to nickle support grids as above, then blocked (incubated) with goat serum diluted 1/10 in a 0.1M phosphate buffered saline with 0.1% glycine prior to incubation with anti-B19 monoclonal antibody. We use 162-2B, an anti-B19 mouse monoclonal antibody of IgM isotope developed by Anderson and his colleagues at the Centers for Disease Control (Anderson et al, 1986). Samples are washed and then incubated with a polyclonal goat anti-mouse IgM (μ chain-specific) antibody conjugated to colloidal gold to enable visualization of the antibody by EM. Samples are then negatively stained with uranyl acetate and carbon coated (Figure 30.2; Naides and Weiner, 1989). Observation of viral particles with specific colloidal gold conjugated antibody allows species identification. B19 virus may be difficult to dis-

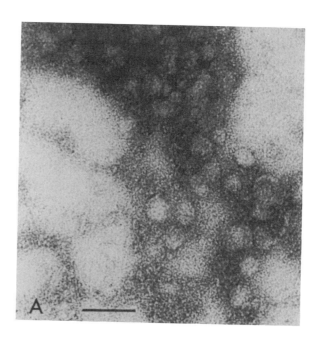

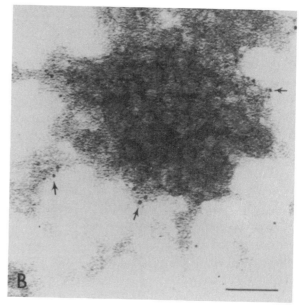

Figure 30.2 (A) Fetal ascites from a hydropic fetus showing viral particles without label, prepared by a pseudoreplica technique, and negatively stained by uranyl acetate. Original magnification, ×100,000; pictured here, ×280,000; internal scale, 50nm. (B) The fetal ascites prepared by pseudoreplica technique and identified as B19 by indirect labelling with colloidal gold (arrow) before negative staining with uranyl acetate. Original magnification, ×50,000; pictured here, ×140,000; internal scale, 100 nm (with permission from Naides and Weiner, 1989).

tinguish from enteroviruses on the basis of morphology alone. B19 particles may exclude uranyl acetate or phosphotungstic acid giving them a "full" appearance, or stain may enter the capsid giving the appearance of an "empty" shell.

30.9.2 Immunoassay

Both radioimmunoassays and enzyme-linked immunoabsorbent assays (ELISA) have been used to detect B19 antigen and specific antibody to the B19 virus. In antigen-capture assays, an anti-human IgM or IgG antibody is allowed to coat a solid phase. In the second step, a serum known to contain either anti-B19 IgM or IgG antibody is then incubated on the plate. Excess antibody is removed by washing. In the third step, the test serum is added to allow capture of B19 antigen should any be present. Detection of captured antigen uses a mouse monoclonal antibody to viral antigen followed by an anti-mouse antibody labelled with ^{125}I (RIA) or peroxidase (ELISA). Cohen and his colleagues first described the antibody-capture assay for anti-B19 IgM antibody in which the serum to be tested for anti-B19 antibody is added in the second step, and a serum known to contain B19 virus is substituted in the third step (Cohen et al, 1983). Detection of the captured antigen indicates the presence of antibodies to B19 virus in the patient's serum.

We employ an antibody-capture ELISA for detection of anti-B19 IgM and IgG antibody developed by Anderson and his colleagues at the Centers for Disease Control. Microtiter plates designed for use in ELISA are coated with goat antihuman IgM or IgG antibody diluted in carbonate/ bicarbonate buffer (Anderson et al, 1986). After overnight incubation, the plates are washed with phosphate buffered saline/ 0.15% Tween 20 (PBS-T) and serum to be tested for antibody is added. After 1.5 hours at 35 °C, the plates are washed with PBS-T and then serum containing high titer B19 virus that contains no endogenous antibody is added to the test wells. A control, antigen-negative, serum is added to parallel wells. After overnight incubation the plates are washed, and monoclonal antibody 162-2B (a mouse IgM) is added to the wells. The monoclonal antibody is diluted in PBS with 0.5% gelatin and 0.15% Tween 20 for the IgG serology, and in PBS with 0.5% gelatin, 0.15% Tween 20, and 2% normal goat serum for the IgM test. After washing, a peroxidase conjugated goat anti-mouse

IgM antibody is added. After incubation for one hour at 35 °C, the wells are washed, then 3,3',5,5'-tetramethylbenzidine dihydrochloride color developer is added to each well in the presence of the hydrogen peroxide, and the reaction stopped with 25 μL 2M H_2SO_4. A serum is considered positive when the value of the patient's serum sample incubated with virus (a serum from a viremic patient) is greater than three standard deviations (SD) above the mean value of controls (all patients' sera samples in an assay incubated with normal serum), and the difference between the values from wells with and without virus for a given test serum is greater than three SD (Bell et al, 1989). We utilize as an antigen source serum containing native B19 virus in the absence of endogenous anti-B19 antibodies. The presence of B19 virus in a prospective antigen source is detected by antigen-capture ELISA and confirmed by direct DNA hybridization methods, or by polymerase chain reaction with B19-specific primers, followed by hybridization of the amplified product using B19 specific probes (Naides et al, 1990). Anderson and his colleagues have demonstrated that the antibody-capture ELISA is highly sensitive and specific (Anderson et al, 1986). Early reports of cross reactivity between anti-B19 and anti-rubella antibodies were based on counter immunoelectrophoresis techniques which have been replaced by RIA and ELISA methods. In the latter, cross reactivity has not been a problem (Cohen and Shirley, 1985; Kurtz and Anderson, 1985; Cohen and Supran, 1987).

In response to the dependence of the laboratory on B19 viremic serum, a number of recombinant antigens have been developed for B19 testing. While these are not yet widely available, they no doubt will provide useful diagnostic reagents. Young and his colleagues were able to express B19 VP1 and VP2 protein in Chinese hamster ovary cells transfected with a B19 plasmid construct. The VP1 and VP2 self-assembled into empty capsids (Kajigaya et al, 1989). In addition, transfected cell lysates have been used as an antigen source. VP1 and VP2 have now been expressed in the baculovirus system as well (Brown et al, 1990). B19 synthetic peptides based upon sequence analysis of the viral capsid gene have been employed as an antigen source as well. Synthetic peptides provide approximately 90% sensitivity and specificity comparing serologic titers using native virus (Fridell et al, 1991). B19 antigen has also been prepared as fusion proteins expressed in *Escherichia coli* (Morinet et al, 1989).

30.9.3 Nucleic Acid Probes

B19 DNA may be detected by hybridization with cDNA probes, riboprobes (synthetic RNA), or synthetic oligonucleotide probes. These B19 specific probes have been used to identify virus by in situ hybridization (Hassam et al, 1990; Schwarz et al, 1991). Anderson and her colleagues, and Clewley first reported detection of B19 viral DNA using molecularly cloned probes that were ^{32}P-labeled. A virus-containing sample was blotted onto nitrocellulose filters using a dot blot manifold. It was then baked and hybridized with virus-specific probe (Anderson et al, 1985b; Clewley, 1985).

Since B19 encapsidates both positive and negative strands in equal numbers, B19 DNA forms double-stranded molecules on extraction and purification (Clewley, 1984). Clewley purified double-stranded B19 DNA then treated it with nuclease S1 to remove hairpin palindromic loops by cutting at exposed single-stranded regions in the termini. A nearly full-length genomic fragment was cloned by homopolymeric tailing after addition of cytidylic acid tails using terminal transferase. An insertion site was produced by adding guanylic acid tails to the cloning vector pBR322 after its linearization by restriction at the Pst I site (Clewley, 1985).

Non-radioactive labels for probes may be used for safety and because they have a long shelf life. For example, Mori and her colleagues used a biotin-labeled DNA probe and streptavidin-alkaline phosphatase conjugate (Mori et al, 1989). A number of investigators have used digoxigenin-labeled probes for detection of B19 (Azzi et al, 1990; Zerbini et al, 1990).

30.9.4 Polymerase Chain Reaction

The polymerase chain reaction (PCR) offers exquisite sensitivity and the ability to detect B19 DNA in an array of clinical specimens. Primers directed against sequences both in the non-structural protein and the viral capsid protein genes have been used. PCR is highly sensitive as demonstrated by Clewley who reported that 60 out of 95 anti-B19 IgM positive serum samples were found to be positive by PCR while only one was positive by dot blot hybridization. PCR was 99% specific in that only one sample in a control panel of 100 sera from individuals with other infections was positive by PCR, as well as by dot blot hybridization. He concluded that the high rate of detection of B19 DNA by PCR represents a slow rate of decay of viral DNA after peak viremia and is not a clinically significant phenomenon (Clewley, 1989). Koch and Adler, using oligonucleotide primers and detection of amplification products on agarose gels, found PCR 10,000 times more sensitive than dot blot hybridization. Southern analysis of amplified product using a radio-labeled oligonucleotide probe complementary to a sequence between the primers was 10^7 times more sensitive than dot blot hybridization. PCR detected B19 DNA in urine, amniotic fluid, pleural fluid, ascites, and leukocyte extracts (Koch and Adler, 1990). Other investigators have now reported the use of PCR to detect B19 in fetal and adult tissues (Clewley, 1989; Salimans et al, 1989; Foto et al, 1990). PCR has been successfully utilized to diagnose persistent B19 infection in immunocompromised patients in whom small amounts of B19 DNA in serum may be detected (Frick-hofen and Young, 1990; Naides et al, 1991b).

30.10 UNUSUAL FEATURES

Simpson and his colleagues reported a small parvovirus-like agent isolated from synovial tissue of a patient with severe rheumatoid arthritis after co-culture with the continuous cell line WI38, then passage through suckling mouse brain. The virus, designated RA-1, elicited a syndrome in neonatal mice that included neurological disturbances, dwarfism, alopecia, blephari-tis, and thoracic spine curvature (Simpson et al, 1984). However, presence of anti-RA-1 antibodies has not been associated with any clinical syndrome in humans. An-ti-B19 antibodies do not react with RA-1 isolates (Stierle et al, 1987). When various clinical specimens from more than 100 hu-man donors, including healthy subjects and patients with rheumatic disorders, were tested for RA-1 sequences by PCR amplifi-cation and Southern analysis, all donors tested positive for RA-1 related DNA se-quences. Therefore, RA-1 is distinct from B19. Reports of fecal parvovirus-like agents described above suggest that human parvoviruses other than B19 exist. How-ever, RA-1 has not been associated with any clinical syndrome to date, nor is it entirely clear that it is a human virus.

30.11 PREVENTION OF AND THERAPY FOR DISEASES DUE TO HUMAN PARVOVIRUS

Parvovirus B19 is ubiquitous. It is therefore difficult to prevent exposure. Community contacts are frequently asymptomatic. However, avoidance of potentially viremic individuals by those at high risk for compli-cations of B19 infection remains a prudent course.

There is no specific anti-viral therapy for parvovirus B19. In general, manage-ment of patients is symptomatic and sup-portive. Patients with aplastic crisis usually require transfusion support during the pe-riod of areticulocytosis. Adults with chronic B19 arthropathy benefit from non-steroidal anti-inflammatory drugs which control symptoms of arthralgia. We have attempted a short course of oral prednisone 5 mg daily in one patient with B19 arthrop-athy without apparent benefit (Naides et al, 1990).

When fetal infection is accompanied by severe anemia, transfusion support is required for fetal survival. Three centers have now reported using fetal transfusion in utero to support anemic fetuses through their aplastic crisis. Fetuses treated in this manner have survived to term and have been born without apparent long-term se-quelae (Peters and Nicolaides, 1990; Soothill, 1990; Sahakian et al, 1991). We reported one case of B19-induced fetal hy-drops without significant anemia. Hydrops in this fetus was due to a viral cardiomyop-athy. Digitalization of the fetus in utero resulted in resolution of ascites and other stigmata of congestive heart failure, but the fetus died despite the good response (Naides and Weiner, 1989). As described above, individuals with immunodeficiency may develop persistent B19 infection be-cause they lack the ability to produce anti-B19 neutralizing antibodies. Infusion of commercial immune serum globulin or im-munoglobulin, has been used successfully to clear persistent infection in patients with congenital immune deficiency, hematologic malignancy, or AIDS (Kurtzman et al, 1988; 1989a; Frickhofen et al, 1990). Intra-venous immunoglobulin is given at a dose of 0.4 grams/kg daily for five or ten days. Therapy results in clearing of B19 viremia and improvement in bone marrow suppression (Frickhofen et al, 1990). As an alternative to the intravenous route, we have given one AIDS patient serum immune globulin intramuscularly with good initial response. This patient was chronically anemic with a monthly transfusion requirement. He had no detectable reticulocytes and the anemia did not respond to erythropoietin. After his initial intramuscular injection of immunoglobulin, the reticulocyte count and he-

moglobin spontaneously increased. His transfusion requirements decreased (Naides et al, 1991b). While intramuscular immune serum globulin has the advantage of ease of administration and significantly decreased costs, efficacy of the intramuscular route compared to intravenous immune serum globulin administration remains to be determined. Specific anti-viral chemotherapy for B19 disease has not been evaluated.

REFERENCES

Ager, E.A., Chin, T.D.Y., and Poland, J.D. 1966. Epidemic erythema infectiosum. N. Engl. J. Med. 275:1326–1331.

Anand, A., Gray, E.S., Brown, T., Clewley, J.P., and Cohen, B.J. 1987. Human parvovirus infection in pregnancy and hydrops fetalis. N. Engl. J. Med. 316:183–186.

Anderson, L.J., Tsou, R.A., Chorba, T.L., Wulff, H., Tattersall, P., and Mortimer, P.P. 1986. Detection of antibodies and antigens of human parvovirus B19 by enzyme-linked immunosorbent assay. J. Clin. Microbiol. 24:522–526.

Anderson, L.J. and Hurwitz, E.S. 1988. Human parvovirus B19 and pregnancy. Clin. Perinatol. 15:273–286.

Anderson, M.J., Jones, S.E., Fisher Hoch, S.P., Lewis, E., Hall, S.M., Bartlett, C.L., Cohen, B.J., Mortimer, P.P., and Pereira, M.S. 1983. Human parvovirus, the cause of erythema infectiosum (fifth disease)? Lancet 1:1378.

Anderson, M.J., Lewis, E., Kidd, I.M., Hall, S.M., and Cohen, B.J. 1984. An outbreak of erythema infectiosum associated with human parvovirus infection. J. Hyg. 93:85–93.

Anderson, M.J., Higgins, P.G., Davis, L.R., Willman, J.S., Jones, S.E., Kidd, I.M., Pattison, J.R., and Tyrrell, D.A. 1985a. Experimental parvoviral infection in humans. J. Infect. Dis. 152:257–265.

Anderson, M.J., Jones, S.E., and Minson, A.C. 1985b. Diagnosis of human parvovirus infection by dot-blot hybridization using cloned viral DNA. J. Med. Virol. 15:163–172.

Anderson, M.J., Hall, S.M., and Serjeant, G.R. 1989. Risks associated with human parvovirus B19 infection. M. M. W. R. 38:81–97.

Andrews, M., Martin, R.W.Y., Duff, A.R., Greig, H.D., and Frost, S.A.H. 1984. Fifth disease: Report of an outbreak. J. R. Coll. Gen. Pract. 34:573–574.

Arnett, F.C., Edworthy, S.M., Bloch, D.A., McShane, D.J., Fries, J.F., Cooper, N.S., Healey, L.A., Kaplan, S.R., Liang, M.H., Luthra, H.S., Medsger, T.A. Jr., Mitchell, D.M., Neustadt, D.H., Pinals, R.S., Schaller, J.G., Sharp, J.T., Wilder, R.L., and Hunder, G.G. 1988. The American Rheumatism Association 1987 revised criteria for the classification of rheumatoid arthritis. Arthritis Rheum. 31:315–324.

Astell, C.R. 1990. In P. Tijssen (ed.), Handbook of Parvoviruses, Volume I. Boca Raton: CRC Press, Inc., 59–79.

Astell, C.R. and Blundell, M.C. 1989. Sequence of the right hand terminal palindrome of the human B19 parvovirus genome has the potential to form a "stem plus arms" structure. Nucleic Acids Res. 17:5857.

Azzi, A., Zakrzewska, K., Gentilomi, G., Musiani, M., and Zerbini, M. 1990. Detection of B19 parvovirus infections by a dot-blot hybridization assay using a digoxigenin-labelled probe. J. Virol. Methods 27:125–133.

Bell, L.M., Naides, S.J., Stoffman, P., Hodinka, R.L., and Plotkin, S.A. 1989. Human parvovirus B19 infection among hospital staff members after contact with infected patients. N. Engl. J. Med. 321:485–491.

Berns, K.I. and Hauswirth, W.W. 1984. In K.I. Berns (ed.), The Parvoviruses. New York: Plenum Press, 1–31.

Bertrand, Y., Lefrere, J.J., Leverger, G., Courouce, A.M., Feo, C., Clark, M., Schaison, G., and Soulier, J.P. 1985. Autoimmune haemolytic anaemia revealed by human parvovirus linked erythroblastopenia. Lancet 2:382–383.

Blundell, M.C., Beard, C., and Astell, C.R. 1987. In vitro identification of a B19 parvovirus promoter. Virology 157:534–538.

Bond, P.R., Caul, E.O., Usher, J., Cohen, B.J., Clewley, J.P., and Field, A.M. 1986. Intrauterine infection with human parvovirus. Lancet 1:448–449.

Bowman, C.A., Cohen, B.J., Norfolk, D.R., and Lacey, C.J. 1990. Red cell aplasia associated with human parvovirus B19 and HIV infection: Failure to respond clinically to intravenous immunoglobulin. AIDS 4:1038–1039.

Brandrup, F. and Larsen, P.O. 1976. Erythema infectiosum (fifth disease). Br. Med. J. 1:47–48.

Brown, C.S., van Bussel, M.J.A.W.M., Wassenaar, A.L.M., van Elsacker-Niele, A.W., Weiland, H.T., and Salimans, M.M.M. 1990. An immunofluorescence assay for the detection of parvovirus B19 IgG and IgM antibodies based on recombinant viral antigen. J. Virol. Methods 29:53–62.

Brown, K.E., Mori, J., Cohen, B.J., and Field, A.M. 1991. In vitro propagation of parvovirus B19 in primary foetal liver culture. J. Gen. Virol. 72:741–745.

Brown, T., Anand, A., Ritchie, L.D., Clewley, J.P., and Reid, T.M. 1984. Intrauterine parvovirus infection associated with hydrops fetalis. Lancet 2:1033–1034.

Burton, P.A. 1986. Intranuclear inclusions in marrow of hydropic fetus due to parvovirus infection. Lancet 2:1155.

Carrington, D., Gilmore, D.H., Whittle, M.J., Aitken, D., Gibson, A.A., Patrick, W.J., Brown, T., Caul, E.O., Field, A.M., and Clewley, J.P. 1987. Maternal serum alpha-fetoprotein—a marker of fetal aplastic crisis

during intrauterine human parvovirus infection. Lancet 1:433–435.

Carter, B.J., Mendelson, E., and Trempe, J.P. 1990. In P. Tijssen (ed.), Handbook of Parvoviruses, Volume I. Boca Raton: CRC Press, Inc., 169–226.

Caul, E.O., Usher, M.J., and Burton, P.A. 1988. Intrauterine infection with human parvovirus B19: A light and electron microscopy study. J. Med. Virol. 24:55–66.

Chorba, T., Coccia, P., Holman, R.C., Tattersall, P., Anderson, L.J., Sudman, J., Young, N.S., Kurczynski, E., Saarinen, U.M., and Moir, R. 1986. The role of parvovirus B19 in aplastic crisis and erythema infectiosum (fifth disease). J. Infect. Dis. 154:383–393.

Chrystie, I.L., Almeida, J.D., and Welch, J. 1990. Electron microscopic detection of human parvovirus (B19) in a patient with HIV infection. J. Med. Virol. 30:249–252.

Clarke, H.C. 1984. Erythema infectiosum: An epidemic with a probable posterythema phase. Can. Med. Assoc. J. 130:603–604.

Clewley, J.P. 1984. Biochemical characterization of a human parvovirus. J. Gen. Virol. 65:241–245.

Clewley, J.P. 1985. Detection of human parvovirus using a molecularly cloned probe. J. Med. Virol. 15:173–181.

Clewley, J.P., Cohen, B.J., and Field, A.M. 1987. Detection of parvovirus B19 DNA, antigen, and particles in the human fetus. J. Med. Virol. 23:367–376.

Clewley, J.P. 1989. Polymerase chain reaction assay of parvovirus B19 DNA in clinical specimens. J. Clin. Microbiol. 27:2647–2651.

Cohen, B.J. 1988. In J.R. Pattison (ed.), Parvoviruses and Human Disease. Boca Raton: CRC Press, Inc., 69–83.

Cohen, B.J. and Shirley, J.A. 1985. Dual infection with rubella and human parvovirus. Lancet 2:662–663.

Cohen, B.J. and Supran, E.M. 1987. IgM serology for rubella and human parvovirus B19. Lancet 1:393.

Cohen, B.J., Mortimer, P.P., and Pereira, M.S. 1983. Diagnostic assays with monoclonal antibodies for the human serum parvovirus-like virus (SPLV). J. Hyg. 91:113–130.

Cohen, B.J., Courouce, A.M., Schwarz, T.F., Okochi, K., and Kurtzman, G.J. 1988. Laboratory infection with parvovirus B19. J. Clin. Pathol. 41:1027–1028.

Cossart, Y.E., Field, A.M., Cant, B., and Widdows, D. 1975. Parvovirus-like particles in human sera. Lancet 1:72–73.

Cotmore, S.F. and Tattersall, P. 1984. Characterization and molecular cloning of a human parvovirus genome. Science 226:1161–1165.

Cotmore, S.F. and Tattersall, P. 1989. A genome-linked copy of the NS-1 polypeptide is located on the outside of infectious parvovirus particles. J. Virol. 63:3902–3911.

Cotmore, S.F., McKie, V.C., Anderson, L.J., Astell, C.R., and Tattersall, P. 1986. Identification of the major structural and nonstructural proteins encoded by human parvovirus B19 and mapping of their genes by procaryotic expression of isolated genomic fragments. J. Virol. 60:548–557.

Cramp, H.E. and Armstrong, B.D.J. 1976. Erythema infectiosum: An outbreak of "slapped cheek" disease in north Devon. Br. Med. J. 1:885–886.

Davidson, R.J., Brown, T., and Wiseman, D. 1984. Human parvovirus infection and aplastic crisis in hereditary spherocytosis. J. Infect. 9:298–300.

de Mayolo, J.A. and Temple, J.D. 1990. Pure red cell aplasia due to parvovirus B19 infection in a man with HIV infection. South. Med. J. 83:1480–1481.

Deiss, V., Tratschin, J.D., Weitz, M., and Siegl, G. 1990. Cloning of the human parvovirus B19 genome and structural analysis of its palindromic termini. Virology 175:247–254.

Duncan, J.R., Potter, C.B., Cappellini, M.D., Kurtz, J.B., Anderson, M.J., and Weatherall, D.J. 1983. Aplastic crisis due to parvovirus infection in pyruvate kinase deficiency. Lancet 2:14–16.

Evans, J.P., Rossiter, M.A., Kumaran, T.O., Marsh, G.W., and Mortimer, P.P. 1984. Human parvovirus aplasia: Case due to cross infection in a ward. Br. Med. J. 288:681.

Faden, H., Gary, G.W. Jr., and Korman, M.

1990. Numbness and tingling of fingers associated with parvovirus B19 infection. J. Infect. Dis. 161:354–355.

Foto, F., Scharosch, L.L., Howard, E.J., and Naides, S.J. 1990. Detection of parvovirus B19-specific DNA sequences in bone marrow aspirates from chronic B19 arthropathy patients (Abstract). Arthritis Rheum. 33: S142.

Frickhofen, N. and Young, N.S. 1989. Persistent parvovirus B19 infections in humans. Microb. Pathog. 7:319–327.

Frickhofen, N. and Young, N.S. 1990. Polymerase chain reaction for detection of parvovirus B19 in immunodeficient patients with anemia. Behring. Inst. Mitt. 85:46–54.

Frickhofen, N., Abkowitz, J.L., Safford, M., Berry, J.M., Antunez de Mayolo, J., Astrow, A., Cohen, R., Halperin, I., King, L., and Mintzer, D. 1990. Persistent B19 parvovirus infection in patients infected with human immunodeficiency virus type 1 (HIV-1): A treatable cause of anemia in AIDS. Ann. Intern. Med. 113:926–933.

Fridell, E., Cohen, B.J., and Wahren, B. 1991. Evaluation of a synthetic-peptide enzyme-linked immunosorbent assay for immunoglobulin M to human parvovirus B19. J. Clin. Microbiol. 29:1376–1381.

Goldman, F., Rotbart, H., Gutierrez, K., and Ambruso, D. 1990. Parvovirus-associated aplastic crisis in a patient with red blood cell glucose-6-phosphate dehydrogenase deficiency. Pediatr. Infect. Dis. J. 9:593–594.

Graeve, J.L., de Alarcon, P.A., and Naides, S.J. 1989. Parvovirus B19 infection in patients receiving cancer chemotherapy: The expanding spectrum of disease. Am. J. Pediatr. Hematol. Oncol. 11:441–444.

Gray, E.S., Anand, A., and Brown, T. 1986. Parvovirus infections in pregnancy. Lancet 1:208.

Gray, E.S., Davidson, R.J., and Anand, A. 1987. Human parvovirus and fetal anaemia. Lancet 1:1144.

Green, D.H., Bellingham, A.J., and Anderson, M.J. 1984. Parvovirus infection in a family associated with aplastic crisis in an affected sibling pair with hereditary spherocytosis. J. Clin. Pathol. 37:1144–1146.

Hanada, T., Koike, K., Takeya, T., Nagasawa, T., Matsunaga, Y., and Takita, H. 1988. Human parvovirus B19-induced transient pancytopenia in a child with hereditary spherocytosis. Br. J. Haematol. 70:113–115.

Hassam, S., Briner, J., Tratschin, J.D., Siegl, G., and Heitz, P.U. 1990. In situ hybridization for the detection of human parvovirus B19 nucleic acid sequences in paraffin-embedded specimens. Virchows Arch. [B]. 59:257–261.

Joseph, P.R. 1986. Incubation period of fifth disease. Lancet 1:1390–1391.

Kajigaya, S., Shimada, T., Fujita, S., and Young, N.S. 1989. A genetically engineered cell line that produces empty capsids of B19 (human) parvovirus. Proc. Natl. Acad. Sci. USA 86:7601–7605.

Kelleher, J.F. Jr., Luban, N.L., Cohen, B.J., and Mortimer, P.P. 1984. Human serum parvovirus as the cause of aplastic crisis in sickle cell disease. Am. J. Dis. Child 138: 401–403.

Kilbourne, E.D., Cerini, C.P., Khan, M.W., Mitchell, J.W. Jr., and Ogra, P.L. 1987. Immunologic response to the influenza virus neuraminidase is influenced by prior experience with the associated viral hemagglutinin. J. Immunol. 138:3010–3013.

Kinney, J.S., Anderson, L.J., Farrar, J., Strikas, R.A., Kumar, M.L., Kliegman, R.M., Sever, J.L., Hurwitz, E.S., and Sikes, R.K. 1988. Risk of adverse outcomes of pregnancy after human parvovirus B19 infection. J. Infect. Dis. 157:663–667.

Klouda, P.T., Corbin, S.A., Bradley, B.A., Cohen, B.J., and Woolf, A.D. 1986. HLA and acute arthritis following human parvovirus infection. Tissue Antigens 28:318–319.

Knisely, A.S. 1990. Parvovirus B19 infection in the fetus. Lancet 336:443.

Knisely, A.S., O'Shea, P.A., McMillan, P., Singer, D.B., and Magid, M.S. 1988. Electron microscopic identification of parvovirus virions in erythroid-line cells in fatal hydrops fetalis. Pediatr. Pathol. 8:163–170.

Knott, P.D., Welply, G.A., and Anderson, M.J. 1984. Serologically proved intrauterine infection with parvovirus. Br. Med. J. 289: 1660.

Koch, W.C. and Adler, S.P. 1990. Detection of human parvovirus B19 DNA by using the polymerase chain reaction. J. Clin. Microbiol. 28:65–69.

Komatsu, N., Nakauchi, H., Miwa, A., Ishihara, T., Eguchi, M., Moroi, M., Okada, M., Sato, Y., Wada, H., Yawata, Y., Suda, T., and Miura, Y. 1991. Establishment and characterization of a human leukemic cell line with megakaryocytic features: Dependency on granulocyte macrophage colony-stimulating factor, interleukin 3, or erythropoietin for growth and survival. Can. Res. 51:341–348.

Kurtz, J.B. and Anderson, M.J. 1985. Cross-reactions in rubella and parvovirus specific IgM tests. Lancet 2:1356.

Kurtzman, G.J., Ozawa, K., Cohen, B., Hanson, G., Oseas, R., and Young, N.S. 1987. Chronic bone marrow failure due to persistent B19 parvovirus infection. N. Engl. J. Med. 317:287–294.

Kurtzman, G.J., Cohen, B., Meyers, P., Amunullah, A., and Young, N.S. 1988. Persistent B19 parvovirus infection as a cause of severe chronic anaemia in children with acute lymphocytic leukaemia. Lancet 2:1159–1162.

Kurtzman, G., Frickhofen, N., Kimball, J., Jenkins, D.W., Nienhuis, A.W., and Young, N.S. 1989a. Pure red-cell aplasia of 10 years' duration due to persistent parvovirus B19 infection and its cure with immunoglobulin therapy. N. Engl. J. Med. 321:519–523.

Kurtzman, G.J., Cohen, B.J., Field, A.M., Oseas, R., Blaese, R.M., and Young, N.S. 1989b. Immune response to B19 parvovirus and an antibody defect in persistent viral infection. J. Clin. Invest. 84:1114–1123.

Lauer, B.A., MacCormack, J.N., and Wilfert, C. 1976. Erythema infectiosum: An elementary school outbreak. Am. J. Dis. Child. 130:252–254.

Lefrere, J.J., Courouce, A.M., Muller, J.Y., Clark, M., and Soulier, J.P. 1985. Human parvovirus and purpura. Lancet 1:730.

Lefrere, J.J., Courouce, A.M., Bertrand, Y., Girot, R., and Soulier, J.P. 1986a. Human parvovirus and aplastic crisis in chronic hemolytic anemias: A study of 24 observations. Am. J. Hematol. 23:271–275.

Lefrere, J.J., Courouce, A.M., Boucheix, C., Chomienne, C., Bernadou, A., and Soulier, J.P. 1986b. Aplastic crisis and erythema infectiosum (fifth disease) revealing a hereditary spherocytosis in a familial human parvovirus infection. Nouv. Rev. Fr. Hematol. 28:7–9.

Lefrere, J.J., Courouce, A.M., Girot, R., Bertrand, Y., and Soulier, J.P. 1986c. Six cases of hereditary spherocytosis revealed by human parvovirus infection. Br. J. Haematol. 62:653–658.

Lefrere, J.J., Courouce, A.M., Girot, R., and Cornu, P. 1986d. Human parvovirus and thalassaemia. J. Infect. 13:45–49.

Lefrere, J.J., Courouce, A.M., Soulier, J.P., Cordier, M.P., Guesne Girault, M.C., Polonovski, C., and Bensman, A. 1986e. Henoch-Schonlein purpura and human parvovirus infection. Pediatrics 78:183–184.

Lefrere, J.J., Girot, R., Courouce, A.M., Maier Redelsperger, M., and Cornu, P. 1986f. Familial human parvovirus infection associated with anemia in siblings with heterozygous beta-thalassemia. J. Infect. Dis. 153: 977–979.

Lefrere, J.J., Courouce, A.M., and Kaplan, C. 1989. Parvovirus and idiopathic thrombocytopenic purpura. Lancet 1:279.

Luzzi, G.A., Kurtz, J.B., and Chapel, H. 1985. Human parvovirus arthropathy and rheumatoid factor. Lancet 1:1218.

Lyon, D.J., Chapman, C.S., Martin, C., Brown, K.E., Clewley, J.P., Flower, A.J., and Mitchell, V.E. 1989. Symptomatic parvovirus B19 infection and heat-treated factor IX concentrate. Lancet 1:1085.

Mabin, D.C. and Chowdhury, V. 1990. Aplastic crisis caused by human parvovirus in two patients with hereditary stomatocytosis. Br. J. Haematol. 76:153–154.

Mansfield, F. 1988. Erythema infectiosum. Slapped face disease. Aust. Fam. Physician 17:737–738.

Matsunaga, Y., Matsukura, T., Yamazaki, S., Sugase, M., and Izumi, R. 1987. Hydrops fetalis caused by intrauterine human parvovirus infection. Jpn. J. Med. Sci. Biol. 40: 165–169.

Metzman, R., Anand, A., DeGiulio, P.A., and

Knisely, A.S. 1989. Hepatic disease associated with intrauterine parvovirus B19 infection in a newborn premature infant. J. Pediatr. Gastroenterol. Nutr. 9:112–114.

Mori, J., Beattie, P., Melton, D.W., Cohen, B.J., and Clewley, J.P. 1987. Structure and mapping of the DNA of human parvovirus B19. J. Gen. Virol. 68:2797–2806.

Mori, J., Field, A.M., Clewley, J.P., and Cohen, B.J. 1989. Dot blot hybridization assay of B19 virus DNA in clinical specimens. J. Clin. Microbiol. 27:459–464.

Morinet, F., D'Auriol, L., Tratschin, J.D., and Galibert, F. 1989. Expression of the human parvovirus B19 protein fused to protein A in *Escherichia coli*: Recognition by IgM and IgG antibodies in human sera. J. Gen. Virol. 70:3091–3097.

Mortimer, P.P., Cohen, B.J., Buckley, M.M., Cradock Watson, J.E., Ridehalgh, M.K., Burkhardt, F., and Schilt, U. 1985a. Human parvovirus and the fetus. Lancet 2:1012.

Mortimer, P.P., Cohen, B.J., Rossiter, M.A., Fairhead, S.M., and Rahman, A.F.M.S. 1985b. Human parvovirus and purpura. Lancet 1:730–731.

Mynott, M.J. 1985. An epidemic of erythema infectiosum in a school. Practitioner 229: 767–768.

Naides, S.J. 1987. Transient liver enzyme abnormalities in acute human parvovirus (HPV) infection (Abstract). Clin. Research 35:859A.

Naides, S.J. and Field, E.H. 1988. Transient rheumatoid factor positivity in acute human parvovirus B19 infection. Arch. Intern. Med. 148:2587–2589.

Naides, S.J. and Weiner, C.P. 1989. Antenatal diagnosis and palliative treatment of non-immune hydrops fetalis secondary to fetal parvovirus B19 infection. Prenat. Diagn. 9:105–114.

Naides, S.J., Cuthbertson, G., Murray, J.C., and Stapleton, J.T. 1988a. Neonatal sequelae of parvovirus B19 infection in utero, (Abstract). 28th ICAAC Proc. 199.

Naides, S.J., Piette, W., Veach, L.A., and Argenyi, Z. 1988b. Human parvovirus B19-induced vesiculopustular skin eruption. Am. J. Med. 84:968–972.

Naides, S.J., Scharosch, L.L., Foto, F., and Howard, E.J. 1990. Rheumatologic manifestations of human parvovirus B19 infection in adults. Initial two-year clinical experience. Arthritis Rheum. 33:1297–1309.

Naides, S.J., Foto, F., Marsh, J.L., Scharosch, L.L., and Howard, E.J. 1991a. Synovial tissue analysis in patients with chronic parvovirus B19 arthropathy, (Abstract). Clin. Research 39:733A.

Naides, S.J., Howard, E.J., Swack, N.S., and Stapleton, J.T. 1991b. Parvovirus B19 infection in HIV-1 infected individuals failing or intolerant to Zidovudine therapy, (Abstract). Seventh International Conference on AIDS, Florence, Italy.

Nunoue, T., Okochi, K., Mortimer, P.P., and Cohen, B.J. 1985. Human parvovirus (B19) and erythema infectiosum. J. Pediatr. 107:38–40.

Okabe, N., Koboyashi, S., Tatsuzawa, O., and Mortimer, P.P. 1984. Detection of antibodies to human parvovirus in erythema infectiosum (fifth disease). Arch. Dis. Child 59:1016–1019.

Oliver, A.R. and Phillips, A.D. 1988. An electron microscopical investigation of faecal small round viruses. J. Med. Virol. 24:211–218.

Ozawa, K. and Young, N. 1987. Characterization of capsid and noncapsid proteins of B19 parvovirus propagated in human erythroid bone marrow cell cultures. J. Virol. 61:2627–2630.

Ozawa, K., Kurtzman, G., and Young, N. 1986. Replication of the B19 parvovirus in human bone marrow cell cultures. Science 233:883–886.

Ozawa, K., Ayub, J., Hao, Y.S., Kurtzman, G., Shimada, T., and Young, N. 1987a. Novel transcription map for the B19 (human) pathogenic parvovirus. J. Virol. 61:2395–2406.

Ozawa, K., Kurtzman, G., and Young, N. 1987b. Productive infection by B19 parvovirus of human erythroid bone marrow cells in vitro. Blood 70:384–391.

Pattison, J.R. (ed.) 1988. Parvoviruses and Human Disease. Boca Raton: CRC Press, Inc., 1–4.

Pattison, J.R., Jones, S.E., Hodgson, J., Davis, L.R., White, J.M., Stroud, C.E., and Murtaza, L. 1981. Parvovirus infections and hypoplastic crisis in sickle-cell anaemia. Lancet 1:664–665.

Paver, W.K., Caul, E.O., Ashley, C.R., and Clarke, S.K. 1973. A small virus in human faeces. Lancet 1:237–239.

Peters, M.T. and Nicolaides, K.H. 1990. Cordocentesis for the diagnosis and treatment of human fetal parvovirus infection. Obstet. Gynecol. 75:501–504.

Plummer, F.A., Hammond, G.W., Forward, K., Sekla, L., Thompson, L.M., Jones, S.E., Kidd, I.M., and Anderson, M.J. 1985. An erythema infectiosum-like illness caused by human parvovirus infection. N. Engl. J. Med. 313:74–79.

Rao, K.R., Patel, A.R., Anderson, M.J., Hodgson, J., Jones, S.E., and Pattison, J.R. 1983. Infection with parvovirus-like virus and aplastic crisis in chronic hemolytic anemia. Ann. Intern. Med. 98:930–932.

Rao, S.P., Miller, S.T., and Cohen, B.J. 1990. Severe anemia due to B19 parvovirus infection in children with acute leukemia in remission. Am. J. Pediatr. Hematol. Oncol. 12:194–197.

Rappaport, E.S., Quick, G., Ransom, D., Helbert, B., and Frankel, L.S. 1989. Aplastic crisis in occult hereditary spherocytosis caused by human parvovirus (HPV B19). South. Med. J. 82:247–251.

Rechavi, G., Vonsover, A., Manor, Y., Mileguir, F., Shpilberg, O., Kende, G., Brok Simoni, F., Mandel, M., Gotlieb Stematski, T., and Ben Bassat, I. 1989. Aplastic crisis due to human B19 parvovirus infection in red cell pyrimidine-5'-nucleotidase deficiency. Acta Haematol. 82:46–49.

Reid, D.M., Reid, T.M., Brown, T., Rennie, J.A., and Eastmond, C.J. 1985. Human parvovirus-associated arthritis: A clinical and laboratory description. Lancet 1:422–425.

Rubinstein, A.I. and Rubinstein, D.B. 1990. Inability of solvent-detergent (S-D) treated factor VIII concentrate to inactivate parvoviruses and non-lipid enveloped non-A, non-B hepatitis virus in factor VIII concentrate: Advantages to using sterilizing 100 °C dry heat treatment. Am. J. Hematol. 35: 142.

Saarinen, U.M., Chorba, T.L., Tattersall, P., Young, N.S., Anderson, L.J., Palmer, E., and Coccia, P.F. 1986. Human parvovirus B19-induced epidemic acute red cell aplasia in patients with hereditary hemolytic anemia. Blood 67:1411–1417.

Sahakian, V., Weiner, C.P., Naides, S.J., Williamson, R.A., and Scharosch, L.L. 1991. Intrauterine transfusion treatment of nonimmune hydrops fetalis secondary to human parvovirus B19 infection. Am. J. Obstet. Gynecol. 164:1090–1091.

Salimans, M.M., van de Rijke, F.M., Raap, A.K., and van Elsacker Niele, A.M. 1989. Detection of parvovirus B19 DNA in fetal tissues by in situ hybridization and polymerase chain reaction. J. Clin. Pathol. 42: 525–530.

Salimans, M.M. 1990. Detection of human parvovirus B19 DNA by dot-hybridization and the polymerase chain reaction: Applications for diagnosis of infections. Behring. Inst. Mitt. 85:39–45.

Sasaki, T., Takahashi, Y., Yoshinaga, K., Sugamura, K., and Shiraishi, H. 1989. An association between human parvovirus B-19 infection and autoantibody production. J. Rheumatol. 16:708–709.

Sato, H., Hirata, J., Furukawa, M., Kuroda, N., Shiraki, H., Maeda, Y., and Okochi, K. 1991a. Identification of the region including the epitope for a monoclonal antibody which can neutralize human parvovirus B19. J. Virol. 65:1667–1672.

Sato, H., Hirata, J., Kuroda, N., Shiraki, H., Maeda, Y., and Okochi, K. 1991b. Identification and mapping of neutralizing epitopes of human parvovirus B19 by using human antibodies. J. Virol. 65:5485–5490.

Schwarz, T.F., Nerlich, A., Hottenträger, B., Jäger, G., Wiest, I., Kantimm, S., Roggendorf, H., Schultz, M., Gloning, K.-P., Schramm, T., Holzgreve, W., and Roggendorf, M. 1991. Parvovirus B19 infection of the fetus: Histology and in situ hybridization. Am. J. Clin. Pathol. 96:121–126.

Semble, E.L., Agudelo, C.A., and Pegram, P.S. 1987. Human parvovirus B19 arthropathy in two adults after contact with childhood erythema infectiosum. Am. J. Med. 83:560–562.

Serjeant, G.R., Topley, J.M., Mason, K., Serjeant, B.E., Pattison, J.R., Jones, S.E., and Mohamed, R. 1981. Outbreak of aplastic crises in sickle cell anaemia associated with parvovirus-like agent. Lancet 2:595–597.

Shade, R.O., Blundell, M.C., Cotmore, S.F., Tattersall, P., and Astell, C.R. 1986. Nucleotide sequence and genome organization of human parvovirus B19 isolated from the serum of a child during aplastic crisis. J. Virol. 58:921–936.

Shimomura, S., Komatsu, N., Frickhofen, N., Anderson, S., Kajigaya, S., and Young, N.S. 1992. First continuous propagation of B19 parvovirus in a cell line. Blood 79:18–24.

Shiraishi, H., Wong, D., Purcell, R.H., Shirachi, R., Kumasaka, E., and Numazaki, Y. 1985. Antibody to human parvovirus in outbreak of erythema infectiosum in Japan. Lancet 1:982–983.

Shiraishi, H., Umetsu, K., Yamamoto, H., Hatakeyama, Y., Yaegashi, N., and Sugamura, K. 1989. Human parvovirus (HPV/B19) infection with purpura. Microbiol. Immunol. 33:369–372.

Shneerson, J.M., Mortimer, P.P., and Vandervelde, E.M. 1980. Febrile illness due to a parvovirus. Br. Med. J. 280:1580.

Silman, A.J. 1988. The 1987 revised American Rheumatism Association criteria for rheumatoid arthritis (editorial). Br. J. Rheumatol. 27:341–343.

Simpson, R.W., McGinty, L., Simon, L., Smith, C.A., Godzeski, C.W., and Boyd, R.J. 1984. Association of parvoviruses with rheumatoid arthritis of humans. Science 223:1425–1428.

Soloninka, C.A., Anderson, M.J., and Laskin, C.A. 1989. Anti-DNA and antilymphocyte antibodies during acute infection with human parvovirus B19. J. Rheumatol. 16:777–781.

Soothill, P. 1990. Intrauterine blood transfusion for non-immune hydrops fetalis due to parvovirus B19 infection. Lancet 336:121–122.

Stierle, G., Brown, K.A., Rainsford, S.G., Smith, C.A., Hamerman, D., Stierle, H.E., and Dumonde, D.C. 1987. Parvovirus associated antigen in the synovial membrane of patients with rheumatoid arthritis. Ann. Rheum. Dis. 46:219–223.

Summerfield, G.P. and Wyatt, G.P. 1985. Human parvovirus infection revealing hereditary spherocytosis. Lancet 2:1070.

Summers, J., Jones, S.E., and Anderson, M.J. 1983. Characterization of the genome of the agent of erythrocyte aplasia permits its classification as a human parvovirus. J. Gen. Virol. 64:2527–2532.

Takahashi, M., Koike, T., Moriyama, Y., Shibata, A., Koike, R., Sanada, M., and Tsukada, T. 1986. Inhibition of erythropoiesis by human parvovirus-containing serum from a patient with hereditary spherocytosis in aplastic crisis. Scand. J. Haematol. 37:118–124.

Takahashi, T., Ozawa, K., Mitani, K., Miyazono, K., Asano, S., and Takaku, F. 1989. B19 parvovirus replicates in erythroid leukemic cells in vitro. J. Infect. Dis. 160:548–549.

Tattersall, P. and Ward, D.C. 1976. Rolling hairpin model for replication of parvovirus and linear chromosomal DNA. Nature 263:106–109.

Tsukada, T., Koike, T., Koike, R., Sanada, M., Takahashi, M., Shibata, A., and Nunoue, T. 1985. Epidemic of aplastic crisis in patients with hereditary spherocytosis in Japan. Lancet 1:1401.

Turton, J., Appleton, H., and Clewley, J.P. 1990. Similarities in nucleotide sequence between serum and faecal human parvovirus DNA. Epidemiol. Infect. 105:197–201.

van Elsacker-Niele, A.W. and Anderson, M.J. 1987. First picture of erythema infectiosum? Lancet 1:229.

van Elsacker-Niele, A.M., Salimans, M.M., Weiland, H.T., Vermey Keers, C., Anderson, M.J., and Versteeg, J. 1989. Fetal pathology in human parvovirus B19 infection. Br. J. Obstet. Gynaecol. 96:768–775.

Vandervelde, E.M., Goffin, C., Megson, B., Mahmood, N., Porter, A., and Cossart, Y.E. 1974. User's guide to some new tests for hepatitis-B antigen. Lancet 2:1066.

Weiland, H.T., Vermey Keers, C., Salimans, M.M., Fleuren, G.J., Verwey, R.A., and Anderson, M.J. 1987. Parvovirus B19 associated with fetal abnormality. Lancet 1:682–683.

West, N.C., Meigh, R.E., Mackie, M., and Anderson, M.J. 1986. Parvovirus infection associated with aplastic crisis in a patient with HEMPAS. J. Clin. Pathol. 39:1019–1020.

White, D.G., Woolf, A.D., Mortimer, P.P., Cohen, B.J., Blake, D.R., and Bacon, P.A. 1985. Human parvovirus arthropathy. Lancet 1:419–421.

Woernle, C.H., Anderson, L.J., Tattersall, P., and Davison, J.M. 1987. Human parvovirus B19 infection during pregnancy. J. Infect. Dis. 156:17–20.

Woolf, A.D., Campion, G.V., Klouda, P.T., Chiswick, A., Cohen, B.J., and Dieppe, P.A. 1987. HLA and the manifestations of human parvovirus B19 infection, (Abstract). Arthritis Rheum. 30:S52.

Woolf, A.D., Campion, G.V., Chishick, A., Wise, S., Cohen, B.J., Klouda, P.T., Caul, O., and Dieppe, P.A. 1989. Clinical manifestations of human parvovirus B19 in adults. Arch. Intern. Med. 149:1153–1156.

Young, N.S., Mortimer, P.P., Moore, J.G., and Humphries, R.K. 1984. Characterization of a virus that causes transient aplastic crisis. J. Clin. Invest. 73:224–230.

Young, N.S., Baranski, B., and Kurtzman, G. 1989. The immune system as mediator of virus-associated bone marrow failure: B19 parvovirus and Epstein-Barr virus. Ann. N.Y. Acad. Sci. 554:75–80.

Zerbini, M., Musiani, M., Venturoli, S., Gallinella, G., Gibellini, D., Gentilomi, G., and La Placa, M. 1990. Rapid screening for B19 parvovirus DNA in clinical specimens with a digoxigenin-labeled DNA hybridization probe. J. Clin. Microbiol. 28:2496–2499.

Zhou, S.Z., Srivastava, C.H., Munshi, N.C., and Srivastava, A. 1991. Parvovirus B19 replication in human cord blood cells: A possible mechanism of virus-induced hydrops fetalis, (Abstract). 4th Parvovirus Workshop. Program and Abstracts, 10.

31

Measles, Mumps, and Rubella

David A. Fuccillo and John L. Sever

31.1 MEASLES VIRUS

Man is the only known natural host for measles (rubeola) virus. The successful isolation of measles virus in human and rhesus monkey kidney tissue cultures was first reported by Enders and Peebles in 1954. During this period of time, there were more than 400,000 cases of measles reported each year in the United States. However, since virtually all children would acquire measles, the true number probably exceeded four million per year. In 1963 both an inactivated and a live attenuated vaccine (Schwartz, 1962) were licensed for use in the United States. The killed vaccine eventually proved less effective and children who received this material were at risk of developing an atypical severe form of the disease when subsequently exposed to live measles virus. In 1967 the inactivated vaccine was no longer used. A live, further attenuated vaccine (Moraten strain) was licensed in 1968 and is the vaccine used currently in the United States. Initially, vaccine was administered to children over 9 months of age but it became apparent that the young infants did not mount an adequate immune response. It is now recom-

mended that vaccination be initiated at age 15 months and a second dose administered before entering kindergarten or first grade.

31.1.1 Characteristics of Virus

Measles virus is an RNA virus and is classified as a member of the paramyxovirus group. The measles virion is spherical with a diameter ranging from 120 to 250 nm (Hail and Martin, 1974). It has an envelope composed of glycoproteins and lipids and bears short surface projections. The envelope encloses an elongated helical nucleocapsid in which protein units are spirally arranged around the nucleic acid.

31.1.2 Clinical Aspects

Measles is a highly contagious, acute biphasic disease with a prominent prodrome preceding the exanthemic phase. Measles is spread through direct contact with infected droplets originating from a cough or sneeze or from contaminated fomites. Susceptible persons intimately exposed to a measles patient have a 99% chance of acquiring the disease. Prior to the use of vaccines, more than 90% of the population had measles before 10 years of age. After an incubation period of nine to eleven days there is an

571

initial three- to four-day prodromal period characterized by fever, cough, coryza, and conjunctivitis. The incubation period in adults may last up to three weeks. Fever occurs 24 hours or less before other symptoms appear and these increase in severity reaching a peak with the appearance of the rash on the fourth to fifth day.

Bluish-white lesions with a red halo, Koplik spots, will appear on the buccal or labial mucosa in 50% to 90% of the cases, two to three days after the onset of the prodrome. These lesions are small, irregular red spots with a bluish-white speck in the center. This lesion, pathognomonic for measles (Koplik, 1896), is on the inner lip or opposite the lower molars. They may be few in number early in the prodrome; however, they increase rapidly to spread over the entire surface of the mucous membranes. A lesion somewhat similar in appearance to Koplik spots has been reported with ECHO-9 (Coxsackie A23) and Coxsackie A16 and A9 virus infections. The rash is first evident behind the ears or on the forehead. The lesions are red macules, 1 to 2 mm in diameter, which become maculopapules over the next three days. By the end of the second day, the trunk and upper extremities are covered with rash and by the third day the lower extremities are affected. The rash resolves in the same sequence, lasting approximately six days. The lesions turn brown and persist for seven to ten days and then are followed by a fine desquamation.

The most frequent complication of measles involves infections of the lower respiratory tract. Croup, bronchitis, bronchiolitis, and, rarely, giant cell interstitial pneumonia may occur. Otitis media is a common bacterial complication of measles. Prior to the advent of antibiotics these complications contributed significantly to a high number of fatalities and significant morbidity. Excluding pneumonia and otitis media, the most frequent serious complication of measles is postinfectious encephali-

tis. It occurs in 0.1% to 0.2% of measles patients during any stage of the illness, although it is most common two to seven days after the onset of the exanthem. Mortality is about 30%, with the same proportion of patients showing other permanent residual damage. Other complications include thrombocytopenic purpura, appendicitis, myocarditis, and mesenteric lymphadenitis (Gershon and Krupman, 1979).

Subacute sclerosing panencephalitis (SSPE), also called Dawson's encephalitis, is a late or "slow virus" complication of measles. The incidence of SSPE was approximately 1:100,000 to 1,000,000 cases, but after the advent of the vaccine there was a dramatic decrease in the frequency of the disease. Subacute sclerosing panencephalitis is a progressive, invariably fatal encephalopathy characterized by personality changes, mental deterioration, involuntary movements, muscular rigidity, and death. It usually begins 4 to 17 years after the patient has recovered from measles. Measles virus has been successfully isolated from brain and lymphoid tissues of SSPE patients (Barbosa et al, 1969; 1971).

Transplacental infections have been associated with some fetal effects. There is an apparent increased frequency of abortions and stillbirths. The teratogenic potential of gestational measles has been neither proved nor refuted (Fuccillo and Sever, 1973; South and Alford, 1980).

Atypical measles can occur in children previously vaccinated with killed measles virus vaccines when they become infected with wild measles (Fulginiti et al, 1967). The disease is characterized by fever, a prodromal period, and subsequent rash. During the prodrome, patients may experience malaise, myalgia, headache, nausea, and vomiting. Symptoms usually last for two to three days and frequently individuals have a sore throat, conjunctivitis, and photophobia along with a nonproductive cough and pneumonia. Chest x-rays often show patchy infiltrates. The rash produced is different from that of typical measles. It can be a mixture of macules, papules, vesicles, and pustules. Frequently, there is a petechial component which begins at the distal extremities and concentrates on the hands, wrists, ankles, and feet and then progresses centrally toward the trunk. Koplik spots have not been reported and the face is rarely involved. Edema often occurs in extremities. The appearance of atypical measles may be confused with Rocky Mountain Spotted Fever.

31.1.3 Laboratory Diagnosis

Diagnosis of a typical case of measles can be made based upon clinical symptoms. However, demonstration of the virus or seroconversion against the virus is necessary to confirm the diagnosis. Best results for isolation of the virus are achieved from specimens taken within the first few days of illness. Measles virus may be isolated from blood, throat, conjunctivae, and urine. Swabs of nasopharyngeal secretions are best for isolation during first four to five days. Virus may be present in urinary sediment for as long as one week. Measles virus grows very slowly in primary human kidney, monkey kidney, and human amnion cell cultures. The virus produces characteristic multinucleate giant cells after seven to ten days incubation (Katz and Enders, 1969). Uninoculated, control cultures should always be compared with inoculated cultures. Suspicious cultures can then be tested by hemadsorption with monkey red blood cells. Hemadsorption with monkey red blood cells but not with non-simian erythrocytes is useful to distinguish the virus from other paramyxoviruses, especially mumps. The most specific and rapid method for identification is the detection of measles antigen present in the multinucleated giant cells using direct immunofluorescence (Nommensen and Dekkers, 1981).

Measles virus is relatively labile to heat. Infectivity is decreased one half when kept at 37 °C for two hours and complete inactivation can be accomplished at 56 °C for 30 minutes (DeJong and Winkler, 1964). It can also be inactivated after exposure to ultraviolet and visible light. The virus is stable from pH 5 to 10.5 with an optimum at pH 7 (Musser and Underwood, 1960). Measles virus is preserved quite readily by storage at −70 °C in a suspension medium containing protein. Infectivity may also be maintained at 4 °C for several months in a protein containing media or in the presence of stabilizers (McAleer et al, 1980).

The most practical method to make a laboratory diagnosis of measles is to obtain acute and convalescent phase sera and demonstrate a greater than fourfold rise in specific antibody. Hemagglutination inhibition (HI), neutralization, and complement-fixation methods have all been employed with the serodiagnosis of measles. Hemagglutination inhibition tests have been the most useful method in the past, having sensitivity and specificity comparable to neutralization. Recently, newer methods such as enzyme-linked immunosorbent assays (ELISA) and radioimmunoassay (RIA) have been developed which have greater sensitivity than the HI test (Boteler et al, 1983; Rice et al, 1983). Antibody appears within one to two days after onset of rash and titers peak ten days to two weeks later. The presence of specific IgM antibody can be used to diagnose recent infection; however, timing the collection of the specimen is especially important for interpreting results, since IgM antibody peaks approximately ten days after the rash develops and becomes undetectable after 30 days.

31.1.4 Control and Prevention

Individuals having an illness compatible with measles should be cared for in such a way that contact with other people or patients is minimal. The communicability of measles virus is extremely high. Therefore, any susceptible individuals who had direct face-to-face contact with the infectious individual should obtain prophylactic treatment. Risk, other than face-to-face, is very low and therefore postexposure prophylaxis is unnecessary. Measles vaccination may prove protection if given within 72 hours of exposure (Centers for Disease Control, 1982a). The Immunization Practices Advisory Committee supports readmission to school of all previously unimmunized children immediately following vaccination (Centers for Disease Control, 1989). Immune globulin, given within six days of exposure, can prevent or modify measles virus infection. It is indicated for susceptible, close contacts of measles patients, particularly if they are under 1 year of age. If immune globulin is used for a child at this age, measles vaccine should be given about 3 months later but at no less than 15 months of age.

After a further attenuated variant of the Edmonston B vaccine was introduced in 1968, the reported cases of measles took a dramatic downward turn. In 1960, the cumulative total number of cases was 399,852 from week 1 to 35. In 1970 the total was 39,365; in 1981, 2,562; in 1982, 1,188; and in 1983, for the same period, the total number of cases was 1,194 (Centers for Disease Control, 1982b). There was hope that 1983 would be the year in which measles would be eliminated from the United States but this goal was not accomplished.

In fact, reported cases increased every year until 1986 when there were 6,282 cases. A small decrease in total cases was reported for 1987 and 1988 but total cases rose again during 1989 to 16,236 cases (Centers for Disease Control, 1990).

Elimination of measles from the United States was not accomplished primarily due to failure of achieving high vaccination coverage among preschool-aged children. Another possibility could be vac-

cine failure. These failures may be primary in that an adequate response to vaccination never developed or secondary where a response developed but was lost over a period of time. Both of these problems should be eliminated by employing the recent recommendation to give a second dose of vaccine to children before they enter school (Centers for Disease Control, 1989).

31.2 MUMPS VIRUS

Mumps virus infection was probably first described around the fifth century B.C. by Hippocrates. The name "mumps" is thought to be derived from the mumbling speech of patients afflicted with this disease. The etiologic agent was identified as virus by Johnson and Goodpasture in 1934. The virus was first isolated in the amniotic cavity of chick embryo in 1945 (Habel, 1945). Buynak and Hilleman, in 1966, developed the first successful live attenuated vaccine by passage of the virus in chick embryo cell cultures (Buynak and Hilleman, 1966).

31.2.1 Characteristics of Virus

Mumps virus is a member of the paramyxovirus group. Virus particles range in size from 85 to 300 nm in diameter (Cantell, 1961). The virus has a single strand of RNA, contains a nucleoprotein core, and has an outer viral envelope. The envelope contains a hemagglutinin, neuraminidase, and a hemolysin.

31.2.2 Clinical Aspects

Man is the only known host and reservoir of the virus. The infection can be either clinically apparent or subclinical. Infection is endemic worldwide, usually affecting the 6 to 10-year-old age group; it occurs predominantly in the spring. The virus usually causes an uncomplicated infection of the salivary glands (e.g., parotid glands) and is manifested by an enlargement of these organs. The parotitis is sudden and may not be preceded by any prodromal symptoms. Swelling of the glands reaches a maximum after 48 hours and they usually remain swollen for a period of seven to ten days. There may be little or no increase in body temperature. Approximately 20% to 30% of postpubital men acquiring mumps develop epididymo-orchitis between one and two weeks following the parotitis. Sterility is not a common sequela of infection since only 1% to 12% of the cases are bilateral.

Another complication of mumps virus infection is meningoencephalitis which has an incidence from 5% to 10%. Encephalitis is one central nervous system (CNS) complication but mumps virus infection has been linked to other rare CNS complications such as transverse myelitis, cerebellar ataxia, poliomyelitis-like syndrome, and Guillain-Barré syndrome. About 5% of adult females with mumps may develop oophoritis. Other complications such as pancreatitis, thyroiditis, neuritis, inflammation of the eye, and inner ear infection can be encountered. There have been reports of diabetes mellitus being associated with mumps but at present information is inconclusive (Sultz et al, 1975; Ratzmann et al, 1984). Other reports that intrauterine mumps can lead to endocardial fibroelastosis have not been confirmed (St. Geme et al, 1966).

Seroepidemiologic surveys have indicated that 80% to 90% of adults have evidence of prior mumps infection. Mumps is transmitted by saliva containing the virus either by direct transfer, air-suspended droplets, or by recently contaminated fomites. Approximately 85% of susceptible contacts can become infected when first exposed and 25% to 40% of the infections may be asymptomatic. The virus is thought to multiply in the upper respiratory tract, then invade the bloodstream, and finally affect the salivary glands and other organs.

About 18 days elapse between the time of exposure and the first detectable enlargement of the salivary glands. The incubation period may range from 14 to 24 days. The period of communicability can be from seven days before the salivary gland involvement until nine days thereafter. The virus is also excreted in the urine for as long as 14 days after onset of illness.

31.2.3 Laboratory Diagnosis

The diagnosis of mumps infection is quite simple when typical parotitis is produced. However, diagnosis by viral isolation or serologic techniques is most useful when the patient presents with an atypical or asymptomatic infection. A viral isolation from the spinal fluid, blood, saliva, and urine confirms the diagnosis of recent mumps infection. Primary monkey kidney cell cultures are the most sensitive hosts for isolating the virus. These may be obtained from rhesus or cynomolgus ceropithecus monkeys. Continuous human cell lines such as HeLa and primary cell cultures of human amnion or human embryonic kidney can also be used for the growth of the virus (Hopps and Parkman, 1979). This virus produces a characteristic cytopathic effect (CPE) with large syncytia. However, some strains may not produce this CPE, and therefore hemadsorption with guinea pig erythrocytes should be performed for identification. Rapid identification of mumps isolates from cell cultures can be accomplished by immunofluorescence staining (Lennette et al, 1975).

The virus is relatively stable and infectivity is changed very little upon storage at 4 °C for several days. The virus is stable when stored for a number of weeks at −20 °C and for months at −70 °C. Stability can be increased with the addition of protein such as 1% bovine albumin, 0.5% gelatin, or 2% serum. Swabs taken for isolation studies can be placed in tubes containing a suitable medium with a protein stabilizer and with antibiotics. Other specimens such as urine and cerebral spinal fluid can be placed on ice where they may be kept for a few hours before inoculation. For longer storage, specimens should be maintained at −70 °C (Cantell, 1961).

Serologic diagnosis of mumps infection can be very important, especially in those cases of meningitis or encephalitis that occur in the absence of parotitis. Serologic methods for diagnosis of mumps infections include complement fixation (CF), HI, neutralization, and, more recently, an ELISA test. The CF and HI procedures are of approximately equal sensitivity. The neutralization test is generally more sensitive in detecting antibodies than is the HI test. Recent studies have shown the ELISA test to be more sensitive detecting antibodies in acute phase sera than the CF and more sensitive than HI for the detection of low levels of antibodies (Leinikki et al, 1979; Popow-Kraupp, 1981). Other studies have shown that the use of ELISA to detect IgM is particularly suitable for early diagnosis of mumps infection with one serum specimen (Nicolai-Scholten et al, 1980; Ukkonen et al, 1980).

31.2.4 Control and Prevention

More than 59 million doses of vaccine have been distributed in the United States since licensure in December 1967. Reported cases of mumps in 1967 were 185,691 and in 1982 were 5,270. This is a 97% decrease. It appears that incidence rates have declined in all age groups by more than 90%. During this 15-year period, the highest reported rate occurred in 5 to 9 year olds followed by children under 5 years old. These two groups accounted for over 30% of all reported cases. Over the five years from 1978 to 1982, this group accounted for about 50% of recorded cases and the risk of infection for 10 to 14 year olds surpassed that for children under 5 years of age. This age-specific change in risk of infection to

mumps is similar to those noted for measles and rubella and results from a vaccination policy oriented toward preschool and elementary school children. Individuals who are neither vaccinated nor infected at a young age eventually will be exposed at an older age and subsequently come down with the infection (Centers for Disease Control, 1983c).

Mumps vaccine is considered one of the safest of the childhood immunizing agents but 19 states still do not require proof of mumps infection or immunization as a condition for school entry (Centers for Disease Control, 1982c). Despite this, the 1982–1983 school entrance survey indicated nationwide mumps vaccine coverage of 95%. This recent increase is attributed to the use of combined measles, mumps, and rubella vaccine. This preparation has been the vaccine of choice for the routine immunization of children 15 months of age or older. A benefit cost analysis was performed on a mumps vaccination program in which mumps was given as part of a measles, mumps, rubella combination. It was estimated that costs associated with mumps were reduced by more than 86% with a benefit-cost ratio of seven to one using reported incidence rates. When rates were corrected for underreporting, the benefit-to-cost ratio was 39 to 1. Since outbreaks of mumps are still potentially possible in the unvaccinated population, considerable medical and economic savings can be realized by including mumps in the immunization process as part of the state compulsory school immunization laws (Koplan and Preblud, 1982).

31.3 RUBELLA

Rubella was first recognized in 1815 when Maton described "a rash liable to be mistaken for Scarletina" (Maton, 1815). More than a century later, Gregg noted a high incidence of congenital malformations, mostly cataracts, and reported the association of these abnormalities with rubella contracted in the first trimester of the pregnancy (Gregg, 1941). In 1962, several groups (Parkman et al, 1962; Weller and Neva, 1962) reported the successful isolation of the rubella virus.

Several live attenuated vaccines were developed and the first was licensed in 1969 (Meyer et al, 1966; Stokes et al, 1967). The HPV-77 and Cendehill vaccine strains were introduced first, whereas the presently used vaccine, Wistar RA 27/3, was licensed in 1979 (Plotkin et al, 1967; Ingalls et al, 1970).

31.3.1 Characteristics of Virus

Acquired rubella, also known as German or three-day measles, is caused by a double-layered, single-stranded RNA virus classified as a togavirus. The virion is a spherical particle with a diameter of 60 to 70 nm having surface projections 6 nm in length. It has a dense central nucleoid of about 30 nm surrounded by an envelope acquired when virus buds out through the cytoplasmic or plasma membrane of cell (Herrmann, 1979).

31.3.2 Clinical Aspects

The disease is quite contagious and is usually transmitted by respiratory secretions. Before rubella immunization, the disease occurred most commonly in childhood although it affected adults as well. The clinical manifestations of rubella are usually mild; in fact, they may be absent in as much as one third of the individuals infected. Catarrhal symptoms are first to be seen, followed by lymphadenopathy involving the posterior auricular, posterior cervical, and postoccipital lymph nodes, and finally the emergence of a maculopapular rash on the face, then on the neck and trunk. A low grade fever is usually present and some individuals may have transient arthralgia and arthritis, which is usually more severe in adults.

The signs and symptoms of rubella may be difficult to distinguish from other rash-associated diseases. For this reason neither the clinical presentation of apparent rubella nor the absence of classic rubella signs and symptoms can be regarded as definitively diagnostic. Acute rubella lasts from three to five days and generally requires little treatment. The incubation period for rubella varies from 10 to 21 days, with 12 to 14 days being typical. Infected individuals are usually contagious for 12 to 15 days, beginning five to seven days before the appearance of a rash. Acquired rubella infection almost always confers permanent immunity to the disease. There is some evidence that reinfection can occur rarely but usually it is asymptomatic and is not associated with fetal infection (Cradock-Watson et al, 1985).

Acquired rubella in a pregnant woman is no more intrinsically dangerous than it is in any other patient but the potential damage to the fetus can be quite disasterous. The point in the gestation cycle at which maternal rubella occurs greatly influences the severity and risk of the congenital rubella syndrome (CRS). If rubella occurs in the first trimester of pregnancy there is a 25% chance of fetal anomalies; the risk is 50% when infection occurs in the first month. The risk becomes less than 10% by the third month and drops to approximately 6% when infection occurs in the fourth or fifth month. After the fifth month, although fetal infections continue to occur, rubella poses no known threat to the fetus (Sever, 1983). Fetal infection is a result of placental infection during the viremic phase of the disease. One study of mortality associated with CRS showed a 13% death rate in the first 18 months of life (Cooper, 1968). Another study showed a 20% mortality rate in the first 18 months of life (Desmond et al, 1967).

The extent of rubella-produced anomalies varies from one neonate to another. Fetal abnormalities include stillbirths, cataracts, cardiovascular defects, mental retardation, microcephaly, encephalitis, hepatosplenomegaly, bone defects, and growth retardation. Severely affected children are likely to have multiple organ involvement. In neonates with CRS, thrombocytopenia purpura is common, as is low birth weight and failure to thrive.

Chronic infection may persist for months or even years. Rubella immunity develops in most children who have had congenital rubella. However, in late childhood, about a third of these children lose antibody and become susceptible to acquired rubella which, if it occurs, follows a typical benign course (Hardy et al, 1969).

In the past few years late manifestations of CRS have been recognized. These are disabilities that do not make their appearance until years after birth. One of the first disabilities to be found was insulin-dependent diabetes mellitus (Menser et al, 1967; Plotkin and Kaye, 1970). Additional studies revealed a 20% incidence of latent or overt diabetes in a study involving 50 older subjects from the 1964 rubella epidemic (Forrest et al, 1971). All patients had a prenatal history of rubella before the 16th week of gestation. All were deaf and most had other rubella-associated defects of the eyes and heart. In a follow-up study, 40% of these CRS patients had developed evidence of overt or latent diabetes (Menser et al, 1974). Other endocrine disorders have been seen in small numbers of survivors of congenital rubella. They include hypothyroidism (Ziring et al, 1975, 1977), hyperthyroidism (Floret et al, 1980), hypoadrenalism (Ziring et al, 1977), and a growth hormone deficiency (Preece et al, 1977).

Ocular consequences of CRS are observed during and after the neonatal period. One study describes 13 patients having glaucoma 3 to 22 years after birth. All had cataracts early in life which had either been removed surgically or had resolved spontaneously. Another group of patients were found to have keratic precipitates without other evidence of acute ocular inflammation (Boger, 1980, 1981).

Another disability associated with congenital rubella is bilateral hearing loss. There has been one report of a CRS child who had numerous audiograms, developed normal speech patterns, and attended to everyday activities until age 10, when signs of progressive deafness began to appear (Desmond et al, 1978).

The last disability associated with CRS is progressive rubella panencephalitis (PRP) (Townsend et al, 1976). This is a "slow virus" manifestation of rubella that is similar to SSPE which is due to measles. Progressive rubella panencephalitis usually appears during the second decade of life. Progressive deterioration of intellectual and motor function occurs with dementia close to the time of death. There is an intense immune response against rubella antigens and high titers of rubella antibody are present in both serum and cerebrospinal fluid (Weller et al, 1964). Virus has been recovered from brain by rescue techniques (Wolinsky, 1978). The pathologic findings are similar to SSPE but without the inclusions and with perivascular deposits (Rosenberg et al, 1981).

Progressive rubella panencephalitis has now been found in two patients with postnatal rubella infection and in nine cases of CRS (Lebon and Lyon, 1974; Weil et al, 1975; Townsend et al, 1975; Wolinsky, 1978). No correlation can be made between the occurrence of PRP and the presence of rubella-associated defects or the severity of neonatal infection.

The pathology produced with congenital rubella appears to result from a chronic viral infection with an inhibition of cell multiplication at critical points in organogenesis. This causes the hypoplastic organ development and other characteristic structural defects seen with this disease (Rawls and Melnick, 1966). The immune response may also contribute to permanent damage in the developing child either by an impaired immunity or by inflicting damage through inflammatory mechanisms (Fuccillo et al, 1974; Rosenberg et al, 1981).

31.3.3 Laboratory Diagnosis

Rubella virus can be readily isolated from a variety of specimens obtained from patients with congenital or acquired infection if care is taken in obtaining and storing specimens. Specimen collection should be as early as possible after the person becomes ill, preferably within three to four days after symptoms appear. Specimens to be inoculated within 48 hours should be stored at 4 °C and transported on ice. If inoculation is delayed, the specimen should be frozen at −70 °C immediately after collection. Most body fluids and tissues such as placental or aborted material contain the virus. Nasopharngeal washings or throat swabs placed in a protein medium containing antibiotics and protein media are particularly useful for the detection of the agent.

Basically, there are two techniques used to detect the isolation of rubella virus—the direct technique and the indirect interference method. Weller and Neva (1962) first introduced the direct method when they described the CPE of rubella virus in primary cultures of human amnion cells (1962). The spectrum of tissue culture systems capable of supporting the growth of rubella with characteristic CPE after adaptation is quite large. These include the continuous rabbit kidney, RK13, LLC-RK-1, vervet monkey kidney, Vero, GMK-AH-1 and BSC-1, rabbit cornea (SIRC), baby hamster kidney, BHK-21, and a number of others. In primary cells, rubella virus very seldom produces CPE and when it is present the changes are slow to appear and difficult to detect (Herrmann, 1979).

The indirect technique is based on the ability of rubella virus to interfere with the replication of enteroviruses such as Echo 11 or Coxsackie A9 in primary African green monkey kidney. A number of other superinfecting viruses and a variety of cell cultures can also be used to detect rubella propagation (Parkman et al, 1964). Each system has been found to have its advantages and disadvantages, but the interference technique in primary African green monkey kidney with a Coxsackie A9 challenge has proven to be a very sensitive technique for detecting the recovery of virus from specimens (Schiff and Sever, 1966).

The variability of the signs and symptoms of rubella and their similarity to those of various other conditions make a clinical history of rubella unreliable and serologic testing essential. Antibody tests are used today to determine immune status and to diagnose rubella infection. These tests have been available for more than 15 years. Hemagglutination inhibition was the first widely used test and is the standard against which other rubella screening and diagnostic tests are measured (Stewart et al, 1967). Other screening assays include neutralization, CF, passive hemagglutination, latex agglutination, and, more recently, ELISA (Vejtorp, 1983). The ELISA test is very sensitive and, in fact, can detect antibody when HI antibody was not found. It was shown that a small group of individuals who initially seroconverted after vaccination and then lost detectable HI antibody still retained antibody when tested by the ELISA method (Buimovici-Klein et al, 1980; Best et al, 1980). Therefore, the presence of any detectable level of rubella antibody or a history of rubella vaccination is accepted presumptive evidence of immunity. To make an accurate diagnosis using serologic testing requires the utilization of tests in the proper relationship to the progress of the disease and, specifically, to the emergence, increase, and reduction of IgG and IgM antibodies. An individual infected with rubella develops both IgG and IgM antibodies along with the appearance of clinical symptoms when they are present. These IgG antibodies increase rapidly for the next 7 to 21 days, IgG levels off, and then remains present and indefinitely protective. Therefore, detection of IgG antibody is useful in indicating immunity but recent rubella infection can only be documented in cases where paired serum specimens, drawn several weeks apart, show a four-fold or greater increase in IgG antibody. Both serum specimens should be run at the same time to eliminate test variation.

IgM antibodies become detectable a few days after onset of symptoms and are at their peak from 8 to 21 days afterward. During the next four to five weeks, IgM drops off until it is no longer present. The presence of IgM antibody on a single sample thus indicates a recent rubella infection and, in most cases, points to an infection that has occurred within the preceding month (Meurman, 1978; Leinikki et al, 1978).

31.3.4 Control and Prevention

Rubella occurred in epidemic proportions at six- to nine-year intervals before the widespread use of the rubella vaccine. More than 20,000 cases of congenital rubella syndrome and an unknown number of stillbirths occurred in the United States as a result of the 1964 epidemic. This unfortunate event stimulated the quest for an effective rubella vaccine. Three strains of live, attenuated rubella vaccine virus were developed and the first was licensed for use in 1969. Since that time there have been no further rubella epidemics in the United States.

The vaccine presently used is Wistar RA 27/3 (Ingalls et al, 1970) which was licensed in 1979. This vaccine has been effective in inducing immunity in 95% or more of vaccinees. Rash and lymphadenopathy in children and arthralgia in adults are side effects but they are not disabling. Less

than 1% of vaccinees report arthritis. There is no indication of increased risk of these reactions in immune patients receiving the vaccine. Rubella vaccine virus has been isolated from products of conception but there has been no demonstration of congenital rubella syndrome (Banatvala et al, 1981). The vaccine is not recommended by Centers for Disease Control (1981; 1983a; 1986) to be given to pregnant women even though studies have shown 1,142 pregnant women who received the vaccine within three months before or three months after conception failed to produce a child damaged by rubella. Vaccine is available in monovalent preparations and in combination with measles and mumps. Doctors have been encouraged to use the triple combination for routine vaccination of all children at 15 months of age.

The incidence of reported rubella has fluctuated slightly over the past several years, although there has been a dramatic downward trend for most of the United States. In 1980 3,904 cases of rubella were reported, an incidence of 1.7 cases per 100,000 of population. In 1981, the incidence was 0.9 per 100,000. In 1982, the incidence was 1.0 per 100,000 and during the first 38 weeks of 1983, 791 cases were reported; this was a 61% decrease from the number reported during the same period in 1982 (Centers for Disease Control, 1983b). In 1986 the incidence had decreased further to 0.25 cases per 100,000.

According to the National Congenital Rubella Syndrome Registry the incidence rates of confirmed congenital rubella have declined since 1979. Fifty-five cases were reported in 1979, 14 in 1980, and nine in both 1981 and 1982. In 1983 there were three cases reported. The goal of rubella vaccination programs has been to prevent congenital rubella infection. Dramatic results have been accomplished; however, congenital rubella still continues at a very low incidence. The vaccination strategy of doctors in the United States in 1969 was aimed at controlling rubella in preschool and young school-aged children, the known reservoirs for rubella transmission. If it were possible to control disease in this population, exposure of susceptible pregnant females to rubella virus would also be decreased. This approach resulted in dramatic declines as previously mentioned. However, this vaccination strategy had less effect on rubella incidence in persons 15 and older. Approximately 10% to 20% of this population continues to be susceptible, similar to that during prevaccine years. Now increased efforts are being made to immunize susceptible junior and senior high school students and to enforce rubella immunization requirements for school entry. Military recruits are also receiving rubella vaccine. Also, physicians are encouraged to test all women of childbearing age and to immunize susceptible women when not pregnant. Until such time that the susceptibility rate of postpubital women is effectively lowered, congenital rubella will continue to occur. It has been estimated that a child with congenital rubella causes an average lifetime expenditure of $221,600 (Centers for Disease Control, 1984).

REFERENCES

Banatvala, J.E., O'Shea, S., Best, J.M., Nicholls, M.V., and Cooper, K. 1981. Transmission of RA27/3 rubella vaccine strain to products of conception (letter). Lancet 1:392.

Barbosa, L.H., Fuccillo, D.A., Sever, J.L., and Zeman, W. 1969. Subacute sclerosing panencephalitis: Isolation of measles virus from a brain biopsy. Nature 221:974.

Barbosa, H.L., Hamilton, R., Wiltig, B., Fuccillo, D.A., Sever, J.L., and Vernon, M.L. 1971. Subacute sclerosing panencephalitis: Isolation of suppressed measles virus from lymph node biopsies. Science 173:840–841.

Best, J.M., Harcourt, G.C., Druce, A., Palmer, S.J., O'Shea, S., and Banatvala, J.E. 1980. Rubella immunity by four different techniques: Result of challenge studies. J. Med. Virol. 5:239–247.

Boger, W.P. III. 1980. Late ocular complications in congenital rubella syndrome. Opthalmology (Rochester) 87:1244–1252.

Boger, W.P. III. 1981. Spontaneous absorption of the lens in the congenital rubella syndrome. Am. J. Ophthalmol. 99:433–434.

Boteler, W.L., Luipersheck, P.M., Fuccillo, D.A., and O'Beirne, A.J. 1983. Enzyme linked immunosorbent assay for detection of measles antibody. J. Clin. Microbiol. 17:814–818.

Buimovici-Klein, E., O'Beirne, A.J., Millian, S.J., and Cooper, L.Z. 1980. Low level of rubella immunity detected by ELISA and specific lymphocyte transformation. Arch. Virol. 66:321–327.

Buynak, E.B. and Hilleman, M.R. 1966. Live attenuated mumps virus vaccine. Proc. Soc. Exp. Biol. Med. 123:768–775.

Cantell, K. 1961. Mumps virus. Adv. Virus Res. 8:123–164.

Centers for Disease Control. 1981. Rubella Prevention. M.M.W.R. 30:37–47.

Centers for Disease Control. 1982a. Measles prevention. M.M.W.R. 31:217–24, 229–31.

Centers for Disease Control. 1982b. Countdown toward the elimination of measles in the U.S. M.M.W.R. 31:447–478.

Centers for Disease Control. 1982c. Mumps vaccine. M.M.W.R. 31:617–620.

Centers for Disease Control. 1983a. Rubella vaccination during pregnancy—United States, 1971–1982. M.M.W.R. 32:430–432.

Centers for Disease Control. 1983b. Rubella and congenital rubella—United States, 1980–83. M.M.W.R. 32:505–510.

Centers for Disease Control. 1983c. Efficacy of mumps vaccine. M.M.W.R. 32:391–398.

Centers for Disease Control. 1984. Rubella and congenital rubella—United States 1983. M.M.W.R. 33:237–247.

Centers for Disease Control. 1986. Rubella vaccination during pregnancy—United States, 1971–1985. M.M.W.R. 35:275–284.

Centers for Disease Control. 1989. Measles Prevention: Recommendations of the Immunization Practices Advisory Committee (ACIP). M.M.W.R. 38:1–17.

Centers for Disease Control. 1990. M.M.W.R. 38:891.

Cooper, L.Z. 1968. Rubella: A preventable cause of birth defects. In D. Bergsma (ed.). Birth Defects (Original Article Series), Vol 4, Intrauterine Infection. New York: The National Foundation.

Cradock-Watson, J.E., Rudehalgh, M.S., Anderson, M.J., and Patison, J.R. 1985. Rubella reinfection and the fetus. Lancet 2:1039.

DeJong, J.G. and Winkler, K.C. 1964. Survival of measles virus in air. Nature 201:1054–1055.

Desmond, M.M., Wilson, G.S., Melnick, J.L., Singer, D.B., Zion, T.E., Rudolph, A.J., Pineda, R.G., Ziai, M.-H., and Blattner, R.J. 1967. Congenital rubella encephalitis. Course and early sequelae. J. Pediatr. 71: 311–331.

Desmond, M.M., Fisher, E.S., Vorderman, A.L., Schaffer, H.G., Andrew, L.P., Zion, T.E., and Catlin, F.I. 1978. The longitudinal course of congenital rubella encephalitis in non-retarded children. J. Pediatr. 93:584–591.

Enders, J.F. and Peebles, T.C. 1954. Propagation in tissue cultures of cytopathogenic agents from patients with measles 1954. Proc. Soc. Exp. Biol. Med. 86:277–286.

Floret, D., Rosenberg, D., Hage, G.N., and Monnet, P. 1980. Case report: Hyperthyroidism, diabetes mellitus and the congenital rubella syndrome. Acta Paediatr. Scand. 69:250–261.

Forrest, J.M., Menser, M.A., and Burgess, J.A. 1971. High frequency of diabetes mellitus in young adults with congenital rubella. Lancet 1:332–334.

Fuccillo, D.A. and Sever, J.L. 1973. Viral teratology. Bact. Rev. 37:19–31.

Fuccillo, D.A., Steele, R.W., Henson, S.A., Vincent, M.M., Hardy, J.B., and Bellanti, J.A. 1974. Impaired cellular immunity to rubella virus in congenital rubella. Infect. Immunology 9:81–84.

Fulginiti, V.A., Eller, J.J., Downie, A.W., and Kempe, C.H. 1967. Altered reactivity to measles virus. JAMA 202:1075–1080.

Gershon, A. and Krupman, S. 1979. Measles virus. In E.H. Lennette and N.J. Schmidt (eds.), Diagnostic Procedures for Viral, Rickettsial and Chlamydial Infections. Washington, D.C.: American Public Health Association.

Gregg, N. 1941. Congenital cataract following german measles in the mother. Trans. Ophthalmol. Soc. Aust. 3:35–46.

Habel, K. 1945. Cultivation of mumps virus in the developing chick embryo and its application to studies of immunity to mumps in man. Public Health Rep. 60:201–212.

Hall, W.W. and Martin, S.J. 1974. The biochemical and biological characteristics of the surface components of measles virus. J. Gen. Virol. 22:363–374.

Hardy, J.B., Sever, J.L., and Gilkeson, M.R. 1969. Declining antibody titers in children with congenital rubella. J. Pediatr. 75:213–220.

Herrmann, K.L. 1979. Rubella virus. In E.H. Lennette and N.J. Schmidt (eds.), Diagnostic Procedures for Viral, Rickettsial and Chlamydia Infections. Washington, D.C.: American Public Health Association.

Hopps, H.E. and Parkman, P.D. 1979. In E.H. Lennette and N.J. Schmidt (eds.), Diagnostic Procedures for Viral, Rickettsial and Chlamydial Infections. Washington, D.C.: American Public Health Association.

Ingalls, T.H., Plotkin, S.A., Philbrook, F.R., and Thompson, R.F. 1970. Immunization of school children with rubella (RA 27/3) vaccine. Lancet 1:99–101.

Johnson, S.C. and Goodpasture, E.W. 1934. An investigation of the etiology of mumps. J. Exp. Med. 59:1–19.

Katz, S.L. and Enders, J.F. 1969. Measles virus. In E.H. Lennette and N.J. Schmidt (eds.), Diagnostic Procedures for Viral and Rickettsial Infections, 4th ed. New York: American Public Health Association, 504–528.

Koplan, J.P. and Preblud, S.R. 1982. A benefit-cost analysis of mumps vaccine. Am. J. Dis. Child. 136:362–4.

Koplik, H. 1896. The diagnosis of the invasion of measles from a study of the exanthemata as it appears on the buccal mucus membrane. Arch. Pediatr. 13:918–922.

Lebon, P. and Lyon, G. 1974. Non-congenital rubella encephalitis (letter). Lancet 2:468.

Leinikki, P.O., Shekarchi, I., Dorsett, P., and Sever, J.L. 1978. Determination of virus-specific IgM antibodies by using ELISA: Elimination of false-positive results protein A-Sepharose absorption and subsequent IgM antibody assay. J. Lab. Clin. Med. 92:849–857.

Leinikki, P.O., Shekarchi, I., Tzan, N., Madden, D.L., Sever, J.L., McLean, A., and Hilleman, M.R. 1979. Evaluation of enzyme-linked immunosorbent assay (ELISA) for mumps virus antibodies. Proc. Soc. Exp. Biol. Med. 160:363–367.

Lennette, D.A., Emmons, R.W., and Lennette, E.H. 1975. Rapid diagnoses of mumps virus infections by immunofluorescence methods. J. Clin. Microbiol. 2:81–84.

Maton, W.G. 1815. Some account of a rash liable to be mistaken for scarlatina. Medical Tr. Roy. Coll. Phys. London 5:149.

McAleer, W.J., Markus, H.Z., McLean, A.A., Buynak, E.B., and Hilleman, M.R. 1980. Stability on storage at various temperatures of live measles, mumps and rubella virus vaccines in new stabilizer. J. Biol. Stand. 8:281–287.

Menser, M.A., Dods, L., and Harley, J.D. 1967. A twenty-five year follow-up of congenital rubella. Lancet 2:1347–1350.

Menser, M.A., Forrest, J.M., Honeyman, M.C., and Burgess, J.A. 1974. Diabetes, HLA antigens, and congenital rubella. Lancet 2:1058–1059.

Meurman, O.H. 1978. Persistence of immunoglobulin G and immunoglobulin M antibodies after postnatal rubella infection determined by solid-phase radioimmunoassay. J. Clin. Microbiol. 7:34–38.

Meyer, H.M., Parkman, P.D., and Panos, T.C. 1966. Attenuated rubella virus: II. Production of an experimental live virus vaccine and clinical trial. N. Engl. J. Med. 275:575–580.

Musser, S.J. and Underwood, G.E. 1960. Studies on measles virus. II. Physical properties and inactivation studies of measles virus. J. Immunol. 85:292–297.

Nicolai-Scholten, M.E., Ziegelmaier, R., Behrens, F., and Hopken, W. 1980. The enzyme-linked immunosorbent assay (ELISA) for determination of IgG and IgM antibodies after infection with mumps virus. Med. Microbiol. Immunol. 168:81–90.

Nommensen, F.E. and Dekkers, N.W. 1981. Detection of measles antigen in conjunctival epithelial lesions staining by lissamine green during measles virus infection. J. Med. Virol. 7:157–162.

Parkman, P.D., Buescher, E.L., and Artenstein, M.S. 1962. Recovery of rubella virus from Army recruits. Proc. Soc. Exp. Biol. Med. 111:225–230.

Parkman, P.D., McCowan, J.M., Mundon, F.K., and Druzd, A.D. 1964. Studies of rubella. I. Properties of the virus. J. Immunol. 93:595–607.

Plotkin, S.A. and Kaye, R. 1970. Diabetes mellitus and congenital rubella. Pediatrics 46:650–651.

Plotkin, S.A., Farquhar, J.D., Katz, S., and Ingalls, T.H. 1967. Discussion of rubella vaccines. Pan. Am. Health Org. Sci. Publ. 147:405–408.

Popow-Kraupp, T. 1981. Enzyme-linked immunosorbent assay (ELISA) for mumps virus antibodies. J. Med. Virol. 8:79–88.

Preece, M.A., Kearney, P.J., and Marshall, W.C. 1977. Growth-hormone deficiency in congenital rubella. Lancet 2:842–844.

Ratzmann, K.P., Strese, J., Witt, S., Berling, H., Keilacker, H., and Michaelis, D. 1984. Mumps infection and insulin-dependent diabetes mellitus (IDDM). Diabetes Care 7:170–173.

Rawls, W.E. and Melnick, J.L. 1966. Rubella virus carrier cultures derived from congenitally infected infants. J. Exp. Med. 123:795–816.

Rice, G.P.A., Casali, P., and Oldstone, M.B.A. 1983. A new solid-phase enzyme-linked immunosorbent assay for specific antibodies to measles virus. J. Infect. Dis. 147:1055–1059.

Rosenberg, H.S., Oppenheimer, E.H., and Esterly, J.R. 1981. Congenital rubella syndrome: The late effects and their relation to early lesions. Perspect. Pediatr. Pathol. 6:183–202.

St. Geme, J.W. Jr., Noren, G.R., and Adams, P. Jr. 1966. Proposed embryopathic relation between mumps virus and primary endocardial fibroelastosis. N. Engl. J. Med. 275: 339–347.

Schiff, G.M. and Sever, J.L. 1966. Rubella: Recent laboratory and clinical advances. Prog. Med. Virol. 8:30–61.

Schwartz, A.J.F. 1962. Preliminary tests of a highly attenuated measles vaccine. Am. J. Dis. Child. 103:386–389.

Sever, J.L. 1983. Virus as teratogens. In E.M. Johnson and D.M. Kochhar (eds.), Handbook of Experimental Pharmacology, Vol. 65. New York: Springer Verlag.

South, M.A. and Alford, C.A. 1980. The immunology of chronic intrauterine infections. In R. Stiehm and V.A. Fulginiti (eds.), Immunologic Disorders in Infants and Children, 2nd ed. Philadelphia: W.B. Saunders, 702–714.

Stewart, G.L., Parkman, P.D., Hopps, H.E., Douglas, R.D., Hamilton, J.P., and Meyer, H.M. Jr. 1967. Rubella-virus hemagglutination inhibition test. N. Engl. J. Med. 276: 554–557.

Stokes, J. Jr., Weibel, R.E., Buynak, E.D., and Hilleman, M.R. 1967. Clinical and laboratory tests of Merck strain live attenuated rubella virus vaccine. Pan Am. Health Org. Sci. Publ. 147:402–405.

Sultz, H.A., Hart, B.A., Zielezny, M., and Schlesinger, E.R. 1975. Is mumps virus an etiologic factor in juvenile diabetes mellitus? J. Pediatr. 86:654–656.

Townsend, J.J., Baringer, J.R., Wolinsky, J.A., Malamud, N., Medick, J.P., Panirch, H.S.,

Scott, R.A.T., Oshiro, L.S., and Cremer, N.E. 1975. Progressive rubella panencephalitis. N. Engl. J. Med. 292:990–993.

Townsend, J.J., Stroop, W.G., Baringer, J.R., Wolinsky, J.H., McKerrow, J.H., and Berg, B.O. 1976. Neuropathy of progressive rubella panencephalitis after childhood rubella. Neurology 32:185–190.

Ukkonen, P., Vaisanen, O., and Penttinen, K. 1980. Enzyme-linked immunosorbent assay for mumps and parainfluenza type 1 immunoglobulin G and immunoglobulin M antibodies. J. Clin. Microbiol. 11:319–323.

Vejtorp, M. 1983. Serodiagnosis of postnatal rubella. Dan. Med. Bull. 30:(2)53–66.

Weil, M.L., Itabashi, H.H., Cremer, N.E., Oshiro, L.S., Lennette, E.H., and Carnay, L. 1975. Chronic progressive panencephalitis due to rubella virus simulating subacute sclerosing panencephalitis. N. Engl. J. Med. 292:994–998.

Weller, T.H., Alford, C.A., and Neva, F.A. 1964. Retrospective diagnosis by serological means of congenitally acquired rubella infection. N. Engl. J. Med. 270:1039–1041.

Weller, T.H. and Neva, F.A. 1962. Propagation in tissue culture of cytopathic agents from patients with rubella-like illness. Proc. Soc. Exp. Biol. Med. 111:215–225.

Wolinsky, J.S. 1978. Progressive rubella panencephalitis. In P.J. Vinken and G.W. Bruyn (eds.), Handbook of Clinical Neurology, Vol. 34. Amsterdam: North-Holland, 331–341.

Ziring, P.R., Fedun, B.A., and Cooper, L.A. 1975. Thyrotoxicosis in congenital rubella. (letter). J. Pediatr. 88:1002.

Ziring, P.R., Gallo, G., Finegold, M., Buimovici-Klein, E., and Ogra, P. 1977. Chronic lymphocytic thyroiditis: Identification of rubella virus antigen in the thyroid of a child with congenital rubella. J. Pediatr. 90:419–420.

32

Human Retroviruses

Fulvia di Marzo Veronese, Paolo Lusso,
Jorg Schüpbach, and Robert C. Gallo

32.1 INTRODUCTION

The search for human retroviruses started many years ago, prompted by suggestive findings in animal model systems. It was well known that a variety of naturally occurring neoplasias and immune dysfunctions in animals are induced by retroviruses. In contrast, evidence for an etiologic role for retroviruses in human neoplasia and in particular in lymphoproliferative disorders was long missing. In most animal leukemias, abundant retroviral replication was observed and the virus could easily be isolated from infected animals. In humans, however, the lack of detectable virus expression in neoplastic cells in vivo and the inability to grow cells of different lineages in vitro represented the major problems encountered.

The discovery of the first human retrovirus, called human T-cell leukemia virus type I (HTLV-I) (Figure 32.1), was made possible by two principal achievements: 1) the ability to specifically detect the retroviral enzyme reverse transcriptase among a variety of cellular DNA polymerases (Sarngadharan et al, 1978) and 2) the ability to continuously grow human T cells in vitro

with the aid of interleukin 2 (IL-2), a cytokine originally described in this laboratory as T-cell growth factor (TCGF) (Morgan et al, 1976; Ruscetti et al, 1977). The body of knowledge gathered by studying HTLV-I was fundamental in the subsequent discovery and characterization of other human retroviruses. To date, two evolutionarily distinct groups of human retroviruses have been identified: the leukemia viruses (HTLV-I and HTLV-II), belonging to the oncovirus subfamily, and the immunodeficiency viruses (HIV-1 and HIV-2), classified in the lentivirus subfamily. In this chapter, we briefly characterize these viruses and give some details about the pathology they produce and the appropriate methods for their detection.

32.2 HISTORY AND CHARACTERIZATION

32.2.1 Human T-Cell Leukemia Virus Types I and II

As mentioned above, successful isolation of these viruses depended on the ability to grow hematopoietic cells of different lineages in vitro. After the discovery of IL-2, this laboratory developed a procedure for

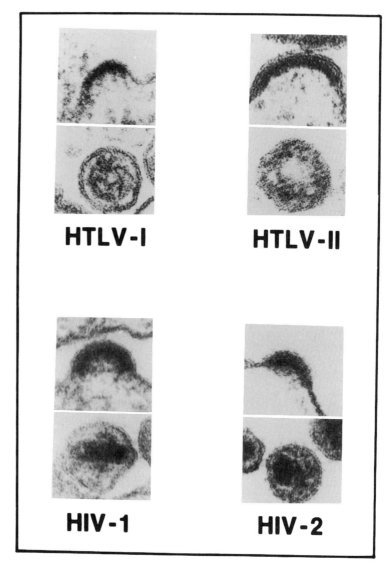

Figure 32.1 Electron microscopy of thin sections of cells producing HTLV-I, HTLV-II, HIV-1 and HIV-2, showing budding and mature viral particles.

culturing normal human mature T lymphocytes (Morgan et al, 1976; Ruscetti et al, 1977). Using the same procedure, it was also possible to grow malignant T cells obtained from patients who were affected by some mature T-cell lymphoproliferative disorders (Poiesz et al, 1980). Peripheral blood cells from these patients were separated on Ficoll-Hypaque gradients and grown in complete medium supplemented with IL-2. The cultures were then monitored for changes in the cellular morphology, and for the production of virus using

electron microscopy (EM) and by measuring viral reverse transcriptase activity in the supernatant fluids. Several human T-cell lines were established from patients with mature T-cell malignancies. In some of these cultures, it was possible to observe viruses with the typical morphology of type C retroviruses.

Human T-cell leukemia viruses were first isolated from cultured cells of two adult black North American patients with an aggressive form of T-cell malignancy (Poiesz et al, 1980a, 1981). These viruses were shown to be very closely related by nucleic acid and antigenic determinant analysis and were eventually called HTLV type I. Human retroviral isolates from patients with mature T-lymphoproliferative disorders were also independently reported by other laboratories. A great number of HTLV-I isolates have been obtained from T-cell malignancies, neurologic disorders, and in some instances from healthy family members of patients around the world (Sarin et al, 1983; Popovic et al, 1984a). In Japan, virus isolates were obtained from patients with a distinct clinical entity, adult T-cell leukemia (ATL), originally recognized in Southwestern Japan (Takatsuki et al, 1979) and showing a pattern of geographic clustering suggestive of the presence of a transmissible agent that is spread horizontally. Such isolates were initially designated adult T-cell leukemia viruses (ATLV). Comparative studies on the various isolates were performed employing immunologic cross-reactivity (Poiesz et al, 1981; Kalyanaraman et al, 1982a; Yoshida et al, 1982; Schüpbach et al, 1983b) and sequence homology, as determined by molecular hybridization (Poiesz et al, 1981; Popovic et al, 1982), cleavage site maps of several restriction endonucleases (Wong-Staal et al, 1983) and nucleotide sequence analyses (Haseltine et al, 1984). The evidence from these studies showed that all the isolates are closely related to or identical with the prototype HTLV-I isolated in the United States.

HTLV-I has been recently associated with human diseases other than cancers. The most conclusive of these associations is with a neurological disorder known as tropical spastic paraparesis (TSP), also known as HTLV-I-associated myelopathy (HAM) (Gessain et al, 1985). The virus isolates obtained from these patients appear to be immunologically and molecularly indistinguishable from those obtained from ATL patients.

In 1982, a retrovirus related to, but distinct from HTLV-I, was isolated in our laboratory from a T-cell line that had been established from a patient with a T-cell variant of hairy cell leukemia. This virus was designated HTLV type II (Kalyanaraman et al, 1982b). HTLV-I and HTLV-II have common biological features. They exhibit a predominant in vivo and in vitro tropism of CD4-positive T cells and can immortalize them in vitro. However, CD8+ T cells can be infected and immortalized in vitro by HTLV-I and II. The receptor for the two viruses is probably the same and has been recently mapped to a locus on human chromosome 17 (Sommerfelt et al, 1988). By electron microscopy, the morphologic appearance of budding particles and mature forms of the two viruses is very similar (Fig. 32.1). In addition, HTLV-I and HTLV-II have a number of cross-reactive antigenic determinants both in the envelope and internal core proteins, for example two monoclonal antibodies against different epitopes of HTLV-I and HTLV-II p24 were unable to discriminate between the two viruses (Sarngadharan et al, 1985). These viruses also share some nucleotide sequence homology and possess reverse transcriptases of similar size and biochemical features. Nevertheless, HTLV-I and HTLV-II are distinct, as demonstrated by core protein (p24) competition radioimmunoassays (Kalyanaraman et al, 1982b). The envelope antigens also are different, as shown by biological assays, such as inhibition of syncytia induction and vesicular stomatitis virus pseudotype neutralization (Clapham et al, 1984). Since 1982, HTLV-II has been identified in several individuals, mostly among intravenous drug users (Lee et al, 1989). An etiological role of HTLV-II in a form of atypical hairy-cell leukemia has been suggested (Rosenblatt et al, 1986, 1988). However, the limited number of patients identified as infected with HTLV-II, in part also as a consequence of poor serologic discrimination from HTLV-I, has hampered a definitive demonstration of an etiological role of HTLV-II in human diseases.

32.2.2 Human Immunodeficiency Virus Types 1 and 2

Human immunodeficiency virus type 1 (HIV-1), initially designated human T-cell lymphotropic virus type-III (HTLV-III) lymphadenopathy-associated virus (LAV), was detected and isolated from patients with acquired immunodeficiency syndrome (AIDS) or with clinical signs that frequently precede full-blown AIDS (Barré-Sinoussi et al, 1983; Gallo et al, 1984; Popovic et al, 1984b; Sarngadharan et al, 1984). In 1986, a second virus, HIV-2, biologically and genetically related to HIV-1, was identified in West Africa and subsequently shown to produce a clinical picture similar to that observed in HIV-1 infection (Clavel et al, 1986). The comparative virulence and pathogenicity of HIV-1 and HIV-2 have yet to be determined.

AIDS has been recognized since 1981 as a distinct new disease entity characterized by severe immune suppression with a dramatic depletion of the CD4-positive subset of mature T lymphocytes, multiple opportunistic infections, and/or neoplasias (Gottlieb et al, 1981; Siegal et al, 1981). Several features of the disease suggested that it could be caused by a transmissible infectious agent, probably a virus. The etiologic agent was filterable through devices that could retain fungi or bacteria and epidemiologic studies indicated that it could be transmitted by blood transfusion or blood products. Furthermore, it was well known that animal retroviruses, such as feline leukemia virus, are able to cause AIDS-like diseases in animals (Jarrett et al, 1964).

Viruses belonging to the new group, HIV, can now be isolated from virtually all infected individuals (Ho et al, 1989; Coombs et al, 1989). These viruses share some properties with HTLV-I and HTLV-II, including tropism for CD4+ T lymphocytes,

Mg^{2+}-dependent reverse transcriptase, a major core protein with a molecular weight of 24 kD (p24), some cross-reactive antigenic determinants, the presence of two essential regulatory gene products whose mRNAs are doubly spliced, and closest relatedness with African primate retroviruses. Unlike the HTLV, however, the individual HIV isolates demonstrate a remarkable level of genetic variability. The main biological difference between HTLV and HIV is the effect seen after transmission to primary human cells in vitro. HTLV-I and HTLV-II can transform some of these cells and propagate indefinitely in culture. In contrast, HIV replicates only transiently since it has strong cytopathogenic effects on susceptible cells, leading to massive cell death. In addition, HIV, similar to other lentiviruses, also has the ability to productively infect primary human monocyte/macrophages, which may constitute the major in vivo reservoir of the infection.

Characterization of HIV-1 and serologic testing of large numbers of patients became possible only after the virus could be propagated in human T-cell lines that have become resistant to the cytopathogenic effects of the virus (Popovic et al, 1984b). In addition to T-cell lines, other human cell lines of B-cell and monocytic origin have been subsequently found to support the long-term growth of both HIV-1 and HIV-2, but do so less efficiently.

32.3 GENOME ORGANIZATION AND PROTEIN PRODUCTS

32.3.1 Human T-Cell Leukemia Virus Types I and II

The molecular cloning of HTLV-I and HTLV-II (Seiki et al, 1982; Manzari et al, 1983; Gelmann et al, 1984) and the complete nucleotide sequence of a DNA clone of HTLV-I (Seiki et al, 1983) allowed detailed studies on their genetic structures (see Figure 32.2). These structures are sim-

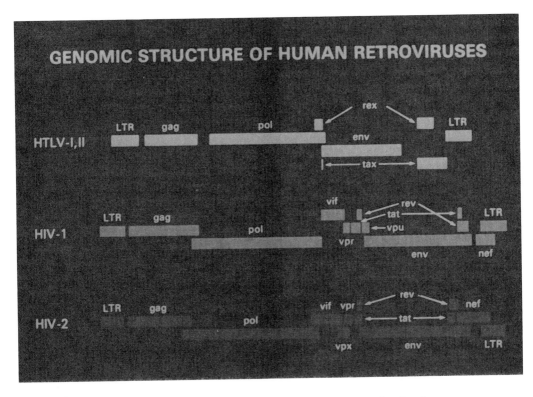

Figure 32.2 Genetic organization of human retroviruses. Refer to text for details.

ilar to that of other replication competent retroviruses, with two long terminal repeat (LTR) elements, one at each end and with three essential structural genes, *gag*, *pol*, and *env*. In addition to the usual complement of genes there is a novel region in HTLV-I and HTLV-II, between the *env* gene and the 3′LTR, not found in other replication competent leukemia viruses (Lee et al, 1984b; Slamon et al, 1984). This region was initially designated as the X region, because no functional assignment could be made (Seiki et al, 1983). Two new genes and their encoded products have now been identified in the X region of HTLV-I and HTLV-II. One is designated *tax* (1 and 2, for HTLV-I and HTLV-II, respectively) and was previously referred to as pX, x, x-*lor*, *tat*-I/*tat*-II or TA1/TA2. This gene encodes a transactivating protein involved in the transcriptional control of viral, and

possibly cellular, genes. The second protein is designated rex and is encoded by a region which partially overlaps with that of the *tax* gene but is in another reading frame. This protein is also essential for HTLV replication and appears to act post-transcriptionally to regulate viral gene expression. The *gag* region is initially translated as a precursor polyprotein Pr53^{gag}, which is then proteolytically processed to form the mature *gag* gene products. These proteins are p19, p24, p15 as mapped in the *gag* region from the amino terminus (Copeland et al, 1983; Hattori et al, 1984). By analogy with other retroviruses, these three proteins are believed to correspond to the matrix (MA), capsid (CA), and nucleocapsid (NC) proteins, respectively. All three gag proteins were purified to homogeneity, their amino acid sequences were determined almost completely (Oroszlan et al,

1982; Copeland et al, 1983), and their immunologic properties characterized (Kalyanaraman et al, 1981, 1984b; Posner et al, 1981; Schüpbach et al, 1983a). Among the gag proteins of HTLV-I and HTLV-II, p24 seems to have the highest degree of antigenic cross-reactivity. A variety of gag-related polypeptides were identified in HTLV-infected cells (Kalyanaraman et al, 1981b; Kalyanaraman et al, 1984b; Schneider et al, 1984). These include molecular species of 70, 55, 37, 29 and 28 kD, as defined by sodium dodecyl sulfate-polyacrylamide gel electrophoretic (SDS-PAGE) analysis. At its 3′ end the *gag* open reading frame overlaps with the 5′ end of the open reading frame encoding the protease. In turn, the 3′ end of the protease open reading frame overlaps with the 5′ end of the *pol* gene. The polymerase region contains the largest open reading frame of the HTLV genome. By analogy with other retroviruses, the 5′ region of the *pol* gene should encode the reverse transcriptase protein and nucleotide sequences found toward the 3′ end of the integrase protein. The HTLV enzymes are larger than other mammalian type C retroviral reverse transcriptases and prefer Mg^{2+} (as divalent cation) instead of Mn^{2+} (Rho et al, 1981). The *env* gene of HTLV-I codes for a primary glycosylated product of 61 to 68 kD (g61-68), the size being dependent on the laboratory and the cell line studied (Lee et al, 1984a; Schneider et al, 1984). In the mature virion, gp61-68 is processed into the external glycoprotein gp46 and the transmembrane protein gp21. The deduced amino acid sequence of the env precursor of HTLV-II shows great similarity with the corresponding protein of HTLV-I (Sodroski et al, 1984).

We already mentioned that two unique genes are found in HTLV, termed *tax* and *rex*, which encode nonstructural proteins translated from a doubly-spliced mRNA containing three exons (Inoue et al, 1987; Kiyokawa et al, 1985; Nagashima et al,

1986; Shima et al, 1986). The HTLV-I and HTLV-II *tax* genes encode proteins of 40 and 37 kD, respectively; both are located in the nucleus of HTLV-infected cells. The *rex* gene encodes proteins of 27 and 21 kD for HTLV-I and 26 and 24 kD for HTLV-II.

32.3.2 Human Immunodeficiency Virus Types 1 and 2

HIV-1 and HIV-2 share several features with the HTLVs, but in most respects they are more closely related to other members of the subfamily of the lentiviruses by genetic and morphologic criteria. HIV is a positive strand RNA virus and produces particles similar in form to the animal type C retroviruses. Electron microscopic examination and analysis of virions reveal a characteristic dense, cylindrical core (Fig. 32.1). This ultrastructure with the cylindrical core is morphologically indistinguishable from that of other lentiviruses such as visna virus, but is quite different from that of HTLV which has a bar- or cone-shaped core.

The genetic structure of HIV-1 and HIV-2, along with the gene products, are shown schematically in Fig. 32.2. Like HTLV, the HIV provirus contains two LTR elements, one at each end, and three structural genes that are essential for virus replication. The *gag* gene encodes a polyprotein precursor Pr53[gag], subsequently cleaved by the viral protease during maturation to give rise to the amino terminal myristoylated protein, p17 (MA), the major core protein p24 (CA) and the nucleocapsid protein p15 (NC), which is further cleaved to generate p7 and p6. In addition to Pr53[gag], alternative gag precursors of 37 and 39 kD in size have been found in HIV-1 infected cells (Veronese et al, 1987). The *pol* gene codes for the protease p10, the two subunits of reverse transcriptase p66/51 and the integrase or endonuclease p31. The *env* gene codes for a primary translational product gp160, which is proteolytically cleaved by cellular enzymes to form the external glycoprotein gp120 and the transmembrane glycoprotein gp41 (Veronese et al, 1985). In addition to these conventional retroviral genes, HIV contains at least seven accessory genes termed: *tat*, *rev*, *nef*, *vif*, *vpr*, *vpu* (HIV-1) or *vpx* (HIV-2) (Wong-Staal, 1990), and *tev* (Felber et al, 1990). Significant levels of HIV gene expression occur only in the presence of the *tat* gene product, the transactivator protein, analogous in function to HTLV tax protein. Tat is 14 kD in size and requires a cis-acting sequence termed the transactivation responsive region (TAR). The *rev* gene product is 19 kD in size and is thought to preferentially stabilize and facilitate the transport of the higher molecular weight RNAs from the nucleus to the cytoplasm (Malim et al, 1989a, 1989b). The HTLV gene *rex* displays a mechanism of action similar to that of *rev* (Hidaka et al, 1988) and could substitute for rev function in complementation studies (Rimsky et al, 1988). Also *rev* activity requires a cis-acting sequence called the rev-responsive element (RRE) (Hadzopoulou-Cladaras et al, 1989; Malim et al, 1989a, 1989b). The *nef* gene product is a 27 kD protein associated with cytoplasmic membranes, and is post-translationally myristoylated and phosphorylated. Since mutants defective in *nef* replicate two to tenfold better than the wild type (Fisher et al, 1986; Luciw et al, 1987), the nef protein is postulated to act as a possible regulator of transcription. The protein product of the *vif* gene corresponds to a 23 kD protein not required for assembly of virions or regulation of viral gene expression. However, viruses containing mutations in *vif* yield only low levels of infectious particles (Fischer et al, 1987; Strebel et al, 1987). The products of *vpr*, *vpu*, and *vpx* are proteins of 18, 16, and 14 kD in size, respectively. None of these proteins appears to be essential for virus replication in T cells in vitro and their function is still unknown (Dedera et al, 1989; Ogawa et al,

1989; Strebel et al, 1988; Cohen et al, 1988; Terwilliger et al, 1989; Henderson et al, 1988; Yu et al, 1988). Tev is a 28 kD protein initiated at the *tat* initiation codon containing the first exon of *tat* at its amino terminus, a small portion of *env* in the middle, and the second exon of *rev* at its carboxy terminus (Benko et al, 1990). This protein displays both tat and rev activities in functional assays. A characteristic feature of the HIV genome is the dramatic degree of heterogeneity among individual isolates. Such hypervariability is unevenly distributed throughout the genome and particularly involves the *env* gene. This feature, which may hamper the development of effective vaccines, is likely a consequence of strong immune selective pressure in vivo.

32.4 PATHOGENESIS OF HUMAN T-CELL LEUKEMIA VIRUS TYPE I INFECTION

The spectrum of human diseases which have been linked to HTLV-I includes a malignancy of T lymphocytes, ATL, and a neurological disorder, TSP or HAM. In many instances, and particularly in the course of ATL, immunodeficiency is often an accompanying manifestation of HTLV-induced disease.

32.4.1 Adult T-Cell Leukemia (ATL)

32.4.1.a Clustering of T-Cell Malignancies and Epidemiology of HTLV-1

A clustering of T-cell malignancies was originally recognized among patients born in the southwestern regions of the Japanese islands (Yodoi et al, 1974; Takatsuki et al, 1977; Uchijama et al, 1977). Serologic studies (Kalyanaraman et al, 1982a; Robert-Guroff et al, 1982b, 1983) subsequently linked these cancers to the newly detected HTLV-I. Independently, Japanese investigators linked a retrovirus, initially designated ATLV, to ATL (Hinuma et al, 1981, 1982; Miyoshi et al, 1981; Yoshida et al, 1982). Later, it was recognized that HTLV-I and ATLV were the same retrovirus (Watanabe et al, 1984). Other clusters of T-cell malignancies indistinguishable from Japanese ATL were later identified in the Caribbean (Costello et al, 1980; Catovsky et al, 1982) and also linked to HTLV-I (Blattner et al, 1982; Schüpbach et al, 1983a). To date, additional clusters or sporadic cases of HTLV-I-associated malignancies have been recognized in the United States, especially in the southeastern regions, in Central and South America, Africa, the Middle East, Italy, India, and some areas of the Far East outside Japan (Blayney et al, 1983a; Fleming et al, 1983; Hunsmann et al, 1983; Biggar et al, 1984; Merino et al, 1984; Saxinger et al, 1984; Manzari et al, 1984; Gallo and Blattner, 1985).

There is now solid evidence that HTLV-I is the etiologic agent of ATL. The virus can be consistently isolated from the neoplastic cells of ATL patients, epidemiologic studies show that ATL occurs only in regions where HTLV-I is endemic and that the incidence of ATL in endemic regions is correlated with the prevalence of antibodies to HTLV-I (Robert-Guroff and Gallo, 1983). Infection by HTLV-I can be demonstrated in all ATL patients by both molecular detection and virus isolation. Moreover, it was shown that the tumor cells of ATL patients are monoclonal in origin and contain one or few copies of the HTLV-I genome with a single (monoclonal) site of proviral integration (Wong-Staal et al, 1983; Yoshida et al, 1984). However, the integration site is not the same in different patients. In vitro models show that HTLV-I has a predominant tropism and immortalizing capability for mature T lymphocytes bearing the CD4 marker, which is the common phenotype of ATL cells (Miyoshi et al, 1981; Yamamoto et al, 1982; Popovic et al,

1983a). However, the pathogenetic mechanism of ATL is still unknown. Unlike retroviruses causing acute leukemia in animals, HTLV-I does not encode for viral oncogenes. In addition, it does not integrate into a specific site of the host cell genome like the chronic (slow) leukemia retroviruses do. Two genes of HTLV-I that may be critical in the leukemogenesis are *tax* and *rex*, which encode for *trans*-acting transcriptional activator proteins which are capable of inducing the expression of several cellular genes. For example, tax induces the expression of IL-2 and its receptor (CD25), which may be implicated in the initial (non-neoplastic) expansion of HTLV-I-infected clones in vivo. The leukemogenesis in ATL is certainly a multi-step progression, since the disease is preceded by a lengthy phase of polyclonal T-cell infection by HTLV-I, often revealed by the spontaneous T-cell proliferation of freshly isolated PBMC from infected individuals. The most likely hypothesis is that the initial polyclonal T-cell proliferation may predispose some T cells to secondary leukemogenic event(s), ultimately leading to the emergence of a neoplastic monoclone (Gallo et al, 1986).

In all geographic regions where clusters of ATL have been found, HTLV-I infection is also present within the healthy population. The seroprevalence and ATL incidence vary greatly among geographic areas that may be only short distances apart (Yamaguchi et al, 1983). Similarly, HTLV-I infection may be restricted to certain ethnic or social groups, such as blacks emigrating from the Caribbean to European countries (Blattner et al, 1982; Greaves et al, 1984; Robert-Guroff et al, 1984; Schaffar-Deshaynes et al, 1984). Family members of ATL patients show a prevalence of antibodies to HTLV-I that is significantly higher than the unrelated control population (Miyamoto et al, 1985). In these instances, HTLV-I transmission occurs mostly through sexual contacts. It is likely that lymphocytes contained in the secretions, rather than cell-free body fluids, are responsible for transmission. This observation is consistent with the low efficiency of cell-free transmission of HTLV-I in vitro (Popovic et al, 1983b). Other routes of virus transmission include blood transfusions (Saxinger and Gallo, 1982; Maeda et al, 1984; Miyamoto et al, 1984; Okochi et al, 1984; Gessain et al, 1990), sharing blood-contaminated needles (Robert-Guroff et al, 1986), and mother-to-child transmission, most likely after birth via lymphocytes contained in breast milk (Tajima et al, 1982; Komura et al, 1983). It is noteworthy that, in contrast to HIV, HTLV-I seems not to be transmissible through fractionated blood products (Epstein and Fricke, 1990). Some investigators have also postulated a role for parasites or other vectors in the transmission of HTLV-I (Tajima et al, 1981, 1983; Merino et al, 1984).

It has been estimated that only about 4% of seropositive persons will eventually develop ATL. Although the minimal incubation time for the development of ATL is not known, circumstantial evidence suggests that it is in the range of years or decades. Cases of ATL were reported to occur in individuals 20 years and longer after emigration from HTLV-I-endemic regions where they probably acquired the infection (Greaves et al, 1984).

32.4.1.b ATL Clinical Course

ATL may manifest itself as a leukemia or, more rarely, as a lymphoma. Usually, the disease is rapidly progressive and the median survival time is less than one year despite chemotherapy (Uchiyama et al, 1977; Hanaoka, 1982). A proportion of patients, however, display a subacute or chronic disease that may eventually progress to an acute clinical course. Sporadically, patients with subacute cases also have undergone spontaneous remission (Kawano et al, 1984).

The malignant cells of ATL display a remarkable pleomorphism, containing typical polylobulated nuclei with prominent nucleoli (Fig. 32.3). They usually express the CD4 antigen (only isolated reports have described HTLV-I-associated CD8+ lymphoproliferations), the 4B4 antigen (CDw29) expressed by "memory" T cells, the transferrin receptor (CD71), and the IL-2 receptor (CD25 or Tac antigen), while they are negative for terminal deoxynucleotide transferase (TdT) (Hattori et al, 1981; Lando et al, 1983; Popovic et al, 1983a; Waldmann et al, 1983). Despite the CD4-positive ("helper") phenotype, in some cases these cells have demonstrated a "suppressor" function in vitro (Takatsuki et al, 1982; Waldmann et al, 1983; Yamada et al, 1983). Chromosomal abnormalities have not been consistently found in ATL. Nevertheless, $6q^-$, $14q^+$, and trisomy 3 or 7 are the most frequent abnormalities observed and may be helpful in selected cases for confirmation of the clonal origin of the neoplastic cells.

Five clinicopathologic patterns of ATL have been described (Table 32.1) (Clark et al, 1986): 1) typical ATL, usually with a rapidly progressive clinical course (average 6 to 8 months); 2) smoldering ATL, showing an indolent clinical course with few circulating malignant cells; skin involvement (see below), lymphadenopathy and/or hepatosplenomegaly may be present; 3) chronic ATL, similar to the smoldering ATL but with a high percentage of circulating malignant cells, skin involvement, lymphodenopathy and/or hepatosplenomegaly; 4) the "blastic" crisis of the smoldering or chronic ATL, which represents the conversion to acute clinical disease with the characteristic features of typical ATL; 5) aleukemic or lymphoma-like ATL, which includes cases indistinguishable from mature non-Hodgkin's T-cell lymphomas in HTLV-I endemic areas, with integrated HTLV-I provirus but no leukemic manifes-

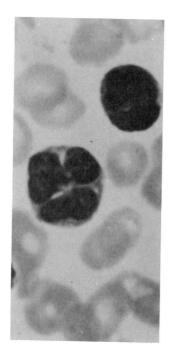
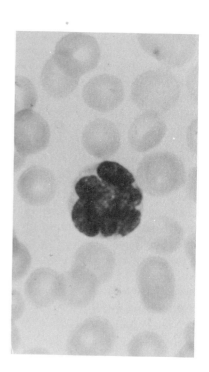

Figure 32.3 Morphology of typical neoplastic cells from the peripheral blood of an ATL patient (Giemsa stain).

tations and sometimes none of the other features of ATL.

In the typical ATL patient, lymphadenopathy and splenomegaly are frequently observed, but there is usually no mediastinal lymph node involvement. A peculiar finding of ATL in about 50% of cases is the presence of diffuse skin lesions consequent to the infiltration of leukemic cells; in contrast, interstitial pneumonia due to neoplastic infiltration is rarely observed. Another possible complication is central nervous

Table 32.1 Clinical Features of ATL Subgroups

Clinical Feature	Acute ATL	Smoldering ATL	Chronic ATL	Non-Hodgkin's Lymphoma-Like[a]
Leukemic Manifestation[b]	++++[c]	−	++++	−
Skin Involvement	++	++++	++	++
Lymphadenopathy	++++	+	+	++++
Hepatosplenomegaly	+++	+	+	++
Bone Marrow Infiltration	+++	−	++	++
Hypercalcemia	+++	−	−	++
Elevated serum LDH	++++	−	+	++++

[a] Aleukemic ATL.

[b] >10,000 WBC/μL and abnormal lymphocytes with features typical of ATLL.

[c] The ratings indicate: ++++ = feature present in 76% to 100% of patients; +++ = 51% to 75%; ++ = 26% to 50%; + = 1% to 25%; − = absent.
(Adapted from Clark et al, 1985)

system involvement. The laboratory findings include leukocytosis with WBC counts over 100,000 in acute ATL, eosinophilia, neutrophilia and hypercalcemia, the latter due to lytic bone lesions in the absence of tumor cell infiltration (Blayney et al, 1983b). Recent data from this laboratory indicate that HTLV-I-transformed cells constitutively produce TNF-α and TNF-β, soluble factors which can induce osteoclast activation and, consequently, osteolysis (Tschachler et al, 1989). The majority of patients show in vitro and in vivo evidence of immune dysfunction and are frequently affected by opportunistic infections.

For the purpose of differential diagnosis, the combination of clinical and pathologic features usually allows the distinction of ATL from other malignancies of mature T cells, [i.e., mycosis fungoides (MF)/Sézary syndrome (SS) and T-cell chronic lymphocytic leukemia (T-CLL)]. However, the distinction may sometimes be difficult (Lennert et al, 1985). Most of the typical features of ATL, with the exception of the skin involvement, are uncommon in MF/SS. In approximately 33% of ATL cases, the leukemic infiltration is restricted to the subcutaneous tissue and the dermis, but does not affect the epidermis as usually seen in MF/SS (Pautrier's microabscesses) (Jaffe et al, 1984). Moreover, the morphology of the polylobulated ATL cells is different from that of typical SS cells. The nuclear convolutions of ATL cells are usually less markedly indented than those of Sézary cells and have a more "lumpy" appearance. The characteristics of ATL that usually distinguish it from T-CLL include its more aggressive clinical course, the higher degree of nuclear pleomorphism with the typical polylobulated nuclei, and the absence of cytoplasmic granules. In cases where a distinction on clinicopathologic grounds is difficult, HTLV-I antibody detection, the demonstration of HTLV-I antigens in cultured T cells, or the direct demonstration of the viral genome in uncul-

tured leukemic cells obviously will help to confirm the diagnosis of ATL. In addition, the presence of the IL-2 receptor (CD25) on malignant CD4+, TdT− T cells strongly indicates an association with HTLV-I.

32.4.2 Tropical Spastic Paraparesis (TSP) or HTLV-I-Associated Myelopathy (HAM)

Although TSP and HAM have been described independently and in different geographic areas, there is now general agreement that they are the same clinical entity. The first evidence for the involvement of HTLV-I in TSP came from the observation that French West-Indian TSP patients were serologically positive for the virus (Gessain et al, 1985). Subsequently, HTLV-I has been identified and isolated from both the peripheral blood and the cerebrospinal fluid (CSF) of TSP patients (Jacobson et al, 1988; Bhagavati et al, 1988). Virus isolates from ATL and TSP patients appear to be genetically indistinguishable (Yoshida et al, 1987). Nevertheless, only in rare instances have TSP and ATL been described in the same patient (Bartholomew et al, 1986; Kawai et al, 1989). At the present time there is little clue concerning the possible mechanism of neurologic disease induction by HTLV-I. Autoimmune mechanisms have been postulated, based on the finding of lymphocytic infiltration in involved tissues and of some favorable clinical responses to corticosteroids. TSP has a wide geographic distribution, including Central and South America, Africa, India, and Japan. The pathologic hallmark of the disease is a myelopathy primarily affecting the pyramidal tracts and, to a lesser extent, the sensory system. The course of the disease is gradual and progressive with no spontaneous remissions. The major clinical manifestations include spastic paraparesis or paraplegia with hyperreflexia and positive Babinski sign, urinary or fecal incontinence, and mild sensory abnormalities. High numbers of lymphocytes, sometimes of atypical morphology, are detectable in the CSF, together with elevated protein levels. Atypical lymphocytes can also be seen in the peripheral blood. Diffuse lesions of the white matter and the paraventricular regions of the brain have been documented using magnetic resonance imaging.

32.4.3 Other Disorders

HTLV-I has also been associated with immunodeficiencies and with neoplastic disorders other than ATL. It is known that patients with ATL frequently present opportunistic infections. Similar to the picture of AIDS, these often include *Pneumocystis carinii* pneumonia, fungal infections by *Candida*, *Aspergillus* or *Cryptococcus*, infections by herpesviruses, such as human cytomegalovirus (hCMV), herpes simplex virus (HSV) or varicella zoster virus (VZV) and parasitic infestations such as strongyloidiasis. Bacterial sepsis, especially with *Klebsiella*, is also frequent. There is extensive evidence for abnormal immune functions in these patients, and in vitro studies have shown that HTLV-I-infected human CD4+ or CD8+ T lymphocytes display impaired function (Mitsuya et al, 1984; Popovic et al, 1984a; Suciu-Foca et al, 1984; Wainberg et al, 1985).

Recent data indicate that HTLV-I/II may be involved as a cofactor in the course of HIV infection. A series of in vitro results suggest that HTLV-I could positively interact with HIV-1; it productively coinfects HIV-1-infected T cells (DeRossi et al, 1986), transactivates the HIV-1 LTR (Siekevitz et al, 1987), and generates viral particles phenotypically mixed with HIV-1, thereby expanding the HIV-1 cellular tropism (Lusso et al, 1990a). In addition, prospective seroepidemiological studies in Trinidad (Bartholomew et al, 1987), in Miami (Page et al, 1990), in Japan (Hattori et al, 1989), and in New Jersey (Weiss et al, 1989) suggest that the progression to full-blown AIDS in more rapid in patients coinfected with HIV-1 and HTLV-I/II than in those infected with HIV-1 alone. However, these studies were not able to control for the time of acquisition of each virus and the small sample size limited the investigators' ability to assess other factors. Age is strongly associated with risk of HTLV-I/II infection and recent data indicate that progression to AIDS is age-dependent (Weiss et al, 1990).

Recently, HTLV-I has also been associated with some cases of polymyositis (Morgan et al, 1989), infectious dermatitis of infants (Mark Kaplan, personal communication), and retinitis (Kabayama et al, 1988).

The direct or indirect involvement of HTLV-I in malignant disorders other than ATL has also been suggested. For example, a role of HTLV-I was proposed in B-cell chronic lymphocytic leukemia (B-CLL), cutaneous T-cell lymphomas (CTCL) including MF and SS, other T-cell non-Hodgkin's lymphoma, T-cell prolymphocytic leukemia, and large granular lymphocytic (LGL) leukemia (Blattner and Gallo, 1985).

32.4.4 Disorders Associated With Human T-Cell Leukemia Virus Type II (HTLV-II)

Similar to HTLV-I, HTLV-II is capable of immortalizing in vitro both CD4+ and CD8+ T lymphocytes. However, its role in human disease and particularly in lymphoproliferative disorders remains to be further elucidated. A limited number of HTLV-II isolates were obtained from patients with atypical hairy-cell leukemia, often of the T-cell variant type (Kalyanaraman et al, 1982b; Rosenblatt et al, 1986). Subsequently, HTLV-II was identified in a T-cell CLL and in a T-cell prolymphocytic leukemia (T-PPL) (Sohn et al, 1986, Cervantes et al, 1986). Other isolates were obtained from patients without evidence of neoplasia, including AIDS patients.

32.5 PATHOGENESIS OF HUMAN IMMUNODEFICIENCY VIRUS (HIV) INFECTION

32.5.1 Epidemiology and Transmission

The following discussion will not distinguish between the clinical consequences of infection by HIV-1 and HIV-2. Although the relative pathogenicity of HIV-2 is not yet fully established and its geographic distribution is much more limited compared to HIV-1, it is nonetheless able to produce full-blown AIDS as well as neurologic disorders similar to those observed in the course of HIV-1 infection.

Infection by HIV is predominantly spread by sexual contact (both homosexual and heterosexual), parenteral inoculation, and perinatal mother-to-child transmission (Friedland and Klein, 1987). So far, HIV has been isolated from blood, semen, vaginal secretions, saliva, lymph nodes, urine, bone marrow, tears, cerebrospinal fluid, and neurologic tissues, (Gallo et al, 1984; Groopman et al, 1984; Ho et al, 1984; Popovic et al, 1984b; Zagury et al, 1984; Salahuddin et al, 1985; Ho et al, 1985b).

Male homosexuals or bisexuals without a history of intravenous drug use account for 63% of all AIDS cases in the United States. The transmission of the virus among homosexuals and heterosexuals is proportional to the number of sexual partners and is facilitated by receptive anal intercourse or other sexual practices (Goedert et al, 1985).

Approximately 25% of the U.S. AIDS cases so far have occurred in parenteral drug addicts and an exceedingly high number of individuals in this population are HIV-1 seropositive. Sharing unsterilized needles or syringes is the major way of viral spread within this group.

In hemophiliacs transfused with clotting factors and in recipients of whole blood, cellular blood components, or plasma, the rate of infection is proportional to the frequency of treatment. In recipients of blood transfusions, the incidence of AIDS is directly linked to the number of units received (Hardy et al, 1985). However, transfusion of a single contaminated unit invariably resulted in the transmission of HIV. Among patients with hemophilia A, about 80% of the recipients of U.S.-produced factor VIII concentrates were seropositive in 1984 (Kitchen et al, 1984; Melbye et al, 1984; Evatt et al, 1984; Koerper et al, 1985), while in patients with hemophilia B, requiring less frequent treatment, the rate of seropositivity was consistently lower. Clotting factor concentrates were obtained from pooled plasma of thousands of commercial donors, which included members of groups at high-risk for AIDS. Subsequently, prolonged heat-treatment of these preparations (e.g., 68 °C for 72 hours) has been extremely effective in preventing the further spread of HIV in hemophiliacs (Koerper et al, 1985; Mössler et al, 1985). Recently developed techniques, such as "dry heating" at 80 °C for 72 hours, have virtually eliminated the presence of infectious HIV, as well as hepatitis B virus (HBV) and C virus (HCV) from clotting factor preparations (Epstein and Fricke, 1990). Accidental exposure to HIV, most frequently by needle-stick injuries, among health-care workers has very rarely resulted in HIV infection, suggesting that the risk of infection is directly proportional to the number of virus particles in the inoculum.

Most cases of pediatric AIDS have occurred in the offspring of mothers belonging to AIDS risk groups. In these cases, the virus may be transmitted in utero across the placental barrier (Lapointe et al, 1985), at the time of delivery by exposure to infected blood or genital secretions, or in the postnatal period, possibly through breast-feeding (Ziegler et al, 1985). In addition, some pediatric cases are due to blood transfusions or the administration of blood-products.

Although the initial cases of HIV infection in the United States were largely confined to homosexual or bisexual males and drug addicts, the virus has since spread to a considerable number of heterosexually active persons and the projections for the second decade of the epidemic indicate that the incidence will rapidly grow among females and young heterosexuals. Heterosexual transmission is particularly frequent through female prostitutes, who may also be drug addicts (Redfield et al, 1985). The spread of clinically silent HIV infection among traditionally non-risk groups may pose important problems of identification of potential carriers, since testing for anti-HIV antibody may not be performed in unsuspected individuals for years. Thus, the epidemiology of HIV infection in the Western world may eventually assume a pattern similar to that observed in third-world countries, particularly in Africa, where HIV probably was prevalent for more than a decade before the outbreak of the epidemic in the United States (Saxinger et al, 1985), and where both the seropositivity and the incidence of AIDS are evenly distributed between males and females (Biggar et al, 1985).

32.5.2 Clinical Manifestations of HIV Infection

The development of accurate serologic tests has made it possible to identify and follow-up HIV infection almost from the onset. In a proportion of cases, primary infection is manifested by an acute illness with flu-like symptoms and diarrhea which can last for a few days or weeks. As occurs with many other retroviruses, HIV infection has a chronic course and may remain clinically silent for prolonged periods of time, in some individuals for over 10 years (asymptomatic carrier status or clinical la-

tency period). Almost invariably, however, a gradual impairment of the immune function occurs, concomitant with the disappearance of the CD4+ T lymphocytic subset, and secondary opportunistic infections or neoplasms develop. Once the diagnosis of AIDS is made, the mean survival time is usually less than one year. However, the natural history of the infection is remarkably variable in different individuals with no clear-cut correlation with any known risk factor. Besides the immune dysfunction and the CD4+ T-cell depletion, HIV may also directly infect neural cells or monocytes/macrophages within the neural tissue and thereby cause several neurologic disorders. In this respect, HIV shows biologic similarity with members of the lentivirus subfamily of retroviruses, most notably with visna virus, which is the cause of slowly progressive, debilitating central nervous system infections in hoofed domestic animals (Gonda et al, 1985).

The spectrum of HIV-related disorders includes: 1) acute primary infection; 2) persistent generalized lympadenopathy (PGL); 3) acquired immunodeficiency syndrome (AIDS); 4) neurologic disorders. Classification systems for HIV-associated illnesses have been proposed by the Centers for Disease Control (CDC) and Walter Reed Army Institute for Research; they are presented in Tables 32.2 and 32.3.

32.5.2.a Acute Primary Infection

Although in many cases seroconversion results from totally asymptomatic primary infection, an acute illness similar to infectious mononucleosis may occur in a proportion of patients (CDC stage I) (Anonymous, 1984; Cooper et al, 1985; Tucker et al, 1985). After an incubation time of a few days to three months, the illness has a sudden onset and may last for two to three weeks, usually followed by complete recovery. The major signs and symptoms include fever, diarrhea, sweats, malaise,

Table 32.2 CDC Classification of HIV Infection

GROUP I	Acute infection
GROUP II	Asymptomatic infection
GROUP III	Persistent generalized lymphadenopathy
GROUP IV	Other disease
Subgroup A	Constitutional disease[a]
Subgroup B	Neurologic disease[b]
Subgroup C	Secondary infectious diseases
Category C-1	Specified secondary infectious diseases listed in the CDC surveillance definition for AIDS[c]
Category C-2	Other specified secondary infectious diseases[d]
Subgroup D	Secondary cancers[e]
Subgroup E	Other conditions[f]

[a] Fever persisting more than one month, involuntary weight loss of greater than 10 percent of baseline, or diarrhea persisting more than one month.

[b] Dementia, myelopathy, or peripheral neuropathy.

[c] *Pneumocystis carinii* pneumonia, chronic cryptosporidiosis, toxoplasmosis, extraintestinal strongyloidiasis, isosporiasis, candidiasis (esophageal, bronchial, or pulmonary), cryptococcosis, histoplasmosis, mycobacterial infection with *Mycobacterium avium* complex or *M. kansasii*, cytomegalovirus infection, chronic mucocutaneous or disseminated herpes simplex virus infection, or progressive multifocal leukoencephalopathy.

[d] Oral hairy leukoplakia, multidermatomal herpes zoster, recurrent *Salmonella* bacteremia, nocardiosis, tuberculosis, or oral candidiasis (thrush).

[e] Kaposi's sarcoma, non-Hodgkin's lymphoma (small, noncleaved lymphoma or immunoblastic sarcoma), or primary lymphoma of the brain.

[f] Other clinical findings or diseases, not classifiable above, that may be attributed to HIV infection or may be indicative of a defect in cell-mediated immunity. Included are patients with chronic lymphoid interstitial pneumonitis.

headache, lethargy, anorexia, nausea, myalgia, arthralgia, sore throat, and an erythematous truncal eruption. Aseptic meningitis with irritability and neck stiffness, neuritis, or even frank encephalopathy, may sometimes dominate the picture. Hematologic abnormalities include transient leukopenia with absolute lymphopenia, thrombocytopenia, and an inversion of the normal CD4+/CD8+ T lymphocyte ratio (normal values >1.5), in most cases due to the relative increase of CD8+ T cells (Cooper et al, 1985). Atypical lymphocytes can be detected in the blood smear. Elevated erythrocyte sedimentation rate is commonly observed.

Seroconversion has been shown to occur in most cases after an interval ranging from one to ten weeks from the onset of the acute illness. IgM antibodies have been detected in the serum of acutely infected patients starting five days after clinical onset (Cooper et al, 1987). In a number of cases who remained seronegative for prolonged periods, the presence of HIV genetic sequences was documented by the use of the polymerase chain reaction (PCR) (Ou et al, 1988).

32.5.2.b Persistent Generalized Lymphadenopathy

The most recent CDC staging of HIV-induced disease defines stage (or group) II

Table 32.3 Walter Reed Classification for HIV Infection

Stage	HIV Antibody or virus detection	Chronic Lymphadenopathy	T helper cells/mm^3	Delayed Hypersensitivity	Thrush	Opportunistic Infections
0	−	−	>400	+	−	−
1	+	−	>400	+	−	−
2	+	+	>400	+	−	−
3	+	±	<400	+	−	−
4	+	±	<400	Partial anergy	−	−
5	+	±	<400	Complete anergy	+	−
6	+	±	<400	Partial or complete anergy	±	+

as the clinically asymptomatic infection with only subtle immunologic abnormalities and stage III as a syndrome called persistent generalized lymphadenopathy (PGL). Lymphadenopathy (lymph node enlargement of 1 cm or greater) must involve at least two or more extrainguinal sites and persist for at least three months in the absence of any identifiable cause other than HIV infection. Usually, the cervical and axillary lymph nodes are involved. Splenomegaly is rarely seen. A minority of the patients have anemia, leukopenia, or thrombocytopenia. The serum immunoglobulin level (mainly IgG and IgA in adults, also IgM in children) is increased polyclonally in about 75% of the patients. Immunologic studies reveal a lowered CD4/CD8 T lymphocyte ratio in almost all patients and cutaneous anergy to a battery of recall skin test antigens. Circulating immune complexes may be present in the serum and the blastogenic responses to T- and B-cell mitogens are generally decreased. HIV-specific antibodies are detected in the serum of virtually all patients.

Biopsy of the soft, nontender, and mobile lymph nodes usually shows follicular hyperplasia with pronounced capillary endothelial cell proliferation. The usual CD4/CD8 T lymphocyte ratio in the lymph node is reversed in the center, mantle, and interfollicular T-cell zone (Ziegler and Abrams, 1985). Molecular hybridization studies showed that only a small fraction of the cells, usually follicular dendritic cells, are productively infected with HIV (Harper et al, 1986; Tenner-Racz et al, 1988).

32.5.2.c Acquired Immunodeficiency Syndrome (AIDS)

Initially, a number of clinical manifestations prodromic to full-blown AIDS, but not fulfilling the definition of AIDS, were referred to as AIDS related complex (ARC), but the designation has been pro-

gressively abandoned. The long-term follow-up of HIV-infected individuals has now permitted the more precise determination of the rate of progression from the asymptomatic state to symptomatic HIV infection and eventually to AIDS. According to different studies, the incidence of full-blown AIDS after four to six years is 25% to 35% (Curran et al, 1988; Eyster et al, 1987). Only 30% or fewer individuals remain completely symptom-free five years post infection. The best prognostic index of disease progression appears to be the absolute number of circulating CD4+ T cells.

Many hypotheses have been formulated to explain the variable course of the disease in different individuals. For example, environmental or constitutional factors may be important to determine the rapidity of the disease progression. Among these potential cofactors are different viruses, parasites, and mycoplasmas. Coinfection with these agents may lead to stimulation of the immune system associated with the release of cytokines, amplification of HIV replication and/or of HIV-induced cytopathogenicity and, in some cases, direct damage to the immune system of the host. It is important to emphasize that the interpretation of any study on the role of cofactors in AIDS is biased by the difficulty in discriminating whether an agent plays a primary role in the disease progression or is opportunistic. In this respect, only the development of appropriate animal model systems will provide definitive evidence for the importance of cofactors.

In the case of mycoplasmas, a putative novel species with atypical characteristics has been detected in several tissues of patients with AIDS. It is called *M. incognitus* (Lo et al, 1989), and its possible pathogenic role remains to be defined. Subsequently, another group proposed that culture-contaminating mycoplasmas may contribute to the cytopathogenic effect of HIV-1 in vitro (Lemaitre et al, 1990). However, it is well established that HIV by itself can kill CD4+ T cells in the absence of concurrent infection. Furthermore, the ubiquitous nature of mycoplasmas and the likelihood of their dissemination in immunocompromised patients complicate the study of their pathological relevance.

Among the viral agents, most herpesviruses [such as HSV, Epstein-Barr virus (EBV), hCMV, and human herpesviurs 6 (HHV-6)], HBV, and HTLV-I are capable of transactivating the HIV-1 LTR in vitro and may thereby activate latent HIV infection or boost its expression in vivo. However, only HHV-6 and HTLV-I are primarily targeted to CD4+ T cells and may productively coinfect these cells along with HIV-1 (DeRossi et al, 1986; Lusso et al, 1989). Coinfection by HHV-6 and HIV-1 also produces an accelerated cell death in CD4+ T-cell cultures. In addition, HHV-6 is able to induce de novo expression of CD4, i.e., the HIV-1 receptor, in normal CD8+ T lymphocytes, thus potentially expanding the range of HIV-1-susceptible cells (Lusso et al, 1991).

Following a variable period of PGL, many clinical signs herald the progression toward AIDS. The lymphadenopathy usually becomes less pronounced as germinal center atrophy replaces the hyperplasia. Progressive fatigue, flu-like illnesses, night sweats, recurrent or chronic fever, and significant weight loss are frequently observed. The patients start developing bacterial, fungal, and viral infections of the skin and mucous membranes, including seborrhoic dermatitis, genital warts, and recurrent genital or oral herpes. Episodes of herpes zoster are also seen, as well as oral candidiasis and hairy leukoplakia. Diarrhea lasting for more than one month can be one of the most prominent symptoms.

The current definition of AIDS is still based principally on clinical criteria (i.e., the secondary effects caused by the underlying immune dysfunction) (Table 32.2.). The primary defect, however, involves the immune system and consists of the deple-

tion of the CD4+ helper/inducer T-lymphocyte subset (Klatzman et al, 1984; Popovic et al, 1984b), as well as the infection of cells of the monocyte/macrophage system (Gartner et al, 1986). The CD4+ T lymphocyte and the monocyte/macrophage represent crucial cells for the development of specific cellular and humoral immune responses. Thus, quantitative and/or qualitative alterations of these cell types have profound repercussions on the immune system as a whole.

Most of the immunologic abnormalities and dysfunctions seen in AIDS patients are related to the deficiency of CD4+ T cells ($<400/mm^3$, normal range 800 to 3600/mm^3). Frequent findings include lymphopenia ($<1000/mm^3$, normal range 1 to 4.8 $\times$ $10^3/mm^3$), lowered CD4/CD8 T lymphocyte subset ratio (<1.0, normal value >1.2), a defect in delayed hypersensitivity manifested by cutaneous anergy, and impaired blastogenic responses of both T and B cells in vitro. Elevated serum immunoglobulin levels (usually IgG and IgA) are almost invariably present as a consequence of polyclonal B cell activation.

Antibodies to HIV are found in virtually all patients with AIDS. In parallel with many other immune functions, however, antibody titers usually decline with the progression of the disease. This decrease in titer predominantly affects anti-gag antibodies in the presence of essentially unchanged titers of antibodies to the env proteins. Concomitantly, a rise in the free p24 antigen levels in plasma is observed.

The spectrum of opportunistic infections that may affect AIDS patients includes a variety of bacterial, fungal, or viral agents, which can involve virtually any part of the patient's body (Table 32.2). The infections most commonly observed at diagnosis are mucocutaneous herpesvirus infections (HSV, VZV, hCMV); oral candidiasis; gastrointestinal hCMV or parasitic (giardia, cryptosporidia, isospora, amoeba) infections, possibly resulting in chronic diarrhea and malabsorption; *Pneumocystis carinii* pneumonia (PCP), present in 60% of the patients and a major cause of death; central nervous system (CNS) infection (by hCMV, toxoplasma, cryptococcus, JC virus); and hCMV retinitis. Pre-neoplastic lesions and neoplasias include hairy oral leukoplakia, oligoclonal B-lymphocytosis, non-Hodgkin's lymphoma (mostly of B-cell origin with particularly frequent extranodal and CNS localization), Hodgkin's disease with a particularly aggressive clinical course, and Kaposi's sarcoma, often multicentric and involving both the skin and the mucous membranes.

32.5.2.d Neurologic Disorders

Apart from immunodeficiency, HIV infection may be the direct or indirect cause of severe neurologic dysfunctions. Primary neurologic disorders (in the absence of secondary infections or tumors) are observed in more than 33% of patients with AIDS while the majority of them display neuropathologic abnormalities at postmortem examination (Snider et al, 1983b). These disorders are generally referred to as AIDS encephalopathy or AIDS dementia complex and are manifested by a progressive cognitive impairment with motor and behavioral anomalies, such as apathy, depression, and personality changes. Spinal cord disease with signs and symptoms of spastic paraparesis and ataxia, pathologically represented by a vacuolar myelopathy, afflicts approximately 20% of patients with AIDS (Petito et al, 1985). Polyneuropathy and other peripheral nerve disorders can also be seen occasionally.

Histopathologically, cerebral atrophy, gliosis with microglial nodules, and focal necrosis are the common features. HIV sequences are detectable in the brain tissue of the majority of AIDS patients, particularly in those with encephalopathy (Shaw et al, 1985). Infectious virus has been transmitted from human brain to chimpanzees

(Gajdusek et al, 1985). HIV was also isolated from cells present in cerebrospinal fluid (CSF) or directly from CSF (Ho et al, 1985). All the isolates of HIV from brain tissue have a pronounced tropism for cells of the monocyte/macrophage lineage (Gendelman et al, 1989). Since in rare cases neurologic signs may be the first or only manifestation of HIV infection, it has been proposed that this peculiar disease presentation could be associated with purely neurotropic HIV isolates.

32.6 LABORATORY DIAGNOSIS OF HUMAN RETROVIRUS INFECTIONS

A variety of tests were developed for the demonstration of infection by HTLV-I, HTLV-II, HIV-1 and HIV-2. They include immunologic tests for the demonstration of virus-specific antibodies or viral proteins in infected cells or tissues, and molecular tests for the demonstration of HIV genetic sequences in nucleic acids isolated from tissues or serum of seropositive individuals.

32.6.1 Tests for the Demonstration of Virus-Specific Antibodies

In dealing with immunologic tests, it should always be kept in mind that positive immunologic tests only indicate relatedness, but never identity of the antigens compared. This restriction is especially important in cases where the individuals tested are not members of risk groups and were not knowingly exposed to the agents tested. However, the problem can be overcome by performing multiple immunoassays such as competition tests, which rule out reactions with host cell antigens, e.g., HLA antigens, and non-immunological tests such as DNA amplification by PCR.

For most kinds of viruses the demonstration of antiviral antibodies usually indicates past but not necessarily ongoing viral infection. In the cases of HTLV and HIV, however, the unambiguous demonstration of specific antiviral antibodies may be considered adequate evidence for ongoing infection, as these retroviruses persist in the body for the lifetime of the infected individual.

32.6.2 Enzyme-Linked Immunosorbent Assays

Enzyme-linked immunosorbent assays (ELISA) were developed to detect antibodies to both HTLV-I and HIV-1 (Saxinger and Gallo, 1983; Sarngadharan et al, 1984). Several commercial kits are now available for these viruses. In these assays, viral antigens (either purified whole virus lysate, single native or recombinant viral proteins, or synthetic peptides representing immunodominant regions) are bound to the wells of microtiter plastic plates or, alternatively, to plastic beads. Human serum or other body fluids to be tested for antibodies are diluted in buffer, added to the wells or tubes and incubated for a fixed period of time. The wells or tubes are then washed and a secondary enzyme-conjugated antibody is added, which is directed against human immunoglobulins. After a brief incubation, the wells or tubes are washed again. The substrate is then added, and a color reaction develops in the wells or tubes where binding of human antibodies and of the enzyme-labeled secondary antibody occurred (see Chapter 9 for a more complete description of ELISA).

The ELISA is ideally suited for large-scale screening of blood supply because of the ease of the assay, its reproducibility, its potential for automation, and its long shelf life. In our experience, the sensitivity and specificity of ELISA for anti-HIV antibody are lower than those of immunoblot assay or radioimmunoprecipitation followed by SDS-PAGE analysis (RIP/SDS-PAGE). This was especially true when ELISA

plates or beads were coated with whole virus lysate as in the first generation ELISA for HTLV-I and HIV-1 detection. To overcome this problem in the second generation of ELISAs for HIV, synthetic peptides or recombinant proteins are used to coat the plates, which increases test reliability. An additional weakness of the ELISA is its high rate of false-positive results with sera improperly stored or transported (Schüpbach et al, 1985). Thus, as we indicated in our earlier publications on the development of a blood test for HIV-1, a positive ELISA result should always be confirmed by an independent method, preferably by immunoblot or RIP/SDS-PAGE.

32.6.3 Immunoblot

The advantages of a properly performed immunoblot are its high sensitivity and its excellent specificity, which greatly surpasses that of the ELISA. Immunoblot allows simultaneous testing for antibodies to a variety of viral antigens, without a need for purifying them first. Several immunoblot kits are now commercially available for both HTLV-I and HIV-1.

The original procedure of Western blot (Towbin et al, 1979) was modified to allow the testing of multiple serum samples for antibodies (Schüpbach et al, 1985; Sarngadharan et al, 1984). Figure 32.4 illustrates the essential steps of the procedure.

1. In the first step, the proteins from detergent-disrupted purified virus (250 mg/gel) are fractionated by electrophoresis on a 12% polyacrylamide slab gel in the presence of SDS according to the method of Laemmli (1970).

2. The protein bands are then transferred (45V, overnight at 4°C) from the gel to a nitrocellulose sheet, as described by Towbin (1979); the sheet is saturated for two hours at 37°C in "BLOTTO" buffer, consisting of phosphate buffered saline (PBS) containing 5% nonfat dry

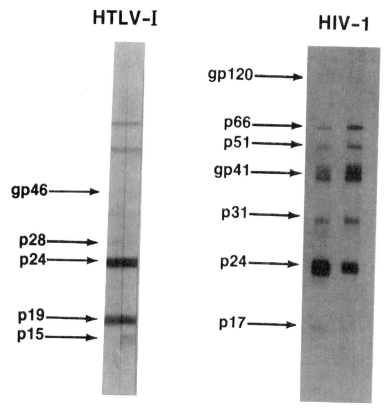

HTLV-I

HIV-1

gp120⟶

p66⟶
p51⟶
gp41⟶

p31⟶

gp46⟶

p28⟶
p24⟶

p24⟶

p19⟶
p15⟶

p17⟶

Figure 32.4 Immunoblot analysis of purified virus from HTLV-I and HIV-1 infected cells. Representative strips obtained testing selected sera from HTLV-I ad HIV-1 infected individuals with the procedure described in the text.

milk and 0.001% Merthiolate (Johnson et al, 1984).

3. The sheet is rinsed with PBS containing 0.05% Tween 20 (PBS-T), placed between two layers of Parafilm and cut to 3 to 5 mm wide strips. These strips can be used directly for testing or stored at −20 °C for several months.

4. For testing, strips are freed of their Parafilm sheaths, placed in plastic tubes containing 2.4 mL BLOTTO and 100 μL normal goat serum (NGS) and incubated for one hour in a horizontal position on a shaker; test serum is then added in a dilution of 1:50 or 1:100 and the strips incubated overnight at 4 °C; the strips are subsequently washed in three changes of PBS-T.

5. A secondary, labeled goat antibody with specificity to human immunoglobulins is added (in BLOTTO containing 100 μL

NGS); excellent results are obtained with affinity-purified and iodinated (^{125}I) antibodies to IgG and IgM, which are used at an activity of 1.25×10^6 cpm per tube. Strips are incubated with labelled antibody for 30 minutes; after a final round of three washings with PBS-T, the strips are dried, mounted on filter paper, and processed by autoradiography.

In addition to iodinated second antibodies, other methods involve the use of horseradish peroxidase or alkaline phosphatase-conjugated antibodies in combination with the appropriate substrate. These procedures are as sensitive as the one using radiolabeled antibodies, and have the further advantage of shortening the length of the assay.

The interpretation of the results is easy when antibodies against a variety of antigens are present and when the bands are

Table 32.4 Frequency of Specific Antigenic Reactivity of 700 HIV Antibody Positive Sera in Virus Immunoblots

HIV Antigen	No. Positive	Percent Positive
gp41	695	99.3
p66/51	551	78.7
p31	523	74.7
p24	453	64.7
p17	275	39.3
gp120	115	16.4

intense. This is generally the case for HIV-1. In contrast, the interpretation of the results with HTLV-I sera and with sera from terminal AIDS patients or early HIV-1 seroconverters is more complex: titers of antibodies may be low and only single bands may be detected. In addition, protein bands of cellular origin may be detected in positions only slightly different from those of viral proteins. This is especially true when poor quality virus is used for the preparation of the strips. The diagnostic potential of immunoblot thus depends largely on the experience of the interpreter. Structural proteins identified as coded for by HTLV-I and clearly visible in immunoblots include the gag proteins p15, p19, p24, and p29 and the env protein p46. Additional bands are often recognized by HTLV-I sera but have not been identified yet as viral proteins. Structural proteins identified as coded for by HIV-1 and clearly visible in immunoblots include the gag proteins p24 and p17, the env proteins gp120 and gp41, and the pol proteins p66/51 and p31. In addition to these antigens that correspond to the mature products, some of the precursor proteins such as gp160, Pr53^{gag}, and Pr39^{gag} may also be present in the antigen mixture and may be "recognized" by the patient's antibodies. Figure 32.4 shows representative immunoblot profiles of HTLV-I and HIV-1 antigens with selected human sera. Table 32.4 summarizes the results obtained with 700 sera analyzed by immu-

noblots using HIV-1 lysate as antigen (De-Vico et al, 1988). Antibodies to *env* gene product gp41 were present in 695 (99%) sera. Next to gp41, the most frequently recognized viral antigens were the *pol* gene products p66 and p51. Antibodies to these proteins were detected in 551 (79%) sera, making the frequency of reactivity to reverse transcriptase greater than that to the *gag* gene proteins p24 and p17, detected in 453 (65%) and 275 (39%) sera, respectively. Antibodies to the integrase protein p31 were present in 75% of the sera. The low percentage of positivity toward the external env glycoprotein gp120 can be explained by technical considerations: the well-documented loss of gp120 during the purification of the virus and the less efficient transfer of high molecular weight proteins to nitrocellulose membranes. Figure 32.5 summarizes the prevalence of serum antibodies to HIV-1 antigens among individuals at different clinical stages of HIV infection. These included asymptomatic, PGL, AIDS, and AIDS associated with Kaposi's sarcoma. The prevalence rate of seropositivity to env gp41 was similar throughout the stages and the one to pol p66/51 did not change significantly. On the contrary, antibodies to gag p24 were seen among fewer AIDS patients than in patients with PGL or in asymptomatic individuals.

32.6.4 Immunofluorescence Tests

Indirect immunofluorescence (IF) tests were among the first tests to be used to detect human retrovirus infections. In fact, most of the serologic studies done initially in Japan were based on the demonstration of an ill-defined "ATLA" (adult T-cell leukemia antigen) (Hinuma et al, 1981), which only later was shown to consist of multiple viral or virus-associated proteins present in HTLV-I-infected cells. The IF tests for human retroviruses, per se, are relatively nonspecific tests since the cells used as targets contain multiple non-viral antigens

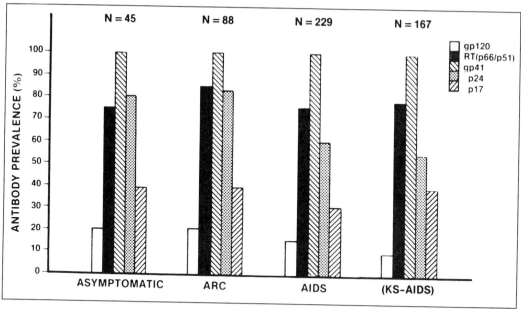

Figure 32.5 Prevalence of serum antibodies to HIV-1 antigens among individuals in different clinical stages of HIV-1 infection.

(mainly HLA antigens and cell-activation markers) potentially reacting with antibodies present in the test sera. Even when uninfected cells of exactly the same type as the virus-producer cells are used as negative controls (e.g., the uninfected H9 cell counterpart of the H9/HIV-1 producer), a positive result on H9/HIV-1 may still be nonspecific, because the reactive antibodies might be directed at a cellular component expressed by H9 cells as a consequence of virus infection. Thus, IF should only be used for general screening, or monitoring, but never as a confirmatory test.

Indirect IF tests using live HTLV-infected cells (membrane IF) were also frequently used. These tests may be very sensitive but suffer from the same problems of low specificity described above.

32.6.5 Agglutination Assays

Particle agglutination assays for the screening of antibodies to HTLV-I and HIV-1 are available as test kits. The assays use a gelatin particle carrier with detergent-disrupted purified virus or recombinant antigens. The HTLV-I and HIV-1-sensitized gelatin particles are agglutinated in the presence of virus-specific antibodies in serum or plasma samples. Unsensitized particles are provided as a negative control. The assay is easy and quick, requiring less than three hours to obtain a result. It is ideal for use in certain settings like hospital emergency rooms, but its sensitivity is not satisfactory for blood screening.

32.6.6 Radioimmunoassays

32.6.6.a Binding Assays (Solid Phase Radioimmunoassays)

The first generation of anti-HTLV antibody tests used were radioimmunoassays (RIA) carried out in our laboratory (Posner et al, 1981; Robert-Guroff et al, 1982b). Whole virus proteins were coupled to wells of microtiter plates, as in the ELISA. An iodinated secondary antibody was used instead of an enzyme conjugate and the activity recovered in the wells was measured in a gamma counter. The procedure appears to offer no advantage over the ELISA.

32.6.6.b Radioimmunoprecipitation Assays (RIPA)

These tests are usually performed with single, purified proteins and are thus highly specific. RIPA were developed for p24 of HTLV-I (Kalyanaraman et al, 1982a) and for the other gag proteins, p19 and p15 (Schüpbach et al, 1983a, 1983b; Kalyanaraman et al, 1984b). RIPA were also developed for p24 of HIV-1 (Kalyanaraman et al, 1984a).

For these tests, purified proteins are iodinated to high activity (10,000 to 20,000 cpm/ng) and 8,000 to 10,000 cpm are incubated with an aliquot of test serum. The total IgG is then precipitated by the addition of a predetermined amount of a goat antiserum to human IgG. The precipitate is washed and counted in a gamma counter. In the presence of virus-specific antibodies, labeled antigen will be found in the precipitate.

The strength of the RIP assay is its specificity, which is a reward for the tedious work and expense of antigen purification.

32.6.6.c Metabolic Labeling of Infected Cells and RIP/SDS-PAGE

In this test, metabolic activity is used to label cellular or viral proteins. Lysates are then prepared and incubated with small aliquots of test serum. The labeled antigens bound in immune complexes are then absorbed to protein A Sepharose and subjected to polyacrylamide gel electrophoresis in the presence of SDS. Following autoradiography, the pattern of bands obtained with the test serum is evaluated in comparison with the patterns obtained with positive and negative control sera.

The sensitivity of this test is higher than that of immunoblot for detection of antibodies to env proteins. This is because some proteins, especially the major envelope glycoproteins, tend to be lost during the virus purification process. In addition, antigens of cellular origin may become much more concentrated in the purified virus and complicate the interpretation of immunoblots. Finally, a significant part of the viral antigens may be present in the form of larger precursors in the cells, while the purified virus used in immunoblots contains mostly mature products. By RIP/SDS-PAGE analysis, sera from HTLV-I-infected subjects clearly identify the env precursor gp61 and the gag precursor p53, along with *gag* gene mature products p15, p19, and the *tax* gene product p40. With HIV-1 the predominant antibodies seen by RIP/SDS-PAGE analysis are usually directed at the env precursor gp160 together with its processed products gp120 and gp41. P24 is always detected when antibodies to these envelope antigens and to its own precursor Pr53[gag] are present (Kitchen et al, 1984; Manca et al, 1987). In contrast, these larger antigens are only infrequently detected by immunoblot with purified virus and p24 may sometimes be the only antigen detected.

RIP/SDS-PAGE assays are used by many investigators and by the American Red Cross as confirmatory tests after an atypical or uninterpretable immunoblot is obtained, but they are not suitable for screening large numbers of sera because they are time-consuming, technically difficult, and expensive.

32.6.6.d Competition Radioimmunoassays

Competition RIA can be used for the detection of either antigens or antibodies. The principle of the test is that the interaction of a solid phase antigen with a relevant radio-labeled antibody is competed out by the previous binding of test serum antibodies. Any degree of antigen specificity may be obtained by the selective use of target antigens or the use of antisera or monoclonals of defined specificity. Tests based on this system were first developed for antibodies specific for p19 of HTLV-I (Robert-Guroff, 1982a) and later for antibodies to the total proteins of HTLV-I and HTLV-II (Tedder et al, 1984) and of HIV-1 (Cheinsong-Popov et al, 1984). These tests appear to have high specificity, as demonstrated by the fact that none of more than 1000 randomly selected blood donors was positive in the HIV-1 assay. The competition RIA may be quite sensitive, provided that the antibodies in the test serum recognize the same epitopes as the labeled reference antibody.

32.6.7 Large Scale Serologic Screening

32.6.7.a Large Scale Screening of Sera for HTLV-I and HTLV-II

The American Red Cross began screening donated blood for antibodies to HTLV-I by late 1988, eight years after the discovery of the virus. This decision was based on several observations about transmissibility, distribution, and pathologic manifestations of this virus. The transmission of HTLV-I by cellular blood products had been described already in 1984 (Okochi et al, 1984). Concerns were further raised in 1986, when a case of HAM/TSP was clearly associated with a transfusion of HTLV-I-infected blood. In 1986, the American Red Cross determined, by a large cross-sectional study, that 0.025% of their blood donors were HTLV-I-positive. In addition, similar studies performed with sera from intravenous drug abusers indicated that HTLV-I infection was endemic in that population. These epidemiological observations, together with the realization that some of the

HTLV-I-associated diseases appear to have an incubation period much shorter than ATL, led to the decision to test all donated blood. Commercial kits were approved in late 1988 for widespread screening by U.S. blood banks. The primary assay consists of an ELISA with purified HTLV-I lysate as antigen. Sera found reactive in this primary assay are then sent to reference laboratories where they are tested by a confirmatory assay like HTLV-I immunoblots. In this assay, sera are considered positive when they react at least with both p24 and either gp46 or gp61. If reactivity to only p24 is found, the serum is further tested by ELISA with recombinant gp21 as antigen. Sera showing an atypical or uninterpretable pattern by all these tests are finally screened by RIP/SDS-PAGE to look for reactivity to the precursor env glycoprotein gp61.

Besides serological detection of HTLV-I and HTLV-II, the ability to distinguish between these two viruses is also needed for an accurate diagnosis, since current commercial methods do not allow such discrimination. Sera from HTLV-II-infected individuals are reactive with commercially available HTLV-I ELISAs and will generally react in RIPA and immunoblot assays to some HTLV-I antigens. Competition RIPA and competition ELISAs (Kalyanaraman et al, 1982b; Robert-Guroff et al, 1986) allow clear-cut discrimination between the two viruses, but these assays are not suitable for large scale screening of blood supplies. Discrimination between HTLV-I and HTLV-II can also be achieved by differential screening using recombinant or synthetic peptides representing antigenic epitopes specific for HTLV-I and HTLV-II, since there is sufficient amino acid divergence in some of the proteins encoded by these two viruses. However, a more specific and sensitive approach is represented by the analysis of the samples by PCR methods since there are differences in the nucleotide sequence between HTLV-I and HTLV-II. By applying PCR and peptide-ELISA techniques, the American Red Cross showed that, unexpectedly, 50% of the blood donors positive for HTLV-I/II and assumed by most investigators to be HTLV-I were instead infected with HTLV-II (Chyang, T. Fang, personal communication).

32.6.7.b Large Scale Screening of Sera for HIV-1 and HIV-2

Blood banks and plasma industries started to screen donated blood for HIV-1 in March 1985, as soon as an antibody screening test became available (Sarngadharan et al, 1984). To date, several commercial kits have been approved for widespread screening. As with screening for HTLV-I antibodies, the primary assay consists of an ELISA with purified HIV-1 lysate as antigen. Sera found reactive in this assay are then sent to the reference laboratories, where they are tested by HIV-1 immunoblot as a confirmatory assay. In this assay, sera are considered positive for HIV-1 if they react with p24 together with gp41 and/or gp120. Batches of blood found positive are discarded and the donors notified. Sera showing an atypical or uninterpretable pattern of reactivity by the above mentioned assays are finally tested by RIP/SDS-PAGE to look for reactivity to the precursor env protein gp160 and the precursor gag protein p53. The combination of these assays virtually eliminated virus transmission by transfusion routes. Nevertheless, the percentage of AIDS cases attributed to transfusion has increased from 1.5% to 2.6%. This is because the probability of developing AIDS is relatively low during the first three years after infection and increases in the following years. Thus, at the present time it is difficult to determine trends in the incidence of HIV infection by looking at trends in the incidence of AIDS. Currently, the most important task in transfusion-as-

sociated AIDS research is the exhaustive investigation of AIDS cases where the only assessed risk was the transfusion of blood pre-screened for HIV. The results of these studies will shed light on the relevance of latent infections, where detectable antibodies to the virus are not present.

As has been discussed for HTLV-I and HTLV-II infection, current serologic methods may not readily distinguish between HIV-1 and HIV-2. Discrimination can be achieved using some of the approaches already described for HTLV-I and HTLV-II. Differential screening by using selected peptides and PCR products easily allows us to distinguish between HIV-1 and HIV-2. In addition, there are commercially available monoclonal antibodies directed against HIV-1 antigens, which do not react with their counterpart antigens from HIV-2. These reagents could represent another serologic approach to differential diagnosis of HIVs by means of antigen capture assays.

32.6.8 Tests for the Demonstration of Viral Proteins

32.6.8.a Antigen Capture Assays

Several antigen ELISA and RIA kits are now available for HIV-1. These easy-to-use kits provide higher sensitivity than reverse transcriptase assays and specifically detect and quantify HIV-1 p24 in infected culture samples, serum, plasma, and other body fluids. The range of detection varies between 6 and 10 pg of p24. The assay usually consists of monoclonal antibodies to HIV-1 p24 coating the plates and either monoclonal antibodies recognizing epitopes on the p24 molecule different from the one recognized by the coating antibody or a polyclonal antibody to detect the captured antigen. Results obtained with the samples are compared to those obtained with an HIV-1 standard curve, calibrated for p24. Antigen assays have been instrumental in quantifying HIV-1 p24 plasma levels, despite problems connected with the presence of p24 complexed with patient antibodies, which potentially compete with the detecting antiserum. HIV-1 p24 antigen capture assays are mostly useful during the window between the time a person becomes infected with HIV and the appearance of detectable antibody. As we already mentioned, HIV-1 seroconversions are usually asymptomatic or subclinical; however a few seroconversions occur between one and ten weeks after the onset of an acute primary infection syndrome. A small number of infected individuals may have a prolonged period of this virus-positive, antibody-negative status, during which viral DNA or RNA can be detected in peripheral blood mononuclear cells (PBMC) by the polymerase chain reaction or viral antigens detected by antigen capture assays. Another application of HIV-1 p24 antigen capture assay is monitoring the decrease of viral load in the serum of patients undergoing antiviral therapy. Recently, HIV-1 gp120 antigen capture assays have become commercially available, but their sensitivity and specificity still have to be assessed. Besides antigen capture assays, viral proteins can be detected by a variety of tests including reverse transcriptase (see below) and direct or indirect IF assays with monoclonal antibodies. Usually, HTLV-I antigens are not detectable in fresh PBMC collected from patients with ATL, TSP, or from asymptomatic HTLV-I carriers, although mRNA for HTLV-I *tax/rex* genes has been detected in these same cells by using reverse transcription followed by PCR (Kinoshita et al, 1989; Gessain et al, 1991). However, after a short-term culture in the presence of IL-2, cells start expressing viral proteins that can be detected by assay for reverse transcriptase, direct or indirect IF, and competition RIP assays. Commercial antigen tests for HTLV-I p24 have become available only recently and their sensitivity and specificity have still to be verified.

32.6.8.b Reverse Transcriptase Assay

The activity of the enzyme reverse transcriptase can be used as a sensitive marker for infection by retroviruses. Even in the absence of demonstrable release of virus particles from cells, reverse transcriptase activity may suggest that retroviral information is being expressed. The reverse transcriptase assay, however, does not allow identification of the type of retrovirus involved.

For the reverse transcriptase assay, cell-free culture supernatant fluids are subjected to polyethylene glycol 6000 precipitation (10% final concentration) in order to concentrate the virus. The samples are placed on ice for at least two hours and centrifuged for 45 minutes at 1000 $\times$ g at 4 °C. The pellet is then resuspended and the virus disrupted using a minimal volume of a buffer containing 25 mM Tris (pH 7.5), 5 mM dithiothreitol (DTT), 0.25 mM EDTA, 0.025% Triton X-100, 5 mM KCl, and 50% glycerol. The reverse transcriptase assay measures the incorporation of tritiated deoxyribonucleotides into acid-insoluble material (Sarngadharan et al, 1978). The reaction mixture in a total volume of 100 μL contains 10 μL of the disrupted virus, 4 μL of 1 M Tris (pH 7.8), 4 μL of 0.2 M DTT, 5 μL of 0.2 M MgCl$_2$, 25 μL of [^{3}H]-TTP (10–20 Ci/mmol), 72 μL of H$_2$O, and 5 μL of the appropriate template [(dT)$\sim_{15}$(A)$_n$ or, as a control, (dT)$\sim_{15}$(dA)$_n$]. The reaction mixture is incubated at 37 °C for one hour and the reaction is terminated by the addition of 50 μg of tRNA and about 3 mL 10% trichloroacetic acid (TCA) containing 0.2 M sodium pyrophosphate. The precipitate is kept on ice for at least ten minutes, then collected on glass microfiber filters presoaked in 5% TCA containing 0.02 M sodium pyrophosphate. The filters are rinsed thoroughly with 5% TCA containing pyrophosphate and finally with 70% ethanol. The washed filters are dried under a heat lamp and the incorporated radioactivity determined using a scintillation spectrometer. Although RNA tumor viruses catalyze endogenous DNA synthesis in the absence of exogenous primer templates, the efficiency of the viral reverse transcriptase is amplified by the addition of several synthetic oligomer-homopolymer hybrid primer templates such as $(dT)\sim_{15}(A)_n$ and $(dG)\sim_{15}(C)_n$. The first of these is also used by certain cellular DNA polymerases like γ polymerase, but a combination of high response with $(dT)\sim_{15}(A)_n$ and poor response with $(dT)\sim_{15}(dA)_n$ in a duplicate sample is indicative of viral reverse transcriptase activity. A ratio between these two activities of four or more is very likely a positive result. $(dG)\sim_{15}(C)_n$ is specific for reverse transcriptase because DNA polymerases cannot use it, but is somehow less efficient than $(dT)\sim_{15}(A)_n$ which is considered the primer-template of choice.

32.6.9 Assays for the Detection of Viral Genomes—PCR Technique

32.6.9.a Detection of HTLV-I and HTLV-II Genomes

Although serological assays for the diagnosis of HTLV-I/II infection are sensitive and specific, direct detection of these viruses would also be desirable to identify the existence of possible seronegative virus carriers. However, a diagnosis of HTLV-I or HTLV-II infection based on genome detection by conventional hybridization techniques is not feasible in asymptomatic subjects, since such a small proportion of cells is infected with the virus. One way to overcome the problem has been to culture patients' cells, allowing virus replication i.e., amplification of the genetic material. The virus can then be detected in the cultured cells by Southern blot analysis, IF tests, or in situ hybridization. However, in vitro culture of cells is time consuming. The newest approach to molecular epidemiol-

ogy is the utilization of an in vitro enzymatic gene amplification technique, called PCR. PCR is used to amplify a region of DNA lying between two regions of known sequence represented by a pair of oligonucleotide primers. These oligonucleotides have different sequences, are complementary to sequences that lie on opposite strands of the template DNA, and are located at both ends of the segment of DNA to be amplified (see Chapter 21 for a detailed description). The template DNA is first denatured by heating in the presence of the two oligonucleotides and the four deoxynucleotide triphosphates. The reaction mixture is then cooled to a temperature that allows the oligonucleotide primers to anneal to their target sequences. The annealed primers are subsequently extended with a thermostable DNA polymerase from *Bacillus thermus aquaticus* (Taq). These cycles of denaturation, annealing, and extension are sequentially repeated many times allowing the exponential amplification of the desired DNA sequence. The amplified genetic material can then be detected by a number of assays, including Southern and dot blots or liquid hybridization techniques (reviewed by Ehrlich et al, 1990). PCR has been successfully applied for the detection of HTLV-I and HTLV-II proviral sequences in patients with TSP/HAM (Ehrlich et al, 1988; Bhagavati et al, 1988; Kwok et al, 1988). In addition, PCR has been extensively applied in epidemiological studies and diagnostic procedures to detect the presence of HTLV-I and HTLV-II genomes in asymptomatic carriers and patients with lymphoproliferative and neurological disorders (Ehrlich et al, 1989a; Ehrlich et al, 1989b; Greenberg et al, 1989). However, PCR's unmatched sensitivity (it can theoretically detect a single molecule of a provirus) carries the threat of false-positive results caused by physical carry-over of amplified DNA into experimental samples. In addition to physical precautions, cautious interpretation of the results is advisable. A diagnosis of HTLV infection should never be made on the basis of a single PCR performed with a single oligonucleotide primer pair, unless these results are substantiated by serological and/or antigenic detection. Additionally, more than one set of primer pairs should produce a positive result, particularly when seronegative individuals are tested.

32.6.9.b Detection of HIV-1 and HIV-2 Genomes

Antibody assays provide a rapid and sensitive means to screen for the presence of antibodies to the HIVs, but these assays provide only indirect markers of HIV infection. Direct virus indentification is especially important for asymptomatic seropositive individuals not belonging to any high-risk group (e.g., laboratory or clinical workers) and required, until recently, the isolation of HIV from lymphocytes of these subjects. As we already mentioned for HTLV, virus isolation is a laborious and time-consuming procedure sometimes leading to inconclusive results. Direct determination of the presence of HIV-1 mRNA by in situ hybridization is also problematic since only 1 in 10,000 PBMC or less from a patient with AIDS is actively expressing the virus (Harper et al, 1986). Therefore, DNA amplification by PCR is now used to complement, and in some instances replace, virus isolation as a routine procedure for determining HIV-1 infection. To make sure that HIV-1 variants can be detected, only highly conserved regions of the viral genome should be targets for amplification. Two oligonucleotide primer pairs have been extensively used for the analysis of HIV-infected samples and both amplify conserved regions of *gag* (Ou et al, 1988, Kwok et al, 1989, Kellogg and Kwok, 1990). However, to avoid problems arising from sequence variation, it would be advisable to utilize multiple primers from different conserved regions of the viral ge-

nome. Detection of the amplified material is better achieved by oligomer hybridization (OH). The combination of PCR and OH is at least twenty-five times more sensitive than Southern blot analysis and can detect as few as eighteen copies of HIV molecules in the presence of 1 μg of genomic DNA (Kwok et al, 1989). Because of the remarkable sensitivity of PCR, the same notes of caution already mentioned above for HTLV are of concern for HIV. HIV-1 sequences were detected by PCR in 100% of DNA specimens from seropositive individuals with positive virus isolation and from none of the DNA samples from seronegative, virus-isolation negative subjects (Ou et al, 1988). Such perfect correlation of PCR with serological assays and virus isolation is not observed when virus-isolation negative, seropositive individuals are tested. A positive PCR in this group does not necessarily mean active infection, since the inability to recover virus from these subjects might be explained by the presence of defective integrated proviruses (Willey et al, 1986). Conversely, a negative PCR in the same group could be due to an insufficient number of infected circulating lymphocytes. In addition to being a valid and less time-consuming alternative to virus isolation for the diagnosis of HIV infection, PCR also represents a formidable research tool to address problems related to HIV latency and heterogeneity.

32.6.10 Isolation Methods

32.6.10.a HTLV-I and HTLV-II

HTLV-I has been mainly isolated from two sources: peripheral blood and bone marrow mononuclear cells. The discovery of IL-2 (or TCGF) allowed the establishment of procedures for long-term growth of mitogen- or antigen-stimulated normal T cells in vitro. When the same procedures are applied to grow T cells from peripheral blood or bone marrow of patients with certain

T-cell neoplasms, prior stimulation with mitogen or antigen is usually not required (Poiesz, 1980a). Heparinized peripheral blood or bone marrow samples are aseptically collected and PBMC isolated by Ficoll-Hypaque gradient centrifugation, washed twice in PBS and resuspended in RPMI 1640 containing 2% L-glutamine, 1% penicillin/streptomycin mixture, 20% fetal bovine serum, and 10 to 100 U/mL IL-2 (or 10% v/v TCFG [from which lectin has been removed]). The cells are then placed into tissue culture flasks at a concentration of 10^6 cells/mL. The flasks must be checked twice weekly by monitoring cell number, viability, and possible morphologic alterations. When the cells reach a concentration of 2×10^6 cells/mL, they should be diluted with an equal volume of medium. After a continuous culture is established in the presence of IL-2, attempts can be periodically made to grow the cells without growth factors. Mitogen-stimulated normal PBMC should also be cultured in parallel as a control. The cell cultures must be monitored at regular intervals for expression and release of virus. Initially, virus expression is detectable by reverse transcriptase assay. Positive cultures should then be examined using specific reagents, such as monoclonal antibodies in IF or RIP/SDS-PAGE assays, which allow the unequivocal detection of specific viral proteins.

An alternative, indirect method for HTLV isolation is the cocultivation of normal activated umbilical cord blood lymphocytes with lethally irradiated lymphocytes from patients (Miyoshi et al, 1981). This procedure results in productive infection and transformation of cord blood T lymphocytes grown in suspension culture. These transformed cord blood cells have morphologic, immunologic, and cytochemical properties similar to the original HTLV-positive tumor cells, but are readily distinguishable by HLA haplotype and chromosomal markers. These cells express HTLV proteins, release virus particles, and express the IL-2 receptor. This method usually results in cell cultures that are not dependent upon addition of IL-2 and production of higher titers of virus.

32.6.10.b HIV-1 and HIV-2

With the techniques now available, HIV can be isolated from virtually all infected patients. The virus has been recovered from several tissue sources as indicated above (see 32.5.1). The procedures established for the isolation of HTLV-I and HTLV-II were modified for the isolation of HIV (Barré-Sinoussi et al, 1983; Popovic et al, 1984b). Since many HIV isolates are cytopathic for CD4+ T lymphocytes, it is necessary to continuously add target cells to the cultures to maintain significant levels of viral replication. However, some HIV isolates are less cytopathic or non-cytopathic (this is observed more frequently for isolates obtained from asymptomatic patients). In addition, HIV isolates with a distinctive tropism for cells of the monocyte/macrophage lineage have been obtained, particularly those derived from the nervous tissue of the patient. To provide sufficient numbers of target cells (both CD4+ T cells and monocytes/macrophages), normal PBMC, previously activated for two to three days with 1 to 5 μg/mL phytohemagglutinin (PHA), are cocultured with the patient cells at day 0, 7, 14, and 21 after establishment of the culture. As observed with other viruses, a way to increase the efficiency of isolation involves removal, from both the patient cells and the normal PBMC, of the CD8+ T-lymphocytic subset, which contains specific and non-specific cytotoxic cells capable of suppressing the spread of HIV within the cell cultures. CD8+ T-cell depletion can be achieved by several techniques, such as panning, complement-mediated cytotoxicity, or immunomagnetic bead rosetting. The latter method has been the most successful in our experience. To isolate HIV strains with tropism for monocytes/macrophages, established monolayers of normal peripheral blood macrophages can be directly cocultured with patient material (e.g., PBMC, fractionated monocytes, or cell-free cerebrospinal fluid) in the presence of 100 U/mL of macrophage-colony stimulating factor. For the best results, the nonadherent cells should be removed after two to three days in culture.

Heparinized peripheral blood and bone marrow samples from HIV-infected patients are collected and processed as mentioned above for HTLV. Cells from lymph nodes and brain biopsy specimens are prepared by mincing tissues to eliminate the stroma and then banding the cells of Ficoll-Hypaque before introducing them into cell cultures. Cell-free plasma is filtered through a 0.45 μm membrane filter and used directly. Saliva samples are diluted to a final volume of 2 mL in complete growth medium, incubated for two hours at 37 °C and centrifuged at 1000 × g for ten minutes at 4 °C. Supernatant fluids are filtered through a 0.45 μm filter and then used for transmission experiments. Semen obtained from AIDS patients can be stored frozen. After thawing, mononuclear cells are banded on Ficoll-Hypaque and seeded in round-bottom tissue culture clusters containing 200 μL of medium. The cells are activated for 24 hours with 2 μg/mL phytohemagglutinin and cultured in the presence of IL-2 and a feeder layer containing 1.5 × 10^5 irradiated allogeneic PBMC.

Permissive cell lines can also be used as target cells for HIV infection. However, the number of isolates capable of adapting to grow in continuous cell lines is low, while almost all of the HIV isolates can be efficiently grown in allogeneic PBMC.

The tests employed to identify new HIV isolates are the reverse transcriptase assay, electron microscopic examination, indirect IF assay using HIV-specific probes, e.g., monoclonal antibodies or hyperimmune sera, and further transmission of the virus to cell cultures which are examined for cytopathic effect and virus release. Primary cells from patients usually start to produce virus within one to three weeks following establishment in culture. As soon as the cells start releasing virus, there is a coincidental reduction of viable cells, especially of those with the helper-inducer (CD4+) phenotype, thus requiring the addition of new target cells to the cultures.

By using high-efficiency virus isolation techniques and end-point-dilution cultures, infectious HIV was precisely quantitated in peripheral blood cells and plasma of infected individuals (Coombs et al, 1989; Ho et al, 1989). In one of these studies the mean plasma titers were 30, 3200, and 3500 tissue-culture-infectious doses (TCID) per mL in patients with asymptomatic infection, PGL, and AIDS respectively, while in PBMC the mean titers in the same groups were 20, 2700, and 2200 TCID per 10^6 cells.

32.7 STABILITY AND SAFETY PRECAUTIONS

Detailed studies were carried out on the stability and inactivation procedures for HIV-1. Similar data are currently not available for HTLV. To mimic some natural and clinical laboratory conditions, HIV-1 in medium supplemented with 50% human plasma was air dried or incubated at different temperatures (room temperature, 37 °C and 56 °C) and then tested for infectivity. Complete inactivation of infectious virus was achieved only after three to seven days in a dried state, 11 to 15 days after exposure to 37 °C, and three to five hours at 56 °C (Resnick et al, 1986). Commonly used chemical disinfectants were also tested for their ability to inactivate HIV-1. Sodium hypochlorite at 0.5% final concentration (a 20-fold dilution of Clorox) and ethanol at 70% concentration completely inactivated the virus within one minute. Quarternary ammonium chloride (A-500) at a 15% concentration required at least ten minutes for complete virus inactivation. The effect of NP40 was also tested as an example of nonionic detergent used in the disruption of HIV-1. Exposure of the virus to 0.5% NP40 resulted in complete inactivation of the virus within one minute. Inactivation of HIV-1 by gamma and ultraviolet irradiation was also studied (Spire et al, 1985). To become noninfectious, HIV-1 required exposure to 2.5×10^5 rad and 5×10^3 J/m^2, respectively. The latter result indicates that HIV-1 is not inactivated by UV irradiation exposures even higher than those normally used under laminar air-flow biosafety hoods or in laboratories. It should be pointed out that these studies were performed with concentrated virus preparations with an exceedingly high number of particles. Nevertheless, these results indicate the need for special precautions in all dealings with this agent. Surfaces stained with possibly infected materials should be decontaminated immediately with 70% alcohol, sodium hypochlorite solution, or chemical germicides approved for use as "hospital disinfectants." In general, biosafety level 2 (BSL-2) standards and special practices, containment equipment, and facilities are recommended for all activities involving known or potentially infectious clinical specimens, body fluids, human tissues, and tissues from animals infected with HIV. Additional primary containment and personnel precautions, such as those described for biosafety level 3 (BSL-3), may be indicated for activities with high potential for droplet or aerosol production and for activities involving concentration of infectious materials. If these recommendations are followed, the risk of job-associated HIV-1 infection is limited (Weiss et al, 1988).

REFERENCES

Allan, J.S., Coligan, J.E., Barin, F., et al. 1985. Major glycoprotein antigens that induce antibodies in AIDS patients are encoded by HTLV-III. Science 228:1091–1094.

Anonymous. 1984. Editorial: Needlestick transmission of HTLV-III from a patient infected in Africa. Lancet 2:1376–1377.

Barré-Sinoussi, F., Chermann, J.C., Rey, F., et al. 1983. Isolation of a T-lymphotropic retrovirus from a patient at risk for acquired immune deficiency syndrome (AIDS). Science 220:868–871.

Batholomew, C., Blattner, W., and Cleghorn, F. 1987. Progression to AIDS in homosexual men co-infected with HIV and HTLV-I in Trinidad. Lancet 2:1469.

Bartholomew, C., Cleghorn, F., Charles, W., et al. 1986. HTLV-I and tropical spastic paraparesis. Lancet 2:99–100.

Benko, D.M., Schwartz, S., Pavlakis, G.N., and Felber, B.K. 1990. A novel human immunodeficiency virus type 1 protein, *tev*, shares sequences with *tat*, *env*, and *rev* proteins. J. Virol. 64:2505–2518.

Bhagavati, S., Ehrlich, G., Kula, R.W., et al. 1988. Detection of human T-cell lymphoma/leukemia virus type I DNA and antigen in spinal fluid and blood of patients with chronic progressive myelopathy. N. Engl. J. Med. 318:1141–1147.

Biggar, R.J., Melbye, M., Kestens, L., et al. 1985. Seroepidemiology of HTLV-III antibodies in a remote population of eastern Zaire. Br. Med. J. 290:808–810.

Biggar, R.J., Saxinger, C., Gardiner, C., et al. 1984. Type-I HTLV antibody in urban and rural Ghana, West Africa. Int. J. Cancer 34:215–219.

Blattner, W.A. and Gallo, R.C. 1985. Human T-cell leukemia/lymphoma viruses: Clinical and epidemiologic features. Curr. Top. Microbiol. Immunol. 115:67–88.

Blattner, W.A., Kalyanaraman, V.S., Robert-Guroff, M., et al. 1982. The human type-C retrovirus, HTLV, in blacks from the Caribbean, and relationship to adult T-cell leukemia/lymphoma. Int. J. Cancer 30:257–264.

Blayney, D.W., Blattner, W.A., Robert-Guroff, M., et al. 1983a. The human T-cell leukemia-lymphoma virus in the southeastern United States. JAMA 250:1048–1052.

Blayney, D.W., Jaffe, E.S., Fisher, R.I., et al. 1983b. The human T-cell leukemia/lymphoma virus, lymphoma, lytic bone lesions, and hypercalcemia. Ann. Intern. Med. 98:144–151.

Catovsky, D., Greaves, M.F., Rose, M., et al. 1982. Adult T-cell lymphoma-leukaemia in blacks from the West Indies. Lancet 1:639–643.

Cervantes, J., Hussain, S., Jensen, F., and Schwartz, J.M. 1986. T-prolymphocytic leukemia associated with human T-cell lymphotropic virus II [Abstract]. Clin. Res. 34:454A.

Cheinsong-Popov, R., Weiss, R.A., Dalgleish, A., et al. 1984. Prevalence of antibody to human T-lymphotropic virus type III in AIDS and AIDS-risk patients in Britain. Lancet 2:477–480.

Clapham, P., Nagy, K., and Weiss, R.A. 1984. Pseudotypes of human T-cell leukemia virus types 1 and 2: Neutralization by patients' sera. Proc. Natl. Acad. Sci. USA 81:2886–2889.

Clark, J.W., Blattner, W.A., and Gallo, R.C. 1986. Human T-cell leukemia viruses and T-cell lymphoid malignancies. In P. Storf and J. Mendelsohn (eds.), Principles of Internal Medicine, 7th ed. New York: McGraw-Hill Co.

Clavel, F., Guetard, D., Brun-Vezinet, F., et al. 1986. Isolation of a new human retrovirus from West African patients with AIDS. Science 233:343–346.

Cohen, E.A., Terwilliger, E.F., Sodroski, J.G., and Haseltine, W.A. 1988. Identification of a protein encoded by the *vpu* gene of HIV-1. Nature 334:532–534.

Coombs, R.W., Collier, A.C., Allain, J.P., et al. 1989. Plasma viremia in human immunodeficiency virus infection. N. Engl. J. Med. 321:1626–1631.

Cooper, D.A., Gold, J., Maclean, P., et al. 1985. Acute AIDS retrovirus infection. Definition of a clinical illness associated with seroconversion. Lancet 1:537–540.

Cooper, D.A., Imrie, A.A., and Penny, R. 1987. Antibody response to human immunodeficiency virus after primary infection. J. Infect. Dis. 155:1113–1118.

Copeland, T.D., Oroszlan, S., Kalyanaraman, V.S., et al. 1983. Complete amino acid sequence of human T-cell leukemia virus structural protein p15. FEBS Lett. 162:390–395.

Costello, C., Catovsky, D., O'Brien, M., et al. 1980. Chronic T-cell leukemias. I. Morphology, cytochemistry and ultrastructure. Leuk. Res. 4:463–476.

Curran, J.W., Jaffe, H.W., Hardy, A.M., et al. 1988. Epidemiology of HIV infection and AIDS in the United States. Science 239:610–616.

Dedera, D., Hu, W., Heyden, N.V., and Ratner, L. 1989. Viral protein R of human immunodeficiency virus types 1 and 2 is dispensable for replication and cytopathogenicity in lymphoid cells. J. Virol. 63:3205–3208.

DeRossi, A., Franchini, G., Aldovini, A., et al. 1986. Differential response to the cytopathic effects of human T-cell lymphotropic virus type III (HTLV-III): Superinfection in T4+ (helper) and T8+ (suppressor) T-cell clones transformed by HTLV-I. Proc. Natl. Acad. Sci. USA 83:4297–4301.

DeVico, A.L., Veronese, F.D., Lee, S.L., et al. 1988. High prevalence of serum antibodies to reverse transcriptase in HIV-1-infected individuals. AIDS Res. Hum. Retroviruses 4:17–22.

Ehrlich, G.D., Greenberg, S., and Abbott, M.A. 1990. Detection of human T-cell lymphoma/leukemia virus. In: M.A. Innis, D.H. Gelfand, and J.J. Sninsky (eds.), PCR Protocols: A Guide to Methods and Applications. Academic Press, 325–336.

Ehrlich, G.D. and Poiesz, B.J. 1988. Clinical and molecular parameters of HTLV-I infection. Clin. Lab. Med. 8:65–84.

Ehrlich, G.D., Davey, F.R., Kirshner, J.J., et al. 1989a. A polyclonal CD4+ and CD8+ lymphocytosis in a patient doubly infected with HTLV-I and HIV-1: A clinical and molecular analysis. Am. J. Hematol. 30:128–139.

Ehrlich, G.D., Glaser, J.B., Abbott, M.A., et al. 1989b. Detection of anti-HTLV-I *tax* antibodies in HTLV-I enzyme-linked immunosorbent assay-negative individuals. Blood 74:1066–1072.

Epstein, J.S. and Fricke, W.A. 1990. Current safety of clotting factor concentrates. Arch. Pathol. Lab. Med. 114:335–340.

Evatt, B.L., Gomperts, E.D., McDougal, J.S., and Ramsey, R.B. 1984. Coincidental appearance of LAV/HTLV-III antibodies in hemophiliacs and the onset of the AIDS epidemic. N. Engl. J. Med. 312:483–486.

Eyster, M.E., Gail, M.H., Ballard, J.O., et al. 1987. Natural history of human immunodeficiency virus infections in hemophiliacs: Effects of T-cell subsets, platelet counts, and age. Ann. Intern. Med. 107:1–6.

Fisher, A.G., Ensoli, B., Ivanoff, L., et al. 1987. The *sor* gene of HIV-1 is required for efficient virus transmission *in vitro*. Science 237:888–893.

Fisher, A.G., Ratner, L., Mitsuya, H., et al. 1986. Infectious mutants of HTLV-III with changes in the 3' region and markedly reduced cytopathic effects. Science 233:655–659.

Fleming, A.F., Yamamoto, N., Bhusnurmath, S.R., et al. 1983. Antibodies to ATLV (HTLV) in Nigerian blood donors and patients with chronic lymphatic leukaemia or lymphoma. Lancet 2:334–335.

Friedland, G.H. and Klein, R.S. 1987. Transmission of the human immunodeficiency virus. N. Engl. J. Med. 317:1125–1135.

Gajdusek, D.C., Amyx, H.L., Gibbs, C.J., et al. 1985. Infection of chimpanzees by human T-lymphotropic retroviruses from brain and other tissues from AIDS patients. Lancet 1:55–56.

Gallo, R.C. 1986. The first human retrovirus. Scientific American 254:88–98.

Gallo, R.C. and Blattner, W.A. 1985. Human T-cell leukemia/lymphoma viruses: ATL and AIDS. In V.T. Devita Jr., S. Hellman, and S.A. Rosenberg (eds.), Important Advances in Oncology. Philadelphia: J.B. Lipincott, Co., pp. 104–138.

Gallo, R.C., Salahuddin, S.Z., Popovic, M., et al. 1984. Frequent detection and isolation of cytopathic retroviruses (HTLV-III) from patients with AIDS and at risk for AIDS. Science 224:500–503.

Gartner, S., Markovits, P., Markovitz, D.M., et al. 1986. The role of mononuclear phagocytes in HTLV-III/LAV infection. Science 233:215–219.

Gelmann, E.P., Franchini, G., Manzari, V., et al. 1984. Molecular cloning of a unique human T-cell leukemia virus (HTLV-II$_{mo}$). Proc. Natl. Acad. Sci. USA 81:993–997.

Gendelman, H.E., Orenstein, J.M., Baca, L.M., et al. 1989. The macrophage in the persistence and pathogenesis of HIV infection. AIDS 1989 3:475–495.

Gessain, A., Louie, A., Gout, O., et al. 1991. Human T-cell leukemia-lymphoma virus type I (HTLV-I) expression in fresh peripheral blood mononuclear cells from patients with tropical spastic paraparesis/HTLV-I-associated myelopathy. J. Virol. 65:1628–1633.

Gessain, A., Barin, F., Vernant, J.C., et al. 1985. Antibodies to human T-lymphotropic virus type-I in patients with tropical spastic paraparesis. Lancet 2:407–409.

Goedert, J.J. and Blattner, W.A. 1985. The epidemiology of AIDS and related conditions. In V.T. DeVita Jr., S. Hellman, and S.A. Rosenberg (eds.), AIDS—Etiology, Diagnosis, Treatment and Prevention. Philadelphia: J.B. Lippincott, Co., pp. 1–30.

Gonda, M.A., Wong-Staal, F., Gallo, R.C., et al. 1985. Sequence homology and morphologic similarity of HTLV-III and visna virus, a pathogenic lentivirus. Science 227:173–177.

Gottlieb, M.S., Schroff, R., Schanker, H.M., et al. 1981. *Pneumocystis carinii* pneumonia and mucosal candidiasis in previously healthy homosexual men: Evidence of a new acquired cellular immunodeficiency. N. Engl. J. Med. 305:1425–1431.

Gout, O., Baulac, M., Gessain, A., et al. 1990. Rapid development of myelopathy after HTLV-I infection acquired by transfusion during cardiac transplantation. N. Engl. J. Med. 322:383–388.

Greaves, M.F., Verbi, W., Tilley, R. et al. 1984. Human T-cell leukemia virus (HTLV) in the

United Kingdom. Int. J. Cancer 33:795–806.

Greenberg, S.J., Ehrlich, G.D., Abbott, M.A., et al. 1989. Detection of sequences homologous to human retroviral DNA in multiple sclerosis by gene amplification. Proc. Natl. Acad. Sci. USA 86:2878–2882.

Groopman, J.E., Salahuddin, S.Z., Sarngadharan, M.G., et al. 1984. HTLV-III in saliva of people with AIDS-related complex and healthy homosexual men at risk for AIDS. Science 226:447–449.

Hadzopoulou-Cladaras, M., Felber, B.K., Cladaras, C., et al. 1989. The *rev(trs/art)* protein of human immunodeficiency virus type 1 affects viral mRNA and protein expression via a *cis*-acting sequence in the *env* region. J. Virol. 63:1265–1274.

Hanaoka, M. 1982. Progress in adult T-cell leukemia research. Acta Pathol. Jpn. (32 suppl.) 1:171–185.

Hardy, A.M., Allen, J.R., Morgan, W.M., and Curran, J.W. 1985. The incidence rate of acquired immunodeficiency syndrome in selected populations. JAMA 253:215–220.

Harper, M.E., Marselle, L.M., Gallo, R.C., and Wong-Staal, F. 1986. Detection of lymphocytes expressing human T-lymphotropic virus type III in lymph nodes and peripheral blood from infected individuals by *in situ* hybridization. Proc. Natl. Acad. Sci. USA 83:772–776.

Haseltine, W.A., Sodroski, J., Patarca, R., et al. 1984. Structure of 3′-terminal region of type-II human T-lymphotropic virus: Evidence for new coding region. Science 225:419–421.

Hattori, S., Kiyokawa, T., Imagawa, K., et al. 1984. Identification of *gag* and *env* gene products of human T-cell leukemia virus (HTLV). Virology 136:338–347.

Hattori, T., Koito, A., Takatsuki, K., et al. 1989. Frequent infection with human T-cell lymphotropic virus type I in patients with AIDS but not in carriers of human immunodeficiency virus type 1. J. Acquir. Immune Defic. Syndr. 2:272–276.

Hattori, T., Uchiyama, T., Toibana, T., et al. 1981. Surface phenotype of Japanese adult T-cell leukemia cells characterized by monoclonal antibodies. Blood 58:645–647.

Henderson, L.E, Sowder, R.C., Copeland, T.D., et al. 1988. Isolation and characterization of a novel protein (X-*orf* product) from SIV and HIV-2. Science 241:199–201.

Hidaka, M., Inoue, J., Yoshida, M., and Seiki, M. 1988. Post-transcriptional regulator (*rex*) of HTLV-I initiates expression of viral structural proteins but suppresses expression of regulatory proteins. EMBO. J. 7:519–523.

Hinuma, Y., Komoda, H., Chosa, T., et al. 1982. Antibodies to adult T-cell leukemia-virus-associated antigen (ATLA) in sera from patients with ATL and controls in Japan: A nationwide sero-epidemiologic study. Int. J. Cancer 29:631–635.

Hinuma, Y., Nagata, K., Hanaoka, M., et al. 1981. Adult T-cell leukemia: Antigen in an ATL cell line and detection of antibodies to the antigen in human sera. Proc. Natl. Acad. Sci. USA 78:6476–6480.

Ho, D.D., Rota, T.R., Schooley, T.R., et al. 1985. Isolation of HTLV-III from cerebrospinal fluid and neural tissues of patients with neurologic syndromes related to the acquired immunodeficiency syndrome. N. Engl. J. Med. 313:1493–1497.

Ho, D.D., Moudgil, T., and Alam, M. 1989. Quantitation of human immunodeficiency virus type 1 in the blood of infected persons. N. Engl. J. Med 321:1621–1625.

Ho, D.D., Schooley, R.T., Rota, T.R., et al. 1984. HTLV-III in the semen and blood of a healthy homosexual man. Science 226:451–453.

Hunsmann, G., Schneider, J., Schmitt, J., and Yamamoto, N. 1983. Detection of serum antibodies to adult T-cell leukemia virus in non-human primates and in people from Africa. Int. J. Cancer 32:329–332.

Inoue, J., Yoshida, M., and Seiki, M. 1987. Transcriptional (p40^X) and post-transcriptional (p27^{XIII}) regulators are required for the expression and replication of human T-cell leukemia virus type I genes. Proc. Natl. Acad. Sci. USA 84:3653–3657.

Jacobson, S., Raine, C.S., Mingioli, E.S., and McFarlin, D.E. 1988. Isolation of an HTLV-I-like retrovirus from patients with tropical spastic paraparesis. Nature 331:540–543.

Jaffe, E.S., Blattner, W.A., Blayney, D.W., et al. 1984. The pathologic spectrum of adult T-cell leukemia/lymphoma in the United States. Human T-cell leukemia/lymphoma virus-associated lymphoid malignancies. Am. J. Surg. Pathol. 8:263–275.

Jarrett, W.F.H., Martin, W.B., Crighton, G.W., et al. 1964. Leukaemia in the cat: Transmission experiments with leukaemia (lymphosarcoma). Nature 202:566–567.

Johnson, D.A., Gautsch, J.W., Sportsman, J.R., and Elder, J.H. 1984. Improved technique utilizing nonfat dry milk for analysis of proteins and nucleic acids transferred to nitrocellulose. Gene Anal. Tech. 1:3–8.

Kabayama, Y., Isashiri, M., Uehara, F., et al. 1988. Ocular disorders associated with adult T-cell leukemia. Jpn. J. Clin. Ophthalmol. 42:139.

Kalyanaraman, V.S., Cabradilla, C.D., Getchell, J.P., et al. 1984a. Antibodies to the core protein of lymphadenopathy-associated virus (LAV) in patients with AIDS. Science 225:321–323.

Kalyanaraman, V.S., Jarvis-Morar, M., Sarngadharan, M.G., and Gallo, R.C. 1984b. Immunological characterization of the low molecular weight *gag* gene proteins p19 and p15 of human T-cell leukemia-lymphoma virus (HTLV) and demonstration of human natural antibodies to them. Virology 132:61–70.

Kalyanaraman, V.S., Sarngadharan, M.G., Nakao, Y., et al. 1982a. Natural antibodies to the structural core protein (p24) of the T-cell leukemia (lymphoma) retrovirus found in sera of leukemia patients in Japan. Proc. Natl. Acad. Sci. USA 79:1653–1657.

Kalyanaraman, V.S., Sarngadharan, M.G., Poisez, B., et al. 1981. Immunological properties of a type C retrovirus isolated from cultured human T-lymphoma cells and comparison to other mammalian retroviruses. J. Virol. 38:906–915.

Kalyanaraman, V.S., Sarngadharan, M.G., Robert-Guroff, M., et al. 1982b. A new subtype of human T-cell leukemia virus (HTLV-II) associated with a T-cell variant of hairy cell leukemia. Science 218:571–573.

Kawai, H., Nishida, Y., Takagi, M., et al. 1989. HTLV-I-associated myelopathy with adult T-cell leukemia. Neurology 39:1129–1131.

Kawano, F., Tsuda, H., Yamaguchi, K., et al. 1984. Unusual clinical courses of adult T-cell leukemia in siblings. Cancer 54:131–134.

Kellogg, D.E. and Kwok, S. 1990. Detection of human immunodeficiency virus. In M.A. Innis, D.H. Gelfand, and J.J. Sninsky (eds), PCR Protocols: A Guide to Methods and Applications. Academic Press, pp. 337–347.

Kinoshita, T., Shimoyama, M., Tobinai, K., et al. 1989. Detection of mRNA for the *tax₁/rex₁* gene of human T-cell leukemia virus type I in fresh peripheral blood mononuclear cells of adult T-cell leukemia patients and viral carriers by using the polymerase chain reaction. Proc. Natl. Acad. Sci. USA 86:5620–5624.

Kitchen, L.W., Barin, F., Sullivan, J.L., et al. 1984. Aetiology of AIDS—antibodies to human T-cell leukaemia virus (type III) in haemophiliacs. Nature 312:367–369.

Kiyokawa, T., Seiki, M., Iwashita, S., et al. 1985. p27^XIII and p21^XIII, proteins encoded by the pX sequence of human T-cell leukemia virus type I. Proc. Natl. Acad. Sci. USA 82:8359–8363.

Klatzmann, D., Barré-Sinoussi, F., Nugeyre, M.T., et al. 1984. Selective tropism of lymphadenopathy-associated virus (LAV) for helper-inducer T-lymphocytes. Science 225:59–63.

Koerper, M.A., Kaminsky, L.S., and Levy, J.A. 1985. Differential prevalence of antibody to AIDS-associated retrovirus in haemophiliacs treated with Factor VIII concentrate virus cryoprecipitate: Recovery of infectious virus. Lancet 1:275.

Komuro, A., Hayami, M., Fujii, H., et al. 1983. Vertical transmission of adult T-cell leukemia virus. Lancet 1:240.

Kwok, S., Kellogg, D., Ehrlich, G., et al. 1988. Characterization of a sequence of human T cell leukemia virus type I from a patient with chronic progressive myelopathy. J. Infect. Dis. 158:1193–1197.

Kwok, S., Mack, D.H., Sninsky, J.J., et al. 1989. Diagnosis of human immunodeficiency virus in seropositive individuals: Enzymatic amplification of HIV viral sequences in peripheral blood mononuclear cells. In P.A. Luciw and K.S. Steimer (eds), HIV Detection by Genetic Engineering Methods. New York: Marcel Dekker, Inc.

Laemmli, U.K. 1970. Cleavage of structural proteins during the assembly of the head of bacteriophage T4. Nature 227:680–685.

Lando, Z., Sarin, P., Megson, M., et al. 1983. Association of human T-cell leukaemia/lymphoma virus with the Tac antigen marker for the human T-cell growth factor receptor. Nature 305:733–736.

Lapointe, N., Michaud, J., Pekovic, D., et al. 1985. Transplacental transmission of HTLV-III virus. N. Engl. J. Med. 312:1325–1326.

Lee, H., Swanson, P., Shorty, V.S., et al. 1989. High rate of HTLV-II infection in seropositive i.v. drug abusers in New Orleans. Science 244:471–475.

Lee, T.H., Coligan, J.E., Homma, T., et al. 1984a. Human T-cell leukemia virus-associated cell membrane antigens: Identity of the major antigens recognized after virus infection. Proc. Natl. Acad. Sci. USA 81:3856–3860.

Lee, T.H., Coligan, J.E., Sodroski, J.G., et al. 1984b. Antigens encoded by the 3'-terminal region of human T-cell leukemia virus: Evidence for a functional gene. Science 226:57–61.

Lemaitre, M., Guetard, D., Henin, Y., et al. 1990. Protective activity of tetracycline analogs against the cytopathic effect of the human immunodeficiency viruses in CEM cells. Res. Virol. 141:5–16.

Lennert, K., Kikuchi, M., Sato, E., et al. 1985. HTLV-positive and -negative T-cell lymphomas. Morphological and immunohistochemical differences between European and HTLV-positive Japanese T-cell lymphomas. Int. J. Cancer 35:65–72.

Lo, S.-C., Dawson, M.S., Wong, D.M., et al. 1989. Identification of *Mycoplasma incognitus* infection in patients with AIDS: An immunohistochemical, in situ hybridization and ultrastructural study. Am. J. Trop. Med. Hyg. 41:601–616.

Luciw, P.A., Cheng-Mayer, C., and Levy, J.A.

1987. Mutational analysis of the human immunodeficiency virus: The *orf*-B region down-regulates virus replication. Proc. Natl. Acad. Sci. USA 84:1434–1438.

Lusso, P., Ensoli, B., Markham, P.D., et al. 1989. Productive dual infection of human CD4+ T lymphocytes by HIV-1 and HHV-6. Nature 337:370–373.

Lusso, P., Lori, F., and Gallo, R.C. 1990. CD4-independent infection by human immunodeficiency virus type 1 after phenotypic mixing with human T-cell leukemia viruses. J. Virol. 64:6134–6137.

Lusso, P., DeMaria, A., Malnati, M., et al. 1991. Induction of CD4 and HIV-1 susceptibility to HIV-1 infection in human CD8+ T lymphocytes by HHV-6. Nature 349:533–535.

Maeda, Y., Furukawa, M., Takahara, T., et al. 1984. Prevalence of possible adult T-cell leukemia virus-carriers among volunteer blood donors in Japan: A nation-wide study. Int. J. Cancer 33:717–720.

Malim, M.H., Böhnlein, S., Hauber, J., and Cullen, B.R. 1989a. Functional dissection of the HIV-1 rev *trans*-activator—Derivation of a *trans*-dominant repressor of rev function. Cell 58:205–214.

Malim, M.H., Hauber, J., Le, S.Y., et al. 1989b. The HIV-1 rev *trans*-activator acts through a structured target sequence to activate nuclear export of unspliced viral mRNA. Nature 338:254–257.

Manca, N., Veronese, F.D., Ho, D.D., et al. 1987. Sequential changes in antibody levels to the *env* and *gag* antigens in human immunodeficiency virus infected subjects. Eur. J. Epidemiol. 3:96–102.

Manzari, V., Fazio, V.M., Martinotti, S., et al. 1984. Human T-cell leukemia/lymphoma virus (HTLV-I) DNA: Detection in Italy in a lymphoma and in a Kaposi sarcoma patient. Int. J. Cancer 34:891–892.

Manzari, V., Wong-Staal, F., Franchini, G., et al. 1983. Human T-cell leukemia-lymphoma virus (HTLV): Cloning of an integrated defective provirus and flanking cellular sequences. Proc. Natl. Acad. Sci. USA 80:1574–1578.

Melbye, M., Froebel, K.S., Madhok, R., et al.

1984. HTLV-III seropositivity in European haemophiliacs exposed to factor VIII concentrate imported from the USA. Lancet 2:1444–1446.

Merino, F., Robert-Guroff, M., Clark, J., et al. 1984. Natural antibodies to human T-cell leukemia/lymphoma virus in healthy Venezuelan populations. Int. J. Cancer 34:501–506.

Mitsuya, H., Guo, H.-G., Megson, M., et al. 1984. Transformation and cytopathogenic effect in an immune human T-cell clone infected by HTLV-I. Science 223:1293–1296.

Miyamoto, K., Tomita, N., Ishii, A., et al. 1984. Transformation of ATLA-negative leukocytes by blood components from anti-ATLA-positive donors *in vitro*. Int. J. Cancer 55:721–725.

Miyamoto, Y., Yamaguchi, K., Nishimura, H., et al. 1985. Familial adult T-cell leukemia. Cancer 55:181–185.

Miyoshi, I., Kubonishi, I., Yoshimoto, S., et al. 1981. Type C virus particles in a cord T-cell line derived by co-cultivating normal human cord leukocytes and human leukaemic T cells. Nature 294:770–771.

Morgan, D.A., Ruscetti, F.W., and Gallo, R.C. 1976. Selective in vitro growth of T lymphocytes from normal human bone marrows. Science 193:1007–1008.

Morgan, O.S., Char, G., Mora, C., Rodgers-Johnson, P. 1989. HTLV-I and polymyositis in Jamaica. Lancet 2:1184–1186.

Mösseler, J., Schimpf, K., Auerswald, G., et al. 1985. Inability of pasteurized factor VIII preparations to induce antibodies to HTLV-III after long-term treatment. Lancet 1:1111.

Nagashima, K., Yoshida, M., and Seiki, M. 1986. A single species of pX mRNA of human T-cell leukemia virus type I encodes trans-activator p40x and two other phosphoproteins. J. Virol. 60:394–399.

Ogawa, K., Shibata, R., Kiyomasu, T., Higuchi, T., et al. 1989. Mutational analysis of the human immunodeficiency virus *vpr* open reading frame. J. Virol. 63:4110–4114.

Okochi, K., Sato, H., and Hinuma, Y. 1984. A retrospective study on transmission of adult T cell leukemia virus by blood transfusion: Seroconversion in recipients. Vox Sang. 46:245–253.

Oroszlan, S., Sarngadharan, M.G., Copeland, T.D., et al. 1982. Primary structure analysis of the major internal protein p24 of human type C T-cell leukemia virus. Proc. Natl. Acad. Sci. USA 79:1291–1294.

Ou, C.Y., Kwok, S., Mitchell, S.W., et al. 1988. DNA amplification for direct detection of HIV-1 in DNA of peripheral blood mononuclear cells. Science 239:295–297.

Page, J.B., Lai, S.H., Chitwood, D.D., et al. 1990. HTLV-I/II seropositivity and death from AIDS among HIV-1 seropositive intravenous drug users. Lancet 335:1439–1441.

Petito, C.K., Navia, B.A., Cho, E.S., et al. 1985. Vacuolar myelopathy pathologically resembling subacute combined degeneration in patients with the acquired immunodeficiency syndrome. N. Engl. J. Med. 312:874–879.

Poiesz, B.J., Ruscetti, F.W., Gazdar, A.F., et al. 1980a. Detection and isolation of type C retrovirus particles from fresh and cultured lymphocytes of a patient with cutaneous T-cell lymphoma. Proc. Natl. Acad. Sci. USA 77:7415–7419.

Poiesz, B.J., Ruscetti, F.W., Mier, J.W., Woods, A.M., et al. 1980b. T-cell lines established from human T-lymphocyte neoplasias by direct response to T-cell growth factor. Proc. Natl. Acad. Sci. USA 77:6815–6819.

Poiesz, B.J., Ruscetti, F.W., Reitz, M.S., et al. 1981. Isolation of a new type C retrovirus (HTLV) in primary uncultured cells of a patient with Sézary T-cell leukaemia. Nature 294:268–271.

Popovic, M., Reitz, M.S., Sarngadharan, M.G., et al. 1982. The virus of Japanese adult T-cell leukaemia is a member of the human T-cell leukaemia virus group. Nature 300:63–66.

Popovic, M., Lange-Wantzin, G., Sarin, P.S., et al. 1983a. Transformation of human umbilical cord blood T cells by human T-cell leukemia/lymphoma virus. Proc. Natl. Acad. Sci. USA 80:5402–5406.

Popovic, M., Sarin, P.S., Robert-Guroff, M., et

al. 1983b. Isolation and transmission of human retrovirus (human T-cell leukemia virus). Science 219:856–859.

Popovic, M., Flomenberg, N., Volkman, D.J., et al. 1984a. Alteration of T-cell functions by infection with HTLV-I or HTLV-II. Science 226:459–462.

Popovic, M., Sarngadharan, M.G., Read, E., and Gallo, R.C. 1984b. Detection, isolation, and continuous production of cytopathic retroviruses (HTLV-III) from patients with AIDS and pre-AIDS. Science 224:497–500.

Posner, L.E., Robert-Guroff, M., Kalyanaraman, V.S., et al. 1981. Natural antibodies to the human T cell lymphoma virus in patients with cutaneous T cell lymphomas. J. Exp. Med. 154:333–346.

Redfield, R.R., Markham, P.D., Salahuddin, S.Z., et al. 1985. Heterosexually acquired HTLV-III/LAV disease (AIDS-related complex and AIDS). Epidemiologic evidence for female-to-male transmission. JAMA 254:2094–2096.

Resnick, L., Veren, K., Salahuddin, S.Z., et al. 1986. Stability and inactivation of HTLV-III/LAV under clinical and laboratory environments. JAMA 255:1887–1891.

Rho, H.M., Poiesz, B., Ruscetti, F.W., and Gallo, R.C. 1981. Characterization of the reverse transcriptase from a new retrovirus (HTLV) produced by a human cutaneous T-cell lymphoma cell line. Virology 112:355–360.

Rimsky, L., Hauber, J., Dukovich, M., et al. 1988. Functional replacement of the HIV-1 rev protein by the HTLV-I rex protein. Nature 335:738–740.

Robert-Guroff, M. and Gallo, R.C. 1983. Establishment of an etiologic relationship between the human T-cell leukemia/lymphoma virus (HTLV) and adult T-cell leukemia. Blut 47:1–12.

Robert-Guroff, M., Coutinho, R.A., Zadelhoff, A.W., et al. 1984. Prevalence of HTLV-specific antibodies in Surinam emigrants to the Netherlands. Leuk. Res. 8:501–504.

Robert-Guroff, M., Fahey, K.A., Maeda, M., et al. 1982a. Identification of HTLV p19 specific natural human antibodies by competition with monoclonal antibody. Virology 122:297–305.

Robert-Guroff, M., Nakao, Y., Notake, K., et al. 1982b. Natural antibodies to human retrovirus HTLV in a cluster of Japanese patients with adult T cell leukemia. Science 215:975–978.

Robert-Guroff, M., Weiss, S.H., Giron, J.A., et al. 1986. Prevalence of antibodies to HTLV-I, -II, and -III in intravenous drug abusers from an AIDS endemic region. JAMA 255:3133–3137.

Rosenblatt, J.D., Giorgi, J.V., Golde, D.W., et al. 1988. Integrated human T-cell leukemia virus II genome in CD8+ T cells from a patient with "atypical" hairy-cell leukemia: Evidence for distinct T and B cell lymphoproliferative disorders. Blood 71:363–369.

Rosenblatt, J.D., Golde, D.W., Wachsman, W., et al. 1986. A second isolate of HTLV-II associated with atypical hairy-cell leukemia. N. Engl. J. Med. 315:372–377.

Ruscetti, F.W., Morgan, D.A., and Gallo, R.C. 1977. Functional and morphologic characterization of human T cells continuously grown *in vitro*. J. Immunol. 119:131–138.

Salahuddin, S.Z., Markham, P.D., Popovic, M., et al. 1985. Isolation of infectious human T-cell leukemia/lymphotropic virus type III (HTLV-III) from patients with acquired immunodeficiency syndrome (AIDS) or AIDS-related complex (ARC) and from healthy carriers: A study of risk groups and tissue sources. Proc. Natl. Acad. Sci. USA 82:5530–5534.

Sarin, P.S., Aoki, T., Shibata, A., et al. 1983. High incidence of human type-C retrovirus (HTLV) in family members of a HTLV-positive Japanese T-cell leukemia patient. Proc. Natl. Acad. Sci. USA 80:2370–2374.

Sarngadharan, M.G., diMarzo-Veronese, F., Lee, S., and Gallo, R.C. 1985. Immunological properties of HTLV-III antigens recognized by sera of patients with AIDS, and AIDS-related complex and asymptomatic carriers of HTLV-III infection. Cancer Res. 45:4574–4577.

Sarngadharan, M.G., Popovic, M., Bruch, L., et al. 1984. Antibodies reactive with human T-lymphotropic retroviruses (HTLV-III) in the sera of patients with AIDS. Science 224:506–508.

Sarngadharan, M.G., Robert-Guroff, M., and

Gallo, R.C. 1978. DNA polymerases of normal and neoplastic mammalian cells. Biochem. Biophys. Acta 516:419–487.

Saxinger, W., Blattner, W.A., Levine, P.H., et al. 1984. Human T-cell leukemia virus (HTLV-I) antibodies in Africa. Science 225: 1473–1476.

Saxinger, W.C. and Gallo, R.C. 1982. Possible risk to recipients of blood from donors carrying serum markers of human T-cell leukaemia virus. Lancet 1:1074.

Saxinger, W.C. and Gallo, R.C. 1983. Methods in laboratory investigation. Application of the indirect enzyme-linked immunosorbent assay microtest to the detection and surveillance of human T cell leukemia/lymphoma virus. Lab. Invest. 49:371–377.

Saxinger, W.C., Levine, P.H., Dean, A.G., et al. 1985. Evidence for exposure to HTLV-III in Uganda before 1973. Science 227: 1036–1038.

Schaffar-Deshayes, L., Chavance, M., Monplaisir, N., et al. 1984. Antibodies to HTLV-I p24 in sera of blood donors, elderly people and patients with hemopoietic diseases in France and in French West Indies. Int. J. Cancer 34:667–670.

Schneider, J., Yamamoto, N., Hinuma, Y., and Hunsmann, G. 1984. Sera from adult T-cell leukemia patients react with envelope and core polypeptides of adult T-cell leukemia virus. Virology 132:1–11.

Schüpbach, J., Haller, O., Vogt, M., et al. 1985. Antibodies to HTLV-III in Swiss patients with AIDS and pre-AIDS and in groups at risk for AIDS. N. Engl. J. Med. 312:265–270.

Schüpbach, J., Kalyanaraman, V.S., Sarngadharan, M.G., et al. 1983a. Antibodies against three purified proteins of the human type C retrovirus, human T-cell leukemia/lymphoma, in adult T-cell leukemia/lymphoma patients and healthy blacks from the Caribbean. Cancer Res. 43:886–891.

Schüpbach, J., Kalyanaraman, V.S., Sarngadharan, M.G., et al. 1983b. Antibodies against three purified structural proteins of the human type-C retrovirus, HTLV, in Japanese adult T-cell leukemia patients, healthy family members, and unrelated normals. Int. J. Cancer 32:583–590.

Seiki, M., Hattori, S., and Yoshida, M. 1982. Human adult T-cell leukemia virus: Molecular cloning of the provirus DNA and the unique terminal structure. Proc. Natl. Acad. Sci. USA 79:6899–6902.

Seiki, M., Hattori, S., Hirayama, Y., and Yoshida, M. 1983. Human adult T-cell leukemia virus: Complete nucleotide sequence of the provirus genome integrated in leukemia cell DNA. Proc. Natl. Acad. Sci. USA 80:3618–3622.

Shaw, G.M., Harper, M.E., Hahn, B.H., et al. 1985. HTLV-III infection in brains of children and adults with AIDS encephalopathy. Science 227:177–182.

Shima, H., Takano, M., Shimotohno, K., and Miwa, M. 1986. Identification of p26^{xb} and p24^{xb} of human T-cell leukemia virus type II. FEBS Lett. 209:289–294.

Seigal, F.P., Lopez, C., Hammer, G.S., et al. 1981. Severe acquired immunodeficiency in male homosexuals, manifested by chronic perianal ulcerative herpes simplex lesions. N. Engl. J. Med. 305:1439–1444.

Siekevitz, M., Josephs, S.F., Dukovich, M., et al. 1987. Activation of the HIV-1 LTR by T-cell mitogens and the trans-activator protein of HTLV-I. Science 238:1575–1578.

Slamon, D.J., Shimotohno, K., Cline, M.J., et al. 1984. Identification of the putative transforming protein of the human T-cell leukemia viruses HTLV-I and HTLV-II. Science 226:61–65.

Snider, W.D., Simpson, D.M., Aronyk, K.E., and Nielsen, S.L. 1983a. Primary lymphoma of the central nervous system associated with acquired immune-deficiency syndrome. N. Engl. J. Med. 308:45.

Snider, W.D., Simpson, D.M., Nielson, S., et al. 1983b. Neurological complications of acquired immune-deficiency syndrome: Analysis of 50 patients. Ann. Neurol. 14: 403–418.

Sodroski, J., Patarca, R., Perkins, D., et al. 1984. Sequence of the envelope glycoprotein gene of type II human T lymphotropic virus. Science 225:421–424.

Sohn, C.C., Blayney, D.W., Misset, J.L., et al. 1986. Leukopenic chronic T cell leukemia mimicking hairy cell leukemia: Association with human retroviruses. Blood 67:949–956.

Sommerfelt, M.A., Williams, B.P., Clapham, P.R., et al. 1989. Human T cell leukemia viruses use a receptor determined by human chromosome 17. Science 242:1557–1559.

Spire, B., Dormont, D., Barré-Sinoussi, F., et al. 1985. Inactivation of lymphadenopathy-associated virus by heat, gamma rays, and ultraviolet light. Lancet 1:188–189.

Strebel, K., Daugherty, D., Clouse, K., et al. 1987. The HIV "A" (*sor*) gene product is essential for virus infectivity. Nature 328:728–730.

Strebel, K., Klimkait, T., and Martin, M.A. 1988. A novel gene of HIV-1, *vpu*, and its 16-kilodalton product. Science 241:1221–1223.

Suciu-Foca, N., Rubinstein, P., Popovic, M., et al. 1984. Reactivity of HTLV-transformed human T-cell lines to MHC class II antigens. Nature 312:275–277.

Tajima, K., Fujita, K., Tsukidate, S., et al. 1983. Seroepidemiological studies on the effects of filarial parasites on infestation of adult T-cell leukemia virus in the Goto Islands, Japan. Gann 74:188–191.

Tajima, K., Tominaga, S., and Suchi, T. 1982. Clinico-epidemiological analysis of adult T-cell leukemia. Gann Monogr. Cancer Res. 28:197–210.

Tajima, K., Tominaga, S., Shimizu, H., and Suchi, T. 1981. A hypothesis of the etiology of adult T-cell leukemia/lymphoma. Gann 72:684–691.

Takatsuki, K., Uchiyama, J., Sagawas, K., and Yodoi, J. 1977. Adult T-cell leukemia in Japan. In S. Seno, F. Takaku, and S. Irino (eds.), Topics in Hematology. Amsterdam: Excerpta Medica, pp. 73–77.

Takatsuki, K., Uchiyama, T., Ueshima, Y., et al. 1982. Adult T-cell leukemia: Proposal as a new disease and cytogenic, phenotypic, and functional studies of leukemia cells. Gann Monogr. Cancer Res. 28:13–22.

Tedder, R.S., Shanson, D.C., Jeffries, D.J., et al. 1984. Low prevalence in the UK of HTLV-I and HTLV-II infection in subjects with AIDS, with extended lymphadenop-athy, and at risk of AIDS. Lancet 2:125–128.

Tenner-Racz, K., Racz, P., Schmidt, H., et al. 1988. Immunohistochemical, electron microscopic and *in situ* hybridization evidence for the involvement of lymphatics in the spread of HIV-1. AIDS 1988 2:299–309.

Terwilliger, E.F., Cohen, E.A., Lu, Y.C., et al. 1989. Functional role of human immunodeficiency virus type 1 *vpu*. Proc. Natl. Acad. Sci. USA 86:5163–5167.

Towbin, H., Staehelin, T., and Gordon, J. 1979. Electrophoretic transfer of proteins from polyacrylamide gels to nitrocellulose sheets: Procedure and some applications. Proc. Natl. Acad. Sci. USA 76:4350–4354.

Tschachler, E., Robert-Guroff, M., Gallo, R.C., and Reitz, M.S. 1989. Human T-lymphotropic virus I-infected T cells constitutively express lymphotoxin *in vitro*. Blood 73:194–201.

Tucker, J., Ludlam, C.A., Craig, A., et al. 1985. HTLV-III infection associated with glandular-fever-like illness in a haemophiliac. Lancet 1:585.

Uchiyama, T., Yodoi, J., Sagawa, K., et al. 1977. Adult T-cell leukemia: Clinical and hematologic features of 16 cases. Blood 50:481–492.

Veronese, F.D., DeVico, A.L., Copeland, T.D., et al. 1985. Characterization of gp41 as the transmembrane protein coded by the HTLV-III/LAV envelope gene. Science 229:1402–1405.

Veronese, F.D., Rahman, R., Copeland, T.D., et al. 1987. Immunological and chemical analysis of p6, the carboxyl-terminal fragment of HIV p15. AIDS Res. Hum. Retroviruses 3:253–264.

Wainberg, M.A., Spira, B., Boushira, M., and Margolese, R.G. 1985. Inhibition by human T-lymphotropic virus (HTLV-I) of T-lymphocyte mitogenesis: Failure of exogenous T-cell growth factor to restore responsiveness to lectin. Immunology 54:1–7.

Waldman, T., Broder, S., Greene, W., et al. 1983. A comparison of the function and phenotype of Sezary T-cells with human T-cell leukemia/lymphoma virus (HTLV)-associated with T-cell leukemia. Clin. Res. 31:5474–5480.

Watanabe, T., Seiki, M., and Yoshida, M. 1984. HTLV type I (U.S. isolate) and ATLV (Japanese isolate) are the same species of human retrovirus. Virology 133:238–241.

Weiss, S.H., French, J., Denny, T.N., et al. Five year follow-up findings from the New Jersey intravenous drug abusers cohort study. Annual Meeting American Public Health Association, New York, 1990.

Weiss, S.H., French, J., Holland, B., et al. HTLV-I/II coinfection is significantly associated with risk for progression to AIDS among HIV+ intravenous drug abusers. Fifth International Conference on AIDS, Montreal 1989, Abstract Th.A.O.23, p75.

Weiss, S.H., Goedert, J.J., Gartner, S., et al. 1988. Risk of human immunodeficiency virus (HIV-1) infection among laboratory workers. Science 239:68–71.

Willey, R.L., Rutledge, R.A., Dias, S., et al. 1986. Identification of conserved and divergent domains within the envelope gene of the acquired immunodeficiency syndrome retrovirus. Proc. Natl. Acad. Sci. USA 83: 5038–5042.

Wong-Staal, F., Hahn, B., Manzari, V., et al. 1983. A survey of human leukaemias for sequences of a human retrovirus. Nature 302:626–628.

Wong-Staal, F. 1990. Human immunodeficiency viruses and their replication. In B.N. Fields et al. (eds.), Virology. New York: Raven Press, pp. 1529–1543.

Yamada, Y. 1983. Phenotypic and functional analysis of leukemic cells from 16 patients with adult T-cell leukemia/lymphoma. Blood 61:192–199.

Yamaguchi, K., Nishimura, H., and Takatsuki, K. 1983. Clinical features of malignant lymphoma and adult T-cell leukemia in Kumamoto. Rinsho Ketsueki 24:1271–1276.

Yamamoto, N., Okada, M., Koyanagi, Y., et al. 1982. Transformation of human leukocytes by cocultivation with an adult T cell leukemia virus producer cell line. Science 217: 737–739.

Yodoi, J., Takatsuki, K., Aoki, N., and Masuda, T. 1974. Chronic lymphocytic leukemia of T-cell origin: Demonstration in two cases by the use of anti-thymocyte membrane antiserum. Acta Haematol. Jap. 37: 289–292.

Yoshida, M., Miyoshi, I., and Hinuma, Y. 1982. Isolation and characterization of retrovirus from cell lines of human adult T-cell leukemia and its implication in the disease. Proc. Natl. Acad. Sci. USA 79:2031–2035.

Yoshida, M., Osame, M., Usuku, K., et al. 1987. Viruses detected in HTLV-I-associated myelopathy and adult T-cell leukaemia are identical on DNA blotting. Lancet 1:1085–1086.

Yoshida, M., Seiki, M., Yamaguchi, K., and Takatsuki, K. 1984. Monoclonal integration of human T-cell leukemia provirus in all primary tumors of adult T-cell leukemia suggests causative role of human T-cell leukemia virus in the disease. Proc. Natl. Acad. Sci. USA 81:2534–2537.

Yu, X.F., Ito, S., Essex, M., Lee, T.H. 1988. A naturally immunogenic virion-associated protein specific for HIV-2 and SIV. Nature 335:262–265.

Zagury, D., Bernard, J., Leibowitch, J., et al. 1984. HTLV-III in cells cultured from semen of two patients with AIDS. Science 226:449–451.

Zeigler, J.B., Cooper, D.A., Johnson, R.O., and Gold, J. 1985. Postnatal transmission of AIDS-associated retrovirus from mother to infant. Lancet 1:896–898.

Ziegler, J.L. and Abrams, D.I. 1985. The AIDS-related complex. In V.T. DeVita, et al. (eds.), AIDS-Etiology, Diagnosis, Treatment and Prevention. Philadelphia: J.B. Lippincott Co., pp. 223–233.

Chlamydia

Julius Schachter

33.1 INTRODUCTION

Although members of the genus *Chlamydia* cause a number of human diseases, in a clinical virology setting the diagnosis most commonly called for is that of *Chlamydia trachomatis* genital infections (Schachter, 1978). The lymphogranuloma venereum serovars of *C. trachomatis* cause a more extensive systemic sexually transmitted disease which is relatively uncommon in the United States. *C. psittaci*, another species, is a very common organism present among domestic mammals and virtually ubiquitous in the avian kingdom, but effects humans only as a zoonosis and this diagnosis is usually established serologically (Schachter and Dawson, 1978). Serology is also usually used to diagnose *C. pneumoniae* infection. This organism, which appears to have no animal hosts, is now being recognized as a common cause of human respiratory disease (Grayston et al, 1990). Currently, serologic procedures provide the only reliable diagnosis, as culture appears to be difficult. Knowledge of *C. pneumoniae* infections is in its infancy and it's uncertain as to how much pressure will be placed on diagnostic laboratories to test for these infections.

During their growth, chlamydia produce characteristic intracytoplasmic inclusions which were first demonstrated in the epithelium of experimentally infected subhuman primates. By 1910 similar studies using Giemsa stain of epithelial cell scrapings had demonstrated that *C. trachomatis* could infect the conjunctivae of adults and newborns, as well as cervical or urethral epithelial cells.

The trachoma agent was first isolated by T'ang and colleagues in China in 1957. The interest in sexually transmitted chlamydial infections was renewed by the reports of Jones and colleagues which documented recovery of the organism from the urethra and cervix of adults in England (Jones, 1964). The introduction of tissue culture procedures for the isolation of *Chlamydia* increased the clinical relevance of its detection, as the earlier yolk sac procedures used for isolation of the organism were very time consuming (Gordon and Quan, 1965). The original isolation procedure which involved irradiation of McCoy cells has now been supplanted by treatment of the tissue culture cells with antimetabolites. Cultural diagnosis can be made in two to seven days after processing of the specimen. Noncultural methods, such as direct fluorescent antibody techniques and enzyme immunoassays, have accelerated diagnosis of chlamydial infections (Stamm, 1988). Some of these tests can be done within one hour or less of specimen collection and all will provide a result in less than 24 hours.

The importance of *C. psittaci* as a human pathogen was made apparent in a pandemic in 1929–1930 (Schachter & Dawson, 1978). Human psittacosis was recognized as a severe respiratory disease which was life threatening until the introduction of the tetracyclines. More recently, *C. pneumoniae*, an organism which was isolated during trachoma surveys and recognized to be an atypical chlamydia when compared to trachoma strains, was found to be an important cause of human respiratory disease (Grayston et al, 1990).

33.2 CHARACTERISTICS OF THE ORGANISM

The chlamydiae are among the more common pathogens found throughout the animal kingdom (Storz, 1971; Page, 1972; Schachter, 1978; Schachter and Dawson, 1978). They are nonmotile, gram-negative, obligate intracellular bacteria. Their unique developmental cycle differentiates them from all other microorganisms (Moulder et al, 1984). They replicate within the cytoplasm of host cells, forming characteristic intracellular inclusions which can be seen using light microscopy. They differ from the viruses by possessing both RNA and DNA and cell walls that are quite similar in structure to those of gram-negative bacteria. They are susceptible to many broad-spectrum antibiotics, possess a number of enzymes, and have a restricted metabolic capacity. None of these metabolic reactions results in the production of energy.

Thus, they have been considered as energy parasites that use the ATP produced by the host cell for their own metabolic requirements (Moulder, 1966).

33.2.1 Growth Cycle

Chlamydiae are ingested by susceptible host cells (Byrne and Moulder, 1978). Chlamydia are capable of directly enhancing their ingestion by the host cells. Following attachment, at specific sites on the surface of the cell, the elementary body (EB) enters the cell in a phagosome where the entire growth cycle is completed. The chlamydiae prevent phagolysosomal fusion. Once the EB (diameter, 0.25 to 0.35 μm) has entered the cell, it reorganizes into a reticulate body (RB) which is larger (0.5 to 1 μm) and contains higher levels of RNA. After approximately eight hours, the RB begins dividing by binary fission. Approximately 18 to 24 hours after infection, these RBs become EBs by a reorganization or condensation process that is poorly understood. The EBs are then released to initiate another cycle of infection. The EBs are specifically adapted for extracellular survival and are the infectious form, whereas the intracellular metabolically active and replicating form, the RB, does not survive well outside the host cell and seems adapted for an intracellular milieu.

33.2.2 Taxonomy

Chlamydiae are presently placed in their own order, the *Chlamydiales*, family *Chlamydiaceae*, with one genus, *Chlamydia* (Moulder et al, 1984). There are three species, *C. trachomatis*, *C. psittaci* and *C. pneumoniae*. *C. trachomatis* includes the organisms causing trachoma, inclusion conjunctivitis, lymphogranuloma venereum (LGV), and other sexually transmitted infections, and some strains producing pneumoniae in rodents. *C. trachomatis* strains are sensitive to sulfonamides. They produce a glycogen-like material while replicating within the cytoplasmic inclusion vacuole, which stains with iodine. *C. psittaci* strains infect many avian species and mammals, producing the diseases psittacosis, ornithosis, feline pneumonitis, bovine abortion, etc. (Storz, 1971; Page, 1972). They are resistant to the action of sulfonamides and produce inclusions which do not stain with iodine. *C. pneumoniae* has characteristics similar to *C. psittaci* but shows little DNA relatedness to the other species. The EBs of *C. pneumoniae* appear to be pear-shaped rather than round, as are the EBs of other *Chlamydia* species.

33.2.3 Antigenic Relationships

The chlamydiae possess group(genus)-specific, species-specific, and type-specific antigens. Most of these are located within the cell wall, but precise structural relationships are not known. The major outer membrane protein (MOMP) contains species, subspecies, and serovar-specific antigens (Caldwell and Schachter, 1982; Stephens et al, 1982). The group antigen, shared by all members of the genus, appears to be a lipopolysaccharide (LPS) complex with a ketodeoxyoctanoic acid as the reactive moiety (Dhir et al, 1971). It is similar to the LPS of some gram-negative bacteria (Nurminen et al, 1983). Species-specific protein antigens have also been identified but have not been characterized (Caldwell and Kuo, 1977). Specific antigens of *C. psittaci* strains can be demonstrated by neutralization tests (Banks et al, 1970). The specific antigens of *C. trachomatis* are best recognized by a microimmunofluorescence (Micro-IF) technique (Wang and Grayston, 1970), although these antigens are also associated with a toxic factor (large numbers of viable chlamydiae may kill mice in less than 24 hours after intravenous inoculation). The MOMP molecule has been cloned and sequenced and species and serovar-specific antigens have been identified

within its variable regions (Stephens et al, 1988).

33.3 PATHOGENESIS

C. trachomatis is almost exclusively a human pathogen (Grayston and Wang, 1975; Schachter, 1978). Serotypes within this species cause trachoma (serotypes A, B, Ba, and C have been associated with endemic trachoma, the most common form of preventable blindness), inclusion conjunctivitis, and LGV (serotypes L1, L2, and L3). Where sexual transmission of *C. trachomatis* strains other than LGV has been studied, serotypes D through K have been found to be the major identifiable cause of nongonococcal urethritis in men and may also cause epididymitis. Proctitis may occur in either sex. In women, cervicitis is a common result of chlamydial infection, and acute salpingitis may occur. The agent in the cervix may be transmitted to the neonate as it passes through the infected birth canal; eye disease, inclusion conjunctivitis of the newborn, and a characteristic chlamydial pneumonia of infants may develop (Beem and Saxon, 1977). Vaginal infection and enteric infection in neonates have been recognized. The organism is essentially a parasite of squamocolumnar epithelial cells (the LGV biovars are more invasive and involve lymphoid cells). Typical of the genus, *C. trachomatis* strains are capable of causing chronic and inapparent infections. Because their growth cycle is approximately 48 hours, the incubation periods are relatively long, generally one to three weeks. *C. trachomatis* causes cell death as a result of its replicative cycle and thus is capable of producing cell damage whenever it persists. However, because there are no toxic manifestations demonstrated nor is there sufficient cell death as a result of replication, it is likely that the majority of the disease manifestations are due to immunopathologic mechanisms or non-specific host responses to the organism or its byproducts. Genus-specific proteins can be found in extracts of EBs and have been implicated as a potential sensitizing antigen capable of inducing hypersensitivity reactions in the eye and skin of previously infected hosts (Watkins et al, 1986). This protein has been identified as a heat shock-protein which shares antigenic epitopes with similar proteins of other bacteria (Morrison et al, 1989).

In the absence of therapy chlamydial infections may persist for years, although symptoms usually abate. Pathogenic mechanisms of *C. pneumoniae* have yet to be elucidated. The same may be said for *C. psittaci*, except that this agent infects cells very efficiently and disease may reflect cytopathic effects.

33.4 INFECTED SITES AND METHODS OF COLLECTION

For cytological studies, impression smears of involved tissues or scrapings of involved epithelial cell sites should be appropriately fixed (methanol may be used for immunofluorescent and Giemsa stains). It is imperative that samples be collected from the involved epithelial cell sites by vigorous swabbing or scraping. This is also true for the isolation attempts. Purulent discharges are inadequate and should be cleaned from the site prior to sampling.

For most *C. trachomatis* infections of humans, the involved mucous membranes should be vigorously swabbed or sampled by scraping. Thus, the conjunctiva for trachoma-inclusion conjunctivitis, the anterior urethra (several centimeters into the urethra) for urethritis, or the cervix at the endocervical canal for cervicitis would be tested. As these strains appear to infect only columnar cells, cervical specimens must be collected at the transitional zone or within the os. Since the organism also can infect the urethra of the female, it may

improve recovery rates if another sample is collected from the urethra and sent to the laboratory for testing in the same tube with the cervical sample. For women with salpingitis the samples may be collected by needle aspiration of the involved fallopian tube or endometrial specimens may yield the agent. Rectal mucosa, nasopharynx, and throat may also be sampled. For infants with pneumonia, swabs may be collected from the posterior nasopharynx or the throat, although nasopharyngeal or tracheobronchial aspirates collected by intubation appear to be a superior source for recovery of the agent.

33.5 STABILITY, STORAGE, AND TRANSPORT

C. trachomatis is not a particularly labile organism and by providing reasonable care in the handling of specimens a minimal loss in infectivity is achieved by simply decreasing the time between specimen collection and processing in the laboratory.

Swabs, scrapings, and small tissue samples should be collected in a special transport medium. Because *Chlamydia* are bacteria, the selection of antibiotics to prevent other bacterial contamination is restricted. Broad-spectrum antibiotics such as tetracyclines, macrolides, or penicillin must be excluded. Aminoglycosides and fungicides are most commonly added to transport media. The chlamydial specimens should be refrigerated if they can be processed within 48 hours after collection. If they can not be processed by this time, they should be frozen at $\geq -60\,°C$.

For isolation in cell culture, swabs and specimens are placed in suspending medium (2SP), consisting of 0.2 M sucrose in 0.02 M phosphate buffer, pH 7.0 to 7.2 supplemented with 5% fetal bovine serum and antibiotics. A sucrose-phosphate-glutamate (SPG) medium also has been commonly used.

Chlamydial media

A. *Growth medium for cells for chlamydial culture*

Eagle's minimum essential medium
in Eagle's salts 10×   50 mL
Fetal bovine serum   50 mL
L-glutamine, 200 mM solution   5 mL
Sterile distilled water, up to   500 mL
Adjust pH to 7.4 with 7.5% sodium bicarbonate.

B. *Medium for isolation of chlamydia in cell cultures*

Growth medium (above)
    containing added:
Vancomycin   50 μg/mL
Gentamicin   10 μg/mL
Amphotericin B   2 μg/mL
Glucose   0.594 mg/mL
Cycloheximide   1–2 μg/mL

This medium may be used as a collection medium by doubling the concentrations of vancomycin and amphotericin B.

C. *Sucrose-phosphate transport medium (2SP)*

Sucrose   68.46 g
K_2HPO_4 2.088 g
$Na_2H\,PO_4$ 1.088 g
Distilled water to   1,000 mL
Adjust pH to 7.0 and autoclave;
    add:
Bovine serum to   5%
Streptomycin   50 μg/mL
Vancomycin   100 μg/mL
Nystatin   25 U/mL

D. *Sucrose-phosphate-glutamine transport medium (SPG)*

Sucrose   75.00 g
KH_2PO_4 0.52 g
Na_2HPO_4 1.22 g
Glutamic acid   0.72 g
Distilled water to   1,000 mL
Adjust pH to 7.4 to 7.6 and autoclave, add antibiotics as above (2SP).

It may be simpler to place the clinical specimen directly into standard tissue culture growth medium containing streptomycin (200 μg/mL) or gentamicin (10 μg/mL), vancomycin (100 μg/mL), and amphotericin B (4 μg/mL).

33.6 NONCULTURE METHODS FOR DETECTING CHLAMYDIAE

There are a number of ways to detect chlamydiae directly in clinical specimens; some of the procedures have been used for many years. For example, direct microscopic examination for organisms in impression smears of infected avian tissues is still a useful procedure. Direct light microscopy on human specimens is less often used as modern techniques have replaced the Giemsa stain. Newer procedures, such as DNA hybridization or the use of polymerase chain reactions have yet to find extensive use in diagnosing chlamydial infection. Though some reagents for these tests are commercially available there is insufficient experience to recommend them at this time. Both enzyme immunoassays (EIA) and direct fluorescent antibody (DFA) procedures are commercially available. These non-culture assays should not be used where diagnosis may be required for legal purposes such as instances of sexual abuse or rape, because of limitations of specificity.

These non-culture tests do offer a number of advantages. They are less technically demanding than the cultural methods, and thus, are more widely available. However, because all non-culture tests have some false-positive reactions, they must be used with care in low prevalence settings. The DFA procedures are best suited for evaluating relatively small numbers of specimens as they can be relatively time consuming and require expensive microscopes and trained microscopists. They will, however, provide laboratories with the assurance that an adequate specimen was collected.

The same is not true for the EIAs (nor for that matter for culture). The EIAs offer an advantage of allowing for batch processing of specimens, and thus, may be better suited for screening purposes in populations with a high prevalence of chlamydial infections.

33.7 DIRECT CYTOLOGIC EXAMINATION

C. trachomatis infections of the conjunctiva, urethra, or cervix can be diagnosed by demonstrating typical intracytoplasmic inclusions, but cytological procedures, i.e., direct detection of chlamydia in the specimen, are usually less sensitive than isolation in cell culture or antigen detection procedures. A Macchiavello stain or modifications, such as the Giminez stain, may be useful in detecting EBs in infected avian species or infected animals, but not humans (Schachter and Dawson, 1978).

Ultimately, microscopy is still used in detecting chlamydiae in isolation attempts. Giemsa stain, direct fluorescent antibody, or other procedures may be suitable.

33.7.1 Direct Fluorescent Antibody Technique

Fluorescein-conjugated monoclonal antibodies are commercially available and are routinely used in some laboratories to identify chlamydial inclusions in infected cell cultures (Stamm et al, 1983). These antibodies may be used directly on clinical specimens as well (Tam et al, 1984). Species-specific antibodies to *C. trachomatis* are recommended for use directly with clinical specimens suspected of containing those organisms. Genus specific antibodies may be used for infected cell cultures but they are not recommended for use with direct detection in clinical specimens because of the irregular staining of the LPS. *C. pneumoniae*-specific monoclonal anti-

bodies appear to be better able to detect that organism in infected cell cultures as the infectivity of the organism appears to be relatively low. The DFA is somewhat less sensitive than isolation in culture but is faster and less expensive and may represent an alternate method of diagnosing chlamydial infections in settings where cell culture isolation is not available.

33.7.2 Iodine Staining Technique

Scrapings are air dried, fixed in absolute methanol, and stained for three to five minutes with Lugol iodine or 5% iodine in 10% potassium iodide. Slides are examined as wet mounts. The matrix of inclusions may appear as a reddish-brown mass recognizable under low magnification. The slides may be decolorized with methanol and restained with Giemsa stain. This technique is the least sensitive cytological procedure and is therefore not recommended for use with clinical specimens. Its speed and simplicity however have made it a popular test for examining *C. trachomatis*-infected cell cultures.

33.7.3 Giemsa Staining Technique

The smear is air dried, fixed with absolute methanol for at least five minutes, air dried and then covered with the diluted Giemsa stain (freshly prepared the same day) for one hour. The slide is then rinsed rapidly in 95% ethyl alcohol to remove excess dye and to enhance differentiation. It is then dried and examined microscopically. Longer staining periods (one to five hours) may be preferable with heavy tissue culture monolayers. EBs stain reddish-purple. The initial bodies are more basophilic, staining bluish, as do most bacteria.

33.8 ENZYME IMMUNOASSAY

Most enzyme immunoassays use polyclonal or monoclonal antibodies to detect chlamydial lipopolysaccharide (LPS) in clinical specimens. This is a soluble antigen and is more readily detected in soluble phase assays than is major outer membrane protein (MOMP). The advantages and disadvantages of the test have been discussed previously. These tests are also generally less sensitive than cell culture performed under "ideal settings" but, as with the FA procedures, there are advantages: specimens do not have to be maintained on ice and can be collected from clinic settings distant from the processing laboratory. The introduction of blocking assays for confirmatory purposes improves the specificity of the EIAs to a stage where they can be theoretically used for screening in low prevalence settings. The sensitivity is still somewhat lower than culture.

33.9 METHODS FOR ISOLATION

The recommended procedures for primary isolation of chlamydiae use culture of the organism in tissue culture cell lines. The most common technique involves inoculation of clinical specimens into cycloheximide-treated McCoy cells (Ripa and Mardh, 1977). Some workers have found human cells to be more sensitive for isolation of both *C. trachomatis* and *C. pneumoniae*. At the moment, HL cells may be the cell line of choice for isolation of *C. pneumoniae* (Cles and Stamm, 1990).

McCoy cells are plated onto 13-mm coverslips contained in 15-mm-diameter (1 dram) disposable glass vials. Cell concentration (approximately 1×10^5 to 2×10^5) is selected to give a light, confluent monolayer after 24 to 48 hours of incubation at 37 °C. The cells should, for optimal results, be used within 24 to 72 hours after reaching confluency. If the laboratory is only passing cells on a sporadic basis, they may then be held at room temperature or in a low (2%) serum medium for at least two weeks prior to inoculation.

The clinical specimens should be shaken with glass beads to disrupt the cells prior to inoculation onto cell cultures. This procedure is safer and more convenient than sonication. Standard inoculation procedure involves removing medium from the cell monolayer and replacing it with the inoculum in a volume of 0.1 to 1 mL. The specimen is then centrifuged onto the cell monolayer at approximately 3,000 × g at 35 °C for one hour. The vials are then held at 37 °C for an additional two hours before the cells are washed or the medium is changed. Tissue culture medium containing 1 to 2 μg of cycloheximide per mL (this must be titrated for each batch) is placed onto the cells two hours after centrifugation. The cells are then incubated at 37 °C for 48 to 72 hours, after which one coverslip is examined for glycogen-positive inclusions by use of iodine, Giemsa, or immunofluorescence staining. Fluorescent antibody staining may allow earlier (at 24 hours postinoculation) detection of the inclusion (Thomas et al, 1977). Use of the above mentioned fluorescein-conjugated monoclonal antibodies represents the most sensitive method for detecting *C. trachomatis* inclusions in cell culture (Stamm et al, 1983). The procedure requires more attention to staining than the iodine staining procedure and is more costly. Giemsa stain is more sensitive than iodine stain, but the microscopic evaluation is more difficult. Slide reading can be facilitated by examining the Giemsa-stained coverslip using dark-field rather than bright-field microscopy (Darougar et al, 1971). The iodine stain is the simplest procedure and the one most commonly used, although it is less sensitive than either of the other two.

If passage of a positive sample or blind passage of a negative sample is desired, serial passage of the sample should occur at 72 to 96 hours after the initial inoculation. The cell monolayer is disrupted by shaking with glass beads on a Vortex mixer. Cell debris is removed by low speed centrifugation and the supernatant fluid is inoculated as above. Approximately 90% of positive specimens that eventually are cell-culture-positive for *Chlamydia* are inclusion-positive in the first passage (i.e., initial inoculation).

For trachoma, inclusion conjunctivitis, and the genital tract infections, the technique for isolation is as described above. In LGV, the aspirated bubo pus is diluted (10^{-1} and 10^{-2}) and treated as above. Second passages are always made because detritus from the inoculum may make it difficult to read the slides.

For laboratories processing large numbers of specimens, it may be convenient to use flat-bottomed 96-well microtiter plates rather than vials for the specimens (Yoder et al, 1981). Cells are plated on coverslips or can be placed directly on the plastic. Processing and incubation are described above, but microscopy is modified to use either long working objectives or an inverted microscope. The use of microtiter plates is less sensitive than the vial technique but offers considerable savings in terms of reagents and time and may be suitable in settings where the majority of specimens to be screened are from symptomatic patients. These patients usually yield higher numbers of chlamydiae and thus minimize the impact of the decreased sensitivity of the test.

33.10 IDENTIFICATION

Since most laboratories use tissue culture isolation systems, the basic procedure for identification of chlamydiae involves demonstration of typical intracytoplasmic inclusions by appropriate (Giemsa or iodine) staining procedures. However, in laboratories initiating work with chlamydiae, it would be prudent to use at least one other parameter for identification of chlamydiae. Fluorescent-antibody staining provides both a morphological and an immunological identification.

C. trachomatis strains may be serotyped by the Micro-IF technique (Wang and Grayston, 1970). Monoclonal antibodies are available to type identify all known serovars.

33.11 SERODIAGNOSIS

The most widely-used serological test for diagnosing chlamydial infections is the genus-specific complement fixation (CF) test. It is useful in diagnosing psittacosis, in which paired sera often show fourfold or greater increases in titer. About one third of patients with other evidence of *C. pneumoniae* infection will also show rising CF titers to *C. psittaci* or *C. pneumoniae*. There are some commercially available serologic tests based on measurement of antibodies reactant to chlamydial inclusions in cell culture or EIA using chlamydial antigens. These tests are often described as "specific" by the manufacturers. However, the inclusions contain LPS and genus-specific cross reaction will occur. At this writing, none of these tests are recommended for routine use. However, these may be useful in diagnosing LGV, for which single-point titers greater than 64 are highly supportive of this clinical diagnosis. With LGV it is difficult to demonstrate rising antibody titers since the nature of the disease results in the patient being seen by the physician after the acute stage. Any titer above 16 is considered significant evidence of exposure to chlamydiae. The CF test is not particularly useful in diagnosing trachoma-inclusion conjunctivitis or the related genital tract infections, and it plays no role in diagnosing neonatal chlamydial infections.

The Micro-IF method is a much more sensitive procedure for measuring antichlamydial antibodies. It may be used in diagnosing psittacosis and *C. pneumoniae*, in which paired sera show rising immunoglobulin G (IgG) titers (and often IgM antibody). With LGV it is again difficult to demonstrate rising titers, but single-point titers in active cases usually have relatively high levels of IgM antibody (32) and IgG levels 1:2,000. Trachoma, inclusion conjunctivitis, and the genital tract infections may be diagnosed by the Micro-IF technique if appropriately timed paired acute and convalescent sera are obtained. However, it is often difficult to demonstrate rising antibody titers, particularly in sexually active populations, since many of these individuals are seen for chronic or repeat infections. The background rate of seroreactors in venereal disease clinics is 60%, making it particularly difficult to demonstrate seroconversion. In general, first attacks of chlamydial urethritis have been regularly associated with seroconversion (Bowie et al, 1977). Individuals with systemic infection (epididymitis, salpingitis) usually have much higher antibody levels than those with superficial infections, and women tend to have higher antibody levels than men.

Serology is particularly useful in diagnosing chlamydial pneumonia in neonates. In this case, high levels of IgM antibody are regularly found in association with disease (Schachter et al, 1982). IgG antibodies are less useful because the infants are being seen at a time when they have considerable levels of circulating maternal IgG since all these infections are acquired from the infected mother, who is seropositive. It takes between six and nine months for maternal antichlamydial antibodies to disappear from the infant. Infants older than that age may be tested for determination of prevalence of chlamydial infection without fear of confounding effects of maternal antibody. Infants with inclusion conjunctivitis or respiratory tract carriage of *Chlamydia* without pneumonia usually have very low levels of IgM antibodies. Thus, a single titer of 32 or greater may support the diagnosis of chlamydial pneumonia.

The Micro-IF technique uses many serotypes of chlamydiae and the procedure as simplified by Wang is recommended (Wang et al, 1975). Since serology is particularly useful in diagnosing neonatal infection, and the IgM antibody responses tend to be markedly specific, the use of single broadly reacting antigen will miss at least 15% to 25% of the infections that can be shown to be due to *Chlamydia* by other procedures, or that would be positive by a multiple antigen Micro-IF. The single-antigen tests may be employed for detection of suspensions of agent propagated in yolk sac or identification of fluorescent inclusions in tissue monolayers. Serotypes within the DEL serogroup are commonly chosen for this purpose.

Research workers should be warned that monotypic A seroreactions, at least in the United States, are liable to be spurious. Long-term longitudinal studies on infants in the laboratory of the author suggest that the appearance of antibodies against type A (and to a lesser extent the crossreacting CJI serotypes) may appear in response to non-chlamydial antigenic stimulus (Schachter et al, 1982). These antibodies are usually transient and do not result in the persistent high levels of IgG antibodies that usually follow chlamydial infections. With *C. pneumoniae* infection seroconversion may take more than four weeks, thus requiring a delay in collection of convalescent sera.

33.11.1 Complement Fixation

The CF test may be performed using either test tubes or a microtiter system. Reagents should be standardized in the tube system, regardless of which test system will be used (see CF chapter in this text). The microtiter systems are most useful in screening large numbers of sera, but it is preferable to retest all positive sera in the tube system. Occasionally, sera giving titers in the four to eight range in the microtiter system are positive at 16 (the significant level) in the tube system. The microtiter system uses standard round-bottom 96-well plates and volumes one tenth of those used in the tube test. The CF test is performed on serum specimens heated at 56°C for 30 minutes (preferably acute and convalescent paired sera tested together). In each test a positive control serum of high titer is included together with a known negative serum. The reagents for the CF test are standardized by the Kolmer technique and include special buffered saline, group antigen, antigen (normal yolk sac) control, the positive serum, the negative serum, guinea pig complement, rabbit anti-sheep hemolysin, and sheep erythrocytes. (The guinea pig complement should be carefully tested for chlamydial antibodies since many herds are enzootically infected with chlamydia, such as the guinea pig inclusion conjunctivitis agent.) The hemolytic system is titrated and the complement units (U) are determined. The standard units used in the test are 4 U of antigen and 2 exact U of complement. The test may be performed either using a water bath at 37°C for two hours or using overnight incubation at 4°C, the former being preferable. Doubling dilutions of the serum (from 1:2) are made in a 0.25-mL volume of saline. The antigen is added at 4 U (0.25 mL), and 2 exact U of complement (0.5 mL) are added. Standard reagent controls are always included. The normal yolk sac control is used at the same dilution as the group antigen. The tubes are shaken well and incubated in a water bath at 37°C for two hours. Then 0.5 mL of sensitized sheep erythrocytes are added and the tubes are placed in a water bath for one hour, after which they are read for hemolysis on a 1+ to 4+ scale (roughly equivalent to 25% to 100% inhibition of erythrocyte lysis). The endpoint of the serum is considered the highest dilution inhibiting at least 50% (2+) hemolysis after a complete inhibition of hemolysis has been observed. It is good practice to shake the tubes to resuspend the settled cells, refrigerate them overnight, and recheck the results the next morning.

All reagents are available commercially, except for the high-titered group antigen. This may be prepared as follows. Yolk sacs of seven-day embryonated eggs are inoculated with *Chlamydia* (e.g., psittacosis isolate 6BC) at a dose estimated to result in death of about 50% of inoculated eggs in five to seven days. Eggs are candled daily, and those dying early are discarded. When the 50% death endpoint is approached, the remaining eggs (recently dead or live) are refrigerated for three to 24 hours. The yolk sacs are then harvested. If examination of random samples shows large numbers of particles, the yolk sacs are pooled. This preparation may be stored at −20°C until processed further. The yolk sacs are ground in a mortar with sterile sand. Beef heart broth (pH 7.0) is added to make a 20% suspension, and the material is cultured to determine if it is free of bacterial contamination. The suspension is placed in a flask containing sterile glass beads and stored at 4°C for three to six weeks but shaken daily. It is then centrifuged at ca. 500 × g to remove coarse particles, transferred to a heavy sterile flask, and steamed at 100°C or immersed in boiling water for 30 minutes. After it has cooled, liquefied phenol is added to 0.5%. The antigen should then be refrigerated for at least one week before being used. It is stable for at least one year if not contaminated and should have an antigen titer of 256 or greater. A similar preparation from uninfected yolk sacs must be prepared for use as a negative control.

33.11.2 Microimmunofluorescence

The Micro-IF test is usually performed using chlamydial organisms grown in yolk sac. Tissue-culture-grown agent may be used, but it may be necessary to concentrate the EBs and to add some normal yolk sac to improve contrast for microscopy.

The individual yolk sacs are selected for high content of EB and pretitrated to give an even distribution of particles for the various serovars. It is generally found that a 1% to 3% (W/V) yolk sac suspension (PBS, pH 7.0) is satisfactory. The antigens may be stored as frozen aliquots; after thawing, they must be well mixed on a Vortex mixer before use. Antigen dots are placed on a slide in a specific pattern using a pen. Each cluster of dots includes all the antigenic types to be tested. The antigen dots are air dried and fixed on slides with acetone (15 minutes at room temperature). Slides may be stored frozen. When thawed for use, they may sweat, but they can be conveniently dried (as can the original antigen dots) with the cool air flow of a hair dryer. The slides have serial dilutions of serum (or tears or exudate) placed on the different antigen clusters. The clusters of dots are placed sufficiently separated to avoid serum from one cluster contaminating another cluster. After the diluted serum samples have been added, the slides are incubated for one-half to one hour in a moist chamber at 37 °C. They are then placed in a buffered saline wash for 5 minutes, followed by a second five-minute wash. The slides are then dried and stained with fluorescein-conjugated anti-human globulin. Conjugates are pretitrated in a known positive system to determine appropriate working dilutions. This reagent may be prepared against any class of globulin being considered (IgA or secretory piece for secretions, IgG or IgM). Counterstains such as bovine serum albumin conjugated with rhodamine may be included. The slides are then washed twice, dried, and examined by standard fluorescence microscopy. Use of a monocular tube is recommended to allow greater precision in determining fluorescence of individual EB particles. The endpoints are read as the dilution giving bright fluorescence clearly associated with the well-distributed EBs throughout the antigen dot. Identification of the type-specific response is based upon dilution differences reflected in the endpoints for different prototype antigens.

For each run of either CF or Micro-IF, known positive and negative sera should always be included. These sera should always duplicate their titers as previously observed within the experimental error (± 1 dilution) of the system.

REFERENCES

Banks, J., Eddie, B., Sung, M., Sugg, N., Schachter, J., and Meyer, K.F. 1970. Plaque reduction technique for demonstrating neutralizing antibodies for Chlamydia. Infect. Immun. 2(4):443–447.

Beem, M.O. and Saxon, E.M. 1977. Respiratory-tract colonization and a distinctive pneumonia syndrome in infants infected with *Chlamydia trachomatis*. N. Engl. J. Med. 296:306–310.

Bowie, W.R., Wang, S.-p., Alexander, E.R., Floyd, J., Forsyth, P., Pollock, H., Tin, J.-S., Buchanan, T., and Holmes, K.K. 1977. Etiology of nongonococcal urethritis: Evidence for *Chlamydia trachomatis* and *Ureaplasma urealyticum*. J. Clin. Invest. 59:735–742.

Byrne, G.I. and Moulder, J.W. 1978. Parasite-specified phagocytosis of *Chlamydia psittaci* and *Chlamydia trachomatis* by L and HeLa cells. Infect. Immun. 19:598–606.

Caldwell, H.D. and Kuo, C.-C. 1977. Serologic diagnosis of lymphogranuloma venereum by counter immunoelectrophoresis with a *Chlamydia trachomatis* protein antigen. J. Immunol. 118:442–445.

Caldwell, H.D. and Schachter, J. 1982. Antigenic analysis of the major outer membrane protein of *Chlamydia*. Infect. Immun. 35(3): 1024–1031.

Cles, L.D. and Stamm, W.E. 1990. Use of HL cells for improved isolation and passage of

Chlamydia pneumoniae. J. Clin. Microbiol. 28(5):938–940.

Darougar, S., Kinnison, J.R., and Jones, B.R. 1971. Simplified irradiated McCoy cell culture for isolation of chlamydiae. In R.L. Nichols (ed.), Trachoma and Related Disorders Caused by Chlamydial Agents. Amsterdam: Excerpta Medica, 63–70.

Dhir, S.P., Kenny, G.E., and Grayston, J.T. 1971. Characterization of the group antigen of *Chlamydia trachomatis.* Infect. Immun. 4:725–730.

Gordon, F.B. and Quan, A.L. 1965. Isolation of the trachoma agent in cell culture. Proc. Soc. Exp. Biol. Med. 118:354.

Grayston, J.T., Campbell, L.A., Kuo, C.-C., Modhorst, C.H., Saikku, P., Thom, D.H., and Wang, S.-p. 1990. A new respiratory tract pathogen: *Chlamydia pneumoniae* strain TWAR. J. Infec. Dis. 161(4):618–625.

Grayston, J.T. and Wang, S.-p. 1975. New knowledge of chlamydiae and the diseases they cause. J. Infect. Dis. 132:87–105.

Jones, B.R. 1964. Ocular syndromes of TRIC virus infection and their possible genital significance. Br. J. Vener. Dis. 40(1):3–18.

Morrison, R.P., Belland, R.J., Lyng, K., and Caldwell, H.D. 1989. Chlamydial disease pathogenesis. The 57-kD chlamydial hypersensitivity antigen is a stress response protein. J. Exp. Med. 170(4):1271–1284.

Moulder, J.W. 1966. The relation of the psittacosis group (chlamydiae) to bacteria and viruses. Ann. Rev. Microbiol. 20:107–130.

Moulder, J.W., Hatch, T.P., Kuo, C.-C., Schachter, J., and Storz, J. 1984. Order II. Chlamydiales Storz and Page 1971, 334. In N.R. Krieg and J.G. Holt (eds.), *Bergey's Manual of Systematic Bacteriology.* Baltimore, Maryland: Williams and Wilkins, 729–739.

Nurminen, M., Leinonen, M., Saikku, P., and Makela, P.H. 1983. The genus-specific antigen of *Chlamydia*: Resemblance to the lipopolysaccharide of enteric bacteria. Science 220(4603):1279–1281.

Page, L.A. 1972. Chlamydiosis (ornithosis). In M.S. Hofstad (ed.), *Diseases of Poultry.*

Ames, Iowa: Iowa State University Press, 414–417.

Ripa, K.T. and Mardh, P.-A. 1977. New simplified culture technique for *Chlamydia trachomatis.* In K.K. Holmes and D. Hobson (eds.), *Nongonococcal Urethritis and Related Infections.* Washington, D.C.: American Society for Microbiology, 323–327.

Schachter, J. 1978. Chlamydial infections. N. Engl. J. Med. 298:428–435, 490–495, 540–549.

Schachter, J. and Dawson C.R. 1978. *Human Chlamydial Infections.* Littleton: Publishing Sciences Group.

Schachter, J., Grossman, M., and Azimi, P.H. 1982. Serology of *Chlamydia trachomatis* in infants. J. Infect. Dis. 146:530–535.

Stamm, W.E. 1988. Diagnosis of *Chlamydia trachomatis* genitourinary infections. Ann. Intern. Med. 108(5):710.

Stamm, W.E., Tam, M., Koester, M., and Cles, L. 1983. Detection of *Chlamydia trachomatis* inclusions in McCoy cell cultures with fluorescein-conjugated monoclonal antibodies. J. Clin. Microbiol. 17(4):666–668.

Stephens, R.S., Tam, M.R., Kuo, C.-C., and Nowinski, R.C. 1982. Monoclonal antibodies to *Chlamydia trachomatis*: Antibody specificities and antigen characterization. J. Immunol. 128:1083–1089.

Stephens, R.S., Wagar, E.A., and Schoolnik, G.K. 1988. High-resolution mapping of serovar-specific and common antigenic determinants of the major outer membrane protein of *Chlamydia trachomatis.* J. Exp. Med. 167(3):817.

Storz, J. 1971. *Chlamydia and Chlamydia-induced Diseases.* Springfield: Charles C. Thomas.

Tam, M.R., Stamm, W.E., Handsfield, H.H., Stephens, R., Kuo, C.-C., Holmes, K.K., Ditzenberger, K., Crieger, M., and Nowinski, R.C. 1984. Culture-independent diagnosis of *Chlamydia trachomatis* using monoclonal antibodies. N. Engl. J. Med. 310(18):1146–1150.

Thomas, B.J., Evans, R.T., Hutchinson, G.R., and Taylor-Robinson, D. 1977. Early detection of chlamydial inclusions combining the use of cycloheximide-treated McCoy cells and immunofluorescence staining. J. Clin. Microbiol. 6(3):285–292.

Wang, S.-p. and Grayston, J.T. 1970. Immunologic relationship between genital TRIC, lymphogranuloma venereum and related organisms in a new microtiter indirect immunofluorescence test. Am. J. Ophthalmol. 70:367–374.

Wang, S.-p., Grayston, J.T., Alexander, E.R., and Holmes, K.K. 1975. Simplified micro-immunofluorescence test with trachoma-lymphogranuloma venereum (*Chlamydia trachomatis*) antigens for use as a screening test for antibody. J. Clin. Microbiol. 1:250–255.

Watkins, N.G., Hadlow, W.J., Moos, A.B., Caldwell, H.D. 1986. Ocular delayed hypersensitivity: A pathogenetic mechanism of chlamydial conjunctivitis in guinea pigs. Proc. Nat. Acad. Sci. USA 83(19):7480–7484.

Yoder, B.L., Stamm, W.E., Koester, M.C., and Alexander, E.R. 1981. Microtest procedure for isolation of *Chlamydia trachomatis*. J. Clin. Microbiol. 13(6):1036–1039.

Section 3

Reference Laboratories

34

Virology Services Offered by the Federal Reference Laboratory Centers for Disease Control

Kenneth L. Herrmann

34.1 INTRODUCTION

A strong collaborative effort among local, state, and federal laboratories provides the foundation for a successful nationwide program for the surveillance, prevention, and control of infectious diseases. Laboratories at each level have distinct responsibilities. As the state public health laboratories provide reference and disease surveillance at the state level, so too does the Centers for Disease Control (CDC) provide reference and disease surveillance at the national level. The effectiveness and timeliness with which each laboratory fulfills its responsibilities greatly influence the success of the nationwide program.

Providing reference and disease surveillance (RDS) in microbiology, hematology, histopathology, and immunology at CDC is the responsibility of the National Center for Infectious Diseases (NCID). These program activities, performed in collaboration with state and other qualified laboratories, constitute an important segment of NCID's mission. *All* RDS speci-

mens, with proper justification and a completed request form, must be submitted to NCID by or through the state public health laboratory and with the knowledge and consent of the state laboratory director or his/her designee. Within the limitations outlined in the following sections, NCID will provide RDS on the following:

- Cultures, serum or cerebrospinal fluid (CSF) samples, transudates, exudates, tissues, or histologic specimens from patients suspected of having an unusual infectious disease and/or other kinds of specimens (e.g., vectors, foods, liquids) that aid in the diagnosis of life-threatening, unusual, or exotic infectious diseases.

- Cultures or serum specimens obtained from patients who have infectious diseases that occur only sporadically or from patients who are involved in outbreaks of diseases caused by organisms for which satisfactory diagnostic reagents are not commercially or widely available.

- Organisms suspected of being unusual pathogens or that are associated with hospital-acquired infections.

- Specimens forwarded to NCID for con-

Material in this chapter is reprinted with modifications from Reference and Disease Surveillance (February 1986) prepared by the Centers for Disease Control, Atlanta, Georgia, with permission.

firmation of quality assurance for test performance.

- Serum specimens or cultures that are clinically important and are sent to NCID for confirmation because the results in state laboratories were atypical, aberrant, or difficult to interpret, or because difficulties were encountered with the reagents used.

- Arthropod and vertebrate specimens necessary for confirmation of zoonotic diseases.

In assigning reference priorities, NCID places strong emphasis on the quality of both the specimen and the accompanying information. Prior consultation on urgent or unusual specimens will enable NCID to be more responsive and efficient. Protecting our nation's health through an effective reference and disease surveillance program requires teamwork, cooperation, and good communication.

34.2 NCID ORGANIZATION

Listed below are the different kinds of laboratories of the National Center for Infectious Diseases. The virology laboratories are located in the NCID Divisions or Programs as shown below. The staff members to contact for consultation are listed by organization in Table 34.1.

Bacteriology Laboratories
  Division of Bacterial and Mycotic Diseases (DBMD)
  Hospital Infections Program (HIP)
  Division of Sexually Transmitted Diseases Laboratory Research (DSTDLR)
  Plague Lab (see Division of Vector-Borne Infectious Diseases (DVBID))
  Lyme Disease Lab (see DVBID)
  Arctic Investigations Program (AIP)

Mycology Laboratories
  Division of Bacterial and Mycotic Diseases (DBMD)

Parasitology Laboratories
  Division of Parasitic Diseases (DPD)

Rickettsial Laboratories
  Division of Viral and Rickettsial Diseases (DVRD)

Virology Laboratories
  Division of Vector-Borne Infectious Diseases (DVBID) (includes arbovirus laboratories)
  Division of Viral and Rickettsial Diseases (DVRD)
  Division of HIV/AIDS
  Arctic Investigations Program (AIP)

34.3 REFERENCE AND DISEASE SURVEILLANCE

Table 34.2 lists the virology RDS activities of NCID. This table is organized by disease category and alphabetized for easy reference. Use Table 34.2 and the footnoted instructions to obtain the shortest possible response time from NCID laboratories.

Reference and disease surveillance activities are not undertaken for diseases for which satisfactory reagents or tests are commercially available or for which there is no public health need. The Associate Director for Laboratory Science, NCID, (Dr. Joseph McDade, phone: 404-639-3967) is available to consult with state health department laboratory directors about firms producing reagents or offering interstate diagnostic services.

34.4 REQUIREMENTS FOR ALL SPECIMENS

34.4.1 General Information

The information provided below is intended as a general guideline for those who wish to send specimens to CDC for evaluation. Although this chapter is comprehensive, special circumstances may arise which are not covered, and the user is encouraged to

Table 34.1 Laboratories of the NCID Handling Virus Specimens

NCID Division/Program		Location	Commercial Phone Number
VECTOR-BORNE INFECTIOUS DISEASES			
Director	Duane J. Gubler, Sc.D.	Ft. Collins, CO	303 221-6428
Assistant Director	Dennis W. Trent, Ph.D.	Ft. Collins, CO	303 221-6420
ARBOVIRUS DISEASES BRANCH			
Chief	Theodore F. Tsai, M.D.	Ft. Collins, CO	303 221-6407
DENGUE BRANCH			
Chief	Gary G. Clark	San Juan, PR	809 749-4400
VIRAL AND RICKETTSIAL DISEASES			
Director	Brian W.J. Mahy, Ph.D., Sc.D.	Atlanta, GA	404 639-3574
Deputy Director	Kenneth L. Herrmann, M.D.	Atlanta, GA	404 639-2293
EPIDEMIOLOGY ACTIVITY			
Chief	Lawrence B. Schonberger, M.D.	Atlanta, GA	404 639-3091
HEPATITIS BRANCH			
Chief	Harold S. Margolis, M.D.	Atlanta, GA	404 639-2340
INFLUENZA BRANCH			
Chief	Nancy J. Cox, Ph.D.	Atlanta, GA	404 639-3591
RESPIRATORY AND ENTERIC VIRUSES BRANCH			
Chief	Larry J. Anderson, M.D.	Atlanta, GA	404 639-3596
RETROVIRUS DISEASES BRANCH			
Chief	Thomas M. Folks, Ph.D.	Atlanta, GA	404 639-1024
SPECIAL PATHOGENS BRANCH			
Chief	Clarence J. Peters, M.D.	Atlanta, GA	404 639-1115
VIRAL EXANTHEMS AND HERPESVIRUS BRANCH			
Chief	William C. Reeves, M.D.	Atlanta, GA	404 639-3532
VIRAL AND RICKETTSIAL ZOONOSES BRANCH			
Chief	James G. Olson, Ph.D.	Atlanta, GA	404 639-1075
HIV/AIDS			
Acting Director	Harold W. Jaffe, M.D.	Atlanta, GA	404 639-2003
LABORATORY INVESTIGATIONS BRANCH			
Chief	Gerald Schochetman, Ph.D.	Atlanta, GA	404 639-1000
ARCTIC INVESTIGATIONS PROGRAM			
Director	Robert B. Wainwright, M.D.	Anchorage, AK	907 271-4011

consult with appropriate staff within NCID in such instances.

1. Etiologic agents should be cultivated and shipped in a medium that will protect and ensure the viability of the microorganism during transit.

2. Only pure cultures of etiologic agents should be sent; mixed cultures cannot be accepted without written justification.

3. Optimum containers for shipping different groups of etiologic agents vary de-

(Continued on page 650)

Table 34.2 Viral Diseases for which Testing is Available at CDC[a]

Disease or Agent[b]	Organizational Unit	Serologic Tests Available	Isolation Specimens	Antigen Detection	
				Tests Available	Specimens Requested
Arboviruses					
Calif. encephalitis (LaCrosse)	DVBID	EIA, HI, CF, NT	Brain, mosquitoes	IIF, EIA	organs, mosquitoes
Colorado tick fever	DVBID	EIA, CF, NT	Blood (unfrozen),[c] ticks	IIF, EIA	blood, ticks
Dengue 1–4	DVBID	EIA, HI, CF, NT	Serum, organs, mosquitoes	IIF, EIA	organs, mosquitoes
Eastern equine encephalitis	DVBID	EIA, HI, CF, NT	Brain, CSF, serum, mosquitoes	IIF, EIA	organs, mosquitoes
Japanese encephalitis	DVBID	EIA, HI, CF, NT	Brain, CSF, serum, mosquitoes	IIF, EIA	organs, mosquitoes
Murray Valley encephalitis	DVBID	EIA, HI, CF, NT	Brain, CSF, serum, mosquitoes	IIF, EIA	organs, mosquitoes
Powassan	DVBID	EIA, HI, CF, NT	Brain, CSF, serum, mosquitoes	IIF, EIA	organs, ticks
St. Louis encephalitis	DVBID	EIA, HI, CF, NT	Brain, CSF, serum, mosquitoes	IIF, EIA	organs, mosquitoes, body fluids
Venez. equine encephalitis	DVBID	EIA, HI, CF, NT	Brain, CSF, throat, serum, mosquitoes	IIF, EIA	organs, mosquitoes
Western equine encephalitis	DVBID	EIA, HI, CF, NT	Brain, CSF, serum, mosquitoes	IIF, EIA	organs, mosquitoes
Yellow fever	DVBID	EIA, HI, CF, NT	Serum, organs, mosquitoes	IIF, EIA, Dot Blot	blood, organs, mosquitoes
Other arboviruses	DVBID	EIA, HI, CF, NT	Serum, arthropods	IIF, EIA	organs, arthropods
Crimean hemorrhagic fever (Congo)[d]	DVRD	IIF	Serum, liver, spleen	IIF	
Kyasanur Forest disease[d]	DVRD	EIA, HI, IIF, CF, NT	Serum, organs, ticks	IIF	organs, ticks
Omsk hemorrhagic fever[d]	DVRD	EIA, HI, IIF, CF, NT	Serum, ticks	IIF	organs, ticks
Tick-borne encephalitis[d]	DVRD	EIA, HI, CF, NT	Brain, CSF, serum, ticks	IIF	organs, ticks
Hantaan (Korean hemorrhagic fever)[d]	DVRD	IIF, HI, EIA	Whole blood		

Arenaviruses				
Junin[d]	DVRD	IIF	Serum, liver, spleen	IIF
Lassa[d]	DVRD	IIF, EIA	Serum, liver, spleen	IIF
Lymphocytic choriomeningitis	DVRD	IIF	Serum, CSF	
Machupo[d]	DVRD	IIF	Serum, liver, spleen	IIF
Marburg/Ebola[d] viruses	DVRD	IIF, WIB	Serum, liver, spleen	
Enteroviruses	DVRD			
Coxsackieviruses[e]	DVRD	NT[f]		
Echoviruses[e]	DVRD	NT[f]		
Enterovirus 70[e]	DVRD	NT[f]		
Polioviruses 1–3[e]	DVRD			
Respiratory viruses				
Adenoviruses[e]	DVRD	HI, CF, NT[f]		EIA
Coronaviruses[e]	DVRD			EIA
Influenza[g]	DVRD	HI, NT		
Mumps[h]	DVRD	HI,		EIA
Parainfluenza[e]	DVRD	HI, NT[f]		EIA
Parvovirus, human[e]	DVRD	EIA		
Resp syncytial virus[e]	DVRD	EIA		
Picornaviruses[e]	DVRD	NT[f]		
Herpesviruses				
Cytomegalovirus[e]	DVRD	EIA	Urine (unfrozen), throat	EIA
Herpes simplex[e]	DVRD	EIA	Vesicular fluid, brain	IIF, EIA
Infectious mononucleosis (Epstein-Barr)[e]	DVRD	IIF, OCH		
Varicella-Zoster	DVRD	IIF[i]	Vesicular fluid, scabs[j]	
Exanthem viruses				
Measles (rubeola)	DVRD	EIA[i]		
Rubella	DVRD	EIA[i]		
Orf-paravaccinia	DVRD	IIF		
Vaccinia[j]	DVRD	HI, IIF, EIA	Vesicular fluid scabs, brain, saliva[h]	

Table 34.2 (*continued*)

Disease or Agent[b]	Organizational Unit	Serologic Tests Available	Isolation Specimens	Antigen Detection	
				Tests Available	Specimens Requested
Miscellaneous viruses					
Rabies	DVRD	RFFIT,[j] DFA	Brain or skin biopsy		
Retroviruses					
HTLV-I/II[k]	DVRD	EIA, WIB, RIP	Blood		
HIV-1/2[l]	DHA	EIA, WIB, RIP	Blood		
Hepatitis A					
Hepatitis A (anti-HAV)	DVRD	EIA[h]			
IgM-anti HAV	DVRD	EIA[h]			
Hepatitis B					
HBsAg	DVRD	EIA[h]			
Anti HBs	DVRD	EIA[h]			
Anti HBc	DVRD	EIA[h]			
IgM anti HBc	DVRD	EIA[h]			
HBeAg	DVRD	RIA[h]			
Anti HBe	DVRD	RIA[h]			
Subtyping HBsAg	DVRD	EIA[h]		Dot-Blot[h]	Liver, biopsy
HBV-DNA Hybridization					
Enterically transmitted non-A non-B hepatitis	DVRD	IIF[m]			
Delta Hepatitis					
Anti Delta	DVRD	EIA[h]			
IgM anti Delta	DVRD	EIA[h]			
Delta antigen	DVRD	EIA[h] IIF		EIA[h]	Blood, liver, biopsy
Viral gastroenteritis					
Rotavirus, human[d]	DVRD	EIA		EIA	
Norwalk virus[d]	DVRD	EIA[h]		EIA	
Adenovirus[d]	DVRD	HI, NT		EIA	

Abbreviations: CDC, Centers for Disease Control; CF, complement fixation; CSF, cerebrospinal fluid; DFA, direct fluorescent antibody test; DHA, Division of HIV/AIDS, Atlanta, GA; DVBID, Division of Vector-Borne Infectious Diseases, Ft. Collins, CO; DVRD, Division of Viral and Rickettsial Diseases, Atlanta, GA; EIA, enzyme immunoassay; HI, hemagglutination inhibition; IIF, indirect immune fluorescence; NT, neutralization; OCH, ox cell hemolysin; PPT, precipitation test; RFFIT, rapid fluorescent focus inhibition test; RIP, radioimmunoprecipitation; RIA, radioimmunoassay; WIB, Western immunoblot.

[a] The absence of a notation in any column indicates that the service is not offered for a given organism.

[b] Other tests or tests for agents not listed may be made by prior consultation and arrangement. All serum specimens for serologic tests should be sterile and shipped frozen or on wet ice. Serum specimens may be sent without refrigeration if delivery is assured within 48 hours. In most instances paired (acute and convalescent) serum specimens are required for serologic diagnosis. Specimens for viral isolation should be shipped frozen on dry ice if they will not arrive within 48 hours of collection; if delivery to CDC will take less than 48 hours, specimens for viral isolation should be shipped chilled by coldpacks or wet ice.

[c] Send specimens on wet ice or cold packs; *do not freeze.*

[d] These viruses are among the most highly pathogenic viruses known. Infections (and deaths) may occur in medical and supportive personnel through close contact with patients and their exudates. Contact Special Pathogens Branch, Division of Viral and Rickettsial Diseases, NCID, CDC, for special instructions before taking, packing, and shipping specimens suspected of containing these agents. DO NOT SUBMIT SPECIMENS FOR TESTING WITHOUT MAKING PRIOR ARRANGEMENTS.

[e] Because of the ubiquitous nature of these viruses, primary diagnosis by serologic testing or virus isolation cannot routinely be offered. These services may be provided by prior consultation and arrangement for outbreaks or cases of unusual public health significance.

[f] Neutralization tests are performed by prior arrangement only when virus is isolated from patient.

[g] Because of the epidemic nature of the virus, primary diagnosis by serologic testing or virus isolation cannot routinely be offered. These services may be provided by individual agreement for outbreaks or cases of unusual public health significance. Reference antigenic analysis of representative or unusual isolates ("strain comparison") is organized and conducted yearly by the WHO Collaborating Center for Influenza at CDC. Isolates should be sent directly to that Center (Bldg. 7, Room 112). The completed WHO data form, *not* CDC form 50.34, is required. The WHO data form may be obtained from the Influenza Branch, DVRD, NCID, CDC.

[h] Prior consultation and arrangements with the appropriate Branch are required before submitting specimens; indicate on the request form that such arrangements were made.

[i] Immune status testing will be performed only by prior consultation and arrangements; indicate on the request form that such arrangements were made.

[j] For information on collecting, packing, and shipping specimens from a person with suspected smallpox, contact the Viral Exanthems and Herpesvirus Branch, DVRD, NCID, CDC.

[k] Laboratory support and collaboration for HTLV-I and HTLV-II, other non-AIDS retroviruses, and diseases associated with these viruses (adult T-cell leukemia/lymphoma, tropical spastic paraparesis, or HTLV-I-associated myelopathy) are available through prior arrangements with the Retrovirus Diseases Branch, DVRD, NCID, CDC.

[l] Laboratory support and collaboration for HIV-1 and HIV-2, and for AIDS, are available through prior arrangements with the Laboratory Investigations Branch, Division of HIV/AIDS, NCID, CDC.

[m] Used at CDC for research purposes; selected specimens from outbreaks may be tested after prior arrangements have been made with the Hepatitis Branch, DVRD, NCID, CDC; tests not routinely performed.

pending on the agent and the distance involved in shipment. In all instances, however, the primary container should be of a durable material that, when properly packaged, is leakproof and can withstand the temperature and pressure variations likely to occur in the air and on the ground during shipment to CDC.

4. Serum specimens for serologic testing should be aseptically separated from whole blood. Contaminated serum specimens are unsuited for almost all purposes. Paired serum specimens are preferred and, in many cases, required. (See specific requirements for various etiologic agents). Generally, the first specimen should be obtained as soon after the onset of illness as possible and refrigerated. The second specimen should be collected two to four weeks later. The optimal interval for collecting the second serum specimen will vary with different infectious agents.

5. When whole blood is sent for isolation of particular viral agents, the blood should be kept cold but not frozen prior to shipment and shipped on wet ice (*not* dry ice); however, whole blood submitted for isolation of suspected rickettsial agents *must* be packed in dry ice and shipped frozen.

6. Slides with tissue sections, blood films, or smears of clinical material should be dry, free of immersion oil, properly labeled, and carefully packed in a slide container. The slide container must be wrapped in absorbent material and placed in a sealed, leakproof container.

34.4.2 Instructions for Identifying Specimens

1. Identify individual specimen tubes or vials by encircling them with adhesive tape which contains typed or printed

patient identification number and other identifying information. *Do not* use ballpoint pens, wax, or other writing instruments that tend to smear.

2. Print patient's identification number, type of specimen, and date the specimen was collected on the specimen label.

34.4.3 Instructions for Packing Specimens

1. Package specimens properly in accordance with the provisions of subparagraph 72.3(a) of Title 42, Code of Federal Regulations (Federal Register, August 20, 1980) to protect both the material while in transit and the personnel who handle the packages.

   - **Never** mail clinical specimens or cultures in glass or plastic Petri plates or similar containers!

   - **Never** enclose dry ice in hermetically sealed containers!

   - Enclose patient history and specimen information using CDC form 50.34 which can be obtained directly from your State health department.

2. Avoid delay at CDC by addressing shipments as follows:

   Data and Specimen Handling Section
   Bldg. 4, Rm. B-35, Mail Stop G12
   National Center for Infectious Diseases
   Centers for Disease Control
   Atlanta, GA 30333

3. Whenever possible, time the shipment to arrive at CDC at the beginning or middle of the work week, *not* just before or on the weekend or a holiday.

34.5   SPECIAL REQUIREMENTS FOR VIRAL DISEASES

Considerable research and development in the rapid diagnosis of viral diseases, together with the recognition of many new viral agents responsible for infectious disease, prompt NCID to encourage state laboratories to request consultation and reference testing for those diseases and agents listed in Table 34.2, as well as for new agents that may not be included.

The Division of Viral and Rickettsial Diseases (DVRD) and the Division of HIV/AIDS (DHA) will provide confirmation of serologic and virologic diagnosis when deemed essential by the state laboratories (after that laboratory has consulted with the DVRD specialty laboratory or the DHA). Highest priority will be given to the diagnosis of a life-threatening condition or to an outbreak of illness of public health importance. Infection with a highly pathogenic virus, such as Lassa or Marburg, is a major health hazard not only for the individual patient, but also for the medical and support personnel involved. In such cases, prompt contact with the Special Pathogens Branch (SPB), DVRD, is important. Laboratory support is also available to evaluate problem specimens and to study the complications of viral vaccines.

The viral disease programs of the Division of Vector-Borne Infectious Diseases (DVBID) are concerned with surveillance of arthropod-borne viral diseases in the United States and offer clinical and public health laboratory support in a) primary or confirmatory diagnosis of arboviral infections; b) identification of arbovirus isolates; and c) serologic and virologic surveillance of arthropods and vertebrates.

Clinical specimens from persons or animals with suspected arboviral diseases *must* be accompanied by the following items of clinical and epidemiologic history: dates and places of travel; yellow fever, Japanese encephalitis, or other relevant immunization (WEE, EEE, VEE virus vaccine in horses and avians); and dates of onset and specimen collections.

34.6 ADDITIONAL SUGGESTIONS FOR VIRAL SPECIMENS

34.6.1 General

When shipping specimens in tubes or vials, use only tight fitting, soft rubber stoppers or leak-proof screw caps; seal well with waterproof tape. Avoid direct contact between specimen container and dry ice to prevent breakage. To prevent thawing of specimens in the event of transit delays, use enough dry ice to last 48 hours beyond expected arrival time.

34.6.2 Specimens for Isolation of Etiologic Agents

Select specimens during the acute, febrile phase of illness. Depending on the disease suspected, the specimen submitted may be nasal or throat washings, sputum, urine, feces, CSF, skin scrapings, aspirate from lesions, blood, serum, or various tissues. Throat, nasal, or rectal swabs must be immersed in an appropriate virus transport fluid to prevent drying. Blood, CSF, and tissues should be handled aseptically. Autopsy tissues from several organs should be placed in separate containers, not pooled. Most specimens deliverable to the laboratory within 24 hours of collection may be sent chilled by using frozen ice packs to maintain the initial condition. If more than 24 hours are required, the specimens should be frozen and kept frozen during shipment. (Specimens suspected of containing cytomegalovirus [CMV] should not be frozen but sent chilled.) Consult with the appropriate laboratory if you are uncertain about the optimum method for shipping a given specimen. When shipment of frozen specimens is not possible, some types of specimens can be preserved in buffered glycerin (50% glycerin/50% PBS pH 7.2), but partial loss of virus generally occurs (complete rickettsial and chlamydial loss will occur). Consultation is required before buffered glycerin is used for transport of specific virus specimens.

34.6.3 Specimens for Serologic Testing

Generally, serum may be shipped unfrozen if drawn aseptically and delivery within 48 hours is assured. *Never* add a preservative to serum that will be used in serologic tests.

34.6.4 Arthropod Specimens (Virus Isolation)

Arthropod specimens must be collected alive, killed by freezing or exposure to ether vapor, sealed in ampules, and shipped on dry ice. Cyanide or chloroform must not be used because these chemicals inactivate viruses and rickettsia.

34.7 REQUIREMENTS FOR SPECIMENS FOR DVRD AND DVBID

34.7.1 Sera and Other Body Fluids

Paired sera (acute and convalescent phase) are preferred for serologic diagnosis, but in some cases a presumptive diagnosis is possible if only a single convalescent serum is available. Consult with the respective laboratory if you have questions in this regard.

The DVBID requests CSF specimens from patients with encephalitis. Newly-introduced antibody capture tests of antibody in CSF offer a rapid and sensitive means of specific diagnosis. Detection of viral antigen in other body fluids is under investigation and other fluids may be accepted for testing if approved. Prearrangements for the latter tests should be made with the DVBID.

In cases of suspected viral hemorrhagic fever, the Special Pathogens Branch, DVRD and the DVBID require acute phase sera both for immediate serologic testing, which may be diagnostic, and for virus isolation. The handling and transport of such specimens must be discussed in ad-

vance with either the SPB, DVRD, or the DVBID, as appropriate.

34.7.2 Tissues

Biopsy or autopsy specimens from patients with suspected arboviral disease or viral hemorrhagic fever may be submitted for viral isolation and/or examination by immunofluorescence for viral antigen. Tissues submitted for immunofluorescence testing should be embedded in appropriate medium for frozen sections, or sent cold for processing by DVBID or DVRD. Formalin-fixed tissues may be suitable for immunohistochemical examination in some instances. Prearrangements for these procedures are required.

34.7.3 Isolates

Viral isolates referred to the DVBID or DVRD for identification and/or characterization should be accompanied by as complete an isolation history as possible. Source of the isolate, date of isolation, host system used for isolation, success of reisolation and passage level, and titer of the material being sent should be given. If available, information concerning sensitivity to lipid solvents, spectrum of sensitive host systems, presence or absence of hemagglutinin activity, and antigenic relationship to other established viruses also should be provided. Prearrangements for these studies must be made with the appropriate laboratory.

34.7.4 Arthropod and Avian Specimens

In the event of a known or presumed arthropod-borne viral encephalitis outbreak, arthropods may be submitted for virus isolation. Contact with appropriate DVBID personnel is required to arrange for testing prior to submission of specimens. Specimens should be carefully put into lipless glass vials, tightly stoppered, and shipped on dry ice. The date(s), method(s), and location(s) of collection should accompany the specimens. Sera from wild birds may also be submitted for antibody tests under similar conditions. Sera should be collected from abundant passerine bird species from the potential outbreak area. Birds may be bled from the jugular vein (0.2 mL) and the blood specimen added to 0.9 mL diluent (buffered saline, pH 7.4 to 7.8), preferably with a protein stabilizer such as fetal bovine serum (5% to 10%) or bovine serum albumin (0.5%). Specimens should be centrifuged at low speed to remove red cells; the supernatant fluid should then be removed, frozen, and shipped on dry ice. As is required for arthropods, prearrangements for testing of avian sera must be secured.

34.8 REFERENCE AND DISEASE SURVEILLANCE REQUEST FORM

34.8.1 Instructions for Completing the Request Form for Reference and Disease Surveillance Support (CDC 50.34 REV. 11-90)

The CDC form is a combined request, specimen information, patient history, and results report form. The upper third of the front side of the form must contain the required information for identification of the specimen that is used to start a computer record on each specimen, used to track a specimen while being tested in NCID laboratories, and used to direct the reports to the proper addressee. *All* of the information requested on this *front third* must be provided. The back of the form is to provide NCID laboratories with essential information about the specimen, the patient, and the assistance requested. The required information for this portion of the form depends on the type of specimen submitted and the assistance requested. The results of tests performed at CDC can be affixed to the lower portion of the front

(once the information has been put into the computer), and this portion will constitute the report that is returned to the state health department laboratory or other authorized sender. Detailed instructions for completing the form are as follows.

34.8.1a Front of Form (Figure 34.1)

Justification: The justification section must be completed and signed by a state health department laboratory representative before a specimen can be accepted by CDC.

Name, address, and phone number of requesting physician or organization: Print or type the name, address, and telephone number of the physician, microbiologist, or organization from which the specimen originated. (Include the person or institution to contact if additional information is needed and to whom the final report will be forwarded by the state laboratory.)

State health department address: Print or stamp the address of the state health department laboratory or agency sending specimen. (This will ensure that reports are directed through the correct state health department.)

State health department number: Print the state health department laboratory number assigned to the specimen, if any (used as a cross-reference for specimen identification).

Date sent to CDC: Print date (month, day, and year) that specimen is shipped to CDC.

Hospital number: For those specimens originating from a hospital, print patient's hospital number. (This is not required by CDC, but it is helpful information for the hospital record office in matching the report to patient's record.)

*Justification must be completed by State health department laboratory before specimen can be accepted by CDC. Please check the first applicable statement and when appropriate complete the statement with the *.*

1. Disease suspected to be of public health importance. Specimen is:
   - (a) ☐ from an outbreak.
   - (b) ☐ from uncommon or exotic disease.
   - (c) ☐ an isolate that cannot be identified, is atypical, shows multiple antibiotic resistance, or from a normally sterile site(s).
   - (d) ☐ from a disease for which reliable diagnostic reagents or expertise are unavailable in State.
2. ☐ Ongoing collaborative CDC/State project.
3. ☐ Confirmation of results requested for quality assurance.

*Prior arrangement for testing has been made. Please bring to the attention of:

(Name) ______

Completed by: ______

Date: ______

Name, Address and Phone Number of Physician or Organization:

(For CDC Use Only) **CDC NUMBER**

UNIT | FY 3-4) | NUMBER (5-10) | SUF (11)

DATE RECEIVED (12-17) Month Day Year

STATE HEALTH DEPARTMENT LABORATORY ADDRESS:

STATE HEALTH DEPT. NO.:

DATE SENT TO CDC: Month Day Year

PATIENT IDENTIFICATION Hospital No.:

NAME: Last (18-37) First (38-47) Middle Initial (48)

BIRTHDATE: (49-54) Month Day Year

SEX: (55)

CLINICAL DIAGNOSIS: (56-57)

ASSOCIATED ILLNESS: (58-59)

DATE OF ONSET (Mo. Da. Yr.) (60-65)

FATAL? (66) ☐YES ☐NO

REVERSE SIDE OF THIS FORM MUST BE COMPLETED

Type Specimen

THIS FORM MUST BE EITHER PRINTED OR TYPED

PLEASE PREPARE A SEPARATE FORM FOR EACH SPECIMEN

D.A.S.H.

0	3

(12-13)

Date Reported

Mo. Day Yr.

(14-19)

Comments:

(40-41)

D	6	5

(198-200)

DEPARTMENT OF HEALTH AND HUMAN SERVICES
Public Health Service
Centers for Disease Control
Center for Infectious Diseases
Atlanta, Georgia 30333

The Centers for Disease Control (CDC), an agency of the Department of Health and Human Services, is authorized to collect this information, including the Social Security number (if applicable), under provisions of the Public Health Service Act, Section 301 (42 U.S.C. 241). Supplying the information is voluntary and there is no penalty for not providing it. The data will be used to increase understanding of disease patterns, develop prevention and control programs, and communicate new knowledge to the health community. Data will become part of CDC Privacy Act system 09-20-0106, "Specimen Handling for Testing and Related Data" and may be disclosed: to appropriate State or local public health departments and cooperating medical authorities to deal with conditions of public health significance; to private contractors assisting CDC in analyzing and refining records; to researchers under certain limited circumstances to conduct further investigations; to organizations to carry out audits and reviews on behalf of HHS; to the Department of Justice in the event of litigation, and to a congressional office assisting individuals in obtaining their records. An accounting of the disclosures that have been made by CDC will be made available to the subject individual upon request. Except for permissible disclosures expressly authorized by the Privacy Act, no other disclosure may be made without the subject individual's written consent.

CDC 50.34 REV. 11-90
(Formerly 3.203)

Figure 34.1

Name: Print last name, first name, and middle initial of patient or other equally appropriate specimen identification (required to track and locate specimens).

Birthdate: Print date (month, day, and year) of patient's birth; age in years, or months if an infant.

Sex: Print an "M" (male) or "F" (female).

Clinical diagnosis: Print patient's clinical diagnosis; if none has been made, indicate why assistance is requested (e.g., possible outbreak, exotic isolate, or possible disease).

Associated illness: Print patient's associated or underlying illness, such as cancer, arthritis, hypertension, immunocompromised, or enter major symptoms.

Date of onset: Print date (month, day, and year) illness started. (This date is critical for the interpretation of serologic results; if uncertain, give approximate date.)

Fatal: Check one box only. (This element has epidemiologic significance).

Type of specimen: Print type of specimen (e.g., serum, CSF, fungus culture) in shaded area.

Exercise good judgment to determine the relevance of these items. Paired sera are required for serologic diagnosis of viral and bacterial diseases and a single serum specimen is required for mycotic and parasitic diseases and for syphilis serology (congenital syphilis excepted). In all instances the date(s) of collection of serum specimens *must* be provided. Immunization history is required when such information relates to the kind of serologic analysis requested (e.g., required for polio, measles). Information on treatment, such as administration of

immune serum or globulin, or antibiotics, is often of great benefit when doing serologic tests or identifying reference cultures. As much relevant epidemiologic data as can be obtained should be provided. History of travel and animal or arthropod contacts are required for those RDS requests in which this kind of information is clearly necessary. If any required item of information is not available after efforts have been made to obtain it, please so indicate.

34.8.1b Back of Form (Figure 34.2)

Previous laboratory results/other clinical information: Include any additional information related to this case or specimen that might be helpful to the laboratorian in determining or selecting appropriate laboratory tests or in interpreting results.

The types of specimens usually sent to CDC laboratories are serum specimens, reference cultures, or clinical specimens. To assist state health department laboratories and others in obtaining the information on the request form that NCID requires, the following tabulation for each of the three types of specimens will serve as a guide.

SERUM SPECIMENS

Required
Laboratory exam requested
Specific agent suspected
Serum information
Immunization
Treatment
Epidemiologic data
Previous lab results

Useful
Clinical information
Signs, symptoms, etc.

REFERENCE CULTURES

Required
Laboratory exam requested
Category of agent suspected

LABORATORY EXAMINATION(S) REQUESTED (31-36)
- ☐ ANtimicrobial Susceptibility ☐ IDentification ☐ SErology (Specify Test) _____
- ☐ HIstology ☐ ISolation ☐ OTher (Specify) _____

CATEGORY OF AGENT SUSPECTED (37) ☐ Bacterial ☐ Viral ☐ Fungal ☐ Rickettsial ☐ Parasitic ☐ Other (Specify) _____

SPECIFIC AGENT SUSPECTED: _____ (38-40) ☐☐☐ OTHER ORGANISM(S) FOUND: _____ (41-46)

ISOLATION ATTEMPTED? (47) ☐ Yes ☐ No   **NO. TIMES ISOLATED (48-49)** _____   **NO. TIMES PASSED (50-51)** _____
SPECIMEN SUBMITTED IS (52): ☐ Original Material ☐ Pure Isolate ☐ Mixed Isolate

DATE SPECIMEN TAKEN (53-58) Mo. Da. Year   **ORIGIN (59-60)** ☐ FOod ☐ ANimal (specify) _____
☐ HUman ☐ SOil ☐ OTher (Specify) _____

SOURCE OF SPECIMEN (61-62): ☐ BLood ☐ GAstric ☐ SErum ☐ SPutum ☐ URine ☐ CSF ☐ HAir ☐ SKin ☐ STool ☐ THroat
☐ WOund (Site) _____ ☐ TIssue (Specify) _____
☐ EXudate (Site) _____ ☐ OTher (Specify) _____

SUBMITTED ON (63-64): ☐ MEdium (Specify) _____ ☐ EGg ☐ TIssue Culture (Type) _____
☐ ANimal (Specify) _____ ☐ OTher (Specify) _____

SERUM INFORMATION: Mo. Da. Yr.
- (65-72) ☐ ACute
- (73-80) ☐ COnvalescent
- (81-88) ☐ S3
- (89-96) ☐ S4
- (97-104) ☐ S5

IMMUNIZATIONS: Mo. Yr.
_____ (105-110)
_____ (111-116)
_____ (117-122)
_____ (123-128)

TREATMENT: Drugs Used ☐ None (129)   Date Begun   Date Completed
Mo. Da. Yr. Mo. Da. Yr.
_____ (130-143)
_____ (144-157)
_____ (158-171)

EPIDEMIOLOGICAL DATA: (172-173)
☐ SIngle Case ☐ SPoradic ☐ COntact ☐ EPidemic ☐ CArrier
- Family Illness _____ (174-175)
- Community Illness _____ (176-177)
- Travel and Residence (Location)   Mo. Yr.
- ☐ Foreign _____ (178-183)
- ☐ USA _____ (184-189)
- Animal Contacts (Species) _____ (190-191)
- Arthropod Contacts: (192) ☐ None ☐ Exposure Only ☐ Bite
- Type of Arthropod _____ (193-194)
- Suspected Source of Infection _____ (195-196)

PREVIOUS LABORATORY RESULTS/OTHER CLINICAL INFORMATION:
(Information supplied should be related to this case and/or specimen (s) and relative to the test (s) requested.

CLINICAL TEST RESULTS: (12-13) 0 2
Sputum and Histological Findings _____
Blood Counts _____   Urine Exams _____
Type Skin Tests Performed   Date   Mo. Da. Yr. Strength Pos. Neg.
_____ (14-21)
_____ (23-30) _____ (22)
_____ (32-39) _____ (31)
_____ (40)

SIGNS AND SYMPTOMS
(48-49) ☐ FEver
Maximum Temperature: _____
Duration: _____ Days (50-53)
(54-55)
(56-57) ☐ CHills

RASH:
- (58-59) ☐ MAculopapular
- (60-61) ☐ HEmorrhagic
- (62-63) ☐ VEsicular
- (64-65) ☐ Erythema Nodosum
- (66-67) ☐ Erythema Marginatum
- (68-69) ☐ OTher _____

RESPIRATORY:
- (70-71) ☐ RHinitis
- (72-73) ☐ PUlmonary
- (74-75) ☐ PHaryngitis
- (76-77) ☐ CAlcifications
- (78-79) ☐ PNeumonia (type) _____
- (80-81) ☐ OTher _____

CARDIOVASCULAR:
- (82-83) ☐ MYocarditis
- (84-85) ☐ PEricarditis
- (86-87) ☐ ENdocarditis
- (88-89) ☐ OTher _____

GASTROINTESTINAL:
- (90-91) ☐ DIarrhea
- (92-93) ☐ BLood
- (94-95) ☐ MUcous
- (96-97) ☐ COnstipation
- (98-99) ☐ ABdominal pain
- (100-101) ☐ VOmiting
- (102-103) ☐ OTher _____

CENTRAL NERVOUS SYSTEM:
- (104-105) ☐ HEadache
- (106-107) ☐ MEningismus
- (108-109) ☐ MIcrocephalus
- (110-111) ☐ HYdrocephalus
- (112-113) ☐ SEizures
- (114-115) ☐ CErebral Calcification
- (116-117) ☐ CHorea
- (118-119) ☐ PAralysis
- (120-121) ☐ OTher _____

MISCELLANEOUS:
- (122-123) ☐ JAundice
- (124-125) ☐ MYalgia
- (126-127) ☐ PLeurodynia
- (128-129) ☐ COnjunctivitis
- (130-131) ☐ CHorioretinitis
- (132-133) ☐ SPlenomegaly
- (134-135) ☐ HEpatomegaly
- (136-137) ☐ LIver Abscess
- (138-139) ☐ LYmphadenopathy
- (140-141) ☐ MUcous Membrane Lesions
- (142-143) ☐ OTher _____

STATE OF ILLNESS:
- (144-145) ☐ SYmptomatic
- (146-147) ☐ ASymptomatic
- (148-149) ☐ SUbacute
- (150-151) ☐ CHronic
- (152-153) ☐ DIsseminated
- (154-155) ☐ LOcalized
- (156-157) ☐ INtraintestinal
- (158-159) ☐ EXtraintestinal
- (160-161) ☐ OTher _____

FOR CDC USE ONLY 0 1 (12-13)   No. Specimens: (16-20) _____   No. Tests: (21-25) _____

TYPE SERVICE: (14-15)
- 01-Reference
- 02-Epid. Aid
- 03-Proficiency Testing
- 04-Special Projects
- _____ - Other

LOCATION CODE: (26-27)

AR Argentina	CM Cameroon	GT Guatemala	NU Nicaragua	SZ Switzerland	
AS Australia	CO Colombia	HA Haiti	NZ New Zeland	TD Trinidad-Tobago	
AU Austria	CS Costa Rica	HO Honduras	PA Paraguay	TH Thailand	
BC Bermuda	CY Cyprus	IN India	PE Peru	TW Taiwan	
BE Belgium	DR Dominican Rep.	IS Israel	PK Pakistan	UK United Kingdom	
BH British Honduras	EC Ecuador	IT Italy	PL Poland	UR Soviet Union	
BL Bolivia	ES El Salvador	IV Ivory Coast	PN Panama	UV Uruguay	
BR Brazil	ET Ethiopia	JM Jamaica	PP New Guinea	VE Venezuela	
CA Canada	FR France	MX Mexico	RP Philippines	VN Vietnam	
CB Cambodia	GE Germany	MY Malaysia	RQ Puerto Rico	VQ Virgin Islands	
CI Chile	GQ Guam	NI Nigeria	SL Sierra Leone		
			SP Spain	_____ Other _____	

SPECIMEN SUBMITTED BY: (28-30)
- 100-Health Dept.
- 200-CDC Clinic
- 205-Proficiency Testing
- 225-CDC Non-clinic
- 301-Army
- 302-Navy
- 303-Air Force
- 307-V.A. Hosp.
- 310-U.S.D.A.
- 323-Indian Hosp.
- 325-NIH
- 400-Foreign
- 402-Peace Corps.
- 550-University
- 606-Physician/Clinic
- _____ - Other

CDC 50.34 REV. 11-90 (BACK)
(Formerly 3.203)

CDC NUMBER | UNIT | FY | NUMBER | SUF. |

Figure 34.2

Specific agent suspected
Kind of specimen
Origin of specimen
Source of specimen
Submitted on what medium
Previous lab results
Biochemical reactions
(can be attached to a separate sheet)

Useful
Isolation attempted
Date specimen taken
Number times isolated
Other clinical information
Clinical test results
Signs, symptoms, etc.
Other organisms found

Epidemiologic data
Treatment

CLINICAL SPECIMENS

Required
Laboratory exam requested
Category of agent suspected
Specific agent suspected
Specimen submitted is
Date specimen taken
Source of specimen
Epidemiologic data
Previous lab results

Useful
Other clinical information
Clinical test results
Signs, symptoms, etc.

State Laboratory Virology Services

Steven Specter and Gerald Lancz

35.1 INTRODUCTION

State public health laboratories, much like the federal laboratories at the Centers for Disease Control (CDC), are charged with providing laboratory diagnosis of viral infections when local services are not available. Thus, most states do not encourage routine use of the state virology laboratories as a primary diagnostic laboratory, preferring to defer to a local laboratory service. Availability of only limited diagnostic services by state laboratories is often necessitated by limited financial support. Therefore, private hospital, commercial, or local public health laboratories are often used for routine primary diagnostic virology services. The state virus laboratory should be utilized for diagnostic problems that go beyond the scope or capability of local laboratories, especially for viruses for which statewide surveillance is performed (e.g., influenza, arboviruses) as well as viruses of epidemiologic significance. Although many state laboratories will accept specimens for routine primary isolation, their budget may be adversely affected if the laboratories are confronted with large numbers of specimens. This may impinge unfavorably on their ability to perform a key function as a center to collate and disseminate information regarding viral diseases to the CDC. Within this context we have listed some of the functions and viral diagnostic services available in state and U.S. territorial public health laboratories.

35.2 SUBMISSION OF SPECIMENS

All but a handful of the states have laboratories that accept specimens for the diagnosis of viral diseases (Table 35.1). In most cases the submission of specimens to the state laboratory is via local public health laboratories. Each state with laboratories that accept specimens for virus isolation and/or identification has its own set of requirements for shipping, type of specimen, etc. In most cases these are described in detail in the literature provided by the appropriate state authority. In many states, conditions for specimen submission are similar; however, specific requirements are imposed by other states. It is advised that anyone who desires to submit a specimen to their state clinical virology laboratory, contact the laboratory head to determine the requirements for submitting clinical ma-

Table 35.1 Virology Services Available in State and Territorial Public Health Laboratories[a]

	Serology	Isolation	Refer Specimens to CDC[b]	Regulate Primary Labs in State	Arbovirus Surveillance	Influenza Surveillance	Rabies Detection	Special Epidemiologic Services
1. Alabama	+	+	+	+[c]	+	+	+	Hepatitis testing
2. Alaska	+	+	No	NA	+	+	+	HIV screening
3. Arizona	+	+	+	+	+	+	+	Measles IgM/IgG, HIV screen, Measles, Rubella, Varicella-zoster immune status screen
4. Arkansas	+	No	+	No	+	+	+	
5. California	+	+	+	+	+	+	+	Many others
6. Colorado	+	No	+	No	+	+	+	Upon request
7. Connecticut	+	+	+	+	No	+	+	Many others
8. Delaware	+	+	+	+[c]	+	+	+	Many others
9. District of Columbia	+	+[d]	+	+[c]	No	+	+	
10. Florida	+	+	+	+[c]	+	+	+	HIV; Many others by request
11. Georgia	+	+	+	+[c]	No	+	+	Programmatic support
12. Guam	+	No	+	No	NI	NI	+	
13. Hawaii	+	+	+	No	NI	+	+	Respiratory virus surveillance
14. Idaho	+[d]	+[d]	+	+	+	+	+	
15. Illinois	+	+	+	+	+	+	NI	Measles
16. Indiana	+	+	+	No	+	+	+	Upon request
17. Iowa	+	+	+	No	+	+	+	Enterovirus surveillance and Human Retrovirus surveillance

State								Comments
18. Kansas	+	+	+	No	NI	+	NI	Enterovirus, Rotavirus, Resp. virus surveillance
19. Kentucky	+	+	+	No	NI	+	+	
20. Louisiana	+	+[e]	+	No	+	+	+[f]	
21. Maine	+	+	+	+	No	+	+	
22. Maryland	+	+	+	+	+	+	+	Vaccine Preventable Dis. and Hepatitis and Human Retrovirus surveillance
23. Massachusetts	+	+	+	No	+	NI	+	Epidemiologic services Hepatitis B, HIV testing, Measles, and Rubella IgM
24. Michigan	+	+	+	+	+	NI	+	
25. Minnesota	+	+	+	No	+	+	+	Measles, Mumps, Rubella, Enteroviruses
26. Mississippi	+	No	+	No	NI	NI	+	Hepatitis B, Measles
27. Missouri	+	+	+	No	+	+	+	Many others
28. Montana	+	+	+	+	+	+	NI	
29. Nebraska	NI	NI	NI	NI	NI	NI	NI	
30. Nevada	+	No	+	+[c]	NI	NI	+	Hepatitis, HIV, Rubella, Measles, others
31. New Hampshire	+	+	+	+[c]	+	+	+	Hepatitis, Measles
32. New Jersey	+	+	+	+[c]	+	+	+	
33. New Mexico	+	+	+	No	+	+	+	Enterovirus, Herpesvirus isol./ident., Adenovirus, Rubeola and Rubella IgM/IgG, Human Retrovirus, Rotavirus and Chlamydia

Table 35.1 (*continued*)

	Serology	Isolation	Refer Specimens to CDC[b]	Regulate Primary Labs in State	Arbovirus Surveillance	Influenza Surveillance	Rabies Detection	Special Epidemiologic Services
34. New York	+	+	+	+	+		+	
35. North Carolina	+	+	+	No	+	+	+	
36. North Dakota	+	+	+	No	+	+	+	
37. Ohio	+	+	+	No	+	+	+	Vaccine preventable disease, Enterovirus and respiratory virus surveillance, Chlamydia
38. Oklahoma	+	+	+	No	NI	+	+	
39. Oregon	+	+	+	+	NI	+	+	Hepatitis B screening and HIV testing certain high risk groups
40. Pennsylvania	+	+	+	+	No	+	+	Measles, Rubella IgM
41. Puerto Rico	+	No	+	No	NI	NI	+	Hepatitis, Rubella, Rubeola
42. Rhode Island	+	No	+	No	NI	NI	+	
43. South Carolina	+	+	+	No	No	+	+	Many others per State Services Manual
44. South Dakota	+	+	+	No	+	+	+	
45. Tennessee	+	+	+	+[c]	No	+	+	
46. Texas	+	+	+	No	+	+	+	Herpes simplex screening 33rd week pregnancy, Rubella screening

							Prenatal Hepatitis B screening, HPV screen and type
47. Utah	+	+	+	No	+	+	+
48. Vermont	No	+e	+	No	No	+	+
49. Virgin Islands	NI	No	NI	No	NI	NI	NI
50. Virginia	+	NI	+	No			
51. Washington	+	+	+	NI	+	+	+
52. West Virginia	+	No	+	No	NI	NI	+
53. Wisconsin	+	+		No	+	+	+
54. Wyoming	No	No	+	No	NI	NI	NI

Abbreviations: NA, not applicable; NI, no information provided; +, yes.

a Information is based on responses to a questionnaire sent to the state laboratories in fall and winter of 1989 to 1990.

b Specimens submitted to CDC refer to those not normally tested in the state laboratories.

c Regulation by state authorities but not through the Public Health Laboratory.

d Herpes simplex virus and influenza only.

e Influenza only

f Rabies direct immunofluorescence of animal brain only.

terial. Some generally accepted requirements include:

1. A good clinical history with a listing of the patient's name and age, date of specimen collection, date of onset of illness, type of specimen and collection site, major clinical symptoms, relevant immunization history, virus(es) for which specimens are to be tested, and physician's name, address, and phone number.
2. Serologic testing for antiviral antibodies generally requires simultaneous submission of acute and convalescent sera, except for special screening studies (e.g., rubella or varicella-zoster immune status).
3. Neonatal serum should be accompanied by a maternal serum.

Many state laboratories request that during an epidemic, only a limited number of specimens be submitted to allow determination of the causative agents, rather than the submission of specimens for each patient seen in the course of the epidemic. Shipping instructions often include conditions for handling *and* packaging specimens as well as prepayment of shipping costs. Most states do not require a fee for testing; however, a few states may have minimal fees. Submission of specimens may be limited, e.g., Georgia, Maryland, and New York have requirements that limit submission of specimens by physicians and/or patients who reside in their state. Several states limit specimen submission to licensed physicians, whereas, others also accept specimens from veterinarians and other legitimate public health services, including hospitals, state agencies, and public or community health laboratories.

Specimens for the detection of rabies virus often have additional requirements for handling, shipping, and clinical history. Many states provide specific instructions as well as a Rabies Investigation Report Form to accompany such specimens.

35.3 SCOPE OF SERVICES

An overview of the services available in each state laboratory is provided in Table 35.1. The diversity of the services offered by different states is apparent. These range from laboratories that offer only serology for select viruses, to those that provide extensive serology and isolation services for many viruses. There are laboratories that a) exclude class IV agents (e.g., lassa fever virus); b) exclude viruses that could be easily tested for in hospital or private laboratories (e.g., herpes simplex); c) include only viruses that have epidemiologic importance (e.g., influenza, arboviruses). A listing of such specific services by state has been avoided here because these services no doubt change periodically. Again, it is recommended that you refer to your state laboratory to determine the extent of services offered. The addresses and telephone numbers for the state laboratories, as of July, 1990, are provided in Table 35.2.

Most laboratories will send specimens to the CDC if they do not handle them on site. This may be limited to class IV specimens or may cover a broad range of viral agents. Also listed in Table 35.1 are some special services offered by the laboratories, including participation in national or international surveillance programs for arboviruses and influenza; detection services for rabies virus, which usually results in rapid reporting in suspected rabies exposure cases; and special epidemiologic services. However, a service that is considered to be special by one state laboratory may be part of the normal service offered by another state laboratory. The listing of special services placed in Table 35.1 is based on information supplied by the laboratory directors.

Regulation of licensure of virology laboratories by the state laboratories is not a common practice. Approximately 24% of the state laboratories indicated that they were involved in regulating primary laboratories. Eight additional state laboratories indicated that they are not involved in the regulation of primary laboratory licensure but that another stage agency performed this function. In some states there is no State Virology Laboratory, whereas, in other states there are no practicing primary virology laboratories that require regulation. It would seem that some regulation is desirable in all states to ensure that standard, accepted practices are used to obtain reliable results. The establishment of some level of regulation within states would promote this standardization.

35.4 TURNAROUND TIME FOR RESULT REPORTING

The bane of viral diagnosis by state laboratories in the past has been the long turnaround time from submission of a specimen by the physician until a report is returned to that physician. Frequently, this was a matter of months in all but emergency cases, as with exposure to rabies virus. Today, the turnaround time for diagnosis of many viral diseases is no longer a significant problem.

Based on the response to our questionnaire, it appears that most laboratories send out a report upon identification of a virus, in some states this may be a telephone report.

Virtually all laboratories indicated that the length of time for reporting results was variable and dependent on the type of testing to be performed, as well as whether a specimen is positive or negative. Serologic results are frequently reported within a few days to one week of receipt of the acute and convalescent sera, however, a few laboratories indicate this may take as long as two to four weeks. Testing for immune status against a particular virus is reported from a few days to four weeks after the receipt of a single serum. Positive isolation of many viruses is reported out within 72 hours, whereas, some isolation (as in the case of cytomegalovirus) may take up to two weeks. Reports on specimens that are negative for virus isolation may be sent out as soon as two weeks or not until six weeks after receipt of the specimen.

ACKNOWLEDGEMENT

We thank the directors of the various state laboratories who supplied information used in this chapter.

Table 35.2 State and Territorial Public Health Laboratories

ALABAMA	ALASKA
Director	Chief
Bureau of Clinical Laboratories	Section of Laboratories
State Department of Public Health	Alaska Division of Public Health
8140 University Drive	Department of Health and Social Services
Montgomery, AL 36130	Pouch H-06-D
FTS Direct and Commercial: (205) 277-8660, ext. 215	Juneau, AK 99811
	FTS Direct and Commercial: (907) 465-3140

Table 35.2 (*continued*)

ARIZONA
Assistant Director
Division of State Laboratory Services
1520 West Adams Street
Phoenix, AZ 85007
FTS Direct and Commercial: (602) 542-1194
FAX (602) 542-1169

ARKANSAS
Director
Division of Public Health Laboratories
4815 West Markham Street
Little Rock, AR 72205-3867
FTS Operator: 740-5011
Commercial: (501) 661-2191

CALIFORNIA
Chief
Division of Laboratories
California Department of Health Services
2151 Berkeley Way
Berkeley, CA 94704
FTS Direct and Commercial: (415) 540-2408

COLORADO
Director
Division of Laboratories
Department of Public Health
4210 East 11th Avenue
Denver, CO 80220
FTS Direct and Commercial: (303) 331-4700

CONNECTICUT
Director of Laboratories
State Department of Health
P.O. Box 1689
Hartford, CT 06101
FTS Direct: 641-5063
Commercial: (203) 566-5102
FAX (203) 566-7813

DELAWARE
Director
Division of Public Health Laboratories
P.O. Box 618
Dover, DE 19903
FTS Direct and Commercial: (302) 736-4714

DISTRICT OF COLUMBIA
Director
Bureau of Laboratories
Department of Human Services
300 Indiana Avenue, NW, Room 6154
Washington, DC 20001
FTS Direct and Commercial: (202) 727-0557

FLORIDA
Chief
Office of Laboratory Services
Department of Health and Rehabilitative Services
P.O. Box 210 (1217 Pearl Street)
Jacksonville, FL 32231
FTS Direct and Commercial: (904) 359-6145

GEORGIA
Director of Laboratories
Georgia Department of Human Resources
47 Trinity Avenue, SW, Room 13-H
Atlanta, GA 30334
Commercial: (404) 656-4850

GUAM
Laboratory Director
Public Health and Social Services
P.O. Box 2816
Agana, Guam 96910

HAWAII
Chief
Laboratories Division
State Department of Health
P.O. Box 3378
Honolulu, HI 96801
FTS Direct and Commercial: (808) 548-6324

IDAHO
Chief
Bureau of Laboratories
Department of Health and Welfare
2220 Old Penitentiary Road
Boise, ID 83712
FTS Direct: 554-2235
Commercial: (208) 334-2235
Evaluation and Specimens:
  Box 640
  Boise, ID 83701

Table 35.2 (*continued*)

ILLINOIS
Chief
Division of Laboratories
Illinois Department of Public Health
535 W. Jefferson, 4th Floor
Springfield, IL 62761
FTS Direct and Commercial: (217) 782-4977

INDIANA
Director
Bureau of Laboratories
State Board of Health
1330 West Michigan Street
Indianapolis, IN 46206-1964
FTS Direct and Commercial: (317) 633-0720

IOWA
Director
State Hygienic Laboratory
University of Iowa
Iowa City, IA 52242
FTS Direct and Commercial: (319) 335-4500

KANSAS
Director
Kansas Health and Environmental Laboratory
Department of Health and Environment
Forbes Building, #740
Topeka, KS 66620
FTS Direct and Commercial: (913) 296-1620

KENTUCKY
Director
Division of Laboratory Services
Department for Health Services
Cabinet for Human Resources
275 East Main Street
Frankfort, KY 40621-0001
FTS Direct and Commercial: (502) 564-4446

LOUISIANA
Director
Division of Laboratory Services
Office of Public Health
Louisiana Department of Health
325 Loyola Avenue, 7th Floor
New Orleans, LA 70112
FTS Direct and Commercial: (504) 568-5375

MAINE
Director
Public Health Laboratory
Department of Human Services
State House - Station No. 12
Augusta, ME 04333
FTS Direct and Commercial: (207) 289-2727
FAX (207) 626-5555

MARYLAND
Director
Laboratories Administration
State Department of Health and Mental
 Hygiene
P.O. Box 2355
Baltimore, MD 21203
FTS Direct and Commercial: (301) 225-6100

MASSACHUSETTS
Director
State Laboratory Institute
Department of Public Health
305 South Street
Jamaica Plain, MA 02130
FTS Direct and Commercial: (617) 522-3700

MICHIGAN
Laboratory Director
Laboratory and Epidemiological Services
 Administration
Michigan Department of Public Health
P.O. Box 30035 - 3500 N. Logan
Lansing, MI 48909
FTS Direct: 253-1381
Commercial: (517) 335-8067

MINNESOTA
Director
Division of Public Health Laboratories
Minnesota Department of Health
P.O. Box 9441
Minneapolis, MN 55440
FTS Direct and Commercial: (612) 623-5200

MISSISSIPPI
Director
Public Health Laboratories
State Board of Health
P.O. Box 1700
Jackson, MS 29215
Commercial: (610) 960-7582

Table 35.2 (*continued*)

MISSOURI
Director
State Public Health Laboratory
Missouri Department of Health
307 W. McCarty
Jefferson City, MO 65101
FTS Direct and Commercial: (314) 751-3334

MONTANA
Director
Public Health Laboratory
State Department of Health and Environmental
    Sciences
Cogswell Building
Helena, MT 59620
FTS Direct: 587-2642
Commercial: (406) 444-2642

NEBRASKA
Director of Laboratories
State Department of Health
P.O. Box 2755
Lincoln, NE 68502
FTS Direct: 541-2122
Commercial: (402) 471-2122

NEVADA
Director
Nevada State Health Laboratory
Department of Human Resources
1660 N. Virginia Street
Reno, NV 89503
Commercial: (702) 789-0335
FAX (702) 789-0460

NEW HAMPSHIRE
Chief
Public Health Laboratories
Division of Public Health
State Laboratory Building
6 Hazen Drive
Concord, NH 03301
FTS Direct: 842-1110, ext. 4657
Commercial: (603) 271-4657

NEW JERSEY
Director
Public Health and Environmental Laboratories
State Department of Health
P.O. Box 1540-CN 360
Trenton, NJ 08625-0360
FTS Direct and Commercial: (609) 292-5605

NEW MEXICO
Director
Scientific Laboratory Division
700 Camino de Salud, NE
Albuquerque, NM 87106
FTS Direct and Commercial: (505) 841-2500

NEW YORK
Director
Wadsworth Center for Laboratories and
    Research
New York State Department of Health
Empire State Plaza
Albany, NY 12201-0509
FTS Direct and Commercial: (518) 474-4180

NORTH CAROLINA
Director
Division of Laboratory Services
Department of Environment, Health and
    Natural Resources
P.O. Box 28047
Raleigh, NC 27611-8047
FTS Direct and Commercial: (919) 733-7834

NORTH DAKOTA
Director
Consolidated Laboratories
State Department of Health and Consolidated
    Laboratories
Box 937
Bismarck, ND 58502-0937
FTS Operator: 783-4011
Commercial: (701) 221-6140
FAX (701) 221-6145

Table 35.2 (*continued*)

OHIO
Chief
Division of Public Health Laboratories
State Department of Health
P.O. Box 2568
Columbus, OH 43216-2568
FTS Direct and Commercial: (614) 421-1078

OKLAHOMA
Chief
Public Health Laboratory Service
State Department of Health
P.O. Box 24106
Oklahoma City, OK 73124
FTS Operator: 736-4011
Commercial: (405) 271-5070

OREGON
Manager-Director
Public Health Laboratory
Department of Human Resources
1717 SW 10th Avenue
Portland, OR 97201
FTS Direct and Commercial: (503) 229-5884

PENNSYLVANIA
Director
Bureau of Laboratories
Pennsylvania Department of Health
P.O. Box 500
Exton, PA 19341
FTS Direct and Commercial: (215) 363-8500
FAX (215) 436-3346

PUERTO RICO
Director, Laboratory Program
Institute of Health Laboratories
Department of Health
Building A - Call Box 70184
San Juan, PR 00920
FTS Direct and Commercial: (809) 764-6945

RHODE ISLAND
Associate Director
Division of Laboratories
Health Laboratory Building
50 Orms Street
Providence, RI 02904
FTS Operator: 838-1000
Commercial: (401) 274-1011

SOUTH CAROLINA
Chief
Bureau of Laboratories
Department of Health and Environmental
    Control
P.O. Box 2202
Columbia, SC 29202
FTS Direct and Commercial: (803) 737-7042

SOUTH DAKOTA
Director
State Health Laboratory
Laboratory Building
Pierre, SD 57501
FTS Operator: 782-7000
Commercial: (605) 773-3368

TENNESSEE
Director
Laboratory Services
Tennessee Department of Health and
    Environment
630 Ben Allen Road
Nashville, TN 37247-0801
FTS Direct and Commercial: (615) 262-6300

TEXAS
Chief
Bureau of Laboratories
Texas Department of Health
1100 West 49th Street
Austin, TX 78756
FTS Operator: 729-4011
Commercial: (512) 458-7318

Table 35.2 (*continued*)

UTAH
Director
Utah State Health Laboratory
44 Medical Drive, Room 207
Salt Lake City, UT 84113
FTS Direct and Commercial: (801) 584-8300

VERMONT
Director
State Public Health Laboratory
Vermont Department of Health
195 Colchester Avenue, P.O. Box 70
Burlington, VT 05402-0070
FTS Direct and Commercial: (802) 863-7335

VIRGIN ISLANDS
Director of Public Health Laboratory
P.O. Box 8585
St. Thomas, VI 00801
FTS Direct and Commercial: (809) 774-5955

VIRGINIA
Director
Bureau of Microbiological Science
Division of Consolidated Laboratory Services
Department of General Services
Commonwealth of Virginia
Box 1877
Richmond, VA 23215
FTS Direct: 936-3756
Commercial: (804) 786-3756

WASHINGTON
Assistant Secretary
Department of Health
Division of Laboratories
1610 N.E. 150th Street
Seattle, WA 98155-7224
FTS Direct and Commercial: (206) 361-2816

WEST VIRGINIA
Director
State Hygienic Laboratory
167 11th Avenue
South Charleston, WV 25303
FTS Direct: 885-3530
Commercial: (304) 348-3530

WISCONSIN
Director
State Laboratory of Hygiene
William D. Stovall Building
465 Henry Mall
Madison, WI 53706
FTS Direct and Commercial: (608) 262-1293

WYOMING
Director
Public Health Laboratory Services
Division of Health and Medical Services
State Office Building
Cheyenne, WY 82001
FTS Operator: 328-1110
Commercial: (307) 777-7431

36

Laboratories Offering Viral Diagnostic Services

Steven Specter and Gerald Lancz

36.1 INTRODUCTION

This chapter identifies laboratories that offer viral diagnostic services, including hospital, university, and commercial laboratories (Table 36.1). We have not included public health laboratories, because the services they provide are generally limited to epidemiology. Because diagnostic services offered by any individual laboratory may change, we have not attempted to indicate the extent of services available. Additionally, laboratories that provide viral diagnostic services may open, close, or discontinue services at any time. This must be considered when reference is made to the information compiled in Table 36.1 for possible submission of specimens, utilization of services, etc.

The information presented was supplied by state public health laboratories and individual clinical virologists, who provided names of laboratories with which they were familiar. In this regard, the editors are especially indebted to Steven Racioppi, President, Microtest, Snellville,

GA. For some states, the state laboratory personnel were unable to supply such a list and it was not possible for us to identify individuals with knowledge of viral diagnostic services. Thus, a failure to list services available in a particular state may reflect either our inability to identify these laboratories or that laboratory services are not offered in that state.

Some commercial laboratories have a national network to process clinical specimens. These laboratories fill the void when there is no local laboratory that performs these services (Table 36.2).

The extent of services available from viral diagnostic laboratories varies widely. Some laboratories offer viral diagnostic and serology services for virtually all common human pathogenic viruses, while others may perform a limited number of tests only (e.g., herpes simplex virus isolation or rubella serology). Individuals should contact the laboratory(ies) listed in their locale to determine how their needs can best be served.

Table 36.1 Laboratories Performing Viral Diagnosis

Alabama
  Birmingham
    Children's High Tower Medical Center
    Medical Laboratory Associates
    University of Alabama Medical Center

Alaska
  Anchorage
    Providence Hospital

Arizona
  Phoenix
    Bolin Laboratories, Inc.
    Cigna Health Plan of Arizona
    Consultants Medical Laboratory
    Damon Clinical Labs
    Good Samaritan Regional Medical Center
    Metpath of Arizona
    National Health Laboratories, Inc.
    Phoenix Baptist Hospital
    Sonora Laboratory Sciences
    St. Joseph's Hospital
    St. Luke's Hospital
  Scottsdale
    Blood Systems of Arizona
  Tucson
    American Red Cross
    Kino Community Hospital
    Tucson Medical Center
    University of Arizona Medical Center
    Yuma Regional Medical Center

Arkansas
  Little Rock
    Arkansas Children's Hospital
    University of Arkansas Medical School

California
  Berkeley
    Kaiser Permenante Medical Group
  Emeryville
    Virolab, Inc.
  Fresno
    Valley Children's Hospital
  Glendale
    Pathology Clinical Laboratories

Long Beach
  Long Beach Medical Center
  Medical Reference Laboratory
  Memorial Hospital Medical Center
Los Angeles
  Cedars Sinai Medical Center
  Children's Hospital of Los Angeles
  CLMG, Inc.
  Specialty Labs
  UCLA
  USC/LAC General Hospital
Newberg Park
  Reference Laboratory
North Hollywood
  Kaiser-Permanente (Southern California)
Orange
  University of California, Irvine
Pasadena
  Immunology Consultants
Sacramento
  Physician's Clinical Laboratory
San Diego
  University of California, San Diego
San Francisco
  Children's Hospital
  Damon Clinical Laboratories
  Davies Medical Center
  Mt. Zion Hospital
Santa Ana
  HCA Laboratory
Stanford
  Stanford University Medical Center
Torrance
  Harbor General Hospital
Van Nuys
  Smith Kline Bioscience

Colorado
  Denver
    Children's Hospital
    St. Joseph's Hospital
    University of Colorado Health Science
      Center

Connecticut
  Danbury
    Danbury Hospital

Table 36.1 (*continued*)

Farmington
  University of Connecticut Medical Center
Hartford
  Hartford Hospital
  St. Francis Hospital
New Britian
  North American Laboratory Group
New Haven
  New Haven Hospital
West Haven
  Veterans Hospital

Delaware
  Newark
    Christiana Hospital
  Wilmington
    Delaware Medical Laboratories
    Medical Center of Delaware

District of Columbia
  Children's Hospital National Medical Center
  Columbia Hospital for Women
  Georgetown University
  George Washington University Hospital
  Providence Lab Associates
  Walter Reed Army Medical Center

Florida
  Ft. Lauderdale
    Holy Cross Hospital
  Gainesville
    Shands Teaching Hospital-UF
  Jacksonville
    Baptist Medical Center
    Mayo Clinic
  Lakeland
    Lakeland Regional Medical Center
  Melbourne
    Holmes Regional Medical Center
  Miami
    Bascom Palmer Eye Institute
    Jackson Memorial Hospital
    Mercy Hospital
    Sekot Laboratories
    University of Miami
  Naples
    Diagnostic Services Inc.

Orlando
  Orlando Regional Medical Center
Palm Beach
  St. Mary's Hospital
Sarasota
  Sarasota Memorial Hospital
St. Petersburg
  All Children's Hospital
Tampa
  SmithKline Bioscience
  St. Joseph's Hospital
  Tampa General Hospital

Georgia
  Atlanta
    Emory University Laboratory
  Augusta
    Humana Hospital
    Medical College of Georgia
    St. Joseph's Hospital
  Doraville
    SmithKline Bioscience
  Ft. Gordon
    Pathology Laboratory

Hawaii
  None Identified

Idaho
  Boise
    St. Luke's Hospital

Illinois
  Chicago
    Children's Memorial Hospital
    Columbus Hospital
    Cook County Hospital
    Illinois Masonic Medical Center Hospital
    Michael Reese Hospital
    Mt. Sinai Hospital
    Northwestern Memorial Hospital
    Rush-Presbyterian St. Luke's Hospital
    University of Chicago Hospital
  Elmhurst
    Memorial Hospital of DuPage County
  Maywood
    Loyola Medical Center Hospital

Table 36.1 (*continued*)

Park Ridge
  Victoria Clinic Reference Laboratory
Peoria
  Mobilab, Inc.
Rockford
  Rockford School of Medicine
Springfield
  Memorial Medical Center

Indiana
  Indianapolis
    Indiana Medical Center
  South Bend
    Notre Dame University

Iowa
  Des Moines
    Iowa Methodist Medical Center
  Iowa City
    University of Iowa Hospital and Clinics
  Sioux City
    St. Luke's Hospital

Kansas
  Kansas City
    Kansas University Medical Center
  Wichita
    Consolidated Biological Laboratory
    St. Francis Hospital
    Wesley Medical Center

Kentucky
  Lexington
    University of Kentucky Medical Center
    VA Medical Center
  Louisville
    American Red Cross
    Humana Hospital—University Louisville
    Jewish Hospital
    National Health Laboratory
    NKC, Inc.
    University Louisville Pediatric Virology
      Laboratory
    Viromed Laboratory
  Madisonville
    Regional Medical Center
  Owensboro
    Western Kentucky Regional Blood Center

Louisiana
  Jefferson
    Alton Oschner Clinic
  New Orleans
    Charity Hospital of Louisiana
    LSU Medical Center
  Shreveport
    LSU Medical Center

Maine
  Bangor
    Affiliated Laboratory, Inc.
  Portland
    Maine Medical Center
    Osteopathic Hospital of Maine

Maryland
  Baltimore
    Greater Baltimore Medical Center
    Johns Hopkins Hospital
    Maryland Medical Laboratories
    Sinai Hospital
    University of Maryland Hospital/Baltimore
    Bethesda Naval Medical Command NIH
  Fort Meade
    Kimbrough Hospital
  Rockville
    NIH Clinical Center
  Walkersville
    MA Bioproducts Laboratory

Massachusetts
  Boston
    Beth Israel Hospital
    Brigham and Women's Hospital
    (Knudsin Lab)
    Children's Hospital
    City of Boston Hospital
    Dana Farber Cancer Institute
    Lahey Clinic Medical Center
    Massachusetts General Hospital
    New York Deaconess Hospital
    Tufts-New England Medical Center
  Salem
    Salem Hospital
  Springfield
    Pro Med
    Valley Medical Labs

Table 36.1 (*continued*)

West Springfield
  Baystate Medical Center
Wilmington
  New England Pathology
Worcester
  University of Massachusetts Medical
    Center

Michigan
  Ann Arbor
    University of Michigan Hospital
  Detroit
    Children's Hospital
    Damon Clinical Laboratories
    Henry Ford Hospital
  Farmington Hills
    Bio-Science Lab
  Grand Rapids
    Continental Clinical Biochemical
    St. Mary's Hospital
  Lansing
    Edward Sparrow Hospital
  Livonia
    Roche Bio-Medical Laboratory
  Troy
    William Beaumont Hospital

Minnesota
  Duluth
    University of Minnesota Med. School
  Minneapolis
    Hennepin County Metropolitan Medical
      Center
    Lufkin Laboratories
    University of Minnesota Hospitals
    Veteran's Hospital
  Minnetanka
    Viromed
  Rochester
    Mayo Clinic
  St. Cloud
    North Central Laboratories
  St. Paul
    Children's Hospital
    Ramsey Medical Center

Mississippi
  Jackson
    University of Mississippi Medical Center

Missouri
  Columbia
    Boyce and Bynum Laboratories
    Veteran's Hospital
  Kansas City
    Children's Mercy Hospital
    North Kansas City Memorial Hospital
  Springfield
    Cox Medical Center
  St. Louis
    Cardinal Glennon Children's Hospital
    Jewish Hospital
    Medical Pediatrics
    St. John's Medical Center
    St. Louis Children's Hospital

Montana
  Missoula
    Community Hospital

Nebraska
  Omaha
    AMI/St. Joseph's Hospital

Nevada
  Las Vegas
    Associated Pathologist's Hospital
    Sunrise Hospital Laboratory
  Reno
    Sierra Nevada Laboratory
    Washoe Medical Center

New Hampshire
  None Identified

New Jersey
  Hackensack
    Hackensack Medical Center
  Long Branch
    Monmouth Medical Center
  Newark
    St. Michael's Hospital

Table 36.1 (*continued*)

New Brunswick
  Middlesex Hospital
  Robert Wood Johnson Medical School
Teterboro
  Metpath

New Mexico
  Albuquerque
    New Mexico Reference Laboratory, Inc.
    Presbyterian Hospital
    Scientific Laboratory Division
    SED Medical Laboratories
    University of New Mexico Medical School

New York
  Albany
    Albany Medical Center
    Wadsworth Center for Labs and Research
  Binghamton
    Binghamton General Hospital
  Bronx
    Montifiore Medical Center
    VA Medical Center
  Brooklyn
    SUNY Downstate Hospital
  Buffalo
    Children's Hospital of Buffalo
    Erie County Laboratory
  East Meadow
    Nassau County Medical Center
  Johnson City
    United Health Services
  Manhasset
    North Shore University Hospital
  New York City
    Bellevue Hospital
    Memorial Sloan Kettering Cancer Center
    Mt. Sinai Hospital
    New York Hospital
    St. Luke's Roosevelt Hospital
  Rochester
    Rochester General Hospital
    University of Rochester-Strong
    Memorial Hospital
  Staten Island
    IBR Consolidated Clinical Lab

Stony Brook
  University Hospital at Stony Brook
Syracuse
  Crouse-Irving Memorial Hospital
  SUNY Health Science Center at Syracuse
Valhalla
  Westchester County Public Health
    Laboratory
Valley Cottage
  Rockland Medilabs, Inc.

North Carolina
  Asheville
    St. Joseph's Hospital
  Burlington
    Roche Biomedical Reference Labs
  Chapel Hill
    Frank Porter Graham Child Development
  Center
    North Carolina Memorial Hospital
    University of North Carolina
  Charlotte
    Charlotte Memorial Hospital
    Presbyterian Hospital
  Durham
    Duke University Medical Center
    Veteran's Hospital
  Greensboro
    Moses Cone Memorial Hospital
    Westly Long Hospital
  Greenville
    East Carolina University Virology
      Laboratory
  Raleigh
    Rex Hospital
  Winston-Salem
    North Carolina Baptist Hospital

North Dakota
  None Identified

Ohio
  Akron
    Akron City Hospital
    Children's Hospital

Table 36.1 (*continued*)

Canton
  Aultman Hospital
  Timkin Mercy Medical Hospital
Cincinnati
  Children's Hospital
  University of Cincinnati Hospital
Cleveland
  Case Western Reserve University Hospital
  Cleveland Clinic Foundation
  Cleveland Metropolitan General Hospital
  St. Luke's Hospital
  St. Vincent's Medical Center
Columbus
  Children's Hospital
  Doctor's Hospital
  Grant Hospital
  Ohio State University Hospital
Dayton
  Children's Hospital Medical Center
Dublin
  Roche Biomedical Labs
Miamisburg
  Diagnostic
  Virology Service
Rootstown
  North East Ohio Universities
Springfield
  Mercy Medical Center
  Springfield Community Hospital
Toledo
  Medical College of Ohio
Youngstown
  Youngstown Hospital

Oklahoma
  Broken Arrow
    Symex Corporation-Vironostics
  Oklahoma City
    HCA Presbyterian Hospital
    Medical Arts Laboratory
    Northwest Laboratories
    Oklahoma Children's Memorial Hospital
    Oklahoma Medical Center
    St. Anthony Hospital
  Tulsa
    St. Francis Hospital
    St. John's Hospital

Oregon
  Bend
    St. Charles Medical Center
  Clakamas
    Kaiser Regional Laboratory
  Coos Bay
    Coastal Medical Laboratory, Inc.
  Corvallis
    Good Samaritan Hospital Laboratory
  Eugene
    Oregon Medical Laboratories
  Medford
    Rouge Valley Medical Center
  Oregon City
    Drs. Haug and Hoffman
  Pendleton
    Interpath Laboratory, PC
  Portland
    Dr. D.H. McGowan, Medical Laboratory
    Emmanuel Hospital/Metro Lab
    Good Samaritan Hospital Laboratory
    Medical Lab-Pathologists Central
  Laboratory
    Oregon Health Sciences University
  Clinical Laboratory
    Physicians Medical Laboratory
    Providence Medical Center
  Roseburg
    Douglas Community Hospital
  The Dalles
    Mid Columbia Medical Center

Pennsylvania
  Abington
    Abington Memorial Hospital
  Allentown
    Health East Laboratories
  Bethlehem
    St. Luke's Hospital
  Chester
    Crozer Chester Medical Center
    Sacred Heart Hospital
  Danville
    Geisinger Medical Center
  Erie
    Associated Clinical Laboratory

Table 36.1 (*continued*)

Harrisburg
  Harrisburg Hospital
Hershey
  Hershey Medical Center
Langhorne
  Delaware Valley Medical Center
Latrobe
  Latrobe Area Hospital
Norristown
  SmithKline Clinical Labs
Philadelphia
  Albert Einstein Medical Center
  Ayer Clinical Laboratory/Pennsylvania
    Hospital
  Children's Hospital of Philadelphia
  Hospital of Medical College of
    Pennsylvania
  Jeanes Hospital
  Lankenau Hospital
  St. Christopher's Hospital for Children
  Temple University Hospital
  Thomas Jefferson University Hospital
Pittsburgh
  Allegheny General Hospital
  Children's Hospital of Pittsburgh
  Clinical Pathology Facility
  Eye and Ear Hospital
  Fulton Medical Laboratory
  Joel Alcoff, M.D.
  Magee Women's Hospital
  Med-Chek Laboratories
  Mercy Hospital
  Presbyterian University Hospital
  St. Francis Hospital
  Western Pennsylvania Hospital
Plymouth Meeting
  Dekalb Laboratories
Sayre
  Robert Packer Hospital
Scranton
  Clinical Laboratories, Inc.
State College
  Centre Community Hospital
Trevose
  Damon Clinical Laboratories

West Chester
  Chester County Hospital
York
  York Hospital

Puerto Rico
  San Juan
    Nichols Institute

Rhode Island
  Providence
    Brown University Medical Center
    Rhode Island Hospital

South Carolina
  Charleston
    Medical University of South Carolina
    Roper Hospital
  Greenville
    Greenville Memorial Hospital

South Dakota
  Sioux Falls
    Veteran's Medical Center
  Vermillion
    University of South Dakota Virology Lab

Tennessee
  Chattanooga
    Allied Labs of Chattanooga
  Johnson City
    East Tennessee State Medical School
  Knoxville
    University of Tennessee
  Memphis
    Baptist Regional Laboratory
    LeBonheur Children's Hospital
    Regional Medical Center
    St. Francis Memorial Hospital
    St. Jude's Children's Research Hospital
  Nashville
    ICL Baptist Hospital
    Meharry Medical College
    National Health Laboratory
    St. Thomas Hospital

Table 36.1 (*continued*)

Texas
  Austin
    Austin Pathology
    Seton Medical Center
  Dallas
    Baylor University Medical Center
    National Pathology Laboratory
    Parkland Memorial Hospital
    University of Texas S.W. Medical Center
  Fort Worth
    Cook-Ft. Worth Children's Hospital
    Harris Medical Laboratory
  Galveston
    Shriner's Burn Institute
    University of Texas Medical Branch
  Houston
    Baylor College of Medicine
    Hermann Hospital
    Influenza Research Center/Baylor College
      of Medicine
    Methodist Hospital
    Microbiology Specialists, Inc.
  Hurst
    Allied Clinical Laboratory
  Lubbock
    Texas Tech University Health Science
      Center
  Richardson
    Viral Diagnostics
  San Antonio
    Brooke Army Medical Center
    Brooks Aero
    Medical Center Hospital/University of
      Texas
    Oak Hills Pathology Labs
    Severans Reference Lab
    Southwest Bioclinical Labs
  Temple
    Scott and White Clinic
  Texarcana
    Doctors Diagnostic Labs
  Victoria
    Lyster Reference Laboratory

Utah
  Salt Lake City

    Association Regional and University
      Pathologists, Inc.
    University of Utah Medical Center

Vermont
  Burlington
    Medical Center Hospital of Vermont

Virginia
  Charlottesville
    University of Virginia
  Fairfax
    American Medical Labs
  Norfolk
    Medical College of Hampton Roads
  Richmond
    Division of Consolidated Lab Services
    Medical College of Virginia
  Vienna
    National Health Labs

Washington
  Seattle
    Children's Hospital Medical Center
    Fred Hutchinson Cancer Research Center
    Laboratory of Pathology for Seattle, Inc.
    Virginia Mason Clinic
  Spokane
    Deaconess Hospital
    Pathology Associates Medical Labs
    Sacred Heart Medical Center

West Virginia
  Charleston
    Charleston Area Medical Center
  Huntington
    Marshall University School of Medicine

Wisconsin
  Green Bay
    Bellin Memorial Hospital
  LaCrosse
    Gunderson Clinic
  Marshfield
    St. Joseph's Hospital

Wyoming
  Jackson
    Intermountain Virology Laboratory

Table 36.2 Commercial Laboratories Which
Offer Viral Diagnostic Services Nationally

Bionetics Lab	Kensington, MD
Bio-Science Lab	Van Nuys, CA
Damon Clinical Labs	Phoenix, AZ
MA Bioproducts	Walkersville, MD
Metpath	Teterboro, NJ
National Health Laboratory	Nashville, TN
Smith Kline Bioscience	Norristown, PA
Specialty Labs	Los Angeles, CA
Virolab, Inc.	Emeryville, CA
Viromed	Minneapolis, MN

Index

Page numbers in italics indicate figures; page numbers followed
by t indicate tables.

A

ABTS. *See* 2.2'Amino-di[3 ethyl-benzthiazoline
    sulfonate (6)]
Acquired immunodeficiency syndrome (AIDS), 588,
    600–602. *See also* Human immunodeficiency
    virus
  cytomegalovirus infections in, 502
    treatment of, 505
  Epstein-Barr virus infections in, 513
  parvovirus infection in, treatment of, 561–562
  varicella-zoster virus infections in, 509
Acyclovir (ACV)
  in Epstein-Barr virus, 513
  in herpes simplex virus infections, 495
  herpes simplex virus 2 plaque formation and, 56
  herpes simplex virus resistant to, 282
  in varicella-zoster virus infections, 509
Adenosine monophosphate, in varicella-zoster virus
    infections, 509–510
Adenovirus(es), 332–334, 333t
  age and susceptibility to, 322t
  CDC testing for, 647t, 648t
  enteric, 375–376
    characteristics of, 375
    clinical manifestations of, 375
    diagnostic methods for, 376
    epidemiology and pathophysiology of, 375
  eye infections and, cytopathology of, 84–85, *85*,
    85t, 86
  in feces, electron microscopy and, 92
  group 1, hemagglutination inhibition test and, 247
  IgM determination and, interpretation of assay
    results and, 271
  immunoelectron microscopy and, 97
  immunofluorescence and, *123*, 124, 125, 125t
  incubation for cell culture, 47
  inoculation and, 34
  isolation of, 59
  laboratory diagnosis of, 324t, 334

radioimmunoassay antibody detection and, 140t
radioimmunoassay antigen detection and, 139t,
  142, 143, 144
respiratory tract infections and, cytopathology of,
  74, 74t, 76
specimens of, 21t, 24t, 26t
  cerebrospinal fluid, 30
  fecal, 29
  ocular, 30
  storage of, 33
  timing of collection of, 27
  tissue, 31
  urine, 29
  types of, 333t, 333–334
  urinary tract infections and, cytopathology of, 77t
Adult T-cell leukemia (ATL)
  clinical course of, 593–595, *594*, 594t
  clustering of T-cell malignancies and epidemiology
    of HTLV-I and, 591–593
  pathogenesis of, 591–595
Adult T-cell leukemia viruses (ATLV), 587
Affinity chromatography, immunoglobulin
    purification by, 162
Agar diffusion method, of negative staining, electron
    microscopy and, 96
Agar gel precipitation (AGP) test
  for antibody assay, poxviruses and, 544
  poxviruses and, 539
Agglutination assays. *See also* Hemagglutination
    inhibition test; Immune adherence
    hemagglutination
  retroviruses and, 606
Agglutinins, naturally occurring, removal of, 245
AGP test. *See* Agar gel precipitation test
Airfuge ultracentrifugation, of negative staining,
    electron microscopy and, 96–97
Alkaline phosphatase conjugates, preparation of, 163
  one-step glutaraldehyde method for, 163

681

Allantoic cavity, virus isolation and
  harvesting, assay, and identification of isolates
    and, 63–64
  inoculation and, 62, *63*
Alphaviruses, radioimmunoassay antibody detection
    and, 139
Alsever's solution, 249
Amantadine (Symmetrel), influenza A virus and, 327
2.2'Amino-di[3 ethyl-benzthiazoline sulfonate (6)]
    (ABTS), purification and storage of, 164
Amniotic fluid, virus isolation and
  harvesting, assay, and identification of isolates
    and, 63–64
  inoculation and, 62, *63*
Amphotericin B, in culture media, 10
Antibodies
  anti-rabies, detection of, 433–434
  for enzyme-linked immunosorbent assay, 161
    determination of optimal dilution of, 167
    purification of, 161–162
  enzyme-linked immunosorbent assay detection of,
    167
Antigen(s)
  for enzyme-linked immunosorbent assay, 159–161
    determination of optimal dilution of, 166–167
  enzyme-linked immunosorbent assay detection of,
    167
Antigen capture assays, retroviruses and, 609–610
Antigenicity, of enteroviruses, 343
Antigen titration, for complement fixation test. *See*
    Complement fixation test, antigen titration for
Antimicrobial agents, in culture media, 10
Antisense technology, in herpes simplex virus
    infections, 496
Antisera
  quality control and, 12–13
  reference, 249
Anti-species conjugate, quality control of, for
    immunofluorescence, 119
Antiviral drug susceptibility testing. *See* Dye-uptake
    (DU) assay
Arboviruses, 443–451. *See also* Colorado tick fever
    virus; Eastern equine encephalitis virus;
    Jamestown Canyon virus; LaCrosse virus;
    Powassan virus; Snowshoe hare virus; St.
    Louis encephalitis virus; Venezuelan equine
    encephalitis virus; Western equine
    encephalitis virus
  CDC testing for, 646*t*
  characteristics of, 444–447, 445*t*
  clinical description of, 448*t*, 448–449
  history of, 443–444
  laboratory diagnosis of, 449–451
    antigen and antibody detection and, 450–451
    specimen collection and, 449
    virus isolation and, 449–450
  pathogenesis of, 447–448
  serologic determination of, 35
  trends in, 451
Arenaviruses, CDC testing for, 647*t*
Arthropod specimens, National Center for Infectious
    Diseases requirements for, 652–653
Astroviruses, 379

characteristics of, 379
detection methods for, 379
epidemiology, pathophysiology, and clinical
    manifestations of, 379
in feces, electron microscopy and, 92
radioimmunoassay antibody detection and, 144
ATL. *See* Adult T-cell leukemia
ATLV. *See* Adult T-cell leukemia viruses
Australia antigen. *See* Hepatitis B virus
Autoclaves, routine maintenance and performance
    checks on, 14
Avian specimens, National Center for Infectious
    Diseases requirements for, 652–653

B
Backup procedures, 15–16
BAL. *See* Bronchoalveolar lavage
Base-pairing, nucleic acid hybridization and, 287
BGM kidney cells. *See* Buffalo green monkey kidney
    cells
Biological safety cabinets, routine maintenance and
    performance checks on, 14
BK virus (BKV), 465
  characteristics of, 465
  laboratory diagnosis of, 469
  pathogenesis and disease potential of
    in immunocompetent hosts, 466
    malignancies and, 468
    pregnancy and, 468
    primary immunodeficiency diseases and, 467–
      468
    in renal and bone marrow transplant recipients,
      467
  urinary tract infections and, cytopathology of, 77*t*,
    77–79, *79, 80*
  urine specimens of, 29
Blood
  electron microscopy and, 93, *95*
  interference assay and, 112
  as source of specimens, 30–31
Bone marrow transplants, polyomaviruses and, 467
Bovine papular stomatitis (BPS) virus, 532
  antigen identification and, 540*t*
Brain, herpes simplex virus in, electron microscopy
    and, 93, *95*
Breda virus, 380–381
Bronchial washes, for immunofluorescence, 118
Bronchoalveolar lavage (BAL)
  for immunofluorescence, 118
  as source of specimens, 28
Buffalo green monkey (BGM) kidney cells, specimen
    processing and, 34
Buffalopox virus, 530–531
  isolation of
    cell cultures and, 537
    chicken embryo chorioallantoic membrane and,
      536
Buffers, for enzyme-linked immunosorbent assay,
    165

C
Calciviruses, 379–380
  characteristics of, 379

diagnostic methods for, 380
epidemiology and clinical manifestations of, 379–380
in feces, electron microscopy and, 92
immunoelectron microscopy and, 97
California encephalitis virus. *See* LaCrosse virus
CAM. *See* Chorioallantoic membrane
Camelpox virus, antigen identification and, 540*t*
Capsaicin, in varicella-zoster virus infections, 510
Cell culture(s), 8, *9*
  continuous cell lines and, 44, 44*t*
  diploid, 44, 44*t*
  enterovirus replication in, 344–345, *346*
  of human B-lymphotrophic virus, 517–518, *518*
  isolation of hepatitis viruses and, 409–410
  neutralization and, 233
  primary cells and, 44, 44*t*
  processing isolates for electron microscopy, 102, *102–104*, 104–105
  susceptibility to viruses and, 44, 45*t*
  virus isolation in. *See* Virus isolation, in cell cultures
Cell lysates, for enzyme-linked immunosorbent assay, 159–160
Centers for Disease Control (CDC), testing by. *See* National Center for Infectious Diseases
Centrifugation culture. *See* Shell vial technique
Centrifuges, routine maintenance and performance checks on, 14
Cerebrospinal fluid (CSF)
  electron microscopy and, 92
  interference assay and, 112
  as source of specimens, 29, 30
Cervical dysplasia, human papillomavirus and, 461–462, 462*t*
CF test. *See* Complement fixation test
Chickenpox. *See* Varicella-zoster virus
Chlamydia, 627–637
  characteristics of, 628–629
    antigenic relationships and, 629
    growth cycle and, 628
    taxonomy and, 628–629
  direct cytologic examination and, 632
    direct fluorescent antibody technique and, 632
    Giemsa staining technique and, 632
    iodine staining technique and, 632
  enzyme immunoassay and, 632–633
  eye infections and, cytopathology of, 85*t*, *86*, 86–87
  identification of, 634
  infected sites and collection methods and, 630
  inoculation and, 34
  isolation methods for, 633–634
  nonculture methods for determining, 631–632
  pathogenesis of, 629–630
  serodiagnosis of, 634–637
    complement fixation and, 635–636
    microimmunofluorescence and, 636–637
  stability, storage, and transport of specimens and, 630–631
Chorioallantoic membrane (CAM), virus isolation and, 62

harvesting, assay, and identification of isolates and, 63–64
inoculation and, 62–63, *63*
poxviruses and, 535–536
Ciliocytophthoria, in respiratory tract infections, 76–77, *77*
CMV. *See* Cytomegalovirus
Cocultivation, viral isolation and, 61
Colorado tick fever (CTF) virus, 444
  CDC testing for, 646*t*
  characteristics of, 445*t*, 446
  clinical features of, 449
  laboratory diagnosis of
    antigen detection and, 450
    specimen collection for, 449
    virus isolation and, 449–450
Color standards, for complement fixation test, 213–214
  hemoglobin solution preparation and, 213
  preparation of, 214, 214*t*
  red blood cell suspension preparation and, 214
Competition radioimmunoassays, retroviruses and, 607–608
Competitive binding radioimmunoassay. *See* Radioimmunoassay, solid phase
Complement fixation (CF) test, 203–223
  for antibody assay, poxviruses and, 544
  antigen titration and, 216–220
    addition of reactants to microtiter plate and, 217–218
    complement preparation and, 218–219
    preparation of hemolysin sensitized sheep erythrocytes and, 219
    reading of test and, 219*t*, 219–220, 220*t*
    setup for, 216–217, *217*
  background of, 204–205
  Chlamydia and, 635–636
  color standards and, 213–214, 214*t*
    preparation of hemoglobin solution and, 213–214
    preparation of red blood cell suspension and, 214
  comparison with other methods, 205–206
  complement titration and, 214–216, 215*t*
    computation of complement volume producing 50% hemolysis and, 215*t*, 215–216, *216*
    preparation of complement dilution and, 214
    preparation of sensitized cells and, 214
  diagnostic serology test and, 220–223
    addition of reactants to microtiter plate and, 221–222
    reading of test and, 222–223
    setup for, 220–221, *221–222*
  hemolysin titration and, 211–213, *213*
    preparation of complement dilution and, 212
    preparation of hemolysin dilution and, 211
    preparation of hemolysin sensitized cells and, 212, 212*t*
  herpes simplex virus and, 490
  interpretation of, 208–209
  mumps virus and, 576
  performance of, 211–223
  poxviruses and, 539
  reagents for, 207–208, 208*t*

Complement fixation (CF) test [*cont.*]
  setup for, 209–210
    equipment and, 209
    reagent preparation and, 209–210
    reagents and, 209
    test outline and, 210
  sheep erythrocyte preparation and, 211
    cell standardization-centrifugation method for, 211
    cell washing and, 211
  specimens for, 207
  technique for, 206–207
  test principles and, 203–204, *204*
Congenital rubella syndrome (CRS), 577–579
Congo virus, CDC testing for, 646*t*
Conjugates, for enzyme-linked immunosorbent assay, 161–164
  determination of optimal concentration of, 165–166
Contamination checklist, 11, 11*t*
Coronaviruses, 334–335, 380
  age and susceptibility to, 322*t*
  in feces, electron microscopy and, 92
  laboratory diagnosis of, 324*t*, 335
Corticosteroids, in Epstein-Barr virus, 513
Cowpox virus, 530
  antigen identification and, 540*t*
  isolation of
    cell cultures and, 537
    chicken embryo chorioallantoic membrane and, 536
Coxsackievirus(es). *See also specific coxsackieviruses*
  asymptomatic infections with, 348
  CDC testing for, 647*t*
  pathogenesis and clinical syndromes and, 350, 351*t*, 352
  radioimmunoassay antibody detection and, 139
  serologic diagnosis of, 357
Coxsackievirus A
  age and susceptibility to, 322*t*
  chronic infections with, 352–353
  discovery of, 342
  identification of, 356
  incubation times of, 353
  isolation of, 59*t*, 60
    in mice, 64
  pathogenesis and clinical syndromes and, 351*t*
  pregnancy and, 353
  serotypes of, 341*t*
  specimens of, dermal, 30
Coxsackievirus B
  age and susceptibility to, 322*t*
  chronic infections with, 352–353
  discovery of, 342
  IgM determination and, interpretation of assay results and, 271
  incubation times of, 353
  isolation of, 59, 59*t*
    in mice, 64
  pathogenesis and clinical syndromes and, 351*t*
  pregnancy and, 353
  radioimmunoassay antibody detection and, 144
  serotypes of, 341*t*

CPE. *See* Cytopathic effects
Crimean hemorrhagic fever virus, CDC testing for, 646*t*
CRS. *See* Congenital rubella syndrome
Culturettes, 31–32
Cytocentrifugation, 72
Cytology, herpes simplex virus and, 488–489
Cytomegalovirus (CMV), 501–507
  biology and pathogenesis of infection and, 502–504, *503*
  CDC testing for, 647*t*
  clinical manifestations of, 504–505
  complement fixation test and, 205–206
  detection by immune adherence hemagglutination, 257
  enzyme immunoassay kits for, 158*t*
  epidemiology of, 504
  history of, 501–502
  immunofluorescence and, 124, 125, 125*t*
  incubation for cell culture, 47
  inoculation and, 34
  isolation of, 58–59
    shell vial technique for, 51, *52*
  laboratory diagnosis of, 505–507, *506*
  nucleic acid hybridization and, 293
  prevention and treatment of, 505
  radioimmunoassay antibody detection and, 140*t*, 142
  reference virus stocks, 12
  respiratory tract infections and, cytopathology of, 73, 74*t*, *75*
  specimens of, 21*t*, 24*t*, 25*t*
    blood, 30
    bronchoalveolar lavage, 28
    cerebrospinal fluid, 30
    dermal, 30
    ocular, 30
    sputum, 28
    timing of collection of, 27
    tissue, 31
    urine, 29
  urinary tract infections and, cytopathology of, 77*t*, 77–78, *78*
Cytomegalovirus hyperimmune globulin, 505
Cytopathic effects (CPE)
  cytomegalovirus and, 506, *506*
  detection in cell cultures, 48, *49–51*
  herpes simplex virus and, 484
  mumps virus and, 575
  varicella-zoster virus and, 510, *510*
  virus infectivity assay by end point of, 54, 54*t*
Cytopathology, 73–87
  of genital infections, 79–84, 81*t*, *81–84*
  of ocular infections, 84–87, *85*, 85*t*, *86*
  of respiratory tract infections, 73–77, 74*t*, *75–77*
  of urinary tract infections, 77*t*, 77–79, *78–80*

D

Dane particle. *See* Hepatitis B virus
Dawson's encephalitis, measles virus and, 572
Denaturation, nucleic acid hybridization and, 287
Dengue virus

CDC testing for, 646*t*
characteristics of, 445*t*
radioimmunoassay and, 144
trends with, 451
Dermal lesions, as source of specimens, 30
  for immunofluorescence, 118
Dexamethasone (DMX), incubation and, 48
DIA. *See* Dot immunobinding assay
Dimethyl sulfoxide (DMSO)
  incubation and, 47–48
  specimen storage and, 34
  in varicella-zoster virus infections, 509–510
Direct application method, of negative staining,
  electron microscopy and, 94
Direct fluorescent antibody technique, *Chlamydia*
  and, 632
Direct immunoelectron microscopy method, 98
Direct interference method, rubella virus and, 579
Direct smears, 72
DMSO. *See* Dimethyl sulfoxide
DMX. *See* Dexamethasone
DNA, problems encountered in using in clinical
  testing, 309–310, 310*t*
DNA probes
  parvovirus and, 560
  preparation of, 290
Dot blots, *288*, *289*, 291, 312–313
  human papillomavirus and, 464*t*
Dot immunobinding assay (DIA), rabies virus and,
  434
Drugs, antiviral. *See also specific drugs*
  susceptibility testing and. *See* Dye-uptake assay
DU. *See* Dye-uptake assay
DVBID. *See* National Center for Infectious
  Diseases, Division of Vector-Borne Infectious
  Diseases of
DVRD. *See* National Center for Infectious Diseases,
  Division of Viral and Rickettsial Diseases of
Dye-uptake (DU) assay
  advantages of, 281–282
  applications of, 281
  disadvantages and technical problems of, 282–283
  history of, 277
  materials for, 277–278
  virus infectivity assays and, 278–279, 279*t*
  virus inhibition assay and, 279–281, 280*t*, 281*t*

     E
Eastern equine encephalitis (EEE) virus, 443
  CDC testing for, 646*t*
  characteristics of, 444–446, 445*t*
  clinical features of, 448*t*
  laboratory diagnosis of, virus isolation and, 449–
    450
ECHO-11 challenge, interference assay and, 113
Echoviruses
  age and susceptibility to, 322*t*
  asymptomatic infections with, 348
  CDC testing for, 647*t*
  incubation times of, 353
  isolation of, 59, 59*t*
  pathogenesis and clinical syndromes and, 351*t*–
    352*t*

serologic diagnosis of, 357
serotypes of, 341*t*
Eczema vaccinatum, vaccinia and, 529
Edmonston B vaccine, 574
EEE virus. *See* Eastern equine encephalitis virus
EIA. *See* Enzyme immunoassay
Electron microscopy (EM), 89–105. *See also*
  Immunoelectron microscopy
  advantages of, 89–90
  limitations of, 90
  negative staining methods and, 93–97
    agar diffusion method of, 96
    airfuge ultracentrifugation method of, 96–97
    direct application method of, 94
    materials for, 93–94
    pseudoreplica method of, 96
    water drop method of, 95–96
  parvovirus and, 557, *558*
  poxviruses and, 537–539
  processing cell culture isolates for, 102, *102–104*,
    104–105
    negative staining and, 104
    thin sectioning and, 104–105
  rotaviruses and, 372
  safety precautions for, 91
  special facilities for, 90
  thin sectioning methods and, 101–102, 104–105
  transmission electron microscope and, 90
  virus detection by
    in blood, 93, *95*
    in cerebrospinal fluid, 92
    in feces, 92, *92*, *94*
    in respiratory tract secretions, 92
    in tissues, 93, *95*
    in urine, 93
    in vesicle fluid and crusts, *91*, 91–92
Electropherotyping, rotaviruses and, 374
ELISA. *See* Enzyme-linked immunosorbent assay
Embryonated eggs
  neutralization and, 233
  virus isolation in. *See* Virus isolation, in
    embryonated eggs
Encephalitis. *See* Eastern equine encephalitis virus;
  Japanese encephalitis virus; Murray Valley
  encephalitis virus; St. Louis encephalitis
  virus; Subacute sclerosing panencephalitis;
  Venezuelan equine encephalitis virus;
  Western equine encephalitis virus
  Dawson's, measles virus and, 572
  due to herpes simplex virus, laboratory diagnosis
    of, 489
  vaccinia and, 529
Enteroviruses, 341–357. *See also* Coxsackievirus A;
  Coxsackievirus B; Coxsackievirus(es);
  Echoviruses; Hepatitis A virus; Poliovirus;
  Rhinovirus
  age and susceptibility to, 322*t*
  antigenicity and neutralization of, 343–344
  asymptomatic infections with, 348–349
  CDC testing for, 647*t*
  discovery of, 341–342
  geographic, seasonal, socioeconomic, sex, and age
    factors and, 347–348, *348*

Enteroviruses [*cont.*]
  hemagglutination inhibition test and, 247
  immune response to, 353–354
  immunoelectron microscopy and, 97
  immunofluorescence and, 125*t*
  incubation times of, 353
  isolation of, 59*t*, 59–60, 60
  laboratory diagnosis of, 355–357
    serologic, 357
    virus isolation and identification and, 355–356
  mode of transmission of, 346–347, 347*t*
  pathogenesis and clinical syndromes and, 349–353, 351*t*–352*t*
  radioimmunoassay for, 138–139
    antibody detection and, 138–139, 140*t*
  reactivity to chemical and physical agents, 343
  replication in cell culture, 344–345, *346*
  specimens of, 21*t*, 23*t*, 25*t*
    blood, 30–31
    cerebrospinal fluid, 30
    fecal, 29
    storage of, 33
    timing of collection of, 20, 27
  storage of, 343
  structure of, 342–343, *343*
  vaccination against, 354–355
Enzyme immunoassay (EIA), 299–300. *See also* Enzyme-linked immunosorbent assay
  Chlamydia and, 632–633
  IgM determination and, 269
  kits for, 158*t*–159*t*
  rabies virus and, 433
Enzyme-linked immunosorbent assay (ELISA), 35, 153–170, 154*t*–156*t*, *157*
  antibodies for, 161
  antigen for, 159–161
  buffers for, 165
  conjugates and substrates for, 161–164
    alkaline phosphatase conjugate preparation and, 163
    antibody purification and, 161–162
    horse radish peroxidase conjugate preparation and, 163–164
    test specimens and, 164–165
  error sources in, 169–170
  herpes simplex virus and, 483*t*, 486–488, 490
  instruments for, 169
  materials for, 157–165
  mumps virus and, 576
  parvovirus and, 558–560
  poxviruses and, 539, 543
  quantitation and, 168–169
  rabies virus and, 433, 434
  retroviruses and, 603
  rotaviruses and, 373–374
  rubella virus and, 579
  solid phase, materials for, 157, 159
  test methods for, 165–168
    antiviral antibody detection and, 167
    antiviral IgM detection and, 167–168
    optimal antibody dilution determination and, 167
    optimal antigen dilution determination and, 166–167

    optimal conjugate concentration determination and, 165–166
    viral antigen detection and, 167
Epidermodysplasia verruciformis, 459
Epifluorescence, 120
Epstein-Barr virus (EBV), 501, 511–515
  biology and pathogenesis of infection and, 511–512
  CDC testing for, 647*t*
  clinical manifestations of, 512–513
  detection by immune adherence hemagglutination, 256, 257, 258
  enzyme immunoassay kits for, 158*t*
  epidemiology of, 512
  history of, 511
  IgM determination and, interpretation of assay results and, 271
  laboratory diagnosis of, 513–515, *514*
  serologic determination of, 35
  specimens of, 24*t*
  treatment of, 513
Errors, sources of, 15–16, 16*t*
  in enzyme-linked immunosorbent assay, 169–170
Erythema multiforme, vaccinia and, 529
Erythrocytes, 249
  for immune adherence hemagglutination, 255–256
  preparation of red blood cell suspension for hemagglutination inhibition test and, 244
  sheep, for complement fixation test, preparation of, 211
Exanthem viruses, CDC testing for, 647*t*
Explant culture, viral isolation and, 61
Eye
  infections of, cytopathology of, 84–87, 85, 85*t*, 86
  as source of specimens, 30

F
FA. *See* Fluorescent antibody test
Facility design and maintenance, 3–5
FAMA. *See* Fluorescent antibody-to-membrane antigen
Fecal specimens, electron microscopy and, 91*t*, 92, *92, 92, 94*
Fetal transfusion, in parvovirus infection, 561
Filtration, 72
Flaviviruses, radioimmunoassay antibody detection and, 141
Fluorescence, 119. *See also* Immunofluorescence; Microimmunofluorescence
Fluorescent antibody (FA) test, rabies virus and, 432
Fluorescent antibody-to-membrane antigen (FAMA), varicella-zoster virus and, 511
Foscarnet
  in cytomegalovirus, 505
  in human B-lymphotrophic virus infections, 517
  in varicella-zoster virus infections, 509
Freezers, routine maintenance and performance checks on, 15
δ agent. *See* Hepatitis D virus

G
Ganciclovir
  in cytomegalovirus, 505

in Epstein-Barr virus, 513
in human B-lymphotrophic virus infections, 517
in varicella-zoster virus infections. *See* Varicella-zoster immune globulin
Gastrointestinal tract viruses, 361–381. *See also* Adenovirus(es), enteric; Norwalk virus; Rotaviruses
treatment and prevention of viral gastroenteritis and, 381
Gel filtration, IgM determination by, 265
Genetic analysis, rotaviruses and, 374–375
Genital tract infections
cytopathology of, 79–84, 81*t*, *81–84*
human papillomavirus and, 460
Gentamicin sulfate, in culture media, 10
Giemsa staining technique, *Chlamydia* and, 632

H

HAd inhibition test. *See* Hemadsorption inhibition test
HAd test. *See* Hemadsorption test
HAI test. *See* Hemagglutination inhibition test
HAM. *See* Human T-cell leukemia virus I, myelopathy associated with
Hanks' balanced salt solution (HBSS), 32
Hantaan virus, CDC testing for, 646*t*
HBLV (human B-lymphotrophic virus). *See* Human herpesvirus-6
HBSS. *See* Hanks' balanced salt solution
HDF cells. *See* Human diploid fibroblast cells
HDRV vaccine. *See* Human diploid rabies virus vaccine
Heat blocks, routine maintenance and performance checks on, 15
Hemadsorption (HAd), detection in cell cultures, 48–49, *52*
Hemadsorption (HAd) inhibition test, 248
methods for, 248
Hemadsorption (HAd) test, 247–248, *248*
materials for, 247
methods for, 247–248
Hemagglutinating viral antigens, 249
Hemagglutination inhibition (HAI) test, 243–247. *See also* Immune adherence hemagglutination
applications of, 246–247
hemagglutinin titration for, 244, *244*
for influenza, 243–246, *246*
kaolin treatment and, 245
KIO treatment and, 245
materials for, 244
measles virus and, 573
mumps virus and, 576
poxviruses and, 540–541, 541*t*
receptor destroying enzyme treatment and, 245
red blood cell suspension preparation for, 244
removal of naturally occurring agglutinins for red blood cells and, 245
serum treatment for, 245
Hemolysin, titration of, for complement fixation test, 211–213, 212*t*
Heparin-MnCl$_2$. *See* HEPES-saline-albumin-gelatin
Hepatitis, 397–417. *See also specific hepatitis viruses*

enzyme immunoassay kits for, 158*t*
laboratory diagnosis of, 409–417
cell culture for virus isolation and, 409–410
histologic, biochemical, and hematologic tests and, 409
serologic markers and immunodiagnosis and, 410–417, *411*
nature and types of, 397
pathogenesis of, 403–407
serologic determination of, 35
Hepatitis A virus (HAV), 398
antibody of, radioimmunoassay for, 136–137, *138*
CDC testing for, 648*t*
characteristics of, 400–401
discovery of, 342
in feces, electron microscopy and, 92
identification of, 356
immune response and, 353–354
pathogenesis and clinical syndromes and, 352, 352*t*, *353*, 403–404
pathology of, 407
serologic diagnosis of, 357
serologic markers and immunodiagnosis in, 410–412, *412*
specimens of, 24*t*
Hepatitis B virus (HBV), 398–399
in blood, electron microscopy and, 93, *95*
CDC testing for, 648*t*
characteristics of, 401–402
immunoelectron microscopy and, 97
indirect radiometric assays for antibodies of, 134–135
one-step procedure, 134, *134*, 136*t*
two-step procedure, 134–135, 136*t*, *137*
nucleic acid hybridization and, 294
pathogenesis of, 403, *404*, 404–405
host factors and, 405
virus and, 405
pathology of, 407
radiometric assays for antigens and antibodies of, 131–134, *132*, *133*
serologic markers and immunodiagnosis in, 412–414, 413*t*, *414*
serologic markers for, 130–131, 131*t*
specimens of, 24*t*
Hepatitis C virus (HCV), 400
characteristics of, 402
pathogenesis of, 403, 406
pathology of, 408–409
radioimmunoassay antibody detection and, 144
serologic markers and immunodiagnosis in, 416–417
specimens of, 24*t*
Western blotting and, 305
Hepatitis D virus (HDV), 399
characteristics of, 402
pathogenesis of, 403, 405–406
pathology of, 407–408
serologic markers and immunodiagnosis in, 414–415, *415*, *416*
Hepatitis E virus (HEV), 399–400
CDC testing for, 648*t*
characteristics of, 402–403

Hepatitis E virus (HEV) [*cont.*]
  pathogenesis of, 403, 406–407
  pathology of, 408–409
  serologic markers and immunodiagnosis in, 415–416
HEPES-saline-albumin-gelatin (HSAG), 249
Herpes simplex virus (HSV), 473–496. *See also specific herpes simplex viruses*
  acyclovir-resistant, 282
  age and susceptibility to, 322*t*
  antigen detection and, 485–488, *487*
  in brain, electron microscopy and, 93, *95*
  CDC testing for, 647*t*
  characteristics of, 473–475, 475*t*
  chorioallantoic membrane pock formation and, 62
  cytology and, 488–489
  detection by immune adherence hemagglutination, 258
  DNA technology and, 488
  encephalitis due to, laboratory diagnosis of, 489
  enzyme immunoassay kits for, 158*t*
  epidemiology of, 479–480
  eye infections and, cytopathology of, 85*t*, 86
  genital tract infections and, cytopathology of, 79–80, *81*, 81*t*, *82*
  immunity and vaccine development and, 491–493, 492*t*
  immunofluorescence and, 117, 122, 124, 125, 125*t*
  immunoperoxidase staining and, 192, 195–197, 199
  inoculation and, 34
  laboratory diagnosis of, 482, 483*t*, 484–489
  latency induced by, 480–482
  medical importance of, 19
  nucleic acid hybridization and, 295
  pathogenesis of, 477*t*, 477–479
  radioimmunoassay and, 142–143
    antibody detection and, 140*t*, 142–143, 144–145
    antigen detection and, 139*t*, 143
  replicative cycle of, 476–477
  respiratory tract infections and, cytopathology of, 74*t*, 74–75, *75*, *76*
  serologic, 489–491
  specimens of, 21*t*, 22*t*, 23*t*, 24*t*, 25*t*, 26*t*
    cerebrospinal fluid, 30
    dermal, 30
    ocular, 30
    sputum, 28
    storage of, 33
    tissue, 31
    transport of, 32, 32*t*, 33
    urine, 29–30
  treatment of, 494*t*, 494–496
  unusual features of, 493–494
  urinary tract infections and, cytopathology of, 77*t*, 77–78, *79*, *80*
  virus isolation and, 484–485
Herpes simplex virus 1 (HSV-1), 473
  characteristics of, 474
  epidemiology of, 480
  isolation of, 57, 58*t*
  latency induced by, 481
  pathogenesis of, 467, 479
  replicative cycle of, 476

Herpes simplex virus 2 (HSV-2), 473
  characteristics of, 474–475, 475*t*
  enzyme-linked immunosorbent assay and, 160
  epidemiology of, 480
  isolation of, 57, 58*t*
  latency induced by, 481
  pathogenesis of, 467, 479
  plaque formation and, 56
  replicative cycle of, 476
  specimens of, cerebrospinal fluid, 30
Herpesviruses. *See also* Cytomegalovirus; Herpes simplex virus; Herpes simplex virus 1; Herpes simplex virus 2; Herpes virus hominis; Herpes zoster virus; Human herpesvirus-6; Varicella-zoster virus
  CDC testing for, 647*t*
  in cerebrospinal fluid, electron microscopy and, 92
  incubation for cell culture, 47–48
  stability of, 31
  in urine, electron microscopy and, 93
Herpes virus hominis, immunofluorescence and, *124*
Herpes zoster virus, eye infections and, cytopathology of, 86
HHV-6. *See* Human herpesvirus-6
HIV. *See* Human immunodeficiency virus; Human immunodeficiency virus 1; Human immunodeficiency virus 2
Horse radish peroxidase (HRP) conjugates
  preparation of, 163–164
    one-step glutaraldehyde method for, 163
    periodate and, 164
    two-step glutaraldehyde method for, 163–164
  purification and storage of, 164
Host factors
  human papillomavirus and, 458
  pathogenesis of hepatitis B and, 405
  polyomaviruses and, 466
Host system, neutralization and, 233
  cell cultures and, 233
  embryonated hens' eggs and, 233
  mice and, 233
HPV. *See* Human papillomavirus
HRP. *See* Horse radish peroxidase
HSAG. *See* HEPES-saline-albumin-gelatin
HSV. *See* Herpes simplex virus; Herpes simplex virus 1; Herpes simplex virus 2
HTLV. *See* Human T-cell leukemia virus; Human T-cell leukemia virus I; Human T-cell leukemia virus II
HTLV-III (human T-cell lymphotrophic virus type-III). *See* Human immunodeficiency virus 1
Human B-lymphotrophic virus (HBLV). *See* Human herpesvirus-6
Human diploid fibroblast (HDF) cells, specimen processing and, 34
Human diploid rabies virus (HDRV) vaccine, 435
Human herpesvirus-6 (HHV-6), 501, 515–518
  biology and pathogenesis of, 515–516
  clinical manifestations of, 516–517
  epidemiology of, 516
  history of, 515
  laboratory diagnosis of, 517–518, *518*
  treatment of, 517

Human immunodeficiency virus (HIV). *See also*
    Acquired immunodeficiency syndrome;
    specific human immunodeficiency viruses
  enzyme immunoassay kits for, 158*t*
  human T-cell leukemia virus I and, 596
  parvovirus infections and, 554–555
  pathophysiology of infection with, 597–602
    clinical manifestations of, 598–602, 599*t*
    epidemiology and transmission and, 597–598
  Western blotting and, 299, 301, 304, 305
Human immunodeficiency virus 1 (HIV-1)
  CDC testing for, 648*t*
  detection of genomes of, 612
  genome organization and protein products of, 590–591
  history and characteristics of, 588
  isolation of, 613–614
  large-scale serologic screening for, 609
  stability and safety precautions and, 614–615
Human immunodeficiency virus 2 (HIV-2)
  CDC testing for, 648*t*
  detection of genomes of, 612
  genome organization and protein products of, 590–591
  history and characteristics of, 588
  isolation of, 613–614
  large-scale serologic screening for, 609
Human monkeypox virus, 528
  antigen identification and, 540*t*
  enzyme-linked immunosorbent assay and, 543
  hemagglutination inhibition test and, 541
  indirect fluorescent antibody test and, 542–543
  isolation of
    cell cultures and, 537
    chicken embryo chorioallantoic membrane and, 536
    radioimmunoassay-adsorption test and, 543–544
    radioimmunoassay and, 543
    stained smears and, 539
Human papillomavirus (HPV), 455–464, 456*t*
  characteristics of, 456–457
  clinical types of warts and, 458–462
    cervical dysplasia and, 461–462, 462*t*
    epidermodysplasia verruciformis and, 459
    genital warts and, 460
    respiratory papilloma and, 460–461
    skin warts and, 458–459
  genital tract infections and, cytopathology of, 81*t*, 81–83, *83*
  laboratory diagnosis of, 462–463
    antigen tests and, 462–463
    identification of viral genotypes and, 463, 464*t*
  nucleic acid hybridization and, 287, 293–294
  pathogenesis and disease potential of, 457–458
    host factors and, 458
    location of lesion and, 458
    virus genotype and, 458, 458*t*
  specimens of, 25*t*
    ocular, 30
    urine, 29
  treatment of, 463–464
Human T-cell leukemia virus (HTLV). *See also*
    specific viruses

Western blotting and, 305
Human T-cell leukemia virus I (HTLV-I), 585, *586*
  in adult T-cell leukemia, 591–595
  CDC testing for, 648*t*
  detection of genomes of, 611–612
  enzyme immunoassay kits for, 158*t*
  genome organization and protein products of, 588–590, *589*
  history and characteristics of, 585–588
  isolation of, 612–613
  large-scale serologic screening for, 608–609
  myelopathy associated with (HAM), 595–596
  pathophysiology of infection with, 591–596
    adult T-cell leukemia and, 591–595
    tropical spastic paraparesis and and HTLV-I-associated myelopathy and, 595–596
Human T-cell leukemia virus II (HTLV-II), 585, *586*
  CDC testing for, 648*t*
  detection of genomes of, 611–612
  genome organization and protein products of, 588–590, *589*
  history and characteristics of, 585–588
  isolation of, 612–613
  large-scale serologic screening for, 608–609
  pathophysiology of infection with, 596
Human T-cell lymphotrophic virus type-III (HTLV-III). *See* Human immunodeficiency virus 1
Hybridization. *See* In situ hybridization; Nucleic acid hybridization

I

IAHA. *See* Immune adherence hemagglutination
Idoxuridine
  in herpes simplex virus infections, 495
  in varicella-zoster virus infections, 509–510
IEM. *See* Immunoelectron microscopy
IF. *See* Immunofluorescence
IFA test. *See* Indirect fluorescent antibody test
IM (infectious mononucleosis). *See* Epstein-Barr virus
Immune adherence hemagglutination (IAHA), 251–259, 253*t*
  advantages and disadvantages of, 258–259
  applications of, 253*t*
  background of, 252
  hemagglutination patterns and, 256
  IgM antibody detection with, 256–257
  microtiter procedure for, 252–255
  rabies virus and, 433–434
  red blood cells for, 255–256
  sensitivity and specificity of, 257–258
Immune globulin
  in measles virus infection, 574
  prophylaxis against enteroviruses with, 355
  rabies, 436
    intravenous, 436
Immune response, enteroviruses and, 353–354
Immune status, assessment of, 35
Immunity, to herpes simplex virus, 491–493, 492*t*
Immunoassays. *See also* Enzyme immunoassay;
    Enzyme-linked immunosorbent assay;
    Radioimmunoassay

Immunoassays [*cont.*]
  membrane, 35
  rotaviruses and, 373*t*, 373–374
Immunoblot. *See also* Western blot
  retroviruses and, 603–605, *604*, 605*t*, *606*
Immunocompetence, specimen collection and, timing
    of, 20, 27
Immunodeficiency diseases. *See also* Acquired
    immunodeficiency syndrome; Human
    immunodeficiency virus; Human
    immunodeficiency virus 1; Human
    immunodeficiency virus 2
  BK virus and, 467–468
Immunodiagnosis, in hepatitis, 410–417, *411*
Immunoelectron microscopy (IEM), 97–101
  direct method of, 98
  immunogold method of, 100, *101*
  Norwalk virus and, 378
  parvovirus and, 557, *558*
  rotaviruses and, 372–373
  routine, 101
  serum-in-agar method of, 98, *99*
  solid phase, 98–100
    materials for, 99–100, *100*
Immunofluorescence (IF), 117–127. *See also*
    Microimmunofluorescence
  applications and future of, 126–127
  applications in detection of antigens in cell culture,·
    124–125, 125*t*
  direct versus indirect, 120
  herpes simplex virus and, 483*t*, 486, 489
  history of, 117
  indirect
    herpes simplex virus and, 485–486
    practical details for, 121–123, *123*, 124
    retroviruses and, 605–606
  microscope and, 120
  monoclonal antibodies and, 120
  physics of, 119–120
  poxviruses and, 539
  quality control of anti-species conjugate for, 119
  quality control of specific antisera for, 118–119
  reagents for, 118
  resolution of technical problems with, 125, 126*t*
  rotaviruses and, 374
  specimen preparation for, 120–121
  specimens for, 117–118
  viruses for which technique is available and, *123*,
    124, *124*, 125*t*
Immunofluorescence, indirect, IgM determination
    and, 268–269
Immunoglobulin(s)
  intravenous, rabies immune globulin and, 436–437
  purification of
    affinity chromatography and, 162
    ammonium sulfate precipitation and, 162
Immunoglobulin, intravenous
  cytomegalovirus and, 505
  in parvovirus infection, 561
Immunoglobulin M (IgM)
  detection by immune adherence hemagglutination,
    256–257
  determination of, 35, 263–270, 263–272, 264*t*

based on chemical inactivation of IgM, 263–264
based on physicochemical separation of IgM,
    265–267
interpretation of assays and, 270–272
solid phase "capture" immunoassays and, 269–
    270, *270*, *271*
solid phase indirect immunoassays and, *267*,
    267–279, *268*
enzyme-linked immunosorbent assay detection of,
    167–168
radioimmunoassay for, 135–137, *138*
Immunogold method, for immunoelectron
    microscopy, 100, *101*
Immunoperoxidase staining, 189–199, *191*, 192*t*. *See*
    also Unlabeled peroxidase anti-peroxidase
    technique
  applications of, 190
  indirect, herpes simplex virus and, 489
Inactivated absorbed rabies vaccine (RDRV), 435–
    436
Incubation, for cell culture, 47–48
Incubators, routine maintenance and performance
    checks on, 14
Indirect fluorescent antibody (IFA) test, poxviruses
    and, 542–543
Indirect interference method, rubella virus and, 579
Infectious mononucleosis (IM). *See* Epstein-Barr
    virus
Infectivity assay, by end point of cytopathic effects,
    54, 54*t*
Influenza virus(es), 323–325, 324*t*, *326*, 327, 327*t*
  age and susceptibility to, 322*t*
  CDC testing for, 647*t*
  complement fixation test and, 206
  hemadsorption of, 48
  hemagglutination inhibition test and, 243–246, *246*
  laboratory diagnosis of, 324*t*, 325, 327
  radioimmunoassay antigen detection and, 139*t*, 143
  respiratory tract infections and, cytopathology of,
    76
  specimens of, 21*t*
    sputum, 28
    tissue, 31
  vaccine against, 327, *327*
Influenza virus A, 323, 324, 324*t*, 325
  immunofluorescence and, *123*, 124, 125, 125*t*
  immunofluorescence detection of, 117
  inoculation and, 34
  isolation of, 60, 62
  prophylaxis of, 327
  tissue specimens of, 31
Influenza virus B, 323, 325
  immunofluorescence and, 124, 125, 125*t*
  isolation of, 60, 62
Influenza virus C, 323, 325
Influenza viruses. *See also specific influenza viruses*
Inoculation, 34–35
  for cell culture, 45–47, *47*
  in embryonated eggs
    amniotic or allantoic cavity inoculation and, 62,
      63
    chorioallantoic membrane inoculation and, 62–
      63, *63*

for interference assay, 112–113
for shell vial technique, 53
virus isolation in mice and, 64, *65*
In situ hybridization, 291–292, *292*
  herpes simplex virus and, 488
  human papillomavirus and, 464*t*
Instruments, quality control and, 13–15
Interference, detection in cell cultures, 49–50
Interference assay, 111–114
  ECHO-11 challenge and neutralization and, 113–114
  history of, 111–112
  pros and cons of, 114
  specimen collection, transport, and storage for, 112
  specimen preparation and inoculation for, 112–113
Interferon, rabies prevention and, 436
Intravenous immunoglobulin
  cytomegalovirus and, 505
  in parvovirus infection, 561
Intravenous rabies immune globulin (IRIG), 436–437
Iodine staining technique, *Chlamydia* and, 632
Isolates, National Center for Infectious Diseases requirements for, 652
Isolation. *See* Virus isolation

J
Jamestown Canyon (JC) virus, 444, 465
  characteristics of, 446–447, 465
  clinical features of, 448*t*
  laboratory diagnosis of, 468, 469
  pathogenesis and disease potential of
    in immunocompetent hosts, 466
    malignancies and, 468
    pregnancy and, 468
    progressive multifocal leukoencephalopathy and, 466–467
    in renal and bone marrow transplant recipients, 467
  specimens of
    tissue, 31
    urine, 29
Japanese encephalitis virus, CDC testing for, 646*t*
Junin virus, CDC testing for, 647*t*

K
Kaolin, 249
Kits, quality control and, 12–13
Korean hemorrhagic fever virus, CDC testing for, 646*t*
Kyasanur Forest disease virus, CDC testing for, 646*t*

L
Laboratories
  hospital, university, and commercial, 671, 672*t*–680*t*
  of National Center for Infectious Diseases. *See* National Center for Infectious Diseases
  state, 659–670
    scope of services offered by, 664, 665*t*–669*t*, 670

submission of specimens to, 659, 660*t*–663*t*, 664
turnaround time for result reporting by, 670
LaCrosse (LAC) virus, 443–444, *444*
  CDC testing for, 646*t*
  characteristics of, 445*t*, 446–447
  clinical features of, 448*t*, 448–449
  laboratory diagnosis of, virus isolation and, 449–450
Lassa virus, CDC testing for, 647*t*
Latency, herpes simplex virus-induced, 480–482
Latency associated transcript (LAT), 481–482
Latex agglutination tests, rotaviruses and, 373–374
LAV (lymphadenopathy-associated virus). *See* Human immunodeficiency virus 1
Leibovitz-Emory medium (LEM), 32
Leukemia. *See* Adult T-cell leukemia; Human T-cell leukemia virus; Human T-cell leukemia virus I; Human T-cell leukemia virus II
Liquid hybridization, 313
Liver, as source of specimens, 31
Lung, as source of specimens, 31
  for immunofluorescence, 118
Lymphadenopathy-associated virus (LAV). *See* Human immunodeficiency virus 1
Lymphocytic choriomeningitis virus, CDC testing for, 647*t*

M
Machupo virus, CDC testing for, 647*t*
Malignancies
  polyomaviruses and, 468
  T-cell, clustering of, in adult T-cell leukemia, 591–595
MC virus. *See* Molluscum contagiosum virus
Measles virus, 571–574
  CDC testing for, 647*t*
  characteristics of, 571
  clinical aspects of, 571–573
  complement fixation test and, 206
  control and prevention of, 573–574
  immunofluorescence and, *124*, 124, 125, 125*t*
  laboratory diagnosis of, 573
  respiratory tract infections and, cytopathology of, 74*t*, 75–76, *76*
  serologic determination of, 35
  specimens of, 22*t*, 23*t*
  urinary tract infections and, cytopathology of, 77*t*
Media, quality control and, *10*, 10–11, 11*t*
Metabolic labeling, retroviruses and, 607
4 Methylumbelliferylphosphate (4MUP), preparation of, 163
Mice
  neutralization and, 233
  virus isolation in. *See* Virus isolation, in mice
Microdiluters, routine maintenance and performance checks on, 15
Microimmunofluorescence, *Chlamydia* and, 636–637
Microscopes
  for immunofluorescence, 120
  routine maintenance and performance checks on, 14–15

Milker's nodule virus, 532
  antigen identification and, 540*t*
  isolation of, cell cultures and, 537
Molluscum contagiosum (MC) virus, 532–533, 535
  antigen identification and, 540*t*
  complement fixation test and, 539
  electron microscopy and, 538–539
  eye infections and, cytopathology of, 85*t*
  genital tract infections and, cytopathology of, 81*t*, 83–84, *84*
  isolation of, cell cultures and, 537
Monoclonal antibodies
  biotinylated, 35
  cytomegalovirus diagnosis and, 506
  for immunofluorescence, 120
Mumps virus, 574–576
  CDC testing for, 647*t*
  in cerebrospinal fluid, electron microscopy and, 92
  characteristics of, 574
  clinical aspects of, 574–575
  control and prevention of, 576
  enzyme immunoassay kits for, 158*t*
  immunofluorescence and, 124, 125*t*
  laboratory diagnosis of, 575–576
  specimens of, 21*t*, 23*t*
    urine, 29
4MUP. *See* 4 Methylumbelliferylphosphate
Murray Valley encephalitis virus, CDC testing for, 646*t*
Myxoviruses
  immunoelectron microscopy and, 97
  isolation of, 57
  stability of, 31

N

NANB (non-A, non-B hepatitis). *See* Hepatitis E virus
Nasal aspirates, for immunofluorescence, 117–118
Nasal washings, as source of specimens, 27
Nasopharyngeal aspirates, as source of specimens, nasopharyngeal swabs compared with, 27–28
Nasopharyngeal swabs
  for immunofluorescence, 117
    preparation of, 120
  as source of specimens, nasopharyngeal aspirates compared with, 27–28
National Center for Infectious Diseases (NCID), 643–657
  Division of HIV/AIDS (DHA) of, 651
  Division of Vector-Borne Infectious Diseases (DVBID) of, 651
    requirements for specimens for, 652–653
  Division of Viral and Rickettsial Diseases (DVRD) of, 651
    requirements for specimens for, 652–653
  organization of, 644, 645*t*
  reference and disease surveillance activities of, 644, 646*t*–649*t*
  reference and disease surveillance request form of, 653–657, *654*, *656*
  requirements for specimens and, 644–645, 650
    packing of specimens and, 650

specimen identification and, 650
  viral diseases and
    arthropod and avian specimens and, 652–653
    arthropod specimens and, 652
    isolates and, 652
    requirements for, 650–651
    sera and other body fluids and, 652
    specimens for isolation of etiologic agents and, 651–652
    specimens for serologic testing and, 652
    tissues and, 652
Neuralgia, post zoster-associated, 509–510
Neurologic disorders. *See* Eastern equine encephalitis virus; Encephalitis; Japanese encephalitis virus; Murray Valley encephalitis virus; St. Louis encephalitis virus; Subacute sclerosing panencephalitis; Venezuelan equine encephalitis virus; Western equine encephalitis virus
  in human immunodeficiency virus infections, 602
Neutralization test (NT), 229–240
  calculation of 50 percent endpoints and, 236
    Karber method for, 236
    Reed-Muench method for, 237*t*, 238*t*
  enteroviruses and, 343–344
  herpes simplex virus and, 483*t*, 490–491
  interference assay and, 113–114
  interpretation of serologic results and, 240
    tests for viral antibodies and, 240
  mumps virus and, 576
  poxviruses and, 541–542
  procedures for, 233–236
    constant antiserum, varying virus and, 234–235
    constant virus
      constant antiserum and, 235–236
      varying serum and, 234
    varying virus, varying antiserum and, 236
  serum pool preparation for, 236–240
  standardization of test materials for, 230–233
    host system and, 233
    serum and, 230, *232*, 233
    virus and, 230, *231*
Norwalk-like agents, specimens of, 24*t*
Norwalk virus, 376–379
  CDC testing for, 648*t*
  characteristics of, 376
  clinical manifestations of, 377*t*, 377–378
  epidemiology of, 376–377
  in feces, electron microscopy and, 92, *92*
  immunoelectron microscopy and, 97
  isolation and identification of, 378–379
    isolation sites and, 378
  pathophysiology of, 377
  radioimmunoassay antigen detection and, 139*t*, 142
  stability of, 378
NT. *See* Neutralization test
Nucleic acid, identification of type of, 56
Nucleic acid hybridization, 285–296, 286*t*
  applications of, 285, 286*t*, 293–295
  base-pairing and denaturation and, 187
  choosing correct test for, 292
  controlling hybridization test and, 287, *288*, 288*t*
  dot blots and, *288*, *289*, 291

probe preparation and, 290–291
pros and cons of, 295–296
signal detection and, 287, *289*, 289–290, *290*
in situ hybridization and, 291–292, *292*
Southern blots and, *290*, 291
specimen handling for, 285–287
Nucleic acid probes, parvovirus and, 560

O

Omsk hemorrhagic fever virus, CDC testing for, 646*t*
OPD. *See* Ortho phenylenediamine
Orf virus, 532
antigen identification and, 540*t*
CDC testing for, 647*t*
isolation of, cell cultures and, 537
Organ culture, viral isolation and, 61
Ortho phenylenediamine (OPD), purification and
storage of, 164
Orthopoxviruses, 533–534, 535. *See also* Buffalopox
virus; Cowpox virus; Human monkeypox
virus; Smallpox virus; Vaccinia virus
agar gel precipitation test and, 539
electron microscopy and, 538
enzyme-linked immunosorbent assay and, 543
immunofluorescence and, 539
indirect fluorescent antibody test and, 542
isolation of
cell cultures and, 536
chicken embryo chorioallantoic membrane and,
535–536
neutralization test and, 541–542
radioimmunoassay and, 543

P

Papanicolaou stain, modified, 72
Papillomavirus. *See* Human papillomavirus
Papovaviruses, 455–469, 456*t*. *See also* BK virus;
Human papillomavirus; Jamestown Canyon
virus; Polyomavirus(es)
immunoelectron microscopy and, 97
in urine, electron microscopy and, 93
Parainfluenza virus (PIV), 327–329, *328*, 329*t*. *See*
also specific parainfluenza viruses
age and susceptibility to, 322*t*
CDC testing for, 647*t*
hemadsorption of, 48
hemagglutination inhibition test and, 246
IgM determination and, interpretation of assay
results and, 271
isolation of, 60–61
laboratory diagnosis of, 324*t*, 329
respiratory tract infections and, cytopathology of,
74*t*, 76
specimens of, 21*t*, 22*t*
tissue, 31
Parainfluenza virus (PIV) 1, 328
immunofluorescence and, *123*, 124, 125, 125*t*
Parainfluenza virus (PIV) 2, 328
immunofluorescence and, 124, 125, 125*t*
isolation of, 61
radioimmunoassay antigen detection and, 139*t*, 143

Parainfluenza virus (PIV) 3, 328
immunofluorescence and, *123*, 124, 125, 125*t*
radioimmunoassay antigen detection and, 139*t*
Parainfluenza virus (PIV) 4, 328
Parainfluenza virus (PIV) 5, 328
Paramyxoviruses
in cerebrospinal fluid, electron microscopy and, 92
isolation of, 57
stability of, 31
Paranitrophenyl phosphate (pNPP), preparation of,
163
Parapoxviruses, 534–535. *See also* Milker's nodule
virus; Orf virus
complement fixation test and, 539
electron microscopy and, 538
enzyme-linked immunosorbent assay and, 543
indirect fluorescent antibody test and, 543
neutralization test and, 542
Parvoviruses, 380, 547–562, *548*
CDC testing for, 647*t*
characteristics of, 548–549
clinical manifestations of, 551–555
isolation of, sites for, 556
laboratory diagnosis of, 557–560
electron microscopy and, 557, *558*
immune electron microscopy and, 557, *558*
immunoassay and, 558–560
nucleic acid probes and, 560
polymerase chain reaction and, 560
pathogenesis of, 549–550
prevention and treatment of of diseases due to,
561–562
stability of, 556
in tissues, 555–556
unusual features of, 561
in vitro propagation of, 556–557
Parvovirus-like agents, 380
Paul-Bunnell antibodies, detection by immune
adherence hemagglutination, 256
PBS. *See* Phosphate buffered saline
PCR. *See* Polymerase chain reaction
Persistent generalized lymphadenopathy (PGL), in
human immunodeficiency virus infections,
599–600
Personnel, 5
Pestiviruses, 380
PGL. *See* Persistent generalized lymphadenopathy
pH meters, routine maintenance and performance
checks on, 15
Phosphate buffered saline (PBS), 248–249
Picornaviruses. *See also* Enteroviruses; Rhinovirus
CDC testing for, 647*t*
Pipettors, routine maintenance and performance
checks on, 15
Plaque formation, 55, 55–56
PML. *See* Progressive multifocal
leukoencephalopathy
pNPP. *See* Paranitrophenyl phosphate
Poliovirus(es)
asymptomatic infections with, 348
CDC testing for, 647*t*
discovery of, 341–342
identification of, 356

Poliovirus(es) [*cont.*]
  incubation times of, 353
  isolation of, 59, 59*t*
  pathogenesis and clinical syndromes and, *349,*
    349–350, 351*t*
  replication in cell culture, 344–345
  serologic diagnosis of, 357
  serotypes of, 341*t*
  vaccination against, 354–355
Polymerase chain reaction (PCR), 35, 285, 309–317
  basic principles of, 310–311
    amplification and, 311, *311*
    DNA polymerase and, 310, *310*
    steps in polymerase chain reaction and, 310–
     311, *311*
  contamination by amplified product and
    prevention of, 314, 314*t*
    in vivo amplified DNA fragments and, 314–315
  cytomegalovirus and, 506–507
  human papillomavirus and, 464*t*
  jumping, 314–315, *315*
  limitations of, 313*t*, 313–315
  nested, 315, *316*
  parvovirus and, 560
  problems encountered using DNA in clinical
    testing and, 309–310, 310*t*
  retroviruses and, detection of viral genomes and,
    611–612
  rotaviruses and, 375
  in vitro amplification and, 311–313
    capability to detect genetic changes and, 313
    enhanced sensitivity and, 311–312, 312*t*
    flexibility of detection methods and, 312–313
    simplifying DNA purification procedures and,
     313
Polyomavirus(es), 464–469. *See also* BK virus;
    Jamestown Canyon virus
  characteristics of, 465
  laboratory diagnosis of, 468–469
  pathogenesis and disease potential of, 466–468
    in immunocompetent hosts, 466
    in immunocompromised hosts, 466
    malignancies and, 468
    pregnancy and, 468
    primary immunodeficiency diseases and, 467–
     468
    progressive multifocal leukoencephalopathy and,
     466–467
    in renal and bone marrow transplant recipients,
     467
  treatment of, 469
  in urine, electron microscopy and, 93
Powassan (POW) virus, 444
  CDC testing for, 646*t*
  characteristics of, 445*t*, 446
  clinical features of, 448*t*
Poxviruses, 527–544
  agar gel precipitation and, 539
  antigens of, 534–535
  chorioallantoic membrane pock formation and, 62
  complement fixation and, 539
  electron microscopy and, 537–539

genital tract infections and, cytopathology of, 83–
    84, *84*
  immunofluorescence and, 539
  isolation of, 535–537
    cell cultures and, 536–537
    chicken embryo chorioallantoic membrane and,
     535–536
  physical characteristics and classification of, 533*t*,
    533–534, *534*
  serodiagnosis of, 540–544
    complement-fixation test and agar gel
     precipitation for antibody assay and, 544
    enzyme-linked immunosorbent assay and, 543
    hemagglutination inhibition test and, 540–541, 541*t*
    indirect fluorescent antibody test and, 542–543
    neutralization test and, 541–542
    radioimmunoassay-adsorption test and, 543–544
    radioimmunoassay and, 543
  specimen collection and handling and, 535
  stained smears and, 539, 540*t*
Pregnancy
  coxsackieviruses and, 353
  polyomaviruses and, 468
  rubella and, 577–579, 580
Primary rabbit kidney cells, specimen processing
    and, 34
Procedure manual, 5
Proficiency testing, 16–17
Progressive multifocal leukoencephalopathy (PML)
  laboratory diagnosis of, 468
  polyomaviruses and, 465, 466–467, 468
  treatment of, 469
Progressive rubella panencephalitis (PRP), 578
Prostaglandin E$_2$, herpes simplex virus-associated
    latency and, 482
Pseudoreplica method, of negative staining, electron
    microscopy and, 96
PXAPX. *See* Unlabeled peroxidase anti-peroxidase
    technique

Q

Quality control, 3–17
  of anti-species conjugate, for immunofluorescence,
    119
  cell culture and, 8, *9*
  facility design and maintenance and, 3–5
  instruments and, 13–15
  media and solutions and, *10*, 10–11, 11*t*
  personnel and, 5
  procedure manual and, 5
  proficiency testing and, 16–17
  reagents, stains, antisera, and kits and, 12–13
  reference virus stocks and, 12, *13*, 14*t*
  of specific antisera, for immunofluorescence, 118–
    119
  statistics and backup procedures and, 15–16, 16*t*
  submission of sera and, 7–8
  submission of smears and, 7
  viral transport media and, 6–7
  water quality and, 11–12

R

Rabies enzyme immunoassay (REIA), 433
Rabies immune globulin (RIG), 436
  intravenous, 436
Rabies tissue culture infection test (RTCIT), 432–433
Rabies virus, 425–438
  CDC testing for, 648*t*
  characteristics of, 426–427, 427*t*
  considerations for clinical virologist with, 434–436
  future prevention of infection by, 436–438
  history of, 425–436
  identification of, 431–434
    detection of anti-rabies antibody and, 433–434
  immunofluorescence and, 117, 125*t*
  isolation and cultivation of, 430–431
  pathogenesis of, 427–430
  specimens of, 23*t*
  unique features of infection by, 437–438
  vaccine against, 434–436
Radioimmunoassay (RIA), 129–145
  adenovirus and, 143
    antigen detection and, 142, 144
  astrovirus antibody detection and, 144
  competition, retroviruses and, 607–608
  competitive binding, 129–130
  coxsackie B virus antibody detection and, 144
  cytomegalovirus antibody detection and, 142
  dengue virus and, 144
  enteroviruses and, 138–139
    antibody detection and, 138–139
  hepatitis B virus and. *See also*
    Radioimmunoassay, indirect, for hepatitis B
      antibodies
    serologic markers and, 130–131
  hepatitis C virus antibody detection and, 144
  herpes simplex virus and, 142–143, 483*t*, 488
    antibody detection and, 142–143, 144–145
    antigen detection and, 143
  IgM determination and, 135–137, *138*, 269
  indirect, for hepatitis B virus antibodies, 134–135
    one-step procedure, 134, *135*, 136*t*
    two-step procedure, 134–135, 136*t*, *137*
  influenza virus antigen detection and, 143
  Norwalk virus and, 378–379
    antigen detection and, 142
  parainfluenza virus 2 and, 143
  parvovirus and, 558–560
  poxviruses and, 543
  respiratory syncytial virus and, 143
  rotavirus and, 141–142
    antibody detection and, 141–142
    antigen detection and, 141
  solid phase, 130, 131*t*
    direct solid phase sandwich assays for hepatitis
      B virus antigens and antibodies, 131–134, *132*, *133*
    retroviruses and, 606
  togavirus and, 139, 141
    anti-alphavirus antibody detection and, 139
    anti-flavivirus antibody detection and, 141
    anti-rubella virus antibody detection and, 141
  varicella-zoster virus antibody detection and, 143, 145

  yellow fever virus and, 145
Radioimmunoassay-adsorption test, poxviruses and, 543–544
Radioimmunoprecipitation assays (RIPA), retroviruses and, 607
Rapid fluorescent focus inhibition test (RFFIT), rabies virus and, 433
RD cells. *See* Rhabdomyosarcoma cells
RDE. *See* Receptor destroying enzyme
RDRV. *See* Inactivated absorbed rabies vaccine
Reagents, quality control and, 12–13
Receptor destroying enzyme (RDE), 249
Record keeping, for cell culture, 8, *9*
Rectal swabs, as source of specimens, 29
Red blood cell suspension, for hemagglutination inhibition test, preparation of, 244
Reference antisera, 249
Reference virus stocks, quality control and, 12, *13*, 14*t*
Refrigerators, routine maintenance and performance checks on, 15
REIA. *See* Rabies enzyme immunoassay
Renal syndrome viruses, detection by immune adherence hemagglutination, 258
Renal transplants, polyomaviruses and, 467
Reoviruses, hemagglutination inhibition test and, 247
Respiratory syncytial virus (RSV), 329–332, 330*t*, *331*, 331*t*
  age and susceptibility to, 322*t*
  CDC testing for, 647*t*
  enzyme immunoassay kits for, 158*t*
  immunofluorescence and, 122, *123*, 124, 125*t*
  isolation of, 61
  laboratory diagnosis of, 324*t*, 332
  radioimmunoassay antigen detection and, 139*t*, 143
  respiratory tract infections and, cytopathology of, 74*t*, 74–75
  specimens of, 21*t*, 22*t*
    collection of, timing of, 20
    inoculation and, 34–35
    sputum, 28
    storage of, 34
  subgroups of, 330
Respiratory tract infections
  cytopathology of, 73–77, 74*t*, 75–77
  human papillomavirus and, 460–461
Respiratory tract secretions, electron microscopy and, 92
Respiratory tract specimens, for interference assay, 112
Respiratory viruses, 321–336, 322*t*–324*t*, *323*. *See also* Adenovirus(es); Coronaviruses; Influenza virus(es); Parainfluenza virus; Respiratory syncytial virus; Rhinovirus; *specific viruses*
  CDC testing for, 647*t*
Restriction endonuclease mapping, herpes simplex virus and, 484–485
Retroviruses, 585–615. *See also* Human immunodeficiency virus; Human immunodeficiency virus 1; Human immunodeficiency virus 2; Human T-cell leukemia virus; Human T-cell leukemia virus I; Human T-cell leukemia virus II

Retroviruses [*cont.*]
 CDC testing for, 648*t*
 laboratory diagnosis of, 602–614
  agglutination assays and, 606
  assays for detection of viral genomes and, 611–612
  enzyme-linked immunosorbent assay and, 603
  immunoblot and, 603–605, *604*, 605*t*, *606*
  indirect immunofluorescence tests and, 605–606
  isolation methods and, 612–614
  large scale serologic screening and, 608–609
  radioimmunoassays and, 606–608
  tests for demonstration of viral proteins and, 609–611
  tests for virus-specific antibodies and, 602–603
 stability and safety precautions and, 614–615
Reverse transcriptase assay, retroviruses and, 610–611
RFFIT. *See* Rapid fluorescent focus inhibition test
Rhabdomyosarcoma (RD) cells, specimen processing and, 34
Rhinovirus, 335–336
 age and susceptibility to, 322*t*
 isolation of, 60
 laboratory diagnosis of, 324*t*, 336
 specimens of
  sputum, 28
  tissue, 31
RIA. *See* Radioimmunoassay
Ribavarin (Virazole), in respiratory syncytial virus, 332
RIG. *See* Rabies immune globulin
RIP/SDS-PAGE, retroviruses and, 607
RNA probes, preparation of, 290–291
Rotators, routine maintenance and performance checks on, 15
Rotaviruses, 361–375
 antigenic characteristics of, 362–364, *365*, 366*t*
 CDC testing for, 648*t*
 clinical manifestations of, 368–370, 369*t*
 diagnostic methods for, 372–375
 enzyme immunoassay kits for, 158*t*
 epidemiology of, 364–366
 in feces, electron microscopy and, 92
 immunoelectron microscopy and, 97
 immunology of, 367
 isolation methods for, 370–372
 isolation sites of, 370
 mixed infection and, 370
 pathophysiology of, 366–367
 radioimmunoassay and
  antibody detection and, 140*t*, 141–142
  antigen detection and, 139*t*, 141
 sequential infection and, 370
 solid phase immunoelectron microscopy and, 98
 specimens of, 24*t*
 stability of, 370
 structure of, 362, *362*, *363*, 364*t*
 transmission of, 367–368
RSV. *See* Respiratory syncytial virus
RTCIT. *See* Rabies tissue culture infection test
Rubella virus, 576–581
 CDC testing for, 647*t*

characteristics of, 577
 clinical aspects of, 577–579
 control and prevention of, 580–581
 enzyme immunoassay kits for, 159*t*
 hemagglutination inhibition test and, 247
 IgM determination and, sucrose density gradient ultracentrifugation and, 265
 immunoelectron microscopy and, 97
 interference and, 49–50
 interference assay for. *See* Interference assay
 laboratory diagnosis of, 579–580
 radioimmunoassay antibody detection and, 140*t*, 141
 serologic determination of, 35
 specimens of, 22*t*, 25*t*
  ocular, 30
Rubeola virus. *See* Measles virus

S

Sabin vaccine, 354
Safety precautions, for electron microscopy, 91
St. Louis encephalitis (SLE) virus, 443
 CDC testing for, 646*t*
 characteristics of, 445*t*, 446
 clinical features of, 448*t*, 449
 radioimmunoassay antibody detection and, 140*t*, 141
 specimens of, 23*t*
Salk vaccine, 354
Sera
 National Center for Infectious Diseases requirements for, 652
 specific, quality control of, for immunofluorescence, 118–119
 submission of, 7
Serodiagnosis, 35
 of *Chlamydia*, 634–637
  complement fixation and, 635–636
  microimmunofluorescence and, 636–637
 of cytomegalovirus, 507
 of Epstein-Barr virus, *514*, 514–515
 of herpes simplex virus, 489–491
 of mumps virus, 576
 of poxviruses. *See* Poxviruses, serodiagnosis of
 of rubella virus, 579–580
 of varicella-zoster virus, 511
Serologic markers, in hepatitis, 410–417, *411*
Serologic screening, for retroviruses, 608–609
Serum immune globulin, in parvovirus infection, 561–562
Serum-in-agar method, for immunoelectron microscopy, 98, *99*
Sheep erythrocytes, for complement fixation test, preparation of, 211
Shell vial technique
 coverslip staining and, 53
 immunofluorescence and, 124–125, 125*t*
 inoculation procedure and, 53
 reading procedure and, 54
 reagents and equipment for, 53
 virus isolation and, 50–54, *52*
Signal detection, nucleic acid hybridization and, 287, *289*, 289–290, *290*

SLE virus. *See* St. Louis encephalitis virus
Smallpox virus, 527–528
  antigen identification and, 540*t*
  complement fixation test and, 539
  hemagglutination inhibition test and, 541
  isolation of
    cell cultures and, 536–537
    chicken embryo chorioallantoic membrane and, 536
    radioimmunoassay and, 543
    vaccinia and, 528–530
  "Small round viruses", in feces, electron microscopy and, 92
Smears, submission of, 7
Snowshoe hare (SSH) virus, 444
  characteristics of, 446–447
Solid phase "capture" immunoassays, IgM determination and, 269–270, *270, 271*
Solid phase immunoelectron microscopy (SPIEM), 98–100, *100*
Solid phase indirect immunoassays, IgM determination and, *267,* 267–269, *268*
Solid phase radioimmunoassays, 130, 131*t*
  retroviruses and, 606
Southern blots, *290,* 291, 300, 313
Southern transfer, human papillomavirus and, 464*t*
Specimen(s), 19–35
  collection of, 20, 27–31
    arboviruses and, 449
    Chlamydia and, 630
    of herpes simplex virus, 486, *487*
    for interference assay, 112
    poxviruses and, 535
    source and, 27–31
    timing of, 20, 27
  for complement fixation test, 207
  for enzyme-linked immunosorbent assay, 164–165
  handling of, poxviruses and, 535
  for immunofluorescence, 117–118
    preparation of, 120–121
  inoculation of, for interference assay, 112–113
  for interference assay
    collection, transport, and storage of, 112
    preparation and inoculation of, 112–113
  for National Center for Infectious Diseases. *See* National Center for Infectious Diseases
  obtaining, 45
  preparation and staining of, 71–73
    cytocentrifugation and filtration and, 72–73
    direct smears and, 72
    for immunofluorescence, 120–121
    for interference assay, 112
  processing, 34–35, 45
    examination and, 35
    inoculation and, 34–35
  selection of, 20, 21*t*–26*t*
  storage of, 33–34
    Chlamydia and, 630–631
    for interference assay, 112
  submission to state laboratory virology services, 659, 660*t*–663*t*, 664
  transport of, 31–33
    Chlamydia and, 630–631

for interference assay, 112
stability of viruses and, 31
swabs and, 31
transport system and, 31–33, 32*t*
Spectrophotometers, routine maintenance and performance checks on, 15
SPG. *See* Sucrose-phosphate-glutamate
SPIEM. *See* Solid phase immunoelectron microscopy
Spleen, as source of specimens, 31
Spot blot, 312–313
  herpes simplex virus and, 483*t*, 488
Sputum, as source of specimens, 28
SSH virus. *See* Snowshoe hare virus
SSPE. *See* Subacute sclerosing panencephalitis
Stained smears, poxviruses and, 539, 540*t*
Staining, negative, electron microscopy and. *See* Electron microscopy, negative staining methods and
Stains, quality control and, 12–13
State laboratory virology services, 659–670
  scope of, 664, 665*t*–669*t*, 670
  submission of specimens to, 659, 660*t*–663*t*, 664
  turnaround time for result reporting by, 670
Statistics, 15–16
Stool, electron microscopy and, 92, *92, 94*
Stool specimens, 29
Stuart's transport medium, 31–32
Subacute sclerosing panencephalitis (SSPE), measles virus and, 572
Substrates, for enzyme-linked immunosorbent assay, 161–164
Sucrose density gradient ultracentrifugation, IgM determination by, 265–266, *266*
Sucrose-phosphate-glutamate (SPG)
  specimen storage and, 34
  specimen transport and, 33
Symmetrel (Amantadine), influenza A virus and, 327

T
Tanapoxviruses, 531, 535
  antigen identification and, 540*t*
  complement fixation test and, 539
  electron microscopy and, 538
  indirect fluorescent antibody test and, 543
  isolation of, cell cultures and, 537
  neutralization test and, 542
  stained smears and, 539
Temperature, specimen storage and, 33–34
Thin sectioning methods, for electron microscopy, 101–102, 104–105
Throat swabs, for immunofluorescence, 117
Throat washings, as source of specimens, 27
Tick-borne encephalitis virus, CDC testing for, 646*t*
Tissue(s)
  electron microscopy and, 93, *95*
  infected, harvesting and processing, virus isolation in mice and, 64–65
  National Center for Infectious Diseases requirements for, 652
  parvovirus B19 in, 555–556
  as source of specimens, 31

Tissue(s) [*cont.*]
  specimens for interference assay and, 112
Togaviruses. *See also* Rubella virus
  hemagglutination inhibition test and, 247
  IgM determination and, interpretation of assay
    results and, 271
  isolation in mice, 65
  radioimmunoassay antibody detection and, 139,
    140*t*, 141
  serologic determination of, 35
  specimens of, cerebrospinal fluid, 30
Transmission electron microscope (TEM), 90
Transporter Tube, 33
Tropical spastic paraparesis (TSP), 595–596
Tryptose phosphate broth (TPB), 33

U

Ultracentrifugation, sucrose density gradient, IgM
    determination by, 265–266, *266*
Unlabeled peroxidase anti-peroxidase (PXAPX)
    technique, 190, *191*, 191–192, 192*t*
  applications of, 192*t*
  materials for, 193
  methods for, 193–194
  results of, 194–195, *196–199*
  staining and, 192–193
Urinary tract infections, cytopathology of, 77*t*, 77–
    79, *78–80*
Urine
  electron microscopy and, 93
  interference assay and, 112
  as source of specimens, 29–30

V

Vaccination
  against cytomegalovirus, 505
  against Eastern equine encephalitis virus, 445–446
  against enteroviruses, 354–355
  against gastrointestinal tract viruses, 381
  against herpes simplex virus, 491–493, 492*t*
  against influenza virus, 327, *327*
  against measles virus, 574
  against mumps virus, 576
  against rabies virus, 434–436
  against rubella virus, 580–581
  against smallpox, vaccinia and, 528–530
  against varicella-zoster virus, 509
  against Venezuelan equine encephalitis virus, 446
  against Western equine encephalitis virus, 446
Vaccinia virus, 528–530
  antigen identification and, 540*t*
  CDC testing for, 647*t*
  isolation of
    cell cultures and, 537
    chicken embryo chorioallantoic membrane and,
      536
  radioimmunoassay-adsorption test and, 543–544
  stained smears and, 539
Vancomycin hydrochloride, in culture media, 10
Varicella-zoster immune globulin (VZIG), 509
Varicella-zoster virus (VZV), 501, 507–511

  age and susceptibility to, 322*t*
  biology and pathogenesis of infection and, 507–508
  CDC testing for, 647*t*
  clinical manifestations of, 508–509
  detection by immune adherence hemagglutination,
    256, 257, 258
  enzyme immunoassay kits for, 159*t*
  epidemiology of, 508
  eye infections and, cytopathology of, 85*t*
  history of, 507
  immunofluorescence and, 124, 125*t*
  incubation for cell culture, 47
  isolation of, 57–58
  laboratory diagnosis of, *510*, 510–511
  prevention and treatment of, 509–510
  radioimmunoassay antibody detection and, 140*t*,
    143, 144–145
  reference virus stocks, 12
  specimens of, 22*t*
    blood, 30
    cerebrospinal fluid, 30
    collection of, timing of, 20
    dermal, 30
    inoculation and, 34
    ocular, 30
    storage of, 34
    timing of collection of, 27
Variola virus
  radioimmunoassay-adsorption test and, 543–544
  stained smears and, 539
Venezuelan equine encephalitis (VEE) virus, 444
  CDC testing for, 646*t*
  characteristics of, 445*t*, 446
  laboratory diagnosis of, virus isolation and, 449–
    450
Vesicle fluid and crusts, electron microscopy and,
    91, 91–92
Vidarabine
  in herpes simplex virus infections, 495
  in varicella-zoster virus infections, 509
Viral transport media (VTM), 6
Virazole (ribavarin), in respiratory syncytial virus,
    332
Virocult, 33
Virus infectivity assays, 278–279, 279*t*
Virus inhibition assay, 279–280, 280*t*, 281*t*
Virus isolation, *43*, 43–65
  of arboviruses, 449–450
  in cell cultures, 44–61
    adenovirus and, 59
    advantages and limitations of, 56–57
    background of, 44
    cell culture types and, 44, 44*t*
    cytomegalovirus and, 58–59
    cytopathic effects and, 48, *49–51*
    enteroviruses and, 59*t*, 59–60
    hemadsorption and, 48–49, *52*
    herpes simplex virus types 1 and 2 and, 57, 58*t*
    influenza virus and, 60
    inoculation and incubation and, 45–48, *47*
    interference and, 49–50
    obtaining and processing specimens and, 45
    parainfluenza virus and, 60–61

plaque formation and, *55*, 55–56
respiratory syncytial virus and, 61
rhinovirus and, 60
shell vial technique and, 50–54, *52*
supplies and equipment needed for, 45, 46*t*
variation in susceptibility to viruses and, 44, 45*t*
varicella-zoster virus and, 57–58
virus infectivity assay by end point of
    cytopathic effect and, 54, 54*t*
virus isolate identification and, 56
Chlamydia and, 633–634
in embryonated eggs, 62–64
    background of, 62
    harvesting, assay, and identification of isolates
        and, 63–64
    inoculation and incubation and, 62–63
    maintenance and source of eggs and, 62
of enteroviruses, 355
explant culture or cocultivation for, 61
of herpes simplex virus, 483*t*, 484–485
in mice, 64–65
    background of, 64
    harvesting and processing infected tissues and,
        64–65
    identification of isolates and, 65
    inoculation and observation for illness and, 64,
        65
organ culture for, 61
of parvovirus, preferred sites for, 556
of poxviruses, 535–537
    cell cultures and, 536–537
    chicken embryo chorioallantoic membrane and,
        535–536
of rabies virus, 430
of retroviruses, 612–614
Virus size, determination of, 56
Virus stocks, reference, quality control and, 12, *13*,
    14*t*

VTM. *See* Viral transport media
VZIG. *See* Varicella-zoster immune globulin
VZV. *See* Varicella-zoster virus

W
Water baths, routine maintenance and performance
    checks on, 15
Water drop method, of negative staining, electron
    microscopy and, 95–96
Water quality, 11–12
WEE virus. *See* Western equine encephalitis virus
Western blot, 299–305
    advantages and disadvantages of, 304–305, *305*
    commercial kits and nitrocellulose strips with
        blotted proteins and, 305
    herpes simplex virus and, 490
    history and principle of, 299–301, *300*
    procedure for, 301–304
        materials and, 302–303
Western equine encephalitis (WEE) virus, 443
    CDC testing for, 646*t*
    characteristics of, 444, 445*t*, 446
    clinical features of, 448*t*, 449
    laboratory diagnosis of, virus isolation and, 449–
        450
    specimens of, 23*t*
Whitepox virus, 531
    antigen identification and, 540*t*
    isolation of
        cell cultures and, 536–537
        chicken embryo chorioallantoic membrane and,
        536
Wistar RA 27/3 vaccine, 580

Y
Yellow fever virus, radioimmunoassay and, 145